Untangling Complex Systems

Untangling Complex Systems

A Grand Challenge for Science

Pier Luigi Gentili

CRC Press
Taylor & Francis Group
Boca Raton London New York

CRC Press is an imprint of the
Taylor & Francis Group, an **informa** business

CRC Press
Taylor & Francis Group
6000 Broken Sound Parkway NW, Suite 300
Boca Raton, FL 33487-2742

First issued in paperback 2019

© 2019 by Taylor & Francis Group, LLC
CRC Press is an imprint of Taylor & Francis Group, an Informa business

No claim to original U.S. Government works

ISBN-13: 978-1-4665-0942-9 (hbk)
ISBN-13: 978-0-367-48562-7 (pbk)

Dedication

To God, my family, all my past, present, and next students, and whoever will read this book with curiosity and open-mindedness.

Contents

Preface

SCOPE AND GENESIS

Complex Systems are natural systems that science is unable to describe exhaustively. Examples of Complex Systems are both unicellular and multicellular living beings; human brains; human immune systems; ecosystems; human societies; the global economy; the climate and geology of our planet. Science cannot predict the behavior of such systems, especially in the long term. Why is it so important to study Complex Systems? Because humanity must tackle compelling challenges that affect Complex Systems. For example, we need to predict catastrophic events, such as earthquakes and volcanic eruptions, to avoid many deaths. We struggle to protect our ecosystems and the environment from climate change and the risk of shrinking biodiversity. We need to find innovative solutions to guarantee a worldwide sustainable economic growth, primarily by focusing on the energy issue. We also need to find creative solutions to ensure stability and justice in our societies. Finally, there are still incurable diseases that must be defeated. I have made a list of what I like to call "Natural Complexity Challenges." To try to win the "Natural Complexity Challenges," we need to understand Complex Systems deeply. But this is not an easy task because Complex Systems are intertwined networks, working in out-of-equilibrium conditions, which exhibit emergent properties, such as self-organization phenomena and chaotic behaviors in time and space. I decided to contribute to the untangling of Complex Systems by writing this book. This book is an account of an amazing scientific and intellectual journey I made to understand Natural Complexity. I have undertaken my trip, equipped with the fundamental principles of physical chemistry, and in particular, the Second Law of Thermodynamics that describes the spontaneous evolution of our universe. Two central questions have guided me:

1. If the Second Law of Thermodynamics is true, how is it possible to observe the spontaneous emergence of order in time and space? Is it possible to violate the Second Law?
2. What are the common features of Complex Systems? When and how do the emergent properties emerge?

To find answers to my questions, I have gone on a *marvelous interdisciplinary journey*. I dealt with many disciplines; particularly, chemistry, biology, physics, economy, and philosophy. Chapter 1 presents an excursus on the evolution of the scientific knowledge and its mutual fruitful relationship with technology. Chapter 2 is a thorough analysis of the Second Law of Thermodynamics and to understand if its violation is feasible. Chapters 3 and 4 present the principles of Non-Equilibrium Thermodynamics. Then, the theory of Non-Linear Dynamics is introduced by the description of the emergence of the temporal order in ecosystems (Chapter 5), economy (Chapter 6), within a living being (Chapter 7), and in a chemical laboratory (Chapter 8). Chapter 9 describes the emergence of spatial order in chemistry, along with examples regarding biology, physics, and geology. Then, Chapter 10 introduces the concept of Chaos in time, whereas Chapter 11 covers Chaos in space by presenting fractals. Chapter 12 offers the typical features of Complex Systems and outlines the link between Natural Complexity and Computational Complexity. Finally, Chapter 13 proposes strategies to try to untangle Complex Systems and win the Complexity Challenges.

PURPOSES

This book has four principal objectives and one hope. First, it traces a new interdisciplinary didactic path in Chemistry. This book is useful for upper-division undergraduate and graduate students in Chemistry, who want to learn the principles and theories regarding Non-Equilibrium Thermodynamics, Non-Linear Dynamics, and Complexity.

Second, the contents I present should boost the spread of interdisciplinary courses in Complex Systems to universities around the world. Teachers of Complexity can choose this textbook when they want to highlight the relevant contribution of chemistry.

Third, this book contributes to the training of a new generation of PhD students and researchers who want to comprehend Complex Systems and win the Complexity Challenges.

Fourth, I want to stimulate public and private funding agencies to sustain interdisciplinary projects on Complex Systems.

This book terminates with a question: "Can we formulate a new scientific theory for understanding and predicting the behavior of Complex Systems?" I hope that someone, also inspired by this text, will contribute to the formulation of that scientific theory that we have been waiting for many years.

ADVICE AND NOTES FOR STUDENTS AND INSTRUCTORS

In every chapter, after the presentation of the theory, I offer a list of key questions and key words that want to help students in fixing the most important concepts that have been presented. Teachers can exploit these lists to check the level of preparation of their students. Then, I suggest books and papers for deepening the knowledge of the proposed content. All the chapters, except the first, offer exercises with solutions. These exercises are useful tools to test the degree of understanding of the theory and the subjects presented in each chapter. My suggestion is that students solve all the exercises by themselves. Some of the exercises require the numerical solution of differential equations. The solutions that I propose have been obtained by using MATLAB software. Students can use any other software they are familiar with.

LIMITATIONS AND APOLOGIES

Regarding the references, I apologize if the reader is upset by any omissions as I did not want the chapters to be exhaustive reviews of all the work done on that particular subject. But rather, some representative and didactic examples are proposed. No implication is intended towards the importance of works cited relative to works not cited. I apologize in advance for those cases where my selection is faulty. The subject of Complexity is amazingly rich and polyhedral, and apologies are offered to those readers whose favorite topics are omitted. Of course, this book is the report of a wonderful personal journey, and it could be enriched by additional content.

Regarding the Figures, I decided to print them in black and white to maintain a low manufacturing price of the book and make it more affordable.

Whoever wants to suggest me either improvements or constructive corrections or share their experience in using this textbook, please send me an e-mail to the following address: pierluigigentili@gmail.com.

MATLAB® is a registered trademark of The MathWorks, Inc. For product information, please contact:

The MathWorks, Inc.
3 Apple Hill Drive
Natick, MA 01760-2098 USA
Tel: 508 647 7000
Fax: 508-647-7001
E-mail: info@mathworks.com
Web: www.mathworks.com

Acknowledgments

First, I want to acknowledge Dr. Lance Wobus, who in Summer 2011, after reading one of my papers related to the field of Natural Computing, sent an e-mail proposing that I write a book on that topic. Despite being aware of the commitment, I accepted his invitation after extending the subject of the book to the content of my teaching activity that regards Complex Systems and the role that Natural Computing can play in the comprehension of Complexity. I am also thankful to Senior Editor Barbara (Glunn) Knott and Editorial Assistant Danielle Zarfati, who trusted in my project and helped me to finalize the publication of this book.

I am grateful all my school teachers and professors at the Chemistry Department of Perugia University, who, along with my family members and friends, contributed to my knowledge and my *forma mentis*. In particular, I like to mention Prof. Giuseppe Cardaci, who taught me the principles of Non-Equilibrium Thermodynamics and made me passionate about that field. I am also grateful to my tutors of degree and PhD thesis in Chemistry, Prof. Giovanna Favaro, Prof. Gian Gaetano Aloisi, Prof. Aldo Romani and Prof. Massimo Olivucci, who helped me to grow as a researcher. The mind of the researcher, including his knowledge, his skills, and his questions, is like a mosaic whose pieces are made from the books and papers he reads, the lectures he attends, and the colleagues he meets and works with. In particular, I acknowledge all my direct collaborators, who are too numerous to be listed here. Among them, I want to mention those who helped me to understand Complex Systems and hosted me in their research groups. They are Prof. Irving R. Epstein, Prof. Milos Dolnik, Prof. Vladimir Vanag (now working at Immanuel Kant Baltic Federal University) of the Brandeis University (MA, USA), Prof. Jean-Claude Micheau of the Université Paul Sabatier-Toulouse III (France), Prof. Peter Bentley of the University College of London (UK), and Prof. Peter Tompa of the Vrije University in Brussels (Belgium). Moreover, I want to mention those who noticeably helped me to develop my research on Natural Computing, who are Prof. Mark B. Heron, Prof. Raimondo Germani, and Prof. Hiroshi Gotoda. I am also grateful to the Santa Fe Institute for providing exciting courses and other educational materials related to the Complex Systems science on the "Complexity Explorer" website. I also want to acknowledge Dr. Veronica Dodero, Dr. Federico Rossi, Dr. Marcello Budroni, Dr. Christophe Coudret and Dr. Otto Hadač for fruitful discussions on some of the themes of this book and for suggesting me significant references. I thank Mr. Danilo Pannacci and Prof. Cristiano Zuccaccia for their sharp questions about some of the subjects of this book, and for the lively discussions that we had during our shared lunchtimes. I also want to acknowledge my past students because the lectures that I gave them and the questions they asked me have been very beneficial for writing this manuscript. I thank Mr. Andrea Nicoziani who helped me to build a Hele-Shaw cell used for an experiment proposed in Chapter 11. I thank Mr. Antonio Maria Cinti for bringing me a sample of malachite and one of agate with traces of periodic precipitations, whose pictures are shown in Chapter 9. I thank Mr. Nicomede Pelliccia for helping me in preparing some pictures. Then, I want to thank all my family, specifically, my parents, who have never stopped encouraging me in my studies and research. My father helped me also to understand the principles of the economy. Finally, I am grateful to God for the gift of life and for infusing me the passion of scrutinizing His Creation. The more I study nature, the more I find it is breathtaking and amazing. I thank God for guiding me in my research and for all the keen scientists who allowed me to meet, so far.

About the Author

Pier Luigi Gentili is a PhD in Chemistry. His research and teaching activities are focused on Complex Systems. He is trusting in Natural Computing as an effective strategy to understand Complex Systems and face the Computational Complexity Challenges. In particular, he is developing the innovative Chemical Artificial Intelligence. He has several collaborations and work experience in many laboratories such as, the "Photochemistry and Photophysics Group" of the University of Perugia (Italy); the "Nonlinear Dynamics Group" of the Brandeis University (USA); the "European Laboratory of Nonlinear Spectroscopy" in Florence (Italy); the "Center for Photochemical Sciences" of the Bowling Green State University (USA); the "Laboratory of Computational Chemistry and Photochemistry" of the University of Siena (Italy).
ORCID: 0000-0003-1092-9190

1 Introduction

A life without research is not worthy of being lived.

Socrates (470–399 BC)

The most beautiful thing we can experience is the Mysterious. It is the source of all true art and science.

Albert Einstein (1879–1955 AD)

1.1 THE NEVER-ENDING JOURNEY TO DISCOVERING THE SECRETS OF NATURE

One of the most precious gifts of our life is the possibility of appreciating the beauty of nature. For example, the bright colors and the peaceful silence of a breathtaking sunset admired on top of a mountain; the variety of perfumes, colors, and shapes of flowers; the magnificence of grand trees (see Figure 1.1); the astonishing vastness of a starry sky. These are just a few examples of a countless number of marvels we can enjoy.

We can scrutinize the beauty of nature simply by using our senses of sight, hearing, smell, taste, and touch. In fact, our senses are "endo-somatic tools" we use to collect information about the outside world. The information collected by the sensory cells is transduced in electrochemical signals that are sent to the brain. Within our brain, such information satisfies our unquenchable "perceptual curiosity" of always experiencing something completely new, and "diversive curiosity" that refers to the relentless desire we must explore and seek new stimulation to avoid boredom (Livio 2017). However, we also have "epistemic curiosity" to satisfy. It represents our "appetite for knowledge." Epistemic curiosity spurs us to get acquainted with natural wonders and understand how they originated.

I think that everybody will agree with me if I say that the beauty of nature resides in its *harmony, organization, functionality, efficiency, variety, complexity....* In other words, the beauty of nature derives from the presence of an inherent logos (λόγος), i.e., a rational logic based on laws and principles that are universal in space and time. The natural marvels have drawn the attention and ignited the curiosities of many men and women in the course of history. This attraction is still active, and it will never cease until the end of life on earth. People who dedicated their lives, entirely or partly, to the study of nature, can be called "Philo-physicists," from the Greek "φίλος-φύσις," which means "fond of nature." Philo-physicists have been gifted with unquenchable epistemic curiosity about nature. They have discovered many natural wonders, unveiled fundamental natural laws, and relentlessly extended human knowledge. The acquisition of new knowledge about nature has often promoted technological developments. Technology is the ensemble of methods and tools helpful to fulfill the natural human will of improving our psycho-physical well-being by solving practical problems. Between science and technology, there exists a reciprocal positive feedback action[1] (see Figure 1.2). In fact, not only does scientific knowledge promotes technological development, but any new technical achievement allows for more time and/or new tools, i.e., "exo-somatic facilities," to deepen our exploration of nature.

---

[1] Feedback occurs when the output of an action becomes the input of another.

FIGURE 1.1   Three examples of natural wonders.

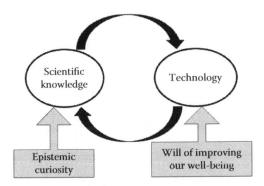

FIGURE 1.2   Reciprocal positive feedback action between Science and Technology. Science is fed mainly by epistemic curiosity, whereas technology by the dream of improving our welfare.

The journey to discovering the secrets of nature, made by humanity, has been punctuated by two revolutionary intellectual events: (1) the birth of philosophy in the ancient Greek colonies, during the sixth century BC, and (2) the use of the experimental method for inquiring about nature, proposed by Galileo Galilei and finalized by Isaac Newton in the seventeenth century AD. These two intellectual revolutions have been two "gateway events" (to use a term coined by Nobel prize Murray Gell-Mann[2]). In fact, they have induced profound and fundamental changes in the human methodology of gaining insights about nature. The two mentioned "gateway events" have been cultural facts, and as such, they have not been abrupt, but gradual outcomes of slow evolutionary intellectual processes. Once across these gateways, scientific inquiry has never been the same again. These two events split out the scientific journey into three main stages (see Figure 1.3). The first stage is the period that preceded the birth of philosophy. It can be named as the "Practical

---

[2] Murray Gell-Mann (1994) coined the term "gateway event" to indicate a change opening up a new kind of a phase space with a huge increase in kinds and levels of complexity in system's dynamics. Once through the gateway, life is never the same again. For example, new technology is often an economic gateway event because it expands the production possibilities frontier.

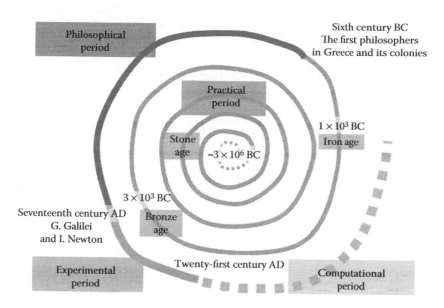

FIGURE 1.3 The humankind journey to discovering the secrets of nature represented as a spiral partitioned in four periods. First, the "Practical Period" started with the appearance of humankind on earth, about 3 million years ago. Then, the birth of philosophy in sixth century BC initiated the "Philosophical Period." The formulation of the experimental method began the "Experimental Period" in the seventeenth century AD. Finally, in the twenty-first century, we are waiting for the next gateway event to enter the "Computational Period" and deeply understand Complexity.

Period" because humans, spurred by their necessities, were particularly ingenious in making arti-facts for solving practical problems. Unconsciously, they obtained the first important achievements in the technological development. The second stage is named as the "Philosophical Period," because authentic Philo-physicists started to investigate nature and human thoughts by using philosophical reasoning. The rigorous and systematic use of experiments as a method of inquiry into nature began only in the seventeenth century AD, opening a new stage that we can name as the "Experimental Period." We are still living it. Three hundred years of productive scientific investigation and aston-ishing technological development have elapsed. Nevertheless, we still experience strong limita-tions on our attempts to exhaustively describe systems, such as the climate and the geology of our planet; the living beings; the human brain; the human immune system; the ecosystems on earth; the human societies and the global economy. These are called Complex Systems. We are aware that the traditional scientific methodologies, the available theories, and computational tools are not enough to deeply understand and predict the behavior of Complex Systems. Therefore, we expect that more efficient algorithms, brand-new computing machines, and probably new methodologies and theories are needed. Do we need to study Complex Systems? Of course, yes. In fact, if we suc-ceed to comprehend Complex Systems, we will surely possess new strategies and effective tools to tackle the Natural Complexity Challenges. The Natural Complexity Challenges are: (1) predicting catastrophic events on our planet (like earthquakes or volcanic eruptions) to save lives; (2) defeat-ing diseases that are still incurable (such as glioblastoma, diabetes, HIV, etc.); (3) protecting our environment and ecosystems from climate change and the risk of shrinking biodiversity; (4) guar-anteeing worldwide sustainable economic growth, primarily by focusing on the energy issue; and (5) ensuring stability in our societies.

After reading this book, it will be evident that for the comprehension of Complex Systems we probably need a new intellectual gateway event. This third gateway event will spark a new stage in the scientific journey to discovering the secrets of nature. It seems plausible that a proper name for the expected next and fourth stage is "Computational period" (see Figure 1.3). In fact, whenever we face the description of Complex Systems, or we tackle the Natural Complexity Challenges, we need to collect and process a vast amount of data, the so-called Big Data (Marx 2013).

In the next three paragraphs of this chapter, I present just a few relevant historical events and achievements in the first three stages of that exciting experience, which is the human journey to discovering the secrets of nature. Then, I outline what we expect in the fourth stage.

1.1.1 THE "PRACTICAL PERIOD"

The first stage of the scientific journey begun, of course, with the appearance of humankind on earth, a few millions of years ago. The early humans had to face many practical problems to survive, such as those of retrieving food supplies, defending themselves in dangerous situations, and resting in safe shelters. It is reasonable to think that, in the beginning, our ancestors used to pick fruit and vegetables for eating and collect available tools made of stones, wood, leaves, bones, and leather for hunting and making shelters and clothing. Then, humankind took a giant step in the development of Physical Technologies by making artifacts. Physical Technologies are the methods and tools for transforming matter, energy, and information from one state into another for specific goals (Beinhocker 2007). Direct evidence of first artifacts is found in the archeological records. The earliest appearances of toolmaking are the crude, flaked-stone hand axes found in Africa and date back to, at least, 3.3 million years ago (Hovers 2015). Over time, tools became lighter, smaller and more heavily modified, suggesting a trend towards greater technological sophistication (Shea 2011). The technical improvements were promoted by careful observation of the surrounding environment; processes of trial and error; serendipity[3] and formulation of inductive rules of thumb. Every breakthrough was presumably transmitted to children and peers, at first by grunts and body language and then by formulating spoken languages. The invention of languages promoted the development of Social Technologies that are the methods and tools for organizing people in pursuit of goals (Beinhocker 2007). In fact, without language, the spread of knowledge and instructions is highly inefficient. Language makes it easier for people to live in large groups, helps the build-up of complex belief systems, establishes laws and theories over several generations (Szathmáry and Számadó 2008). It was crucial to teach how to spark a fire by friction of adequately selected materials, grow a plant from a seed, or establish symbiotic relationships with dogs, sheep, goats, and other animals. The domestication of plants and animals triggered the first agricultural revolution during the Neolithic period of the Stone Age.[4] It determined the transition from hunting and gathering to agriculture and settlement. The Neolithic revolution favored the development of sedentary societies based on villages and towns, which radically modified their surrounding environment using irrigation and food storage technologies. The size of human groupings rose significantly. Cooperation began to extend beyond clans of family members. The Neolithic revolution determined consistent improvement in both Physical and Social Technologies and brought about the production of a surplus of foodstuffs. The food in excess was stored and carried by stone and wicker containers first, then ceramic vessels. The first ceramic containers were produced by drying clay under the sun; then the fire was used to "cook" them. Clay was more functional than the other materials. Our ingenious ancestors devised the wheel to build better vessels more quickly. The principle of the wheel was also applied to the cart, and this novelty made transport and transfer of supplies easier. There is evidence that around 5000 BC, not only stones but even metals, such as gold and copper, found

---

[3] The term serendipity come from the Persian fairy tale *The Three Princes of Serendip*, whose heroes "were always making discoveries, by accidents and sagacity, of things they were not in quest of". An inquisitive human mind can turn accidents into discoveries.

[4] The Stone Age is the first period of a Three-Age System proposed by the Danish archeologist Christian Jürgensen Thomsen (1788–1865) for classifying ancient societies and prehistoric stages of progress. After the Stone Age, the Bronze and Iron Ages followed.

in their native states, were used to obtain beautiful jewels and tools (Brostow and Hagg Lobland 2017). The malleability of gold allowed it to be formed into very thin sheets, and its early uses were only decorative (Boyle 1987). Initially, copper was chipped into small pieces, then hammered and ground with techniques similar to those used for bones and stones. However, when copper is hammered, it becomes brittle, and it easily breaks. Maybe, by serendipity or after many attempts, someone found out the solution to the fragility of copper by annealing it, or dropping pieces of it in a campfire and then hammering them. Maybe, again by serendipity, ancient potters, whose clay firing furnaces could reach very high temperatures (around 1100°C–1200°C), discovered that from certain green friable stones (i.e., malachite) copper smelting was possible. This event initiated the metallurgical technique. When different ores were blended in the smelting process, it created alloys, such as bronze, which flowed more smoothly than pure copper, was stronger after forming, and was easy to cast. Bronze was much more useful than copper as farm implements and weapons. A further improvement was gained around 1000 BC with the introduction and manipulation of iron.

As you notice from Figure 1.3, the first stage of the human scientific journey was very long, lasting a few millions of years. During this extended period, humankind developed the written language. The need of writing came from trade, particularly active since the 3000 BC, in the so-called Fertile Crescent.[5] The trade of goods also promoted the formulation of the first mathematical principles. In fact, we know that Egyptians started to make some operations of calculus in the 2nd millennium BC, for measuring amounts of foodstuffs and partitioning goods among people (Merzbach and Boyer 2011). Similarly, geometry was formulated for facing practical goals, such as measuring fields after the periodic flooding of Nile or designing and building pyramids. Babylonians gained some astronomical insights always for practical purposes, such as traveling across deserts and seas and dispelling fears by making predictions and horoscopes.

1.1.2 The "Philosophical Period"

A revolutionary intellectual event occurred in the sixth century BC in Greece: the birth of philosophy. The word philosophy derives from the Greek "φίλος σωφία" that means "love of wisdom." It refers to the attempts at answering foundational questions about the universe and the role of human beings in it through reflection. In fact, the philosophical answers are searched by reasoning, i.e., by using rational logic (λόγος) of the human mind (Snell 1982). The first philosophers were active in the oriental Greek colonies, in Asia Minor; then, in occidental Greek colonies, i.e., in the South of Italy, and finally, they spread also in the motherland. In Greece and its colonies, the right cultural, religious, socio-economic, and political conditions existed for the birth of philosophy (Reale 1987). The Homeric poems, *The Iliad* and *The Odyssey*, along with the poems written by Hesiod, such as the *Theogony* and the *Works and Days*, and the gnomic poetry, all composed between the late eighth and the sixth century BC, have been the cultural roots of the Greek philosophy. They molded some of the intellectual and spiritual pillars of ancient Greece (see Figure 1.4). These pillars are (Jaeger 1965): (1) love of harmony and right proportion; (2) etiological vision of reality, consisting of the search for the causes of any event; (3) the will of having a global vision of reality; and (4) justice as the supreme value. Other spiritual pillars were built on the religious beliefs of ancient Greece. When we talk about Greek religion, we must distinguish the public religion from the mystery religions, such as the Orphism (Reale 1987). In the public religion, gods are natural forces or human features embodied in idealized human beings. For example, Poseidon was the god of the Sea; Zeus was the god of sky and thunder; Athena was the goddess of wisdom and Aphrodite was that of love.

---

[5] The Fertile Crescent is a term proposed by the archaeologist James Henry Breasted at the University of Chicago in 1906. It refers to a crescent-shaped region located in Western Asia, including Mesopotamia, Syria, Jordan, Israel, Lebanon, the West Bank, and Egypt. It was a region so fertile to be defined the cradle of civilization. It saw the development of many of the earliest human civilizations and the birth of writing and wheel.

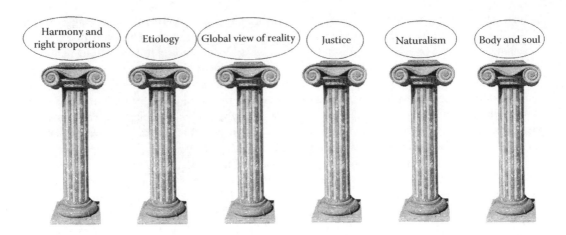

FIGURE 1.4   Intellectual and spiritual pillars inspiring Greek philosophy.

The Greek public religion was a form of "naturalism." It was spurring humans to follow their nature and was considering the bodily death as the final stage of the existence for every person. The public religion was not considered as satisfactory by many Greeks, who turned to the mystery religions. One of these was Orphism that introduced the relevant concept of the human soul as divine and immortal. Human souls were doomed to live in a "grievous circle" of successive bodily lives. The release from the "grievous circle" was guaranteed by an ascetic way of life (repressing some instinctive natural tendencies) along with secret initiation rites. Greek religion did not have holy books, conceived as the product of divine revelation. The poets were the vehicles of diffusion for their religious beliefs and a sacerdotal class, keeper of dogmas, did not exist. This lack of dogmas and their guardians left wide berth for free intellectual inquiry. The ancient Greeks, especially those living in the colonies, also experienced important political freedom. In the seventh and sixth centuries BC, Greece had a considerable socio-economic transformation. It changed from a country based predominantly on agriculture to one where the craft industry and commerce developed to an ever-increasing extent. The new class of tradesmen and artisans gradually gained considerable economic strength and opposed the centralization of the political power, which was in the hands of the landed aristocracy. The old aristocratic forms of government were transformed into new republican forms. The City-States were established and consolidated; the Greek man was led to feel like a citizen.

In synthesis, the cultural, religious, socio-economic, and political environments of Greece in the seventh and sixth centuries BC created the right conditions for the birth of philosophy, which is a great revolution in the history of human thinking. The first philosophers had the merit of raising "Really Big Questions" (RBQs)[6] mainly regarding nature and the origin ($\alpha\rho\chi\acute{\eta}$) of everything. They were named "natural philosophers" due to their interests. They endeavored to answer their RBQs through rational and logical methods, particularly by intuition and induction. Their intuitions were mere intellectual formulations of principles. Their inductions were based on all the information gathered by their senses[7] along with the practical and technical knowledge stored by their ancestors. For sure, some of the first philosophical RBQs were also arisen by our ancestors, during the "Practical Period." Answers were found through the myth, exploiting fantasy and religious beliefs. On the other hand, philosophers proposed solutions to their RBQs, trusting just in their minds,

---

[6] The expression "Really Big Questions" was coined by the physicist John A. Wheeler (1911–2008), who, in his career, formulated questions so broad and important to involve both physics and philosophy.

[7] Some philosophers, such as Melisso from Samo (fifth century BC), denied any validity to the information gathered by senses, trusting only in mind.

through reasoning purified from any superstition and prejudice. In some cases, they suggested solutions and answers, which appear astonishing even nowadays. In fact, they are still valid in their essence. For instance, Pythagoras and his disciples, between the sixth and fifth century BC, realized that the key to understanding nature is mathematics. Everything in the universe is harmony and number. A mathematical regularity exists everywhere.[8] Empedocles and Anaxagoras (fifth century BC) formulated what is known today as the principle of conservation of mass, stating that nothing comes from nothing, and nothing can be utterly destroyed. Leucippus and Democritus (fifth to fourth century BC) guessed that everything is composed of atoms and between atoms lies empty space; atoms are indestructible; atoms have been and will always be in motion. Moreover, there is a huge number of atoms and kinds of atoms.

With the birth of philosophy, the human journey to discovering the secrets of nature undertook a second stage named the "Philosophical Period" (see Figure 1.3). Within the scope of philosophical mentality, i.e., within the scope of its etiological rationalism (prone to search for the cause of everything by reasoning), the knowledge of nature specialized, thus separating in different disciplines. These disciplines were mathematics, geometry, astronomy, geography, and medicine. The philosophers developed mainly theoretical aspects of the various scientific disciplines, overlooking their technical and practical aspects.[9] Through induction and intuitions, they formulated postulates and axioms of the different scientific disciplines. For example, Euclid wrote the *Elements* (around the 300 BC), grounding geometry and mathematics. Archimedes (third century BC) wrote *the Equilibrium of planes*, providing the theoretical basis of statics. Ptolemy (second century AD) composed a comprehensive treatise on astronomy of his time, titled the *Almagest*, and Galen (third century AD) wrote the *On the Elements according to Hippocrates* exerting an important influence over the theory of medicine until the mid-seventeenth century AD. The scientific knowledge maintained its speculative-theoretical facet during the first part of Middle Ages. From the mid-eighth century AD until the mid-thirteenth century, the Arabic culture, fed by both the ancient Greek and Latin texts along with the Chinese and Indian intellectual sources, blossomed into its Golden Age. In the Islamic Golden Age, there were some Muslim Philo-physicists who boosted the scientific inquiry. A famous example is Ibn al-Haytham (also known under the name of Alhazen, 965–1040 AD) whose main contribution was that of placing, for the first time, a particular emphasis on experiments. Experiments are the ultimate ways for choosing between scientific theories that are under debate.[10] Through this brand-new approach, he made significant advances in the field of optics. The contact with the Islamic culture favored an intellectual revitalization of Europe. The revival started in the twelfth century and was sealed by the birth of the first universities. In medieval universities, students were learning the liberal arts, which were grammar, rhetoric, and logic (called the Trivium) along with mathematics, geometry, music, and astronomy (called the Quadrivium). These disciplines were fundamental for a free citizen to study, and they were mandatory to gain access to the higher faculties, that is law, medicine, and theology. Among the many important scholastics, some were distinguished for their thinking about nature. Albertus Magnus (twelfth century), Robert Grosseteste (twelfth century), Roger Bacon (thirteenth century), and finally William of Ockham (fourteenth century), all partly influenced by the Islamic culture, triggered a paradigm shift in the scientific inquiry. They underlined

---

[8] You may ask yourself if mathematics is either an invented ensemble of tools to be continuously improved or something real to be discovered. In my view, the best answer was proposed by Aristotle: the mathematical objects are neither real nor unreal. They exist in potentiality in nature, and our reason catches them by abstraction. The mathematical tools exist just inside our minds.

[9] Of course, medicine could not be just an intellectual discipline because it always had to cure people of illnesses. However, in the "Philosophical Period," medicine improved also from a theoretical point of view with the introduction of etiological explanations of diseases.

[10] Some historians have described Ibn al-Haytham as a pioneer or "the first scientist" of the contemporary scientific method. He established the experiments as proofs of scientific hypotheses. As Gorini (2003) said, "his investigations were based not on abstract theories, but on experimental evidence and his experiments were systematic and repeatable."

the importance of mathematics for understanding nature and stressed that the proof of any scientific acquaintance should come from real experiments. This adhesion to the concrete evidence steered William of Ockham to refuse any metaphysical hypostatization, i.e., abstraction of concepts such as space, time, motion, et cetera. He introduced the principle of parsimony, well known as Ockham's Razor: *Entia non sunt multiplicanda praeter necessitatem* ("entities must not be multiplied beyond necessity"). This principle recommends Philo-physicists to not to postulate unnecessary entities and, among competing hypotheses, to select the one that makes the fewest new assumptions until evidence is presented to prove it false. During the first part of Renaissance (on the whole spanning about two centuries, the fifteenth and the sixteenth), there was the polymath Leonardo da Vinci (1452–1519) who went on the new way traced by Ockham and the other scholastics cited earlier. Leonardo insisted on the idea that just mathematics allows for the interpretation of the mechanical and necessary order of nature. Moreover, he weeded out any animistic, mystic, and spiritual forces from empirical events. Finally, he gave a significant contribution to the next scientific revolution by bringing mechanical[11] and liberal arts to the same level of cultural dignity.

1.1.3 The "Experimental Period"

Two thousand, two hundred years elapsed before witnessing the second gateway event in the human journey to discovering the secrets of nature. This second gateway event was the formulation and application of a mature experimental method. It took place in the Scientific Academies that blossomed in Italy during the seventeenth century and then spread to the rest of Europe.[12] In the Scientific Academies, figures as diverse as natural philosophers and craftsmen started to collaborate by merging theory and real experiments. Together, they devised "exo-somatic" tools bringing great benefits. In fact, instruments (1) extend the frontiers of human knowledge about nature, otherwise delimited by the investigating power of our senses. (2) They avoid misunderstandings which could sometimes derive from a blind trust on our sensorial perceptions. Finally, (3) they gain objective, reproducible and universally valid responses from nature. With the instruments in hand, natural philosophers and craftsmen could establish highly constructive dialogues with nature. They were asking nature if it obeys their hypothesized theories and laws. If nature repeatedly and unequivocally assented, the laws and theories were validated; otherwise, new ideas and models were needed. The experimental methodology was first theorized and applied by Galileo Galilei (1564–1642) and then supported and completed by Isaac Newton (1642–1727). Thanks to the great contributions of Galilei and Newton, the "Experimental Period" begun (see Figure 1.3). According to the Hegel's dialectic,[13] this third period can be conceived as the synthesis of the two previous stages: the "Practical Period," which was the thesis, and the "Philosophical Period," which was the antithesis (see Figure 1.5).

During the "Experimental Period," theory and practice walked hand in hand. Usually, natural philosophers used to formulate a question and a possible answer. Then, they, along with artisans, were designing experiments by devising suitable and reliable facilities. To collect unequivocal and reproducible answers from nature about the validity of their hypothesis, the team of authentic Philo-physicists used to "purify" the phenomenon, which they wanted to analyze, by isolating it from the rest of the world. Moreover, natural philosophers started to describe the natural phenomena by the

---

[11] Mechanical arts are activities requiring manual skills rather than only mental abilities.

[12] The first academy focused exclusively on scientific knowledge was the Accademia dei Lincei founded in Rome in 1603. In 1657, Prince Leopoldo of Tuscany, student and friend of G. Galilei, founded the Accademia del Cimento in Florence. In 1662, Charles II of England created the Royal Society of London for the Improvement of Natural Knowledge, whose Isaac Newton was first member and, then, secretary. This academy promoted the publication of the *Philosophical Transactions*, which is the first example of periodic journal regarding scientific subjects published in Europe. Under the reign of Louis XIV, the minister Colbert founded the Académie Royale des Sciences. Many other academies were born during the eighteenth century.

[13] Georg W. F. Hegel (1770–1831) was a German philosopher.

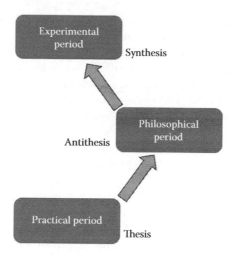

FIGURE 1.5 The first three stages of the humankind journey to discovering the secrets of nature analyzed through the lens of Hegel's dialectic.

universal language of mathematics and geometry. As Galilei stated in his book titled *The Assayer* (1623 AD), the universe "stands continually open to our gaze, but it cannot be understood unless one first learns to comprehend the language and interpret the characters in which it is written. It is written in the language of mathematics, and its characters are triangles, circles, and other geometrical figures, without which it is humanly impossible to understand a single word of it; without these, one is wandering around in a dark labyrinth."

The new methodology to understand the secrets and the marvels of nature, proposed in the Academies, brought about revolutionary discoveries. During the seventeenth century, the first relevant results were achieved in astronomy. Nicolaus Copernicus (1473–1543), Tycho Brahe (1546–1601), Johannes Kepler (1571–1630), Galileo Galilei were provided with accurate sextants,[14] quadrants, armillary spheres, and telescopes to study our Solar System. They discovered that the Solar System is heliocentric and not geocentric. Moreover, the orbits of the planets are elliptical and not circular, as believed before. Finally, the heavenly bodies comply with the same physical laws as the terrestrial bodies. Therefore, the planets can stay in their orbits without being fixed to solid spheres, just because they interact through the gravitational force, as Newton inferred. Newton (1687) wrote a book titled *Philosophiae Naturalis Principia Mathematica*, which is considered "as one of the masterpieces in the history of science."[15] In the *Principia*, Newton laid out the foundations of what is nowadays known as the "Classical Physics." He formulated the laws governing the physical behavior of macroscopic bodies. Moreover, he invented calculus to rigorously describe change and motion, through new mathematical notions such as infinitesimal, derivative, integral, and limit. Finally, he formulated the four "Rules of Reasoning in Philosophy,"[16] which became the foundations of two important "epistemological pillars." Epistemological pillars are platonic ideas guiding the interpretation of natural phenomena and the formulation of axioms and postulates. The first epistemological pillar is "Simplicity:" "Nature loves Simplicity." Therefore, the truth is always to be found in simplicity. It resembles the Ockham's Razor. The idea of a "Simple Nature" inspired

---

[14] Sextants and quadrants are instruments to measure angles; the armillary spheres were models of the solar system to demonstrate how it works.

[15] Assessment extracted from "Reading the Principia: The Debate on Newton's Mathematical Methods for Natural Philosophy from 1687 to 1736" by N. Guicciardini.

[16] Remember that with the term "Philosophy," Newton meant what we nowadays define "Science." The term "scientist" was coined by William Whewell (1834) to indicate all those figures who dedicated their lives to the study of nature by using the austere and rigorous experimental method (since the seventeenth century AD).

the reductionist approach in the scientific inquiry. Such an approach consists in describing a natural system by decomposing it in its constituents and studying their properties, singularly. Finally, the picture of the entire system can be reconstructed as a simple sum of the features of its elements.

The second epistemological pillar is "Uniformity:" "Nature is Uniform." The natural laws, which are valid *hic et nunc* ("here and now"), are true always and everywhere in the universe: they are "Universal." The idea of Uniformity, also known as Uniformitarianism, is at the core of any scientific discipline, but in particular of geology. In fact, as proposed by the Scottish geologist James Hutton (1726–1797), the same natural laws that rule the processes in the universe now, have always been in action in the past and everywhere in the universe.

During the eighteenth and nineteenth centuries, the classical mechanics, formulated first by Newton, was further developed and improved to such an extent that the reliance on it was almost absolute. The confidence in the simple laws of classical physics favored the establishment of two further epistemological pillars: the "Determinism" and the "Mechanism" (see Figure 1.6). The Determinism is well epitomized by the statement written by Pierre-Simon Laplace[17] in his *A Philosophical Essay on Probabilities* (1814 AD): "We may regard the present state of the universe as the effect of its past and the cause of its future. An intellect which at a certain moment would know all forces that set nature in motion, and all positions of all items of which nature is composed, if this intellect were also vast enough to submit these data to analysis, it would embrace in a single formula the movements of the greatest bodies of the universe and those of the tiniest atom; for such an intellect, nothing would be uncertain, and the future just like the past would be present before its eyes." In other words, Laplace was advocating that since the natural laws are deterministic and known, if we could determine position and momentum of every particle in the universe at a specific moment in time, we would be able to predict any subsequent event. The future is potentially predictable.

The fourth pillar, the Mechanism, sustains that everything in the universe, either inanimate or animate, behaves like a machine. Also "vital" phenomena, like passion, memory, and imagination "follow from the mere arrangement of the machine's organs every bit as naturally as the movements

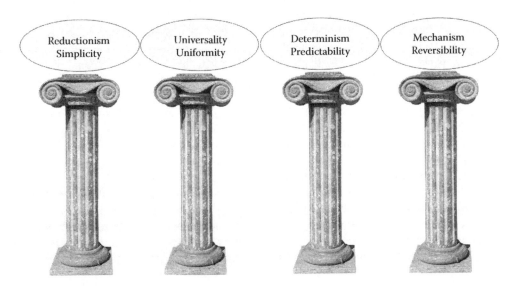

FIGURE 1.6   Four epistemological pillars guiding and inspiring the scientific inquiry during the Experimental Period.

---

[17] Pierre-Simon, marquis de Laplace (1749–1827) was a French mathematician and astronomer whose work was crucial for the development of mathematical astronomy and statistics. He formulated the Laplace's equation and the Laplace's transform. The widely used Laplacian differential operator is also named after him.

of a clock or other automaton follow from the arrangement of its counter-weights and wheels," as René Descartes stated in his *Treatise of Man* in 1664. All activities and qualities of bodies are reduced to quantitative realities, i.e., mass and motion. From this idea, a picture of a universe hosting just reversible transformations emerges. All motions are reversible: when a moving object has covered the distance from A to B, we at once imagine that it can go back over the path from B to A. If, therefore, everything that happens is motion, it is clear that any event in nature should occasionally retrace its march.

The picture of a reversible universe faded during the Industrial Revolution, in the nineteenth century AD. During this century, there was a change from an agrarian, handicraft economy to one dominated by industry and machinery manufacture. The entrepreneurs of that time tried in any way to optimize the functioning of their thermally or electrically powered machines. They clashed with the impossibility of converting all the available heat into work, and they could not avoid the degradation of part of the mechanical or electrical energy into heat. The industrial processes like the most events in the Universe we stumble across are irreversible: energy is continuously squandered into heat. In the original scientific method proposed by Galileo and Newton and consisting in a "sublimation" of empirical phenomena before their mathematical description and interpretation, the irreversible feature of natural events was simply ignored. To interpret the empirical irreversibility, a new theory was formulated: The Thermodynamics. Its Second principle introduces a new state variable, Entropy,[18] and it asserts that the Entropy of the Universe increases relentlessly due to the irreversible processes. After the formulation of Thermodynamics, the epistemological pillar of Mechanism was still maintained, but in a reviewed form. It is still valid that the Universe looks like a machine, but in most of cases, it works irreversibly due to frictions and other energy dissipation processes.

The existence of an "Arrow of Time" in nature was corroborated by biology with the development of the Theory of Evolution. Such theory is attributed to Charles Darwin (1809–1882), but it was "in the air" in the nineteenth century (Mitchell 2009). Most likely, life on earth comes from one form, called LUCA, i.e., the Last Universal Common Ancestor, which evolved in the myriad life forms we know, nowadays. Since "there is a frequently recurring struggle for existence, it follows that any being, if it varies however slightly in any manner profitable to itself, under the complex and sometimes varying conditions of life, will have a better chance of surviving, and thus be naturally selected. From the strong principle of inheritance, any selected variety will tend to propagate its new and modified form" (as stated by Darwin in his famous book,[19] *On the Origin of Species*, published in 1859). Evolutionary processes, governed by natural selection, have given rise to the diversity of species. Life on earth has its history: the simple early forms transformed in always more various, beautiful and marvelous ones (paraphrasing Darwin) along the millennia.

In the first half of the twentieth century AD, new scientific theories blossomed. They brought about a reworking of two of the pillars shown in Figure 1.6. The first new theory was formulated by Albert Einstein[20] (1879–1955). The Theory of Relativity (Einstein 1916) describes the behavior of bodies moving at very high speed, close to that of light propagating in empty space at 3×10^8 m/s (that means at 1.1×10^9 Km/h). According to the Relativity theory, the description of mechanical

---

[18] The new function Entropy was introduced by Rudolf Clausius (1822–1888), a German physicist and mathematician, in 1865. Entropy derives from the Greek "έν τροπή," having the meaning "in transformation" or "in change."

[19] The full title of the Darwin's book was *On the origin of Species by Means of Natural Selection, or the Preservation of Favoured Races in the Struggle for Life*. In this book, Darwin exposed his ideas inferred after his long scientific expedition around the world made on board of the *Beagle*. The content of Darwin's book was as much revolutionary as the *De Revolutionibus orbium coelestium* by Copernicus. Copernicus reorganized the spatial order of things and conferred to the earth, and hence to human beings, a position that is not the center of "Universe." On the other hand, Darwin reorganized the temporal order of life on earth and overthrew human being from his supposed prominent position.

[20] Albert Einstein may be considered one of the utmost Philo-physicists of the twentieth century. Time Magazine has defined him as the "Person of the 20th century," "unfathomably profound-the genius among the geniuses."

phenomena depends on the properties of the reference system. Although the natural laws are still universal, time is no longer uniform and absolute as Newton thought. Its flow depends on (1) the velocity of the reference system[21] and (2) the strength of the gravitational field. Even the lengths of objects change in relation to the motion[22] and have been predicted to contract in gravitational fields. As soon as the Theory of Relativity was experimentally confirmed, the pillar of Universality was permanently weakened with respect to the power it had before: the features of space and time are context dependent. A universal space and a universal time do not exist. What exists is a spacetime that can be curved by gravitational fields.

The second revolutionary scientific theory formulated in the first half of twentieth century was Quantum Mechanics. It describes the behavior of microscopic bodies, such as molecules, atoms, and subatomic particles moving at ordinary velocities. Its foundations were established by many scientists: Max Planck (1858–1947), Niels Bohr (1885–1962), Werner Heisenberg (1901–1976), Louis de Broglie (1892–1987), Erwin Schrödinger (1887–1961), Max Born (1882–1970), Albert Einstein, John von Neumann (1903–1957), Paul Dirac (1902–1984), Wolfgang Pauli (1900–1958), David Hilbert (1862–1943), among others. Monumental discoveries in its early development were the concept of "Quantization" of physical variables and the "Uncertainty Principle." The latter claims a fundamental limit on the accuracy with which certain pairs of physical properties of a particle can be simultaneously determined. For example, position and momentum. It means that it is impossible to simultaneously measure the present position while also determining the future motion of a particle with unlimited accuracy. This Principle struck a hard blow to the epistemological pillar of Determinism and Predictability. Laplace's dream of predicting the future through the laws of physics and the determination of the instantaneous state of all the particles in the Universe was shattered irreparably. In the natural world, particles normally remain in an uncertain, non-deterministic "blurred" path and their evolution can be predicted just in probabilistic terms.

The dream of predicting natural phenomena was shattered even earlier, at the end of nineteenth century, and for macroscopic bodies, by Henri Poincaré (1854–1912).[23] He found out that a system as simple as that constituted by three orbiting planets exhibits a dynamic that is aperiodic and extremely sensitive to the initial conditions. In essence, Poincaré was the first Philo-physicist to experience Deterministic Chaos. In fact, Deterministic Chaos appears when a deterministic system (whose dynamics can be described by differential equations) exhibits aperiodic behavior that depends sensitively on the initial conditions, thereby rendering long-term prediction impossible. "If we knew exactly the laws of nature and the situation of the universe at the initial moment, we could predict exactly the situation of that same universe at a succeeding moment. But even if it were the case that the natural laws had no longer any secret for us, we could still only know the initial situation approximately," as Poincaré (1908) stated in his essay "Science and Method." In fact, the instrumental determination of the value of any variable is affected by a certainly unavoidable uncertainty. Therefore, as Poincaré continues, "it may happen that small differences in the initial conditions produce very great ones in the final phenomena. A small error in the former will produce an enormous error in the latter. Prediction becomes impossible, and we have the fortuitous phenomenon." With the advent of the electronic computers in the 1950s, the solution of nonlinear differential equations, even much more complex than that formulated by Poincaré, could be determined numerically and in a reasonable time. Therefore, many computational experiments were performed. One of these was particularly relevant. It was carried out by the meteorologist

---

[21] The duration of a phenomenon is longer in a moving system than in one at rest; moreover, the simultaneity of two events depends on the motion of the observer.

[22] At speed close to that of light, objects appear to be smaller than what they appear as when stationary or moving at ordinary speeds.

[23] Henri Poincaré was a French mathematician, theoretical physicist, engineer, and philosopher of science. He was indeed a polymath and was the first person to discover a chaotic deterministic system which laid the foundations of modern chaos theory.

Edward Lorenz (1917–2008). He formulated a simplified model of the dynamics of the terrestrial atmosphere to make weather forecasts. He described the dynamic of the terrestrial atmosphere by a system of three non-linear differential equations and re-discovered deterministic chaos. Lorenz found that if he started his simulations from two slightly different initial conditions, the resulting dynamical behavior would soon become completely different. This result implies that the weather is intrinsically unpredictable: tiny uncertainties in defining the initial conditions of the atmosphere are amplified rapidly, eventually leading to embarrassing forecasts. It was coined the "Butterfly Effect" to refer to this idea of sensitive dependence on the initial conditions for the dynamics of non-linear chaotic systems.

The first half of twentieth century was dramatic not only for natural sciences but also for mathematics. In the year 1900, at the International Congress of Mathematicians in Paris, the German mathematician David Hilbert (1862–1943) made a list of unsolved problems in mathematics. The most important ones were those regarding mathematics itself and what can be proved by using mathematics. They can be summarized in three fundamental questions (Mitchell 2009): (1) Is mathematics complete? (2) Is mathematics consistent? (3) Is every statement in mathematics decidable? In the nineteenth century, mathematics was dominated by the axiomatic methodology. According to the axiomatic approach, any branch of mathematics must start from the formulation of a series of fundamental assumptions, the axioms, and then generate all relevant statements by logical deduction. Therefore, the concept of "truth" is reduced to the concept of "provable from the axioms." Of course, the success of this methodology depends on the ability to formulate "good axioms." A mathematical statement is acceptable as axiom if it is rather simple and elementary enough to be considered "obviously true" (Devlin 2002). Mathematics is complete if every mathematical statement can be proved or disproved from a given finite set of axioms. Mathematics is consistent if only the true statements can be proved. Finally, every statement in mathematics is decidable if there is a definite procedure that tells us in finite time whether a statement is true or false. Until 1930, the three fundamental questions mentioned earlier remained unanswered, but Hilbert was confident that the answers would be three "yes." In fact, Hilbert and many others were convinced they were on the verge of discovering an automatic way to prove or disprove any mathematical statement. However, Hilbert, along with the entire community of mathematicians and philosophers, remained astounded when a twenty-five-year-old mathematician named Kurt Gödel (1906–1978) presented a proof of the so-called Incompleteness Theorem.[24] This theorem states that if mathematics is consistent, then it is incomplete and there are true statements that cannot be proved. If it were inconsistent, then there would be false statements that could be proved, and the entire building of mathematics would crash down.[25] Therefore, the answer to question (1) is "no." But the surprises had not ended, here. In fact, in 1935, the twenty-three-year-old Alan Turing (1912–1954), a graduate student at Cambridge, demonstrated that the answer to question (3) mentioned earlier, was "no," again. Turing invented the computing machine that could solve any computational problem for which an algorithm could be devised. He found that there are mathematical problems that cannot be solved by his machine (known as the Turing Machine), and it follows that any other automatic computer cannot solve these problems. They are problems for which algorithms cannot be written, even in principle. The example studied by Turing was the problem of predicting whether the Turing Machine, once it is set in motion, will ever finish its calculation and halt. By analyzing this problem, Turing was able to demonstrate that there can be no general procedure for telling whether mathematical propositions

---

[24] Gödel used arithmetic to demonstrate his "Incompleteness Theorem."

[25] The proof given by Gödel is complicated, but it can be intuitively understood by saying (Mitchell 2009): "This statement is not provable." Let us call this sentence "statement S." Now, let us suppose that "statement S" could indeed be proved. But then it would be false because it states that it cannot be proved. That would mean that a false statement could be proved, and mathematics would be inconsistent. If we assume that "statement S" cannot be proved, then it would mean that "statement S" is true. The result is that mathematics is consistent, but it would be incomplete because there is a true statement that cannot be proved. Therefore, mathematics is either inconsistent or incomplete.

are true or false. A result of Turing's work was to partition all possible mathematical problems into two sets. One set contains all those problems for which algorithms can never be written; they are unsolvable. The other set includes all those problems that can be solved by algorithms (Lewis and Papadimitriou 1978). Some of the solvable problems are tractable, but others are intractable because they cannot be solved accurately and in a reasonable time.

Just as Quantum Mechanics and Chaos Theory shattered Laplace's dream of the unlimited predictive power of science, Gödel's and Turing's results shattered Hilbert's dream of the unlimited computing power of mathematics.

In the last decades of the twentieth century, among the epistemological pillars depicted in Figure 1.6, the only one that was not debunked by the evolution of the scientific theories was that sustaining the Simplicity of nature, which inspired the reductionist approach. The reason is the kind of systems that had been investigated until then. In the seventeenth, eighteenth, and nineteenth centuries, scientists faced problems of "simplicity," involving two or a few more variables. In the first half of the twentieth century, scientists also faced problems of the so-called "disorganized complexity" involving billions of variables (Weaver 1948). They developed powerful techniques of probability theory and statistical mechanics to rationalize the behavior of "disorganized complex" systems.[26] In the second half of the twentieth century, scientists turned their attention to problems of "organized complexity." Organized Complexity is a peculiarity of systems that are in out-of-equilibrium conditions and are constituted of many strongly interacting elements.[27] Such systems exhibit emergent properties. Emergent properties are characteristics not directly traceable to the system's components, but rather to how these parts interact together within the entire system as a whole. For instance, a living organism can be dissected in its elementary constituents, biopolymers, and other molecules, but in dissecting it, we kill it, and we cannot learn what gives it life. The out-of-equilibrium systems having emergent properties are examples of what we nowadays refer to as Natural Complex Systems, which are the subject of this book. To fully comprehend how a Complex System works, it is necessary not only to break it into pieces but also to examine it in its entirety. Therefore, even the apparently robust epistemological pillar of reductionism sways when we strive to describe Complex Systems. A Complex System is like an outstanding piece of music or a brilliant work of art. If we listen to a Beethoven's symphony or we look at Raphael's fresco, we can try to understand their beauties by decomposing them in their constitutive elements, which are musical notes and colors, respectively. If the single constituents are taken apart, they are not masterpieces. Only when they are organized uniquely by the musician in time (symphony) or by the painter in space (fresco), musical notes and colors become attractive. The same is true for Complex Systems. Moreover, Complex Systems exhibit as either stationary, periodic, chaotic or stochastic dynamics. Whatever the kind of dynamics, it is usually extremely sensitive to initial and contour conditions. This property imposes substantial limitations on the reproducibility of experiments. Many of the experiments on Complex Systems are historical events that are not reproducible. We can take a beautiful image from the essay "Of clouds and clocks" written by the philosopher Karl Popper (1972) and say that science, in the past, had been occupied with clocks, i.e., simple, deterministic systems having reproducible behaviors. Now, instead, science has to deal with clouds, i.e., Complex Systems, having unique and hardly replicable behaviors (Figure 1.7).

Complex Systems are everywhere: in physics, chemistry, biology, geology, economy, sociology, and so on. To try to describe them, we collect a gargantuan amount of data by monitoring accurately and continuously their dynamics. If we want to handle the Big Data, it is urgent to improve the performances of our electronic computers, contrive more powerful computing machines with brand-new architectures, and formulate more efficient algorithms. Probably, we need a new methodology

---

[26] An example is a gas consisting of an Avogadro's number of particles.

[27] When we say, "strong interactions" we mean "non-linear interactions." In fact, any Natural Complex System is characterized by non-linear feedback loops among its constitutive elements. The consequence is that any Natural Complex System is extremely sensitive to small changes.

FIGURE 1.7 Simple Systems are epitomized by clocks; Complex Systems by clouds.

of scientific inquiry and a new theory, as well. In other words, we expect a third gateway event that should usher us in a new scientific era that will likely be called the "Computational Period."

1.1.4 THE "COMPUTATIONAL PERIOD"

The experimental investigation of Complex Systems presents two significant hurdles. First, experiments are often not reproducible. Second, many of the experiments require long periods of time. Fortunately, there are computational experiments to get around these hurdles. Today, people working in all areas of Complexity use computers to make experiments, test hypothesis, and gain insights when no other routes are feasible. "Really efficient high-speed computing devices may, in the field of nonlinear partial differential equations as well as in many other fields, which are now difficult or entirely denied access, provide us with those heuristic hints which are needed in all parts of mathematics for genuine progress" (von Neumann 1966).

A promising strategy to deal with Complex Systems and face the Complexity Challenges presented in paragraph 1.1 is the interdisciplinary research area of Natural Computing. The fundamental idea is that every natural causal phenomenon is a kind of computation because information is encoded through the states of natural systems. Therefore, every element of the universe works like a computing machine that processes information. Accordingly, the epistemological pillar of Mechanism (see Figure 1.6) assumes another aspect: the universe is a vast computational machine (Lloyd 2006). This new idea brings to the forefront the Information variable. As Murray Gell-Mann stated, "although Complex Systems differ widely in their physical attributes, they resemble one another in the way they handle information. That common feature is perhaps the best starting point for exploring how they operate." We need a new theory. This new theory should allow us to predict the emergent properties of Complex Systems. Through it, it should become evident why living beings have the unique feature of using energy and matter to encode and process information and exploit this information to make decisions. The new theory will probably include the two fundamental attributes of information: the quantity, already rationalized by Claude Shannon (1948), and the quality, i.e., its semantics. It is not an easy task to rationalize semantic information because it is context-dependent. During the industrial age, in the eighteenth and nineteenth centuries AD, entrepreneurs and scientists were engaged in optimizing the performances of thermal machines, and the laws of "Thermodynamics" have been formulated. Nowadays, scientists, philosophers, entrepreneurs, and politicians are all engaged in trying to win the Complexity Challenges, and the laws of an "Info-dynamics" might be developed. Then, perhaps, Complex Systems would become more understandable and their investigation less daunting.

1.2 WHAT IS SCIENCE, TODAY?

If we want to contribute to the formulation of new scientific methodologies and theories, it is fundamental to think how science evolves. If we consider the "Experimental Period," we may say that scientific knowledge has grown like a tree. The principal macroscopic components of a tree are roots, trunk, branches, and leaves. Roots absorb water and nutrients from the soil and feed the rest of the tree. Leaves harvest two ingredients, light from the Sun and carbon dioxide from the atmosphere, to synthesize carbohydrates and permit the tree to grow. In the "Tree of Scientific Knowledge" (see Figure 1.8), the roots are represented by Mathematics. The trunk is made up of Physics and Chemistry. The branches are the other scientific disciplines, which are grounded in Mathematics, Physics, and Chemistry. For instance, Geology that deals with the properties and the transformations of the Earth; Astronomy, whose interests are the properties of the extra-terrestrial objects and phenomena. Another relevant branch of the "Tree of Scientific Knowledge" is Biology, which is the study of that fantastic phenomenon that is life on earth. Medicine and Veterinary Science ramify from Biology, and they focus on the physical health of humans and animals, respectively. Psychology studies human mind and behavior. Social Sciences, based on Biology and Psychology, concern about the meta-biological aspects of human behavior. They include Anthropology, Sociology, and Economics. Anthropology is focused on the definition of human life and its origin. Sociology studies social structures, their organizations, the rules, and processes binding people and their institutions. Economics concerns the exploitation of scarce resources and deals with the organization of productive activities and interchange of goods to best meet individual and collective needs. Based mainly on Chemistry, Biology, and Social Sciences, Agricultural science deals with the use of plants, fungi, and animals for food, fiber, and other products used to sustain life. The application of physical, chemical, social, and economic knowledge to design and build structures, machines, devices, systems, and processes is the scope of Engineering.

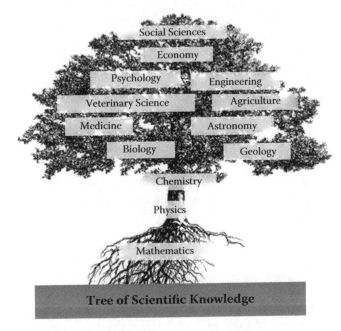

FIGURE 1.8   Schematic structure of the "Tree of Scientific Knowledge."

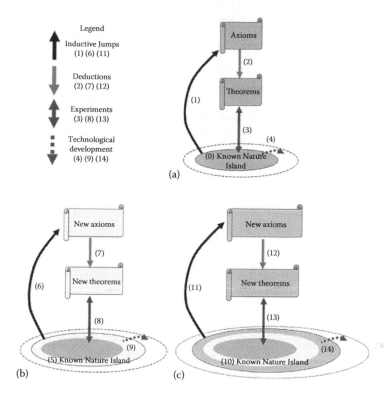

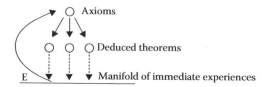

FIGURE 1.9   The mechanism of growth of scientific knowledge.

In the "Tree of Scientific Knowledge," the leaves represent the multitudes of men and women who have been dedicating their lives to the growth of scientific knowledge. Their nutrients are curiosity and the will of improving human welfare, along with the rigorous principles of logic and the acquired scientific notions.

The mechanism of growth of scientific knowledge is represented schematically in Figure 1.9.

Scientists collect data about natural phenomena by using instruments. The sensitivity and resolution of the available facilities define the boundaries of what can be observed from what remains unexplored; they outline the extension of the "Known Nature Island" (see graph [a] in Figure 1.9). From the investigation of the detectable natural phenomena, just a few ingenious scientists can make inductive jumps (represented by the arrow labeled as [1] in the graph [a] of Figure 1.9) and formulate axioms and postulates. Axioms and postulates[28] are the fundamental principles and laws of scientific knowledge. They are created by induction, i.e., by the free invention of the human mind.[29]

---

[28] Both axioms and postulates are assumed to be true without any proof or demonstration. Usually, axioms are evident statements, whereas postulates are not necessarily evident; they are accepted as far as they originate theorems having predicting power.

[29] Albert Einstein defined the inductive jumps as "free creation of human mind" in a letter sent to his lifelong friend Maurice Solovine, in 1952. In this letter, Einstein offered a remarkable description of his idea about the scientific methodology. He drew a diagram in which the functions of induction, deduction, and experience were clearly laid out (see the sketch below and von Bayer 2004).

The first examples of known intuitive figures were Euclid, who formulated the axioms of geometry, and some naturalist philosophers, such as Anaxagoras and Empedocles, who put the principle of mass conservation forward. Further brilliant intuitive minds were Newton and Leibniz (giving birth to the Calculus), Clausius (formulating the definition of Entropy and the Second Law of Thermodynamics to interpret the irreversibility of nature), Darwin (who, after his long trip around the world on board of the *Beagle*, formulated the law of Natural Selection as principle that rules biological evolution), Planck (advancing the idea of Quantization to understand phenomena of the microscopic world), Schrödinger (formulating the postulates of Quantum Mechanics), Einstein (developing the Relativistic theory), and the Polish mathematician Benoit Mandelbrot (1924–2010) (proposing Fractal geometry to describe the shapes of natural objects such as coasts, trees, clouds, ferns, et cetera).

After the inductive jump (1), the next step in the evolution of scientific knowledge is the deduction of theorems and propositions from axioms and postulates (see arrow [2] in the graph [a]). Scientists who deduce theorems are usually named as "theoreticians." The validity of theorems and hence, implicitly, of axioms must be proved by designing and performing suitable experiments. The theoretical predictions must be compared with the results of real experiments (see the arrow labeled as [3] in the graph [a] of Figure 1.9). Experiments are carried out by those scientists we call "experimentalists." The main reasoning strategy of experimentalists is *abduction* consisting of finding out the causes of phenomena, after detecting the effects and considering the rules. The acquisition of new scientific knowledge usually promotes technological development. New technologies often allow deepening the observation of nature. Therefore, science and technology extend the borders of "Known Nature Island" (see step [4] in the graph [a]). In a larger "Known Nature Island" (see graph [b] of Figure 1.9), new phenomena and features become evident. Their interpretation often requires the formulation of new axioms through brand-new inductive jumps (step [6] in the graph [b]). The novel axioms allow the deduction of new theorems and propositions (step [7]). Then, they need to be experimentally confirmed (step [8]). As seen before, the acquisition of new knowledge induces further technological development. More powerful technologies extend our observational capabilities. The "Known Nature Island" becomes even more extensive (step [9]) and new surprising details pop up. For their interpretation, we require a new cycle of intuitions, deductions, and abductions[30] accompanied by a permanent dialogue with nature through experiments (graph [c] of Figure 1.9). And so on, indefinitely.

Now, at the beginning of the twenty-first century, we are scrutinizing the behavior of Complex Systems, hoping that someone will make a fruitful inductive jump and will formulate those axioms that we are still waiting. Such an inductive jump is not an easy task because it requires an interdisciplinary knowledge. In fact, Natural Complexity is a subject that involves all the disciplines depicted in Figure 1.8. Such disciplines are now strongly linked as nodes of a network and grow through mutual connections.

1.3   PURPOSE AND CONTENTS OF THIS BOOK

To favor the awaited inductive jumps in the field of Complexity, we need to prepare the new generations of students to have interdisciplinary interests and knowledge. This book is born with this purpose. It has been written by; I dare say, a Philo-physicist with a background in chemistry but

---

[30] The three types of scientific reasoning about a causal event ($Cause \xrightarrow{Rule} Effect$) are:

Induction: Finding the rule when cause and effect are known.

Deduction: Finding the effect when cause and rule are known.

Abduction: Finding the cause when the rule and effect are known

with multidisciplinary interests. It has been written for upper-division undergraduate, graduate, and PhD students in chemistry, who should acquire an interdisciplinary outlook, but also for students in interdisciplinary courses who want to deepen the role of chemistry in the field of Complexity. To me, the writing of this book has been like a marvelous journey. I have had the luck of telling the most breathtaking moments of my trip every year when I teach the subject of Complexity to my students. As it occurs in our daily life, we undertake an exciting, fruitful, and unforgettable journey whenever one or more queries guide us. In my case, after noticing that in the outside world there are both self-organizing and chaotic phenomena, I have written this book with the intention of finding answers to the two following Really Big Questions:

1. If the Second Law of Thermodynamics is true, how is it possible to observe the spontaneous emergence of order in time and/or space? Is it possible to violate the Second Law?
2. What are the features of the Complex Systems? When and how do the emergent properties emerge? Can we untangle Complex Systems?

The next three chapters give answers to the first RBQ. Chapter 2 is a thorough analysis of the Second Law. Chapters 3 and 4 present the theory of non-equilibrium thermodynamics. Then, the theory of non-linear dynamics is introduced by the description of the emergence of temporal order in ecosystems (Chapter 5), economy (Chapter 6), within a living being (Chapter 7), and in a chemical laboratory (Chapter 8). Chapter 9 describes the emergence of order in space through phenomena such as Turing structures, chemical waves, and periodic precipitations by presenting examples in chemistry, biology, physics, and geology. Then, Chapter 10 introduces the concept of Chaos in time, whereas Chapter 11 covers the concept of Chaos in space, by explaining fractals. In Chapters 12 and 13, we are ready to give answers to the second RBQ. Chapter 12 shows the intimate relation between Natural Complexity and Computational Complexity. It outlines the Complexity Challenges that must be won. Moreover, it describes the main characteristics of Complex Systems. Chapter 13 proposes strategies to untangle Complex Systems and try to win the Complexity Challenges. Lastly, this book is completed by five appendices that are useful tools for students and researchers. Appendices A and C face the numerical solution of differential equations and the Fourier transform of time series, respectively. Appendix B is a short introduction to the Maximum Entropy Method. The last two Appendices give instructions on how to deal with experimental (D) and computational (E) errors.

Finally, with this book, I want to achieve another goal. I want to stimulate public and private funding agencies to sustain interdisciplinary projects and to not be afraid of providing money to Dionysian scientists engaged in "Untangling Complexity." As the Hungarian biochemist Albert Szent-Györgyi (who won the Nobel Prize in Physiology or Medicine in 1937) said in his funny letter to *Science* (1972), scientists can be divided into Apollonians and Dionysians. An Apollonian "tends to develop established lines to perfection, while a Dionysian rather relies on intuition and is more likely to open new, unexpected alleys of research." The Apollonian sees the future lines of his research and has no difficulty writing a systematic and classical project. Not so the Dionysian. He knows only the direction in which he wants to go; he wants to explore the unknown, and he has no idea what is going to discover. In fact, intuition, which guides a Dionysian, is "a sort of subconscious reasoning, only the end result of which becomes conscious." Therefore, "defining the unknown or writing down the subconscious is a contradiction in absurdum." Hopefully, this book will be useful for preparing the next generations of both Dionysian and Apollonian scientists. The former will make the awaited inductive jumps for the formulations of axioms regarding Complexity. The latter will deduce systematically all the theorems and propositions necessary to predict the behavior of Complex Systems and will perform the experiments for confirming the new theories. Together, they will contribute to finding innovative solutions to the Complexity Challenges.

1.4 KEY QUESTIONS

- What promotes the development of science and technology?
- What are the two gateway events in the journey of humankind to discovering nature?
- What are the principal achievements of the "Practical Period"?
- Why was philosophy born in ancient Greece?
- Which are the intellectual and spiritual pillars inspiring philosophy?
- Who paved the way for the experimental method?
- When and where did the experimental method sprout?
- What are the four epistemological pillars established by the formulation of Classical Physics?
- Which theories induced to debunk the four epistemological pillars?
- How does the scientific knowledge evolve?
- What are Natural Complex Systems?
- What are the "Natural Complexity Challenges"?

1.5 KEY WORDS

Philo-physicists; Physical Technology and Social Technology; Gateway Events; Philosophy; Etiology; Ockham's razor; Reductionism; Uniformitarianism; Determinism; Mechanism; Completeness and Consistency of Mathematics; Solvability and Tractability of a Computational Problem; Induction; Deduction; Abduction.

1.6 HINTS FOR FURTHER READING

- More information about the origin of languages can be found in the book by Cavalli-Sforza (2001) and the book by Carruthers and Chamberlain (2000).
- Norbert Wiener (1894–1964) was convinced that to understand Complex Systems it is necessary to focus on the concepts of information, communication, feedback, control, and purpose or "teleology." He is considered the originator of cybernetics that is the discipline of control and communication theory (Wiener 1948).
- Among other attempts at formulating a general theory for Complex Systems, it is worthwhile mentioning the General System Theory by the biologist Ludwig von Bertalanffy (1969), Synergetics by the physicist Hermann Hacken (2004), Self-Organized Criticality by the physicist Per Bak (Bak et al. 1987) and the theory of Local Activity by the electrical engineer Leon Chua (Mainzer and Chua 2013).
- Examples of books on Complexity, which include philosophical considerations, are those by Prigogine and Stengers (1984), Mainzer (2007), Capra and Luisi (2014).

2 Reversibility or Irreversibility? That Is the Question!

> If you do not ask me what time is, I know it. If you ask me, I do not know.
>
> **Augustine of Hippo (354–430 AD)**

> Life can only be understood backwards, but it must be lived forwards
>
> **Søren Kierkegaard (1813–1855 AD)**

2.1 INTRODUCTION

A "Really Big Question" (RBQ) in philosophy is the one spoken by Hamlet in the homonymous play written by William Shakespeare (1564–1616 AD):

"To be or not to be? That is the question."

Analogously, a "Really Big Question" in science is the following:

"Reversibility or Irreversibility? That is the question."

In other words, are the transformations in nature reversible or irreversible?

If we reflect on our life, we do not doubt to answer that the events are irreversible. Suffice to think about our unrepeatable experiences of everyday life and to consider the unequivocal existence of an arrow of time in the evolution of the universe and the history of the humankind. However, if we study the fundamental laws of physics, we discover that most of them tell us that transformations in nature are reversible. Physics can be divided into four main disciplines:

- *Classical physics* that deals with the mechanics of macroscopic bodies having speeds much smaller than that of light (being about $c \approx 3 \times 10^8$ m/s)
- *Quantum physics*, whose purpose is the description of the mechanics of microscopic bodies
- *Relativistic physics*, concerning the motion of both macroscopic and microscopic entities whose velocities approach that of light
- *Thermodynamics*, dealing with the transfer of energy from one place to another and from one form to another

All these different disciplines of physics are based on three Conservation Laws:

1. Conservation of Mass
2. Conservation of Energy
3. Conservation of Charge

The famous Einstein's formula,

$$E = mc^2 \qquad [2.1]$$

reveals that the Conservation of Mass is included in the Conservation of Energy.[1] Classical, Quantum, and Relativistic physics present only reversible processes, i.e., transformations wherein it is possible to go from one state A to another state B, but also from B to A, without limitations. The time-reversibility or, we could even say, the time-reversal symmetry of physics is a consequence of the Conservation Laws.

The puzzle is that we continuously ascertain the irreversibility. If a glass falls from our hands, it shatters irreversibly; wear and tear of objects are irreversible; aging and dying are exhibited by living beings, but also by inanimate objects such as the stars in the universe.

Thermodynamics is the only discipline that does not ignore the irreversibility observed in nature and tries to explain it. As Ilya Prigogine (Nobel laureate in chemistry in 1977 for his work on dissipative structures and irreversibility) and Kondepudi (1998) claimed: "Science has no final formulation. And it is moving away from a static geometrical picture towards a description in which evolution and history play essential roles. For this new description of nature, thermodynamics is basic." Even Einstein (Schilpp 1979) had high regard for Thermodynamics: "A theory is more impressive the greater the simplicity of its premises is, the more different kinds of things it relates, and the more extended its area of applicability. Therefore, the deep impression which classical thermodynamics made upon me. It is the only physical theory of universal content concerning which I am convinced that, within the framework of the applicability of its basic concepts, it will never be overthrown."

2.2 THE THERMODYNAMIC APPROACH

According to the thermodynamic approach, the description of a natural phenomenon requires the partition of the universe into two parts: a "system" that is the theatre of the transformation, and what surrounds the system, which is called "environment."

The system can interact or not with the environment; it depends on the properties of its boundaries. In fact, a system can be

- Open when it exchanges matter and energy with its environment
- Closed when it exchanges energy but not matter with the environment
- Adiabatic when it transfers neither matter nor heat
- Isolated when it exchanges neither matter nor energy

At every instant, the thermodynamic state of a system can be described as determining the values of macroscopic variables such as temperature T, pressure P, volume V, surface Ar, chemical composition. When we define the chemical composition of a system, we need to specify the compounds that are present (labeled as $i = 1,\ldots, n$) and their concentrations: χ_i ($\forall i = 1,\ldots,n$). If there are the same values of T, P, and $\chi_i \forall i = 1,\ldots,n$, in every point of the system and they are constant over time, then it means that the system is at equilibrium. Otherwise, it is out of equilibrium. Table 2.1 reports two states, labeled as A and B, regarding four different systems. Read it, carefully. Now, I ask you: "Is it possible that state A precedes state B for all the systems? Is it possible the opposite, i.e., state B precedes state A?"

According to the First Principle of Thermodynamics, which applies the conservation of energy, it is possible that state A comes first, but also the opposite, i.e., state B precedes state A. However, our daily experience induces us to admit that states A surely come first for all the four systems listed in Table 2.1. State A evolves spontaneously to state B, but not vice versa. It means that there is an arrow of time: A→B. The direction of the spontaneous evolution can be predicted merely by considering the thermodynamic properties of the two states involved in the transformation. Therefore, there will be a state function distinguishing between spontaneous and nonspontaneous changes.

---

[1] Unequivocal experimental proofs of the mass-energy equivalence come from nuclear reactions.

TABLE 2.1

Examples of Four Systems in Two Different States: A and B

System	State A	State B
Two identical pieces of Al, labeled as 1 and 2, are maintained in an isolated container	The two pieces have different temperatures: $T_1 = 30°C$, $T_2 = 40°C$	The two pieces have the same temperature: $T_1 = T_2 = 35°C$
An isolated container is divided into two identical volumes: $V_1 = V_2 = V$	V_1 contains pure water; V_2 contains water with sugar at $C = 1$ M	V_1 and V_2 contain water and sugar at $C = 0.5$ M
A reactor having isolating walls	Water and Na are inside the reactor at T_A	NaOH, $H_{2(g)}$, and water are inside the reactor at $T_B > T_A$
A computer	N bits[a] are stored in memory	(N/2) bits are stored in the memory and (N/2) bits are present as (N/2) $kTln2$ of thermal energy

[a] One bit (Binary unIT) is the basic unit of extent of information.

This state function is entropy, which is introduced in the Second Principle of Thermodynamics. There are different definitions of entropy (its symbol is S). I present some of them in the next three paragraphs.

2.2.1 THE CLASSICAL DEFINITION OF ENTROPY

During the Industrial Revolution, the first definition of entropy was formulated by the German physicist Rudolf Clausius in the nineteenth century. At that time, entrepreneurs and scientists were striving to optimize the efficiencies of heat engines. A heat engine (see Figure 2.1) is a machine that transforms heat into work. The engine takes heat q_h from a hot reservoir. Part of this heat is transformed into work (w). Another part ($q_h - w = q_c$) is squandered as heat in a cold reservoir. The efficiency of the thermal machine is

$$\eta = \frac{w}{q_h} = \frac{q_h - q_c}{q_h}.$$

[2.2]

Equation [2.2] does not take into account the possible reduction of the efficiency due to frictions or other defects functioning in the engine.

If there is a thermal gradient between the two reservoirs (Figure 2.1), then it is still possible to carry out work. William Thomson (Thomson 1852) extrapolated this evidence to a universal scale.

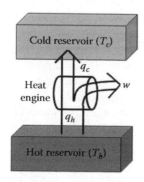

FIGURE 2.1 Scheme illustrating the working principle of a thermal machine.

He proposed the idea of the heat death of the Universe. The Universe is asymptotically approaching a state where all energy is evenly distributed, and its temperature is uniform. In this state, it is not possible to perform any kind of work and any activity is suppressed.

According to Clausius' definition, the entropy of a system increases if it absorbs heat, and the lower the temperature, the larger the increment of S:

$$dS_{sys} = \left(\frac{dq}{T} \right)_{rev}.$$ [2.3]

In [2.3], dq is the differential heat absorbed by the system in a reversible (*rev*) manner. Any reversible transformation is an ideal process because it is a time sequence of equilibrium states. Only extremely slow transformations can approach the idea of thermodynamic reversibility. A reversible transformation looks like the slow motion of a sloth. It is convenient to refer to reversible transformations because we can describe them by using the principle of equilibrium thermodynamics.

The entropy of the universe increases in any irreversible or spontaneous transformation, whereas it remains constant in any reversible process. In mathematical terms, this is stated as:

$$dS_{sys} + dS_{env} = dS_{univ} \geq 0.$$ [2.4]

The inequality [2.4] is the mathematical formulation of the Second Principle of Thermodynamics. The entropy of the Universe (which is an isolated system) unrelentingly grows due to irreversible processes, whereas it does not change if only reversible events occur. An isolated system evolves towards the equilibrium where the entropy of the system is maximized. If this system is not perturbed anymore, it will not change its macroscopic properties over time: it has exhausted all its capacity for change.

As far as the four systems of Table 2.1 are concerned, the spontaneous transformations are those from state A to state B. They are the following ones: (1) the flow of heat from the warmer to the colder piece of Al; (2) the diffusion of sugar from the concentrated solution to pure water; (3) the partial degradation of chemical energy into heat; (4) the degradation of potential energy of bits into heat. Each of these four processes determines a maximization of the entropy of the universe. The reverse transformations, B→A, are not allowed, and in fact, they are in contradiction with the Second Principle of Thermodynamics.

2.2.2   THE STATISTICAL DEFINITION OF ENTROPY

In the second half of the nineteenth century, James C. Maxwell (Scottish mathematician and physicist), Ludwig Boltzmann (Austrian mathematician and physicist), and Josiah Gibbs (American engineer, chemist, and physicist) interpreted the laws of classical thermodynamics through the laws of statistical mechanics applied to gases. In statistical mechanics, physical variables, referred to macroscopic systems, are related to the properties of their constitutive elements, i.e., atoms and molecules. For instance, the temperature of a gas is a macroscopic parameter related to the kinetic energy of the atoms. If the atoms move faster in average, i.e., their average kinetic energy increases, then the temperature of the gas rises. Statistical mechanics combines probability theory (essential for dealing with large populations) with the laws of classical or quantum mechanics.

The features of an ensemble of many atoms or molecules in an isolated system are described through the concept of *microstates*. The definition of entropy according to the statistical thermodynamics is

$$S = -k_B \sum_i p_i (ln p_i),$$ [2.5]

where:

k_B is Boltzmann's constant

p_i is the probability of the i-th microstate and the sum is extended to all the microstates

If a system is in a macroscopic state, which corresponds to Ω all equally probable microstates, the definition of statistical entropy becomes

$$S = k_B ln\Omega.$$ [2.6]

From equation [2.6], it is evident that an increase of entropy means a rise in the number of the accessible microstates, Ω. A microstate is defined by specifying the distribution of structural units (1) in the available space and (2) among the accessible energetic levels. Therefore, the total entropy is the sum of two contributions (Denbigh 1981): *configurational entropy* associated with the structural and spatial disorder, and *thermal entropy* associated with the distribution of the structural units among the accessible energetic levels (see Figure 2.2). A spontaneous process does not necessarily imply a growth of the structural and spatial disorder. For example, the crystallization of an under-cooled liquid inside an adiabatic vessel is a spontaneous process. It gives rise to the solid phase, which is spatially more ordered than the liquid one. Apparently, it violates the Second Law of Thermodynamics. However, the solution to what seems like a paradox comes from the consideration that the crystallization is an exothermic process that releases heat. Therefore, the structural elements of the crystals have more thermal energy available, and the number of accessible energetic states increases. In other words, the decrease of *configurational entropy* is counterbalanced by the larger increase of *thermal entropy*.

We can make another example to understand that entropy is not merely a synonym of disorder, but it is related to all the degrees of freedom of the structural units. The second example is the chromosome segregation phenomenon during the bacterial cell replication (Jun and Wright 2010). If we look at Figure 2.3, we may ask ourselves: how is it possible to have segregation of chromosomes if the driving force is the increase of disorder? In fact, from studying the behavior of gases, we know that two species of particles that are initially separated by a wall, mix perfectly when the wall is removed (see Figure 2.3a and b). We expect that when two chromosomes are intermingled together (see Figure 2.3c), they maintain this state because the entropy is maximized. Actually, this is not true. The two chromosomes spontaneously segregate, and the final state is that labeled as d in Figure 2.3. This event can be understood if we imagine long polymer chains as random walks that

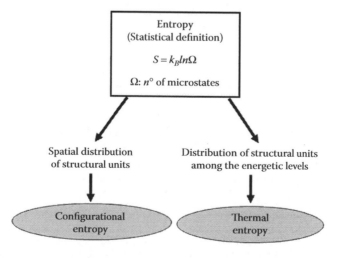

FIGURE 2.2   The two contributions to the statistical entropy.

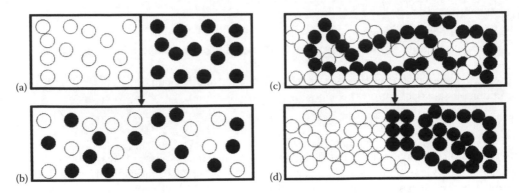

FIGURE 2.3   Free mixing of two gases: (a) and (b) are the initial and the final states, respectively. Segregation of two polymers: (c) and (d) are the initial and final states, respectively.

cannot cross their own path (the so-called "self-avoiding random walk"). Finding intermixed but self-avoiding conformations is a tough task. Intermixed chains have fewer conformational degrees of freedom than the ones that are segregated. Therefore, the driving force of chromosome segregation is the maximization of entropy that measures the total degrees of freedom of a system.

According to the Second Law of Thermodynamics, the value of Ω remains constant, or rises in an isolated system, when it hosts a reversible or an irreversible transformation, respectively. The system evolves spontaneously towards the state hiding the largest number of microstates (Ω_f):

$$S_f - S_i = k_B ln\left(\frac{\Omega_f}{\Omega_i}\right). \qquad [2.7]$$

When the number of possible microstates increases due to irreversible transformations, we lose information about the state of the system at the molecular level. To avoid this loss of knowledge, we should know the position and the momentum of every particle at a certain instant of time, and then predict their evolution in time, knowing the laws that describe their dynamics. However, such detailed knowledge is out of our reach, because when we try to retrieve information about the microscopic world, we must account for the Indetermination Principle of Heisenberg.[2] From the statistical definition of entropy, it is evident that the Second Principle of Thermodynamics comes from our incapacity to know the accurate mechanical state of our system at the microscopic level. This unavoidable uncertainty prevents us from transforming heat entirely into work without causing modifications in any other system. In fact, heat is a "disordered" form of energy, whereas work is a more "ordered" form of energy.

Finally, the statistical interpretation of the Second Principle shows us that irreversibility is a matter of probability. It is not impossible that our glass, fallen from our table and shattered in many pieces on the floor, goes back intact to its initial place. It is only a matter of probability and hence a matter of time. If we waited enough time, we could witness the unlikely event of fragments of our glass recomposing themselves on the table. The instant in which this may happen is completely unpredictable. In other words, the statistical interpretation of the Second Principle tells us that it is not impossible that an isolated system moves away from its equilibrium state. It is a matter of probability, and such probability largely decreases with the growth of the number of particles constituting the system itself.

---

[2] According to the Indetermination Principle of Heisenberg, it is impossible to determine accurately and simultaneously position and momentum of a particle.

2.2.3 THE LOGICAL DEFINITION OF ENTROPY

In the twentieth century, with the revolution of wireless telecommunications and the appearance of the first computers, a new era began: The Information Age. In fact, communication devices and computers receive, store, process, and send information.

In 1948, the American engineer and mathematician Claude Shannon wrote a paper titled "A Mathematical Theory of Communication," which is a founding work for the field of Information Theory. In this key paper, Shannon proposed concepts useful to theorize and devise communication systems. The communication of information involves five essential elements (see Figure 2.4). The primary component is the *information source* producing a message. The message is transformed into a signal by the *transmitter*. The signal is a series of symbols and is sent by the transmitter through a medium which is named as the *communication channel*. A *receiver* collects the signal crossing the communication channel and reconstructs the message intended to be received by the *destination*.

The signal consists of a series of symbols, conveying a certain amount of information. If the signal is encoded in binary digits, 0s and 1s, the basic unit amount of information becomes the bit. To distinguish between two symbols, we need to receive one bit. To distinguish among 2^n symbols, we need to retrieve n bits (see Table 2.2).

The quantity of information (I) collected by the receiver can be quantified by determining the difference between the uncertainty before (H_{bef}) and after (H_{aft}) receiving the signal:

$$I = H_{bef} - H_{aft}.$$ [2.8]

The uncertainty is given by the equation [2.9]:

$$H = -\sum_{i=1}^{M} p_i \log_2 p_i,$$ [2.9]

where:
  M is the number of symbols
  p_i is the probability of i-th symbol

FIGURE 2.4 Schematic representation of a communication system.

TABLE 2.2
Relation Between States and Bits

2 states: 1 bit	4 states: 2 bits	8 states: 3 bits
State A: 0	State A: 00	State A: 000
State B: 1	State B: 01	State B: 001
	State C: 10	State C: 010
	State D: 11	State D: 100
		State E: 110
		State F: 101
		State G: 011
		State H: 111

If the M symbols are all equally likely before receiving the signal, it derives that

$$p_i = \frac{1}{M}$$ [2.10]

$$H_{bef} = \log_2 M.$$ [2.11]

If, after receiving the signal, all the symbols are perfectly known, then

$$H_{aft} = 0$$ [2.12]

$$I = \log_2 M.$$ [2.13]

Equation [2.13] represents the maximum value of the information we can get in this case.[3] On the other hand, if there is noise over the communication channel, perturbing the transmission of the signal, $H_{aft} \neq 0$, and $I < \log_2 M$.

Once the information is received, it can be processed by using a computer. Current computers are based on electronic circuits and the so-called Von Neumann architecture that was proposed for the first time in the mid-forties of the twentieth century (Burks et al. 1963). Four main elements constitute a computer (see Figure 2.5). First, a memory for storing information. Second, a Central Processing Unit (CPU) processes the information. CPU consists of two subunits: The Arithmetic Logic Unit (ALU), which performs arithmetic and logical operations, and the Control Unit (CU), which extracts instructions from memory, decoding and executing them. The third element of a computer is the Information Exchanger (IE) allowing for the transfer of information in and out of the computer; it consists of a screen, keyboards, et cetera. Finally, the fourth element is the data and instructions Bus, working as a communication channel and binding the other three components of the computer by conveying information among them. Electronic computers are general purpose computing machines because the instructions to make computation are stored into the memory as sequences of bits into

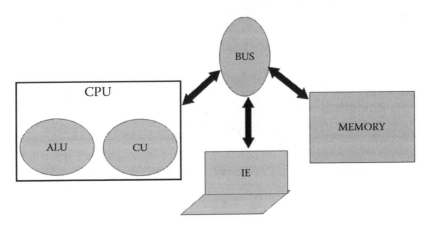

FIGURE 2.5   Schematic structure of a computer having the Von Neumann architecture.

---

[3] If the M symbols are not equally probable, $I < \log_2 M$.

the software. The information inside the computer is encoded in electrical signals, and transistors act as the basic switching elements of the CPU. One of the main tasks of Information Technology is that of always devising more powerful computers, capable of processing larger amounts of information at increasingly higher speeds but at lower power levels, volume, and price.

Computers are physicochemical systems. The laws of physics and chemistry dictate what they can and cannot do. The amount of information that a physical system can store and process is related to the number of distinct physical states that are accessible to the system. A system with Ω accessible states can register $\log_2\Omega$ bits of information. If we keep in mind the statistical definition of entropy (equation [2.6]), it is evident that information and entropy are intimately linked. The laws of thermodynamics play key roles in computation. When the information of one bit goes into observable degrees of freedom of the computer, such as another bit, then it has been moved and not erased; but if it goes into unobservable degrees of freedom such as microscopic random motion of molecules, it results in an increase of entropy of $k_B ln2$ (Lloyd 2000). The rate at which a computer can process information is limited by the maximum number of distinct states that the system can pass through, per a unit of time. It has been demonstrated (Margolus and Levitin 1998) that an isolated quantum system with average energy E takes time at least $\Delta t = \pi\hbar/2E$ to evolve to a distinct (orthogonal) state. From this relation, it derives that if E_l is the energy allocated in the l-th logic gate of a computer, the total number of logic operations performed per second is equal to the sum over all logic gates of the operations per second per gate:

$$\sum_l \frac{1}{\Delta t_l} \le \sum_l \frac{2E_l}{\pi\hbar} = \frac{2E_{tot}}{\pi\hbar}. \tag{2.14}$$

In other words, the rate at which a computer can compute is limited by the total amount of energy available: E_{tot}.

The theory of information shows that information is strictly related to the statistical definition of entropy (compare equation [2.9] with equation [2.5]). The analogy between H and S is not simply formal, but also semantic. In fact, if a macroscopic system is in a certain thermodynamic state, entailing Ω_i equally probable microstates, the uncertainty we have about it is proportional to the logarithm in base two of Ω_i. If this system undergoes an irreversible transformation, it will end up in a new state, entailing a larger number (Ω_f) of equally likely microstates. After the spontaneous event, the information (equation [2.8]) we have about the system, is

$$I = \log_2\left(\frac{\Omega_i}{\Omega_f}\right) \tag{2.15}$$

This amount of I is negative: after an irreversible transformation, we have lost information about the microscopic state of the system. Note that if we multiply equation [2.15] by $-k_B ln2$, we obtain equation [2.7], that is the statistical definition of ΔS. Shannon confirmed, through his theory of communication, that an increase of entropy corresponds to a loss of information about the microscopic state of the system. The strict link between the Second Principle of Thermodynamics and the Mathematical Theory of Communication is confirmed by the law of "Diminishing Information" (Kåhre 2002). This law states that if we consider the transmission of information in a chain $A{\rightarrow}B{\rightarrow}C$, the information that C has about A ($I(C@A)$) is less or at most equal to the information B has about A ($I(B@A)$):

$$I\left(C@A\right) \le I\left(B@A\right) \tag{2.16}$$

Compared to direct reception, an intermediary can only decrease the amount of information. As in any effort to produce work, and likewise in any attempt at sending information, there are unavoidable sources of loss.

2.3   AN EXHAUSTING FIGHT AGAINST ENTROPY

So far, we have learned that spontaneous processes bring about an increase in the entropy in our universe. The growth of entropy means dissipation of energy and/or loss of information. Now, we ask ourselves:

"Is it possible to violate the Second Principle of Thermodynamics?"

This question was also raised by Isaac Asimov in his original story titled *The Last Question* published in 1956. In this science-fiction short story, Asimov plays to predict the human history for the next several trillion years. He imagines humans always devising more powerful computing machines. One of these, called Multivac, helps humankind to solve the energy issue by unveiling how to exploit solar energy. Armed with this discovery, humans colonize the Universe in the coming millennia, because they can harvest and exploit the energy of the stars. However, one great challenge remains to be won. How can the net amount of entropy of the Universe be massively decreased? They pose this question to their sharpest computers, but the only answer they receive is:

"THERE IS INSUFFICIENT DATA FOR A MEANINGFUL ANSWER."

Actually, Maxwell, Boltzmann, and Gibbs found an answer in their theory of statistical thermodynamics. In fact, they told us that it is possible to witness a decrease in entropy. It is just a matter of waiting for the right time; in other words, it is just a matter of probability. The more spontaneous a process, the less probable its reverse, and the longer the time we need to wait, on average. Aware of this, but not satisfied by this random way of violating the Second Principle of Thermodynamics, we are wondering if it is possible to break it whenever we want, without waiting for biblical times. Over the years, some "thought experiments" (the so-called "Gedankenexperiments") have been proposed to understand if and how it is possible to decrease the entropy of the universe.

2.3.1   THE MAXWELL'S DEMON

The first valuable thought experiment dates back to James Maxwell. In his book titled *Theory of Heat*, Maxwell (1902) states that "the second law of thermodynamics is undoubtedly true as long as we can deal with bodies only in mass and have no power of perceiving or handling the separate molecules of which they are made up." Therefore, the violation of the second law is in the power of "a being whose facilities are so sharpened that he can follow every molecule in its course." Let us suppose a vessel full of air at equilibrium, i.e., having uniform temperature and pressure. Each molecule, inside the vessel, has its own velocity and kinetic energy. There is a well-known distribution of velocities.[4] Let us imagine partitioning the vessel into two halves, A and B, through a division having a small hole. If the being, able to see the single molecules, slowly opens and closes the small hole indefinitely through a frictionless slide so as to allow only the swiftest molecules to pass from A to B, then only the slowest ones will pass from B to A, "he will thus, without expenditure of work, raise the temperature of B and lower that of A, in contradiction to the second law of thermodynamics". This being was dubbed Maxwell's Demon due to his ability to overthrow the laws of nature. Maxwell concluded that since humans cannot see and manipulate single molecules and atoms without the expenditure of work, they will not violate the second principle. This is not a satisfying and complete exorcism, because it leaves open the possibility of devising mechanisms, tools, or artificial beings, which can carry out transformations reducing the entropy of the universe.

2.3.2   A FIRST MECHANICAL ATTEMPT

An example of a device which, at first sight, seems to be successful in our attempt of violating the Second Principle of Thermodynamics is depicted in Figure 2.6.

---

[4] The ensemble of molecules or atoms of an ideal gas, which do not interact each other except when they elastically collide, exhibits the Maxwell-Boltzmann distribution of speeds. This distribution depends on the temperature.

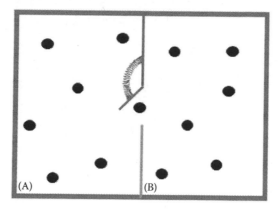

FIGURE 2.6 Device consisting of two rooms (A and B) separated by a spring-loaded trapdoor containing molecules of an ideal gas (black spheres).

Two chambers (A and B), initially containing gas at the same temperature and pressure, are in contact through a tiny hole closed by a spring-loaded trapdoor. The door swings open in only one direction, allowing molecules from B to A, but not vice versa. When a molecule in room B impinges on the trapdoor with sufficient energy, it opens the door and comes into A. The repetition of this event eventually accumulates molecules in A at the expense of B, creating an inequality of pressure. This would mean that the system abandons its original equilibrium state, in contradiction with the Second Law of Thermodynamics. If so, the trapdoor works like a unidirectional valve. Is this feasible? It is not! It was Marian Smoluchowski,[5] who, in 1912, revealed why a device such as that in Figure 2.6 could not create the gradient of pressure. The energy needed to open the trapdoor should be comparable with the average kinetic energy of the single molecules. When the trapdoor is repeatedly hit, it eventually stores a content of energy equal to the order to that of the molecules striking it. Therefore, it will jitter, opening and closing randomly. When the trapdoor maintains open, it cannot operate as a valve, since molecules can pass freely in both directions. Moreover, the probability that a molecule impinging on the door goes from B to A, will be counterbalanced by the probability that the trapdoor pushes a molecule from A to B. If a device initially has uniform temperature and pressure (being at equilibrium), then ability of the trapdoor to play as a unidirectional valve is null.

2.3.3 Another Mechanical Attempt

Another Gedankenexperiment for the violation of the Second Law through a mechanical device was proposed by Richard Phillips Feynman.[6] Feynman designed a device that is depicted in Figure 2.7. It consists of an axle having vanes at one end, and a ratchet and a pawl system at the other. The vanes are maintained within a chamber containing a gas at temperature T_A. The ratchet and the pawl system are contained within another box full of gas at temperature T_B. The idea is to exploit the random thermal motion of the molecules of the gas contained in the box at T_A to do work and lift a load through a wheel. The ratchet and the pawl system must rectify the random motion of the vanes. The ratchet and pawl system must have a spring because the pawl must return after coming off a tooth.

---

[5] Marian Smoluchowski (Vorderbrühl, 1872–1917) was a Polish scientist who gave significant contributions in the field of statistical physics and kinetic theory.

[6] Richard Phillips Feynman was an American theoretical physicist (New York City 1918–Los Angeles 1988). He is known for his contribution to the theory of quantum electrodynamics, which was awarded a Nobel Prize for, in 1965. He opened a new vista on nanotechnology with his famous lecture titled "There's Plenty of Room at the Bottom."

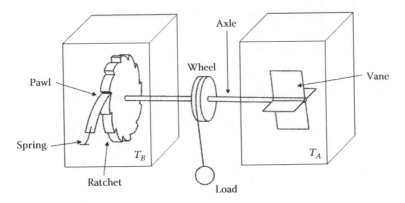

FIGURE 2.7   Structure of Feynman's device for the violation of the Second Law of Thermodynamics.

Unfortunately, this device does not allow us to violate the Second Law of Thermodynamics. In fact, when the two boxes are at the same temperature ($T_A = T_B$), the molecules of gas contained in the two chambers have the same thermal energy. Such thermal energy is transferred to the mechanical parts of the device. Therefore, the pawl and the ratchet have Brownian motion, as the vanes do. It is correct that this Brownian motion will, every once in a while, lift the pawl up and over a tooth just at the moment when the vanes are turning the axle backward. The net result is nothing if the two gases are at the same temperature. In fact, the wheel will do a lot of jiggling this way and that way, but it will not lift the load. The situation is analogous to Smoluchowski's trapdoor. It is impossible that a system at equilibrium in this universe spontaneously moves away from it without the intervention of any external force. Feynman demonstrated that only if T_A is maintained higher than T_B, the device can rectify thermal motion in useful mechanical work (Feynman et al. 1963).

2.3.4   THE INVOLVEMENT OF ARTIFICIAL INTELLIGENCE: A "THOUGHT EXPERIMENT"

From the previous "thought experiments," it seems evident that a mechanical device is not enough to violate the Second Law of Thermodynamics. Somehow, we need to introduce Maxwell's Demon. The role of Maxwell's Demon can be played by a system of Artificial Intelligence (AI). Such an AI system should be able to monitor and manipulate single molecules.

The first proposal of such a system came from the physicist Leo Szilard,[7] who published a paper (in the *Zeitschrift für Physik* in 1929), whose title, translated in English, reads: "On the decrease of entropy in a thermodynamic system by the intervention of intelligent beings" (Szilard 1964). Szilard devised a machine whose components (see Figure 2.8) are a cylinder having its two ends blocked by two pistons and containing just a molecule, a movable partition, which can be inserted in the middle of the cylinder to trap the molecule in one of the two halves, a detector to see which half the molecule is in and a memory to record that information.[8] The cylinder and the molecule inside it are thermally connected with a heat reservoir at temperature T.

---

[7] Leo Szilard (Budapest 1898–1964) was a Hungarian physicist, who participated in the Manhattan Project for building the atomic bomb, during World War II.

[8] The machine presented here corresponds to that described by Bennett (1987). It is slightly different from that proposed by Leo Szilard in his original paper, appeared in 1929 in Zeitschrift für Physik.

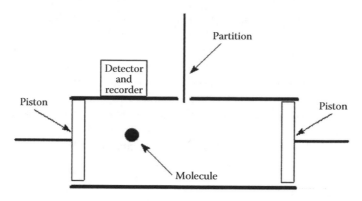

FIGURE 2.8 Structure of the Szilard's machine.

Szilard's machine works by making a cycle consisting of six steps.

First step: The movable partition is inserted inside the cylinder, and the molecule is trapped in one half of the cylinder. The work needed to perform this first step is null because it is made in the void, in the absence of any friction.

Second step: The detector determines the position of the molecule. This information is recorded in the memory as either L, if the molecule is on the left part, or R, if it is on the right part. The two cognitive states, L and R, correspond to two distinct physical states of the memory.

Third step: After knowing the position of the molecule, the piston on the side which does not contain it, is pushed in until it reaches the partition. This operation does not require work because it is a compression of empty space.

Fourth step: The partition is removed.

Fifth step: The molecule repeatedly hits the piston that has just been advanced and drives the piston back to its original position. The molecule does mechanical work on the piston by exploiting the heat of the surrounding thermal bath. Since the molecule behaves like an ideal gas at uniform T, its internal energy does not change:

$$\Delta U = q + w = 0 \qquad [2.17]$$

$$q = -w = k_B T ln2. \qquad [2.18]$$

Sixth step: The content of the memory is erased, and it returns to its original blank state.

After these six steps, the machine is again in its initial state. After the entire cycle, it seems that the Szilard's machine has violated the Second Law of Thermodynamics because it has converted heat completely in work. By repeating plenty of times the cycle, a relevant amount of heat can be seemingly converted in work.

There are two steps of the cycle that produce appreciable amounts of entropy. First of all, the second step is the act of measurement. The same Szilard was aware of this aspect, which was clearly stated by Leon Brillouin[9] and Denis Gabor[10] some years later. If we use one photon to locate the

---

[9] Leon Brillouin (Sèvres 1889–1969) was a French physicist who contributed to quantum mechanics, solid state physics, radio-wave in the atmosphere, and information theory.

[10] Denis Gabor (Budapest 1900–1979) was an electrical engineer and physicist, most notable for inventing holography, which is a kind of photography without lenses: an image of an object is captured as an interference pattern between the light scattered from the object and a reference beam. This recorded interference pattern enables to view a three-dimensional image of the object. Gabor received the Nobel Prize in Physics in 1971 for his invention of holography.

molecule, we waste the energy of that photon (that is $E = h\nu$). The frequency of this photon must be larger than the frequencies of the thermal photons[11] emitted by the molecule and the inner walls of the cylinder, which are at the same temperature. If the probe-photon had the same frequency as the thermal photons, it would be confused with them. Since the frequency of the probe-photon must be high, the waste of energy in the second step would be larger than the gain in the fifth step ($h\nu > k_B T ln2$). Charles Bennett[12] (1987) has shown that the measurement of the position of the molecule does not have to involve necessarily a photon. It may be carried out by a device operating in a reversible manner and hence it may be not thermodynamically costly. It is the sixth step that produces entropy inevitably, as the IBM physicist Rolf Landauer[13] (1961) revealed. Landauer started from the premise that distinct logical states of a computer must be represented by distinct physical states of computer's hardware. When we erase the content of the memory in the sixth step, we carry out an irreversible transformation, since the information stored as "noble" energy (that is as electrical or potential energy) is dissipated as heat. The stored information ($I = \log_2 2 = 1$ bit) is wasted as heat ($q = k_B T ln2$). The sixth step produces entropy: $S = k_B ln2$. To avoid this detrimental step, we should dispose of an unlimited memory. The trouble is that without the sixth step, the overall remaining set of operations would not be a real cycle. Moreover, although we can sometimes trust in very large memories, unlimited ones do not exist. Therefore, in the moment we decide to polish the memory of our device from all old data, we produce an appreciable amount of heat and hence of entropy. In conclusion, it seems clear that even the device designed by Szilard, based on the involvement of artificial intelligence, cannot violate the second law of thermodynamics.

TRY EXERCISE 2.1

2.3.5 THE EMBODIMENT OF MAXWELL'S DEMON: A "REAL EXPERIMENT"

Since the early 1990s, the microscopic techniques have improved so much that we are now able to detect single molecules and atoms.[14] The observation of the microscopic world stirs up emotions similar to those felt after diving into the deep ocean wherein completely new and amazing scenarios are unveiled. The discovery of a new world spurs us to monitor and govern it. In fact, microscopists are getting used to manipulating microscopic bodies, as if they were many Maxwell's Demons. For example, recently, the physicist Masaki Sano and his colleagues of University of Tokyo (Toyabe et al. 2010), exploiting an upright microscope equipped with a high-speed camera, tried to verify if the detection and manipulation of tiny objects entail wasting energy. They observed and manipulated the motion of a colloidal polystyrene bead (having hundreds of nanometers in diameter) immersed in an electric field having a "spiral-staircase-like" profile of potential energy. The bead could rotate either clockwise or counterclockwise. However, the team created a "spiral staircase" that was harder to mount in the counterclockwise direction than to descend in the clockwise direction. When the bead was left alone, it had some thermal energy available, and it moved in random

---

[11] As we will learn in Chapter 12, any material at uniform temperature T is in equilibrium with its thermal radiation, which is emitted and absorbed by the same object. The spectrum and the internal energy of thermal radiation depend only on temperature.

[12] Charles Bennett (born New York in 1943) is currently working at IBM research as an IBM fellow.

[13] Rolf Landauer (Stuttgart 1927–1999) was an IBM physicist and he is best known for having elevated the physics of information processing to a serious field of scientific inquiry. His main inheritance is the belief that "information is physical" and cannot be understood outside its physical embodiment. He formulated a principle, bearing his name, stating that only irreversible computations, which are those discarding information, produce entropy. It is feasible to design reversible computing and energy-saving circuits. For the most recent developments of Landauer's Principle, read Chapter 13.

[14] Traditional optical microscopes allow detecting details of objects up to the Abbe's diffraction limit. The Abbe's diffraction limit corresponds to about 200 nm when blue light is used as a probe. Nowadays, there are microscopic techniques which bypass the Abbe's limit. Examples are the (1) Two-photon Excitation Microscopy, (2) Stimulated Emission Depletion (STED) Microscopy, (3) Near-Field Microscopy, (4) Atomic Force Microscopy (AFM), (5) Scanning Tunnelling Microscopy, and the (6) Electron Microscopy.

directions. More often, it turned clockwise, i.e., it went "downstairs," although from time to time, it also went "upstairs" moving anticlockwise. To play as Maxwell's Demon, Sano and his team were looking over the bead, and when it randomly turned counterclockwise, they quickly adjusted the potential to hinder a turn back clockwise. It was like closing the hole of the device described in paragraph 2.3.1 when a hot molecule was let to pass from room A to room B by the demon. In the experiment of Sano's team, the bead has been forced to keep climbing "upstairs" in seeming violation of the Second Law of Thermodynamics. Energy was consumed by the observer and his equipment to monitor the bead and switch the voltage when needed. This experiment shows that a source of energy dissipation can be the actions played by artificial intelligent agents that, like the Maxwell's demon, detect and manipulate microscopic bodies.

2.3.6 THE SURPRISING BEHAVIOR OF SMALL SYSTEMS

As we have ascertained in the previous paragraph, the recent refinement of the micromanipulation technology has opened new opportunities to shed light on the unique properties of microscopic systems (Grier 2003). It is now possible to perturb single small systems, driving them away from the equilibrium and observe their subsequent responses. Whereas the thermodynamic state of a macroscopic system is defined by specifying parameters such as temperature, pressure, and concentrations of the chemical species, the state of a microscopic system is determined by measuring the so-called "controlled parameters" (Bustamante et al. 2005). Examples of controlled parameters are the elongation of a macromolecule, the force acting on a colloidal particle (having linear dimensions in the range between a few units and hundreds of nanometers) and its position in space with respect to a reference's system. As a small system is pushed out of equilibrium, its dynamics are effectively random, because it is soaked into a thermal bath where unpredictable fluctuations become relevant. The behavior of the small system is described by the *Fluctuation Theorem,* formulated in the mid-1990s within the theory of non-equilibrium statistical mechanics (Evans and Searles 2002). The state of a small system is represented by a point in its phase space. The phase space is a multidimensional reference system whose coordinates are the controlled parameters needed to define the states of the small system. When a small system evolves in time, it describes a trajectory in its phase space. Every trajectory is characterized by a value of the "entropy production" P^*:

$$P^* = \frac{dq}{Tdt},$$ [2.19]

which is the ratio of the rate at which the system exchanges heat with the bath over its temperature. The Fluctuation Theorem defines the ratio between the probability that the small system traces a trajectory generating a positive entropy production P^* (i.e., $\Pr(P^*)$) and the probability that the same small system follows the respective anti-trajectory generating a negative entropy production $-P^*$ (whose probability is $\Pr(-P^*)$). Its formulation for arbitrary averaging times is the following:

$$\frac{\Pr(P^*)}{\Pr(-P^*)} = e^{\frac{P^*(\Delta t)}{k_B}}.$$ [2.20][15]

---

[15] The Fluctuation theorem presents different formulations. A formulation refers to non-equilibrium steady-state fluctuations ([2.21]) whereas the other [2.20] refers to transient fluctuations.

$$\lim_{t \to \infty} \frac{k_B}{t} ln\left(\frac{\Pr(P^*)}{\Pr(-P^*)}\right) = P^*$$ [2.21]

The Fluctuation Theorem gives a precise mathematical expression for the probability of violating the Second Law of Thermodynamics. The probability of observing an event reducing entropy is inversely proportional to the exponential of the product $P^*(\Delta t)$. In simple words, it means that the probability of consuming entropy is negligible when we deal with events occurring over long-time scales (Δt large) and involving macroscopic systems producing large entropy production values.[16] We may rewrite equation [2.20] in the following form:

$$\Pr\left(P^*\right) = \Pr\left(-P^*\right) e^{\frac{P^*(\Delta t)}{k_B}} \qquad [2.22]$$

Reminding that $\Pr(P^*) + \Pr(-P^*) = 1$, it derives that:

$$\Pr\left(-P^*\right) = \frac{1}{1 + e^{P^*(\Delta t)/k_B}} \qquad [2.23]$$

$$\Pr\left(P^*\right) = \frac{1}{1 + e^{-P^*(\Delta t)/k_B}}. \qquad [2.24]$$

When we are at equilibrium, Δt is 0, and $\Pr(-P^*) = \Pr(P^*) = 0.5$, in agreement with the postulate of microscopic reversibility. In fact, at equilibrium, a transformation of a microscopic system tracing a trajectory in its phase space is counterbalanced by the corresponding anti-trajectory. The anti-trajectory is overlapped to the trajectory, but it is crossed in the opposite direction and with equal probability.

From equations [2.23] and [2.24], we see that outside of equilibrium $\Pr(-P^*)$ decays from 0.5 to 0, and the larger the system, the faster the decay. On the other hand, $\Pr(P^*)$ grows from 0.5 to 1 (see Figure 2.9).

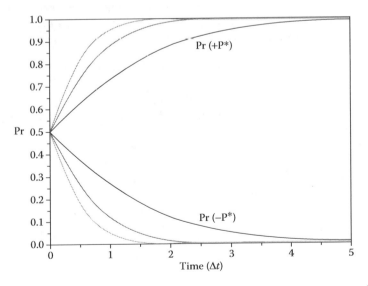

FIGURE 2.9 Trends of the probabilities that the entropy production is positive ($\Pr(+P^*)$) and negative ($\Pr(-P^*)$) expressed by equations [2.24] and [2.23], respectively. The three curves in the graph have been traced for three values of P* that grows going from the black to the gray up to the light gray one.

---

[16] Note that the entropy production is an extensive variable: its values depend on the extent of the system.

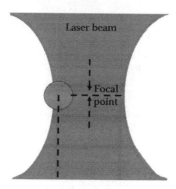

FIGURE 2.10 Sketch representing the latex particle trapped by the laser beam. The laser beam draws the particle towards its focal point by a harmonic force.

Fluctuation Theorem has been proved experimentally. An important empirical demonstration has been offered by the team of Denis Evans at the Australian National University in Canberra (Wang et al. 2002). Evans and his colleagues followed the trajectory of a 6.3 μm diameter latex particle contained in a glass sample cell filled with water and put on the stage of an inverted microscope.[17] The latex particle was trapped within a laser beam having 980 nm as the wavelength (see Figure 2.10).

The infrared (IR) rays of the laser beam are refracted at different intensity over the volume of the sphere and draw the particle towards the focus of the beam, i.e., the region of highest intensity. The force exerted by the laser beam (F_{opt}) was of the order of pico-Newton (10^{-12} N) and was assumed to be harmonic near the focal point. It was defined by the following equation:

$$F_{opt} = -k\left(x_{\Delta t} - x_0\right), \qquad [2.25]$$

where in k is the trapping constant which was tuned to about 1×10^{-5} pN/Å by adjusting the laser power. The team first recorded x_0 which represents the position of the particle in the absence of any stage translation (averaging for a minimum of 2 sec). Then, the stage was translated at a constant velocity ($v = 1.25$ μm/s), and the position of the particle was recorded at different time intervals, $x_{\Delta t}$. The entropy produced along a trajectory of duration Δt was estimated by the following equation:

$$\Delta S = T^{-1}\int_0^{\Delta t} vF_{opt}dt, \qquad [2.26]$$

where T is the temperature of the heat sink surrounding the system. The determination of ΔS has been repeated over 500 trajectories and for different Δt. They found that when Δt was 10^{-2} sec, the trajectories were distributed nearly symmetrically around $\Delta S = 0$, that is with entropy-consuming and entropy-producing trajectories equally probable. At longer times, the entropy-consuming trajectories occurred less often, and the mean value of ΔS shifted towards positive numbers. For Δt greater than a few seconds, the entropy-consuming trajectories could not be detected at all.

The Fluctuation Theorem has received further deft empirical proofs (see for instance the work by Dillenschneider and Lutz (2009), who proved Landauer's Principle in the limit of long erasure

---

[17] An inverted microscope is an optical microscope "upside down". In fact, its light source and condenser (a lens concentrating light from the illumination source) are on the top, above the stage, pointing down, while the objectives are below the stage, pointing up.

cycles for a system of a single colloidal bead 2 μm in diameter), and it is now accepted as a valid generalization of the laws of thermodynamics to small systems (Bustamante et al. 2005). Its content is revolutionary, because small systems, such as the molecular machines we encounter in biology and nanotechnology, manifest striking behaviors if compared with the workings of the macroscopic world. Biological and nanotech machines can transform heat in work over short time scales when the work performed during a cycle is comparable to the thermal energy available for each degree of freedom. In other words, the nano-machines spend some time working in "reverse mode." It is as if a car could transform its products of combustion and the released heat into fuel and oxygen to go ahead.

TRY EXERCISE 2.2

2.3.7   THERE IS STILL AN OPEN QUESTION

In the last paragraph, we have discovered that small systems (having dimensions not larger than a few μm) can break the second law of thermodynamics over short time scales (less than tens of milli-sec). It means that heat can be transformed entirely into work only in the microscopic world and over a limited lapse of time.

Now a further question arises. How can we conciliate the Second Law of Thermodynamics with the theory of evolution? The theory of evolution asserts that life on earth evolved from simple forms (unicellular organisms) towards more and more organized and complex species (multi-cellular living beings). Biological evolution requires very long-time scales, being of the order of billions of years. Moreover, how is it possible that after fertilizing an egg, a multi-cellular organism, which is highly organized in both space and time, emerges? It seems that amazing phenomena, such as biological evolution and the birth of any living being, are in sharp contrast with the Second Law of Thermodynamics. This antinomy cannot be solved by invoking the Fluctuation Theorem, because it predicts that only microscopic systems can decrease entropy, and just for short time scales. To address this apparent conundrum, we must go ahead on our journey and learn the principle of non-equilibrium thermodynamics. So, let us plunge into Chapter 3.

2.4   KEY QUESTIONS

- What are the fundamental Conservation Laws?
- What kind of system can we encounter in Thermodynamics?
- How many definitions of entropy do you know?
- What happens to entropy in an irreversible transformation?
- What is a reversible transformation?
- What happens to entropy in a reversible transformation?
- What are the two contributions to statistical entropy?
- What is the connection between information and entropy?
- Is it possible to violate the Second Law of Thermodynamics?
- What is the role of Maxwell's demon?
- Why did the mechanical attempts at violating the Second Principle fail?
- What is the significant concept proposed by Landauer in the analysis of the Szilard's machine?
- What does the Fluctuation Theorem tell us?

2.5   KEY WORDS

Conservation Laws; Irreversibility and Reversibility; Configurational and Thermal Entropy; Degrees of freedom; Uncertainty and Information; Fluctuations.

2.6   HINTS FOR FURTHER READING

To deepen the relationship between thermodynamic entropy and information, read Brillouin (1956), Leff and Rex (1990), and Maruyama et al. (2009). To meditate more on the Maxwell's demon, read Maddox (2002).

To collect more information about optical trapping, read Ashkin (1997).

A pleasant book to meditate on the rise of complexity in nature is that written by Chaisson (2002).

2.7   EXERCISES

2.1. If we erase 8 kilobytes in the memory of our computer working at 300 K, how much heat do we generate according to Landauer's Principle? How much entropy do we generate? Remember that one byte is a unit of information that consists of eight bits.

2.2. In the experiment of Evans and his team, the latex particle trapped by the laser beam feels a harmonic potential; it is like a ball bound to a spring following Hooke's Law (see Figure 2.11).

After $\Delta t = 10^{-2}$ sec, the minimum value of ΔS obtained was $-6k_B$. Knowing that the force constant k of the spring was 0.1 pN/μm and the speed of the stage v was -1.25 μm/s, which was the value of $\Delta x = x_{\Delta t} - x_0$? Assume that the temperature of water's bath wherein the latex particle was immersed was 300 K.

2.8   SOLUTIONS TO THE EXERCISES

2.1. According to Landauer's Principle, any logically irreversible transformation of classical information is necessarily accompanied by the dissipation of at least $k_B T ln2$ of heat per lost bit. Since we erase 64000 bits, the heat produced will be $64000 k_B T ln2 = 1.8 \times 10^{-16}$ J. The entropy is $S = q/T = 6.1 \times 10^{-19}$ J/K. Note that in current silicon-based digital circuits, the energy dissipation per logic operation is about a factor of 1000 greater than the ultimate Landauer's limit.

2.2. After applying the laser beam and maintaining constant its power, the internal energy of the particle is

$$dU = dq + dw = dq - F_{opt}dx = TdS - F_{opt}dx$$

The entropy production P becomes:

$$\frac{P^*}{k_B} = \frac{dq}{k_B T dt} = \frac{F_{opt}dx}{k_B T dt} = \frac{F_{opt}vdt}{k_B T dt}$$

$$P^* = \frac{F_{opt}v}{T}$$

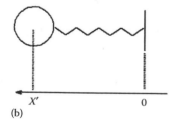 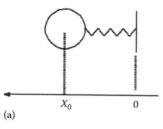 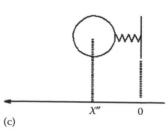

(b)   (a)   (c)

FIGURE 2.11   The latex particle trapped by the laser beam behaves like a ball bound to a spring. In (a), (b), and (c), the spring is at equilibrium, stretched and compressed, respectively.

$$dS = \frac{F_{opt}v}{T}\,dt$$

$$\Delta S = \frac{1}{T}\int_0^{\Delta t} F_{opt}v\,dt$$

$$\Delta S = \frac{F_{opt}v\Delta t}{T} = \frac{(-k\Delta x)v\Delta t}{T}$$

Introducing the values of variables, we obtain $\Delta x = -19.9\ \mu m$. Note that within the same time interval Δt, the displacement due to the movement of the stage is $-0.0125\ \mu m$. This result means that the latex particle exploits the thermal energy to do work and go further away from the focal point of the laser.

3 Out-of-Equilibrium Thermodynamics

Nature, to be commanded, must be obeyed.

Francis Bacon (1561–1626 AD)

Science is built up with facts, as a house is with stones. But a collection of facts is no more a science than a heap of stones is a house.

Jules Henri Poincaré (1854–1912 AD)

3.1 INTRODUCTION

Out-of-equilibrium systems are widespread. The Universe, although it is an isolated system, is very far-from-equilibrium because it is expanding and its cosmic background thermal radiation, of about 3 Kelvin (K), is not in thermodynamic equilibrium with the matter concentrated in the galaxies. Each star is out-of-equilibrium; in fact, nuclear reactions of fusion occur in their cores, which have higher temperatures and different compositions than their outer shells. The same is true for each planet, particularly for Earth. The core of our planet is melted with a higher temperature and pressure than its surface. Moreover, the influx of electromagnetic radiation coming from the sun maintains our planet far-from-equilibrium. Also, humans, like every other living being on Earth, are out-of-equilibrium. Every living creature is an open system exchanging matter and energy with its surrounding. Even in our scientific laboratories, out-of-equilibrium systems are typical. Open systems are constantly far-from-equilibrium. On the other hand, we may assume that closed, adiabatic and isolated systems reach an equilibrium state when they are not perturbed, and the effect of the gravitational field of the earth can be overlooked.

Out-of-equilibrium systems exhibit the power of self-organizing. How is it possible? Do they violate the Second Law of Thermodynamics?

3.2 DEFINITION OF THE ENTROPY CHANGE FOR AN OUT-OF-EQUILIBRIUM SYSTEM

To define the entropy change (dS) for an out-of-equilibrium system in the time interval dt, we distinguish two contributions (Prigogine 1968):

$$dS = d_i S + d_e S \tag{3.1}$$

wherein $d_i S$ is the change of entropy due to processes occurring inside the system as if it were isolated, whereas $d_e S$ is the change of entropy due to the exchange of energy and/or matter with the environment. The Second Law of Thermodynamics allows us to state that

$$d_i S \geq 0 \tag{3.2}$$

The term d_iS is positive when it refers to irreversible events, whereas it is null for reversible transformations. On the other hand, d_eS can be positive, negative, and null. Equations [3.1] and [3.2] hold for any "macroscopic portion" of the universe. A portion of the universe is considered "macroscopic" if it contains a number of structural units (for instance molecules) so large to allow to overlook microscopic fluctuations. The fluctuations are due to the random motion of the structural units and their interactions with the surrounding environment. For an extensive thermodynamic variable X, the fluctuations become negligible if they determine a change δX, which is very small compared to the value of X. The value of X is proportional to the number of structural units, N, whereas δX, in agreement with the law of large numbers, is proportional to $N^{1/2}$ (being the root-mean-square). Therefore, the relative effect of the microscopic fluctuations ($\delta X / X$) will be inversely proportional to the square root of the number of particles:

$$\frac{\delta X}{X} = \frac{1}{\sqrt{N}} \qquad\qquad [3.3]$$

If we want to overlook the fluctuations, we must consider "macroscopic portions" of the universe containing many structural units (N). However, when we deal with an out-of-equilibrium system, distinct parts of it may have a different temperature, pressure, chemical composition, et cetera. When we describe the behavior of the system, it is convenient to partition it into smaller subsystems that are in internal equilibrium, i.e., have uniform values of their intensive thermodynamic variables. This approach is called "approximation of local equilibrium."

In Figure 3.1, there is a schematic representation of a generic out-of-equilibrium system. Inside this system, there is not a unique value of T, P, and a uniform distribution of the chemicals. The intensive variables, T, P, and μ_k (the chemical potential of the k-th species) depend on the spatial coordinates (x, y, z) and the time (t): $T(x,y,z,t), P(x,y,z,t), \mu_k(x,y,z,t)$. To describe the evolution of the entire system, we partition it in a certain number of smaller "macroscopic" regions (those labeled by different capital letters in Figure 3.1), within which the conditions of internal equilibrium are verified. The extensive variables of these subsystems can be transformed into densities, dividing them by the volume of each portion. Instead of having the total internal energy U, the total entropy S, and the total moles of k, and so on, we will have the density of internal energy ($u = U/V$), the density of entropy ($s = S/V$), the molar concentration of k ($C_k = n_k/v$), which are intensive variables that depend on time and spatial coordinates: $u(x,y,z,t), s(x,y,z,t), C_k(x,y,z,t)$.

Within each subsystem of Figure 3.1, it is possible to apply equations [3.1] and [3.2]. This approach is known as "local formulation" of the Second Principle of Thermodynamics. For the entire system, we can write

$$d_iS_{tot} = d_iS_A + d_iS_B + \ldots + d_iS_H \geq 0 \qquad\qquad [3.4]$$

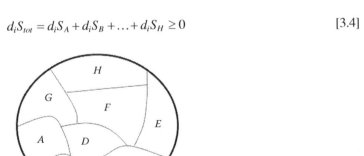

FIGURE 3.1 Partition of the out-of-equilibrium system in eight macroscopic subsystems. The partition must be performed in such a way that each subsystem is in internal equilibrium, and the microscopic fluctuations can be overlooked.

The terms d_iS will be positive or null for each subsystem. It may happen that inside the same subsystem there are transformations reducing entropy, which are "coupled" with transformations producing entropy. The overall internal balance will determine an entropy growth in agreement with the Second Law of Thermodynamics. It may also occur that two or more subsystems are combined by a common transformation: in some of them, the internal entropy decreases, whereas into the others increases of a larger amount, such that the second principle is not violated.

When we apply the approximation of local equilibrium, we overlook the flows between pairs of subsystems. Such flows are generated by the inevitable gradients that are present between subsystems. They are negligible as far as the overall system is not so far from equilibrium condition. On the other hand, if the system is very far-from-equilibrium, it is not fair to overlook the gradients and the respective fluxes. In these situations, we need to apply the theory of "Extended Thermodynamics" (Jou et al. 2010).

Finally, the equations [3.1] and [3.2], which we have introduced at the beginning of this paragraph, are postulates that are indirectly confirmed by their power of describing the behavior of out-of-equilibrium systems. To become familiar with them, we are now dealing with some examples of irreversible transformations, the first of which is the conduction of heat.

3.2.1 HEAT CONDUCTION

Imagine having a system consisting of two metallic cylinders, each having a certain temperature (Figure 3.2): T_A for the block A and T_B for block B, with $T_A \neq T_B$. If the surrounding environment has another T_{env}, different from those of the two metals, we must consider three distinct flows of heat:[1]

1. Either the heat flow from A to B ($d_iq_B > 0$) or the heat flow from B to A ($d_iq_A > 0$)
2. The heat flow from the environment to A ($d_eq_A > 0$)
3. The heat flow from the environment to B ($d_eq_B > 0$)

The overall entropy change for the system is given by

$$dS_{tot} = dS_A + dS_B \tag{3.5}$$

$$dS_{tot} = d_iS_A + d_eS_A + d_iS_B + d_eS_B \tag{3.6}$$

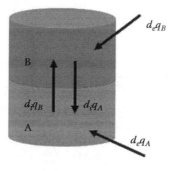

FIGURE 3.2 A system consisting of two cylinders, A and B, at two different uniform temperatures, T_A and T_B, which exchange heat between them and the environment.

---

[1] In this book, we choose the convention according to which dq is positive when the heat is absorbed by the system, and dw is positive when the work is performed on the system.

$$dS_{tot} = \frac{d_i q_A}{T_A} + \frac{d_e q_A}{T_A} + \frac{d_i q_B}{T_B} + \frac{d_e q_B}{T_B} \qquad [3.7]$$

From the energy conservation law, it derives that $d_i q_A = -d_i q_B = dq$. Therefore,

$$d_i S_{tot} = dq \left(\frac{1}{T_A} - \frac{1}{T_B} \right) \geq 0 \qquad [3.8]$$

From equation [3.8], it derives that if A is warmer than B ($T_A > T_B$), heat is released from A to B, then $dq < 0$. On the other hand, if $T_B > T_A$, the warmer B will release heat to the colder A, then $dq > 0$. Of course, if A and B are at the same T, there is no net exchange of heat, then $dq = 0$. If we divide both terms of equation [3.8] by the unit of time, we achieve the definition of entropy production P^* (remember equation [2.19]):

$$P^* = \frac{d_i S_{tot}}{dt} = \frac{dq}{dt} \left(\frac{1}{T_A} - \frac{1}{T_B} \right) \qquad [3.9]$$

It is evident that the entropy production corresponds to the product between two distinct terms: (1) the flow of heat, which is the rate of the irreversible process; and (2) $(1/T_A - 1/T_B)$, which is the cause, or the "force" of the irreversible process since its sign rules the direction of the flow.

   If the thermal gradient is not maintained, as the heat flows from the warmer to the colder block, the difference between the two temperatures eventually vanishes. As soon as the two subsystems have the same T, the system is at equilibrium, and there is no net exchange of heat between the two blocks.

3.2.2 CHEMICAL REACTIONS

Another example of an out-of-equilibrium system is schematically depicted in Figure 3.3. It is an open system exchanging energy and matter with its environment, performing mechanical work (because the boundaries are assumed to be flexible) and hosting a reaction in its interior. The internal energy change is

$$dU = TdS - PdV + \sum_k \mu_k d_i n_k + \sum_k \mu_k d_e n_k \qquad [3.10]$$

wherein $d_i n_k$ and $d_e n_k$ are the changes in the moles of the k-th species due to the internal chemical reaction and the influx of matter, respectively.

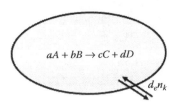

FIGURE 3.3 Scheme of an open system hosting a chemical reaction and exchanging matter with the environment.

The last two terms of equation [3.10] represent chemical work. The symbol μ_k is the chemical potential of the k-th species.

BOX 3.1   CHEMICAL POTENTIAL

The chemical potential for a k-th species is

$$\mu_k = \left(\frac{\partial G}{\partial n_k} \right)_{T,P,n_{j\neq k}}$$

It represents how the Gibbs free energy of the system varies when the number of moles of k-th species changes at T, P constant and when the moles of all the other species present in the system do not change. In the case of just one species, μ_k is the molar free energy at specific T and P, which is the chemical energy available to carry out work. The equation that links the chemical potential of a species to its concentration or pressure is

$$\mu_k = \mu_k^0 + RT ln C_k$$

$$\mu_k = \mu_k^0 + RT ln \left(P_k / P^0 \right)$$

wherein μ_k^0 is the chemical potential in standard conditions (i.e., at $P^0 = 1$ atm and temperature T), and C_k and P_k are the concentration and pressure of k-th species, respectively. The previous equations are valid when the solid, liquid or gaseous solutions show ideal behaviors. In the case of real solutions, the concentration must be substituted by the activity $a_k = \gamma_k C_k$ and the pressure by the fugacity $f_k = \gamma_k P_k$ in the definition of μ_k. The coefficients γ_k of activity or fugacity quantify how much the real solution deviates from the ideal case.

The changes in the moles of the different species involved in a chemical reaction can be expressed as $d_i n_k = \nu_k d\xi$, where ν_k is the stoichiometric coefficient of the k-th compound (being positive for a product and negative for a reagent), whereas $d\xi$ is the change in the extent of reaction. Therefore:

$$dw = \sum_k \nu_k \mu_k d\xi + \sum_k \mu_k d_e n_k \qquad [3.11]$$

The total variation of entropy for the system of Figure 3.3 can be split in its two contributions, $d_i S$ and $d_e S$, according to equation [3.1], and we obtain:

$$d_e S = \frac{dU}{T} + \frac{pdV}{T} - \frac{1}{T} \sum_k \mu_k d_e n_k \qquad [3.12]$$

$$d_i S = -\frac{1}{T} \sum_k \nu_k \mu_k d\xi \qquad [3.13]$$

The definition of the chemical work associated with the chemical reaction can be further simplified, introducing the concept of "chemical affinity" A (proposed for the first time by Théophile De Donder):[2]

$$A = -\sum_k \nu_k \mu_k \qquad [3.14]$$

---

[2] Théophile De Donder (Bruxelles 1872–1957) was a mathematician and physicist, founder of the Belgian school of thermodynamics.

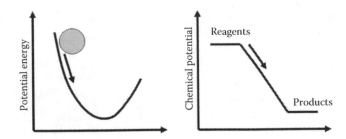

FIGURE 3.4 Analogy between a ball sliding over a potential well and a chemical reaction wherein the reagents at high chemical potentials convert to the products at lower potentials.

The chemical affinity A is the difference between the sum of the chemical potentials of the reagents and the sum of the chemical potentials of the products; wherein every chemical potential is multiplied by its stoichiometric coefficient. When $A > 0$ (it means the sum of the chemical potentials of the reagents is larger than that of the products), the reagents spontaneously transform to the products. The species at higher potentials transform into other species having lower potentials. This behavior is similar to the phenomenon of a ball that slides along a well from a point at high potential energy to the minimum of the well (see Figure 3.4).

Introducing the definition of A, from equation [3.14], into equation [3.13], we obtain:

$$d_i S = \frac{A}{T} d\xi \qquad [3.15]$$

Therefore, the entropy production for the system of Figure 3.3 is

$$P^* = \frac{d_i S}{dt} = \frac{A}{T} \frac{d\xi}{dt} \geq 0 \qquad [3.16]$$

Analogously to what we learnt in the case of heat conduction, the entropy production in [3.16] results the product of two terms: (1) $(d\xi/dt) = v$, which represents the overall rate of the reaction, and can be conceived as a "flow" of matter; (2) A/T, which rules the direction and the extent of the flow of matter, i.e., it is the force generating the flow. From equation [3.16], it derives that if $A > 0$, then $v > 0$, i.e., the reagents convert to the products. On the other hand, if $A < 0$, then $v < 0$, i.e., the products transform into the reagents. Finally, if $A = 0$, then $v = 0$. This latter condition corresponds to the state of chemical equilibrium, wherein the rate of reagents-to-products transformation is equal to the rate of the products-to-reagents transformation.

If more than one chemical reaction proceeds inside the system, the overall entropy production will be the sum of the contributions of each reaction:

$$P^* = \frac{d_i S}{dt} = \sum_j \frac{A_j}{T} \frac{d\xi_j}{dt} \geq 0 \qquad [3.17]$$

Usually, each reaction gives a positive contribution to the entropy production. However, there are exceptions. For instance, in case of two simultaneous reactions occurring inside the same system, it may happen that:

$$\frac{A_1}{T} \frac{d\xi_1}{dt} > 0 \text{ and } \frac{A_2}{T} \frac{d\xi_2}{dt} < 0, \text{ but } \frac{A_1}{T} \frac{d\xi_1}{dt} + \frac{A_2}{T} \frac{d\xi_2}{dt} > 0 \qquad [3.18]$$

Equation [3.18] shows that one reaction goes unexpectedly uphill, whereas the other goes downhill. However, the sum of the entropy production due to the two reactions is positive. Situations similar to that of equation [3.18] are possible when the two reactions are coupled, i.e., they have two or more species in common.[3]

3.2.3 DIFFUSION

Let us imagine having a tank divided into two portions, A and B, containing only the k-th species but at two different concentrations: $C_{k,A}$ in A and $C_{k,B}$ in B (see Figure 3.5).

If the system is closed, delimited by rigid walls, and at a uniform temperature, the only irreversible process occurring inside it is the diffusion of k-th molecules. The internal change of entropy will be given by:

$$d_iS = -\frac{\mu_{k,A}d_in_{k,A}}{T} - \frac{\mu_{k,B}d_in_{k,B}}{T}$$
[3.19]

The Conservation Law of mass allows us to write: $-d_in_{k,A} = d_in_{k,B} = d_in_k$. Therefore, the entropy production is

$$\frac{d_iS}{dt} = \frac{(\mu_{k,A} - \mu_{k,B})}{T}\frac{d_in_k}{dt}$$
[3.20]

The entropy production, as shown in equations [3.20], is again the product of two terms: (1) the rate of exchange of k-th moles, $\left(\frac{d_in_k}{dt}\right)$; (2) the gradient of chemical potential divided by the temperature, $\frac{(\mu_{k,A}-\mu_{k,B})}{T}$, which drives the direction and the extent of flow of matter. If the gradient is positive, that is $\mu_{k,A} > \mu_{k,B}$, it means that $C_{k,A} > C_{k,B}$ and the k-th molecules diffuse from the A to the B region. On the other hand, if $\mu_{k,A} < \mu_{k,B}$, the k-th molecules diffuse from the B to the A compartment. When the chemical potential of the k-th species is uniform, there is no net diffusion, and the system is at the equilibrium.

3.2.4 MIGRATION

When a chemical species is within a conservative vector field,[4] its molar Gibbs free energy is expressed through the extended chemical potential, $\bar{\mu}_k$:

$$\bar{\mu}_k = \mu_k + \bar{\tau}_k\psi = \mu_k^0 + RTlnC_k + \bar{\tau}_k\psi$$
[3.21]

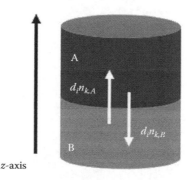

z-axis

FIGURE 3.5 Tank divided into two portions (A and B) containing the k-th species at two different concentrations.

---

[3] Coupled reactions are important in biochemistry, where the system hosting the reactions could be a cell or an organelle.
[4] A conservative vector field has the property that the value of the line integral from one point to another is path independent; it depends just on the position of the starting and final points enclosed into the conservative vector field.

where μ_k^0 is the chemical potential in standard conditions, and $\overline{\tau}_k\psi$ represents the molar potential energy of the k-th species immersed inside the conservative vector field of potential ψ. In the case of charged atoms or molecules within an electric field ($\vec{E}$), we have

$$\overline{\tau}_k\psi = z_k F\psi \text{ with } \vec{E} = -\vec{\nabla}\psi \qquad [3.22]$$

where z_k is the charge of the k-th molecule, and F is the Faraday constant (i.e., the charge per mole of electrons: 96,485 C/mol).

In the case of dipolar molecules embedded in an electric field, we have

$$\overline{\tau}_k\psi = -N_{Av}p_k E\cos\theta \qquad [3.23]$$

wherein N_{Av} is the Avogadro number, p_k is the dipole moment of the k-th molecule and θ is the angle included between the direction of the electric field and that of the dipole moment.

In case of a gravitational field generating a force $\vec{F}_G$ per unit of mass, we have

$$\overline{\tau}_k\psi = \left(MW_k\right)\psi \text{ with } \vec{F}_G = -\vec{\nabla}\psi \qquad [3.24]$$

where MW_k is the molecular weight of k-th molecule.

In the previous paragraph we learned that when inside a system there is a gradient of chemical potential due to an inhomogeneous distribution of a compound, there will be diffusion of its molecules. The diffusion tends to reduce the gradient until it becomes null. When it is null, the system is at equilibrium. If this system is immersed inside a vector field exerting a force on the k-th species, the k-th extended chemical potential will not be uniform. The vector field pushes the system out-of-equilibrium. For a matter of simplicity, let us imagine that the vector field generates forces along just one spatial coordinate: the z-axis. The extended chemical potential will be a function of the z coordinate, and between two points separated by an infinitesimal distance dz, there will be a gradient: $\overline{\mu}_k\left(z_0\right) - \overline{\mu}_k\left(z_0 + dz\right)$ (see Figure 3.6a). This gradient induces the migration, i.e., an ordered motion of molecules of k-th species along the direction of the vector field.

The internal energy variation inside the volume $dV = (Ar)dz$ will be

$$\frac{dU}{dV} = du = T\frac{dS}{dV} + \left(\overline{\mu}_k\left(z_0 + dz\right) - \overline{\mu}_k\left(z_0\right)\right)dC_k \qquad [3.25]$$

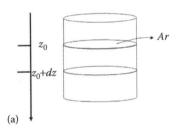

(a)

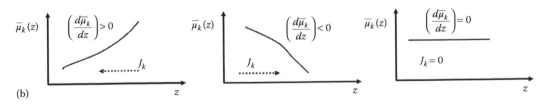

(b)

FIGURE 3.6 Migration of a species contained in a cylinder of section Ar in the presence of a vector field directed along the z-axis (graph a). The relationship between the gradient of the extended chemical potential $\left(\frac{d\overline{\mu}_k}{dz}\right)$ and the flow of matter J_k (graphs labeled as b).

The term $\bar{\mu}_k\left(z_0 + dz\right)$ can be expressed through the Taylor series expansion arrested at the first-order term:

$$\bar{\mu}_k\left(z_0 + dz\right) \approx \bar{\mu}_k\left(z_0\right) + \left(\frac{d\bar{\mu}_k}{dz}\right)_{z_0} dz + \dots \qquad [3.26]$$

It derives that the entropy production per unit of volume is

$$\frac{d_i S}{dVdt} = \frac{P^*}{dV} = p^* = -\frac{1}{T}\left(\left(\frac{d\bar{\mu}_k}{dz}\right)_{z_0} dz\right)\frac{dC_k}{dt}$$

$$= -\frac{1}{T}\left(\frac{d\bar{\mu}_k}{dz}\right)_{z_0}\left(\frac{dn_k}{(Ar)dt}\right) = -\frac{1}{T}\left(\frac{d\bar{\mu}_k}{dz}\right)_{z_0} J_k\left(z\right) \geq 0 \qquad [3.27]$$

wherein $J_k(z)$ is the flow of the k-th species along the z-axis. The entropy production per unit of volume is the product of a flow ($J_k(z)$) and the gradient of extended chemical potential divided by T. The latter term represents the cause of the flow; it is also defined as the thermodynamic force responsible of $J_k(z)$. Based on equation [3.27], we can claim that when $\left(\frac{d\bar{\mu}_k}{dz}\right)_{z_0}$ is positive, the flow is towards the negative direction of z; when $\left(\frac{d\bar{\mu}_k}{dz}\right)_{z_0}$ is negative, the flow is towards the positive direction of z (see figure 3.6b). Finally, when $\left(\frac{d\bar{\mu}_k}{dz}\right)_{z_0}$ is null, the flow of migration is null. If our system starts from a uniform distribution of k-th species, whereby μ_k does not depend on z, introducing equation [3.21] into equation [3.27], we achieve that

$$p^* = -\frac{\bar{\tau}_k}{T}\left(\frac{d\psi}{dz}\right)_{z_0} J_k \geq 0 \qquad [3.28]$$

The term $-\left(\frac{d\psi}{dz}\right)_{z_0}$ represents the force due to the vector field. The flow of migration ceases when the extended chemical potential has the same value everywhere into space.

Equation [3.28] can be applied to define the entropy production for the process of electrical conduction inside a mono-dimensional electrical wire (extended along the z-axis). The electrical wire is a conductor wherein we can assume that the electron density and the temperature have spatially uniform values. Therefore, the electron chemical potential is constant along z, and the entropy production per unit of length reduces to:

$$p^* = \frac{F}{T}\left(\frac{d\psi}{dz}\right)_{z_0}\left(\frac{dn_e}{(Ar)dt}\right) = \frac{E_z}{T}\left(\frac{I}{Ar}\right) \geq 0 \qquad [3.29]$$

where E_z is the electrical field along z, and I is the electrical current.

If we integrate equation [3.29] along the entire length of the wire (L), we obtain

$$P^* = \frac{d_i S}{dt} = \left(\frac{1}{T}\right)\int_0^L E_z I dz = \frac{(\Delta V)I}{T} \geq 0 \qquad [3.30]$$

The term ΔV represents the potential difference across the entire wire: if $\Delta V > 0$, the electrons flow towards the positive direction of z. On the other hand, if $\Delta V < 0$, the electrons migrate in the opposite direction. Finally, if $\Delta V = 0$, there is no electron migration.

TRY EXERCISES 3.1 AND 3.2

3.2.5 Generalization

The thermodynamic analysis of heat conduction, chemical reaction, diffusion, and migration, has revealed that entropy production is always the product of two terms: a gradient (divided by T) and a flow. The gradient, divided by T, is a thermodynamic force (F). It drives a thermodynamic flow (J). The force is the cause of the flow. In the case of more than one force and hence of more than one flow, the entropy production is given by the summation

$$\frac{d_iS}{dt} = \sum_k F_k J_k \geq 0 \qquad [3.31]$$

If at least one force is not null, the system is out-of-equilibrium. On the other hand, when all the forces are null, the system is at the equilibrium. At the equilibrium, all the thermodynamic flows vanish. Remember that when a system evolves up to reach an equilibrium state, the entropy of the universe (i.e., the entropy of the system and environment) reaches a maximum, whereas the Gibbs free energy of the system reaches a minimum

TRY EXERCISE 3.3

The out-of-equilibrium systems can operate in two distinct regimes: (1) in the linear regime, when they are not so far-from-equilibrium, and (2) in the non-linear regime when they are very far-from-equilibrium. In the next paragraphs of this chapter, we are going to discover the properties of out-of-equilibrium systems in the two different regimes.

3.3 NON-EQUILIBRIUM THERMODYNAMICS IN LINEAR REGIME

When a system is out-of-equilibrium, but it is close to the equilibrium, the relationships linking the flows to the forces are linear. This statement grounds on few empirical laws, such as the Fourier's Law for heat conduction, Ohm's Law for electrical conduction, Fick's Law for diffusion, and the Poiseuille's Law for the laminar flow of fluids.

3.3.1 Fourier's Law: The Law of Heat Conduction

In the first half of the nineteenth century, the French mathematician and physicist Joseph Fourier formulated the empirical law of heat conduction, named after his surname as Fourier's Law:

$$\vec{J}_q = \frac{\partial q}{\partial t \partial (Ar)} = -k\vec{\nabla}T \qquad [3.32]$$

The heat, which flows per unit time and per unit of area (Ar), the latter being disposed perpendicular to the direction of the flow, is directly proportional to the gradient of temperature:

$$\vec{\nabla}T = \hat{i}\frac{\partial T}{\partial x} + \hat{j}\frac{\partial T}{\partial y} + \hat{k}\frac{\partial T}{\partial z} \qquad [3.33]$$

where $\hat{i}, \hat{j}, \hat{k}$ are the unit vectors of the three Cartesian coordinates. The higher the temperature gradient, the larger the heat flow. The proportionality constant, k, represents the thermal conductivity, which is a peculiar property of every material. It is larger in metals than in non-metallic materials. In metals, the valence electrons are not tightly bound to the reticular ions. Therefore, they move quite easily. The thermal conductivity can be increased by cooling down the metal because at low temperature the random thermal motion of electrons is damped.

In case of anisotropic solids, we do not have just one k, but many. Fourier's law for such systems has the following form:

$$
\begin{aligned}
J_{q,x} &= -k_{xx}\frac{\partial T}{\partial x} & -k_{xy}\frac{\partial T}{\partial y} & -k_{xz}\frac{\partial T}{\partial z} \\
J_{q,y} &= -k_{yx}\frac{\partial T}{\partial x} & -k_{yy}\frac{\partial T}{\partial y} & -k_{yz}\frac{\partial T}{\partial z} \\
J_{q,z} &= -k_{zx}\frac{\partial T}{\partial x} & -k_{zy}\frac{\partial T}{\partial y} & -k_{zz}\frac{\partial T}{\partial z}
\end{aligned}
\tag{3.34}
$$

The thermal conductivity is a tensor. If we consider the expression of the thermal force proposed in equation [3.9], i.e., as $\vec{\nabla}\left(\frac{1}{T}\right)$, equation [3.32] can be rearranged in the following linear form:[5]

$$
\vec{J}_q = \frac{\partial q}{\partial t \partial (Ar)} = kT^2 \vec{\nabla}\left(\frac{1}{T}\right) = L_q \vec{\nabla}\left(\frac{1}{T}\right)
\tag{3.35}
$$

where $L_q = kT^2$ is named as the phenomenological coefficient.

TRY EXERCISE 3.4

3.3.2 Ohm's Law: The Law of Electrical Conduction

In the first half of the nineteenth century, the German mathematician and physicist Georg Simon Ohm, drawing inspiration from Fourier's work on heat conduction, formulated the empirical law of electrical conduction. When an electric potential difference (ΔV) is applied between two points inside a conductor, the charges start to migrate orderly. The unidirectional migration of charges corresponds to electrical current. The intensity of the current is related to ΔV through the following equation:

$$
\Delta V = V_i - V_f = RI
\tag{3.36}
$$

The constant of proportionality R is the resistance of the conductor. For a conductor of length l and cross-sectional area Ar (see Figure 3.7), the resistance R is proportional to $\rho \cdot l/Ar$, where ρ is the resistivity. Introducing the definition of R in equation [3.36], we obtain

$$
\frac{\Delta V}{\rho l} = \sigma_e E = \frac{I}{(Ar)} = J_e
\tag{3.37}
$$

wherein $\sigma_e = 1/\rho$ is the electrical conductivity and J_e is the flow of charges.

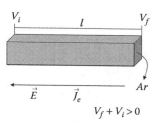

FIGURE 3.7 Block of a conductor of length l and section Ar crossed by a flow of electric current ($\vec{J}_e$) generated by the electric field $\vec{E}$. Note that $\vec{J}_e$ and $\vec{E}$ have the same direction and orientation.

---

[5] Note that $d\left(\dfrac{1}{T}\right) = -\dfrac{1}{T^2}dT$.

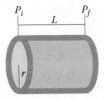

FIGURE 3.8 A tube of length L and radius r crossed by the fluid that flows due to a pressure gradient.

3.3.3 POISEUILLE'S LAW: THE LAW OF LAMINAR FLOW OF FLUIDS

The flow of fluid in parallel layers with no disruption between layers, termed *laminar flow*, is described by a law that is formally equivalent to that describing an electrical conduction. Instead of having the migration of charges due to an electrical potential difference, we have the non-turbulent movement of liquid due to a pressure gradient. The flow of the volume of liquid (dV/dt) is linearly dependent on the pressure gradient (ΔP), and the proportionality constant is the inverse of the resistance to flow R', which depends on the length L of the tube (see Figure 3.8), the fourth power of its radius r, and the viscosity η of the fluid:

$$\frac{dV}{dt} = \frac{1}{R'}\left(P_i - P_f\right) = \frac{\pi r^4}{8\eta L}\Delta P \qquad [3.38]$$

This law was formulated by the French physicist and physiologist Jean Leonard Poiseuille in the first half of the nineteenth century.

The physical meaning of viscosity η can be understood if we consider layers of liquid in contact with each other moving at different speeds (see Figure 3.9). If we apply a shear force on layer 1 of the liquid along the x-direction, a linear momentum will propagate along the z-axis from one layer to the other. The flow of linear momentum along z is linearly dependent on the gradient of v_x along the z-axis:

$$\vec{J}_{p_z} = -\eta\,\hat{k}\,\frac{dv_x}{dz} \qquad [3.39]$$

The proportionality constant is the viscosity, whose unit is the poise in the centimeter–gram–second (cgs) system, and the Poiseuille in the SI (International System of Unit). 1 poiseuille = 10 poise.

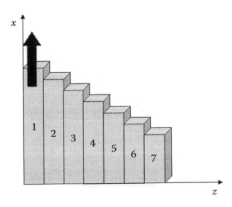

FIGURE 3.9 Laminar flow of a fluid partitioned in ideal seven layers. A shear force (represented by the black arrow) along the x-axis is applied to layer 1. A linear momentum propagates along the positive direction of z.

TRY EXERCISE 3.5

3.3.4 FICK'S LAW: THE LAW OF DIFFUSION

In the second half of the nineteenth century, the German physiologist Adolf Fick, investigating the fluxes of salts diffusing between two reservoirs through tubes of water, formulated a law that governs the transport of mass through any diffusive means (in the absence of bulk fluid motion). The flow of moles of the k-th solute ($\vec{J}_k$) depends on the gradient of its concentration ($\vec{\nabla}C_k$):[6]

$$\vec{J}_k = -D_k \vec{\nabla} C_k \qquad [3.40]$$

The constant of proportionality is the diffusion coefficient D_k. If we consider an ideal solution, wherein the intermolecular forces are not relevant, the chemical potential of the k-th component can be expressed as $\mu_k = \mu_k^0 + RT\ln C_k$. Knowing that $d\ln C_k = \left(\frac{1}{C_k}\right)dC_k$, we can rearrange equation [3.40] and write the Fick's law as a function of the thermodynamic force ($-\vec{\nabla}\mu_k/T$):

$$\vec{J}_k = -D_k C_k \vec{\nabla}\ln C_k = -\frac{D_k C_k}{RT}\vec{\nabla}\mu_k = -L_{k,d}\frac{\vec{\nabla}\mu_k}{T} \qquad [3.41]$$

wherein

$$L_{k,d} = \frac{D_k C_k}{R} \qquad [3.42]$$

is the phenomenological coefficient for the diffusion.

In the presence of a conservative vector field exerting a force on the k-th species, equation [3.41] can be rewritten in terms of the extended chemical potential $\bar{\mu}_k$ (see equation [3.21]). The flow of the k-th moles due to its concentration gradient and the vector field is given by two distinct contributions:

$$\vec{J}_k = -L_k \frac{\vec{\nabla}\bar{\mu}_k}{T} = -L_k R\vec{\nabla}\ln C_k - \frac{L_k \bar{\tau}_k}{T}\vec{\nabla}\psi \qquad [3.43]$$

The first contribution is the diffusion, which is described by equation [3.40]. The second contribution represents the migration. The vector field exerts a force drawing the k-th molecules towards the region where the potential is lower. Against this drift motion proceeding at velocity $\vec{v}_k$, the surrounding medium exerts a frictional force (see Figure 3.10), given by $\vec{F}_{fr} = -\gamma_k \vec{v}_k$ (γ_k is the frictional coefficient).

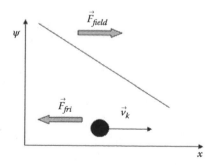

FIGURE 3.10 A conservative field exerts a force ($\vec{F}_{field}$) drawing a molecule (black circle) in the region where the potential (ψ) decreases. A frictional force, having opposite direction, hampers this motion.

---

[6] The concentration C_k is expressed in moles/volume units.

When the two forces balance, we obtain

$$\vec{F}_{field} = -\vec{F}_{fr} \tag{3.44}$$

$$-\tau_k \vec{\nabla}\psi = \gamma_k \vec{v}_k \tag{3.45}^7$$

The migration produces a flow given by:

$$\vec{J}_{k,field} = \vec{v}_k C_k = -\frac{\tau_k C_k}{\gamma_k}\vec{\nabla}\psi \tag{3.46}$$

The ratio τ_k/γ_k is named as the mobility of k, driven by a force field in a specific medium. By merging equations [3.46] and [3.43], we obtain another definition of the phenomenological coefficient L_k:

$$L_k = \frac{TC_k}{N_{Av}\gamma_k} \tag{3.47}$$

Combining the two definitions of L_k (equations [3.42] and [3.47]), we achieve an expression of the diffusion coefficient, known as Stokes–Einstein relation:

$$D_k = \frac{k_B T}{\gamma_k} \tag{3.48}$$

Note that both the diffusion coefficient and the mobility are inversely proportional to the friction coefficient. The Irish physicist Stokes (1819–1903) derived an expression for the frictional coefficient valid in case of small spherical molecules (having radius r_k) moving inside a continuous viscous fluid (with viscosity η):

$$\gamma_k = 6\pi\eta r_k \tag{3.49}$$

Combining equations [3.48] and [3.49], we obtain a relationship linking the microscopic parameters D_k and r_k with the macroscopic properties T and η.

TRY EXERCISE 3.6 AND 3.7

3.3.5 GENERALIZATION: SYMMETRY PRINCIPLE AND ONSAGER RECIPROCAL RELATIONS

In the previous paragraphs, we have ascertained that the phenomena of heat and electrical current conduction, laminar flow in fluids, diffusion, and migration, are all based on linear relationships between the flows and the forces: $J \propto LF$. One force may give rise to more than one flow. For example, a thermal gradient in a solution can promote heat flow and diffusion: this phenomenon is known as thermal diffusion, or Soret effect (see Section 3.3.8). On the other hand, a chemical potential gradient drives a matter flow and a heat flow, known as the Dufour effect (Section 3.3.8). It may occur that when a thermal gradient is applied between two metal junctions, it induces not only a thermal flow but also a flow of charges: this is the so-called Seebeck effect (see Section 3.3.6). On the other hand, when an electrical potential difference is applied between the two metal junctions, both a flow of charges and a thermal flow are originated according to the phenomenon known as Peltier effect (see Section 3.3.6). These are examples of cross effects in the linear regime. In case of cross effects, we can write the following relations:

---

[7] Note that $\tau_k N_{Av} = \overline{\tau}_k$.

$$J_j = \sum_k L_{jk} F_k \qquad [3.50]$$

$$\frac{d_i S}{dt} = \sum_j J_j F_j = \sum_{j,k} L_{jk} F_k F_j \geq 0 \qquad [3.51]$$

Equation [3.50] shows that the j-th flow is linearly dependent on k forces. The terms L_{jk} are named as "proper phenomenological coefficients" when $j = k$, whereas they are named as "cross phenomenological coefficients" when $j \neq k$.

Now a question arises: Is it possible to observe any combination of flows and forces in linear regime? To answer this question, we need (1) the "symmetry principle," and (2) the Onsager reciprocal relations.

1. The "symmetry principle" was formulated by the French physicist Pierre Curie (1894) before working on radioactivity with his wife, Marie. The symmetry principle states that when certain causes produce certain effects, the symmetry elements of the causes must be found in the produced effect. In other words, when certain effects show a certain asymmetry, such asymmetry must be found also in the causes that generated them. The symmetry principle asserts that the effects have the same or larger number of symmetry elements than their causes. In other words, macroscopic causes have a degree of symmetry always equal or lower than the effects they induce.[8] From this principle, we can infer that an isotropic force can originate an isotropic flow, but not an anisotropic flow. In other words, a scalar force gives rise just to a scalar flow (see Figure 3.11).

   An example of scalar force is A/T, i.e., the force for a reaction performed inside a thermo-stated well-stirred tank reactor. Such a force has a spherical symmetry, meaning it has all the symmetry elements and it is perfectly isotropic. It cannot originate an anisotropic flow, like a unidirectional thermal flow. Therefore, if we consider a system where a chemical reaction close to the equilibrium[9] is going on and the heat flows in one spatial direction, we can write

$$v = L_{cc} \frac{A}{T} + L_{cq} \vec{\nabla}\left(\frac{1}{T}\right)$$

$$\vec{J}_q = L_{qc} \frac{A}{T} + L_{qq} \vec{\nabla}\left(\frac{1}{T}\right) \qquad [3.52]$$

---

[8] The properties of symmetry for flows and forces may be described in terms of the presence of symmetry elements and their associated symmetry operations.

Symmetry Element	Symmetry Operation	Symbol
Identity	Identity	E
Proper axis	Rotation by $(360/n)°$	C_n
Plane	Reflection	σ
Inversion center	Inversion of a point x,y,z to $-x,-y,-z$	i
Improper axis	Rotation by $(360/n)°$ followed by reflection in a plane perpendicular to the rotation axis	S_n

[9] In this chapter, we will learn that the relation between the chemical flow and force is non-linear unless we are close to equilibrium.

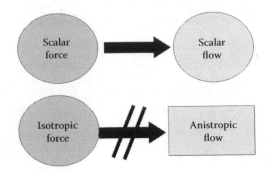

FIGURE 3.11   Schematic representation of the symmetry principle proposed by Pierre Curie.

The symmetry principle allows us to add that $L_{qc} = 0$. The scalar chemical force cannot be the cause of the thermal flow that is a vector. To know if the thermal gradient can be a cause of the chemical flow v, we need the Onsager reciprocal relations.

2. The "reciprocal relations" were first noticed by William Thomson (Lord Kelvin) in the nineteenth century, and then theoretically demonstrated by Lars Onsager in the first half of the twentieth century:[10]

$$L_{jk} = L_{kj} \qquad\qquad [3.53]$$

If we use equation [3.53] for the case indicated by equation [3.52], we infer that $L_{cq} = L_{qc} = 0$. This result can be generalized. If we consider a system hosting three different kinds of flows and forces of tensorial (*ten*),[11] vectorial (*vec*), and scalar (*sc*) features, respectively, we can expect to write the following equations in linear regime:

$$J_{ten} = L_{ten,ten}F_{ten} + L_{ten,vec}F_{vec} + L_{ten,sc}F_{sc}$$

$$J_{vec} = L_{vec,ten}F_{ten} + L_{vec,vec}F_{vec} + L_{vec,sc}F_{sc} \qquad [3.54]$$

$$J_{sc} = L_{sc,ten}F_{ten} + L_{sc,vec}F_{vec} + L_{sc,sc}F_{sc}$$

The symmetry principle along with the Onsager reciprocal relations allow us to infer that $L_{ten,vec} = 0 = L_{vec,ten}$; $L_{ten,sc} = 0 = L_{sc,ten}$; $L_{vec,sc} = 0 = L_{sc,vec}$. Therefore,

$$J_{ten} = L_{ten,ten}F_{ten}$$

$$J_{vec} = L_{vec,vec}F_{vec} \qquad [3.55]$$

$$J_{sc} = L_{sc,sc}F_{sc}$$

---

[10] The demonstration of the "reciprocal relations" by Onsager is based on the principle of detailed balance or microscopic reversibility that is valid for systems at equilibrium. For the interested reader, I recommend the original papers published by Onsager (1931a, 1931b) in the *Physical Reviews* journal.

[11] To have an idea of what a tensorial force is, imagine a tensorial force applied in one point along the *x*-axis and directed towards the positive direction of *x*. Such tensorial force can induce effects not only along *x*-axis, but also along *y*-axis and *z*-axis.

It derives that only forces and flows having the same degree of symmetry can interact mutually. In a system where there are tensorial, vectorial and scalar forces and flows, the total entropy production will be given by

$$\frac{d_i S}{dt} = \sum_{sc} J_{sc} F_{sc} + \sum_{vec} J_{vec} F_{vec} + \sum_{ten} J_{ten} F_{ten}$$

[3.56]

and for each term

$$\sum_{sc} J_{sc} F_{sc} \geq 0$$

$$\sum_{vec} J_{vec} F_{vec} \geq 0$$

$$\sum_{ten} J_{ten} F_{ten} \geq 0$$

[3.57]

3.3.6 AN EXPERIMENTAL PROOF OF THE RECIPROCAL RELATIONS

An experimental proof of the validity of the reciprocal relations (Miller 1960) regarding cross phenomenological coefficients is offered by the thermoelectric phenomena.

Suppose to have a thermocouple, i.e., two rods of different metals soldered together and to have the two soldered points at two different temperatures (see Figure 3.12). If the temperature gradient ΔT is maintained constant, it gives rise to an electromotive force (*emf*). Both a thermal and an electrical flow cross the two rods when the circuit is closed. This phenomenon is known as the Seebeck effect after the German physicist who discovered it in 1821. The *emf* is measured by a potentiometer when no current is permitted to flow:

$$emf = -\int_{x_1}^{x_2} \left(\frac{d\varphi}{dx} \right) dx$$

[3.58]

where φ is the electrical potential. Its derivative, with respect to T

$$\frac{d(emf)}{dT} = -\frac{d\varphi}{dT}$$

[3.59]

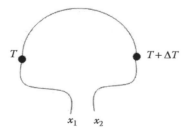

FIGURE 3.12 Sketch of a thermocouple. A potentiometer (in the case of Seebeck effect) or a battery (in the case of Peltier effect) are bound to the circuit at x_1 and x_2.

is called the thermoelectric power. Its value is small, of the order of a few $\mu V/K$. It can be exploited just to devise thermometers. However, arranging several thermocouples in a series, a thermopile is obtained that generates an appreciable *emf*.

The reverse of the Seebeck effect, which is called the Peltier effect (after the French physicist who discovered it in 1834), occurs when an electric potential is applied to the two junctions, kept at the same temperature. An electrical current will pass through the two rods and heat will be produced at one of the two junctions and absorbed at the other.[12] The ratio between the heat current (I_q) and the electrical current (I) is defined as Peltier heat (Π):

$$\Pi = \frac{I_q}{I} \qquad [3.60]$$

The Peltier effect can be used as a refrigerator.

For simplicity, we assume that our circuit is mono-dimensional. The entropy production per unit of length generated by the two irreversible processes, i.e., heat and electrical conduction, is given by:

$$p^* = -I_q \frac{1}{T^2}\left(\frac{dT}{dx}\right) - I\frac{1}{T}\left(\frac{d\varphi}{dx}\right) \qquad [3.61]$$

The linear relations for the two rods soldered together are

$$
\begin{aligned}
I_q &= \quad -L_{qq}\frac{1}{T^2}\left(\frac{dT}{dx}\right) \quad -L_{qe}\frac{1}{T}\left(\frac{d\varphi}{dx}\right) \\
I &= \quad -L_{eq}\frac{1}{T^2}\left(\frac{dT}{dx}\right) \quad -L_{ee}\frac{1}{T}\left(\frac{d\varphi}{dx}\right)
\end{aligned}
\qquad [3.62]
$$

If we want to observe the Seebeck effect, the electrical current must be null. From the second relation in equation [3.62], we derive that

$$-\frac{d\varphi}{dT} = \frac{d\left(emf\right)}{dT} = \frac{1}{T}\frac{L_{eq}}{L_{ee}} \qquad [3.63]$$

If we want to detect the Peltier effect, the junctions of the thermocouple are maintained at the same temperature (i.e., $\left(\frac{dT}{dx}\right) = 0$) while electric current passes through the rods. This condition causes a heat current from one junction to the other. The Peltier heat is given by

$$\Pi = \frac{I_q}{I} = \frac{L_{qe}}{L_{ee}} \qquad [3.64]$$

Few values of the Peltier heat per unit of temperature are reported in Table 3.1.

The ratio

$$\frac{\Pi}{T}\frac{dT}{d\left(emf\right)} = \frac{L_{qe}}{L_{eq}} \qquad [3.65]$$

---

[12] The thermoelectric phenomena ground on the involvement of two kinds of metals. Due to the two different crystal structures, electrons pass across the boundary more easily in one direction than in the other.

TABLE 3.1

Experimental Data Regarding the Peltier and Seebeck Effects and the Proof of the Reciprocal Relations. More Extended Data Can Be Found in Miller (1960)

Thermocouple	T (K)	$\dfrac{\Pi}{T}$ ($\mu V/K$)	$\dfrac{d(emf)}{dT}$ ($\mu V/K$)	$\dfrac{\Pi}{T}\dfrac{dT}{d(emf)}$
Fe-Ni	289	33.1	31.2	1.06
Fe-Al	273	11.0	11.5	0.956
Zn-Ni	290	22.1	22.0	1.00
Zn-Bi	273	25.4	25.1	1.01
Cu-Ni	287	20.2	20.7	0.976
Cu-Fe	273	−10.16	−10.15	1.000

allows us to verify the reciprocal relations. In Table 3.1, there are values of the ratio experimentally determined for a number of thermocouples. The ratio L_{qe}/L_{eq} is remarkably close to 1 leaving no doubt that the Onsager reciprocal relations are verified within the experimental error.

3.3.7 CROSS-DIFFUSION

The phenomenon of cross-diffusion offers another experimental proof of the validity of the Onsager's reciprocal relations. Cross-diffusion occurs when a flux of one species is induced by the concentration gradient of another species. For simultaneous diffusion of two or more species under isothermal conditions, a "generalized Fick's law" holds:

$$J_k = -\sum_i D_{k,i} \nabla C_i \qquad [3.66]$$

Examples of systems that exhibit cross-diffusion are strong-electrolytes, micelles, microemulsions, and mixtures containing molecules of significantly different sizes, for example, proteins and salts (Vanag and Epstein 2009). Three phenomena can be responsible for cross-diffusion: electrostatic interactions, excluded volume effects, and formation of aggregates or complexes. Cross-diffusion can play a role in phenomena of pattern formation in out-of-equilibrium conditions (Rossi et al. 2011; Budroni et al. 2015).

The cross-diffusion coefficients can be both positive and negative. When $D_{k,i}$ is positive, it means that species k tends to diffuse towards lower values of C_i (generating a co-flux). On the other hand, when $D_{k,i}$ is negative, species k diffuses towards larger values of C_i as if species i attracted species k (generating a counter-flux).

To verify the Onsager reciprocal relations through cross-diffusion, we need a system with at least three components. The entropy production per unit of volume for a system where isothermal diffusion of three components occurs is:

$$p^* = -\frac{J_1}{T}\nabla\mu_1 - \frac{J_2}{T}\nabla\mu_2 - \frac{J_3}{T}\nabla\mu_3 \qquad [3.67]$$

where the subscript 1, 2, and 3 refer to solute 1, solute 2, and solvent 3, respectively; the J_i terms (with $i = 1, 2, 3$) refer to the flows and the $\nabla\mu_i$ terms (with $i = 1, 2, 3$) refer to chemical potential gradients. Reminding that the total volume V_{TOT} of the system is defined by

$$V_{TOT} = n_1\bar{V}_1 + n_2\bar{V}_2 + n_3\bar{V}_3 \qquad [3.68]$$

where $\bar{V}_i$ is the partial molar volume of the i-th component, in the absence of volume flow, we may write:

$$0 = J_1\bar{V}_1 + J_2\bar{V}_2 + J_3\bar{V}_3 \qquad [3.69]$$

When temperature and pressure are constant, the Gibbs–Duhem relation [3.70] holds:

$$0 = n_1 d\mu_1 + n_2 d\mu_2 + n_3 d\mu_3 \qquad [3.70]$$

Since $d\mu_k = (\nabla\mu_k)dr$ (where r is a generic spatial coordinate), then equation [3.70] can be re-written in the following form:

$$0 = n_1\nabla\mu_1 + n_2\nabla\mu_2 + n_3\nabla\mu_3 \qquad [3.71]$$

Using equations [3.69] and [3.71], the terms J_3 and $\nabla\mu_3$ can be eliminated from the expression [3.67] that defines the entropy production. Therefore, we obtain

$$p^* = -\frac{J_1}{T}\left(\nabla\mu_1 + \frac{n_1}{n_3}\frac{\bar{V}_1}{\bar{V}_3}\nabla\mu_1 + \frac{n_2}{n_3}\frac{\bar{V}_1}{\bar{V}_3}\nabla\mu_2\right) - \frac{J_2}{T}\left(\nabla\mu_2 + \frac{n_2}{n_3}\frac{\bar{V}_2}{\bar{V}_3}\nabla\mu_2 + \frac{n_1}{n_3}\frac{\bar{V}_2}{\bar{V}_3}\nabla\mu_1\right) \qquad [3.72]$$

Equation [3.72] can be formally expressed in compact way as

$$p^* = J_1 Y_1 + J_2 Y_2 \qquad [3.73]$$

The corresponding linear relations take the form

$$J_1 = L_{11}Y_1 + L_{12}Y_2$$
$$J_2 = L_{21}Y_1 + L_{22}Y_2 \qquad [3.74]$$

To verify the Onsager's reciprocal relation, we must express the phenomenological coefficient L_{ik} as a function of the measured diffusion coefficients D_{ik} of the generalized Fick's law:

$$J_1 = -D_{11}\left(\frac{\partial C_1}{\partial r}\right) - D_{12}\left(\frac{\partial C_2}{\partial r}\right)$$
$$J_2 = -D_{21}\left(\frac{\partial C_1}{\partial r}\right) - D_{22}\left(\frac{\partial C_2}{\partial r}\right) \qquad [3.75]$$

From the latter two equations, it is evident that if J_1 is constant due to a constant concentration gradient for 1, and $J_2 = 0$, we have, anyway, a concentration gradient also for C_2.

The forces Y_1 and Y_2 of the equations [3.74] can be expressed in terms of the concentration gradients of species 1 and 2, according to the following equations:

$$\nabla\mu_1 = \left(\frac{\partial\mu_1}{\partial C_1}\right)\left(\frac{\partial C_1}{\partial r}\right) + \left(\frac{\partial\mu_1}{\partial C_2}\right)\left(\frac{\partial C_2}{\partial r}\right)$$
$$\nabla\mu_2 = \left(\frac{\partial\mu_2}{\partial C_2}\right)\left(\frac{\partial C_2}{\partial r}\right) + \left(\frac{\partial\mu_2}{\partial C_1}\right)\left(\frac{\partial C_1}{\partial r}\right) \qquad [3.76]$$

Introducing equations [3.76] in [3.72], and after fixing

$$a = -\frac{1}{T}\left[\left(1 + \frac{n_1\bar{V}_1}{n_3\bar{V}_3}\right)\left(\frac{\partial\mu_1}{\partial C_1}\right) + \left(\frac{n_2\bar{V}_1}{n_3\bar{V}_3}\right)\left(\frac{\partial\mu_2}{\partial C_1}\right)\right] \qquad [3.77]$$

$$c = -\frac{1}{T}\left[\left(1 + \frac{n_1\bar{V}_1}{n_3\bar{V}_3}\right)\left(\frac{\partial\mu_1}{\partial C_2}\right) + \left(\frac{n_2\bar{V}_1}{n_3\bar{V}_3}\right)\left(\frac{\partial\mu_2}{\partial C_2}\right)\right] \qquad [3.78]$$

$$b = -\frac{1}{T}\left[\left(\frac{n_1\bar{V}_2}{n_3\bar{V}_3}\right)\left(\frac{\partial\mu_1}{\partial C_1}\right) + \left(1 + \frac{n_2\bar{V}_2}{n_3\bar{V}_3}\right)\left(\frac{\partial\mu_2}{\partial C_1}\right)\right] \qquad [3.79]$$

$$d = -\frac{1}{T}\left[\left(\frac{n_1\bar{V}_2}{n_3\bar{V}_3}\right)\left(\frac{\partial\mu_1}{\partial C_2}\right) + \left(1 + \frac{n_2\bar{V}_2}{n_3\bar{V}_3}\right)\left(\frac{\partial\mu_2}{\partial C_2}\right)\right] \qquad [3.80]$$

the equations [3.74] become

$$\begin{aligned} J_1 &= -\left(aL_{11} + bL_{12}\right)\left(\frac{\partial C_1}{\partial r}\right) \quad -\left(cL_{11} + dL_{12}\right)\left(\frac{\partial C_2}{\partial r}\right) \\ J_2 &= -\left(aL_{21} + bL_{22}\right)\left(\frac{\partial C_1}{\partial r}\right) \quad -\left(cL_{21} + dL_{22}\right)\left(\frac{\partial C_2}{\partial r}\right) \end{aligned} \qquad [3.81]$$

Comparing equations [3.81] with [3.76], we can obtain the expressions of the phenomenological coefficients as functions of the diffusion coefficients:

$$L_{11} = \frac{dD_{11} - bD_{12}}{\left(ad - bc\right)} \qquad [3.82]$$

$$L_{12} = \frac{aD_{12} - cD_{11}}{\left(ad - bc\right)} \qquad [3.83]$$

$$L_{21} = \frac{dD_{21} - bD_{22}}{\left(ad - bc\right)} \qquad [3.84]$$

$$L_{22} = \frac{aD_{22} - cD_{21}}{\left(ad - bc\right)} \qquad [3.85]$$

From equations [3.83] and [3.84], it derives that the Onsager's reciprocal relation is confirmed if

$$aD_{12} + bD_{22} = cD_{11} + dD_{21} \qquad [3.86]$$

being $\left(ad - bc\right) \neq 0$. Different ternary systems have been experimentally investigated. They are listed in Miller (1960) and in Kondepudi and Prigogine (1998). The relation $L_{12} = L_{21}$ holds within the experimental error.

3.3.8 THERMAL DIFFUSION

With thermal diffusion, we mean two phenomena. The first phenomenon is the relative motion of the components of a gaseous mixture or a liquid solution when there is a thermal gradient; it is named as Soret effect. The second phenomenon is a heat flow driven by concentration gradients, which is named as the Dufour effect. For both effects, the entropy production per unit volume can be written as

$$p^* = J_u\nabla\left(\frac{1}{T}\right) - \sum_k J_k\nabla\left(\frac{\mu_k}{T}\right) \qquad [3.87]$$

This definition of entropy production can be rearranged as

$$p^* = \left(J_u - \sum_k J_k \mu_k \right) \nabla \left(\frac{1}{T} \right) - \sum_k J_k \frac{1}{T} \nabla \mu_k \qquad [3.88]$$

The separation between the thermal and the concentration gradients is not complete because $\nabla \mu_k$ contains the gradient of T, as well. In fact, μ_k is a function of C_k, T, and pressure, P. The extended definition of $\nabla \mu_k$ can be achieved from the equation as follows:

$$d\mu_k = \left(\frac{\partial \mu_k}{\partial n_k} \right)_{T,P} dn_k + \left(\frac{\partial \mu_k}{\partial T} \right)_{n_k,P} dT + \left(\frac{\partial \mu_k}{\partial P} \right)_{T,n_k} dP \qquad [3.89]$$

The second term of equation [3.89] can be re-written as

$$\left(\frac{\partial \mu_k}{\partial T} \right)_{n_k,P} = \frac{\partial}{\partial T} \left(\frac{\partial G}{\partial n_k} \right)_{P,T} = \left(\frac{\partial}{\partial n_k} \left(\frac{\partial G}{\partial T} \right) \right)_{P,T} = -\left(\frac{\partial S}{\partial n_k} \right)_{P,T} \qquad [3.90]$$

The last term of equation [3.90] is the partial molar entropy of species k. The variation of the variables $x = \mu_k$, n_k, T and, P with respect to the spatial coordinates r, can be written as $dx = (\nabla x)dr$. Introducing this latter definition in equation [3.89], it derives that

$$\nabla \mu_k = \left(\frac{\partial \mu_k}{\partial n_k} \right)_{T,P} \nabla n_k - S_k \nabla T + \left(\frac{\partial \mu_k}{\partial P} \right)_{T,n_k} \nabla P \qquad [3.91]$$

If we assume that our system is in mechanical equilibrium, $\nabla P = 0$. Introducing [3.91] in [3.88], after rearranging the term $-S_k \nabla T$ in $S_k T^2 \nabla \left(\frac{1}{T} \right)$, we obtain:

$$p^* = \left(J_u - \sum_k J_k (\mu_k + TS_k) \right) \nabla \left(\frac{1}{T} \right) - \sum_k J_k \frac{1}{T} \left(\frac{\partial \mu_k}{\partial n_k} \right)_{T,P} \nabla n_k \qquad [3.92]$$

The term $(\mu_k + TS_k) = H_k$ is the partial molar enthalpy, that is $\left(\frac{\partial H}{\partial n_k} \right)_{P,T}$. Therefore, from equation [3.92], it derives that a heat flow that includes a matter flow is given by

$$J_q = J_u - \sum_k J_k H_k \qquad [3.93]$$

If we express the entropy production per unit volume as function of J_q that is called the reduced heat flow, we have

$$p^* = J_q \nabla \left(\frac{1}{T} \right) - \sum_k J_k \frac{1}{T} \left(\frac{\partial \mu_k}{\partial n_k} \right)_{T,P} \nabla n_k \qquad [3.94]$$

Let us imagine having a two-component system. From the Gibbs-Duhem relation [3.70], it derives that

$$n_1 \left(\nabla \mu_1 \right)_{P,T} + n_2 \left(\nabla \mu_2 \right)_{P,T} = n_1 \left(\frac{\partial \mu_1}{\partial n_1} \right)_{T,P} \nabla n_1 + n_2 \left(\frac{\partial \mu_2}{\partial n_2} \right)_{T,P} \nabla n_2 = 0 \qquad [3.95]$$

Moreover, in the absence of volume flow, we may write, analogously to equation [3.69], the following equation:

$$0 = J_1 \bar{V}_1 + J_2 \bar{V}_2 \qquad [3.96]$$

Introducing equations [3.95] and [3.96] into [3.94], the definition of entropy production becomes

$$p^* = J_q \nabla \left(\frac{1}{T} \right) - \frac{J_1}{T} \left(1 + \frac{n_1 \bar{V}_1}{n_2 \bar{V}_2} \right) \left(\frac{\partial \mu_1}{\partial n_1} \right)_{T,P} \nabla n_1 \qquad [3.97]$$

We are in the presence of two independent forces $\left(\nabla \left(\frac{1}{T} \right) \right.$ and $-\frac{1}{T} \left(1 + \frac{n_1 \bar{V}_1}{n_2 \bar{V}_2} \right) \left(\frac{\partial \mu_1}{\partial n_1} \right)_{T,P} \nabla n_1$ or $-\frac{1}{T} \left(1 + \frac{n_2 \bar{V}_2}{n_1 \bar{V}_1} \right) \left(\frac{\partial \mu_2}{\partial n_2} \right)_{T,P} \nabla n_2 \right)$ and two independent flows (J_q and J_1 or J_2).

The phenomenological linear laws are:

$$
\begin{aligned}
J_q &= L_{qq} \nabla \left(\frac{1}{T} \right) & -L_{q1} \frac{1}{T} \left(1 + \frac{n_1 \bar{V}_1}{n_2 \bar{V}_2} \right) \left(\frac{\partial \mu_1}{\partial n_1} \right)_{T,P} \nabla n_1 \\
J_1 &= L_{1q} \nabla \left(\frac{1}{T} \right) & -L_{11} \frac{1}{T} \left(1 + \frac{n_1 \bar{V}_1}{n_2 \bar{V}_2} \right) \left(\frac{\partial \mu_1}{\partial n_1} \right)_{T,P} \nabla n_1
\end{aligned}
\qquad [3.98]
$$

The term $L_{1q} \nabla \left(\frac{1}{T} \right) = -\frac{L_{1q}}{T^2} \nabla T$ represents the Soret effect. It is also expressed as $-C_1 D_S \nabla T$, where D_S is the coefficient of thermal diffusion. The ratio

$$c_S = \left(\frac{D_S}{D_1} \right) \qquad [3.99]$$

(where D_1 is the ordinary diffusion coefficient for species 1) is called the Soret coefficient, that has the dimension of $[T]^{-1}$. A thermal gradient applied to a solution contained in a closed vessel gives rise to a concentration gradient. The value of the concentration gradient can be obtained from the second equation in [3.98], setting $J_1 = 0$:

$$\nabla C_1 = -C_1 \frac{D_S}{D_1} \nabla T = -C_1 c_S \nabla T \qquad [3.100]$$

The Soret coefficient is generally small, of the order of 10^{-3} or $10^{-2} \, K^{-1}$.

The cross term $-L_{q1} \frac{1}{T} \left(1 + \frac{n_1 \bar{V}_1}{n_2 \bar{V}_2} \right) \left(\frac{\partial \mu_1}{\partial n_1} \right)_{T,P} \nabla n_1$ in first equation of [3.98] represents the Dufour effect, that is a heat flow generated by a concentration gradient. The Dufour effect is expressed as $-C_1 D_D \nabla n_1 = -L_{q1} \frac{1}{T} \left(1 + \frac{n_1 \bar{V}_1}{n_2 \bar{V}_2} \right) \left(\frac{\partial \mu_1}{\partial n_1} \right)_{T,P} \nabla n_1$ where D_D is the Dufour coefficient. The Onsager reciprocal relation $L_{q1} = L_{1q}$ imposes that

$$\frac{D_D}{D_S} = T \left(1 + \frac{n_1 \bar{V}_1}{n_2 \bar{V}_2} \right) \left(\frac{\partial \mu_1}{\partial n_1} \right)_{T,P} \qquad [3.101]$$

Equation [3.101] has been verified in ideal gas mixtures and liquid isotope solutions (Würger 2014).

3.4 EVOLUTION OF OUT-OF-EQUILIBRIUM SYSTEMS IN LINEAR REGIME

A system is maintained out-of-equilibrium when at least one force is not null. If we fix the value of the force, how does the system evolve when it works in linear regime?

We are now considering two examples: heat conduction and diffusion. We will notice that in the linear regime, the system evolves to a stationary state that has time-independent properties.

3.4.1 THE CASE OF HEAT CONDUCTION

Suppose to have a system of length L and rectangular section Ar (see Figure 3.13), which is in physical contact with two heat reservoirs at its ends. One reservoir is at high temperature (T_h), and the other is colder, at T_c. If we assume that the temperature gradient is not large, the heat will flow from the hot to the cold bath by conduction along the x-axis.

The entropy production per unit length and per unit of area due to the conduction process will be given by the product of the flow and the force:

$$p^*(x) = J_q \frac{d}{dx}\left(\frac{1}{T}\right) = -\frac{J_q}{T^2}\frac{dT}{dx} \qquad [3.102]$$

If we consider a portion of the system having an infinitesimal width along x-axis (see Figure 3.13), the net heat variation (according to the Conservation Law of Energy) is given by

$$\frac{dq}{dt} = J_q(x_0)(Ar) - J_q(x_0 + dx)(Ar) \qquad [3.103]$$

where $J_q(x_0)$ is the heat flow entering through Ar in x_0, whereas $J_q(x_0 + dx)$ is the heat flow getting out through Ar in $x_0 + dx$. We can express $J_q(x_0 + dx)$ as a Taylor series and obtain

$$\frac{dq}{dt} \approx J_q(x_0)(Ar) - \left[J_q(x_0) + \left(\frac{\partial J_q}{\partial x}\right)_{x_0} dx\right](Ar) = -\left(\frac{\partial J_q}{\partial x}\right)_{x_0}(Ar)dx \qquad [3.104]$$

If we introduce the Fourier's law [3.32] into equation [3.104], we achieve

$$\frac{dq}{dt} = k\left(\frac{\partial^2 T}{\partial x^2}\right)_{x_0}(Ar)(dx) \qquad [3.105]$$

If we employ the definition of the heat capacity at constant pressure $(C_p = dq/dT)$, finally we obtain

$$\frac{\partial T}{\partial t} = \frac{k(Ar)(dx)}{C_p}\left(\frac{\partial^2 T}{\partial x^2}\right)_{x_0} \qquad [3.106]$$

If the temperatures of the two reservoirs are kept constant, the system will evolve up to reach a time-independent stationary state, wherein the derivative of T with respect to time becomes null. The term $\left(\frac{\partial T}{\partial t}\right)$ is null when the second derivative of T with respect to x is null (see equation [3.106]). The latter condition is true when T is a linear function of x. In this condition, i.e., at the stationary state, J_q is constant and, therefore,

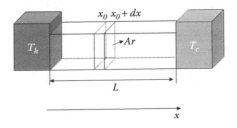

FIGURE 3.13 A system of length L is in physical contact with two reservoirs: one hot at T_h and the other cold at T_c. A portion of the system of section Ar and infinitesimal width (dx) is shown.

$$J_q \int_0^L dx = -k \int_{T_h}^{T_c} dT \qquad\qquad [3.107]$$

$$J_q L = -k\left(T_c - T_h\right) \qquad\qquad [3.108]$$

From equation [3.108] it is evident that the higher the thermal gradient, the larger the conductivity, the stronger the heat flow.

The total entropy production per unit of area is obtained by integrating along the entire length L of the system

$$p^* = \frac{d_i S}{dt} = \int_0^L -\frac{J_q}{T^2}\left(\frac{dT}{dx}\right)dx = -\int_{T_h}^{T_c} J_q\left(\frac{1}{T^2}\right)dT = J_q\left(\frac{1}{T_c} - \frac{1}{T_h}\right) \geq 0 \qquad\qquad [3.109]$$

At the stationary state, the positive entropy production is counterbalanced by the entropy that is exchanged with the environment:

$$\frac{d_e S}{dt} = \left(\frac{J_q}{T_h} - \frac{J_q}{T_c}\right)(Ar) = -\frac{d_i S}{dt} \qquad\qquad [3.110]$$

At the stationary state, the total entropy of the system is constant: it is produced inside and released into the environment.

TRY EXERCISES 3.8 AND 3.9

3.4.2 THE CASE OF DIFFUSION

Suppose to have two reservoirs containing compound k at two different concentrations: $C_{1,k}$ in reservoir 1 and $C_{2,k}$ in reservoir 2 (with $C_{1,k} > C_{2,k}$). The two reservoirs are kept in contact through a third vessel of length L along the x-axis and section Ar (see Figure 3.14a). Species k diffuses from

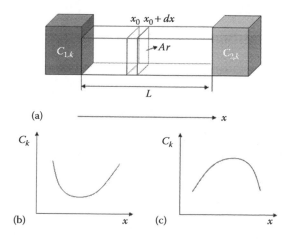

FIGURE 3.14 (a) Diffusion of the species k into a vessel of length L and in between two reservoirs containing k at concentrations of $C_{1,k}$ and $C_{2,k}$, respectively. Two unstable profiles of $C_k(x)$: (b) convex (positive curvature) and (c) concave (negative curvature).

reservoir 1 to 2. The variation of the number of k moles in a portion of the vessel having volume (Ar) (dx), according to the Conservation Law of Mass, is given by

$$\frac{\partial n_k}{\partial t} = J_k(x_0)(Ar) - J_k(x_0 + dx)(Ar) \approx J_k(x_0)(Ar) - \left[J_k(x_0) + \left(\frac{\partial J_k}{\partial x} \right)_{x_0} dx \right](Ar) \quad [3.111]$$

$$\frac{\partial n_k}{\partial t} \approx -\left(\frac{\partial J_k}{\partial x} \right)_{x_0} (Ar)dx \qquad [3.112]$$

It derives that

$$\frac{\partial C_k}{\partial t} \approx -\left(\frac{\partial J_k}{\partial x} \right)_{x_0} \qquad [3.113]$$

Introducing equation [3.40], known as the "first Fick's law," into equation [3.113], we achieve

$$\frac{\partial C_k}{\partial t} = D_k \frac{\partial^2 C_k}{\partial x^2} \qquad [3.114]$$

Equation [3.114] is termed as "diffusion equation" or "second Fick's law." It is formally equivalent to equation [3.106] formulated for heat conduction. It reveals that the variation of C_k on time is directly proportional to the curvature or concavity of the function $C_k(x)$. In particular, if $C_k(x)$ has a positive curvature, i.e., it is convex (sometimes referred to as concave up, meaning that the tangent line lies below the graph of the function, see Figure 3.14b), C_k will grow until it annihilates the trough. If $C_k(x)$ has a negative curvature, i.e., it is concave (sometimes referred to as concave down, meaning that the tangent line lies above the graph of the function, see Figure 3.14c), C_k will decrease up to level the hunch. Finally, when C_k is uniformly distributed or linearly distributed along x, it does not change over time.

If we fix the concentrations of species k into the two reservoirs at two different values, the system will evolve towards a steady state, wherein the diffusion flow of k is constant over time, and the k concentration depends linearly on x:

$$J_k \int_0^x dx = -D_k \int_{C_{1,k}}^{C_k(x)} dC_k \qquad [3.115]$$

$$C_k(x) = C_{1,k} - \frac{J_k}{D_k} x \qquad [3.116]$$

The entropy production per unit length and per unit of area due to the diffusion process will be given by the product of the flow and the force

$$p^*(x) = -J_k \frac{d}{dx}\left(\frac{\mu_k}{T} \right) \qquad [3.117]$$

At uniform T and for an ideal solution

$$p^*(x) = -\frac{J_k}{T} \frac{d\mu_k}{dx} = -RJ_k \frac{d\ln C_k}{dx} \qquad [3.118]$$

If we separate the variables, x, and lnC_k, and integrate between 0 and L, we obtain that the total entropy production $P*$ is

$$p^* = \int_0^L p^*(x)\,dx = -RJ_k \int_{C_{1,k}}^{C_{2,k}} dlnC_k = RJ_k ln\left(\frac{C_{1,k}}{C_{2,k}}\right) \qquad [3.119]$$

Since $C_{1,k} > C_{2,k}$, $P*$ is positive. At the stationary state, the entropy produced inside the system is released outside, and, in fact,

$$\frac{d_eS}{dt} = -RJ_k lnC_{1,k} + RJ_k lnC_{2,k} = -\frac{d_iS}{dt} \qquad [3.120]$$

As in the case of heat conduction, at the stationary state, the total entropy of the system is constant: it is produced inside and released into the environment.

TRY EXERCISES 3.10 AND 3.11

3.5 THE THEOREM OF MINIMUM ENTROPY PRODUCTION IN LINEAR REGIME

In Section 3.4, we have ascertained that a system, working out-of-equilibrium and in the linear regime, evolves spontaneously towards a stationary state wherein the variables generating the forces become linear functions of the spatial coordinates, whereas the flows become constant over time. When such stationary states are reached, the entropy production of the system is minimized. This statement is known as the Minimum Entropy Production Theorem (Nicolis and Prigogine 1977). Now, we demonstrate this theorem in two situations: (1) in the case of a single force and flow and (2) in the case of many forces and flows.

3.5.1 A SINGLE FORCE AND FLOW

For the system described in Section 3.4.2, consisting of a vessel of length L (along the x coordinate) and having two reservoirs at its extremes, containing the k-th species at concentration $C_{1,k}$, and $C_{2,k}$ respectively, the entropy production $P*$ due to the diffusion of the k-th species is

$$P^* = -\int_0^L J_k \left(\frac{\partial \mu_k / T}{\partial x}\right) dx \qquad [3.121]$$

If we insert equation [3.41] in [3.121], we obtain

$$P^* = \int_0^L L_{k,d} \left(\frac{\partial \mu_k / T}{\partial x}\right)^2 dx \qquad [3.122]$$

To find the minimum of $P*$, we exploit the Euler–Lagrange equation,[13] assuming that $f = \mu_k / T$ and $\dot{f} = \partial(\mu_k / T)/\partial x$:

$$\frac{d}{dx}\left(2L_{k,d}\dot{f}\right) = 0 \qquad [3.123]$$

---

[13] In the calculus of variation, the Euler–Lagrange equation is useful for solving optimization problems. If In is an integral of the form $In = \int_a^b \Lambda(x, f(x), \dot{f}(x))\,dx$, where the integrand Λ is a function of x, f and its derivative with respect to x ($\dot{f}(x) = (\partial f/\partial x)$), it is extremized when it is verified the Euler–Lagrange differential equation $\frac{d}{dx}\frac{\partial\Lambda}{\partial\dot{f}} = \frac{\partial\Lambda}{\partial f}$.

It means that $2L_{k,d}\dot{f}$ is a constant (symbol *cost.*),

$$\frac{2L_{k,d}}{T}\frac{\partial \mu_k}{\partial x} = 2L_{k,d}R\frac{\partial lnC_k}{\partial x} = \frac{2L_{k,d}R}{C_k}\frac{\partial C_k}{\partial x} = cost \qquad [3.124]$$

Keeping in mind that $L_k = \frac{C_k D_k}{R}$, then the final result is

$$2D_k\frac{\partial C_k}{\partial x} = cost \qquad [3.125]$$

This means that J_k is constant and it confirms that P^* minimizes when C_k is a linear function of x.

TRY EXERCISE 3.12

3.5.2 The Case of More Than One Force and One Flow

Imagine having a system with two forces (F_1 and F_2) and two flows (J_1 and J_2), that are coupled. An example can be the thermal diffusion, where F_1 is the thermal gradient, whereas F_2 is the concentration gradient. The total entropy production will be

$$P^* = \int \left(J_1 F_1 + J_2 F_2\right) dV \qquad [3.126]$$

Assume that only F_1 is maintained at a fixed value, whereas F_2 is free to change over time. The system will evolve to a stationary state characterized by a null value for J_2 and a value different from zero for J_1. This stationary state minimizes the entropy production of the system. Since we are in linear regime

$$J_1 = L_{11}F_1 + L_{12}F_2$$
$$J_2 = L_{21}F_2 + L_{22}F_2 \qquad [3.127]$$

Exploiting the Onsager's reciprocal relation $L_{12} = L_{21}$, and inserting the definitions of the flows [3.127] into [3.126], we obtain

$$P^* = \int \left(L_{11}F_1^2 + 2L_{12}F_1 F_2 + L_{22}F_2^2\right) dV \qquad [3.128]$$

The function linking P^* to the two forces is represented in Figure 3.15: It is an elliptic paraboloid.

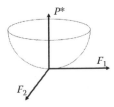

FIGURE 3.15 Dependence of P^* on the two forces F_1 and F_2 in linear regime and when the Onsager's reciprocal relations are applicable. Note that the equilibrium state for this system is located in a unique point of the graph, i.e., in the origin of the axes, where both F_1 and F_2 are null.

To find the minimum for P^* with respect to F_2, being F_1 fixed, we calculate $\frac{\partial P^*}{\partial F_2}$:

$$\frac{\partial P^*}{\partial F_2} = \int \left(2L_{12}F_1 + 2L_{22}F_2 \right) dV = 0 \tag{3.129}$$

In fact, the integrand is $2J_2$. The derivative of P^* with respect to F_2 is null when J_2 is null. This condition represents a minimum of P^* because

$$\frac{\partial^2 P^*}{\partial F_2^2} = \int 2L_{22} dV > 0 \tag{3.130}$$

In the case of thermal diffusion, if the thermal gradient is fixed, the system evolves up to reach the stationary state characterized by constant heat flow and the absence of the flow of matter.

The result achieved can be generalized to an arbitrary number of flows and forces. When we have a system with n forces, and n flows, if just the first $m < n$ forces are fixed, the system will evolve to a stationary state where the flows $J_{m+1}, J_{m+2}, \ldots, J_n$ are null, and the entropy production of the system is minimized.

In linear regime, a system changes over time in such a way that

$$\frac{dP^*}{dt} = \frac{d}{dt} \left(\frac{d_i S}{dt} \right) \le 0 \tag{3.131}$$

The time derivative of P^* becomes null at the stationary state, which corresponds to a minimum for P^*. Since P^* minimizes, the stationary state in linear regime is stable. In fact, the function P^* looks like the free energy G for a system that evolves towards an equilibrium state.

3.6 EVOLUTION OF OUT-OF-EQUILIBRIUM SYSTEMS IN NONLINEAR REGIME

When strong thermodynamic forces are present, they bring a system very far from equilibrium. These strong forces give rise to flows, which are in nonlinear relation with their causes. It is in the nonlinear regime that a system can exhibit several amazing dynamical evolutions, which depend on the initial and contour conditions, as we will discover in the rest of this book. Now, we focus on chemical reactions.

3.6.1 CHEMICAL REACTIONS

Let us consider an elementary chemical reaction that occurs as it is written in equation [3.132]. It involves four species, A, B, C, and D that react in a system showing ideal behavior; a, b, c, and d are the stoichiometric coefficients.[14]

$$aA + bB = cC + dD \tag{3.132}$$

The rate of the forward elementary step is

$$v_f = k_f \left[A \right]^a \left[B \right]^b \tag{3.133}$$

---

[14] For an ideal chemical system, the activity or the fugacity of a species is equivalent to its concentration or pressure, respectively. See also Box 3.1 in this chapter.

whereas the rate of the backward step is

$$v_b = k_b \left[C \right]^c \left[D \right]^d \tag{3.134}$$

and k_f and k_b are the kinetic constants of the forward and backward elementary steps, respectively. The net reaction rate is

$$v = v_f - v_b = k_f \left[A \right]^a \left[B \right]^b \left\{ 1 - \frac{k_b \left[C \right]^c \left[D \right]^d}{k_f \left[A \right]^a \left[B \right]^b} \right\} \tag{3.135}$$

When the reaction is at equilibrium, $v_f = v_b$ and $v = 0$, i.e.,

$$\frac{k_f}{k_b} = \frac{\left[C \right]^c_{eq} \left[D \right]^d_{eq}}{\left[A \right]^a_{eq} \left[B \right]^b_{eq}} = K \tag{3.136}$$

K is the equilibrium constant. Therefore,

$$v = v_f \left\{ 1 - \frac{Q}{K} \right\} \tag{3.137}$$

where Q is the reaction quotient:

$$Q = \frac{\left[C \right]^c \left[D \right]^d}{\left[A \right]^a \left[B \right]^b} \tag{3.138}$$

The chemical affinity A of the elementary reaction is

$$A = a\mu_A + b\mu_B - c\mu_C - d\mu_D \tag{3.139}$$

$$A = a\mu_A^0 + b\mu_B^0 - c\mu_C^0 - d\mu_D^0 + RTln\left(\frac{\left[A \right]^a \left[B \right]^b}{\left[C \right]^c \left[D \right]^d} \right) \tag{3.140}$$

$$A = RTlnK - RTlnQ \tag{3.141}$$

From equation [3.141], it derives that

$$\frac{Q}{K} = e^{-\frac{A}{RT}} \tag{3.142}$$

Introducing the latter equation in [3.137], we obtain

$$v = v_f \left\{ 1 - e^{-\frac{A}{RT}} \right\} \tag{3.143}$$

From equation [3.143], we infer that the chemical flow—the rate of the reaction—is in a nonlinear relationship with the chemical force, which has the ratio (A/T). However, equation [3.143] is not a pure thermodynamic definition of the rate because the kinetic term v_f appears on the right side of [3.143]. When $A \gg RT$, $v \approx v_f$, and v_f, like any other reaction rate, may depend on some non-thermodynamic factors, such as the presence of a catalyst or an inhibitor, etc.

3.6.2 The Glansdorff-Prigogine Stability Criterion

How do systems evolve when they work in nonlinear regime?

So far, we have learned that, in general, for an out-of-equilibrium system, the entropy production is $P^* = \sum_k J_k F_k$. The second principle of thermodynamics lets us state that $P^* \geq 0$. P^* is a function of both the flows and the forces. An infinitesimal variation of P^*, dP^*, can be due to changes in both flows and forces:

$$dP^* = \sum_k J_k dF_k + \sum_k F_k dJ_k = d_F P^* + d_J P^* \qquad [3.144]$$

In linear regime, $J_k = \sum_j L_{k,j} F_j$ and $L_{k,j} = L_{j,k}$. Therefore,

$$d_F P^* = \sum_k \sum_j L_{k,j} F_j dF_k = \sum_k \sum_j F_j d(L_{j,k} F_k) = \sum_j F_j dJ_j - d_J P^* \qquad [3.145]$$

From Section 3.5, we know that $dP^* \leq 0$. We, finally, infer that

$$d_F P^* = d_J P^* = \frac{1}{2} dP^* \leq 0 \qquad [3.146]$$

Equation [3.146] is strictly valid only in linear regime (de Groot and Mazur 1984).

When a system works in nonlinear regime, its evolution is described by the following inequality

$$d_F P^* \leq 0 \qquad [3.147]$$

Equation [3.147], known as Glansdorff-Prigogine stability criterion, becomes the universal principle of evolution for out-of-equilibrium systems, which is verified when the contour conditions do not change over time (Glansdorff and Prigogine 1954). Let us prove the relation [3.147] in the case of chemical transformations. Imagine having a system wherein ρ chemical reactions occur. This system is in physical contact with reservoirs, each containing a specific compound at a constant temperature, pressure, and chemical potential. The reactor is separated from the reservoirs through semi-permeable membranes. These membranes allow the transfer of compounds contained into the reservoirs, but not of the others. For each species k inside the reactor, one of the following two conditions are true: either its chemical potential is constant and fixed by the composition of its reservoir (μ_k constant) or its molecules cannot cross any membrane ($d_e n_k = 0$). The reactor is an open system that is maintained far from equilibrium. The equation defining the time variation of the moles for the k-th species is

$$\frac{dn_k}{dt} = \frac{d_e n_k}{dt} + \sum_\rho v_{k,\rho} v_\rho \qquad [3.148]$$

If we multiply both terms of equation [3.148] for $\dot{\mu}_k$, i.e., the time derivative of the k-th chemical potential, we obtain

$$\dot{\mu}_k \frac{dn_k}{dt} = \dot{\mu}_k \frac{d_e n_k}{dt} + \sum_\rho v_{k,\rho} \dot{\mu}_k v_\rho \qquad [3.149]$$

The first term on the right side of equation [3.149] is null, because either the k-th molecules do not cross the membranes $\left(\frac{d_e n_k}{dt} = 0\right)$ or their chemical potential is constant over time $(\dot{\mu}_k = 0)$. Summing up the relations of the type [3.149] for each chemical component, we achieve

$$\sum_k \dot{\mu}_k \frac{dn_k}{dt} = \sum_k \sum_\rho \nu_{k,\rho} \dot{\mu}_k v_\rho \qquad [3.150]$$

Reminding the definition of the affinity for the ρ-th reaction, $A_\rho = -\sum_k \nu_{k,\rho} \mu_k$, we can rearrange the previous equation, [3.150] in the following form

$$\sum_k \sum_{k'} \left(\frac{d\mu_k}{dn_{k'}}\right)\left(\frac{dn_{k'}}{dt}\right)\left(\frac{dn_k}{dt}\right) = -\sum_\rho \left(\frac{dA_\rho}{dt}\right) v_\rho \qquad [3.151]$$

When T and P are constant, and it is possible to apply the approximation of local equilibrium,[15] from equation [3.151], it results that $-\sum_\rho \left(\frac{dA_\rho}{dt}\right) v_\rho \geq 0$.[16] Therefore,

$$\frac{1}{T} \sum_\rho \left(\frac{dA_\rho}{dt}\right) v_\rho = \frac{d_F P^*}{dt} \leq 0 \qquad [3.152]$$

The relation $d_F P^* \leq 0$ is formally equivalent to the relation $dG \leq 0$ that is true for system evolving towards the equilibrium. However, P^* is not a state function as G, whereby $d_F P^*$ is not an exact differential.[17] Its value depends on the path followed by the system. Equation [3.147] tells us that $d_F P^*$ wanes until it becomes null at the stationary state, but it does not tell anything about how the system evolves and what stationary state will be reached.

3.7 THE CHEMICAL TRANSFORMATIONS AND THE LINEAR REGIME

We have found that the relationship between a chemical flow (v) and its chemical force (A/T) for an elementary step is exponential (see equation [3.143]). However, there are situations wherein their relationship becomes linear. For instance, close to equilibrium, the chemical affinity of a reaction is small. Therefore, the term $e^{-\frac{A}{RT}}$ developed in Maclaurin series[18] becomes $\left(1 - \frac{A}{RT} + \ldots\right)$. Introducing this contribution in equation [3.143], we obtain:

$$v \approx v_f \left(\frac{A}{RT}\right) = L_c \frac{A}{T} \qquad [3.153]$$

---

[15] In the case of chemical reactions, the approximation of local equilibrium is a reasonable assumption if the reaction rates are appreciably slow. When the reaction goes on quite slowly, the distributions of the molecular velocities are close to those expected from the kinetic theory of Maxwell and Boltzmann. The approximation generally fails when the activation energy for a reaction is small, comparable to the available thermal energy RT.

[16] When we can apply the approximation of local equilibrium, the term $\sum_k \sum_{k'} \left(\frac{d\mu_k}{dn_{k'}}\right)\left(\frac{dn_{k'}}{dt}\right)\left(\frac{dn_k}{dt}\right)$ is always ≥ 0. In fact, if k and k' are both reagents, $\left(\frac{dn_k}{dt}\right)$ and $\left(\frac{dn_{k'}}{dt}\right)$ are ≤ 0, whereas $\left(\frac{d\mu_k}{dn_{k'}}\right)$ is ≥ 0; if one is a reagent, for example k, and the other, k', is a product, $\left(\frac{dn_k}{dt}\right) \leq 0$, $\left(\frac{dn_{k'}}{dt}\right) \geq 0$ and $\left(\frac{d\mu_k}{dn_{k'}}\right) \leq 0$.

[17] A differential dG is exact if $\int dG$ is path-independent.

[18] A Maclaurin series is a Taylor series expansion of a function having 0 as accumulation point: $f(x) = f(0) + \frac{x}{1!} \dot{f}(0) + \frac{x^2}{2!} \ddot{f}(0) + \frac{x^3}{3!} \dddot{f}(0) + \ldots$

3.7.1 ONSAGER'S RECIPROCAL RELATIONS FOR CHEMICAL REACTIONS

When a chemical reaction is close to the equilibrium, the relationship between the rate and the chemical force is linear (see equation [3.153]), and the phenomenological coefficient is $L_c = (v_f/R)$. In the linear regime, it is possible to verify the Onsager's reciprocal relations. Let us consider a set of chemical reactions that are not completely independent, such as the following ones:

$$
\begin{aligned}
&(1) \quad X + Y \rightarrow W \\
&(2) \quad W + Y \rightarrow Z \\
&(3) \quad X + 2Y \rightarrow Z
\end{aligned}
\qquad [3.154]
$$

Note that the third reaction is the sum of the previous two. We can express the affinity of the third reaction as the sum of the affinities of the first two reactions: $A_3 = A_1 + A_2$.

The entropy production is

$$
P^* = v_1 \frac{A_1}{T} + v_2 \frac{A_2}{T} + v_3 \frac{A_3}{T}
\qquad [3.155]
$$

If we introduce the equation $A_3 = A_1 + A_2$ into equation [3.155], we obtain

$$
P^* = (v_1 + v_3)\frac{A_1}{T} + (v_2 + v_3)\frac{A_2}{T} = v_1' \frac{A_1}{T} + v_2' \frac{A_2}{T}
\qquad [3.156]
$$

In linear regime, i.e., close to equilibrium, we can write

$$
\begin{aligned}
v_1' &= L_{11}\frac{A_1}{T} + L_{12}\frac{A_2}{T} \\
v_2' &= L_{21}\frac{A_1}{T} + L_{22}\frac{A_2}{T}
\end{aligned}
\qquad [3.157]
$$

Moreover, we know that $v_1 = v_{1f}\frac{A_1}{RT}$, $v_2 = v_{2f}\frac{A_2}{RT}$, and $v_3 = v_{3f}\frac{A_3}{RT}$. Therefore, after a few simple mathematical steps, we obtain

$$
\begin{aligned}
(v_{1f} + v_{3f})\frac{A_1}{RT} + v_{3f}\frac{A_2}{RT} &= L_{11}\frac{A_1}{T} + L_{12}\frac{A_2}{T} \\
(v_{3f})\frac{A_1}{RT} + (v_{2f} + v_{3f})\frac{A_2}{RT} &= L_{21}\frac{A_1}{T} + L_{22}\frac{A_2}{T}
\end{aligned}
\qquad [3.158]
$$

From equations [3.158], we confirm the Onsager's reciprocal relations $L_{12} = (v_{3f}/R) = L_{21}$.

3.7.2 A PARTICULAR CASE

We have discovered that a linear relationship between the chemical flow (v) and the chemical force (A/T) is valid for an elementary step when it is close to equilibrium. A linear relationship between v and A/T can also be verified for a reaction that, although it has a large affinity, occurs through many elementary steps and each one characterized by small values of their affinities[19] (Kondepudi and Prigogine 1998).

---

[19] An example could be a polymerization chain reaction, when the initiation step is over, and the steps of chain propagation are taking place.

Suppose that the overall reaction is $X \to Z$.
X converts to Z through n elementary steps:

$$X \overset{1}{\to} Y \overset{2}{\to} M \overset{3}{\to} N \overset{4}{\to} \ldots \overset{n}{\to} Z \qquad [3.159]$$

For each elementary step we can write $v_1 = v_{1f} \frac{A_1}{RT}$, $v_2 = v_{2f} \frac{A_2}{RT}$, and so on, $v_n = v_{nf} \frac{A_n}{RT}$.

At the stationary state, it results that the rate of the overall reaction is equal to the rates of each elementary process: $v = v_1 = v_2 = \ldots = v_n$.

The affinity of the reaction is the sum of the affinities of each elementary step: $A = A_1 + A_2 + \ldots + A_n$.

It derives that

$$A = v\frac{RT}{v_{1f}} + v\frac{RT}{v_{2f}} + \ldots + v\frac{RT}{v_{nf}} = vRT\left(\frac{1}{v_{1f}} + \frac{1}{v_{2f}} + \ldots + \frac{1}{v_{nf}}\right) \qquad [3.160]$$

If we pose $\frac{1}{v_{f,eff}} = \left(\frac{1}{v_{1f}} + \frac{1}{v_{2f}} + \ldots + \frac{1}{v_{nf}}\right)$, finally we achieve a linear relationship between the rate and the chemical force of the overall reaction:

$$v = v_{f,eff}\frac{A}{RT} \qquad [3.161]$$

although $\frac{A}{RT} \gg 1$.

3.8   THE EVOLUTION OF CHEMICAL REACTIONS IN OPEN SYSTEMS

Any chemical reaction carried out in a closed vessel evolves spontaneously towards the equilibrium state, which is globally stable. A chemical reaction exhibits an entirely different evolution when it is carried out in an open system. Examples of natural open systems are cells and living beings. In the laboratory, an open system is any reactor that allows us to continuously pump fresh reactants in and suck reacted solution out to maintain a constant volume. If the mixture inside the reactor is also stirred (mechanically or magnetically), such an apparatus is named as Continuous-flow Stirred Tank Reactor (CSTR) (see Figure 3.16). When we run a reaction inside a CSTR, it is necessary to fix the flow rate (F_0, measured in mLs^{-1}). Knowing the volume (V) of the reactor, it is possible to define the residence time (τ_0) as the ratio

$$\tau_0 = \frac{V}{F_0} \qquad [3.162]$$

The residence time is the average time that an atom spends inside the reactor.

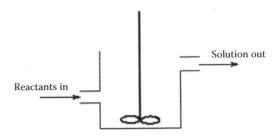

FIGURE 3.16   Scheme of a CSTR.

When a reaction runs in a CSTR, it evolves towards states that can be either stable or unstable or oscillatory. Another way to maintain a chemical reaction far from equilibrium is by irradiation with, for instance, ultraviolet (UV) or visible electromagnetic waves.

3.8.1 THE MONO-DIMENSIONAL CASE

Let us start from the simplest chemical case, for instance, an irreversible reaction following a first-order kinetic law: $X \rightarrow Y$. If k_f is the kinetic constant of the reaction, $[X]_0$ the concentration of X that is introduced inside the reactor, and k_0 is the reciprocal of the residence time, the time variation of $[X]$ is described by the following linear differential equation

$$\frac{d[X]}{dt} = f([X]) = k_0[X]_0 - [X](k_0 + k_f)$$ [3.163]

By fixing the contour conditions (the flow rate, the volume of the reactor, $[X]_0$, the temperature and pressure), the chemical system evolves towards a stationary state, wherein $[X]$ becomes constant over time and equal to

$$[X]_{ss} = \frac{k_0[X]_0}{(k_0 + k_f)}$$ [3.164]

We ask us about the stability of this stationary state. To answer this question, we choose a graphical method (Strogatz 1994). The first derivative of $[X]$ can be partitioned in two contributions: the rate of injection of X, given by $k_0[X]_0$, and the rate of X disappearance, i.e., $(k_0 + k_f)[X]$. If we plot the two rates versus $[X]$, we obtain Figure 3.17.

The two rates are represented by two straight lines, intersecting in one fixed point that represents the stationary state. When the system is on the left with respect to the fixed point, i.e., $[X] < [X]_{ss}$, the rate of injection is larger than the rate of consumption and $[X]$ increases up to reach the stationary state. On the other hand, if the initial conditions of the system are represented by any point where $[X] > [X]_{ss}$, the rate of consumption will be higher than that of injection and $[X]$ will decrease up to reach the $[X]_{ss}$ value. This phenomenology demonstrates that the fixed point is a stable stationary state. A more quantitative measure of stability may be obtained by an analytical method known as linear stability analysis. Let $x = [X] - [X]_{ss}$ be a small perturbation away from the stationary state. To see if the perturbation grows or decays, we write the differential equation [3.163] for x, using the Taylor expansion having $[X]_{ss}$ as accumulation point:

$$\dot{x} = \frac{dx}{dt} = f([X]_{ss}) + x\dot{f}([X]_{ss}) + \dots$$ [3.165]

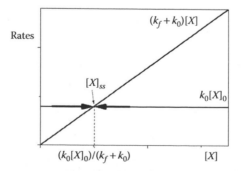

FIGURE 3.17 Rates of injection and consumption for X versus its concentration.

When $\dot{f}\left([X]_{ss}\right) \neq 0$, we can neglect the higher order terms in the expansion. Since $f\left([X]_{ss}\right) = 0$ in equation [3.165] reduces to

$$\dot{x} \approx x\dot{f}\left([X]_{ss}\right) \tag{3.166}$$

which is a linear differential equation in x and it is called the linearization about $[X]_{ss}$. The solution of equation [3.166] will be a function of the type

$$x(t) = x(t_i) e^{\dot{f}\left([X]_{ss}\right)(t-t_i)} \tag{3.167}$$

Equation [3.167] shows that the perturbation x grows exponentially if $\dot{f}\left([X]_{ss}\right) > 0$ and decays exponentially if $\dot{f}\left([X]_{ss}\right) < 0$. In other words, the sign of $\dot{f}\left([X]_{ss}\right)$ determines the stability of the stationary state and the magnitude of $\dot{f}\left([X]_{ss}\right)$ rules the time required for $[X]$ to vary in the neighborhood of $[X]_{ss}$.

In the case of the irreversible $X \rightarrow Y$ reaction described by equation [3.163], we observe that $\dot{f}\left([X]_{ss}\right) = -\left(k_0 + k_f\right) < 0$. This result confirms that its stationary state is stable.

TRY EXERCISE 3.13

3.8.2 THE BI-DIMENSIONAL CASE

Let us consider a generic reaction mechanism of the type:

$$\begin{aligned} A + B &\xrightarrow{k_1} X \\ X + A &\xrightarrow{k_2} Y \\ Y &\xrightarrow{k_3} Z \end{aligned} \tag{3.168}$$

Let us assume that the inflow of A and B and the outflow of Z maintain constant the concentrations of the reagents and the product. $[X]$ and $[Y]$ are the variables. The dynamics of the system is described by two differential equations[20] of the type

$$\begin{aligned} \frac{d[X]}{dt} &= k_1[A][B] - k_2[X][A] - k_0[X] = f\left([X],[A],[B],k_1,k_2,k_0\right) \\ \frac{d[Y]}{dt} &= k_2[A][X] - k_3[Y] - k_0[Y] = g\left([X],[A],[Y],k_3,k_2,k_0\right) \end{aligned} \tag{3.169}$$

The chemical system will evolve towards stationary states that are the solutions of the system of equations as follows:

$$\begin{aligned} \frac{d[X]}{dt} &= f\left([X],[A],[B],k_1,k_2,k_0\right) = 0 \\ \frac{d[Y]}{dt} &= g\left([X],[A],[Y],k_3,k_2,k_0\right) = 0 \end{aligned} \tag{3.170}$$

---

[20] A ordinary differential equation is non-linear when it contains (I) power functions and/or (II) product functions and/or (III) transcendent functions.

To test the properties of the solutions, we can exploit the linear stability analysis that we learned for the mono-dimensional case in the previous paragraph. If $[X]_{ss}$ and $[Y]_{ss}$ represent the concentrations of the stationary state, let us consider the effects of small perturbations: $x = [X] - [X]_{ss}$ and $y = [Y] - [Y]_{ss}$. Introducing x and y into the system [3.170] and expanding the functions f and g in the Taylor series with $[X]_{ss}$ and $[Y]_{ss}$ as accumulation points, we obtain

$$\frac{dx}{dt} = \dot{x} \cong f\left([X]_{ss},[Y]_{ss}\right) + \left(\frac{\partial f}{\partial [X]}\right)_{ss} x + \left(\frac{\partial f}{\partial [Y]}\right)_{ss} y + \ldots$$

$$\frac{dy}{dt} = \dot{y} \cong g\left([X]_{ss},[Y]_{ss}\right) + \left(\frac{\partial g}{\partial [X]}\right)_{ss} x + \left(\frac{\partial g}{\partial [Y]}\right)_{ss} y + \ldots$$

[3.171]

In the system [3.171], the second and higher order terms of the Taylor expansion are neglected, because the perturbations are assumed to be small. The terms $f\left([X]_{ss},[Y]_{ss}\right)$ and $g\left([X]_{ss},[Y]_{ss}\right)$ are null. The derivatives are estimated in correspondence of the stationary points, and they are constant. Therefore, the system [3.171] reduces to a system of linear differential equations. Solutions of these linear equations are exponentials of the form $x = v_{i,1}e^{\lambda_i t}$, $y = v_{i,2}e^{\lambda_i t}$, and any linear combination of them. This result is evident if we consider, at first, two linear differential equations that are uncoupled, like:

$$\frac{dx}{dt} = \lambda_1 x$$

$$\frac{dy}{dt} = \lambda_2 y$$

[3.172]

They are uncoupled because there is no y in the x-equation and vice-versa. In this simple situation, the two differential equations can be solved separately. The solutions are $x = x_0 e^{\lambda_1 t}$ and $y = y_0 e^{\lambda_2 t}$. The x- and y-axes represent the directions of the trajectories for the x and y variables, respectively, as time goes to infinity ($t \to \pm\infty$). In other words, the x- and y-axes contain the straight-line trajectories: a trajectory starting from one point of the two axes, stays on that axis forever. For the general case, we want to find the analog of these straight-line trajectories.

Introducing the solutions $x = v_{i,1}e^{\lambda_i t}$ and $y = v_{i,2}e^{\lambda_i t}$ into equation [3.171] and dividing by the non-zero scalar factor $e^{\lambda_i t}$, we achieve

$$\lambda_i \begin{pmatrix} v_{i,1} \\ v_{i,2} \end{pmatrix} = \begin{pmatrix} \left(\dfrac{\partial f}{\partial [X]}\right)_{ss} & \left(\dfrac{\partial f}{\partial [Y]}\right)_{ss} \\ \left(\dfrac{\partial g}{\partial [X]}\right)_{ss} & \left(\dfrac{\partial g}{\partial [Y]}\right)_{ss} \end{pmatrix} \begin{pmatrix} v_{i,1} \\ v_{i,2} \end{pmatrix}$$

[3.173]

In matrix form, equation [3.173] becomes

$$\lambda_i \bar{v}_i = J \bar{v}_i$$

[3.174]

where J is the Jacobian matrix whose terms are the derivatives calculated at the fixed point; $\bar{v}_i$ is the 2×1 vector whose terms are $v_{i,1}$ and $v_{i,2}$. This vector represents the direction of the phase space, which has trajectories with straight-lines, and is called the eigenvector of J. Linear equations such as [3.174] have nontrivial solutions (i.e., solutions different from that with $v_{i,1}$ and $v_{i,2}$ both null) only when λ is an eigenvalue of the matrix J. The eigenvalues of J can be determined by solving the characteristic equation

$$det\left(J - \lambda I\right) = det\begin{pmatrix} \left(\dfrac{\partial f}{\partial [X]}\right)_{ss} - \lambda & \left(\dfrac{\partial f}{\partial [Y]}\right)_{ss} \\ \left(\dfrac{\partial g}{\partial [X]}\right)_{ss} & \left(\dfrac{\partial g}{\partial [Y]}\right)_{ss} - \lambda \end{pmatrix} = 0 \qquad [3.175]$$

wherein I is the 2×2 identity matrix. If we expand the determinant, we have

$$\lambda^2 - \lambda tr\left(J\right) + det\left(J\right) = 0 \qquad [3.176]$$

where $tr(J)$ and $det(J)$ are the trace (that is the sum of its diagonal elements) and the determinant of the Jacobian,[21] respectively. Equation [3.176] is a quadratic equation in λ having two solutions that are functions of the elements of the Jacobian:

$$\lambda_1 = \frac{tr\left(J\right)}{2} + \frac{\sqrt{\left(tr\left(J\right)\right)^2 - 4det\left(J\right)}}{2} \qquad [3.177]$$

$$\lambda_2 = \frac{tr\left(J\right)}{2} - \frac{\sqrt{\left(tr\left(J\right)\right)^2 - 4det\left(J\right)}}{2} \qquad [3.178]$$

For the two eigenvalues λ_1 and λ_2, we will have two eigenvectors, $\bar{v}_1$ and $\bar{v}_2$, and the general solutions will be of the type

$$\begin{aligned} x\left(t\right) &= c_1 v_{1,1} e^{\lambda_1 t} + c_2 v_{2,1} e^{\lambda_2 t} \\ y\left(t\right) &= c_1 v_{1,2} e^{\lambda_1 t} + c_2 v_{2,2} e^{\lambda_2 t} \end{aligned} \qquad [3.179]$$

wherein the values of the parameters c_1 and c_2 depend on the initial conditions.[22] The eigenvalues λ_1 and λ_2 can be either real or pure imaginary or complex numbers. The eigenvalues are related with the trace and the determinant of J through the following equations:

$$\begin{aligned} \lambda_1 + \lambda_2 &= tr\left(J\right) \\ \lambda_1 \lambda_2 &= det\left(J\right) \end{aligned} \qquad [3.180]$$

All the possible combinations of solutions for the eigenvalues can be expressed as functions of $tr(J)$ and $det(J)$ as shown in Table 3.2, Figure 3.18 and explained as follows.

Points in the region (a):

$$tr\left(J\right) < 0, \, det\left(J\right) > 0 \text{ and the discriminant } \Delta = \left[\left(tr\left(J\right)\right)^2 - 4det\left(J\right)\right] > 0$$

---

[21] The trace of the Jacobian is the sum of its diagonal elements, i.e., $tr\left(J\right) = \left(\frac{\partial f}{\partial [x]}\right)_{ss} + \left(\frac{\partial g}{\partial [y]}\right)_{ss}$. The determinant of J is $det\left(J\right) = \left(\frac{\partial f}{\partial [x]}\right)_{ss}\left(\frac{\partial g}{\partial [y]}\right)_{ss} - \left(\frac{\partial f}{\partial [y]}\right)_{ss}\left(\frac{\partial g}{\partial [x]}\right)_{ss}$.

[22] You can look up any text on Linear Algebra to refresh your memory on the solutions of systems of linear equations, their existence, and uniqueness.

TABLE 3.2

List of All the Possible Solutions of Linear Differential Equations

Region	$tr(J)$	$det(J)$	Δ[a]	Roots	Name
(a)	<0	>0	>0	$\lambda_{1,2} < 0; \lambda_1 \neq \lambda_2$	Stable node
(b)	<0	>0	=0	$\lambda_{1,2} < 0; \lambda_1 = \lambda_2$	Stable star
(c)	<0	>0	<0	$\lambda_{1,2} = a \pm ib; a < 0$	Stable spiral
(d)	=0	>0	<0	$\lambda_{1,2} = \pm i\sqrt{4det(J)}$	Center
(e)	>0	>0	<0	$\lambda_{1,2} = a \pm ib; a > 0$	Unstable spiral
(f)	>0	>0	=0	$\lambda_{1,2} > 0; \lambda_1 = \lambda_2$	Unstable star
(g)	>0	>0	>0	$\lambda_{1,2} > 0; \lambda_1 \neq \lambda_2$	Unstable node
(h)		<0	>0	$\lambda_1 > 0; \lambda_2 < 0$	Saddle node
(l_1)	>0	=0	$=[tr(J)]^2$	$\lambda_1 = 0; \lambda_2 > 0$	Lines of unstable fixed points
($l_2 = 0$)	=0	=0	=0	$\lambda_1 = 0; \lambda_2 = 0$	Plane of fixed points
(l_3)	<0	=0	$=[tr(J)]^2$	$\lambda_1 = 0; \lambda_2 < 0$	Lines of stable fixed points

[a] Δ is the discriminant.

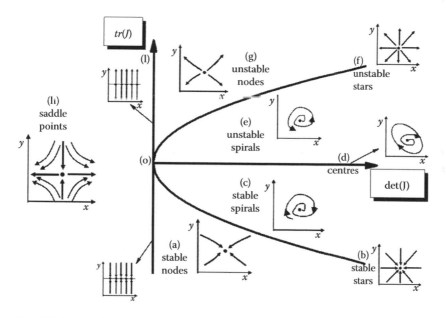

FIGURE 3.18 All the possible solutions of the quadratic equations [3.176] grouped in different regions and curves.

For any point in this region the roots, λ_1 and λ_2, are two different negative real numbers. This means that the stationary states represented by the points in the region (a) are stable. When a system is perturbed, it will restore the initial state with exponential speed. The relative stationary state is termed a *stable node*. Points on the branch (b) of the curve are $tr(J) = \sqrt{4det(J)}$:

$$tr(J) < 0, det(J) > 0 \text{ and } \Delta = \left[(tr(J))^2 - 4\,det(J)\right] = 0$$

For any point along the branch (b) of the parabola, we have two equal roots that are negative real numbers. This result means that after the perturbations, x and y decrease exponentially with the same speed. The stationary state is stable and is defined as a *stable star*.

Points in the region (c):

$$tr(J) < 0, det(J) > 0 \text{ and } \Delta = \left[(tr(J))^2 - 4 \, det(J) \right] < 0$$

For the points in the region (c), the two roots are two complex conjugates of the form $\lambda_{1,2} = a \pm ib$ with the real term a being negative. Reminding that $e^{a \pm ib} = e^a (\cos(b) \pm isen(b))$, it is evident that after the perturbation, the system will restore the initial stationary state after oscillations. These kinds of stationary states are called *stable spirals*.

The points of the regions (a), (b), and (c) are *attractors*.

Points on the straight line (d)

$$tr(J) = 0, det(J) > 0 \text{ and } \Delta = \left[(tr(J))^2 - 4 \, det(J) \right] < 0$$

The solutions are $\lambda_{1,2} = \pm i \sqrt{4 \, det(J)}$, i.e., they are purely imaginary. In these cases, when the system is removed from the stationary state, it enters a limit cycle or periodic orbit: it starts to oscillate. The corresponding fixed points are named as *centers*.

Points in the region (e):

$$tr(J) > 0, det(J) > 0 \text{ and } \Delta = \left[(tr(J))^2 - 4 \, det(J) \right] < 0$$

In this region, the roots are two complex numbers of the type $\lambda_{1,2} = a \pm ib$ with a positive real a. In this case, the perturbation grows exponentially by oscillating. The fixed points of the region (e) are termed *unstable spirals*.

Points on the branch (f) of the curve are $tr(J) = \sqrt{4 \, det(J)}$:

$$tr(J) > 0, det(J) > 0 \text{ and } \Delta = \left[(tr(J))^2 - 4 \, det(J) \right] = 0$$

For any point along the branch (f) of the parabola, we have two equal roots that are positive real numbers. This result means that the effects of the perturbations x and y grow exponentially with the same speed. The initial stationary state is unstable and is defined as an *unstable star*.

Points in the region (g):

$$tr(J) > 0, det(J) > 0 \text{ and the discriminant } \Delta = \left[(tr(J))^2 - 4 \, det(J) \right] > 0$$

For any point in this region the roots, λ_1 and λ_2, are two different positive real numbers. This result means that the many stationary states represented by the points in the region (g) are unstable. When a system is perturbed, it will go away from the initial state. These stationary states are termed *unstable nodes*.

The points of the regions (e), (f), and (g) are *repellers*.

Points in the region (h)

$$det(J) < 0, \Delta = \left[(tr(J))^2 - 4 \, det(J) \right] > 0, \text{ and } \Delta > tr(J)$$

The roots λ_1 and λ_2 are two real numbers, one positive and the other negative. The stationary state is stable if the perturbation involves just the eigenvector with the negative eigenvalue λ_i, whereas it is unstable if it involves the other eigenvector with the positive value of the eigenvalue. A stationary state of this type is called a *saddle point* because its vector field in the phase space resembles a riding saddle.

Points on the straight line (l)

$$det(J) = 0 \text{ and } \Delta = \left[(tr(J))^2\right] > 0$$

If we exclude the origin (o), for the stationary states represented by the points along the straight line (l), an eigenvalue (let us say λ_1) is null, whereas the other (λ_2) is a positive (for the portion of the line above the origin (o)) or a negative (for the portion of the line below (o)) real number. This situation means that there is a *line of fixed points* that are stable when λ_2 is negative, but unstable when λ_2 is positive. In the origin (o) of the graph, both eigenvalues are null, and we have a *plane of fixed points*.

Now, the most curious readers would ask themselves: "is it always correct to overlook the quadratic and the higher order terms in the expansion [3.171]?" The answer is yes, as long as we deal with fixed points that are not those of the borderline cases, such as stars, centers, degenerate nodes, and non-isolated fixed points. The latter can be altered by small nonlinear terms. However, stars, degenerate nodes, and non-isolated fixed points do not change their stability. For instance, an unstable star due to non-linear terms may transform into an unstable node or an unstable spiral, but not into a stable spiral. This possibility is plausible if we retake a look at the stability diagram (Figure 3.18). The centers are much more delicate because they lie at the edge between the stability and the instability regions. Small non-linear terms may transform a center into a stable or unstable spiral.

TRY EXERCISES 3.14 AND 3.15

3.8.3 THE MULTI-DIMENSIONAL CASE

The linear stability analysis presented in the mono- and bi-dimensional cases can be generalized to systems with more than two variables

$$\frac{dx_i}{dt} = \dot{x}_i = f_i(x_1, x_2, \ldots, x_n) \text{ with } i = 1, 2, \ldots, n \qquad [3.181]$$

The character of a steady state is determined by the eigenvalues of the $n \times n$ Jacobian matrix, whose elements are of the type

$$J_{ij} = \left(\frac{\partial f_i}{\partial x_j}\right)_{ss} \qquad [3.182]$$

If all the eigenvalues have negative real parts, the stationary state is stable; if just a single eigenvalue has a positive real part, the stationary state has an intrinsic instability.

3.9 KEY QUESTIONS

- Which equation defines the entropy change in the thermodynamics of out-of-equilibrium systems?
- What is the entropy production?
- How many regimes do we distinguish in non-equilibrium thermodynamics?
- Make examples of irreversible processes in the linear regime.
- What are the rules governing the relationships among forces and flows in linear regime?

- How does an out-of-equilibrium system in the linear regime evolve?
- Explain the Theorem of Minimum Entropy Production.
- What is the relationship between the chemical flow and the chemical force for an elementary reaction?
- What does the Glansdorff-Prigogine stability criterion tell?
- Present the linear stability analysis for a chemical system wherein a reaction with two variables occurs.

3.10 KEY WORDS

Entropy production; Flow and Force; Symmetry principle and Onsager reciprocal relations; Linear stability analysis; Attractors and repellers.

3.11 HINTS FOR FURTHER READING

A landmark paper about the non-equilibrium thermodynamics is the Nobel lecture given by Prigogine in December 8, 1977 and published in *Science* in 1978.

In this book, I use MATLAB for the numerical solution of differential equations and for plotting the results. Alternatively, there exists a software package called SageMath, which is free and can be downloaded from the website: http://www.sagemath.org/. SageMath is similar to the scientific computing environment called Python.

Finally, in October 2017, there was a special issue of *Chaos: An Interdisciplinary Journal of Nonlinear Science* celebrating the 100th birthday of Ilya Prigogine: It is volume 27, issue 10.

3.12 EXERCISES

3.1. Imagine having a solution of tryptophan (whose molecular formula is $C_{11}H_{12}N_2O_2$) dissolved in water and contained inside a cuvette, at a temperature of 300 K. The height of the solution is 3 cm. Determine the ratio of the concentration of tryptophan between the top and the bottom of the solution, knowing that tryptophan molecules feel the terrestrial gravitational field ($g = 9.8$ m/sec^2). How much is the same ratio when we consider a solution of Human Serum Albumin (its molecular mass weight is 67,000 gmol^{-1}) in water?

3.2. The cellular membrane is made of a phospholipid bilayer and acts as a barrier that prevents the intracellular fluid from mixing with the extracellular fluid. These two solutions have different concentrations of their ions and, hence, different electrical potentials. In particular, the inside solution is at a more negative potential than the external solution. If the concentration of K^+ is 5 mmol/L outside the cell, and 140 mmol/L inside the cell, how much is the potassium transmembrane potential?

3.3. Imagine that a system hosts a spontaneous chemical reaction. The dependence of Gibbs free energy of the system (G) and the entropy of the universe (S_{univ}) on the extent of the reaction ξ are shown in Figure 3.19. Predict the shape of the first and second derivatives of G and S_{univ} with respect to ξ.

3.4. Determine the physical dimensions of the thermal conductivity k by making a dimensional analysis of Fourier's Law (equation [3.32])

3.5. Determine the physical dimensions of viscosity η by using equations [3.38] and [3.39].

3.6. Considering Fick's Law (equation [3.40]), which are the physical dimensions of the diffusion coefficient D_k?

3.7. Estimate the ratio between the mobility and the diffusion coefficient for sodium cation at 298 K in the following two situations: (1) in the presence of a gravitational field; (2) in the presence of an electric field.

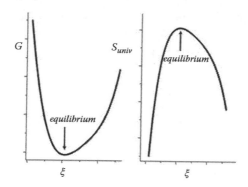

FIGURE 3.19 Trend of G and S_{univ} as function of the extent of reaction.

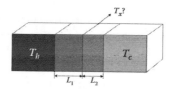

FIGURE 3.20 Heat conduction between two reservoirs and across two blocks.

3.8. Let us imagine having two blocks of two different materials, 1 and 2, with two different thermal conductivities (k_1 and k_2, respectively), two different lengths (L_1 and L_2), and the same section (Ar). Block 1 is in physical contact with a hot reservoir (being at T_h) on one side and with block 2 on the other side. Block 2 is in physical contact with block 1 on one side and with a cold reservoir (being at T_c) on the other side (see Figure 3.20). Determine the temperature T_x at the surface in conjunction between 1 and 2, when the entire system reaches the stationary state.

3.9. Suppose that the function describing the trend of a body temperature along the spatial coordinate x is initially (at time $t = 0$) $T = 273 + 10e^{-\frac{x}{5}}$ (see Figure 3.21). The body has a length of 10 in the x unit. Imagine fixing the values of T at the extremes of the body at the following values: $T(0) = 283.00$ K and $T(10) = 274.35$ K. How does T evolve in time along x? Which is the function describing $T(x)$ at the stationary state, after waiting for an infinite amount of time?

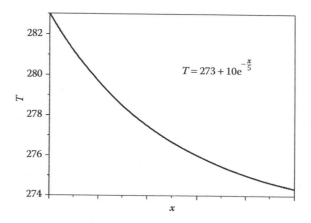

$$T = 273 + 10e^{-\frac{x}{5}}$$

FIGURE 3.21 A trend of T as a function of the spatial coordinate x.

3.10. Imagine having sodium ions at the bottom of a container full of water and having area Ar and height h (along the z-axis). To describe the diffusion of the sodium ions from the bottom to the top of the container, it is necessary to solve the diffusion equation [3.114].

1. Verify that the following Gaussian function

$$C(z,t) = \frac{n_0}{(Ar)(\pi Dt)^{1/2}} e^{-\frac{z^2}{4Dt}}$$

   is the solution of equation [3.114] for a system having n_0 moles of Na^+ and being D the diffusion coefficient of sodium ions.
2. Determine the expression that defines the average distance covered by sodium ions as a function of time.
3. Calculate the one-dimensional average distance covered by sodium ions after 1 minute, 1 hour, and 1 day, knowing that $D = 1.33 \cdot 10^{-5}\,cm^2/sec$ in water at 298 K.

3.11. Determine the equation that defines the one-dimensional mean square distance traveled by a molecule through diffusion. Then, calculate the one-dimensional mean square distance covered by sodium ions, having $D = 1.33 \cdot 10^{-5}\,cm^2/sec$ in water at 298 K, and a protein, having $D = 100\,\mu m^2/sec$ in water at 298 K, after 1 minute, 1 hour and 1 day.

3.12. Demonstrate the validity of Minimum Entropy Production Theorem in the case of heat conduction for a system such as that described in Section 3.4.1.

3.13. Investigate the stability of the stationary states for the following reaction, by using the graphical and the linear stability methods.

$$X + Y \underset{k_{-1}}{\overset{k_1}{\rightleftarrows}} 2X$$

Assume that the reaction is performed in a CSTR where the only Y is injected and sucked out. The concentration of Y introduced into the reactor $[Y]_0$ is so large that it can be conceived constant. Let $\tau_0 = 1/k_0$ be its residence time into the CSTR.

3.14. Apply the linear stability analysis to the chemical reaction described by equation [3.168].

3.15. *The linear stability analysis to predict the course of a love story.*[23] True love is the noblest and most exciting feeling a human can ever experience in his/her life. However, it may become a source of pain as well when it is not mutual and when it fades. We may experience love at first sight, or love may take time to grow. Sometimes the first impression is crucial, sometimes not. Love is the subject of many stories in the literature: the most well-known love story is that of Romeo and Juliette. Other notorious stories are those of Orpheus and Eurydice, Paris and Helen, Tristan and Isolde, Eloise and Abelard, or Dante and Beatrice. Everyone has lived or is living a love story, either concrete or virtual. We all know that a lover asks himself/herself if the counterpart likes him/her; if his/her feeling and actions fit well with those of his/her partner or not; if the relationship with his/her partner will be everlasting or destined to fail. We may try to answer these queries through a model that looks like a chemical mechanism. In fact, such a model describes a love relation between a male and a female by an ensemble of sentimental transformations conceived as they were chemical reactions. Let us get started by indicating the two main characters of the relation: a male (P) and a female (M). P and M stand for Pierre and Marie Curie: this couple is a prototype of a successful love story in the scientific environment. Marie was a hard-working student at the Sorbonne University of Paris in the early 1890s.

---

[23] This is a form of agent-based modeling that we will know in Chapter 13.

She was spending every spare hour reading in the library or the laboratory. The industrious student caught the attention of Pierre Curie, director of one of the laboratories where Marie was working. Let us indicate the feeling of Pierre for Marie with F_P and with I_M Pierre's thought about Marie. F_P is a variable that can assume positive or negative values. It will be positive when Pierre likes Marie, whereas it will be negative when Pierre dislikes Marie. Physiologically, such feelings correspond to the release of excitatory (positive) or inhibitory (negative) neurotransmitters. F_P can also be null if Pierre is indifferent towards Marie and no neurotransmitter is released within his brain at the sight of Marie or thinking about her. Similarly, the feeling of Marie for Pierre will be indicated by F_M and Marie's thought about Pierre with I_P. To describe the evolution of the relationship between Pierre and Marie, we may use the following set of "sentimental" transformations:

$$F_P + I_M \xrightarrow{k_1} n_1 F_P + I_M$$
$$P + F_M \xrightarrow{k_2} n_2 F_P + F_M$$
$$F_M + I_P \xrightarrow{k_3} n_3 F_M + I_P$$
$$M + F_P \xrightarrow{k_4} n_4 F_M + F_P$$

The first and the third processes represent how F_P and F_M evolve spontaneously into the brains of Pierre and Marie when they think about Marie and Pierre, respectively. The second transformation describes how Pierre reacts when he gets acquainted with the feeling of Marie for him, by interpreting her direct or indirect messages. Vice versa, the fourth process describes how Marie reacts when she gets acquainted with Pierre's feeling for her, by interpreting his direct or indirect messages. The terms n_1, n_2, n_3, and n_4 look like stoichiometric coefficients. They can be any, positive or negative, real numbers. For this exercise, let us assume that n_1 and n_3 may be 2, 1 or 0. If n_1 and n_3 are equal to 2, it means that he/she likes her/him and their feelings of love grow autocatalytically; if n_1 and n_3 are 1, it means that their sensations neither grow nor decrease; if they are 0, it means their feelings fade. Note that Pierre's thought about Marie (I_M) and Marie's thought about Pierre (I_P) play as if they were catalysts of the first and third transformation, respectively. Their action is positive when $n_1 = n_3 = 2$, null when $n_1 = n_3 = 1$, and inhibitory when $n_1 = n_3 = 0$.

If we look at the second and fourth processes, we notice that F_M, in the second, and F_P, in the fourth, play as catalysts, as well. If n_2 is equal to 1, it means that love messages sent by Marie to Pierre work as efficient catalysts to increase his love for her. If n_4 is equal to 1, it means that love messages sent by Pierre to Marie work as effective catalysts to boost her love for him.

If n_2 or n_4 is equal to 0, it means that love messages sent by Marie/Pierre do not work at all. Finally, when n_2 or n_4 is equal to -1, it means that love messages sent by Marie/Pierre play as inhibitors of his/her love for her/him. In Table 3.3, there is a summary of all possible values and meanings of the four coefficients.

Note that the values of products $k_1 I_M$, $k_3 I_P$, $k_2 P$, and $k_4 M$ define the time scale of the dynamics of the love story. They may have day^{-1} as the unit if Pierre and Marie think about and encounter each other, every day. In case Pierre and Marie meet only once a week, $k_1 I_M$, $k_3 I_P$, $k_2 D$, and $k_4 B$ will have week^{-1} as units. Now, try to predict the dynamics of the love story between Pierre and Marie. Pierre fell in love with Marie and wooed her. On the other hand, it seems that Marie was a cautious lover. Can you predict what happened? Write the ordinary differential equations for the four transformations listed earlier, find the steady state solution, and use the linear stability analysis to predict the course of the events. You are invited to imagine the evolution of the Pierre and Marie's relationship, but also you would like to predict the dynamics of your own love story and other love affairs that you know.

TABLE 3.3

Values of the Stoichiometric Coefficients for the Love Story

Coefficients	2	1	0	−1
$n_1; n_3$	Positive	Null	Negative	
$n_2; n_4$		Positive	Null	Negative

3.13   SOLUTIONS TO THE EXERCISES

3.1.   When the system is at equilibrium the extended chemical potential at height h_1 must be equal to the extended chemical potential at height h_2:

$$\bar{\mu}_k\left(h_1\right) = \mu_k^0 + RTlnC_k\left(h_1\right) + PM_kgh_1 = \bar{\mu}_k\left(h_2\right) = \mu_k^0 + RTlnC_k\left(h_2\right) + PM_kgh_2$$

If we rearrange the equation, we obtain:

$$\frac{C_k\left(h_2\right)}{C_k\left(h_1\right)} = e^{-\frac{PM_kg}{RT}\left(h_2 - h_1\right)}$$

This ratio is equal to 0.999976 in the case of tryptophan, whereas it is 0.99213 in case of the protein when $h_2 - h_1$ is equal to 3 cm. This result underlines that the solutions are always more concentrated at the bottom than at the top due to the gravitational field. The concentration gradient that originates inside the solution is the "equilibrium" condition in the presence of the gravitational field. It requires time to be reached. The solution must decant. Finally, note that the higher the molecular weight of the solute, the more pronounced is the concentration gradient.

3.2.   At "equilibrium," the electrochemical potential of the monovalent cations must be equivalent inside and outside the cell:

$$\bar{\mu}_{K^+}\left(in\right) = \mu_{K^+}^0 + RTlnC_{K^+}\left(in\right) + F\psi_{in} = \bar{\mu}_{K^+}\left(out\right) = \mu_{K^+}^0 + RTlnC_{K^+}\left(out\right) + F\psi_{out}$$

It derives that

$$\psi_{in} - \psi_{out} = \frac{RT}{F}ln\left(\frac{C_k\left(out\right)}{C_k\left(in\right)}\right) = -86mV$$

3.3.   The Gibbs free energy reaches a minimum at the equilibrium. On the left of the minimum, the slope is always negative. When we are very far from equilibrium the slope is large and negative, then it becomes smaller until it is null at the equilibrium. On the right of the minimum, the slope is positive. Close to the equilibrium the slope is small, and it grows when we get far from the minimum (plot a of Figure 3.22). The second derivative of G with respect to ξ is always positive and has a minimum in correspondence of the equilibrium point (plot b). The first and second derivatives of S_{univ} show opposite trends compared to those of G (see plots c and d) (Figure 3.22).

3.4.   Any mathematical equation has physical meaning if the term on the left of the equal sign has the same physical dimensions of the term appearing on the right of the equal sign.

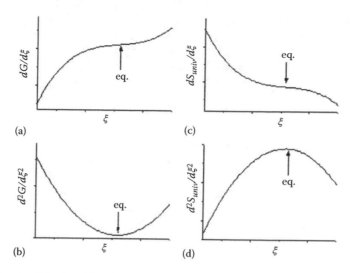

FIGURE 3.22 Trends of the first (a) and second (b) derivative of G with respect to ξ, and trends of the first (c) and second (d) derivatives of S_{univ} with respect to ξ.

Therefore, from equation [3.32], we can infer that

$$dimensions(k) = \frac{dimensions(J_q)}{dimensions(\nabla T)} = \frac{[E]}{[t][L]^2} \frac{[L]}{[T]}$$

[E] represents the energy, [L] is the length, [t] the time, [T] the temperature. The unit of measurements for k are cal/s·cm·K in cgs system. Remember that the cgs system uses cm for length, gram for weight, and second for time.

The dimensions of k can be expressed just by fundamental physical quantities, as

$$dimensions(k) = \frac{[M][L]}{[t]^3[T]}$$

3.5. From equation [3.38], we obtain:

$$\eta = \frac{\pi r^4}{8L} \frac{dt}{dV} \Delta P$$

Therefore, the dimensions of η are pressure per time, i.e., $[M]/[t][L]$. We confirm the result if we consider equation [3.39]:

$$\eta \propto \frac{J_{p_z}}{dv_x/dz} = \left(\frac{[M]}{[t][L]} \right)$$

3.6. From the Fick's Law, we can write:

$$dimensions(D_k) = \frac{dimensions(J_{k,dif})}{dimensions(\nabla C_k)} = \frac{[mol]}{[t][L]^2} \frac{[L]^4}{[mol]} = \frac{[L]^2}{[t]}$$

Note that the value of D_k depends on the compound and the medium where it is estimated. For instance, at 298 K the diffusion coefficient of carbon dioxide (CO_2) is 0.16 cm²/sec in the air, whereas it is $1.6 \cdot 10^{-5}$ cm²/sec in water.

3.7. The ratio between the mobility and the diffusion coefficient is equal to

$$\frac{\tau_k}{\gamma_k}\frac{\gamma_k}{k_BT} = \frac{\tau_k}{k_BT}$$

In the presence of an electric field, the ratio becomes $(z_k e / k_B T)$ (wherein $e = 1.6 \times 10^{-19}$ C), whereas in the presence of a gravitational field it is $(m_k / k_B T)$. In the case of Na⁺, $z_k = +1$ and $m_k = MW(Na)/N_{Av}$. Therefore, the ratio is 38.9 C/J in the presence of an electric field, whereas it is 9.3×10^{-6} Kg/J in the case of the gravitational field. The values of the ratio show that migration is important in the presence of an electric field, whereas it is often negligible when it is due to the terrestrial gravitational field.

3.8. At the stationary state the heat flow J_q is constant. Therefore, we can write

$$\int_0^{L_1}\left(\frac{dq}{dt}\right)dx = -\int_{T_h}^{T_x}k_1(Ar)dT = \int_{L_1}^{L_1+L_2}\left(\frac{dq}{dt}\right)dx = -\int_{T_x}^{T_c}k_2(Ar)dT$$

Finally, it derives that

$$k_1\left(\frac{T_x-T_h}{L_1}\right) = k_2\left(\frac{T_c-T_x}{L_2}\right)$$

$$T_x = \frac{k_2T_cL_1 + k_1T_hL_2}{L_2k_1 + L_1k_2}$$

3.9. To determine how $T(x)$ evolves in time we must use equation [3.106]

$$\frac{dT}{dt} \propto \left(\frac{d^2T}{dx^2}\right) = \left(\frac{10}{5^2}\right)e^{-\frac{x}{5}}$$

The smaller the x coordinate, the faster the T variation.

At the stationary state, J_q is constant and given by equation [3.107]. The temperature will be a linear function of x (see Figure 3.23)

$$\int_0^x J_q dx = -\int_{283.00}^{T(x)} kdT$$

$$\int_0^L J_q dx = -\int_{283.00}^{274.35} kdT$$

Finally, $T(x) = 283.00 - \dfrac{(283.00 - 274.35)}{10}x$

3.10. (1) To verify that the given function is the solution of the second Fick's law, it is necessary to calculate its first derivative with respect to time and compare the result with its second derivative with respect to z. (2) The fraction of sodium ions which has covered a distance dz after a time interval t is given by

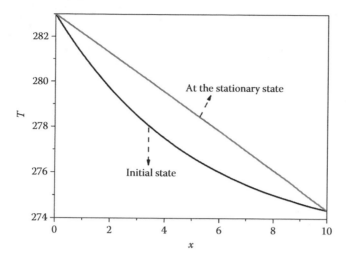

FIGURE 3.23 Evolution of the trend $T(x)$.

$$\frac{C(z,t)(Ar)dz}{n_0}$$

The average distance covered by sodium ions will be

$$\langle z \rangle - \int_0^\infty z \frac{C(z,t)}{n_0}(Ar)dz = \frac{1}{(\pi Dt)^{1/2}} \int_0^\infty z e^{-\frac{z^2}{4Dt}} dz = 2\left(\frac{Dt}{\pi}\right)^{1/2}$$

reminding that $\int z e^{-az^2} dz = -\frac{e^{-az^2}}{2a} + cost.$

(3) The average distances covered by sodium ions after 1 minute, 1 hour, and 1 day are 319 μm, 2.47 mm, and 1.2 cm, respectively.

3.11. The one-dimensional mean square distance traveled by diffusion can be found by integrating $z^2 C(z,t)/n_0$:

$$\langle z^2 \rangle = \int_0^\infty z^2 \frac{C(z,t)}{n_0}(Ar)dz = \frac{1}{(\pi Dt)^{1/2}} \int_0^\infty z^2 e^{-\frac{z^2}{4Dt}} dz = 2(Dt)$$

The three-dimensional mean square distance traveled by diffusion will be three times the mean square distance defined for one dimension: $6(Dt)$.

The second part of the exercise requires comparing quantitatively the root-mean-square distances traveled by two species having diffusion coefficients that differ for one order of magnitude: sodium ion, and a much heavier and bulkier protein. The results of the calculations are listed in Table 3.4. The root-mean-square distance traveled by the sodium ion is ≈3.6 times longer than that travelled by the protein.

When the proteins move within a cell, their diffusion can be slowed down up to tens or even hundreds-fold. Such significant reductions of D are due to molecular crowding, which is particularly pronounced within the nucleus of a eukaryotic cell. In fact, the nucleus contains the highest macromolecular densities in a cell (Bancaud et al. 2009). Molecular crowding hinders diffusion, induces volume exclusion and shifts any binding reaction towards bound states.

TABLE 3.4

Root-Mean-Square Distance Traveled by the Sodium Ion and a Protein

	$\left(\sqrt{\langle z^2 \rangle} = \sqrt{2Dt}\right)_{Na}$	$\left(\sqrt{\langle z^2 \rangle} = \sqrt{2Dt}\right)_{Protein}$
1 minute	0.04 cm	0.011 cm
1 hour	0.31 cm	0.085 cm
1 day	1.5 cm	0.416 cm

3.12. The entropy production for the system described in Section 3.4.1 is

$$P^* = \int_0^L J_q \left(\frac{\partial}{\partial x}\frac{1}{T}\right) dx = \int_0^L L_q \left(\frac{\partial}{\partial x}\frac{1}{T}\right)^2 dx$$

To determine when P^* is minimized, we apply the Euler–Lagrange differential equation (see note 13), and we obtain

$$\frac{d}{dx}\left(2L_q \dot{f}\right) = 0$$

It derives that

$$2L_q \left(\frac{\partial}{\partial x}\frac{1}{T}\right) = -\frac{2L_q}{T^2}\left(\frac{\partial T}{\partial x}\right) = 2k\left(\frac{\partial T}{\partial x}\right) = cost$$

This result means that P^* is minimized at the stationary state when J_q is constant, and T is a linear function of the spatial coordinate x.

3.13. The direct reaction is an example of autocatalysis. When the concentration of X is high, the back reaction becomes relevant. Applying the law of mass action[24] for the two elementary steps, we obtain:

$$\frac{d[X]}{dt} = -k_1[X][Y] + 2k_1[X][Y] - 2k_{-1}[X]^2 + k_{-1}[X]^2 = k_1[X][Y] - k_{-1}[X]^2$$

$$\frac{d[Y]}{dt} = k_0[Y]_0 - k_0[Y] - k_1[X][Y] + k_{-1}[X]^2$$

Since $[Y]_0$ is very high, $[Y]$ is constant over time: $[Y]_t \approx [Y]_0$.

At the stationary state, the production rate of X is equal to its consumption rate (see Figure 3.24):

$$k_1[X][Y] = k_{-1}[X]^2$$

---

[24] Remember that the law of mass action derives from the idea that the rate of a reaction depends on the probability of collision between the reactant molecules.

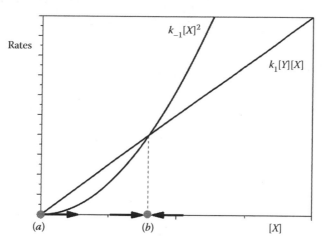

FIGURE 3.24 Rates of production and consumption of X versus [X].

We consider only positive values of [X] since negative concentrations do not have any physical meaning. The straight line, representing the production rate of X, intersects the parabola, representing the rate of X consumption, only in two points: point (a) where $[X]_a = 0$, and point (b) where $[X]_b = (k_1[Y]_0/k_{-1})$. The first solution, $[X]_a = 0$, is unstable because if $[X] > [X]_a$ the rate of X production is larger than that of its consumption and a small perturbation pushes the system away from $[X]_a$. The second solution, $[X]_b$, represents a stable stationary state. In fact, if $[X] < [X]_b$, the rate of X production is larger than that of its consumption and [X] heightens up to reach $[X]_b$. On the other hand, if $[X] > [X]_b$, the rate of X depletion is faster than that of its formation and [X] wanes up to the $[X]_b$ value.

Confirmation of these results can be achieved by applying the linear stability analysis. The derivative $\dot{f}\left([X]_{ss}\right)$ is equal to $k_1[Y]_0 - 2k_{-1}[X]_{ss}$.

When $[X]_{ss} = 0$, $\dot{f}\left([X]_{ss}\right) = k_1[Y]_0 > 0$. A small perturbation on [X] grows exponentially.

When $[X]_{ss} = (k_1[Y]_0/k_{-1})$, $\dot{f}\left([X]_{ss}\right) = -k_1[Y]_0 < 0$. If [X] is perturbed slightly from $[X]_b$, the disturbance will decay monotonically and $[X] \to [X]_b$ as $t \to \infty$.

It is worthwhile noticing that the differential equation written for [X] is formally equivalent to the logistic equation which models the dynamics of growth for a population of organisms, represented by X, in the presence of a constant amount of food represented by Y.

3.14. At the stationary state the concentrations of X and Y are:

$$[X]_{ss} = \frac{k_1[A][B]}{k_2[A] + k_0}$$

$$[Y]_{ss} = \frac{k_2[A]}{(k_3 + k_0)}\left(\frac{k_1[A][B]}{k_2[A] + k_0}\right)$$

Which are the stability properties of this stationary state?

The Jacobian is $J = \begin{pmatrix} -k_2[A] - k_0 & 0 \\ k_2[A] & -k_3 - k_0 \end{pmatrix}$.

The eigenvalues of J can be found solving the characteristic equation [3.175]. The trace and the determinant of J are

$$tr(J) = -k_2[A] - 2k_0 - k_3$$

$$det(J) = (k_2[A] + k_0)(k_3 + k_0)$$

The discriminant $\Delta = \left[(tr(J))^2 - 4det(J)\right] = (k_2[A] - k_3)^2$.

Since $tr(J) < 0$, $det(J) > 0$, when $k_2[A] > k_3$, $\Delta > 0$ and the stationary state is a stable node. When $k_2[A] = k_3$, $\Delta = 0$, and the stationary state is a stable star. Finally, if $k_2[A] < k_3$, $\Delta < 0$ and the stationary state is stable spiral.

3.15. If we sum the four sentimental transformations describing the dynamics of the love story between Pierre and Marie, we obtain that the overall process is

$$P + M + F_P + F_M \rightarrow (n_1 + n_2)F_P + (n_3 + n_4)F_M$$

The ordinary differential equations describing how the feelings of Pierre for Marie and that of Marie for Pierre evolve over time are

$$f = \frac{dF_P}{dt} = (n_1 - 1)k_1 I_M F_P + n_2 k_2 P F_M = (n_1 - 1)k_1' F_P + n_2 k_2' F_M$$

$$g = \frac{dF_M}{dt} = (n_3 - 1)k_3 I_P F_M + n_4 k_4 M F_P = (n_3 - 1)k_3' F_M + n_4 k_4' F_P$$

The steady state solution is $(F_{P,s}, F_{M,s}) = (0,0)$. We may imagine it corresponds to a preliminary situation where Pierre and Marie do not know each other because they have never met, before. Now, imagine that one day they meet each other, casually. They have their first impact on each other and prove their first reciprocal emotions. Their first meeting is like a perturbation to the steady state solution $(F_{P,s}, F_{M,s}) = (0,0)$. To predict the course of the love story, we use the linear stability analysis. The Jacobian is

$$J = \begin{pmatrix} \left(\frac{\partial f}{\partial F_P}\right)_s & \left(\frac{\partial f}{\partial F_M}\right)_s \\ \left(\frac{\partial g}{\partial F_P}\right)_s & \left(\frac{\partial g}{\partial F_M}\right)_s \end{pmatrix} = \begin{pmatrix} (n_1 - 1)k_1' & n_2 k_2' \\ n_4 k_4' & (n_3 - 1)k_3' \end{pmatrix}$$

The trace of the Jacobian is $tr(J) = (n_1 - 1)k_1' + (n_3 - 1)k_3'$ and the determinant is $det(J) = (n_1 - 1)(n_3 - 1)k_1' k_3' - n_2 n_4 k_2' k_4'$. Now, we can play with our imagination and choose different combinations of the four coefficients n_1, n_2, n_3, n_4 and the four kinetic constants: k_1', k_2', k_3', k_4'. The general solutions are bi-exponential functions of the type:

$$F_P = c_1 v_{1,P} e^{\lambda_1 t} + c_2 v_{2,P} e^{\lambda_2 t}$$
$$F_M = c_1 v_{1,M} e^{\lambda_1 t} + c_2 v_{2,M} e^{\lambda_2 t}$$

where λ_1 and λ_2 are the eigenvalues, and $\bar{v}_1 = \begin{pmatrix} v_{1,P} \\ v_{1,M} \end{pmatrix}$ and $\bar{v}_2 = \begin{pmatrix} v_{2,P} \\ v_{2,M} \end{pmatrix}$ are the eigenvectors. The values of the coefficients c_1 and c_2 depend on the initial conditions. Let us examine a few cases.

Case 1. Two identical passionate lovers:

$n_1 = 2$, $k_1' = 2$, $n_2 = 1$, $k_2' = 1$, $n_3 = 2$, $k_3' = 2$, $n_4 = 1$, $k_4' = 1$.

$tr(J) = 4$; $det(J) = 3$.

Eigenvalues: $\lambda_1 = 3$; $\lambda_2 = 1$.

Eigenvectors: $\bar{v}_1 = \begin{pmatrix} 1 \\ 1 \end{pmatrix}$; $\bar{v}_2 = \begin{pmatrix} 1 \\ -1 \end{pmatrix}$.

If the initial conditions are $F_P(0) = +1$, $F_M(0) = +1$, we can calculate the values of the coefficients c_1 and c_2:

$$F_P(0) = +1 = c_1 + c_2$$
$$F_M(0) = +1 = c_1 - c_2$$

It derives that $c_1 = 1$ and $c_2 = 0$. This means that the love between Pierre and Marie will grow exponentially towards complete bliss: $F_P(t) = F_M(t) = e^{3t}$. It is worthwhile noticing that for such a couple if the initial impression is negative, the relationship is designated to fail. For instance, if the initial conditions are $F_P(0) = -0.1$, $F_M(0) = -0.1$, it derives that $c_1 = -0.1$ and $c_2 = 0$. Therefore, the feeling of dislike will grow exponentially.

Case 2. Two identical lovers who are less passionate than those presented in case 1:
$n_1 = 2, k_1' = 1, n_2 = 1, k_2' = 1, n_3 = 2, k_3' = 1, n_4 = 1, k_4' = 1$.
$tr(J) = 2$; $det(J) = 0$.
Eigenvalues: $\lambda_1 = 2$, $\lambda_2 = 0$.
Eigenvectors: $\bar{v}_1 = \begin{pmatrix} 1 \\ 1 \end{pmatrix}$; $\bar{v}_2 = \begin{pmatrix} 1 \\ -1 \end{pmatrix}$.

If initially, the two lovers have a good mutual feeling ($F_P(0) = +0.1$, $F_M(0) = +0.1$), their love is destined to grow exponentially. However, their love will grow slower than that in case 1. In fact, $F_P(t) = F_M(t) = e^{2t}$.

Case 3. One lover is passionate; the other is cautious:
$n_1 = 2, k_1' = 1, n_2 = 1, k_2' = 1, n_3 = 1, k_3' = 1, n_4 = 1, k_4' = 1$.
$tr(J) = 1$; $det(J) = -1$.
Eigenvalues: $\lambda_{1,2} = \dfrac{1 \pm \sqrt{5}}{2}$.
Eigenvectors: $\bar{v}_1 = \begin{pmatrix} 1 \\ (\sqrt{5}-1)/2 \end{pmatrix}$; $\bar{v}_2 = \begin{pmatrix} 1 \\ -(\sqrt{5}+1)/2 \end{pmatrix}$.

The general solution is

$$F_P = c_1 e^{\lambda_1 t} + c_2 e^{\lambda_2 t}$$

$$F_M = c_1 \left(\frac{\sqrt{5}-1}{2} \right) e^{\lambda_1 t} - c_2 \left(\frac{\sqrt{5}+1}{2} \right) e^{\lambda_2 t}.$$

The dynamics of the love relation is strongly dependent on the feelings the protagonists have at their first encounter. In other words, the evolution is strongly dependent on initial conditions. Let us imagine that both Pierre and Marie had good feelings: $F_P(0) = F_M(0) = +1$. It derives that

$$+1 = c_1 + c_2$$

$$+1 = c_1 \left(\frac{\sqrt{5}-1}{2} \right) - c_2 \left(\frac{\sqrt{5}+1}{2} \right)$$

Finally, $c_1 = \left(\sqrt{5} + 3/2\sqrt{5} \right) > 0$, $c_2 = \left(\sqrt{5} - 3/2\sqrt{5} \right) < 0$. This result means that their love is expected to grow exponentially, as occurred in reality. On the other hand, if their first feelings were not good at all, (for instance $F_P(0) = F_M(0) = -1$), then their relationship would

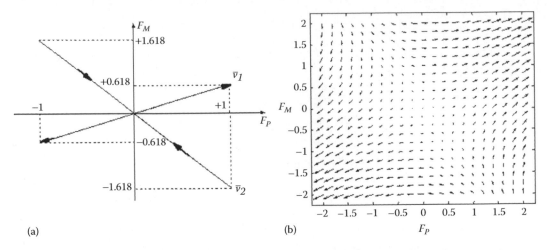

(a)   (b)

FIGURE 3.25   Plot of the eigenvectors $\bar{v}_1$ and $\bar{v}_2$ in (a) and plot of the phase portrait in (b) for case 3.

have been destined to fail. In fact, we would have $c_1 < 0$, and both partners would finish hating each other. There is also a third possibility. The initial condition is a point that lies on the eigenvector $\bar{v}_2$ (see Figure 3.25). For instance, $F_P(0) = +1$ and $F_M(0) = -\left(\sqrt{5}+1\right)/2$.

In such a case, it derives that $c_1 = 0$ and $c_2 = 1$. The love relationship fades to indifference. In Figure 3.25, there is the phase portrait for the system of case 3. Each vector represents the evolution from a particular initial condition or point of the phase space.

Case 4. Both protagonists are cautious lovers. However, one gets excited by positive messages coming from the partner, whereas the other gets discouraged:
$n_1 = 1$, $k_1' = 1$, $n_2 = 1$, $k_2' = 1$, $n_3 = 1$, $k_3' = 1$, $n_4 = -1$, $k_4' = 1$.
$tr(J) = 0$; $det(J) = 1$.
Eigenvalues: $\lambda_{1,2} = \pm i$.
Eigenvectors: $\bar{v}_1 = \begin{pmatrix} 1 \\ -i \end{pmatrix}$; $\bar{v}_2 = \begin{pmatrix} 1 \\ i \end{pmatrix}$.

The love story is oscillatory, as shown in Figure 3.26. The more Pierre shows interest in Marie, the more Marie wants to run away. But when Pierre gets tired and shows discouragement, Marie starts to find him strangely attractive. Her messages of appreciation trigger a warm up in Pierre, who echoes her sentiments. However, the growing attention by Pierre makes Marie more cautious, and the relationship oscillates. The outcome of this link is a

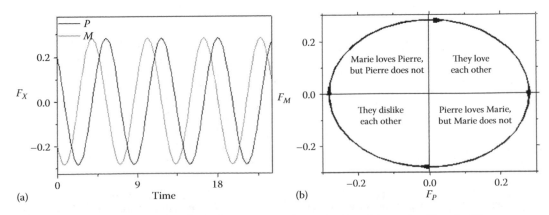

(a)   (b)

FIGURE 3.26   Oscillatory trends of both F_P and F_M over time in (a) and the cycle of love in (b).

never-ending cycle of love and hate as depicted in Figure 3.26. At least, for a quarter of the entire period, they love each other.

Case 5. The two protagonists do not like each other. However, one lover gets excited when the other sends signs of even small interest, whereas the other lover gets excited when the partner looks indifferent.

$n_1 = 0, k_1' = 0.2, n_2 = 1, k_2' = 1, n_3 = 0, k_3' = 0.2, n_4 = -1, k_4' = 1.$

$tr(J) = -0.4; det(J) = 1.04.$

Eigenvalues: $\lambda_1 = -0.2 + i; \lambda_2 = -0.2 - i.$

Eigenvectors: $\bar{v}_1 = \begin{pmatrix} 1 \\ i \end{pmatrix}; \bar{v}_2 = \begin{pmatrix} 1 \\ -i \end{pmatrix}.$

The relationship wanes after few oscillations like shown in Figure 3.27.

Of course, we may consider many more situations. To draw the phase portraits for the love affairs, you may use the following MATLAB code:

```
[y₁, y₂] = meshgrid(-2:0.2:2, -2:0.2:2);
n₁ = 2;
n₂ = 1;
n₃ = 0;
n₄ = -1;
k₁ = 1;
k₂ = 1;
k₃ = 0.5;
k₄ = 0.5;
y₁dot = (n₁ - 1)*k₁*y₁ + n₂*k₂*y₂;
y₂dot = (n₃ - 1)*k₃*y₂ + n₄*k₄*y₁;
quiver(y₁,y₂,y₁dot, y₂dot)
```

To calculate the time evolution of Pierre's and Marie's feelings, you may use the following function file in MATLAB:

```
function dy = love(t, y)
dy = zeros(2,1);
n₁ = 0;
n₂ = 1;
n₃ = 0;
n₄ = -1;
k₁ = 0.2;
```

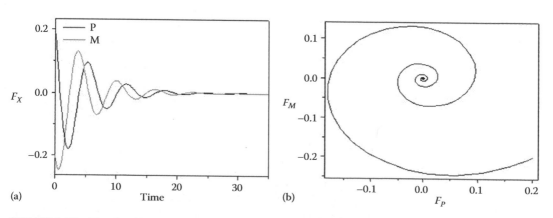

(a) (b)

FIGURE 3.27 Trends of both F_P and F_M over time in (a) and the damped cycles in (b).

```
k₂ = 1;
k₃ = 0.2;
k₄ = 1;
dy(1) = (n₁ - 1)*k₁*y(1)+n₂*k₂*y(2);
dy(2) = (n₃ - 1)*k₃*y(2)+n₄*k₄*y(1);
```

The command line to calculate the evolution in the time window [0 250] and from the initial conditions $F_P(0) = -0.2$, $F_M(0) = 0.2$, will be of the type:

```
[t, y] = ode45('love', [0 250], [-0.2 0.2]);
```

4 An Amazing Scientific Voyage
From Equilibrium up to Self-Organization through Bifurcations

Ye were not form'd to live the life of brutes but virtue to pursue and knowledge high.

Odysseus in 26 Canto of the *Inferno* by Dante

4.1 INTRODUCTION

All the macroscopic transformations that occur in our universe abide by the Second Law of Thermodynamics. It tells us that the total entropy of the Universe either grows or remains constant if either a spontaneous or a reversible process occurs, respectively. The entropy of the Universe is the sum of two contributions: the entropy of the system (S_{sys}) wherein the transformation takes place, and that of the environment (S_{env}). An irreversible or spontaneous transformation ceases when it reaches the equilibrium. When a system arrives at the equilibrium, the entropy of the Universe is maximized:

$$\frac{dS_{tot}}{dt} = \frac{dS_{sys}}{dt} + \frac{dS_{env}}{dt} = 0 \tag{4.1}$$

The entropy of the system consists of two contributions:

$$\frac{dS_{sys}}{dt} = \frac{d_i S_{sys}}{dt} + \frac{d_e S_{sys}}{dt} = 0 \tag{4.2}$$

with both $d_i S_{sys}/dt = 0$ and $d_e S_{sys}/dt = -dS_{env}/dt = 0$.

After reaching the equilibrium, the number of all the possible microscopic configurations are maximized, and our information about the microscopic properties of the system is minimized. This loss of knowledge is usually expressed by saying that disorder has increased. If the system is isolated, the disorder regards only the system. On the other hand, if the system is closed, the disorder does not necessarily store entirely inside of it. In fact, a closed system that reaches the equilibrium, can be the theater of self-assembly phenomena, like the formation of crystals, block copolymer assemblies, and ordered supramolecular structures, such as micelles and micro-emulsions, and others (for a list of many examples of self-assembled systems see Grzybowski et al. 2009). At equilibrium, all the forces are null, and the system stays indefinitely in that state unless it is temporarily perturbed. Even if the system is temporarily pushed away from the equilibrium, it spontaneously recovers its initial state, because the equilibrium is stable. At the equilibrium, the Gibbs free energy of the system[1] (see Figure 4.1) is at its minimum, and whenever the system is moved away from it,

---

[1] We consider Gibbs free energy G when temperature, pressure, and the number of particles are maintained constant. When temperature, volume, and the number of particles are held constant, the Helmholtz free energy reaches a minimum at the equilibrium. By the way, at the equilibrium, the power to do work is minimized.

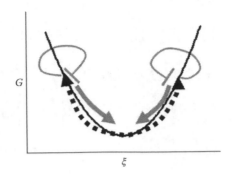

FIGURE 4.1   Profile of the Gibbs free energy (G) as a function of a hypothetical coordinate ξ for either an isolated or an adiabatic or a closed system. The black arrows represent temporary perturbations pushing the system out-of-equilibrium; the slight grey curves represent the deviation-counteracting feedback actions driving the system back to equilibrium (see tick grey arrows).

deviation-counteracting feedback actions draw the system back to the minimum as if it were a ball on a well of potential energy.

Around us, we find many examples of ordered structures in space and/or time. Think about phenomena like convection, the seasonal cycles, all forms of life, a flock of birds or a school of fishes, and so on. A natural question raises; are all the ordered structures we find in nature due to self-assembling phenomena? No, indeed. In fact, most of the ordered structures emerge in out-of-equilibrium conditions. The Fluctuation Theorem tells us that microscopic systems working in out-of-equilibrium conditions may violate the Second Law of Thermodynamics but for short timescales. Fluctuation Theorem does not answer our question exhaustively because even macroscopic systems may give rise to ordered structures over long timescales. Suffice it to think about the evolution of a fertilized egg that originates an extremely ordered living being.

We are in out-of-equilibrium conditions when at least one force is maintained not null. If we fix the value of the force, the system evolves to a stationary state. At the stationary state, the system becomes a dissipative structure, because it produces entropy inward, but it discharges that entropy outward:

$$\frac{dS_{sys}}{dt} = \frac{d_iS_{sys}}{dt} + \frac{d_eS_{sys}}{dt} = 0 \qquad [4.3]$$

with $d_iS_{sys}/dt > 0$ and $d_eS_{sys}/dt < 0$. Of course, the system self-organizes at the expense of the entropy of the surrounding environment, which increases:

$$\frac{dS_{env}}{dt} > 0. \qquad [4.4]$$

When the force is weak, we are close to equilibrium; the emerging flow depends linearly on the force. In the linear regime, we can encounter not only stable states (e.g., in phenomena such as diffusion, heat conduction and electrical conduction), but also unstable (exponential growth) and oscillatory states (in electrical circuits). When the force is so strong that we are very far-from-equilibrium, we abandon the linear regime, and we enter the non-linear regime. In the non-linear regime, we can have not only stable, unstable and oscillatory states but also aperiodic states that are extremely sensitive to initial conditions (see Table 4.1).

TABLE 4.1

All the Possible States in Three Distinct Regimes: (A) at Equilibrium or Out-of-Equilibrium, in (B) Linear or (C) Non-Linear Regime

States Regimes	Stable	Unstable	Oscillatory	Aperiodic
(A) Equilibrium	X			
(B) Out-of-equilibrium, linear	X	X	X	
(C) Out-of-equilibrium, non-linear	X	X	X	X

For non-equilibrium systems, we still lack a general criterion of evolution; there are few principles in an evolutionary landscape that is still forming.[2] The type of evolution and the principle governing the evolution are extremely sensitive to the kind of constraints and variable parameters. For instance, in the previous chapter, we have seen that when the constraints are k fixed forces F_j (with $j = 1, 2, ..., k$), and the variables are the remaining $k + 1, k + 2, ..., n$ forces, the Theorem of the Minimum Entropy Production is the criterion dictating the type of evolution. The system will reach a stationary state wherein the flows corresponding to the constrained forces reach constant values, whereas the unconstrained forces ($j = k + 1, k + 2, ..., n$) adjust to make their corresponding flows zero. Such stationary states are stable because they minimize the entropy production. The symmetry properties of the stationary state depend on those of the force: if the force is isotropic, the flow will be isotropic; if the force is anisotropic, being an n-th order tensor (with $n \geq 1$), the flow may be anisotropic too, in agreement with the Curie's principle and the Onsager's reciprocal relations. If the system in its stationary state is temporarily perturbed, it is pushed away from it. But, left to itself, the system will spontaneously restore the initial stationary state due to deviation-counteracting feedback actions. The value of the flow in the stationary state depends on the strength of the force. The stronger the force, the larger the flow. As long as we are in the linear regime and the presence of stable stationary states, we run across the so-called thermodynamic branch in the diagram of stability (see Figure 4.2). The transition from the linear to the non-linear regime is marked by a point. As soon as we reach this point, having F_c and J_c as coordinates in the plot of Figure 4.2, the thermodynamic branch is no longer stable, and new solutions of stationary state coexist. Such pivotal point is called bifurcation because from there usually two new paths open. Often, we cannot predict the path that will be followed and the state that will exist at the end, because the choice, made by the system, between two possible routes may be a matter of tiny and hardly detectable fluctuations. It is tough to have any general principle predicting the type of evolution in very-far-from-equilibrium conditions. In fact, the Second Law of Thermodynamics is a statistical law in character, rather than being a dynamical law. The consequence is that it tells nothing about the average rate in which the

---

[2] The study of the behavior of Entropy Production P^* for out-of-equilibrium systems becomes relevant for a perspective going "Beyond the Second Law" as shown in a recent book by Dewar et al. (2014). For non-equilibrium systems, there is not just one criterion of evolution, but there are few principles in a landscape that is still forming. Such landscape is somewhat fragmented and the book by Dewar et al. tries to find connections and construct bridges among isolated principles. The authors rightly suggest that to make sense of such landscape, one must identify three key aspects of each principle: (1) Which are the variables? (2) Which are the constraints? (3) Which entropy function is being maximized or minimized? With this interpretative key on hand, it is not so puzzling to find out that besides the Theorem of Minimum Entropy Production formulated by Prigogine and presented in Chapter 3, there is also the Maximum Entropy Production Principle, formulated by Ziegler, which is applicable when the constraints are the fixed n forces and the variables are the n fluxes (Martyushev and Seleznev 2006).

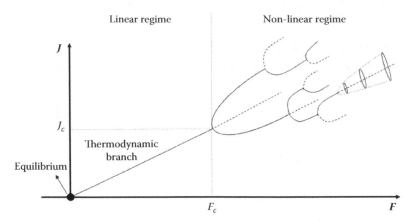

FIGURE 4.2 Sketch representing the evolution of a generic system from the linear to the non-linear regime. The origin of the graph represents the stable equilibrium condition wherein $F = 0 = J$.

entropy of the Universe grows, and also about the probability of statistical fluctuations in energy and mass flow for which, at least momentarily and locally, $(dS_{tot}/dt) < 0$.

The only expression we have seen for the non-linear regime is the Glansdorff-Prigogine stability criterion, mentioned in Chapter 3. It asserts that in the stationary state, $P*$ is minimized only with respect to the forces: $d_F P* \leq 0$. However, this criterion cannot be applied so widely because it requires the validity of the local equilibrium approximation, and it does not tell us anything about how the system evolves and the stationary state that will be reached among many. Therefore, working in a non-linear regime, out-of-equilibrium systems evolve most often unpredictably. In the case of chemical reactions, the final state can be either stationary, periodic, or aperiodic. Such states can be either stable or unstable. The reason why the non-linear regime is so wealthy with surprises is that in very-far-from-equilibrium conditions there are not only deviation-counteracting but also deviation-amplifying feedback actions. It is just the fierce fight between negative and positive feedback actions, which gives rise to the self-organization phenomena. Self-organization, sometimes called "Dynamic Self-Assembling"[3] (Whitesides and Grzybowski 2002), is a symmetry breaking phenomenon because it consists in the spontaneous emergence of order out of less ordered conditions. The order can regard either the time, the space, or both types of coordinates. In non-linear regime, the Curie's principle loses its hold. In fact, the effects may have fewer symmetry elements than their causes. In other words, the flows may be more ordered (or asymmetric) than their forces. Self-organized systems have also been called "dissipative structures" because, to maintain their order, they require a constant supply of energy and matter that is dissipated into heat.

The non-linear regime holds another great surprise: Chaos! Chaos must not be confused with the disorder. Disorder increases at the equilibrium when the entropy is maximized. Chaos is, instead, a synonym of unpredictability. A dynamic is chaotic when it is aperiodic and susceptible to initial conditions. Undetectable differences in the initial conditions may have a considerable impact on the temporal evolution of the system. We will learn more about Chaos in Chapter 10.

The nonlinear regime is breathtaking. Whereas the properties of systems working in the linear regime are precisely equal to the sum of its parts, in non-linear regime the principle of superposition fails spectacularly.

The scrutiny of the behavior of a system moving from the equilibrium to non-equilibrium conditions, crossing the linear regime and diving in the non-linear regime is a marvelous journey. It is

---

[3] Dynamic self-assembly or self-organization refers to the emergence of order in out-of-equilibrium conditions. On the other hand, "Static Self-Assembly" or "Equilibrium Self-Assembly" refers to the appearance of stable spatially ordered structures in equilibrium conditions.

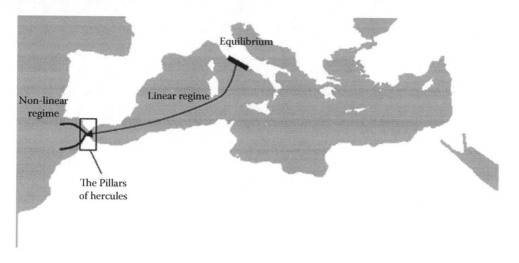

FIGURE 4.3   Journey of Odysseus from the coasts of Italy to the Pillars of Hercules. The Pillars of Hercules are like a bifurcation from the Mediterranean Sea to the Ocean.

comparable to the epic journey set out by Odysseus who, according to Dante in his 26 Canto of the *Inferno* of *Divine Comedy*, wanted to discover the mysteries beyond the Pillars of Hercules (see Figure 4.3). He convinced his terrorized crew to follow him by saying "Ye were not form'd to live the life of brutes but virtue to pursue and knowledge high." Maintaining an equilibrium condition is like staying at the seashore. Working in the linear regime is like sailing the Mediterranean Sea. Arriving at the Pillars of Hercules is like reaching the first bifurcation. Beyond the bifurcation, there is the oceanic non-linear regime reserving uncountable surprises.

A bifurcation point marks the transition from the linear to the non-linear regime. After entering the non-linear regime, we may find many other bifurcations.

4.2   BIFURCATIONS

If, after modifying the contour conditions, we detect changes in the number and/or stability of the fixed-point solutions for a differential equation (e.g., we assist to a change in the topology of the attractor), we have found a bifurcation. The simplest bifurcations of fixed-point solutions, which depend on a single control parameter λ and involve just one variable, are of three types (Strogatz 1994):

$$\text{saddle-node bifurcation: } \frac{dx}{dt} = \lambda - x^2 \qquad\qquad [4.5]$$

$$\text{trans-critical bifurcation: } \frac{dx}{dt} = \lambda x - x^2 \qquad\qquad [4.6]$$

$$\text{pitchfork bifurcation: } \frac{dx}{dt} = \lambda x - x^3 \qquad\qquad [4.7]$$

4.2.1   SADDLE-NODE BIFURCATION

The prototypical example of a saddle-node bifurcation is given by the first-order system [4.5].

The fixed points are $x_0 = \pm\sqrt{\lambda}$. When $\lambda < 0$, there are no fixed points. When $\lambda = 0$, we have one fixed point that is $x_0 = 0$. When $\lambda > 0$, we have two fixed points: one positive and the other negative. Through the linear stability analysis, we define the properties of the fixed points.

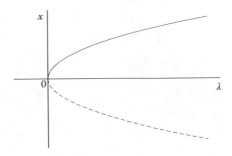

FIGURE 4.4   Bifurcation diagram for the saddle-node case.

$$f(x) = \frac{dx}{dt} = f(x_0) + (x - x_0)\dot{f}(x_0) + \ldots \tag{4.8}$$

Since $f(x_0) = 0$, equation [4.8] can be rearranged in

$$\int \frac{dx}{(x - x_0)} = \dot{f}(x_0)\int dt \tag{4.9}$$

After integrating, we obtain:

$$x = x_0 + e^{\dot{f}(x_0)t} = x_0 + e^{-2x_0 t} \tag{4.10}$$

The solution $x_0 = 0$ is unstable like the fixed points $x_0 = -\sqrt{\lambda}$. On the other hand, the fixed points $x_0 = +\sqrt{\lambda}$ represent stable solutions. Figure 4.4 is the bifurcation diagram, and it shows all the possible solutions of fixed points. The solid line represents the stable solutions, whereas the dashed line represents the unstable fixed points.

TRY EXERCISE 4.1

4.2.2   TRANS-CRITICAL BIFURCATION

The normal form of a differential equation originating a trans-critical bifurcation is equation [4.6]. We have two fixed points: $x_0' = 0$ and $x_0'' = \lambda$. If we apply the linear stability analysis (equation [4.9]), we find that

$$x = e^{\lambda t} \tag{4.11}$$

for $x_0' = 0$, and

$$x = \left(\lambda - \frac{e^{-\lambda t}}{\lambda}\right) \tag{4.12}$$

for $x_0'' = \lambda$. From equation [4.11], it is evident that the fixed point $x_0' = 0$ is stable when $\lambda < 0$ and unstable when $\lambda \geq 0$. From equation [4.12], it is clear that the fixed point $x_0'' = \lambda$ is unstable when $\lambda \leq 0$, whereas it is stable when $\lambda > 0$. Therefore, the bifurcation diagram looks like the plot of Figure 4.5.

If we compare Figure 4.5 with Figure 4.4, we notice that in the trans-critical bifurcation diagram the two fixed points do not disappear after the bifurcation, but they swap their stability.

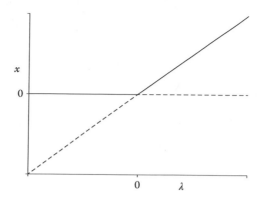

FIGURE 4.5 Bifurcation diagram for the trans-critical bifurcation.

4.2.2.1 From a Lamp to a Laser: An Example of Trans-Critical Bifurcation

A concrete example of trans-critical bifurcation is offered by a simplified model of the laser (Haken 1983; Milonni and Eberly 2010). The word LASER is an acronym that stands for Light Amplification by Stimulated Emission of Radiation. The electromagnetic radiation emitted by a laser has properties that are very different from those of the radiation emitted by a lamp. In fact, the laser radiation is highly monochromatic, polarized, highly coherent (in space and time), highly directional and it has high brightness.[4] The production of laser radiation requires three fundamental ingredients. The first ingredient is the phenomenon of stimulated emission of radiation. For an atomic or molecular system with two levels, having energies E_1 and E_2, respectively, the absorption of a photon, having energy equal to $h\nu = E_1 - E_2$, determines a jump from the lower to the higher level; the spontaneous decay from level 2 to 1 causes the emission of a photon whose energy is $h\nu = E_2 - E_1$. When the system is in its highest level 2, it may also collide elastically with a photon $h\nu$ and decay to the lowest level by emitting a second photon that has the same energy and is in phase with the incident one. The latter is known as a stimulated emission process (see Figure 4.6).

Usually, at room temperature, in the absence of any perturbation, the number of molecules or atoms at the lowest level (1) is much more abundant than the number of units staying in the higher level (2) However, the second ingredient required to generate laser radiation is the promotion of an inversion of the population of the two levels, like shown in Figure 4.7. Such population inversion is feasible by pumping the active materials correctly,[5] either optically or electrically.

The last ingredient to generate laser radiation is the optical cavity. It is a linear device with one relatively long optical axis (see Figure 4.8). It is delimited by two mirrors perpendicular to the

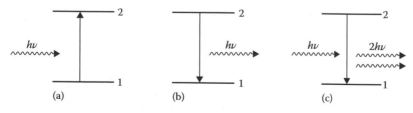

FIGURE 4.6 Schematic representation of the phenomena of absorption of a photon (a), spontaneous emission of a photon (b), and stimulated emission (c).

---

[4] The brightness of radiation represents its power per unit area, unit bandwidth, and steradian, or just its spectral intensity per steradian.

[5] Pumping a system means releasing a lot of energy to the system.

FIGURE 4.7 Population inversion promoted by an efficient pumping.

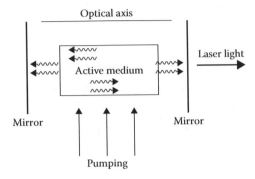

FIGURE 4.8 Sketch of an optical cavity.

optical axis and embedding the active medium emitting the radiation. One mirror is completely reflecting, whereas the other is only partially reflecting. Only the radiation that propagates along the optical axis of the cavity is reflected many times, and it is amplified by a lot of passages through the active medium. The spatial separation d between the two mirrors and the phenomenon of constructive interference select the wavelength (λ) of the laser radiation. It is

$$\lambda = \frac{2d}{m} \tag{4.13}$$

where $m = 1, 2, \ldots$ is a positive integer.

The generation of laser radiation may be described through the formulation of a system of two differential equations: one regarding the number of photons, n, and the other regarding the number of atoms or molecules in level 2: N_2. The system is

$$\frac{dn}{dt} = B_{21}N_2 n - Tn$$

$$\frac{dN_2}{dt} = P_u - B_{21}N_2 n - A_{21}N_2 \tag{4.14}$$

In [4.14], B_{21} represents the probability of the stimulated emission; T is the transmittance of the partially reflecting mirror of the optical cavity; A_{21} represents the probability of the spontaneous emission; P_u is the pump strength. Let us suppose that N_2 relaxes much more quickly than n. Then, $(dN_2/dt) = 0$, and

$$N_2 = \frac{P_u}{B_{21}n + A_{21}} \tag{4.15}$$

If we insert equation [4.15] into the first differential equation of the system [4.14], we obtain:

$$\frac{dn}{dt} = \frac{B_{21}nP_u}{B_{21}n + A_{21}} - Tn$$

[4.16]

We may rearrange equation [4.16] in the following form:

$$\frac{dn}{dt} = \frac{\left(B_{21}P_u - A_{21}T\right)n - TB_{21}n^2}{B_{21}n + A_{21}}$$

[4.17]

In the latter form, the denominator is always positive, and the numerator looks like the differential equation originating a trans-critical bifurcation, i.e., [4.6]. We have two fixed points: $n_0' = 0$ and $n_0'' = \left(B_{21}P_u - A_{21}T\right)/TB_{21}$. The solution $n_0' = 0$ is stable when $B_{21}P_u < A_{21}T$. This condition corresponds to the case wherein we do not have stimulated emission, and the device behaves like a lamp. When $B_{21}P_u$ becomes larger than $A_{21}T$, the system crosses a bifurcation and $n_0'' = \left(B_{21}P_u - A_{21}T\right)/TB_{21}$ becomes the stable solution. The device becomes a laser. The critical value of pumping is given by $\left(P_u\right)_c = \left(A_{21}T/B_{21}\right)$: the pumping must be particularly strong when A_{21} and T are large.

4.2.3 PITCHFORK BIFURCATION

If we determine the fixed points of equation [4.7] and we study their stability, we find a third type of bifurcation, called pitchfork bifurcation. The equation $x\left(\lambda - x^2\right)$ is null when either $x_0' = 0$ or $x_0'' = +\sqrt{\lambda}$ or $x_0''' = -\sqrt{\lambda}$. If we look just for real solutions of the variable x, when λ is negative or null, the fixed point is only the first: $x_0' = 0$. On the other hand, when λ is positive, all the three solutions are possible. We define the properties of the fixed points through the linear stability analysis (see equations [4.8] and [4.9]).

$$\int \frac{dx}{\left(x - x_0\right)} = \dot{f}\left(x_0\right)\int dt = \left(\lambda - 3x_0^2\right)\int dt$$

[4.18]

After integrating, we obtain:

$$x = x_0 + e^{\dot{f}(x_0)t} = x_0 + e^{\left(\lambda - 3x_0^2\right)t}$$

[4.19]

From equation [4.19], it is evident that the fixed point $x_0' = 0$ is stable when λ is negative, whereas it is unstable when $\lambda \geq 0$. When λ is positive, both $x_0'' = +\sqrt{\lambda}$ and $x_0''' = -\sqrt{\lambda}$ are stable solutions. The plot of all the solutions as a function of λ is shown in Figure 4.9. Due to its shape, the graph has been termed the pitchfork bifurcation diagram.

Figure 4.9 shows that if λ moves from negative to positive values, the solution $x_0' = 0$ changes from stable to unstable. At $\lambda = 0$, the system will evolve either towards the positive branch $x_0'' = +\sqrt{\lambda}$ or the negative branch $x_0''' = -\sqrt{\lambda}$. It is not possible to predict the path that will be traced because small, unpredictable fluctuations will push the system towards either one or the other route.

TRY EXERCISE 4.2

4.2.3.1 Chiral Symmetry Breaking

The concept of pitchfork bifurcation has been recently invoked to understand why chiral biomolecules in living beings are present only as one enantiomer.

First of all, let us remember the meaning of chiral species. Any molecule whose structure is not superimposable with that of its mirror image is defined chiral (the word chiral derives from the ancient Greek "χείρ", which means "hand": our right hand is not superimposable to our left hand).

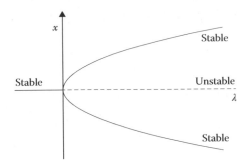

FIGURE 4.9 Pitchfork bifurcation diagram.

Mirror-image structures of a chiral molecule are called enantiomers. The two enantiomeric structures are usually distinguished by the sense of the rotation they give to a linearly polarized light crossing their solutions. An enantiomer will be levogyre (L or (–) as symbols), whereas the other will be dextrogyre (D or (+)). The absolute value of the rotation angle is the same for a couple of enantiomers. When a chiral molecule is synthesized from achiral compounds and, in the absence of any physical "force" having chiral character, for example, in the absence of any chiral interaction (Kondepudi and Nelson 1984), it is obtained as racemate. We have a racemate when the two enantiomers are in an equal amount (it is said that the enantiomeric excess is null).[6] Two examples of natural chiral molecules are shown in Figure 4.10.

What is surprising is that all the chiral amino acids present in living beings exist only as L(–) enantiomers, and the sugar molecules in DNA and RNA (2-deoxyribose in DNA and ribose in RNA) are present just in the D(+) configuration. Wrong-handed amino acids would disrupt the α-helix structure in proteins, and if just one wrong-handed sugar monomer were present, DNA could not be stabilized in a helix. Therefore, a Really Big Question arises:

"How did the biomolecular asymmetry originate?"

A good model for the spontaneous generation of chiral asymmetry has been proposed and confirmed experimentally by Kondepudi and Asakura (2001). It is based on a kinetic mechanism

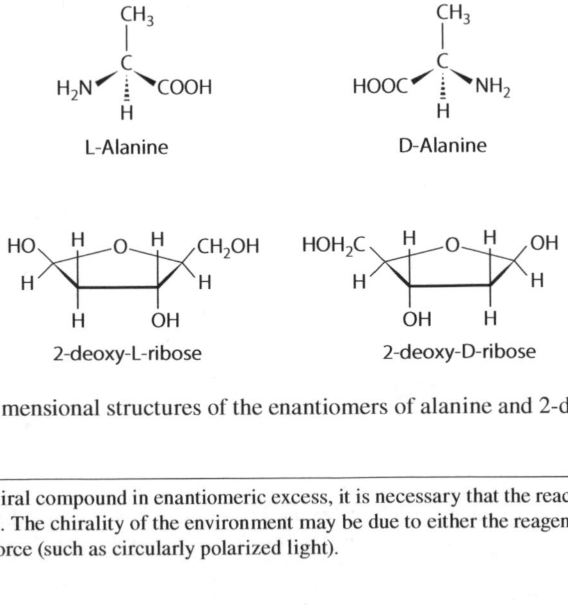

CH₃ structures and sugar structures:

L-Alanine   D-Alanine

2-deoxy-L-ribose   2-deoxy-D-ribose

FIGURE 4.10   Three-dimensional structures of the enantiomers of alanine and 2-deoxyribose.

---

[6] If we want to produce a chiral compound in enantiomeric excess, it is necessary that the reaction takes place in an asymmetric chiral environment. The chirality of the environment may be due to either the reagents or the solvent or the presence of a chiral physical force (such as circularly polarized light).

[4.20–4.24] involving a chiral autocatalytic species X, existing as a couple of enantiomers (X_L and X_D), produced from achiral reactants, S and T.

$$S + T \underset{k_{-1L}}{\overset{k_{1L}}{\rightleftharpoons}} X_L \qquad [4.20]$$

$$S + T \underset{k_{-1D}}{\overset{k_{1D}}{\rightleftharpoons}} X_D \qquad [4.21]$$

$$S + T + X_L \underset{k_{-2L}}{\overset{k_{2L}}{\rightleftharpoons}} 2X_L \qquad [4.22]$$

$$S + T + X_D \underset{k_{-2D}}{\overset{k_{2D}}{\rightleftharpoons}} 2X \qquad [4.23]$$

$$X_L + X_D \overset{k_3}{\longrightarrow} P \qquad [4.24]$$

In the first two steps of the mechanism ([4.20] and [4.21]), the two enantiomeric forms are produced directly from the achiral substrates S and T, whereas in the third and fourth steps ([4.22] and [4.23]), they are produced autocatalytically. Finally ([4.24]), X_L and X_D combine to produce P. The chemical system can be maintained far from equilibrium with a constant inflow of S and T and a constant outflow of P.

The dynamics of the system has a rather simple mathematical description in terms of the variables $\alpha = ([X_L] - [X_D])/2$ and $\lambda = [S][T]$ in the vicinity of the critical point λ_c (Kondepudi and Nelson 1984) at which asymmetric states emerge. It is modeled by the following differential equation:

$$\frac{d\alpha}{dt} = -A\alpha^3 + B(\lambda - \lambda_c)\alpha \qquad [4.25]$$

wherein A and B are functions of the kinetic rate constants of the mechanism [4.20–4.24]. Since A and B are positive, when $\lambda \leq \lambda_c$, the only stationary state solution is $\alpha = 0$. When $\lambda > \lambda_c$, there are three solutions of stationary state: $\alpha_1 = 0$, $\alpha_{2,3} = \pm [B(\lambda - \lambda_c)/A]^{1/2}$. The plot of the stationary state solutions is shown in Figure 4.11.

By applying the linear stability analysis, we can get insight on the stability of the solutions. If $x = (\alpha - \alpha_S)$ represents a small perturbation to the stationary state α_S, its evolution over time is given by:

$$x_t = x_{t_i} e^{[\dot{f}(\alpha_S)](t - t_i)} \qquad [4.26]$$

with

$$\dot{f}(\alpha_S) = \left[-3A\alpha_S^2 + B(\lambda - \lambda_c) \right] \qquad [4.27]$$

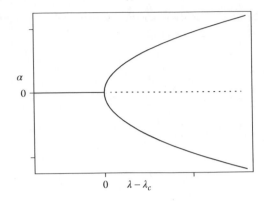

FIGURE 4.11 Bifurcation plot for the system described by the equation [4.25]. All the solutions represent stable stationary states except for those represented by the dotted line ($\alpha = 0$ when $\lambda > \lambda_c$).

When $\lambda \leq \lambda_c$, the solution $\alpha_S = 0$ is stable because the perturbation decays exponentially. When $\lambda > \lambda_c$, the solution $\alpha_1 = 0$ becomes unstable, whereas the solutions $\alpha_{2,3} = \pm\left[B\left(\lambda - \lambda_c\right)/A\right]^{1/2}$ is stable. This result means that by changing the value of the parameter $(\lambda - \lambda_c)$ from negative to positive, the thermodynamic branch defined by the solutions $\alpha_S = 0$ changes its character from stable to unstable at the point having coordinates $\alpha_S = 0$ and $\lambda = \lambda_c$. This point of the graph is a pitchfork bifurcation point (see Figure 4.11). When the system is on the bifurcation point, small random thermal fluctuations can push it towards new stationary states. The system will move towards either the stationary state described by the solution α_2 or that associated with α_3, unpredictably. The solutions α_2 and α_3 represent asymmetric states in which one enantiomer dominates over the other ($[X_L] \neq [X_D]$). Such stationary states are said to have a broken symmetry.

When there is a chiral bias, the evolution of the system is different. The chiral interaction brings about a small gap in the energies of X_L and X_D and those of the transition states, $X_L^{\#}$ and $X_D^{\#}$, involved in the elementary steps producing the chiral species in mechanism [4.20–4.24]. If ΔE is the energy gap between $X_L^{\#}$ and $X_D^{\#}$, the kinetic constants k_{iL} and k_{iD} will be not anymore equal. According to the Arrhenius equation, their ratio becomes

$$\frac{k_{iL}}{k_{iD}} = e^g \text{ where } g = \frac{\Delta E}{RT} \tag{4.28}$$

The chiral interactions may be intrinsic, such as the electroweak force acting within the molecule, or extrinsic, that is due to external electric, magnetic, gravitational and centrifugal fields. A few calculations (Kondepudi and Nelson 1984 and 1985) show that g amounts to 10^{-11}–10^{-15} for the electroweak interaction, whereas it is less than 10^{-19} for ordinary extrinsic forces. In the presence of the small asymmetric factor g, equation [4.25] becomes

$$\frac{d\alpha}{dt} = -A\alpha^3 + B\left(\lambda - \lambda_c\right)\alpha + Cg \tag{4.29}$$

Here C is a function of rate constants. The stationary state solutions for the system described by equation [4.29] are shown in Figure 4.12.

When λ is below or equal to λ_c, there is only one solution that represents a stable stationary state characterized by a small value of α, with small asymmetry. The chiral asymmetry suddenly grows as λ passes through the critical point λ_c. As λ increases, the system will tend to stay on the upper branch: in other words, a small initial asymmetry due to a chiral bias is amplified because the chemical system has autocatalytic nature. When λ is larger than λ_c, three solutions representing three different stationary states are possible (see Figure 4.12). Only the upper and lower solutions are stable

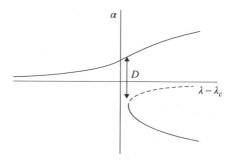

FIGURE 4.12 Bifurcation plot for the system described by equation [4.29] when there is a chiral bias. The points of the continuous lines represent solutions of stable stationary states, whereas the points of the dashed curve represent unstable solutions. D represents the separation between two stable states in the proximity of the critical point λ_c.

stationary states. The system evolves maintaining the upper branch unless it suffers a sufficiently strong perturbation, which knocks it to the lower branch. These perturbations can be due to random fluctuations, which are represented by the introduction of another term in the right part of equation [4.29]: $\sqrt{\varepsilon} f(t)$ where $f(t)$ is assumed to be a normalized Gaussian white noise (Kondepudi and Nelson 1984) with a root-mean-square value $\sqrt{\varepsilon}$.[7] The random fluctuations can play a role in the proximity of the critical point when the separation D (see Figure 4.12) between the upper and lower branches is not large enough. Many different scenarios have been proposed for the possible origin of biomolecular handedness (Cline 1996). Although we are not sure how the biomolecular asymmetry originated, this case study offers the opportunity to make three important considerations.

First, a theory of hierarchical chiral asymmetry (Kondepudi 2003) can be formulated, based on the sensitivity of chiral molecular symmetry breaking on small chiral biases. The breaking of chiral symmetry at one spatial level is the source of a small chiral bias for the next higher spatial level. For example, the parity-violating electroweak asymmetry influences a breaking asymmetry at the molecular level. On the other hand, molecular asymmetry is the origin of morphological asymmetries, such as microtubules in cells.[8]

Second, the theory of bifurcation introduces the concept of "history" in physics and chemistry, unequivocally (Prigogine 1968). If the system, whose bifurcation diagram is shown in Figure 4.13, was initially in state i and that at the end is in state f after a change in the contour conditions from λ_1 to λ_2, then the knowledge of the bifurcation diagram allows us to track the evolution of the system across the bifurcation points a and b.

Third, a system, whose dynamical evolution involves bifurcation points, can be sensitive to fluctuations. Although a system obeys to deterministic laws, when it is close to a bifurcation point, its evolution can be remarkably influenced by random fluctuations, determining the path it will follow.

TRY EXERCISE 4.3

---

[7] The root-mean-square of a set of N values $(x_1, x_2, ..., x_N)$ is given by the following equation:

$$\bar{x}_{rms} = \sqrt{\frac{1}{N}\left(x_1^2 + x_2^2 + ... + x_N^2\right)}.$$

For a continuous function $f(t)$ defined over the time interval $[t_0, t_f]$, the root-mean-square is:

$$\bar{f}_{rms}(t) = \sqrt{\frac{1}{(t_f - t_0)} \int_{t_0}^{t_f} [f(t)]^2 \, dt}.$$

[8] For more information about microtubules in cells, see Chapter 9.

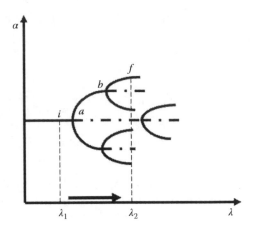

FIGURE 4.13   Hypothetic system with many bifurcations.

4.2.4   Hopf Bifurcations

So far, we have studied bifurcations for one-dimensional systems. As we move up from one-dimensional to two-dimensional systems, we still find that fixed points can be created, destroyed, or destabilized by varying the "contour conditions." In the case of two-dimensional systems, we can encounter also limit cycles. The ensemble of all possible solutions for a two-dimensional system is illustrated in Figure 3.18, and a synthetic view is also reported in the Figure 4.14. It is evident that if the contour conditions determine a negative trace and a positive determinant of the Jacobian (see paragraph 3.8.2 in Chapter 3), the system is in a stable fixed point. In fact, the fixed point is located in the right bottom part of the stability graph (covered by a white patch in Figure 4.14).

If we change the contour conditions so much that either $tr(J)$ changes from negative to positive or $det(J)$ changes from positive to negative values or both variations occur, the fixed point is not any more stable. Imagine starting from a fixed point that is a stable spiral. If $tr(J)$ becomes positive, the fixed point transforms from a stable spiral into an unstable one. If the unstable fixed point is surrounded by a stable limit cycle, we say that we have found a supercritical Hopf bifurcation (see Figure 4.15a). On the other hand, a subcritical Hopf bifurcation occurs when a stable fixed point is encompassed by an unstable limit cycle (see Figure 4.15b). As we reach the bifurcation point, the fixed point becomes unstable. The trajectories must jump to a distant attractor, which may be another fixed point or another limit cycle or goes to infinity. In three or more dimensions, this distant attractor can be "strange" because it can be chaotic, as we will see for the Lorenz's equations in

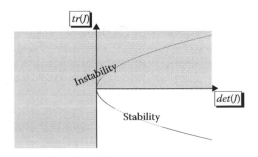

FIGURE 4.14   Plot of $tr(J)$ versus $det(J)$ distinguishing the stable and unstable solutions for a two-dimensional system.

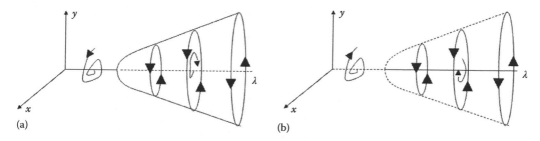

FIGURE 4.15 Representation of the supercritical (a) and subcritical (b) Hopf bifurcations for a two-dimensional system whose variables are x and y, and the control parameter is λ. A dashed line represents an unstable set of fixed points, whereas a continuous line represents a set of stable fixed points.

Chapter 10. Examples of supercritical and subcritical Hopf bifurcations may be found in the oscillating chemical reactions presented in Chapter 8.[9]

TRY EXERCISE 4.4

In Chapter 9, we will encounter a further type of bifurcation in the two-dimensional system: the Turing bifurcation. The Turing bifurcation occurs when a stable fixed-point transforms in a saddle point because the modification of the contour conditions determines the change of the sign for $det(J)$ from positive to negative.

In the next chapters, we are going to discover how out-of-equilibrium systems give rise to ordered phenomena in time. This is true in ecology (Chapter 5), in the economy (Chapter 6), in biology (Chapter 7), and bio/chemical laboratories (Chapter 8). The order in space and time emerging far from equilibrium will be presented in Chapter 9. In Chapter 10, we will know that in non-linear regime even chaotic dynamics can be observed. Chaotic dynamics produce structures that cannot be described with the traditional Euclidean geometry, but with a new geometry: Fractal geometry (see Chapter 11). In Chapter 12, we will learn the features of Complex Systems, and finally, in Chapter 13, we will discover the promising strategies for facing the Complexity Challenges.

4.3   KEY QUESTIONS

- Which are the criteria of evolution for out-of-equilibrium systems?
- What is a bifurcation?
- Present the model that describes the transition from a lamp to a laser.
- What is a plausible hypothesis about the origin of biomolecular asymmetry?

4.4   KEY WORDS

Static and dynamic self-assembly; Self-organization; Dissipative structures; Saddle-node bifurcations; Trans-critical bifurcation; Supercritical and Subcritical Pitchfork bifurcation; Supercritical and Subcritical Hopf bifurcations.

4.5   HINT FOR FURTHER READING

Other examples of bifurcations can be found in the enjoyable book by Strogatz (1994).

---

[9] In the presence of a subcritical Hopf bifurcation, a system may exhibit hysteresis. An example is given by Strogatz (1994) in his delightful book, on page 252. In the range of coexistence of stable fixed point and limit cycle, the system will remain stationary or will oscillate periodically, depending upon its history. Small perturbations to either the steady or the oscillatory state decay, but larger ones can cause a transition from stationary to periodic behavior or vice versa.

4.6   EXERCISES

4.1. For the differential equation $dx/dt = \lambda + x^2$, repeat the linear stability analysis presented in paragraph 4.2.1 and build its bifurcation diagram. Which are the differences if compared to the diagram of Figure 4.4?

4.2. Find the fixed points for the differential equation $dx/dt = \lambda x + x^3$.

4.3. Verify graphically that the differential equation [4.29] has just one fixed point when $(\lambda - \lambda_c) \leq 0$, whereas it has three solutions when $(\lambda - \lambda_c) > 0$.

4.4. The Brusselator is a hypothetical mechanism for an oscillating chemical reaction. It does not describe any particular reaction. It was proposed by Ilya Prigogine and his coworkers at the Université Libre de Bruxelles (Prigogine and Lefever 1968). The mechanism is the following:

$$A \xrightarrow{k_1} X$$

$$B + X \xrightarrow{k_2} Y + D$$

$$2X + Y \xrightarrow{k_3} 3X$$

$$X \xrightarrow{k_4} E$$

If we assume that A and B are held constant, D and E do not participate in any further reactions; the only variables are X and Y. If we write the differential equations for X and Y and we transform in dimensionless form, we get:

$$\frac{dx}{d\tau} = a - bx + x^2 y - x$$

$$\frac{dy}{d\tau} = bx - x^2 y$$

where:

$$x = X \sqrt{k_3/k_4}$$

$$y = Y \sqrt{k_3/k_4}$$

$$\tau = k_4 t$$

$$a = A \left(\frac{k_1}{k_4} \right) \sqrt{k_3/k_4}$$

$$b = k_2 B/k_4$$

Find out the steady state solution and discuss its stability as a function of the values of a and b. Do you find a Hopf bifurcation? Is it supercritical or subcritical?

4.7   SOLUTIONS TO THE EXERCISES

4.1. The solutions of fixed point are $x_0 = \pm\sqrt{-\lambda}$. Real solutions are those with $\lambda \leq 0$. According to the linear stability analysis, $x = x_0 + e^{f(x_0)t} = x_0 + e^{+2x_0 t}$. The solution $x_0 = 0$ and $x_0 = +\sqrt{-\lambda}$ are unstable, whereas the solutions $x_0 = -\sqrt{-\lambda}$ are stable. It derives that the bifurcation diagram is that shown in Figure 4.16.

4.2. The fixed points are: $x_0' = 0$, $x_0'' = +\sqrt{-\lambda}$, and $x_0''' = -\sqrt{-\lambda}$. When $\lambda \leq 0$, all the three fixed points are real solutions. When λ is positive, only $x_0' = 0$ is a possible real solution. Through the linear stability analysis, we define the properties of the fixed points.

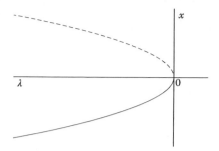

FIGURE 4.16 Saddle-node bifurcation diagram.

$$\int \frac{dx}{(x - x_0)} = \dot{f}(x_0) \int dt = (\lambda + 3x_0^2) \int dt$$

After integrating, we obtain: $x = x_0 + e^{\dot{f}(x_0)t} = x_0 + e^{(\lambda + 3x_0^2)t}$. From the latter equation, it is evident that $x_0' = 0$ is stable when λ is negative, whereas it becomes unstable when $\lambda \geq 0$. The other two solutions, $x_0'' = +\sqrt{-\lambda}$ and $x_0''' = -\sqrt{-\lambda}$, which are possible only when λ is negative, are unstable. The corresponding bifurcation diagram looks like the plot in Figure 4.17.

If we compare this graph with that of Figure 4.9, we notice that now the pitchfork is inverted, and two backward-bending branches of unstable fixed points bifurcate from the origin when $\lambda = 0$. When our system reaches the bifurcation point after following the stable branch defined by the solution $x_0' = 0$, for a further increase of λ, it blows-up: $x(t) \rightarrow \pm\infty$.

The diagram of this exercise is called subcritical pitchfork bifurcation, whereas that shown in Figure 4.9 is named as supercritical pitchfork bifurcation.

4.3. To find the fixed points of the equation $d\alpha/dt = -A\alpha^3 + B(\lambda - \lambda_c)\alpha + Cg$ graphically, we plot the functions $y_1 = B(\lambda - \lambda_c)\alpha - A\alpha^3$ and $y_2 = -Cg$ to find the intersections, which represent the fixed points. When $(\lambda - \lambda_c) \leq 0$, the cubic function, y_1, is monotonically decreasing and intersects the horizontal line at one point at positive α. When $(\lambda - \lambda_c) > 0$, the intersection points can be one, two or three depending on the value of $y_2 = -Cg$. We obtain two solutions when the horizontal line $(y_2 = -Cg)$ is tangent to the minimum of the cubic function. In this case, we are at the bifurcation point. To determine the values of α at the bifurcation point, we search for the minimum of the cubic function. The derivative of y_1 is $\dot{y}_1 = B(\lambda - \lambda_c) - 3A\alpha^2$; it is null when $\alpha = \pm\sqrt{B(\lambda - \lambda_c)/3A}$. The negative solution corresponds to the minimum, whereas the positive solution corresponds to the maximum of the cubic function (Figure 4.18).

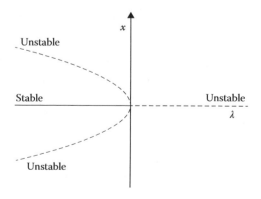

FIGURE 4.17 Subcritical pitchfork bifurcation diagram.

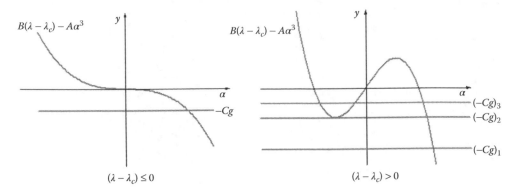

FIGURE 4.18 Graphical determination of the fixed points for the differential equation [4.29], for $(\lambda - \lambda_c) \leq 0$ on the left, and for $(\lambda - \lambda_c) > 0$ on the right.

4.4. The fixed point is $(x_0, y_0) = (a, b/a)$. The Jacobian J is equal to

$$J = \begin{pmatrix} b-1 & a^2 \\ -b & -a^2 \end{pmatrix}.$$

The trace is $tr(J) = b - 1 - a^2$, and the determinant is $\det(J) = a^2$. The determinant is always positive. The characteristic equation is: $\lambda^2 + \lambda(a^2 + 1 - b) + a^2 = 0$. Therefore, the eigenvalues are: $\lambda_{1,2} = \left(-(a^2 + 1 - b) \pm \sqrt{(a^2 + 1 - b)^2 - 4a^2}\right)/2$.

Whenever $b < 1 + a^2$, the trace is negative, and if the discriminant is negative, the eigenvalues are complex numbers with a negative real part. In this case, the fixed point is a stable focus. If we change the contour conditions and we increase the value of b, when it results $b = 1 + a^2$, the trace becomes null, and the eigenvalues become imaginary numbers. The stable focus transforms in a limit cycle. Whenever $b > 1 + a^2$ the trace is positive, and the fixed point becomes unstable. This result means that we have found a Hopf bifurcation. To decide if it is supercritical or subcritical, we must perform simple numerical experiments. In Figure 4.19, there are three examples by fixing a equal to 1 and changing b from 1.9 in (a), to 2 in (b) and 2.1 in (c).

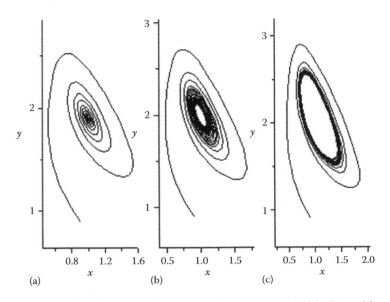

FIGURE 4.19 Dynamics of the Brusselator for $a = 1$ and $b = 1.9$ in (a), $b = 2$ in (b), and $b = 2.1$ in (c).

The dynamics shown in Figure 4.19a demonstrates that the fixed point is a stable focus. In Figure 4.19b, when $b = 1 + a^2$, the system converges to a small limit cycle. By increasing the value of b and maintaining constant the value of a, it is evident that the limit cycle expands. The Hopf bifurcation is supercritical. To do the numerical experiment, you may use MATLAB and a function file like the one reported as follows.

```
function dy = Brusselator(t, y)
dy = zeros(2,1)
a = 1
b = 2.1
dy(1) = a-b*y(1) + (y(1)^2)*y(2)-y(1)
dy(2) = b*y(1)-(y(1)^2)*y(2)
```

The script should look like:

$$[t, y] = ode45(``Brusselator,'' \; [0 \; 750], \; [0.9 \; 0.9])$$

5 The Emergence of Temporal Order in the Ecosystems

The first law of ecology is that everything is related to everything else.

Barry Commoner (1917–2012 AD)

5.1 INTRODUCTION

Both terrestrial and aquatic ecosystems are networks involving interactions among living beings and their environment. The reductionist approach brings us to isolate interactions between pairs of elements. For instance, a recurrent type of simplified interaction in ecosystems is that involving one kind of predator and one kind of prey. The predator is an organism that eats the prey. Some examples are lynx and hare in North America; lion and zebra, leopard and impala in the African Savannah; bear and fish, cat and rat, least weasel and bank vole in Europe; largemouth bass and mosquito fish in the freshwaters of North America, and many more. Predator and prey make efforts to survive in the same ecosystem. The dynamical relationship between predator and prey is one of the dominant themes in ecology.

5.2 PREDATOR-PREY RELATIONSHIP: THE LOTKA-VOLTERRA MODEL

The first plausible model to describe the predator-prey interaction was proposed independently by Lotka (1925) and Volterra (1926).[1] The Lotka-Volterra model involves three elementary steps reported in [5.1], wherein H is the prey, C is the predator, F is the food of prey, and D is the carcass of the predator.

$$
\begin{aligned}
F + H &\xrightarrow{k_1} 2H \\
H + C &\xrightarrow{k_2} 2C \\
C &\xrightarrow{k_3} D
\end{aligned}
\qquad [5.1]
$$

Quite often, H is an herbivore eating, for instance, grass (F) with a rate proportional to k_1. If so, C is a carnivore eating H with a pace proportional to k_2 and dying with a rate which depends on the value of k_3. The overall process is $F \to D$ attainable by summing the three elementary steps of the "mechanism" [5.1].[2] In the long run, the fight for survival between predator and prey reduces, chemically speaking, to a conversion of the food (F) of the prey in the carcass (D) of the predator.

---

[1] Alfred J. Lotka (Lwów, Austria-Hungary, 1880–1949) had American parents and received an international and interdisciplinary education. In fact, he got a Bachelor of Science at the University of Birmingham (England), he graduated at Leipzig University; he received a Master of Arts at Cornell University and a Doctor of Science title at Birmingham University. He is best known for his proposal of the predator-prey model, developed simultaneously and independently of Vito Volterra. Volterra (Ancona, Italy, 1860–1940) was an eminent Italian mathematician known for his work on theoretical ecology. He is also remembered because when Benito Mussolini came to power in 1922 and established the Fascist regime, he was one of only twelve out of 1250 professors who refused to take a mandatory oath of allegiance to the fascist government. As a result of his refusal, he was compelled to resign his university post, and during the following years, he lived mainly abroad.

[2] I am using a terminology that is peculiar to chemical kinetics.

The differential equations describing the time-variations of H and C are

$$\frac{dH}{dt} = -k_1FH + 2k_1FH - k_2HC = k_1FH - k_2HC$$

$$\frac{dC}{dt} = -k_2HC + 2k_2HC - k_3C = k_2HC - k_3C$$

[5.2]

The mechanism [5.1] implies that in the absence of C, H grows exponentially. In fact, the first differential equation of [5.2] reduces to $\left(dH/dt = k_1FH\right)$. On the other hand, in the absence of H, C declines exponentially. In fact, the second differential equation of [5.2] becomes $\left(dC/dt = -k_3C\right)$.

A key feature of the system [5.1] is the involvement of two autocatalytic processes, which are the first and the second steps. It means that the rate of growth of a species is not constant, but it increases with the population of that species. When the amount of F is constant over time, and the kinetic constants are fixed, the overall system will evolve towards a steady state in which the net variation of all species is zero. In the steady state, the rate of prey reproduction is balanced by the rate of prey consumption; at the same time, the death frequency for C is counterbalanced by its birth rate. The characteristics of the steady state can be found by setting the two time-derivatives of the system [5.2] equal to zero. It derives that

$$H_{ss} = \left(\frac{k_3}{k_2}\right)$$

$$C_{ss} = \frac{\left(k_1F\right)}{k_2}.$$

[5.3]

The solutions [5.3] suggest that the steady-state concentration of herbivores will be high if mortality rate of carnivores is larger than their birth rate. On the other hand, the steady-state concentration of C will be relevant when the growth rate of the prey population (k_1F) is high compared to k_2. To get further insight on the properties of this steady state, we can apply the linear stability analysis, which we learned in Chapter 3. We slightly perturb the steady state and trace how it reacts to the small perturbations: $h = H - H_{ss}$ and $c = C - C_{ss}$. Introducing h and c into the system of two differential equations [5.2], and expanding them in the Taylor series with H_{ss} and C_{ss} as accumulation points, we obtain

$$\frac{dh}{dt} = \left(k_1F - k_2C_{ss}\right)h - k_2H_{ss}c$$

$$\frac{dc}{dt} = k_2C_{ss}h + \left(k_2H_{ss} - k_3\right)c.$$

[5.4]

Introducing the steady-state solution in [5.4] and expressing the output in matrix form, we have

$$\begin{pmatrix} \dfrac{dh}{dt} \\ \dfrac{dc}{dt} \end{pmatrix} = \begin{pmatrix} 0 & -k_3 \\ k_1F & 0 \end{pmatrix} \begin{pmatrix} h \\ c \end{pmatrix}.$$

[5.5]

The Jacobian $J = \begin{pmatrix} 0 & -k_3 \\ k_1F & 0 \end{pmatrix}$ has $tr\left(J\right) = 0$ and $det\left(J\right) = k_1Fk_3$. If we look back at Figure 3.18, we notice that we are along line (d), i.e., the line of centers. In fact, by solving the characteristic equation [5.6]

$$det\left(J - \lambda I\right) = det\begin{pmatrix} -\lambda & -k_3 \\ k_1 F & -\lambda \end{pmatrix} = \lambda^2 + k_1 F k_3 = 0 \qquad [5.6]$$

we obtain eigenvalues that are purely imaginary numbers:

$$\lambda_{1,2} = \pm i\sqrt{k_1 F k_3}. \qquad [5.7]$$

This result means that a small perturbation to the steady state [5.3] pushes the system to a stable limit cycle, wherein the number of herbivores and carnivores oscillates. The period of the oscillation is

$$T = \frac{2\pi}{\sqrt{k_1 F k_3}} \qquad [5.8]$$

In the literature, there are many examples of cyclic variations of animal populations (Korpimäki and Krebs 1996). For instance, in the Boreal Zone of North America and some parts of Siberia, the number of hares and their mammalian and avian predators changes cyclically with periods of 9–10 years. The periodic variation of the number of snowshoe hares and Canadian lynxes was recorded by the fur-trading Hudson Bay Company from 1821 to 1934 (Elton and Nicholson 1942). It is possible to use the mechanism [5.1] to interpret their cycles because the hares feed only on a variety of vegetables, and lynxes are essentially single-prey oriented. The proposed values for the constants are: $k_1 F = 0.5471$ year^{-1}, $k_2 = 0.0281$ year^{-1}, $k_3 = 0.8439$ year^{-1}. Based on the values of these kinetic constants, the trend in the number of the two species is periodic as shown in Figure 5.1 for different initial conditions.[3]

The number of both predator and prey oscillates with a period of about nine years, and the peak in the population of predator always follows the maximum in the prey population. An increase in the prey population triggers growth in the predator number. On the other hand, the predator growth leads to a decrease in the prey population. Since there are only two variables, we can picture the evolution of the system in a two-dimensional phase space. The coordinates of the phase space are C and H. In the phase-plane we plot out, at least qualitatively, the vector field. It indicates the direction of evolution in phase space for any value of H and C. It is necessary to attach to each point in the phase space a couple of vectors representing $\dot{H} = \left(dH/dt\right)$ and $\dot{C} = \left(dC/dt\right)$. The evolution of the system will be indicated by the addition of the two vectors. The two curves on which $\dot{H}$ and $\dot{C}$ are null are called H-nullcline and C-nullcline, respectively. The intersections of the two nullclines constitute the steady states of the system, because, at any such points, both derivatives $\dot{H}$ and $\dot{C}$ are null. The time derivative of the variable corresponding to the nullcline changes sign as we cross the nullcline. The nullclines are two straight lines intersecting at one point (see Figure 5.2). From the definitions of $\dot{H}$ and $\dot{C}$ given in [5.2], we can establish their signs in the different regions of the phase plane (try to verify the signs that are indicated in Figure 5.2). From the phase plane of Figure 5.2, we have both the steady state and a qualitative picture of the variation of the number of predator and prey, if the system is perturbed and pushed away from the steady state.

---

[3] It is possible to use MATLAB to solve the Lotka-Volterra system of the two differential equations. The file that can be used is *lv.m*:

```
function yprime = lv(t,y)
%LV: Includes Lotka-Volterra equations
a = 0.5471;b = 0.0281;c = 0.0281;r = 0.8439
yprime = [a*y(1)-b*y(1)*y(2);-r*y(2)+c*y(1)*y(2)]
```

The system can be solved over a time interval of 50 years and be starting from different initial conditions:

```
>>[t,y]=ode45(@lv,[0 50],[H₀;C₀])
```

The MATLAB command ode45 is an implementation of the fourth order Runge-Kutta method to solve differential equations. More information can be found in the Help function of MATLAB software.

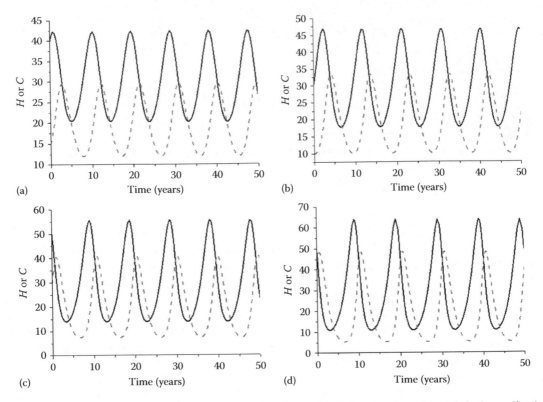

(a)   Time (years)
(b)   Time (years)
(c)   Time (years)
(d)   Time (years)

FIGURE 5.1 Oscillations of the number of prey (continuous black lines) and predator (dashed grey lines) for different initial conditions: (a) $H_0 = 40$, $C_0 = 15$; (b) $H_0 = 30$, $C_0 = 10$; (c) $H_0 = 50$, $C_0 = 30$; (d) $H_0 = 50$, $C_0 = 40$.

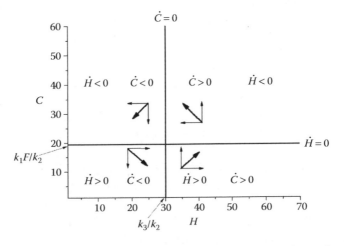

FIGURE 5.2 The evolution of the Lotka-Volterra model in the phase plane. $\dot{H}$ and $\dot{C}$ are the derivatives of prey and predator populations.

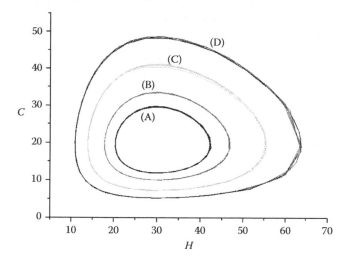

FIGURE 5.3 Trajectories for the simplest Lotka-Volterra model obtained at different initial conditions: (A) $H_0 = 40$, $C_0 = 15$; (B) $H_0 = 30$, $C_0 = 10$; (C) $H_0 = 50$, $C_0 = 30$; (D) $H_0 = 50$, $C_0 = 40$.

The flow in phase plane (see bold arrows) confirms that if the system of carnivores and herbivores is initially in the steady state and is perturbed, it starts to rotate around that point in a closed path. In other words, the plot of the number of carnivores as a function of the number of herbivores gives rise to trajectories, which are closed orbits traced counterclockwise (see Figures 5.2 and 5.3). In Figure 5.3, different closed orbits are shown. They have been obtained for different initial values, C_0 and H_0 (see the caption of Figure 5.3).

5.3   ENTROPY PRODUCTION IN THE LOTKA-VOLTERRA MODEL

Any ecosystem is an open system, exchanging energy and matter with the surrounding environment. To depict the ecosystem from a thermodynamic point of view, we may imagine as if it were an open chemical reactor with constant volume and uniform temperature. For simplicity, we assume that H and C are uniformly distributed in the total volume V. Therefore, we may define the chemical potentials for both H and C:

$$\mu_C = \mu_C^0 + RTln(C)$$
$$\mu_H = \mu_H^0 + RTln(H).$$

[5.9]

If we move from the steady state (*ss*) to another arbitrary state (*as*), the chemical affinity for reaction ρ will change:

$$\Delta A_\rho = A_\rho - A_{\rho,ss} = -\sum_\gamma v_{\gamma,\rho}\mu_\gamma + \sum_\gamma v_{\gamma,\rho}\mu_{\gamma,ss} = -RT\sum_\gamma v_{\gamma,\rho}ln\left(\frac{\Gamma}{\Gamma_{ss}}\right).$$

[5.10]

In equation [5.10], γ is a species, whose concentrations are Γ_{ss} and Γ at the steady state and in the other arbitrary state, respectively. Within the domain of validity of the so-called local equilibrium condition, the corresponding entropy change will be:

$$\Delta S = \int_{ss}^{as}\sum_\rho\left(\frac{\Delta A_\rho}{T}\right)d\xi_\rho$$

[5.11]

The degree of advancement of reaction ρ can be expressed as

$$d\xi_\rho = \frac{V}{\nu_{\gamma,\rho}}(d\Gamma)_\rho.$$ [5.12]

Inserting equations [5.10] and [5.12] into equation [5.11], we obtain

$$\Delta S = -RV \int_{ss}^{as} \sum_{\rho,\gamma} \ln\left(\frac{\Gamma}{\Gamma_{ss}}\right)(d\Gamma)_\rho.$$ [5.13]

The local entropy production will be

$$\frac{P^*}{RV} = -\sum_{\rho,\gamma} \ln\left(\frac{\Gamma}{\Gamma_{ss}}\right)_\rho \left(\frac{d\Gamma}{dt}\right)_\rho.$$ [5.14]

For the Lotka-Volterra kinetic model, the local entropy production is

$$\frac{P^*}{RV} = -\left\{\left(\frac{dH}{dt}\right)\ln\left(\frac{H}{H_{ss}}\right) + \left(\frac{dC}{dt}\right)\ln\left(\frac{C}{C_{ss}}\right)\right\} = -\left\{v_H \ln\left(\frac{H}{H_{ss}}\right) + v_C \ln\left(\frac{C}{C_{ss}}\right)\right\}.$$ [5.15]

The time average of $\left(P^*/RV\right)$ over a cycle of period T is equal to zero (Ishida and Matsumoto 1975). Within one cycle, entropy production assumes both positive and negative values, and its average is zero. This behavior is shown in Figure 5.4 for a cycle of Lotka-Volterra model, describing the periodic variations of snowshoe hares and Canadian lynxes: in the first part of the cycle, entropy

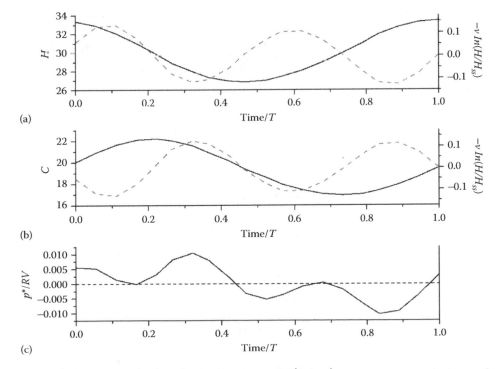

FIGURE 5.4 Trends of H (continuous black trace) and $-v_H \ln\left(H/H_{ss}\right)$ (grey dashed trace) in (a); trends of C (continuous black trace) and $-v_C \ln\left(C/C_{ss}\right)$ (dashed grey trace) in (b), and trend of $\left(P^*/RV\right)$ within a cycle in (c).

production is produced, and it is counterbalanced by the entropy production that is consumed in the second half.

This behavior reminds us that of coupled chemical reactions. Having negative values of entropy production is possible. However, if we consider the sum of all the contributions, the net result is either a positive or a total null entropy production.

5.4 MORE ABOUT PREDATOR-PREY RELATIONSHIPS

The Lotka-Volterra model studied so far accounts for the existence of cyclic variations in populations of predators and prey. However, it has some unsatisfactory features. For example, according to the steps in mechanism [5.1], in the absence of C, the population of H should grow exponentially, and this is not realistic. A model closer to reality describes the growth of H, in the absence of C, like that of a population in the presence of limited food resources.[4]

TRY EXERCISES 5.1 AND 5.2

The Lotka-Volterra model has other weak points. For example, it assumes that the rate of prey consumption is a linear function of the prey density; every predator assault has a positive outcome; the prey assimilated by the predators are all turned into new predators, and all the species are uniformly distributed in the environment. These are rough assumptions of what we can encounter in real ecosystems.

A more general system of differential equations (Harrison 1979) for the predator-prey interaction has the following shape:

$$
\begin{aligned}
\frac{dH}{dt} &= h(H) - f(H)a(C) \\
\frac{dC}{dt} &= g(H)b(C) - c(C)
\end{aligned}
\qquad [5.16]
$$

In the first differential equation of [5.16], the term $h(H)$ represents the intrinsic growth rate of the prey including all factors except predation; the term $\left[-f(H)a(C)\right]$ is the rate of prey consumption per predator, and $f(H)$ is called the functional response of the predator. In the second equation appearing in equations [5.16], the term $-c(C)$ is the intrinsic rate of decrease of the predator, whereas the term $g(H)b(C)$ represents the predator rate of increase in the presence of H and $g(H)$ is called the numerical response of the predator. Holling (1959),[5] studying the predation of small mammals on pine sawflies, identified three types of functional responses. Type I is linear with H, as in Lotka-Volterra model and the example of exercise 5.1 (see Figure 5.5). It is verified with passive predators like spiders. The number of flies captured in spider net is proportional to fly density.

Type II describes a situation in which the number of prey consumed per predator initially rises almost linearly as the density of prey increases, but then it bends and finally levels off (see Figure 5.5). Type II is a valid model when a predator has to accomplish two distinct actions to reach its final goal:

- Capturing a prey, which means wandering, detecting, chasing, catching and subduing it;
- Handling the prey, which means killing, eating, and digesting it.[6]

---

[4] For more information, see also the Logistic Map in Chapter 10. For the moment, try exercises 5.1 and 5.2.

[5] Crawford Stanley Holling (born in 1930 in Theresa, New York) is one of the conceptual founders of ecological economics, which is referred to as both an interdisciplinary field of academic research that aims to address the interdependence and coevolution of human economies and natural ecosystems over time and space.

[6] A further refinement of Type II functional response is presented by Jeschke et al. (2002). In their paper, the authors introduce digestion of the prey by a predator as a process influencing the hunger level of predators and hence the probability of searching for new prey. It emerges that the asymptotic maximum predation rate (i.e., asymptotic maximum number of prey eaten per unit time, for prey density approaching infinity) is determined by either the handling time or the digestion time.

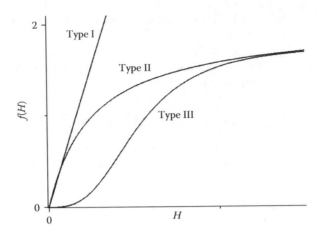

FIGURE 5.5   The three types of functional responses $f(H)$ as functions of prey density.

The mechanistic model for Type II functional response is

$$C + H \underset{k_{-c}}{\overset{k_c}{\rightleftarrows}} (CH) \overset{k_h}{\longrightarrow} 2C, \qquad [5.17]$$

where:
   C captures H with a rate constant k_c,
   (CH) denotes the prey captured by the predator.

H can escape before being killed and eaten by C, with a rate constant k_{-c}. The rate constant of handling is k_h. The expression of the functional response can be achieved by writing how H and (CH) change over time according to the elementary steps of [5.17]:

$$\frac{dH}{dt} = -k_c HC + k_{-c}(CH)$$
$$\frac{d(CH)}{dt} = k_c HC - k_{-c}(CH) - k_h(CH). \qquad [5.18]$$

If we apply the steady-state approximation for (CH) and we exploit the mass balance equation $C_0 = C + (CH)$, where C_0 is the total density of predators, we obtain:

$$(CH) = \frac{k_c HC}{k_{-c} + k_h} = \frac{k_c H\left[C_0 - (CH)\right]}{k_{-c} + k_h}. \qquad [5.19]$$

The definition of (CH) becomes

$$(CH) = \frac{k_c HC_0}{k_{-c} + k_h + k_c H}. \qquad [5.20]$$

By inserting equation [5.20] and the mass balance for a predator into the first differential equation of system [5.18], we achieve:

$$\frac{dH}{dt} = -k_c H\left[C_0 - (CH)\right] + k_{-c}(CH) = -\frac{k_h H C_0}{\dfrac{k_{-c}+k_h}{k_c}+H}.$$ [5.21]

The Type II functional response will be

$$f_{II}(H) = \frac{k_h H}{\dfrac{k_{-c}+k_h}{k_c}+H} = \frac{k_h H}{K+H}.$$ [5.22]

It is a rectangular hyperbolic function having a concave down graph (see Figure 5.5). When H is low and much smaller than K, the functional response increases linearly with H, that is $f_{II}(H) \approx k_h H/K$. When the prey density is high and much larger than K, $f_{II}(H) \approx k_h$. In other words, at high densities, prey is easy to capture, and the rate of their consumption is almost exclusively affected by the time required for pursuing the "handling stage." When a Type II functional response is combined with a numerical response proportional to the rate of prey consumption, the system of two differential equations describing the evolution of the number of prey and predator is:

$$\frac{dH}{dt} = k_1 F\left(1 - \frac{H}{K_1 F}\right)H - \frac{k_h H C}{K+H} = \dot{H}$$

$$\frac{dC}{dt} = \frac{k_h' H C}{K+H} - k_3 C = \dot{C}$$ [5.23]

In equation [5.23], the predators convert the predation into offspring with an efficiency $k_h' \neq k_h$ and experience density-independent mortality at a rate k_3. $K_1 = (k_1/k_{-1})$.

The evolution of the prey and predator system can be depicted in a phase space defined by C and H (see Figure 5.6). The equations representing the null-clines $\dot{H} = 0$ and $\dot{C} = 0$ are a concave-down parabola and a vertical line, respectively:

$$C = \frac{k_1 F K}{k_h} + \frac{k_1 F}{k_h}\left(1 - \frac{K}{K_1 F}\right)H - \frac{k_{-1}}{k_h}H^2$$ [5.24]

$$H = \frac{k_3 K}{k_h' - k_3}.$$ [5.25]

The parabola intercepts the C axis at $C = k_1 F K/k_h$, and the H axis at $H = K_1 F$; its vertex[7] is at $H = (K_1 F - K)/2$ (see Figure 5.6 and equation [5.24]). If $(k_3 K/(k_h' - k_3)) \leq K_1 F$, there is a stationary state represented by the intersection between the parabola and the vertical line. When the vertical isocline is to the right of the vertex of the parabola, that is when

$$\frac{k_3 K}{k_h' - k_3} > \frac{(K_1 F - K)}{2},$$ [5.26]

---

[7] The vertex of the parabola can be found by calculating the derivative of C with respect to H and determining when it is null: $dC/dH = k_1 F/k_h\left(1 - K/K_1 F\right) - 2Hk_{-1}/k_h = 0$.

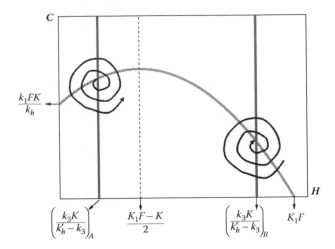

FIGURE 5.6   Phase-plane for a Type II functional response and a numerical response proportional to the rate of prey consumption. The spiraling arrows represent dynamics in the neighborhood of stationary states indicated by the intersections of straight lines (A and B) with the parabola. The vertical dashed line indicates the vertex of the parabola. All the intersection points on the left of the vertex represent unstable solutions.

The corresponding stationary states are stable. Any oscillations in predator and prey densities will eventually be damped out, and both species will coexist at the levels H_{ss} and C_{ss} (defined by the intersection points).

When the vertical isocline is to the left of the vertex or intersects the vertex, the stationary states are unstable. A prey-predator system that starts close to a stationary state exhibits diverging oscillations that spiral out toward a limit cycle. The existence of the limit cycle can be shown by using the Poincaré-Bendixson theorem[8] (Strogatz 1994). This theorem says that if a trajectory is confined to a closed, bounded region that contains no fixed points, then the trajectory must eventually approach a closed orbit. This type of closed curve corresponds to sustained oscillations in time, with unique amplitude and frequency, irrespective of initial conditions. It should be distinguished from oscillations of the Lotka-Volterra type, which present an infinity of amplitudes and frequencies depending on the initial conditions. Numerical solutions of [5.23] show that the further the stationary state is to the left of the vertex of the prey isocline, the faster the oscillations diverge and the larger the amplitude of the limit cycle. Once the trajectory is close to the limit cycle, the system exhibits continued oscillations. Whether the predator and prey can coexist oscillating in their number, depends on how close these oscillations come to the boundaries $C = 0$ and $H = 0$ (Harrison 1995).

TRY EXERCISES 5.3, 5.4 AND 5.5

Finally, there is the Type III functional response, which has a sigmoid shape curve (see Figure 5.5). When the prey density is pretty low, the rate of predation is really slow. At intermediate densities, it grows sharply, and at very high densities, saturation occurs. A functional response having a Type III shape emerges when (1) the ecosystem is heterogeneous, and the prey is not uniformly distributed across the space, or (2) the predator has to do practice to become competent in catching and killing prey.

---

[8] The Poincaré-Bendixson theorem applies only in two-dimensional systems. In higher dimensional systems ($n \geq 3$) trajectories may wander around forever in a bounded region without settling down to a fixed point or a closed orbit. In some cases, the trajectories are attracted to a complex geometric object called a strange attractor. A strange attractor is a fractal object (see Chapter 10) on which the motion is chaotic because it is aperiodic and sensitive to tiny changes in initial conditions. The Poincaré-Bendixson theorem excludes chaos in two-dimensional phase space.

1. When a kind of prey is distributed unevenly in patches, the predator may encounter one of such wealthy areas and have many options ($n > 1$) to chase: all of them in proximity. For example, a kind of prey is browsing only in a few leaves of a tree. When its predator lands in one of those leaves, which are wealthy in prey, it has many chances to catch and eat its favorite food, up to complete satiation. If the predator fails to find his preferred kind of prey after searching in many leaves, it may switch to retrieve another type of prey, just to avoid dying of starvation. Guppy is an example of switching predator. Usually, it hunts fruit flies on the water's surface. When fruit flies decrease in number, guppies change from feeding on the fruit flies to feeding on tubificids living in deep water (Murdoch 1977).

2. Type III functional response is the only type of functional response for which prey mortality can increase with prey density raising from very low values. This property accounts for a natural improvement of a predator's hunting efficiency as prey density increases. For example, many predators respond to chemicals emitted by prey (the so-called kairomones) and increase their activity. Polyphagous vertebrate predators (e.g., birds) can switch to the most abundant prey species by learning to recognize it visually. When a predator finds a type of prey infrequently, it has no experience to develop the best ways to capture and kill that species of prey. For instance, many felines kill prey larger than themselves by inserting canines between vertebrae and disarticulating the spine. They must learn exactly where to bite and how to kill big prey before it injures them.

A mechanism modeling the Type III functional response is

$$C + nH \xrightleftharpoons[k_{-c}]{k_c} (CH_n) \xrightarrow{k_h} 2C, \qquad [5.27]$$

where:

n is a number larger than 1 and the kinetic constants k_c, k_{-c}, k_h have the same meaning encountered in [5.17]

The differential equations describing how H and (CH_n) change over time are

$$\frac{dH}{dt} = -nk_c CH^n + nk_{-c}(CH_n)$$
$$\frac{d(CH_n)}{dt} = k_c CH^n - (k_{-c} + k_h)(CH_n) \qquad [5.28]$$

If we apply the steady-state approximation for (CH_n) and the mass balance for the predator ($C_0 = C + (CH_n)$), we obtain

$$(CH_n) = \frac{k_c C_0 H^n}{k_{-c} + k_h + k_c H^n}. \qquad [5.29]$$

Introducing this definition of (CH_n) in the first differential equation of the system [5.28], we obtain that

$$\frac{1}{n}\frac{dH}{dt} = -\frac{k_c k_h C_0 H^n}{k_{-c} + k_h + k_c H^n} = -\frac{k_h C_0 H^n}{\dfrac{k_{-c} + k_h}{k_c} + H^n} = -\frac{k_h C_0 H^n}{K + H^n}. \qquad [5.30]$$

Type III functional response will be

$$f_{III}(H) = \frac{k_h H^n}{K + H^n}.$$ [5.31]

Equation [5.31] confirms that Type III functional response is a sigmoid function. When the prey density is very low, $f_{III}(H)$ is a function of H to the power of n. On the other hand, when H is high enough to be much larger than K, $f_{III}(H)$ becomes constant and roughly equal to the rate of handling, k_h.

When a Type III functional response with $n = 2$ is combined with a numerical response proportional to the rate of prey consumption, the system of two differential equations describing the evolution of the number of prey and predator is:

$$\frac{dH}{dt} = k_1 F \left(1 - \frac{H}{K_1 F}\right) H - \frac{k_h H^2 C}{K + H^2} = \dot{H}$$

$$\frac{dC}{dt} = \frac{k_h' H^2 C}{K + H^2} - k_3 C = \dot{C}$$ [5.32]

In equation [5.32], k_h' represents the predator's efficiency at turning prey into predator offspring, whereas k_3 is its mortality rate. The phase space for such system is shown in Figure 5.7. The grey curve represents the nullcline $\dot{H} = 0$, whereas the continuous black vertical lines are three possible values of the nullcline $\dot{C} = 0$. The intersection points between the two nullclines, $\dot{H} = 0$ and $\dot{C} = 0$, represent stationary states. Their stability changes depending on the values of the rate constants. We distinguish three regions in the phase space. Stationary states in the region labeled as (a) are stable like those in the region (c). On the other hand, the stationary states of the intermediate region (b) are unstable, and when the system is pushed away from them, it evolves to a limit cycle.

TRY EXERCISE 5.6

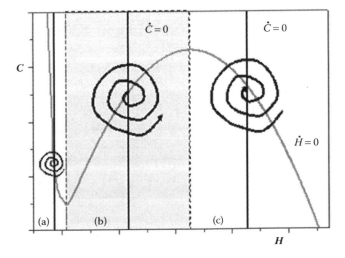

FIGURE 5.7 Phase-plane for a Type III functional response and a numerical response proportional to the rate of prey consumption. The intersection points between the two nullclines in regions (a) and (c) are stable, whereas those in (b) are unstable and the system evolves up to a limit cycle. The spiraling arrows represent the dynamics.

5.5 OTHER RELATIONSHIPS WITHIN AN ECOSYSTEM

Within an ecosystem, we not only have predator-prey interplays, but also many more symbiotic[9] relationships. A list is in Table 5.1. The type of relationship depends on the mutual effects exerting on both protagonists, reminding that an effect can be either positive (+), or negative (−), or null (0).

The antagonism relationship may occur among species and also among members of the same species. Antagonists may compete for the same type of food, living space, and mate. Antagonism may cause the death of some competitors in fighting or when deprived of food for a long time. The severity of antagonism depends on the extent of the similarity of resource requirements of different organisms and the shortage of supply in the habitat.

Parasitism is a relationship in which one species, the parasite, is always benefited at the cost of the other species, the host. A parasite derives nourishment from the host and, in many cases, it finds protection and living space on or inside the host. Usually, the host is bigger than the parasite. Bacteria and fungi are the most important parasites of both plants and animals. Humans use antibiotics to defeat several bacterial parasites. Plants themselves may be parasites either on other plants or animals.

Amensalism is a relationship wherein an organism is inhibited or destroyed whereas the other is unaffected. The classical example is that of the bread mold *Penicillium* that secretes penicillin and destructs bacteria. Many other examples are offered by plants that secrete chemicals to cause harm to other plants.

Neutralism describes a relationship between two species where the health of one species has no effect whatsoever on that of the other. Neutralism is challenging to detect and prove and is often used to describe situations wherein interplays are insignificant.

In commensalism, one organism draws benefits whereas the other receives neither benefits nor harms. Examples of commensalism are offered by epiphytes. All epiphytes use trees only for attachment and derive moisture and nutrients from the air, rain, and sometimes from debris accumulating around them. Sessile invertebrates that grow on plants or other animals represent many commensals.

Finally, a relationship is defined mutualistic when it is favorable to both partners. Mutualism may be facultative when the species involved can exist independently. Otherwise, it may be obligatory, when the relationship is imperative to the existence of one or both the individuals. For example, each lichen is a mutualistic association between a fungus and an alga. The alga photosynthesizes food for itself as well as for the fungus. The fungus, in turn, furnishes water and carbon dioxide. Pollination of flowers by insects is another manifestation of mutualism. Humans engage in mutualisms with other species, including their gut flora without which they would not be able to digest food efficiently.

TABLE 5.1

List of Possible Relationships between Two Species within an Ecosystem

Effect on Species 1	Effect on Species 2	Relationship
−	−	Antagonism
−	+	Parasitism
−	0	Amensalism
0	0	Neutralism
+	0	Commensalism
+	+	Mutualism

---

[9] Symbiosis derives from the Greek συν, i.e., together, and βιοσις, i.e., life. In this book, it is considered in its broadest sense, referring to a close and prolonged association between two or more organisms of different or the same species. In some text, you may find the use of the term symbiosis only for the mutualistic relationship.

5.6 MATHEMATICAL MODELING OF SYMBIOTIC RELATIONSHIPS

The general model to describe any symbiotic relationship between two species A and B, where F is the available food, can be the following one:

$$A + F \underset{k_{-1}}{\overset{k_1}{\rightleftarrows}} 2A$$

$$B + F \underset{k_{-2}}{\overset{k_2}{\rightleftarrows}} 2B \qquad [5.33]$$

$$A + B \xrightarrow{k_{12}} n_A A + B$$

$$A + B \xrightarrow{k_{21}} n_B B + A$$

Note that in the third process, B acts as a catalyst, being present among both the reagents and the products. The same is true for A in the fourth process. The type of relationship that mechanism [5.33] represents depends on the value of the coefficients n_A and n_B. All the possible combinations are reported in Table 5.2.

5.6.1 ANTAGONISM

In an antagonistic relationship between species A and B, the coefficients n_A and n_B appearing in [5.33] are both equal to 0. Therefore, mechanism [5.33] becomes:

$$A + F \underset{k_{-1}}{\overset{k_1}{\rightleftarrows}} 2A$$

$$B + F \underset{k_{-2}}{\overset{k_2}{\rightleftarrows}} 2B \qquad [5.34]$$

$$A + B \xrightarrow{k_{12}} B$$

$$A + B \xrightarrow{k_{21}} A$$

TABLE 5.2

List of All Possible Combinations of the Coefficients n_A and n_B Appearing in the Scheme [5.33]

n_A	n_B	Relationship
0	0	Antagonism
0	2	Parasitism
0	1	Amensalism
1	1	Neutralism
2	1	Commensalism
2	2	Mutualism

Both species, A and B, have logistic growth in the absence of the other (Murray 2002). According to [5.34], the differential equations describing how the number of elements of species A and B change over time are:

$$\frac{dA}{dt} = k_1 FA - k_{-1}A^2 - k_{12}AB = k_1 FA\left(1 - \frac{A}{K_A} - \frac{k_{12}}{k_{-1}K_A}B\right)$$

$$\frac{dB}{dt} = k_2 FB - k_{-2}B^2 - k_{21}AB = k_2 FB\left(1 - \frac{B}{K_B} - \frac{k_{21}}{k_{-2}K_B}A\right)$$

[5.35]

In [5.35], $K_A = (k_1 F/k_{-1})$ and $K_B = (k_2 F/k_{-2})$. To make easier the treatment of the system [5.35], we nondimensionalize it, introducing three dimensionless variables: $a = A_C A$; $b = B_C B$ and $\tau = t_C t$. Inserting these new variables within [5.35], we obtain:

$$\frac{da}{d\tau} = \frac{k_1 F}{t_C}a\left(1 - \frac{a}{K_A A_C} - \frac{k_{12}}{k_{-1}K_A}\frac{b}{B_C}\right)$$

$$\frac{db}{d\tau} = \frac{k_2 F}{t_C}b\left(1 - \frac{b}{K_B B_C} - \frac{k_{21}}{k_{-2}K_B}\frac{a}{A_C}\right)$$

[5.36]

The next step is to choose the non-dimensionalizing constants. If we fix $A_C = (1/K_A)$, $B_C = (1/K_B)$ and $t_C = k_1 F$, we obtain

$$\frac{da}{d\tau} = a\left(1 - a - \frac{k_{12}K_B}{k_{-1}K_A}b\right)$$

$$\frac{db}{d\tau} = \frac{k_2 F}{k_1 F}b\left(1 - b - \frac{k_{21}K_A}{k_{-2}K_B}a\right).$$

[5.37]

Finally, fixing $\rho = (k_2/k_1)$, $r_{12} = (k_{12}K_B/k_{-1}K_A)$, and $r_{21} = (k_{21}K_A/k_{-2}K_B)$, the two differential equations in dimensionless form become:

$$\frac{da}{d\tau} = a(1 - a - r_{12}b) = f$$

$$\frac{db}{d\tau} = \rho b(1 - b - r_{21}a) = g$$

[5.38]

The possible steady states are $(a_{ss}, b_{ss}) = (0,0); (1,0); (0,1); ((1 - r_{12})/(1 - r_{12}r_{21}), (1 - r_{21})/(1 - r_{12}r_{21}))$.
Their stability can be inferred by linear analysis. The Jacobian is

$$J = \begin{pmatrix} \left(\frac{\partial f}{\partial a}\right)_{ss} & \left(\frac{\partial f}{\partial b}\right)_{ss} \\ \left(\frac{\partial g}{\partial a}\right)_{ss} & \left(\frac{\partial g}{\partial b}\right)_{ss} \end{pmatrix} = \begin{pmatrix} 1 - 2a_{ss} - r_{12}b_{ss} & -r_{12}a_{ss} \\ -\rho r_{21}b_{ss} & \rho - 2\rho b_{ss} - \rho r_{21}a_{ss} \end{pmatrix}.$$

[5.39]

The steady state $(0,0)$ is unstable. In fact,

$$J = \begin{pmatrix} 1 & 0 \\ 0 & \rho \end{pmatrix}, tr(J) = 1 + \rho > 0, det(J) = \rho > 0$$

[5.40]

For the steady state $(1,0)$, we have

$$J = \begin{pmatrix} -1 & -r_{12} \\ 0 & \rho - \rho r_{21} \end{pmatrix}, tr(J) = \rho(1 - r_{21}) - 1, det(J) = -\rho(1 - r_{21}) \qquad [5.41]$$

If $r_{21} > 1$, $det(J) > 0$, $tr(J) < 0$. Hence, the steady state is stable. If $r_{21} < 1$, $det(J) < 0$; therefore, the steady state is a saddle point.

For the steady state $(0,1)$, we have

$$J = \begin{pmatrix} 1 - r_{12} & 0 \\ -\rho r_{21} & -\rho \end{pmatrix}, tr(J) = 1 - r_{12} - \rho, det(J) = -\rho(1 - r_{12}) \qquad [5.42]$$

If $r_{12} > 1$, $det(J) > 0$, $tr(J) < 0$. Hence, the steady state is stable. If $r_{12} < 1$, $det(J) < 0$; therefore, the steady state is a saddle point.

For the last steady state,

$$J = \begin{pmatrix} \dfrac{r_{12} - 1}{(1 - r_{12}r_{21})} & \dfrac{r_{12}(r_{12} - 1)}{(1 - r_{12}r_{21})} \\ \dfrac{\rho r_{21}(r_{21} - 1)}{(1 - r_{12}r_{21})} & \dfrac{\rho(r_{21} - 1)}{(1 - r_{12}r_{21})} \end{pmatrix}$$

$$tr(J) = \frac{(r_{12} - 1) + \rho(r_{21} - 1)}{(1 - r_{12}r_{21})} \qquad [5.43]$$

$$det(J) = \frac{\rho(1 - r_{12}r_{21})(r_{21} - 1)(r_{12} - 1)}{(1 - r_{12}r_{21})^2}$$

We distinguish four situations: the first (i) is when both r_{12} and r_{21} are smaller than 1; the second (ii) is when both r_{12} and r_{21} are larger than 1; the third (iii) when $r_{12} > 1$ and $r_{21} < 1$; the fourth (iv) when $r_{12} < 1$ and $r_{21} > 1$. They are graphically represented in Figure 5.8.

In case (i), the steady states $(1,0)$ and $(0,1)$ are saddle points, whereas the steady state represented by the intersection point between the two nullclines, $f = 0$ and $g = 0$, is stable because $tr(J) < 0$, whereas $det(J) > 0$. In the latter steady state, both species coexist. Such situation can occur when two species, having practically the same carrying capacities $(K_A \approx K_B)$,[10] show small interspecific competition parameters, that is $(k_{12}/k_{-1}) < 1$ and $(k_{21}/k_{-2}) < 1$.

In situation (ii), the steady states $(1,0)$ and $(0,1)$ are stable, whereas the steady state, represented by the intersection point between the two nullclines, $f = 0$ and $g = 0$, is a saddle point since it has $det(J) < 0$ and $tr(J) < 0$. In such situation, if the two species have similar carrying capacity, then they have also large ratios (k_{12}/k_{-1}) and (k_{21}/k_{-2}). The competition is strong. It is difficult to predict who ultimately wins out. It depends crucially on the starting condition. Each of the two stable steady states, $(1,0)$ and $(0,1)$, has a domain of attraction. The eigenvector representing the stability direction of the saddle point divides the $a-b$ space into regions: R1 and R2. If the initial condition lies on R1, then eventually species B dies out, and a becomes equal to 1: species A reaches its carrying capacity. On the other hand, if the initial condition lies on R2, then species A will become extinct, and B will reach its carrying capacity K_B. We expect to see the extinction of one species even if the

---

[10] More information about the carrying capacity is presented in Chapter 10, within the paragraph dedicated to the Logistic Map.

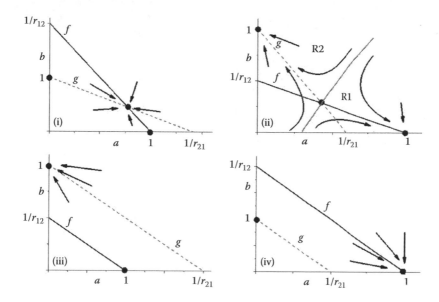

FIGURE 5.8 Phase diagrams for antagonistic relationships between species A and B. In (i), both r_{12} and r_{21} are less than 1; in (ii), both r_{12} and r_{21} are larger than 1; in (iii), $r_{12} > 1$ and $r_{21} < 1$; in (iv), $r_{12} < 1$ and $r_{21} > 1$. The black dots represent the possible steady states. In (ii), the continuous grey curve represents the stable eigenvector of the saddle point and divides the positive quadrant of the a–b space into two regions: R1 and R2.

initial condition lies on the separatrix because random fluctuations will move the system out of the separatrix (i.e., the eigenvector represented as a continuous grey curve in Figure 5.8[ii]) and one of the two species will disappear.

In situations (iii) and (iv), one species is stronger than the other. In (iii), $r_{12} > 1$ and $r_{21} < 1$, and species B finishes to dominate, whereas the other disappears. In (iv), $r_{12} < 1$ and $r_{21} > 1$, and it is species A to win and species B to die. In all cases, except (i), the competition finishes excluding one species.

5.6.2 MUTUALISM

In a mutualistic relationship, the coefficients n_A and n_B appearing in [5.33], are both equal to 2. The differential equations describing how the number of individuals of species A and B change over time are as follows:

$$\frac{dA}{dt} = k_1 FA - k_{-1} A^2 + k_{12} AB = k_1 FA \left(1 - \frac{A}{K_A} + \frac{k_{12}}{k_{-1} K_A} B \right)$$

$$\frac{dB}{dt} = k_2 FB - k_{-2} B^2 + k_{21} AB = k_2 FB \left(1 - \frac{B}{K_B} - \frac{k_{21}}{k_{-2} K_B} A \right)$$

[5.44]

To make easier the treatment of the system [5.44], we nondimensionalize it as we did in the case of the antagonistic relationship. We introduce three dimensionless variables: $a = A_C A$; $b = B_C B$ and $\tau = t_C t$, and we define the non-dimensionalizing constants $A_C = \left(1/K_A \right)$, $B_C = \left(1/K_B \right)$ and $t_C = k_1 F$. We obtain:

$$\frac{da}{d\tau} = a \left(1 - a + \frac{k_{12} K_B}{k_{-1} K_A} b \right)$$

$$\frac{db}{d\tau} = \frac{k_2 F}{k_1 F} b \left(1 - b + \frac{k_{21} K_A}{k_{-2} K_B} a \right)$$

[5.45]

Finally, fixing $\rho = (k_2/k_1)$, $r_{12} = (k_{12}K_B/k_{-1}K_A)$, and $r_{21} = (k_{21}K_A/k_{-2}K_B)$, the two differential equations assume their final dimensionless form:

$$\frac{da}{d\tau} = a(1-a+r_{12}b) = f$$

$$\frac{db}{d\tau} = \rho b(1-b+r_{21}a) = g$$

[5.46]

The possible steady-state solutions are: $(a_{ss}, b_{ss}) = (0,0); (1,0); (0,1); ((1+r_{12})/(1-r_{12}r_{21}), (1+r_{21})/(1-r_{12}r_{21}))$. As usual, their stability can be inferred by the linear analysis method. The Jacobian is

$$J = \begin{pmatrix} \left(\frac{\partial f}{\partial a}\right)_{ss} & \left(\frac{\partial f}{\partial b}\right)_{ss} \\ \left(\frac{\partial g}{\partial a}\right)_{ss} & \left(\frac{\partial g}{\partial b}\right)_{ss} \end{pmatrix} = \begin{pmatrix} 1-2a_{ss}+r_{12}b_{ss} & r_{12}a_{ss} \\ \rho r_{21}b_{ss} & \rho - 2\rho b_{ss} + \rho r_{21}a_{ss} \end{pmatrix}$$

[5.47]

The steady state (0,0) is unstable, because both $tr(J)$ and $det(J)$ are positive.

The steady state (1,0) is a saddle point. In fact,

$$J = \begin{pmatrix} -1 & r_{12} \\ 0 & \rho(1+\rho r_{21}) \end{pmatrix}, tr(J) = \rho(1+r_{21})-1, det(J) = -\rho(1+r_{21}) < 0$$

[5.48]

The steady state (0,1) is also a saddle point. In fact,

$$J = \begin{pmatrix} 1+r_{12} & 0 \\ \rho r_{21} & -\rho \end{pmatrix}, tr(J) = (1+r_{12})-\rho, det(J) = -\rho(1+r_{12}) < 0$$

[5.49]

For the last steady state, we have

$$J = \begin{pmatrix} \dfrac{1+r_{12}}{(r_{12}r_{21}-1)} & \dfrac{r_{12}(r_{12}+1)}{(1-r_{12}r_{21})} \\ \dfrac{\rho r_{21}(r_{21}+1)}{(1-r_{12}r_{21})} & \dfrac{\rho(r_{21}+1)}{(r_{12}r_{21}-1)} \end{pmatrix}$$

$$tr(J) = \frac{(1+r_{12})+\rho(1+r_{21})}{(r_{12}r_{21}-1)}$$

[5.50]

$$det(J) = \frac{\rho(1+r_{12})(1+r_{21}) - \rho r_{12}r_{21}(1+r_{12})(1+r_{21})}{(r_{12}r_{21}-1)^2}$$

When $r_{12}r_{21} < 1$, $det(J) > 0$ and $tr(J) < 0$. Therefore, the fourth steady state is stable. When $r_{12}r_{21} > 1$, the fourth steady state does not lie in the positive quadrant of the $a-b$ space and therefore, it does not have physical meaning.

In synthesis, when $r_{12}r_{21} > 1$, the mutualism has large values for the kinetic constants k_{12} and k_{21} in the case of comparable carrying capacities. In such case, we have only three possible steady states that are either unstable or saddle points. The populations of the two species grow unboundedly

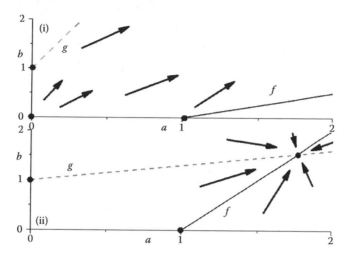

FIGURE 5.9 Possible scenarios for a mutualistic relationship. Case (i) requires $r_{12}r_{21} > 1$; case (ii) holds when $r_{12}r_{21} < 1$.

to infinity (see case (i) in Figure 5.9). On the other hand, when $r_{12}r_{21} < 1$, the mutualism is less strong. We have four steady states. The fourth one represents a stable solution, and all trajectories of its phase space tend to $(a_{ss}, b_{ss}) = (>1, >1)$. In such steady state (see case (ii) in Figure 5.9), the populations of the two species are larger than their maximum values achievable with a neutral relationship.[11]

To see the dynamics of the populations of species A and B in the case of the other symbiotic relationships, *try exercises 5.7, 5.8, 5.9, and 5.10.*

5.7   KEY QUESTIONS

- Describe the Lotka-Volterra model for the predator and prey relationship.
- Is the predator-prey relationship always oscillatory? Describe the dynamics in the case of Type I, II and III functional responses.
- What does the Poincaré-Bendixson theorem state?
- Make examples of symbiotic relationships.
- Which are the possible outcomes of an antagonistic relationship?
- Why should human relationships be mutualistic?

5.8   KEY WORDS

Predator and prey; Functional and numerical responses; Symbiotic relationships.

5.9   HINTS FOR FURTHER READING

Besides *Mathematical Biology* by Murray (2002), additional books on mathematical modeling of ecology are *A Biologist's Guide to Mathematical Modeling in Ecology and Evolution* by Otto and Day (2007) and *Elements of Mathematical Ecology* by Kot (2001).

---

[11] Ideally, any human relationship should be mutualistic. It is the only one that guarantees a reciprocal growth.

5.10 EXERCISES

5.1. Apply the linear stability analysis to the improved Lotka-Volterra model where the growth of the prey population follows the "logistic model" when there are not predators:

$$F + H \underset{k_{-1}}{\overset{k_1}{\rightleftarrows}} 2H$$

$$H + C \xrightarrow{k_2} 2C$$

$$C \xrightarrow{k_3} D$$

5.2. Apply the phase-space analysis to the dynamical system of exercise (5.1).

5.3. Solve the system of two differential equations [5.23] based on Type II functional response and a numerical response proportional to the rate of prey consumption, using the following values for the parameters (extracted from Harrison [1995] and regarding unicellular Paramecium Aurelia and its unicellular predator Didinium nasutum):

$$K_1F = 898; \ k_1F = 1.85; \ k_h = 25.5; \ K = 284.1; \ k_h' = 12.4; \ k_3 = 2.07.$$

Initial conditions: $H_0 = 56$ and $C_0 = 23$.

5.4. According to the results presented by Harrison (1995), when the prey Paramecium and the predator Didinium are grown in a cerophyl medium thickened by methyl-cellulose, the searching effectiveness of the Didinium dramatically reduces. Therefore, k_c becomes much smaller and K much larger. The reduction of the nutrient supply for Paramecium may hinder the growth rate k_1F and diminish the carrying capacity K_1F. For the following sets of parameters $K_1F = 400; \ k_1F = 0.82; \ k_h = 25.5; \ K = 350; \ k_h' = 12.4; \ k_3 = 2.07; \ H_0 = 87; \ C_0 = 10$, determine the stationary state and its stability.

5.5. Consider the system of two differential equations [5.23], involving a Type II functional response, and the values of the constants used in exercise 5.3:

$$K_1F = 898; \ k_1F = 1.85; \ k_h = 25.5; \ K = 284.1; \ k_h' = 12.4; \ k_3 = 2.07.$$

Try to verify the Poincaré-Bendixson theorem by constructing a trapping region TR, i.e., a closed set, such that the vector field points inward everywhere on its boundary.

5.6. Apply the linear stability analysis to confirm that for the system of two differential equations [5.32], there are three regions of stability as shown in Figure 5.7.

5.7. Predict the dynamics of a neutral relationship between two species, A and B, considering the general scheme [5.33].

5.8. Predict the dynamics of parasitism of species B towards species A. Considering the general scheme [5.33], fix $n_A = 0$ and $n_B = 2$.

5.9. Predict the dynamics of amensalism between two species, A and B, assuming that $n_A = 0$ and $n_B = 1$ in the general scheme [5.33].

5.10. Predict the dynamics of commensalism involving two species, A and B, assuming that $n_A = 2$ and $n_B = 1$ in the general scheme [5.33].

5.11 SOLUTIONS TO THE EXERCISES

5.1. The differential equations are

$$\frac{dH}{dt} = k_1 FH - k_{-1}H^2 - k_2 HC$$

$$\frac{dC}{dt} = k_2 HC - k_3 C$$

The solutions of stationary states are:

$H_{ss} = k_3/k_2$ and $C_{ss} = (k_2 k_1 F - k_{-1}k_3)/k_2^2$.

The Jacobian is:

$$J = \begin{pmatrix} -\dfrac{k_{-1}k_3}{k_2} & -k_3 \\ k_1 F - \dfrac{k_{-1}k_3}{k_2} & 0 \end{pmatrix}.$$

$tr(J) = -k_{-1}k_3/k_2$ is always negative.

$$\det(J) = k_1 F k_3 - \frac{k_{-1}k_3^2}{k_2} = \frac{k_1 F k_3 k_2 - k_{-1}k_3^2}{k_2}.$$

The eigenvalues are:

$$\lambda_{1,2} = \frac{tr(J) \pm \sqrt{(tr(J))^2 - 4\det(J)}}{2} = \frac{-\dfrac{k_{-1}k_3}{k_2} \pm \sqrt{\left(\dfrac{k_{-1}k_3}{k_2}\right)^2 - 4\left(k_1 F k_3 - \dfrac{k_{-1}k_3^2}{k_2}\right)}}{2}$$

The determinant is positive if $k_1 F > k_{-1}k_3/k_2$. If it is positive also the discriminant (Δ), the eigenvalues are two real negative roots, and the stationary state is a stable node. When $\Delta = 0$, the stationary state is a stable star. Finally, if $\Delta < 0$, the roots are two complex numbers having a negative real term. Therefore, the stationary state is a stable spiral, and the predator and prey populations exhibit damped oscillations with the predator oscillations lagging in phase behind the prey.

The determinant is null when $k_1 F = k_{-1}k_3/k_2$. In this case, one eigenvalue is a negative real number, and the other is 0: the system admits a line of stable fixed points. When the determinant is negative, i.e., when $k_1 F < k_{-1}k_3/k_2$, the discriminant is positive, and the roots are real, but one positive and the other negative. In this case, the stationary state is a saddle point.

5.2. In the system of two differential equations of exercise 5.1 there are only two variables, H, and C:

$$\frac{dH}{dt} = k_1 FH - k_{-1}H^2 - k_2 HC$$

$$\frac{dC}{dt} = k_2 HC - k_3 C$$

We can represent the evolution of the system in a phase space having C and H as coordinates. The equations representing the null-clines are:

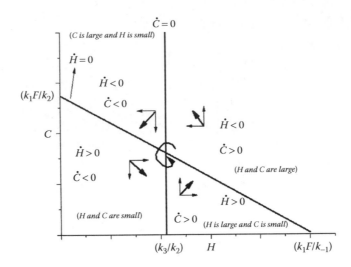

FIGURE 5.10   Phase plane for the "modified" Lotka-Volterra model when the $det(J)$ is positive.

$\dot{H} = 0$ when $C = (k_1 F - k_{-1} H)/k_2$ representing a straight line with $(k_1 F/k_2)$ as intercept and $-(k_{-1}/k_2)$ as slope.

$\dot{C} = 0$ when $H = (k_3/k_2)$ representing a straight line parallel to the C axis.

When $k_1 F > k_{-1} k_3/k_2$, i.e., $det(J) > 0$, the phase space appears like that depicted in Figure 5.10.

The intersection of the two nullclines is the stationary state that in the linear analysis of stability we discovered it is stable (see exercise 5.1). This result is confirmed qualitatively by the vector field shown in Figure 5.10. The signs of $\dot{H}$ and $\dot{C}$ in the different regions of the phase space can be inferred by their definitions and from the magnitude of the variables H and C in respect to their values H_{ss} and C_{ss}.

When $k_1 F = k_{-1} k_3/k_2$, i.e., $det(J) = 0$, the phase space appears as the graph on top of Figure 5.11.

The intersection of the two straight lines is the stationary state where $H_{ss} = k_3/k_2 = k_1 F/k_{-1}$ and $C_{ss} = 0$. The vector field confirms it is a stable stationary state.

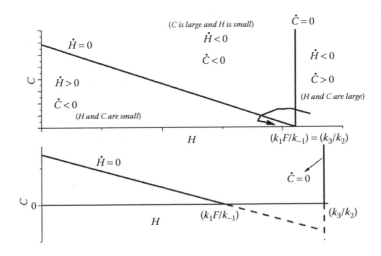

FIGURE 5.11   Phase plane for the "modified" Lotka-Volterra model when the $det(J)$ is null (graph on top) and negative (graph at the bottom).

When $k_1F < (k_3k_{-1}/k_2)$, i.e., $det(J) < 0$, the intersection between the two straight lines occurs for negative values of C and H, and it has no physical meaning.

5.3. The vertical isocline $H = (k_3K/(k'_h - k_3)) = 56.9$ is less than the vertex of the parabola $((K_1F - K)/2) = 306.95$. Therefore, the steady state is unstable. The numerical integration of the system of differential equation [5.23] can be obtained by using MATLAB and the following function file:

```
function dy=Iifr(t, y)
dy=zeros(2,1)
K₁F=898
k₁F=1.85
k_h=25.5
K=284.1
k_h'=12.4
k₃=2.07
dy(1)=k₁F*(1-y(1)/K₁F)*y(1)-(kh*y(1)*y(2))/(K+y(1))
dy(2)=(k_h'*y(1)*y(2))/(K+y(1))-k₃*y(2)
```

The script files should look like the following:

```
[t, y]=ode45('Iifr',[0 100], [56 23])
plot(t, y( :,1),t, y( :,2))
2ove('time')
2ove('prey, predator')

[t, y]=ode45('Iifr',[0 100], [56 23])
plot(y(:,1),y(:,2))
xlabel('prey')
ylabel('predator')
```

If we plot the results, we have: (Figure 5.12)

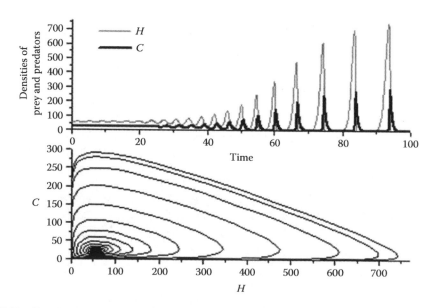

FIGURE 5.12 Trends of the densities of prey and predators over time (graph on top) and diverging oscillations represented in the predator-prey phase plane (graph at the bottom).

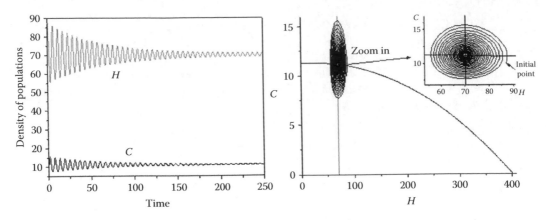

FIGURE 5.13 Trends of the densities of prey and predators over time (graph on the left) and converging oscillations represented in the predator-prey phase plane (graph on the right).

The predator and prey trace a diverging spiral. The numbers of the predators and prey approach the extinction condition. The initial condition is not stable.

5.4. The vertical isocline $H = \left(k_3 K / \left(k'_h - k_3\right)\right) = 70.1$ is larger than the vertex of the parabola $\left(\left(K_1 F - K\right)/2\right) = 25$. Therefore, the steady-state solution is stable. The stationary state corresponds to the intersection point between the parabola and vertical isoclines: $H_{ss} = 70.1$ and $C_{ss} = 11.1$. Using MATLAB and the function files presented in exercise 5.3, we obtain the results shown in Figure 5.13.

From the initial conditions $H_0 = 87$ and $C_0 = 10$, the number of predators and prey oscillates up to reach the stable stationary state.

5.5. The Poincaré-Bendixson theorem assumes that if

1. The phase plane is differentiable in its every point
2. TR is a closed subset of the phase plane
3. TR does contain any fixed point
4. There exists a trajectory J that starts inside TR and remains in TR for all future time; then, either J is a closed orbit, or it spirals towards a closed orbit as $t \to \infty$. In both cases, TR contains a limit cycle

The prey-predator system described by the two differential equations [5.23] verifies the first condition of the Poincaré-Bendixson theorem. To test if also the other three conditions are true, we generate the phase portrait by computing $(\dot{H}, \dot{C})$ at time 0 on a grid over the range of values for H and C we are interested in. If we use MATLAB, the function file should look like this:

```
f=@(t, Y) [1.85*(1-Y(1)/898)*Y(1)-(25.5*Y(1)*Y(2))/(284.1+Y(1));
(12.4*Y(1)*Y(2))/(284.1+Y(1))-2.07*Y(2)]
y1=linspace(0,100,20)
y2=linspace(0,100,20)
[x, y]=meshgrid(y1,y2)
size(x)
size(y)
u=zeros(size(x))
```

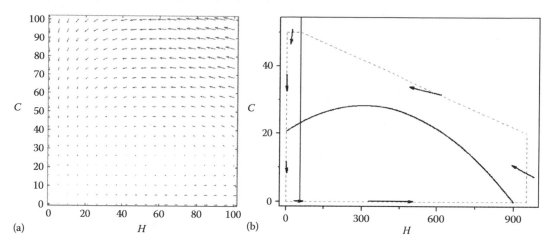

FIGURE 5.14 Phase portrait over the range of values (0, 100) for both H and C and at time 0 (graph a); Trapping Region in the predator-prey phase space (graph b).

```
v=zeros(size(x))
t=0
for I=1:numel(x)
  Yprime=f(t, [x(i); y(i)])
  u(i)=Yprime(1)
  v(i)=Yprime(2)
end
quiver(x, y,u, v,'r'); figure(gcf)
xlabel('H')
ylabel('C')
axis tight equal
```

The result is shown in the left panel (a) of Figure 5.14. The derivatives are plotted as vectors over the range $0 < H < 100$, and $0 < C < 100$, creating a grid of 20×20 points. If we calculate the phase portrait over a broader range, we discover that a trapping region exists, and it is shown in the right panel (b) of Figure 5.14. The region bounded by the dashed lines in (b) is a TR because all the vectors on the boundaries point into the box. The intersection point between the vertical line and the parabola is a fixed point. But it is unstable, and it respells all the trajectories far from itself. However, these trajectories maintain confined inside TR because there is a closed orbit or limit cycle, as confirmed by the arrows, all projected inwards.

5.6. If we consider the system of two differential equations [5.32],

$$\dot{C} = 0 \text{ if } H_{ss} = \sqrt{\frac{k_3 K}{k_h' - k_3}}, \text{ and } \dot{H} = 0 \text{ if } C_{ss} = \frac{k_1 F}{k_h H_{ss}} \left(K + H_{ss}^2 \right) \left(1 - \frac{H_{ss}}{K_1 F} \right).$$

If we want to give meaningful biological values to predator and prey, C and H must be positive.

$$H > 0 \text{ if } k_h' > k_3, \text{ and } C > 0 \text{ if } H_{ss} < K_1 F, \text{ i.e., if } \sqrt{\frac{k_3 K}{k_h' - k_3}} < K_1 F.$$

The Jacobian or the so-called linear stability matrix is:

$$J = \begin{pmatrix} \left(\dfrac{\partial \dot{H}}{\partial H}\right)_C & \left(\dfrac{\partial \dot{H}}{\partial C}\right)_H \\[3mm] \left(\dfrac{\partial \dot{C}}{\partial H}\right)_C & \left(\dfrac{\partial \dot{C}}{\partial C}\right)_H \end{pmatrix}$$

$$= \begin{pmatrix} k_1 F\left(1 - \dfrac{H_{ss}}{K_1 F}\right) - \dfrac{k_1 F H_{ss}}{K_1 F} - \dfrac{2k_h H_{ss} C_{ss}}{K + H_{ss}^2} + \dfrac{2k_h H_{ss}^3 C_{ss}}{\left(K + H_{ss}^2\right)^2} & -\dfrac{k_h H_{ss}^2}{K + H_{ss}^2} \\[5mm] \dfrac{2k_h' K H_{ss} C_{ss}}{\left(K + H_{ss}^2\right)^2} & \dfrac{k_h' H_{ss}^2}{K + H_{ss}^2} - k_3 \end{pmatrix}$$

If λ_1 and λ_2 are the eigenvalues of J, then

$$\lambda_1 + \lambda_2 = tr\left(J\right) = -\dfrac{k_1 F H_{ss}}{K_1 F} - \dfrac{k_h H_{ss} C_{ss}}{K + H_{ss}^2} + \dfrac{2k_h H_{ss}^3 C_{ss}}{\left(K + H_{ss}^2\right)^2}$$

$$\lambda_1 \lambda_2 = det\left(J\right) = \dfrac{k_h H_{ss}^2}{K + H_{ss}^2} \dfrac{2k_h' K H_{ss} C_{ss}}{\left(K + H_{ss}^2\right)^2}$$

We see that $\lambda_1 \lambda_2 = det\left(J\right) > 0$. This result means that the eigenvalues have real parts of the same sign. Using the expression of the nullcline $\dot{H} = 0$, we can rearrange the sum of the eigenvalues and obtain:

$$\lambda_1 + \lambda_2 = -\dfrac{k_1 F H_{ss}}{K_1 F} - k_1 F\left(1 - \dfrac{H_{ss}}{K_1 F}\right) + 2k_1 F\left(1 - \dfrac{H_{ss}}{K_1 F}\right)\left(\dfrac{H_{ss}^2}{K + H_{ss}^2}\right)$$

After tiding up the last equation, we have

$$\lambda_1 + \lambda_2 = -k_1 F + 2k_1 F\left(1 - \dfrac{H_{ss}}{K_1 F}\right)\left(\dfrac{H_{ss}^2}{K + H_{ss}^2}\right)$$

Inserting the expression of H_{ss}, we obtain

$$\lambda_1 + \lambda_2 = k_1 F\left(\dfrac{2k_3}{k_h'} - 1 - \dfrac{2k_3}{k_h' K_1 F}\sqrt{\dfrac{k_3 K}{k_h' - k_3}}\right)$$

When $k_h' > 2k_3$, i.e., when the rate constant of conversion of prey into a predator is at least two times the rate constant predator's mortality, the sum of the eigenvalues is negative, and the stationary states are stable. When $k_h' < 2k_3$, but $\left(\left(2k_3/k_h'\right)-1\right) < \left|\left(2k_3/k_h' K_1 F\right)\sqrt{k_3 K/\left(k_h'-k_3\right)}\right|$, $\lambda_1 + \lambda_2 < 0$ and the stationary states are stable. Finally, when $\left(\left(2k_3/k_h'\right)-1\right) > \left|\left(2k_3/k_h' K_1 F\right)\sqrt{k_3 K/\left(k_h'-k_3\right)}\right|$, $\lambda_1 + \lambda_2 > 0$ and the stationary states are unstable.

5.7. In the case of neutralism, the coefficients n_A and n_B of the scheme [5.33], are both equal to 1. The differential equations that describe the dynamics of the two populations are:

$$\frac{dA}{dt} = k_1 FA\left(1 - \frac{A}{K_A}\right)$$

$$\frac{dB}{dt} = k_2 FB\left(1 - \frac{B}{K_B}\right)$$

with $K_A = \left(k_1 F / k_{-1}\right)$ and $K_B = \left(k_2 F / k_{-2}\right)$

As we did in paragraph 5.6, we nondimensionalize the two differential equations if we introduce the variables $a = A_C A$, $b = B_C B$, $\tau = t_C t$, and the constants $A_C = 1/K_A$, $B_C = 1/K_B$, $t_C = k_1 F$, $\rho = k_2/k_1$. The two differential equations become

$$\frac{da}{d\tau} = a(1-a) = f$$

$$\frac{db}{d\tau} = \rho b(1-b) = g$$

The steady-state solutions are: $(a_{ss}, b_{ss}) = (0, 0); (1, 0); (0, 1); (1, 1)$. The stability of the steady states can be inferred after calculating the trace and determinant of the Jacobian:

$$J = \begin{pmatrix} 1 - 2a_{ss} & 0 \\ 0 & \rho(1 - 2b_{ss}) \end{pmatrix}$$

The result is shown in the Figure 5.15. The (0,0) steady state is an unstable solution; the (1,0) and (0,1) solutions are saddle points; the (1,1) is a stable steady state. All the trajectories converge on (1,1) that corresponds to abundances of the two populations equal to their carrying capacity. The two populations do not affect each other in their growth.

5.8. The differential equations describing the parasitism are:

$$\frac{dA}{dt} = k_1 FA\left(1 - \frac{k_{-1}}{k_1 F}A - \frac{k_{12}}{k_1 F}B\right) = k_1 FA\left(1 - \frac{A}{K_A} - \frac{k_{12}}{k_{-1}K_A}B\right)$$

$$\frac{dB}{dt} = k_2 FB\left(1 - \frac{k_{-2}}{k_2 F}B + \frac{k_{21}}{k_2 F}A\right) = k_2 FB\left(1 - \frac{B}{K_B} + \frac{k_{21}}{k_{-2}K_B}A\right)$$

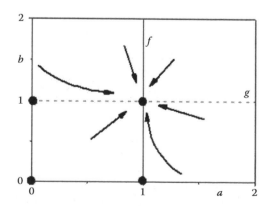

FIGURE 5.15 Phase space and spontaneous evolution in the case of a neutral relationship.

It is convenient to nondimensionalize the two differential equations. We fix $a = A_C A$, $b = B_C B$, $\tau = t_C t$, as usual. Moreover, we introduce the constants $A_C = 1/K_A$, $B_C = 1/K_B$, $t_C = k_1 F$, $\rho = k_2/k_1$, and $r_{12} = ((k_{12}K_B)/(k_{-1}K_A))$, $r_{21} = ((k_{21}K_A)/(k_{-2}K_B))$. The final appearance of the two differential equations is

$$\frac{da}{d\tau} = a(1 - a - r_{12}b) = f$$

$$\frac{db}{d\tau} = \rho b(1 - b + r_{21}a) = g$$

The steady-state solutions are: $(a_{ss}, b_{ss}) = (0, 0); (1, 0); (0, 1); (1 - r_{12})/(1 + r_{12}r_{21}), (1 + r_{21})/(1 + r_{12}r_{21})$. The latter admits positive values for variable a only when $r_{12} < 1$.

We use the linear stability analysis to determine the character of the solutions. The Jacobian is

$$J = \begin{pmatrix} 1 - 2a_{ss} - r_{12}b_{ss} & -r_{12}a_{ss} \\ \rho r_{21}b_{ss} & \rho - 2\rho b_{ss} + \rho r_{21}a_{ss} \end{pmatrix}$$

It derives that the steady state $(0, 0)$ is unstable because $det(J) = \rho > 0$ and $tr(J) = 1 + \rho > 0$.

The steady state $(1, 0)$ is a saddle point. In fact, $det(J) = -\rho(1 + r_{21}) < 0$.

The solution $(0, 1)$ is a saddle point when $r_{12} < 1$, because $det(J) = -\rho(1 - r_{12}) < 0$. On the other hand, it represents a stable steady state when $r_{12} > 1$, because $det(J) > 0$ and $tr(J) = -\rho + (1 - r_{12}) < 0$.

Finally, the last steady-state solution gives

$$det(J) = -\frac{\rho(1 + r_{21})(r_{12} - 1)}{(1 + r_{12}r_{21})^2} - \frac{\rho r_{12}r_{21}(1 + r_{21})(r_{12} - 1)}{(1 + r_{12}r_{21})^2}$$

$$tr(J) = \frac{(r_{12} - 1)}{(1 + r_{12}r_{21})} - \frac{\rho(1 + r_{21})}{(1 + r_{12}r_{21})}$$

When $r_{12} < 1$, $det(J) > 0$ and $tr(J) < 0$. Therefore, we have a stable steady state.

The overall situation is depicted in Figure 5.16. Plot (i) refers to the case $r_{12} < 1$, where we have four possible solutions and the stable one corresponds to a system where population A is not entirely extinct whereas population B is more abundant than its carrying capacity. Plot (ii) describes the evolution when $r_{21} > 1$. In this case, the rate constant k_{21} is so large (assuming that the carrying capacities of A and B are roughly the same) that A dies out and population B converges to a value that corresponds to its carrying capacity. It is beneficial for both if B does not deplete the resources of population A completely.

5.9.   The differential equations describing the amensalism are:

$$\frac{dA}{dt} = k_1 FA\left(1 - \frac{k_{-1}}{k_1 F}A - \frac{k_{12}}{k_1 F}B\right) = k_1 FA\left(1 - \frac{A}{K_A} - \frac{k_{12}}{k_{-1}K_A}B\right)$$

$$\frac{dB}{dt} = k_2 FB\left(1 - \frac{k_{-2}}{k_2 F}B\right) = k_2 FB\left(1 - \frac{B}{K_B}\right)$$

If we nondimensionalize them, by using the variables $a = A_C A$, $b = B_C B$, $\tau = t_C t$, and the constants $A_C = 1/K_A$, $B_C = 1/K_B$, $t_C = k_1 F$, $\rho = k_2/k_1$, and $r_{12} = k_{12}/k_{-1}K_B/K_A$, the equations become:

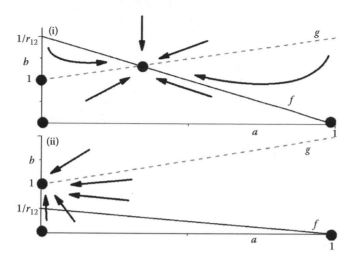

FIGURE 5.16 Phase space in the case of parasitism when $r_{12} < 1$ in (i) (upper graph), and when $r_{21} > 1$ in (ii) (lower graph).

$$\frac{da}{d\tau} = a\left(1 - a - r_{12}b\right) = f$$

$$\frac{db}{d\tau} = \rho b\left(1 - b\right) = g$$

The solutions of steady state are. $(a_{ss}, b_{ss}) = (0, 0);\ (1, 0);\ (0, 1);\ (1 - r_{12}, 1)$. To gain insight on their stability, we use the linear stability analysis. The Jacobian is

$$J = \begin{pmatrix} 1 - 2a_{ss} - r_{12}b_{ss} & -r_{12}a_{ss} \\ 0 & \rho - 2\rho b_{ss} \end{pmatrix}$$

The steady state (0, 0) is unstable because $det\left(J\right) = \rho > 0$ and $tr\left(J\right) = 1 + \rho > 0$.
The steady state (1, 0) is a saddle point; in fact, $det\left(J\right) = -\rho < 0$.

The steady state (0, 1) is a saddle point when $r_{12} < 1$; in fact, $det\left(J\right) = -\rho\left(1 - r_{12}\right) < 0$. On the other hand, it is a stable solution when $r_{12} > 1$. In fact, $det\left(J\right) > 0$ and $tr\left(J\right) = 1 - r_{12} - \rho < 0$.

When $r_{12} < 1$, the only stable solution is $\left(1 - r_{12}, 1\right)$, that has $det\left(J\right) = \rho\left(1 + 3r_{12}\right) > 0$ and $tr\left(J\right) = -1 - 3r_{12} - \rho < 0$. In synthesis, the possible dynamical scenarios are shown in Figure 5.17. Plot (i) refers to the case of $r_{12} < 1$, whereas plot (ii) to $r_{12} > 1$. Assuming that the two populations have similar carrying capacities when the ratio (k_{12}/k_{-1}) is small, the final steady state has $a_{ss} < K_A$ and $b_{ss} = 1$. On the other hand, when the ratio (k_{12}/k_{-1}) is large, population A extinguishes, and B reaches its carrying capacity.

5.10. The differential equations describing the commensalism are:

$$\frac{dA}{dt} = k_1 FA\left(1 - \frac{k_{-1}}{k_1 F}A + \frac{k_{12}}{k_1 F}B\right) = k_1 FA\left(1 - \frac{A}{K_A} + \frac{k_{12}}{k_{-1}K_A}B\right)$$

$$\frac{dB}{dt} = k_2 FB\left(1 - \frac{k_{-2}}{k_2 F}B\right) = k_2 FB\left(1 - \frac{B}{K_B}\right)$$

If we nondimensionalize them, by using the variables $a = A_C A$, $b = B_C B$, $\tau = t_C t$, and the constants $A_C = 1/K_A$, $B_C = 1/K_B$, $t_C = k_1 F$, $\rho = k_2 / k_1$, and $r_{12} = \left(\left(k_{12}K_B\right)/\left(k_{-1}K_A\right)\right)$, the equations become:

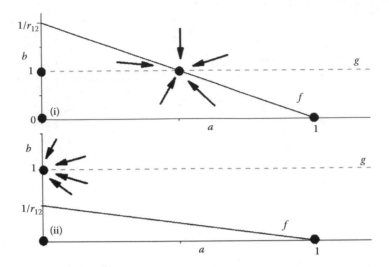

FIGURE 5.17 Phase space in the case of amensalism, when $r_{12} < 1$ in (i) (upper graph) and when $r_{12} > 1$ in (ii) (lower graph).

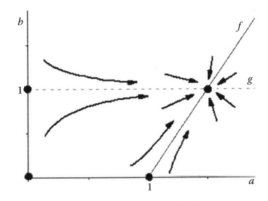

FIGURE 5.18 Phase space in the case of commensalism.

$$\frac{da}{d\tau} = a(1 - a + r_{12}b) = f$$

$$\frac{db}{d\tau} = \rho b(1 - b) = g$$

The steady-state solutions are: $(a_{ss}, b_{ss}) = (0, 0); (1, 0); (0, 1); (1 + r_{12}, 1)$. To gain insight on their stability, we use the linear stability analysis. The Jacobian is:

$$J = \begin{pmatrix} 1 - 2a_{ss} + r_{12}b_{ss} & r_{12}a_{ss} \\ 0 & \rho - 2\rho b_{ss} \end{pmatrix}$$

The solution $(0, 0)$ represents an unstable steady state: in fact, $tr(J) = 1 + \rho > 0$ and $det(J) = \rho > 0$. The solution $(1, 0)$ represents a saddle point because $det(J) = -\rho$ is negative. Even the solution $(0, 1)$ is a saddle point. In fact, $det(J) = -\rho(1 + r_{12}) < 0$. Finally, the solution $(1 + r_{12}, 1)$ is a stable steady state because $det(J) = \rho(1 + r_{12}) > 0$ and $tr(J) = -1 - r_{12} - \rho < 0$. The overall behavior is shown in Figure 5.18. The relationship of commensalism favors population A, whose abundance at the stable steady state is larger than K_A and is not damaging for population B that reaches its own carrying capacity.

6 The Emergence of Temporal Order in the Economy

> …the ideas of economists and political philosophers, both when they are right and when they are wrong, are more powerful than is commonly understood. Indeed, the world is ruled by little else.
>
> **John Maynard Keynes (1883–1946 AD)**

6.1 INTRODUCTION

The economy is strictly bound to the ecology. In Chapter 5, we learned that ecology studies the relationships among living beings within their environment, and how these relationships evolve. Each living-being strives to survive and reproduce. On the other hand, the economy studies how human beings, organized in societies, strive to reach their psycho-physical well-being, by exploiting natural resources, either as they are or after transforming them into goods and services through work. The terms economy and ecology share the Greek etymological root *οίκος*, which means "house." The word ecology was coined by the interdisciplinary figure of Ernst Haeckel[1] with the meaning of "study about the house, dwelling place or habitation of wildlife." The word economy means literally "household management." In the fourth century BC, Xenophon wrote the *Oeconomicus*. In the *Oeconomicus*, the philosopher Socrates discusses the meaning of wealth. He identifies wealth not merely with possessions, but also with usefulness and well-being. The wise Socrates pinpoints in moderation and hard work the two necessary ingredients to succeed in household management. Such ideas are still valid. However, in the last two hundred years or so, the social, political and economic scenarios have changed dramatically. Therefore, a new discipline has been born, political economy, with a meaning outlined by Adam Smith[2] in his book *The Wealth of Nations* (1776). Political economy is the science that studies how single human beings and societies choose to exploit scarce resources to produce various types of goods and services and distribute them for use among people. Since then, economists have been grappling with two fundamental questions: how wealth can be created and how wealth should be allocated. The ultimate answer seems to be that economy is a complex adaptive system (Beinhocker 2007). In economy there is a huge number of stocks and flows dynamically connected in an elaborate web of positive and negative feedback relationships; such feedback relationships have delays and operate at different timescales. The economic systems

---

[1] Ernst Haeckel (1834, Potsdam–1919 AD) was a German biologist, naturalist, philosopher, professor of comparative anatomy, and artist. He coined many words commonly used today in science, such as ecology, phylum, and phylogeny. As artist fond of the amazing forms of life on earth, Haeckel published a book titled *Art Forms in Nature* (1974) wherein he depicted 100 illustrations of various organisms, many of which were first discovered by Haeckel himself.

[2] Adam Smith (1723, Kirkcaldy–1790, Edinburgh) was a Scottish moral philosopher and the pioneer of political economy. His book titled *The Wealth of Nations* is considered the first modern work on economics.

evolve as ecosystems do. It is not by chance that Darwin (1859) proposed the concept of evolution in biology after his *Beagle* expedition and after reading *An Essay on the Principle of Population, as It Affects Future Improvements of Society* written by the English economist Thomas Malthus in 1798. In his book, Malthus presented the economy as a struggle between population growth and human productivity. Nowadays, the human population is so abundant (more than 7 billion), and human activities to produce goods and services are so invasive, widespread, and proceeding at such high pace that they affect the stability of many ecosystems in our planet. Some scientists even believe we are living in a human-dominated geological epoch, called Anthropocene (Lewis and Maslin, 2015). Since the beginning of Anthropocene, humans have had the power of influencing the climate and the integrity of any ecosystem.[3] Therefore, it is clear, that now more than ever, economy and ecology are strictly intertwined. The tight relationship between economy and ecology has been sanctioned by the Romanian-American mathematician and economist Georgescu-Roegen, who, in 1971, published his "epoch-making" contribution titled *The Entropy Law and the Economic Process*. Georgescu-Roegen's book contributed to merge economics and ecology and gave birth to a new discipline: Ecological Economics. The fusion has undoubtedly promoted the recent meta-morphosis of political economy from linear to circular productive processes. Businesspeople and governments are striving to design circular economic processes, with the aim of minimizing waste. The circular economy turns goods that are at the end of their service into resources (Stahel 2016) because the economic systems should work as if they were healthy ecosystems. In any healthy ecosystem, resources are exploited cyclically, and nothing is dumped: the waste of plants and animals become food for fungi and bacteria, and fungi and bacteria repay by feeding plants and animals with minerals. In a balanced ecosystem, nothing is useless. The same should happen in the economy.

As the ecosystems have characteristic dynamics and laws, so does the economy. The main characters and variables of the productive processes reveal dynamics that are formally close to those found in ecology because both disciplines describe far-from-equilibrium systems.

In the next three paragraphs, I will present the main laws and phenomena involved in the economic processes. Then, we will discover that phenomena of temporal self-organization can emerge even in economy. In fact, in the economy, we encounter business cycles that can be interpreted as if they were due to predator-prey-like relationships.

6.2 THE ECONOMIC PROBLEM

When economists analyze the origin and the allocation of wealth at a small scale, focusing on individuals or tiny groups of people or single companies or single productive sectors, they deal with microeconomic systems. On the other hand, when economists study the origin and allocation of wealth at the holistic level, focusing on entire nations and their GDPs (Gross Domestic Products), they deal with macroeconomic systems. Whatever is the level of analysis, the main characters of the economy are humans with their psycho-physical demands and requests. Humans strive to fulfill all their vital demands and satisfy all their reasonable requests by exploiting limited and sometimes scarce resources.

In any economic system, humans may be grouped in two sets: producers and consumers. Producers are those who make goods or provide services. Consumers are those who buy and pay for goods and services. Every human is a consumer. But whoever has a job, he/she is also a producer. The interplay

---

[3] The beginning of the Industrial Revolution, in the late eighteenth century, has most commonly been suggested as the start of the Anthropocene. The researchers Lewis and Maslin demonstrate that Anthropocene may date back to the seventeenth century when the Europeans arrived in the Americas. Colonization of the New World led to the deaths of about 50 million indigenous people, most within a few decades of the seventeenth century due to smallpox. The abrupt near-cessation of farming across the continent and the subsequent re-growth of Latin American forests and other vegetation caused a significant drop of carbon dioxide from the atmosphere.

between producers and consumers allows finding the solution to the economic problem of WHAT, HOW, and FOR WHOM to produce goods and services (Samuelson and Nordhaus 2004).[4] The strategy of solving the economic problem of how to make the best use of limited resources depends on the features of the economic environment. A free-market economic environment finds solutions that are appreciably different from those offered by a command economy and from those determined by a mixed economy. In a free market, what to produce is determined by consumers, how to produce is determined by producers, and who gets the products depends on the purchasing power of the consumers. At the heart of a free market economy, there is the pursuit of self-interest by both producers and consumers. When a government decides what, how and for whom to produce, we have a command economy. In between the two extremes, there are also mixed economies that merge market forces and governmental planning to try to determine the best solutions to the economic problems. In a mixed economy, what to produce is fixed partly by consumers' preferences and partly by the government; how to produce is chosen partly by the producers seeking their profits and partly by the government that, hopefully, looks for social justice. For whom to produce is determined partly by purchasing power and partly by government preferences. In practice, all economies are mixed, although they differ in the balance between public and private sectors.

6.3   LINEAR AND CIRCULAR ECONOMY

Whatever is the kind of economy, three factors allow producing goods and provide services (symbol GS). They are (1) land and natural resources (symbol R), (2) capital (symbol C) including tools, equipment, and factories, (3) labor (symbol L) including skills, risks, and efforts of entrepreneurs.

$$GS = f\left(R, C, L\right) \tag{6.1}$$

Productive processes transform natural resources in goods and services by increasing (a) their Helmholtz free energy A that measures the maximum obtainable work,[5] and/or (b) their content of information I (see Figure 6.1). The value added (symbol VA) to natural resources is the increment of value, represented by the sum $(A+I)_{added}$, that the natural resources undergo by labor. The value added is transformed into the well-being of the buyer. However, when goods and services cease to be useful because they are not used to do further work or give the required information, they are at the end of their lifetime. They become waste, and they are usually dumped because they have no more or a little A+I value.

The scheme sketched in Figure 6.1 represents how the linear economy works: it transforms natural resources into waste, but also into the physical and immaterial psychic well-being of humans. The linear economy is driven by the ambition of relentless exponential growth: the GDP of a nation must always increase. Commonly it is said that a linear economy is affected by the "bigger-better-faster-safer" syndrome, which means progress, emotion and fashion to satisfy the endless human thirst for well-being. Companies make profits if they contrive, produce, and sell appealing and cheap goods or services. Unfortunately, the growth-mania of the linear economy determines an unavoidable fast depletion of natural resources. In fact, our planet, Earth, is like a spaceship (Georgescu-Roegen 1976). The "spaceship earth" is embedded in the gravitational fields of our closest star and moon; its fuels are the flow of electromagnetic energy coming from the sun and the thermal energy produced by the decays of unstable nuclei under the terrestrial crust.

---

[4] The economic problem was outlined by Paul Anthony Samuelson (1915–2009 AD) who was the first American to win the Nobel Prize in Economic Sciences in 1970. As the same Samuelson declared, in the age of specialization, he thinks of himself as the last "generalist" in economics, with interests that ranged from mathematical economics to financial journalism.

[5] From Clausius' inequality $dS \geq dq/T$ and the first principle of Thermodynamics $dU = dq + dw$, it derives that $-dw \leq -dU + TdS$. The work carried out by the system is $dw' = -dw$ and $-dA = -dU + TdS$ a T constant. Therefore, $dw' \leq -dA$. The maximum work that can be performed is given by $-dA$.

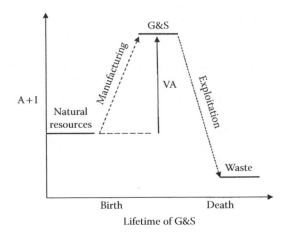

FIGURE 6.1 Scheme describing how the sum of Helmholtz free energy (A) and information (I) of natural resources change when they are transformed in goods and services (*GS*) through manufacturing, and finally in waste after exploitation.

Within the "spaceship earth" there are stocks of mineral resources, fossil fuels, and all other chemicals that are finite. In a thermodynamic sense, the earth may be conceived, approximately, as a closed system.[6] We might delay the exhaustion of natural resources by choosing a strategy of de-growth, which corresponds to negative growth. This radical program proposed by Georgescu-Roegen (1976) wanted to downsize the world economy to the point where it makes use of a very minimum of exhaustible resources. Such proposal has remained unexplored in societies grounding their economy in the accumulation of capital and mass consumption. A valid alternative is the strategy of sustainable growth. Economic growth is sustainable when growth can be maintained without exhausting natural resources and creating environmental problems, especially for future generation. In the last decades, societies have become aware that sustainable growth is feasible if economy mutates from linear to circular. A circular economy works as if it were an ecosystem. Any ecosystem grounds on solar energy (see Figure 6.2). Plants, algae, and phytoplankton feed

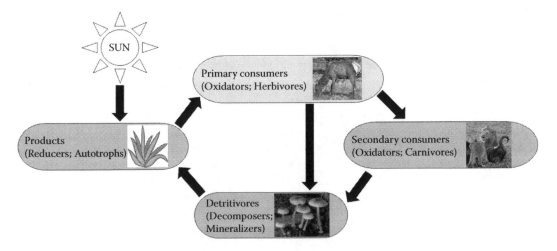

FIGURE 6.2 Schematic structure of an ecosystem fueled by solar radiation.

---

[6] Rigorously, the earth is an open system because compounds may escape from the terrestrial atmosphere and because nuclear particles and meteorites can enter our atmosphere.

FIGURE 6.3   Sketch of the lifetime of Goods and Services in a circular economy.

on solar radiation. They photosynthesize carbohydrates from CO_2 and water. They use carbohydrates as chemical fuel for themselves and for the primary consumers that are the herbivores. The herbivores are the food of carnivores. Both plants and herbivores and carnivores produce dead organic matter through their metabolism. Their waste products feed the detritivores that are fungi and bacteria. The detritivores decompose waste products more effectively and release products that feed the producers. Nothing is wasted.

The same should occur in a circular economy. Circular-economy business models turn old goods into as-new resources by recycling the materials, but at the same time foster reuse and extend service life through repair, remanufacture, upgrades, and retrofits (see Figure 6.3). Waste is nothing but a spring of resources to be harvested. The ultimate goals are industrial loops to turn outputs from one manufacturer into inputs for another. Virgin materials may undergo many cycles of manufacturing. The consumption of virgin materials' stocks is reduced, and the generation of waste is shrunk, too.

6.4   THE LAW OF SUPPLY AND DEMAND

In economy, the value of Helmholtz free energy A and information I of a good or service is expressed through its price. Who determines the prices of goods and services? The protagonists of the economic processes determine the price, i.e., producers and consumers who meet in the marketplace. The number of products the consumers are willing to buy depends on their prices. If the price of a product is low, consumers are encouraged to buy it in a significant amount; on the other hand, if the price of a product is high, few consumers are interested in it (see the Demand curve in Figure 6.4). Producers decide by reasoning oppositely. When a product is sold at high

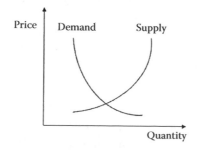

FIGURE 6.4   The curves of demand and supply and their intersection representing the best price for a product or service.

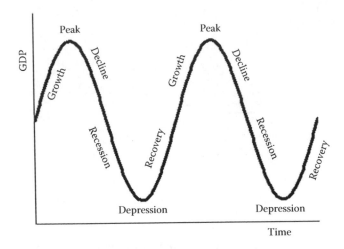

FIGURE 6.5   Oscillations of GDP over time in business cycles.

price, they are strongly willing to manufacture it, whereas, whenever it has a low price, they are not encouraged to invest on it (see the Supply curve in Figure 6.4). The right price for a product is that corresponding to the intersection point of the demand and supply curves. However, the price of a product does not remain constant over time because the economy is a complex adaptive system. In fact, fashion changes people's habits; technology evolves; the availability of natural resources decreases, and so on. The demand and supply curves shift continuously. For example, when the salary of consumers increases, the demand curve shifts towards the upright part of the graph in Figure 6.4. The technological progress allows producers to manufacture much more commodities by cutting costs, and the supply curve moves towards the down-right part of the same graph. When there is limited supply with respect to the demand, producers increase production and prices rise. As soon as the prices rise, consumers decrease consumption. On the other hand, if there is too much supply for the available demand, then prices fall, producers decrease production, and consumers increase consumption.

In macroeconomics, a graph like that of Figure 6.4 is still valid, but it refers to the aggregate demand and the aggregate supply. The aggregate demand is the total amount of money spent by all consumers to buy goods and services, whereas the aggregate supply is the total amount of money spent by all the producers to manufacture commodities, offer services and including their profits. Producers and consumers determine a circular flow of money. The GDP of a nation can be measured by counting either the money spent by the consumers to buy goods and services or the money spent by producers to manufacture commodities, offer services and store profits. The GDP does not remain constant, but moves up and down, periodically (see Figure 6.5). Usually, it is possible to distinguish the phases of growth, peak, decline, recession, depression, recovery that repeat cyclically. In fact, business analysts talk about economic cycles, and they strive to predict them.

6.5   THE BUSINESS CYCLES

For centuries, economists in both the United States and Europe regarded economic downturns as "diseases" that had to be treated; it followed, then, that economies characterized by growth and affluence were regarded as "healthy" economies. By the end of the nineteenth century, however, many economists had begun to recognize that economies were cyclical by their very nature, and studies increasingly turned to determine which factors were primarily responsible for shaping the direction and disposition of national, regional, and industry-specific economies. Today, economists, corporate executives, and business owners cite several factors as particularly important in shaping the complexion of business environments.

There are two main approaches to explaining business cycles (Samuelson and Nordhaus 2004). The first approach invokes exogenous shocks as the leading causes of business cycles. They might be technological innovations, wars, natural catastrophic events, political decisions, and so on. The second approach looks for endogenous factors. In fact, as we have already said in paragraph 6.1, an economic system is an example of an out-of-equilibrium system that can self-organize similarly to ecosystems.

6.5.1 GOODWIN'S PREDATOR-PREY MODEL

Goodwin (1967) presented a model where workers and capitalists play like the predators and preys of the Lotka and Volterra model and give rise to oscillations. It was Karl Marx, who, in his famous book *Das Capital* (1867), formulated the original idea. Marx believed that profits push capitalists to make investments and expand production. The consequent growth of activities provokes excess labor demand. Hence, wages rise. But the rise of the wages squeezes the profit rate of capitalists, slowly eroding the basis for accelerated accumulation. Therefore, the price of labor falls with the needs of the self-expansion of capital. When the capital starts to re-grow, labor demand rises too, again. Then, the entire cycle repeats itself. In Goodwin's model, the total output, Y, of the macro-economy is partitioned between wage-earning workers and profit-earning capitalists. If w represents a single wage and L is the amount of labor employed, the wage bill is wL. The ratio (wL/Y) will be the wage share. The profit of the capitalists will be

$$P = Y - wL. \tag{6.2}$$

The ratio P/Y is the profit share. Of course,

$$\frac{wL}{Y} + \frac{P}{Y} = \frac{w}{\lambda} + \frac{P}{Y} = 1, \tag{6.3}$$

where $\lambda = Y/L$ is the labor productivity. It grows at the positive rate θ:

$$\frac{d\lambda}{dt} = \theta\lambda \tag{6.4}$$

$$\lambda = \lambda_0 e^{\theta t} \tag{6.5}$$

We assume that workers consume all their wages, whereas capitalists invest all their incomes. If K is the capital, $(K/Y) = \kappa$ is the capital-output ratio, which is constant. The capital growth rate g_K is given by:

$$g_K = \left(\frac{dK}{dt}\right)\frac{1}{K} = \frac{P}{K} = \left(1 - \frac{w}{\lambda}\right)\frac{Y}{K} = \left(1 - \frac{w}{\lambda}\right)\frac{1}{\kappa} = (1-v)\frac{1}{\kappa} \tag{6.6}$$

It is evident that g_K increases if the wage share $v = w/\lambda = wL/Y$ decreases.

The growth rate of the labor g_L is

$$g_L = \left(\frac{dL}{dt}\right)\frac{1}{L} = \frac{1}{\lambda}\frac{dY}{dt}\frac{1}{L} + \frac{Y}{L}\frac{d}{dt}\left(\frac{1}{\lambda}\right) = \frac{1}{K}\frac{dK}{dt} - \theta = g_K - \theta. \tag{6.7}^{[7]}$$

From equation [6.7], it is evident that g_L is high when g_K is high and when θ is small.

---

[7] Equation [6.7] requires some steps to be verified. $g_L = (dL/dt)(1/L) = (1/\lambda)(dY/dt)(1/L) + (Y/L)(d/dt)(1/\lambda)$. But $Y = K/v$ and $L = Y/\lambda$. Therefore, $g_L = (dK/dt)(1/K) + (K\lambda/Yv)(d/dt)(1/\lambda) = g_K - \lambda(\theta/\lambda_0 e^{\theta t}) = g_K - \theta$.

If N is the supply of workers and it grows exponentially

$$\frac{dN}{dt} = nN \qquad [6.8]$$

$$N = N_0 e^{nt}, \qquad [6.9]$$

then, the employment rate $\mu = L/N$ decreases exponentially:

$$\mu = \frac{Y}{\lambda N} = \frac{Y}{\lambda_0 N_0 e^{(\theta+n)t}} \qquad [6.10]$$

Its relative variation on time g_μ is

$$g_\mu = \frac{1}{\mu}\frac{d\mu}{dt} = \frac{1}{\mu N}\frac{dL}{dt} + \frac{L}{\mu}\frac{d}{dt}\left(\frac{1}{N}\right) = g_L - n = g_K - (\theta+n) = (1-v)\frac{1}{\kappa} - (\theta+n). \qquad [6.11]$$

Equation [6.11] tells us that the employment growth rate is high if the capital growth rate is high.
    The wage growth rate g_v will be

$$g_v = \frac{1}{v}\frac{dv}{dt} = \frac{1}{v\lambda}\frac{dw}{dt} + \frac{w}{v}\frac{d}{dt}\left(\frac{1}{\lambda}\right) = g_w - g_\lambda. \qquad [6.12]$$

Goodwin assumed that the wage rate g_w can be approximated by the following linear relation:

$$g_w = -\alpha + \beta\mu. \qquad [6.13]$$

It is known as Phillips curve where α and β are positive constants. Phillips straight line tells that wage rate rises as employment increases. From the definition of λ, we infer that $g_\lambda = \theta$. Therefore,

$$g_v = (-\alpha + \beta\mu) - \theta. \qquad [6.14]$$

Merging equations [6.11] and [6.14], we obtain a system of two differential non-linear equations describing how the employment rate μ and wage share v change mutually over time:

$$\frac{d\mu}{dt} = (1-v)\frac{\mu}{\kappa} - (\theta+n)\mu \qquad [6.15]$$

$$\frac{dv}{dt} = (-\alpha + \beta\mu)v - \theta v \qquad [6.16]$$

The system of equations [6.15] and [6.16] is formally equivalent to the Lotka-Volterra equations describing the predator-prey dynamics in a natural ecosystem. This is evident if we arrange [6.15] and [6.16] in the following form:

$$\frac{d\mu}{dt} = \left(\frac{1}{\kappa} - \theta - n\right)\mu - \frac{\mu v}{\kappa} \qquad [6.17]$$

$$\frac{dv}{dt} = \beta\mu v - (\alpha + \theta)v \qquad [6.18]$$

The system of equations [6.17] and [6.18] has two fixed points, namely the trivial fixed point at the origin and

$$v_{ss} = 1 - (\theta + n)\kappa \text{ and } \mu_{ss} = \frac{(\alpha + \theta)}{\beta}. \qquad [6.19]$$

The Jacobian evaluated at the non-trivial fixed point is

$$J = \begin{pmatrix} 0 & -\dfrac{(\alpha + \theta)}{\beta\kappa} \\ \beta(1 - (\theta + n)\kappa) & 0 \end{pmatrix}. \qquad [6.20]$$

The trace of J is null, and its determinant is positive. The eigenvalues are imaginary numbers. The values of the two variables, μ, and v, oscillate: when μ is very high, v starts to grow, but when v increases, μ drops (see Figure 6.6a). μ and v trace closed orbits (see Figure 6.6b). In [6.17 and 6.18], the employment rate μ serves as the prey while the wage rate v acts as the predator. Equation [6.17] implies that in the absence of wage rate, the employment rate grows exponentially at rate $(1/\kappa - \theta - n)$ because labor does not cost. On the other hand, equation [6.18] suggests us that in the absence of employment rate, the wage rate decreases exponentially at the rate $(\alpha + \theta)$.

Goodwin's theoretical model has attracted much interest since its publication. It is appealing because it is simple. It has been found an adequate model at the qualitative level. Proofs underlying Goodwin's model have been collected (Harvie 2000; Molina and Medina 2010). They encourage its theoretical development to more complex models of business cycles in the capitalist economies.

TRY EXERCISE 6.1

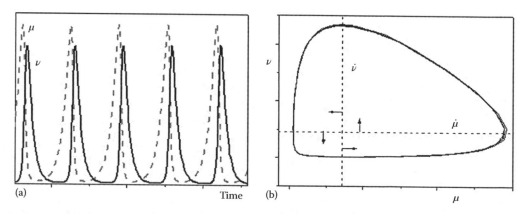

FIGURE 6.6 Oscillations of the "prey-predators" variables μ (dashed gray trace) and v (continuous black trace) in (a), and a closed orbit traced by them in (a). In (b), $\dot{\mu}$ and $\dot{v}$ represent the null-clines, and their intersection is the steady-state solution. The closed orbit is outlined counterclockwise.

6.5.2 THE MULTIPLIER AND ACCELERATOR MODEL

The American economist Paul Samuelson (1939) developed another famous macroeconomic model to interpret the business cycles. It is based on the concepts of multiplier and accelerator.

The theory of multiplier describes how an increase in the level of investment I raises income or output Y:

$$Y_t - Y_{t-1} = \frac{I_t - I_{t-1}}{(1-c)},$$ [6.21]

where c is the marginal propensity to consume.[8] The increase in the income induces growth in investment through the accelerator:

$$I_t = \kappa \left(Y_t - Y_{t-1} \right).$$ [6.22]

In [6.22], κ is the capital-output ratio (K/Y). Thus, the multiplier implies that investment increases output, whereas the accelerator implies that growth in output induces increases in investment. There is a reciprocal positive feedback action between income and investments. It is analogous to the positive feedback relation between science and technology.[9]

A national income at time t, Y_t, can be expressed as the sum of two terms, assuming away government and foreign sector:

$$Y_t = C_t + I_t,$$ [6.23]

where:
  C_t is consumption expenditure
  I_t is private investment.

The consumption depends on past income Y_{t-1} and autonomous consumption c_0:

$$C_t = c_0 + cY_{t-1}.$$ [6.24]

Investment is taken to be a function of the change in consumption and the autonomous investment I_0:

$$I_t = I_0 + \kappa \left(C_t - C_{t-1} \right).$$ [6.25]

Inserting equations [6.24] and [6.25] in [6.23], we obtain the definition [6.26] of how changes in income are dependent on the values of marginal propensity to consume (c_0) and the capital-output ratio κ:

$$Y_t = c_0 + cY_{t-1} + I_0 + c\kappa \left(Y_{t-1} - Y_{t-2} \right).$$ [6.26]

---

[8] The marginal propensity to consume c is the derivative of the consumption function C with respect to the income Y: $c = \left(dC/dY \right)$.
[9] Another action that exerts a positive feedback effect is the tax cut. In fact, a tax cut promotes investments by the producers and outgoings by the consumers.

Equation [6.26] can be rearranged as a nonhomogeneous second order linear difference equation (see Box 6.1 in this chapter):

$$Y_t - c(1+\kappa)Y_{t-1} + c\kappa Y_{t-2} = (c_0 + I_0).$$ [6.27]

The particular "steady-state" solution can be easily determined by fixing $Y_t = Y_{t-1} = Y_{t-2} = Y_{ss}$:

$$Y_{ss} = \frac{c_0 + I_0}{1 - c}.$$ [6.28]

The solution of the homogeneous equation is of the type $Y_h = M_1 r_1^t + M_2 r_2^t$, where M_1 and M_2 are arbitrary constants (that can be determined by fixing particular initial conditions, Y_0 and Y_1), and r_1 and r_2 are the two roots of the characteristic equation

$$r^2 - c(1+\kappa)r + c\kappa = 0.$$ [6.29]

The overall solution is $Y_{tot} = Y_{ss} + Y_h$. If we want to know the dynamics of the system, it is necessary to calculate the roots of equation [6.29]:

$$r_{1,2} = \frac{c(1+\kappa) \pm \sqrt{c^2(1+\kappa)^2 - 4c\kappa}}{2}.$$ [6.30]

We may distinguish five kinds of solutions that correspond to five types of dynamical scenarios (see Figure 6.7). To outline the five possible solutions, it is important to consider the sign of the discriminant $\Delta = \left(c^2(1+\kappa)^2 - 4c\kappa\right)$ and the conditions for stable roots (see Box 6.1 in this chapter), which are

---

BOX 6.1   SOLUTION OF A NONHOMOGENEOUS SECOND ORDER LINEAR DIFFERENCE EQUATION

A nonhomogeneous second order linear difference equation may take the form $y_t + a y_{t-1} + b y_{t-2} = k$, where k is a constant for all t. Its overall solution is the sum of two contributions: $y_{tot} = y_s + y_h$. The first contribution, y_s, is obtained by setting $y_t = y_{t-1} = y_{t-2}$. It derives that $y_s = k/(1+a+b)$. The second term y_h represents the solution of the homogeneous equation: $y_t + a y_{t-1} + b y_{t-2} = 0$. It has the general form $y_h = M_1 r_1^t + M_2 r_2^t$, where M_1 and M_2 are arbitrary constants to be defined, and r_1 and r_2 are the two roots of the characteristic equation $r^2 + ar + b = 0$. The two roots are given by $r_{1,2} = \left(-a \pm \sqrt{a^2 - 4b}\right)/2$. We distinguish three cases. First, when $a^2 > 4b$. The characteristic equation has two distinct real roots, and the general solution is $y_h = M_1 r_1^t + M_2 r_2^t$. Second, when $a^2 = 4b$. The characteristic equation has a single root, $r = -a/2$, and the solution is $y_h = (M_1 + M_2 t) r^t$. Third, when $a^2 < 4b$. The characteristic equation has roots that are complex numbers, and the solution is $y_h = M_1 r^t \cos(\theta t) + M_2 r^t \sin(\theta t)$. The overall solution $y_{tot} = y_s + y_h$ is stable if and only if the modulus of each root of the characteristic equation is less than 1. If the characteristic equation has real roots, then the modulus of each root is its absolute value. Therefore, for stability, we need the absolute values of each root to be less than 1, or $\left|(-a + \sqrt{a^2 - 4b})/2\right| < 1$ and $\left|(-a - \sqrt{a^2 - 4b})/2\right| > 1$. If the characteristic equation has complex roots, then the modulus of each root is b. For stability, we need $b < 1$. It can be demonstrated (Elaydi 2005) that in terms of the coefficients appearing into the second order linear difference equation, the solution is stable if and only if $|a| < 1 + b$ and $b < 1$.

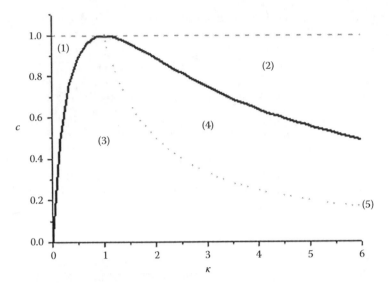

FIGURE 6.7   The output of Samuelson's model. The c vs. κ plot is partitioned in five regions by the discriminant Δ of equation [6.30] represented by the continuous black curve, and the stability conditions for the roots of the characteristic equation [6.29], represented by the dashed gray straight line ($c = 1$) and the dotted gray curve ($c = 1/\kappa$).

$\left|(1+\kappa)c\right| < (1+\kappa c)$ and $\kappa c < 1$. The discriminant can be rearranged in the following form $c = 4\kappa/(1+\kappa)^2$, and it is plotted as a continuous black curve in Figure 6.7. The conditions for stability reduce to the relations $c < 1$ and $c < 1/\kappa$. The functions $c = 1$ and $c = 1/\kappa$ are plotted as dashed gray straight line and dotted gray curve, respectively (see Figure 6.7). Merging the three functions, the graph of marginal propensity to consume (c) versus the capital-output ratio (κ) is partitioned into five regions.

Region (1) embeds positive real roots that are stable solutions. Therefore, in (1), the income, or the GDP, moves from the original steady-state upward or downward at a decreasing rate and finally reaches a new steady-state. In region (2), we encounter positive real roots of c and κ that cause the system to diverge and explode from the original steady state, at increasing rate. In regions (3), (4), and (5) that are below the function $c = 4\kappa/(1+\kappa)^2$ (having a maximum at $c = 1$ and $\kappa = 1$), the roots are complex numbers. The points in (3) represent stable solutions. Therefore, change in investment or consumption will give rise to damped oscillations in income or GDP, until the original state is restored. The solutions in (4) are combinations of c and κ values that are relatively high and correspond to such values of multiplier and accelerator that bring about explosive cycles, that is, the oscillations of income or GDP with greater and greater amplitude. Finally, when the values of c and κ lie over curve labeled as (5), they generate oscillations of constant amplitude in income. It is evident that only a precise constellation of c and κ values yield constant cycles. All the other combinations bring to either complete stability or complete instability, whether monotonic or oscillating. Thus, the multiplier-accelerator model is incomplete as a theory of permanent business cycles, although it has been a great advance in understanding the dynamics of macroeconomy.

TRY EXERCISE 6.2

6.5.3   OTHER MODELS

Other models to elucidate business cycles have been proposed. For example, the real business cycles theory. It supposes that business cycles are always triggered by an exogenous cause, such as new revolutionary technology, geopolitical event, war, natural disaster, and so on (Plosser 1989), and they are responses to the changes in real markets' conditions. Another theory is that of credit and debt cycles, which attributes business cycles to the dynamics of over-borrowing by businesses during the periods of economic booms, followed by the inevitable economic slowdown that brings to a debt crisis and a recession (Fisher 1933). There is also the political cycles theory (Nordhaus 1975) that attributes business cycles to political administrators. For example, soon after an election, new politicians impose austerity to reduce inflation, which increases prices of goods and services. The inflation reflects a reduction in the purchasing power per unit of money. The electors soon take the bitter medicine against inflation. Hopefully, they have time to forget its savor before the new election, a few years later. In fact, around one year before the new vote, politicians, who like to be re-elected, boost the economy by reducing taxes, increasing public investments, and persuading the central banks to maintain the interest rates low. The economic activity of many capitalistic democracies has an electoral rhythm.

6.5.4 THE REAL BUSINESS CYCLES

The real business cycles are not perfectly regular. But they are not perfectly random, either. The economic time series data, whether they refer to the GDP, or unemployment, or inflation, wiggle with many characteristic frequencies (an example is the GDP change data shown in Figure 6.8a, whose Fourier Transform is in b).

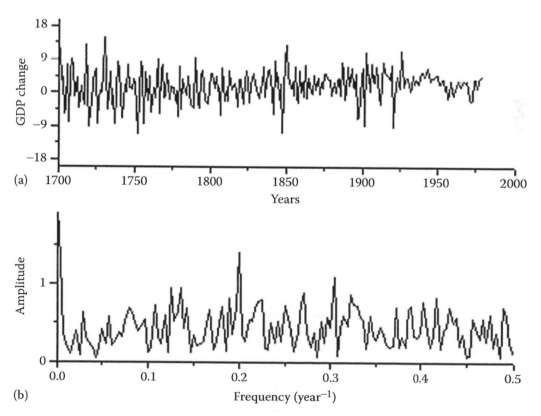

FIGURE 6.8 The trend of the GDP change for Sweden from 1700 up to 2000 in (a) and its Fourier Transform in (b). The source of the GDP data is the Economics Web Institute at the link http://www.economicswebinstitute.org/ecdata.htm.

Nowadays, scientists are aware that the not-quite-regular, not-quite-random economic time series are emergent phenomena typical of complex adaptive systems (Beinhocker 2007). They have three causes: two endogenous and one exogenous. The endogenous are, first, the behavior of the players in the economic system; second, the structure of the economic system, like the market, the production chains, the supply chains, and so on. The third cause is any exogenous input into the economic system, such as the technological changes. All these factors put together are responsible for the not-quite-regular and not-quite-random ups and downs observed in the real business cycles.

TRY EXERCISE 6.3

6.6   KEY QUESTIONS

- Why are ecology and economy related disciplines?
- What is the "economic problem" and how is it possible to solve it?
- Which are the factors that allow to produce goods and provide services?
- What distinguishes a linear and a circular economy?
- Explain the Law of Supply and Demand.
- How is it possible to measure the GDP of a nation?
- Which are the causes of business cycles?
- Describe the Goodwin's predator-prey model.
- Describe the multiplier and accelerator model.

6.7   KEY WORDS

Economy and political economy; Ecological Economics; Micro- and Macro-economic systems; Producers and Consumers; Economic environment; Aggregate demand and supply; Employment rate; Wage share; Marginal propensity to consume; Capital-output ratio.

6.8   HINTS FOR FURTHER READING

If you want to know more about models of business cycles, you can visit the website of the Institute for New Economic Thinking. On this webpage (https://www.ineteconomics.org/), there is a link to "The History of Economic Thought Website." In the "Essays and Surveys," there is a contribution dedicated completely to "Business Cycle Theory" (http://www.hetwebsite.net/het/essays/cycle/cyclecont.htm).

6.9   EXERCISES

6.1.   Goodwin's model into practice: solve numerically the system of two differential equations [6.17] and [6.18] for the conditions listed in Table 6.1.
You may calculate the time evolution of the two variables μ (employment rate) and ν (wage share) in the time interval [0 250] and for the initial conditions $\mu(t=0) = 0.4$ and $\nu(t=0) = 0.6$.

6.2.   The multiplier and accelerator model into practice: solve the nonhomogeneous second order linear difference equation [6.27] for the values of parameters c_0, I_0, c, κ, and the initial conditions listed in Table 6.2. To find the solutions, read carefully Box 6.1 of this chapter. These data have been extracted from Samuelson (1939).

TABLE 6.1

Values of the Capital-Output Ratio (κ), the Labor Productivity (λ), the Rate of Workers Supply (n), the Slope of the Wage Rate (β) and the Intercept of the Wage Rate (α) for Four Different Cases, Labeled as A, B, C, and D, Respectively

Case	κ	θ	n	β	α
(A)	0.5	0.04	0.003	0.52	2.54
(B)	0.5	0.4	0.003	0.52	2.54
(C)	1	0.4	0.003	0.52	2.54
(D)	0.5	0.4	0.003	1	2.54

TABLE 6.2

Values of the Autonomous Consumption (c_0), the Autonomous Investment (I_0), the Marginal Propensity to Consume (c), the Capital-output Ratio (κ), the Two Initial Conditions of the Output Y ($Y(t = 0)$ and $Y(t = 1)$)

Case	$c_0 + I_0$	c	κ	$Y(t = 0)$	$Y(t = 1)$
A	1	0.5	0	1.00	
B	1	0.5	2	1.00	2.50
C	1	0.6	2	1.00	2.80
D	1	0.8	4	1.00	5.00

6.3. The data in the Table 6.3 refers to the percent annual GDP growth for the two strongest economies in the world: that of the USA and that of China (data extracted from the World Bank website). By using the Fourier Transform (see Appendix C), determine and compare the most important frequencies in the two GDP trends (Table 6.3).

6.10 SOLUTIONS TO THE EXERCISES

6.1. To solve numerically the equations [6.17] and [6.18] you may use MATLAB. The script file should look like the following: $[t, y] = \text{ode45}(\text{'Goodwin'},[0\ 250],\ [0.4\ 0.6])$;

The outputs are reported in Figures 6.9 and 6.10.

When the labor productivity increases ten times as from condition (A) to (B), the amplitude of the oscillations grows, as well. In particular, the excursion of employment rate μ raises. If now, at high labor productivity, the capital-output ratio doubles (condition [C]), the period of the oscillations becomes longer. Finally, if the capital-output ratio is low, but the slope β of the wage rate doubles, the amplitude of the oscillations shrinks significantly.

6.2. The complete solution of equation [6.27] is the sum of two contributions: $Y_{tot} = Y_s + Y_h$. The term Y_s is obtained by applying equation [6.28]. The general form for Y_h is $M_1 r_1^t + M_2 r_2^t$, where r_1 and r_2 are the two roots of the characteristic equation [6.29], whereas M_1 and M_2 are arbitrary constants to be defined after knowing the initial conditions.

TABLE 6.3

Data of the Percent Annual GDP Growth for the USA and China in the Period Included between 1961 and 2014

Year	GDP Growth for the USA	GDP Growth for China
1961	2,3	−27,3
1962	6,1	−5,6
1963	4,4	10,2
1964	5,8	18,3
1965	6,4	17
1966	6,5	10,7
1967	2,5	−5,7
1968	4,8	−4,1
1969	3,1	16,9
1970	3,2	19,4
1971	3,3	7
1972	5,3	3,8
1973	5,6	7,9
1974	−0,5	2,3
1975	−0,2	8,7
1976	5,4	−1,6
1977	4,6	7,6
1978	5,6	11,9
1979	3,2	7,6
1980	−0,2	7,8
1981	2,6	5,2
1982	−1,9	9
1983	4,6	10,8
1984	7,3	15,2
1985	4,2	13,6
1986	3,5	8,9
1987	3,5	11,7
1988	4,2	11,3
1989	3,7	4,2
1990	1,9	3,9
1991	−0,1	9,3
1992	3,6	14,3
1993	2,7	13,9
1994	4	13,1
1995	2,7	11
1996	3,8	9,9
1997	4,5	9,2
1998	4,4	7,9
1999	4,7	7,6
2000	4,1	8,4
2001	1	8,3
2002	1,8	9,1
2003	2,8	10
2004	3,8	10,1
2005	3,3	11,4

(*Continued*)

TABLE 6.3 (*Continued*)

Data of the Percent Annual GDP Growth for the USA and China in the Period Included between 1961 and 2014

Year	GDP Growth for the USA	GDP Growth for China
2006	2,7	12,7
2007	1,8	14,2
2008	−0,3	9,6
2009	−2,8	9,2
2010	2,5	10,6
2011	1,6	9,5
2012	2,2	7,8
2013	1,5	7,7
2014	2,4	7,3

Source: World Bank website: (http://www.worldbank.org/)

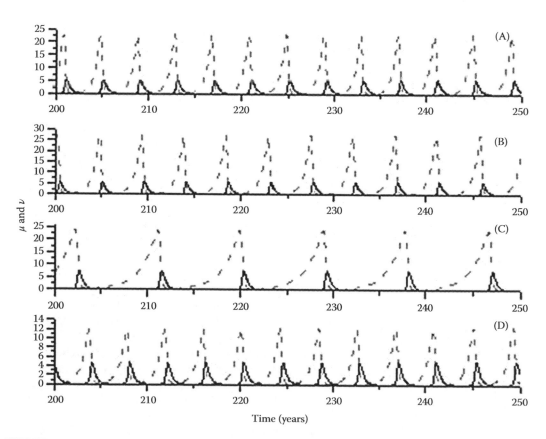

FIGURE 6.9 Trends of employment rate μ (dashed gray traces) and wage share ν (continuous black trace) for the four conditions A, B, C, and D listed in Table 6.1.

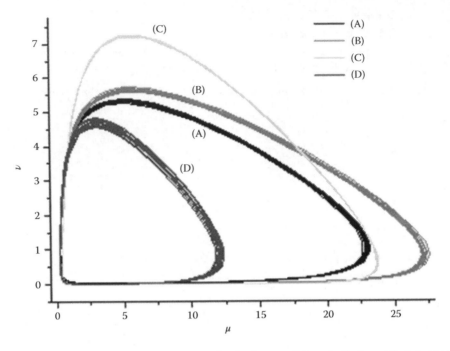

FIGURE 6.10 Limit cycles in the ν-μ phase space for the four conditions A, B, C, and D, listed in Table 6.1 of the exercise.

Case (A): the marginal propensity to consume $c = (dC/dY) = 0.5$ and the capital-output ratio is $\kappa = (K/Y) = 0$. The accelerator does not work. We are in the region (1) of Figure 6.7. The general solution is $Y = 2 - (0.5)^t$, because $Y_s = 2$, and the roots are $r_1 = 0.5$, $r_2 = 0$. With a constant level of governmental expenditure through time, the national income approaches asymptotically the value $Y_s = 2$ (see graph (A) in Figure 6.11).

Case (B): the marginal propensity to consume $c = (dC/dY) = 0.5$ and $\kappa = 2$. Both the multiplier and the accelerator are in action. $Y_s = 2$. The roots of the characteristic equation are $r_{1,2} = 3/4 \pm i\sqrt{7}/4$. The general solution of the homogeneous part of the equation [6.27] is $Y_h = c_1(3/4 + i\sqrt{7}/4)^t + c_2(3/4 - i\sqrt{7}/4)^t = c_1(\alpha + i\beta)^t + c_2(\alpha - i\beta)^t$. In polar coordinates $\alpha = r\cos\theta$, $\beta = r\sin\theta$, $r = \sqrt{\alpha^2 + \beta^2}$, $\theta = \tan^{-1}(\beta/\alpha)$. It derives that $Y_h = c_1(r\cos\theta + ir\sin\theta)^t + c_2(r\cos\theta - ir\sin\theta)^t$. By using De Moivre's theorem $[r(\cos\theta + i\sin\theta)]^t = r^t(\cos(t\theta) + i\sin(t\theta))$, the solution becomes $Y_h = r^t[M_1\cos(t\theta) + M_2\sin(t\theta)]$, with $M_1 = c_1 + c_2$ and $M_2 = i(c_1 - c_2)$. In our case, the general solution is $Y_{tot} = 2 + (1)^t[M_1\cos(41.41t) + M_2\sin(41.41t)]$. From the initial conditions, we can infer the values of the two coefficients, $M_1 = -1$, $M_2 = 1.89$.

This case (B) gives rise to regular oscillations (see graph (B) in Figure 6.11). In fact, it lies on the curve (5) of Figure 6.7.

Case (C): What happens when the marginal propensity to consume is slightly larger than in case (B)? $c = (dC/dY) = 0.6$, and the capital-output ratio is $\kappa = (K/Y) = 2$. $Y_s = 2.5$. The roots of the characteristic equation are $r_{1,2} = 0.90 \pm i0.62$. In polar coordinates, the general solution becomes $Y_{tot} = 2.5 + (1.09)^t[M_1\cos(34.56t) + M_2\sin(34.56t)]$. From the initial conditions, we can infer the values of the two coefficients, $M_1 = -1.5$, $M_2 = 2.65$. Case (C) that lies in the region (4) of Figure 6.7 originates explosive oscillations of income (see graph (C) in Figure 6.11).

Case (D): What happens when both the marginal propensity to consume and the capital-output ratio are high? We have $c = (dC/dY) = 0.8$ and $\kappa = (K/Y) = 4$. $Y_s = 5$.

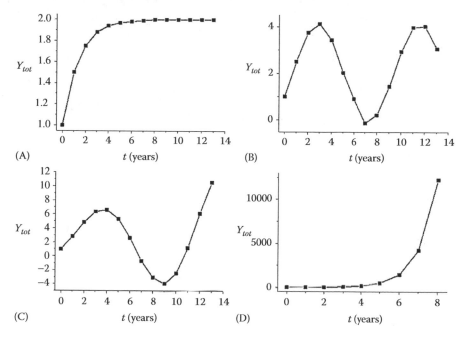

FIGURE 6.11 Outputs of the multiplier and accelerator model in the four situations, labeled as A, B, C, and D, corresponding to the four conditions A, B, C, and D listed in Table 6.2.

The roots of the characteristic equation are $\eta_1 = 2.90$ und $\eta_2 = 1.10$. The general solution is $Y_{tot} = 5 + M_1(2.90)^t + M_2(1.10)^t$. The values of the coefficients M_1 and M_2 are determined from the initial conditions. The final expression is $Y_{tot} = 5 + 2.45(2.90)^t - 6.45(1.10)^t$. The national income explodes, literally, without oscillations if the population is inclined to spend and if the capital-output ratio is high. A $\kappa = 4$ means that four units of capital are required to produce one unit of output. In general, the lower the capital-output ratio, the higher the productive capacity of capital, and the lower will be the investment according to the accelerator.

6.3. Figure 6.12 shows the trend of the percent annual GDP growth for the USA and China. In the last twenty years or so, the percent annual GDP growth of China is impressive. Nevertheless, the gap of GDP per capita between the two countries is still huge according to the data in the World Bank website (http://data.worldbank.org/). For example, in 2015, the GDP per capita was 55,836.80 US$ in the USA and seven times smaller in China (7,924.70 US$).

The two trends of percent annual GDP growth appear different. However, if we look at their Fourier spectra (see Figure 6.13), we find that, in both, the main frequency is 0.148 years⁻¹. It corresponds to almost seven years. This result could suggest that the two economies are somehow linked or experience analogous cycles.

The French economist Clement Juglar (1862) was one of the first to develop a theory of business cycles. He identified cycles of seven to eleven years' long that were attributed to credit oscillations. Other recurrent cycles have been, then, proposed or discovered. For instance, the shortest is the Kitchin (1923) cycle of three to five years, generated by the time lags in information movements affecting the making of commercial firms. Another kind of cycle is the Kuznets swing having a period of 15–25 years attributed to demographic processes, in particular with immigrant inflows and the changes in construction intensity they cause (Kuznets 1930). Finally, the longest type of cycle, ranging between 45 and 60 years,

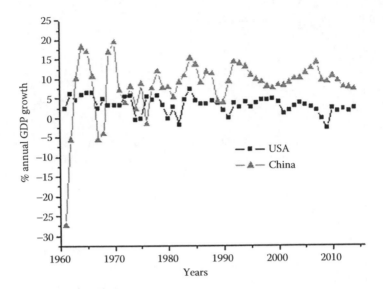

FIGURE 6.12   Percent annual GDP growth for the USA and China in the period included between 1961 and 2014.

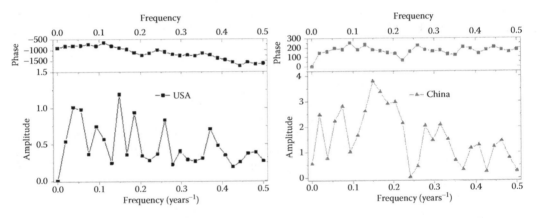

FIGURE 6.13   Fourier spectra of the percent annual GDP growth for the USA (on the left) and China (on the right).

has been proposed by the Soviet economist Nikolai Kondratiev (1935) and was attributed to fundamental technological innovations that spark revolutions and create leading industrial or commercial sectors. Many cycle theorists (Korotayev and Tsirel, 2010) have proposed five cycles so far:

1. The Industrial Revolution: steam engines and industrialization (~1780 ~1830).
2. The Age of steel, railways, heavy engineering (~1830 ~1880).
3. Electricity and chemical industry (~1880 ~1930).
4. The petrochemical industry, automobile, mass production (~1930 ~1970).
5. Information technology (~1970 ~2010).

Now, we are undertaking a new cycle. Its features are still *in fieri*, but they are sometimes connected to nano- and bio-technologies.

7 The Emergence of Temporal Order within a Living Being

A living being is like "a house with clocks in every room and every corner, yet in one way or another, they work in an organized way."

Derk-Jan Dijk (1958 AD)

7.1 INTRODUCTION

A cell is the basic unit of any living being. Within a biological cell, there are tens of thousands of different biochemical species (Reactome website). These species interact in a highly organized manner, both spatially and temporally. A cell looks like a "molecular ecosystem." Within this "microscopic ecosystem," proteins play important roles. In fact, the term protein derives from the Greek "$\pi\rho\omega\tau\varepsilon\tilde{\iota}o\varsigma$" meaning "holding the first place" (Vickery 1950). There are thousands of different kinds of proteins in a human cell. Proteins are the workhorses of the cell (Lodish et al. 2000). Distinct sets of proteins characterize cells belonging to different tissues. Many of the proteins within cells are enzymes, which are the catalysts of the biochemical reactions.[1] Other proteins allow cells to move and do work, maintain internal cell rigidity, and transport molecules across membranes. Reflecting their numerous functions, proteins come in many shapes and sizes. Proteins are formed from only 20 different L-α-amino-acids. The definition of a protein structure requires the specification of hierarchical features. First, it is necessary to know the sequence of amino acids (the so-called primary structure). Then, we need to specify the presence of alpha helices and/or beta sheets (the secondary structure). Furthermore, it is necessary to describe the folding of alpha helices and/or beta sheets (the tertiary structure) forming globules. Finally, we define the arrangement of the globules or subunits (the quaternary structure).[2] This chapter shows examples of the roles played by the proteins in metabolic, signaling, and epigenetic cellular events. We will discover that proteins can participate in phenomena of temporal self-organization.

7.2 METABOLIC EVENTS

Every second, an astonishing number of chemical reactions proceeds inside the cells. Many of these reactions provide the energy for vital processes and produce the substances necessary for growth, internal structuring and reproduction. They constitute the ensemble of processes known as metabolism. In the relatively low-temperature environment of the cell, most of the reactions are unlikely to occur. Fortunately, there are the enzymes (E). Enzymes are highly specific catalysts, due to their structure. The many enzymatic reactions occurring in a living cell must be balanced to keep the cell healthy. The regulation of the enzymatic reaction ultimately requires the control of the enzymatic activity and the control of the enzyme amount.

The mechanism of action of an enzyme is usually described by either the Michaelis-Menten or the Hill equation. These equations look like the Type II and Type III functional responses,

---

[1] The term enzyme was coined by the German physiologist Wilhelm Kühne from the Greek "$\dot{\varepsilon}\nu\zeta\nu\mu o\nu$" which means "in leaven" when he discovered the possibility of inducing the process of fermentation also in the absence of living beings.

[2] You can consult any book on biochemistry to see a picture of the hierarchical structure of a protein.

respectively, which we encountered in Chapter 5. In fact, often, proteins play the roles of predators for their substrates that behave as preys.[3]

7.2.1 MICHAELIS-MENTEN KINETICS

Usually, proteins have one site where they host selectively a substrate S, forming a complex (ES):

$$E + S \underset{k_{-c}}{\overset{k_c}{\rightleftarrows}} (ES) \overset{k_h}{\longrightarrow} E + P. \tag{7.1}$$

Note the formal link between equation [7.1] and the mechanism modeling the Type II functional response (equation [5.17]). S represents the prey. The complex (ES) transforms S into the product P.

The rate of product formation is

$$\frac{d[P]}{dt} = k_h \big[(ES)\big]. \tag{7.2}$$

The rate of change in the concentration of the complex (ES) is

$$\frac{d\big[(ES)\big]}{dt} = k_c [E][S] - k_{-c} \big[(ES)\big] - k_h \big[(ES)\big]. \tag{7.3}$$

If we indicate with C_E the analytical concentration of E (i.e., the sum $[E]+[(ES)]$), and we apply the steady-state condition for $[(ES)]$, we obtain

$$\big[(ES)\big] = \frac{C_E [S]}{[S] + \dfrac{k_h + k_{-c}}{k_c}} = \frac{C_E [S]}{[S] + K_M}. \tag{7.4}$$

The quantity K_M is called the Michaelis constant, after the German enzymologist Leonor Michaelis (1875–1949). Introducing the steady-state concentration of $[(ES)]$ (equation [7.4]) into equation [7.2], we achieve the well-known Michaelis-Menten formula, representing the rate of product formation as a function of the analytical concentration of E, K_M, and $[S]$:

$$\frac{d[P]}{dt} = \frac{k_h C_E [S]}{[S] + K_M}. \tag{7.5}$$

The amount $(1/C_E)(d[P]/dt) = (k_h[S]/([S]+K_M))$ is formally equivalent to the Type II functional response of equation [5.22]. The rate of product formation (equation [7.5]) is a hyperbolic function of $[S]$. When $[S]$ is small compared to K_M, the rate is a linear function of $[S]$. On the other hand, when $[S]$ is much larger than K_M, the rate becomes independent of $[S]$ and reaches its maximum value, which is given by $(k_h C_E)$. The rate of the reaction cannot be any faster than when every enzyme molecule is in a complex with a substrate molecule, i.e., when $C_E = [(ES)]$. The ratio of the maximum rate of reaction to the enzyme analytical concentration is known as the turnover number k_h, i.e., the number of substrate's moles converted to product per unit of time.

---

[3] An excellent example is in our immune system wherein antibodies are predators and antigens are preys.

7.2.2   Hill Kinetics

There exist also enzymes having more than one binding site. Their mechanism of action is like that of equation [7.6]:

$$E + nS \underset{k_{-c}}{\overset{k_c}{\rightleftarrows}} (ES_n) \overset{k_h}{\longrightarrow} ES_{n-1} + P. \qquad [7.6]$$

Note the formal link between equation [7.6] and the mechanism modeling the Type III functional response (equation 5.27).

The rate of change for $[(ES_n)]$ is

$$\frac{d[(ES_n)]}{dt} = k_c [E][S]^n - (k_{-c} + k_h)[(ES_n)]. \qquad [7.7]$$

If C_E is the analytical concentration of the enzyme (i.e., $C_E = [E] + [(ES_n)]$), applying the steady state approximation to ES_n, we obtain:

$$k_c [S]^n \left(C_E - [(ES_n)]\right) = (k_{-c} + k_h)[(ES_n)]. \qquad [7.8]$$

If we rearrange equation [7.8], we achieve

$$[(ES_n)] = \frac{[S]^n C_E}{[S]^n + \dfrac{k_h + k_{-c}}{k_c}} = \frac{[S]^n C_E}{[S]^n + K}. \qquad [7.9]$$

The ratio $[ES_n]/C_E$ is known as Hill equation (Haynie 2001). The rate of product P formation will be

$$\frac{d[P]}{dt} = \frac{k_h C_E [S]^n}{[S]^n + K}. \qquad [7.10]$$

The amount $(1/C_E)(d[P]/dt) = (k_h[S]^n/([S]^n + K))$ is formally equivalent to the Type III functional response of equation [5.31].

The enzymes having more than one site may have an exciting feature: that of showing indirect interactions between the distinct binding sites. In such cases, the enzymes are defined allosteric. This term was introduced by Jacques Monod and François Jacob to characterize the end-product (L-isoleucine) inhibition of the enzyme L-threonine deaminase (Cui and Karplus 2008). L-isoleucine does not compete with the reactant L-threonine in binding at the catalytic site; it instead binds at a regulatory site, inhibiting the reaction. The term "allosteric" comes from the Greek αλλος-στερεος meaning "other-space" and refers to the influence the binding at one site has on the binding at a remote location in the same macromolecule. The ligand that brings about the allosteric regulation of the binding of another ligand is called effector, or modulator.[4] If the effectors are identical to the substrate molecule, we speak about "homotropic interactions" between the ligands and the enzyme. If the binding of a substrate to the first site of the protein facilitates binding to the second, and so on, the interaction is defined as "positive cooperativity." On the other hand, if the binding of the

---

[4] The control of enzyme activity can be achieved by modification of the enzyme molecules through either covalent change (such as phosphorylation and hydrolysis) or non-covalent conformational change.

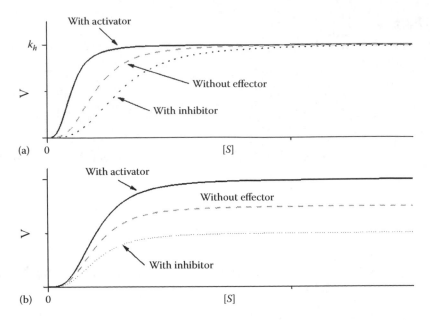

FIGURE 7.1 Rate profiles for allosteric enzymes in case of competitive ligands (a) and uncompetitive ligands (b).

substrate molecule to the first site inhibits the binding to the second, and so on, the interaction is defined as "negative cooperativity." When the effectors are ligands having a structure different from that of the substrate, we speak about "heterotropic interactions." In heterotropic interactions, the effectors can play as either activators or inhibitors of enzymatic activity. Depending on how ligands affect the enzymatic activity, their actions are also typified as "competitive" or "uncompetitive" concerning the substrate binding to the enzyme. Ligands are defined as competitive when their binding changes the K constant of equation [7.10] (in fact, they are also named as "K systems;" see Figure 7.1a). When the binding of a ligand changes the maximum velocity of the enzymatic reaction (i.e., k_h in equation [7.10]), they are referred to as uncompetitive or "V systems" (where V stands for Velocity; see Figure 7.1b) (Hammes and Wu 1974).

7.2.3 THE NONLINEARITY OF ALLOSTERIC ENZYMES

Many proteins turn their activities on and off by the nonlinear allosteric effects. Examples of heterotropic interactions are offered by the allosteric enzyme Aspartate Transcarbamylase (ATCase) catalyzing the reaction shown in Figure 7.2, which is the first step in pyrimidine biosynthesis.

ATCase has catalytic and regulatory sites. The aspartic acid (R_2) binds to the catalytic site of ATCase and in the presence of a saturating concentration of carbamoyl phosphate (R_1) produces N-carbamoyl aspartate (P_1) and phosphate (P_2). This reaction is the first step in the biosynthesis of

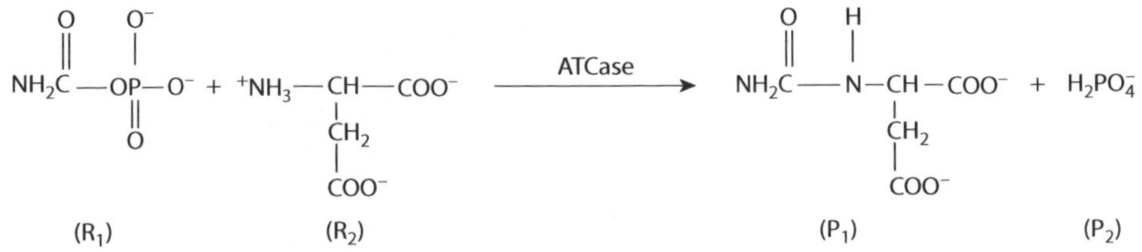

FIGURE 7.2 The first step of pyrimidine synthesis.

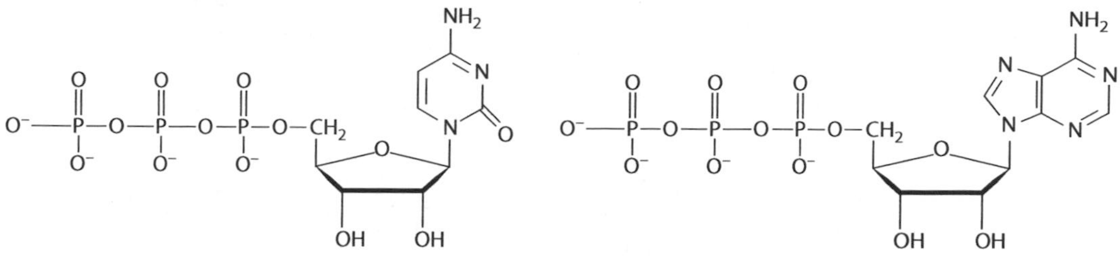

FIGURE 7.3 Structures of cytidine triphosphate (CTP) and adenosine triphosphate (ATP).

molecules such as cytosine, thymine, and uracil, which contain the pyrimidine unit and are some of the nitrogenous bases of DNA and RNA. The velocity of the enzymatic reaction as a function of the concentration of aspartate shows a sigmoidal dependence (Hammes and Wu 1974). Along the synthetic pathway leading to pyrimidines as final products, cytidine triphosphate (CTP) is produced. CTP competes with the structurally similar adenosine triphosphate (ATP) (see Figure 7.3) in binding to the regulatory sites of ATCase.

When the concentration of CTP is high, it overcomes ATP, and it preferentially associates with ATCase exerting an inhibiting effect on its catalytic activity, such as negative feedback in the synthesis of N-carbamoyl aspartate (P_1). On the other hand, when the concentration of CTP is not high, ATP wins the competition in binding to the regulatory sites of ATCase. When it is ATP to bind to ATCase, the final effect is the opposite: ATP activates the catalytic power of ATCase. CTP and ATP are K systems (see Figure 7.1a), because they alter the value of K and not that of k_h.

An example of an allosteric protein exhibiting positive cooperativity in homotropic interactions is the blood protein hemoglobin (Hb). Hb plays a vital role in the transfer of O_2 from the lungs to other cells of the body where it is combustive and participates in reactions releasing thermal energy. Hb is a tetramer, consisting of four subunits each having an oxygen-binding site, due to the presence of iron atom coordinated with a porphyrin ring. When an oxygen molecule binds to a site of hemoglobin, it increases the affinity of the other sites for oxygen (Haynie 2001). In 1965, Monod along with Wyman and Changeux proposed a model (known as MWC model) for allostery based on a few assumptions. Allosteric proteins are oligomers made of identical monomers in a symmetric arrangement. Each monomer has two folded tertiary conformations (T and R, indicated by squares and circles, respectively, in Figure 7.4). All monomers are in either one conformation or the other. When a ligand binds to a monomer of the protein, all monomers undergo the R-to-T or the T-to-R transition simultaneously. This model is represented by the structures enclosed by dashed lines. There are no "hybrid" quaternary structures with some subunits in R state and some in T state. There exists another model described for the first time by Pauling in 1935 for explaining the positive cooperativity of hemoglobin. It was then taken up by D. E. Koshland, G. Némethy and D. Filmer (the so-called KNF model) in 1966. Its basic assumptions are that the two conformational states (T and R) are available to each subunit; only the subunit to which the ligand is bound changes its conformation, and the ligand-induced conformational change in one subunit alters its interactions with neighboring subunits. It is a "sequential" model. A ligand binding induces a change in conformation of an individual subunit without all the rest of the subunits switching conformations. KNF model is represented by the structures enclosed by the grey patch, across the diagonal in Figure 7.4, wherein "hybrid" R/T multimers exist.

Both the Pauling-KNF and the MWC models are limiting cases of a more complicated situation. In fact, proteins are relatively soft polymers and, consequently, have significant structural fluctuations at room temperature. The static view of the structure of molecules has to be replaced by a dynamic picture (Motlagh et al. 2014). Allosteric proteins and biopolymers, in general, are made of semi-rigid domains and subunits, which can move relative to each other, so that "everything that living things

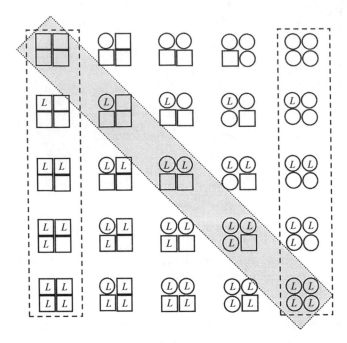

FIGURE 7.4 Models for allostery in the case of a four-subunit enzyme. The squares and the circles designate the T and R conformations of the subunits, respectively. L is the ligand. The Pauling-KNF model, represented by the elements along the diagonal, considers tertiary structural changes; the MWC model, represented by the elements of the right and left columns, considers only quaternary structural changes.

do can be understood in terms of jigglings and wigglings of atoms," as Richard Feynman foresaw in 1963 in his *Lectures in Physics* (Feynman et al. 1963). Recent advances in structural biology indicate that the categories by which protein structures are described (primary, secondary, tertiary, and quaternary hierarchical structural levels) do not encompass the full spectrum. Proteins exist as complex statistical ensembles of conformers, and they unfold and fold continuously in localized regions (Goodey and Benkovic 2008). Many proteins or regions of proteins are structurally disordered in their native, functional state and they cannot be adequately described by a single equilibrium 3D structure, but as a multiplicity of states (Tompa 2002). This multiplicity of states has been named as "super-tertiary structure" (Tompa 2012) in the case of single proteins and as "fuzziness" in the case of protein-protein complexes (Tompa and Fuxreiter 2007). The benefits are flexibility and adaptability in recognition, binding promiscuity, all features that confer an exceptionally plastic behavior in response to the need for the cell. The study of super-tertiary structure and fuzziness can be accomplished by NMR, X-ray crystallography, electron microscopy, atomic force microscopy, light-scattering, fluorescence spectroscopy complemented by advanced computational modeling, and data analysis. In time-resolved spectroscopies, the Maximum Entropy Method (MEM) is a powerful data analysis technique for detecting the multiplicity of molecular states and their dynamics. For example, in an ensemble of hemoglobins, each protein can display its dynamics of oxygen binding. As a consequence, the protein ensemble should be properly described by a distribution function of rate constants $f(k)$. The term $f(k)dk$ represents the fraction of molecules reacting with the rate $k \pm dk$. Information on the rate constant spectrum can be extracted from experiments by recording the kinetics $N(t)$ and expressing it as a weighted infinite sum of exponentials (Brochon 1994):

$$N(t) = \int_{0}^{\infty} f(k)e^{-kt}dk. \qquad [7.11]$$

The recovery of the shapes of the distributions of kinetic constants $f(k)$ is a nontrivial problem in numerical analysis. It is known as the inversion of the Laplace transform, which is notoriously ill-condition (see Appendix B). Methods involving the minimization of chi-squared (χ^2) are commonly used. It is routine to fit $N(t)$ to one- or two-exponential functions by iterative least-squares techniques. When two exponentials fail to fit, three are invoked. When also a three-exponential function fails, one is tempted to try four. Three and four exponentials will provide a good fit with visually random residuals and a good χ^2. However, the parameters obtained from these fits are, in general, meaningless if the data arise from a true distribution of kinetic constants. In these cases, it is suitable to reckon on the MEM. MEM gives the most probable distribution of exponentials, $f(k)$, fitting $N(t)$ according to equation [7.11], by maximizing the function

$$Q = \lambda S - C, \tag{7.12}$$

where S is the Shannon-Jaynes[5] entropy function

$$S = \sum_i \left[f(k_i) - m_i - f(k_i) log\left(\frac{f(k_i)}{m_i} \right) \right] \tag{7.13}$$

formulated for a large but discrete set of rate constants ($i = 1..., n$ with $n \to \infty$), where the ensemble of m_i values is a "prior" distribution used to incorporate prior knowledge into the solution; λ is a Lagrange multiplier and C, in the case of normally distributed noise, takes the familiar form

$$C = \chi^2 = \left\langle \frac{\left(N_{calc} - N_{exp} \right)^2}{\sigma^2} \right\rangle, \tag{7.14}$$

where σ^2 is the variance of data points.

One example of the application of MEM is shown in Figure 7.5 reporting two-lifetime distributions obtained to fit the luminescence decay kinetics of two distinct samples of tryptophan.

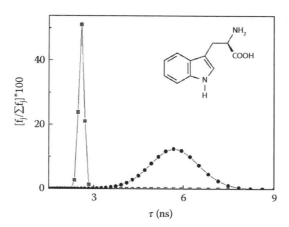

FIGURE 7.5 Fluorescence lifetime distributions for Tryptophan as a single molecule (grey data) and bound to Albumin (black data).

---

[5] Edwin Thompson Jaynes (1922, Waterloo, Iowa–1998) wrote extensively on statistical mechanics and foundations of probability and statistical inference. In 1957, Jaynes proposed the idea of maximum entropy for inferring distributions from data, starting from the observation that the statistical entropy looks the same as Shannon's information uncertainty. According to Jaynes, the "least biased" guess at distribution is to find the distribution of highest entropy.

The distribution in grey is relative to isolated tryptophan molecules in phosphate buffer solution; the distribution in black is relative to the tryptophan as a residue bound to the protein Albumin. The fluorophore tryptophan exhibits a much broader distribution of lifetimes within the macromolecule since it experiences many more micro-environment, due to the many conformers of Albumin (Gentili et al. 2008).

7.2.4 GLYCOLYSIS

The nonlinear behavior of allosteric enzymes can give rise to oscillations. An example appears in glycolysis. Glycolysis is a fundamental metabolic process because it produces chemical energy under the shape of ATP and NADH (Nicotinamide Adenine Dinucleotide) by breaking down glucose ($C_6H_{12}O_6$) into pyruvate ($C_3H_3O_3^-$). A key role in glycolysis is played by the allosteric enzyme Phosphofructokinase (PFK), having distinct regulatory and catalytic sites and exhibiting heterotropic interactions (Schirmer and Evans 1990). PFK is involved in the step of the glycolysis where a phosphate group is transferred from ATP to fructose-6-phosphate (F-6-P, see Figure 7.6).

ATP can bind to both the catalytic and the regulatory site of PFK. When it binds to the catalytic site, it plays as a substrate and phosphorylates F-6-P. On the other hand, when the concentration of the cellular energy vector ATP is high, the cell does not need to convert further glucose into pyruvate. Therefore, a regulatory process of negative feedback takes place: ATP binds to the regulatory site of PFK and lowers the affinity of the enzyme for its other substrate, i.e., F-6-P. ATP inhibits the reaction shown in Figure 7.6. Even fructose-1,6-bisphosphate (F-1,6-bP), i.e., the product of the reaction, binds to the effector site of PFK. Unlike ATP, F-1,6-bP actives the enzyme. F-1,6-bP participates in an autocatalytic step. The autocatalysis is evident if we look at the elementary steps shown in Figure 7.7. In such elementary steps, PFK(ATP) represents the complex between PFK with ATP in its active site. PFK[(ATP)+(F-6-bP)] is the complex of PFK that has both ATP and F-6-P in its active site. (F-1,6-bP)PFK[(ATP)+(F-6-P)] is the enzyme having ATP and F-6-P in its active site and the product (F-1,6-bP) in its regulatory site.

Such autocatalysis favors oscillations that have been observed to have periods of the order of several minutes (Das and Busse 1985). The steps involving PFK are represented schematically in Figure 7.8.

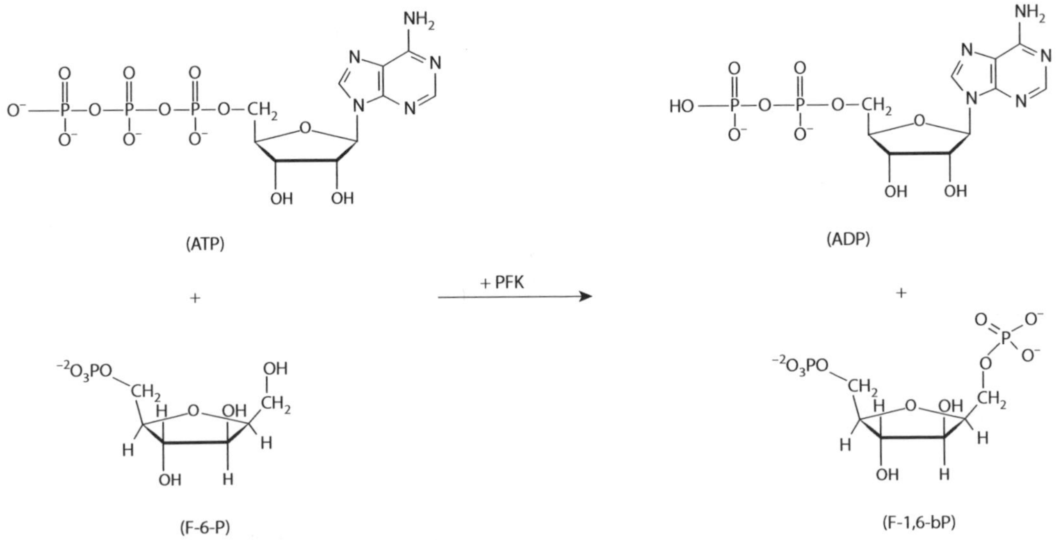

FIGURE 7.6   Reaction of phosphorylation of fructose-6-phosphate (F-6-P) catalyzed by PFK.

$$F\text{-}6\text{-}P + PFK\ (ATP) \rightleftharpoons PFK\ [(ATP) + (F\text{-}6\text{-}P)]$$

$$F\text{-}1,6\text{-}bP + PFK\ [(ATP) + (F\text{-}6\text{-}P)] \rightleftharpoons (F\text{-}1,6\text{-}bP)\ PFK\ [(ATP) + (F\text{-}6\text{-}P)]$$

$$(F\text{-}1,6\text{-}bP)\ PFK\ [(ATP) + (F\text{-}6\text{-}P)] \longrightarrow PFK + 2\ (F\text{-}1,6\text{-}bP) + ADP$$

$$F\text{-}6\text{-}P + PFK\ (ATP) + F\text{-}1,6\text{-}bP \longrightarrow PFK + 2\ (F\text{-}1,6\text{-}bP) + ADP$$

FIGURE 7.7 The autocatalysis of F-1,6-bP.

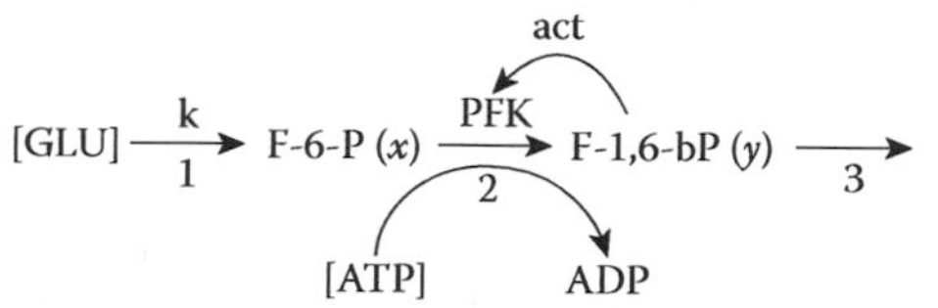

FIGURE 7.8 Scheme of the glycolytic steps involving PFK. [GLU] is the concentration of glucose, which is assumed to be fixed as that of [ATP]. F-6-P and F-1,6-bP are the *x* and *y* variables, respectively, appearing in equation [7.15].

The general differential equations for the mechanism of Figure 7.8 are

$$\frac{dx}{dt} = v_1 - v_2(x, y) = X(x, y)$$

$$\frac{dy}{dt} = v_2(x, y) - v_3(y) = Y(x, y)$$

[7.15]

where:

x is F-6-P

y is F-1,6-bP.

The rate of F-6-P formation is v_1; $v_2(x, y)$ is the rate of the autocatalytic step; $v_3(y)$ is the rate of F-1,6-bP consumption. The collection of the signs of the four partial derivatives (X_x, X_y, Y_x, Y_y) is referred to as the character of the reaction scheme. Thus, it is easy to determine that $X_x = -v_{2|x}$, $X_y = -v_{2|y}$, $Y_x = v_{2|x}$ and $Y_y = v_{2|y} - v_{3|y}$. It derives that the determinant of the Jacobian is always positive:

$$det(J) = X_x Y_y - X_y Y_x = -v_{2|x}\left(v_{2|y} - v_{3|y}\right) + v_{2|x} v_{2|y} = v_{2|x} v_{3|y} > 0.$$

[7.16]

The trace is

$$tr(J) = X_x + Y_y = -v_{2|x} + v_{2|y} - v_{3|y}.$$

[7.17]

We have stationary oscillations when $tr(J) = 0$, i.e., when $Y_y = -X_x$. In other words, it is required that the self-coupling terms are equal in absolute value but with opposite character.

The character of the oscillatory reaction may be qualitatively expressed by a net flux diagram where the large grey arrows represent the collected sets of pathways that produce and remove the chemicals (see Figure 7.9).

When the contour conditions impose an inhibitory self-coupling term also to the variable *y* (i.e., $Y_y = v_{2|y} - v_{3|y} < 0$), the trace of *J* becomes negative, and glycolysis exhibits damped oscillations. If, on the other hand, $|Y_y| > |X_x|$, the trace of *J* becomes positive and the oscillations become sustained.

TRY EXERCISE 7.1

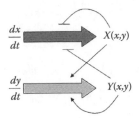

FIGURE 7.9 Net diagram flux for a system of two variables giving rise to stationary oscillations. A traditional arrow, like →, means activation, whereas the symbol ––| means inhibition.

7.3 CELLULAR SIGNALING PROCESSES

All living organisms struggle to survive in an ever-changing environment. They need to collect information about the outside world. They regulate their internal machinery for adapting to the external environment. The fundamental internal control mechanisms are mediated by conformational changes of proteins receiving either physical (like electromagnetic waves, mechanical forces, and thermal energy) or chemical stimuli. Such conformational changes of proteins trigger a cascade of enzymatic reactions, called signal transduction pathway, which ultimately leads to a regulatory or sensing effect. In fact, molecular signaling processes operate within the cell in metabolic regulation, among cells in hormonal and neural signaling, and between an organism and the environment in sensory reception.

7.3.1 THE SIMPLEST SIGNAL TRANSDUCTION SYSTEM

The most straightforward model of signaling event (Ferrell and Xiong 2001) is shown in Figure 7.10.

In Figure 7.10, the receptive signaling protein, R, is converted to its activated form $R*$ by a stimulus enzyme S. Moreover, $R*$ is turned back to the original state R through an inactivating enzyme P. For instance, S may be a protein kinase that transfers phosphate groups from ATP to R, producing a phosphorylated R. The latter is its activated state, $R*$. The addition of a phosphate group to an amino acid residue usually turns a hydrophobic portion of a protein into a polar and extremely hydrophilic one. In this way, it can introduce a conformational change in the structure of the protein via interaction with other hydrophobic and hydrophilic residues in the protein. P will be a phosphatase, which removes the phosphate groups and reverts the signaling protein to its inactivated state, R.

The differential equation describing how the concentration of $R*$ changes over time will be

$$\frac{d\left[R^{*}\right]}{dt} = k_{1}\left[S\right]\left[R\right] - k_{-1}\left[P\right]\left[R^{*}\right]$$

[7.18]

where k_{1} and k_{-1} are the kinetic constants of the forward and backward reactions, respectively. Introducing the mass balance for R ($[R]_{tot} = [R] + [R^{*}]$) in [7.18] and indicating the ratio $[R^{*}]/[R]_{tot}$ as $\chi_{R^{*}}$, we obtain

$$\frac{d\chi_{R^{*}}}{dt} = k_{1}\left[S\right]\left(1 - \chi_{R^{*}}\right) - k_{-1}\left[P\right]\chi_{R^{*}}$$

[7.19]

$$R \underset{+P}{\overset{+S}{\rightleftharpoons}} R^{*}$$

FIGURE 7.10 The most straightforward transduction mechanism.

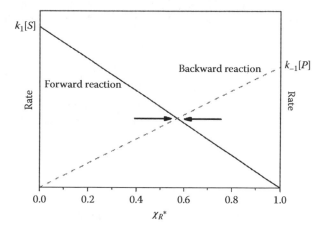

FIGURE 7.11   Rate balance plot. The rate of the forward reaction (continuous black straight line) and the rate of the backward reaction (grey dashed straight line) are plotted as a function of χ_{R^*}. The intersection point represents the stable steady state.

If we plot the rates of the forward and backward reactions as a function of χ_{R^*}, we have two straight lines (see Figure 7.11). Their intersection point represents the steady state of the signaling system. Such steady state is stable. In fact, if it is perturbed by an increase of χ_{R^*}, the back-reaction rate is larger than the forward rate, and χ_{R^*} decreases until it reaches the $(\chi_{R^*})_{ss}$ value. Likewise, if χ_{R^*} is decreased, the forward rate becomes higher, and the system will return to the initial steady state. As the values of $k_1[S]$ and $k_{-1}[P]$ are fixed, there is only one steady state. This result means that the system is monostable.

TRY EXERCISE 7.2

7.3.2   SIGNAL TRANSDUCTION SYSTEMS WITH POSITIVE FEEDBACK

A slightly more complicated signal transduction system involves a positive feedback loop (Figure 7.12).

The positive feedback action may occur when R^* increases the activity of S, or when R is phosphorylated not only by S but also by R^* itself in the process of autophosphorylation.

The differential equation, describing how the concentration of R^* changes over time, contains one term more if compared to the previous case—the rate of feedback. If we assume that the feedback action is based on an autophosphorylation process

$$R + R^* \xrightarrow{\ k_f\ } 2R^* \tag{7.20}$$

we have:

$$\frac{d\left[R^*\right]}{dt} = k_1\left[S\right]\left[R\right] + k_f\left[R\right]\left[R^*\right] - k_{-1}\left[P\right]\left[R^*\right] \tag{7.21}$$

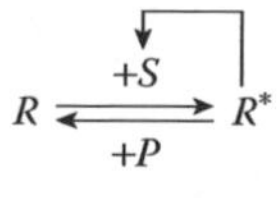

$$R \underset{+P}{\overset{+S}{\rightleftarrows}} R^*$$

FIGURE 7.12   Scheme of signal transduction with feedback.

Introducing the mass balance for R ($[R]_{tot} = [R] + [R^*]$) in [7.21] and indicating the ratio $[R^*]/[R]_{tot}$ as χ_{R^*}, we obtain:

$$\frac{d\chi_{R^*}}{dt} = k_1[S] + \chi_{R^*}\left(k_f[R]_{tot} - k_1[S]\right) - k_f[R]_{tot}\,\chi_{R^*}^2 - k_{-1}[P]\chi_{R^*} \qquad [7.22]$$

The rate of forward reaction expressed as a function of χ_{R^*} is an inverted parabola (see Figure 7.13), whereas the rate of backward reaction is a straight line. When the term $k_1[S]$ is not negligible, the two curves intersect at only one point. In other words, we have just one steady state; the system is monostable. Such steady state is stable, as you can easily infer by looking at Figure 7.13.

TRY EXERCISE 7.3

A more refined model for signal transduction with linear feedback includes a Michaelis-Menten interaction between R^* and P. The mechanism is shown in Figure 7.14.

The differential equation relative to χ_{R^*} becomes:

$$\frac{d\chi_{R^*}}{dt} = k_1[S] + \chi_{R^*}\left(k_f[R]_{tot} - k_1[S]\right) - k_f[R]_{tot}\,\chi_{R^*}^2 - \frac{k_{h2}C_P\chi_{R^*}}{\left(\chi_{R^*}[R]_{tot} + K_{M2}\right)} \qquad [7.23]$$

where:

    K_{M2} is the Michaelis-Menten constant
    C_P is the total concentration of P.

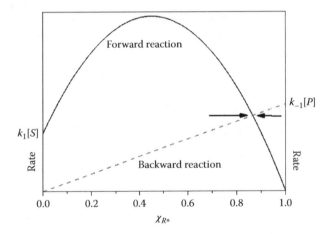

FIGURE 7.13 Rate balance plot for a signaling system with linear feedback. The rate of forward reaction (continuous black curve) and the rate of backward reaction (grey dashed straight line) are plotted as a function of χ_{R^*}. The intersection point represents the stable steady state.

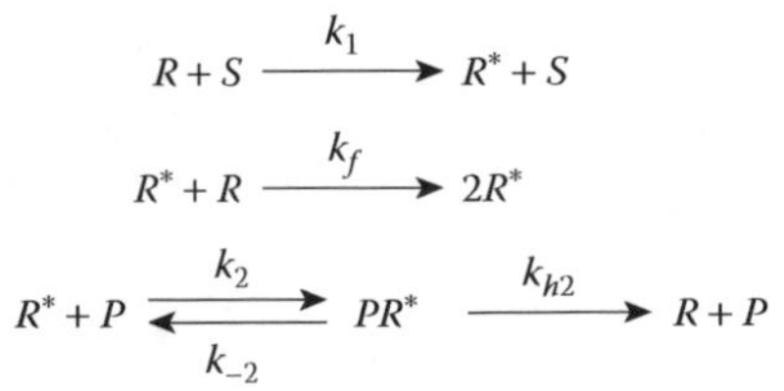

$$R + S \xrightarrow{\ k_1\ } R^* + S$$

$$R^* + R \xrightarrow{\ k_f\ } 2R^*$$

$$R^* + P \underset{k_{-2}}{\overset{k_2}{\rightleftarrows}} PR^* \xrightarrow{\ k_{h2}\ } R + P$$

FIGURE 7.14 Scheme of signal transduction with feedback and a Michaelis-Menten interaction between R^* and P.

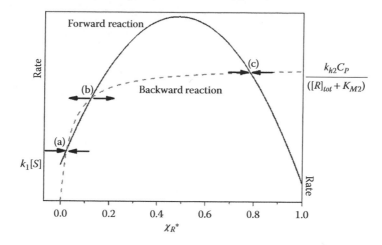

FIGURE 7.15 Rate balance plot for a signaling system with linear feedback and a Michaelis-Menten backward reaction. The rate of forward reaction (continuous black curve) and the rate of backward reaction (grey dashed curve) are plotted as a function of χ_{R^*}. The intersection points (a), (b) and (c) represent the steady states.

The analysis of the rate balance plot (see Figure 7.15) shows that when the stimulus level is low, and the rate of the back reaction is high, the latter can do more than keep up with the first increments of feedback. For higher χ_{R^*}, the back reaction is overwhelmed by the feedback since it continues to rise. Finally, for very high χ_{R^*} values, the back reaction turns to be the fastest step. It is evident that the forward and backward rate curves intersect at three points. Such points are three stationary states: those relative to the lowest and highest values of χ_{R^*} (labeled as (a) and (c), respectively) are stable, whereas the intermediate one, (b), is unstable, as you can easily infer by looking at the rate balance plot. The signaling system described by the mechanism of Figure 7.14 is bi-stable. If the system is anywhere to the left of the (b) state, it will settle into the (a) state; if it is anywhere to the right of the (b) state, it will settle into the (c) state.

TRY EXERCISE 7.4

In exercise 7.4, we observed that the system of Figure 7.14 is bi-stable only for a limited range of the kinetic parameters. For example, if the term $k_1[S]$ is very high, the system will become monostable.

Another condition favorable for having a bi-stable signaling system is a feedback action that is a sigmoidal function of $[R^*]$. Such sigmoidal feedback action, called ultrasensitive, is based on a nonlinear Hill equation relationship (see equation [7.10]) between the amount of R^* produced and the rate of production of more R^*.[6] The differential equation describing how χ_{R^*} changes over time becomes

$$\frac{d\chi_{R^*}}{dt} = k_1[S](1-\chi_{R^*}) + \frac{k_h[R^*]^n}{[R^*]^n + K}(1-\chi_{R^*}) - k_{-1}[P]\chi_{R^*} \qquad [7.24]$$

The rate balance plot reveals that for properly balanced forward and backward reactions, the signaling system is bi-stable (see Figure 7.16). In fact, the curves representing the rates of the forward and backward reactions cross in three points: (a'), (b'), and (c'). The extreme fixed points, (a') and (c'),

---

[6] The sigmoidal feedback action, also known as ultra-sensitivity, can be obtained by many mechanisms. For more information about ultra-sensitivity, go to the "Hints for further reading" (paragraph 7.9).

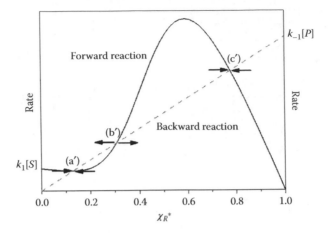

FIGURE 7.16 Rate balance plot for a signaling system with sigmoidal feedback and linear back reaction. The rate of forward reaction (continuous black curve) and the rate of backward reaction (grey dashed straight line) are plotted as a function of χ_{R^*}. The intersection points ([a'], [b'] and [c']) represent the steady states.

are stable, whereas the intermediate fixed point, (b'), is unstable, as it can be easily inferred by analyzing the rate balance plot.

The latter two models of signaling systems have in common the property to be bi-stable. They can reside in either the (a) ([a']) stable state or the (c) ([c']) stable state of Figures 7.15 and 7.16. Does it exist a mechanism for such signaling systems to make a transition from one state to the other? One way to send the system from the (a) ([a']) state to the (c) ([c']) state is to increase the stimulus S continuously. This possibility is demonstrated graphically for the signaling system with ultrasensitive sigmoidal feedback in Figure 7.17. Figure 7.17a is a rate-balance plot showing a family of five curves for the forward reaction, each corresponding to a different level of the continuously variable stimulus. The lowest curve corresponds to the case of no stimulus; the other curves correspond to successive increments of stimulus. Let us suppose that the stimulus is initially zero, and the system starts in the (a') state, having $[R^*]$ equal to zero. It would take a mighty perturbation to drive the system out of the (a') state, past the unstable fixed point (b'), and close to the (c') state, as we can

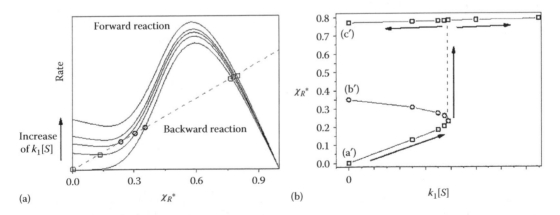

FIGURE 7.17 Effect of the amount of stimulus on the stability of the signaling system with sigmoidal feedback. In (a) we have the rate balance plot with five curves for the forward reaction plotted for different amounts of the stimulus. The intersection points between the forward and back reactions are indicated by squares (for the stable fixed points) and circles (for the unstable fixed points). In (b) the stimulus-response curve is plotted: the squares are relative the stable fixed points of the (a') and (c') type; the circles are the unstable (b') fixed points.

imagine by looking at Figure 7.17b where the χ_{R^*} values of the fixed points are plotted as function of the arbitrary amount of stimulus. Adding stimulus to the system, the curve of the forward reaction skews upward. This change of the curve shifts the (a') state slightly upward and the unstable fixed point (b') slightly down. The (a') and (b') states approach each other. On the other hand, the (c') state shifts upward. These shifts of the fixed points continue by adding more stimulus. There exists a value of the stimulus where the curve of the forward reaction and the straight line of the back-reaction cross in two points: one corresponding to the overlap of the (a') and (b') states and the other corresponding to the stable (c') state. For any further increment of the stimulus, the (a') and (b') states do not longer exist. In such situation, the system leaves the (a') state and reaches the (c') state. What happens if we now lower the stimulus back down? The coexistence of the three states will reappear. But there will be no driving force for the system to leave the (c') state and make the transition back to the (a') state. The (c') state is stable, and the system is stuck in it. Thus, the behavior of the signaling system is different when the stimulus is rising from the one we get when the stimulus is falling (see the black arrows in Figure 7.17b). The system exhibits path dependence that is called hysteresis. For our signaling system, the hysteresis is so substantial that the transition from the (a') state to the (c') state results irreversible.

The hysteresis avoids that the signaling system repeatedly switches back and forth between two chemical states ([a'] and [c']) representing two distinct logical states. This mechanism is suitable for a type of biochemical memory. Unless something happens to fundamentally break the sigmoidal positive feedback or the linear positive feedback combined with a saturable back reaction, systems like those of Figures 7.16 and 7.15 can remain in the (c') and (c) states, indefinitely. Bi-stable cell signaling systems appear to be responsible for the digital, all-or-none character of various cell fate induction processes, and progression from one discrete phase of the cell cycle to another. This mechanism is used by cells to "remember" that they are differentiated long after the differentiation stimulus has been withdrawn, and even long after all of the protein molecules that make up the feedback loop have been replaced by new protein molecules (Xiong and Ferrell 2003; Lisman 1985).

7.4 EPIGENETIC EVENTS

A cell may be described as a room having as many light switches as there are genes. For instance, the human cell is a room with around 22,000–25,000 switches (Ayarpadikannan et al. 2015). A strategy a cell uses to adapt to many conditions is based on altering its genes expression without necessarily changing its DNA sequence, for example, through the so-called "epigenetic" events. The cell can respond to the environmental signals by synthesizing appropriate proteins. For instance, let us assume that Y is the protein synthesized in response to the signal S. For the synthesis of Y, it is necessary that proteins, known as transcription factors (TFs), bind the regulatory region of DNA that precedes the gene coding the target protein Y. Each transcription factor (TF) can bind the DNA to regulate the rate at which the target genes are read. The Hill mechanism describes the association of a transcription factor TF_i to the promoter site of DNA:

$$n_i(TF_i) + DNA \rightleftarrows (DNA)(TF_i)_{n_i} \qquad [7.25]$$

The relative amount (θ) of DNA linked to TF_i will be:

$$\theta = \frac{\left[(DNA)(TF_i)_{n_i}\right]}{[DNA] + \left[(DNA)(TF_i)_{n_i}\right]} = \frac{[TF_i]^{n_i}}{[TF_i]^{n_i} + K_D} = \frac{[TF_i]^{n_i}}{[TF_i]^{n_i} + K_i^{n_i}} \qquad [7.26]$$

where:
  K_D is the thermodynamic dissociation constant for the equilibrium [7.25]
  $K_D = K_i^{n_i}$.

The activated genes are transcribed into messenger RNA (mRNA) through the enzyme called RNA polymerase (RNAp). In a process called transcription, each base pair of the gene is read sequentially, and a corresponding nucleotide is added to the growing single-stranded mRNA molecule. Once prepared, the mRNA must then be decoded to synthesize a protein, in a process called translation. The first step of the translation is the binding of the ribosome to the mRNA strand. Once bound to the mRNA, the ribosome reads the nucleotide sequence of the mRNA and catalyzes the reaction that builds the protein Y one amino acid at a time. Y will act on the environment in response to signal S (Alon 2007). Of course, for each of the above steps, there are other sub-steps. For instance, during transcription, polymerase must read from single-stranded DNA. This process requires that the double-stranded DNA must first be unwound before polymerase can function. These processes consist of other sub-sub-steps, and so on. Due to the large number of steps, a reasonable approximation of the overall dynamics is the system of two differential equations describing the synthesis of protein Y as

$$
\frac{d[\mathrm{mRNA}]}{dt} = k_{tr} \prod_i \left(\frac{[TF_i]^{n_i}}{[TF_i]^{n_i} + K_i^{n_i}} \right) - k_d [\mathrm{mRNA}]
$$

$$
\frac{d[Y]}{dt} = k_{tl} [\mathrm{mRNA}] - k_h E_T \left(\frac{[Y]}{K_M + [Y]} \right)
$$

[7.27]

In [7.27], k_{tr} and k_{tl} are the kinetic constants of RNA transcription and translation, respectively. The kinetic constant k_d is relative to the degradation processes of RNA. Transcription factors, TF_i, can act as either activators ($n_i > 0$) or inhibitors ($n_i < 0$), increasing or reducing the transcription rate of a gene. The second term of the differential equation relative to Y represents its rate of degradation. The term E represents a protease that degrades Y; its total concentration is E_T. Its turnover rate is k_h, and its Michaelis constant is K_M. Note that because binding of TF_i to DNA is fast and often takes seconds to reach the equilibrium, whereas mRNA or protein production take minutes to hours, we can assume the concentration of bound TF_i are those of equilibrium. In Figure 7.18, a scheme of the protein synthesis mechanism described by equation [7.27] is drawn.

TRY EXERCISE 7.5

The Y protein concentration will reach a steady state value without any oscillations, corresponding to a homeostatic condition.

The presence of some form of feedback in the kinetics of protein synthesis can give rise to oscillations. For example, protein Y may inhibit the expression of its gene (see Figure 7.19a). An example is PER protein. It rules certain cellular and high-level biological rhythms having periods of about 24 hours, the so-called circadian rhythms (from the Latin *circa dies*, meaning "approximately a day"), by inhibiting the transcription factors of its own PER gene (Novák and Tyson 2008).

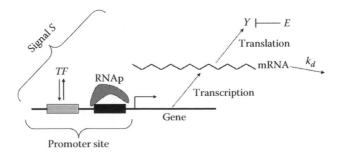

FIGURE 7.18 Scheme of the synthesis of protein Y, where the protagonists are transcription factors (TF_i), RNA polymerase (RNAp), messenger RNA (mRNA), and protease E.

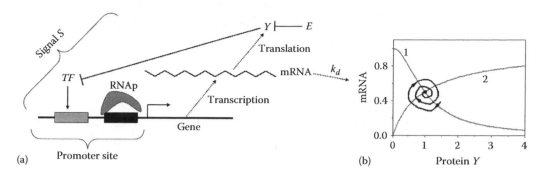

(a) Promoter site   (b)   Protein Y

FIGURE 7.19 Synthesis of protein Y with a negative feedback effect on its transcription factor (TF): scheme of the mechanism in (a) and representative solution[7] of the equations [7.28] in (b).

Knowing that PER protein (symbol Y) binds its transcription factor according to the following chemical equation $nY + TF = Y_n(TF)$ and that protease E degrades PER protein, the differential equations describing the dynamics of the system are:

$$\frac{d[\text{mRNA}]}{dt} = k_{tr}C_{TF}\left(\frac{K^n}{K^n + [Y]^n}\right) - k_d[\text{mRNA}]$$

$$\frac{d[Y]}{dt} = k_{tl}[\text{mRNA}] - k_h E_T\left(\frac{[Y]}{K_M + [Y]}\right)$$

[7.28]

C_{TF} is the total concentration of TF and $(K^n/K^n + [Y]^n)$ represents the fraction of TF not bound to Y. $K^n = K_D$, where K_D is the dissociation constant for the binding process of Y to TF. The other terms have the same meaning seen in equation [7.27].

The nullclines of the non-linear differential equations [7.28] are plotted in Figure 7.19b. The nullcline labeled as 1 is the locus of points where the rate of mRNA synthesis is exactly balanced by the rate of mRNA degradation. On the other hand, the nullcline labeled as 2 is the locus of points where the rate of protein synthesis is balanced by the rate of its degradation. The intersection point of the two nullclines represents the steady state.

The Jacobian of system [7.28] is

$$J = \begin{pmatrix} -k_d & -\dfrac{nk_{tr}C_{TF}K^n[Y]_{ss}^{n-1}}{\left(K^n + [Y]_{ss}^n\right)^2} \\[4ex] k_{tl} & -\dfrac{k_h E_T K_M}{\left(K_M + [Y]_{ss}\right)^2} \end{pmatrix}$$

[7.29]

Its trace is always negative; its determinant is always positive. When the discriminant, that is $\Delta = (tr(J))^2 - 4det(J) = (k_d - k_h E_T K_M/(K_M + [Y]_{ss})^2)^2 - 4nk_{tl}k_{tr}C_{TF}K^n[Y]_{ss}^{n-1}/(K^n + [Y]_{ss}^n)^2$, is negative, the steady state is a stable spiral. Although the system [7.28] may exhibit damped oscillations on the way to the steady state, sustained oscillations in this simple gene regulatory circuit are impossible, and Y reaches a homeostatic steady state.

---

[7] The representative solution for the system of differential equations [7.28] plotted in Figure 7.19b has been obtained with the following parameter values: $k_{tr} = k_d = 0.1$ min^{-1}, $C_{TF} = 1$, $K = 1$, $n = 2$, $K_M = 1$, $k_{tl} = 1$ min^{-1}, $k_h = 1$, $E_T = 1$ (Novák and Tyson 2008).

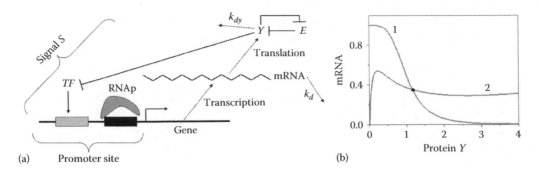

(a)   Promoter site   (b)

FIGURE 7.20 Synthesis of protein Y with a negative feedback effect on its transcription factor (TF) and a negative feedback effect on its degradation process by protease E: scheme of the mechanism in (a) and representative solution[8] of the equations [7.31] in (b).

Undamped oscillations emerge when the negative feedback effect of Y on its gene expression is accompanied by a positive feedback effect, such as the binding of Y to an allosteric site of protease E inhibiting the activity of E (see Figure 7.20a):

$$2Y + E \xrightleftharpoons{} Y_2 E \quad \text{with } K_A = \left([Y_2 E]/[E][Y]^2\right). \tag{7.30}$$

The differential equations describing the dynamics of the system are:

$$\frac{d[\text{mRNA}]}{dt} = k_{tr} C_{TF}\left(\frac{K^n}{K^n + [Y]^n}\right) - k_d[\text{mRNA}]$$

$$\frac{d[Y]}{dt} = k_{tl}[\text{mRNA}] - k_{dY}[Y] - k_h E_T\left(\frac{[Y]}{K_M + [Y] + K_M K_A [Y]^2}\right) \tag{7.31}$$

In these equations, k_{dY} is the rate constant for an alternative pathway of Y degradation. The nullclines of the non-linear differential equations [7.31] are plotted in Figure 7.20b. The nullcline labeled as 1 is the locus of points where the rate of mRNA synthesis is exactly balanced by the rate of mRNA degradation. The nullcline labeled as 2 is the locus of points where the rate of protein synthesis is balanced by the rate of its degradation. The term of positive feedback gives rise to a kink in the nullcline. The intersection point of the two nullclines represents the steady state.

The Jacobian of the system [7.31] is

$$J = \begin{pmatrix} -k_d & -\dfrac{n k_{tr} C_{TF} K^n [Y]_{ss}^{n-1}}{\left(K^n + [Y]_{ss}^n\right)^2} \\[4ex] k_{tl} & \left(-k_{dY} - \dfrac{k_h E_T K_M \left(1 - K_A [Y]_{ss}^2\right)}{\left(K_M + [Y]_{ss} + K_M K_A [Y]_{ss}^2\right)^2}\right) \end{pmatrix} \tag{7.32}$$

---

[8] The representative solution for the system of differential equations [7.31] plotted in Figure 7.20b has been obtained with the following parameter values: $k_{tr} = k_d = k_{dY} = 0.05$ min^{-1}, $C_{TF} = 1$, $K = 1$, $n = 4$, $K_M = 0.1$, $k_{tl} = 1$ min^{-1}, $k_h = 1$, $E_T = 1$, $K_M K_A = 2$ (Novák and Tyson, 2008).

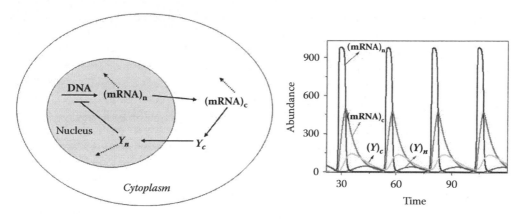

FIGURE 7.21 Mechanism of Y protein synthesis taking into account transport of macromolecules between the nucleus and the cytoplasm (a). Plot of the sustained oscillations for mRNA and Y in cytoplasm and nucleus (b).[9]

Jacobian [7.32] differs from that [7.29] in the Y_y term. When K_A is small, the positive feedback effect is negligible, and the system does not show sustained oscillations. When K_A is sufficiently large, and the negative feedback loop is sufficiently nonlinear (i.e., the n coefficient is sufficiently large), $tr(J)$ becomes null and $det(J) > 0$. Therefore, the system exhibits a limit cycle, and the net diagram flux for the system becomes qualitatively equivalent to that depicted in Figure 7.9 for glycolysis. The mechanism of negative feedback on gene expression combined with inhibition of protein degradation has been suggested as a possible source of circadian rhythms in the reaction network that governs expression of the PER gene in fruit flies (Novák and Tyson 2008). Circadian rhythms in eukaryotic cells[10] have also been modeled by invoking a mechanism involving three or more components and only a negative feedback loop on gene expression (Novák and Tyson 2008). The components may be mRNA that is synthesized in the nucleus (mRNA$_n$) and transported into the cytoplasm (mRNA$_c$) where it gets translated into protein Y_c, which is translocated into the nucleus (Y_n) and represses the activity of its gene (see Figure 7.21a). The four-variable negative-feedback loop oscillates as naturally as a pendulum (see Figure 7.21b).

7.5 BIOLOGICAL RHYTHMS

From the biochemical examples investigated in this chapter, it is evident that living organisms have a natural tendency to exhibit a mixture of homeostatic and oscillatory behaviors: the more complex the biological system, the greater the possibilities. In fact, as far as the oscillatory behavior is concerned, outputs of various frequencies, phases, and amplitudes are possible. Multicellular organisms have cells that specialize in different functions depending on the organ they belong. Each type

---

[9] The results plotted in Figure 7.21b have been achieved assuming that mRNA is synthesized in the nucleus (xn) and transported into the cytoplasm (xc), where it gets translated into protein (yc), which is translocated into the nucleus (yn). If we impose that eps = Vnuc/Vcyt; half-life of mRNA in nucleus = 0.693/kdxn; half-life of protein in cytoplasm = 0.693/kdyc, the differential equations are:

dxn/dt = kdxn*(sig/(1 + yn^p)—xn)—kexport*xn
dxc/dt = eps*kexport*xn—kdxc*xc
dyc/dt = kdyc*(xc − yc)—eps*kimport*yc
dyn/dt = kimport*yc—kdyn*yn/(Km + yn)

The constants appearing in the four differential equations have the following values: sig = 1000, p = 2, kdxn = 10, kexport = 0.2, kdxc = 0.2, eps = 1, kdyn = 8, kdyc = 0.1, Km = 0.1, kimport = 0.1 (Novák and Tyson 2008). Initial conditions: xn = 1.0, xc = 30.0, yc = 200.0, yn = 120.0.

[10] A eukaryote is an organism whose cells contain a nucleus and other organelles enclosed within membranes.

of cell triggers specific "clock genes." In fact, each living being appears as "a house with clocks in every room and every corner; but, yet in one way or another, they work in an organized way," as rightly alleged by Derk-Jan Dijk, director of the Surrey Sleep Research Centre in Guildford (UK). All the periodic processes occurring at either the cellular, organ, or physiological system levels span more than ten orders of magnitude of timescale, from hundredths of seconds up to tens of years[11] (Goldbeter, 2017). They can be sorted out into three time-domains or rhythms: the (1) circadian, (2) ultradian, and (3) infradian rhythms. The circadian processes are ruled by the periodic revolution of the earth around its axis, requiring 24 hours, and determining a significant daily variation in the intensity and spectral composition of the electromagnetic radiation coming from the sun. The infradian processes depend on the slower revolution of the moon around our planet (requiring about one month), or the even slower revolution of the earth around the sun (requiring about one year). Finally, there are organs and physiological systems that have ultradian rhythms, meaning they work cyclically, with periods shorter than 24 hours. A list of ultradian, circadian, and infradian processes found in human beings, is reported in Table 7.1.

Among the ultradian periodic processes, individual neurons detain the record of the fastest cycles possible. In fact, many nerve cells generate impulses of transmembrane depolarization, followed by hyperpolarization, in around 1 millisecond or so (for more details, see Chapter 9). Neurons use such impulses as signals that must travel very fast for effective communication. In the human hearth, the electrical impulses last much longer than in nerve cells. In fact, the heart rate is about 60 beats per minute, so there is a whole second for the impulse to occur and for the heart to recover before next beat. Respiratory rhythm requires the cooperation of many nerve cells, peripheral and central, and it has a frequency of about tenths of Hz. Connected with respiration, there is also the nasal cycle (Atanasov 2014). Our nose holds two parallel breathing passages that are divided by the septum. Our two nostrils shift their workload back and forth in a delicate dance called the nasal cycle. At any moment, most of the air we inhale travels through just one nostril, while a much smaller amount seeps in through the other. This difference is important because odor-causing chemicals vary in the amount of time they take to dissolve through the mucus that lines our nasal cavity. The two nasal cavities are like two chromatographic columns separating and detecting many odor-molecules. Chemicals that dissolve quickly have the strongest effect in a fast-moving airstream that spreads them out over as many odor receptors as possible. But compounds that dissolve slowly are more accessible to smell in a slow-moving airstream. Due to the circadian oscillations of environmental

---

TABLE 7.1

Periodic Processes in Human Beings at the Organ and Physiological System Levels Sorted out According to Their Periods (T) or Frequencies (v)

Ultradian Rhythms ($T < 24$ h)	Circadian Rhythms ($T \approx 24$ h)	Infradian Rhythms ($T > 24$ h)
Nerve cell ($v \approx 1000$ Hz)	Sleep-wake; Hunger	Human ovulation (circalunar)
Heart beating ($v \approx 1$ Hz)	Bowel movements	
Respiration ($v \approx 0.3$ Hz)	Body temperature	Seasonal mutations (circannual)
Nasal cycle ($T \approx 2.8$ h)	Blood pressure	

---

[11] Magicicada species spend most of their 13-year or 17-year lives underground feeding on the roots of trees in the United States. After 13 or 17 years, mature cicada nymphs emerge in the springtime, synchronously and in tremendous numbers. The adults are active for about four to six weeks, and they reproduce. Then, the lifecycle is complete, the eggs have been laid, and the adult cicadas disappear for another 13 or 17 years.

light and temperature, many signaling and cellular metabolic processes, having intrinsic periods of about 24 hours, tend to entrain with the day-night cycle. There is a central pacemaker that gives human body a sense of the time of day: the suprachiasmatic nucleus, a group of 10,000 or so neurons in the hypothalamus, located at the very bottom of the brain, deep within the skull. It is the blue light absorbed by melanopsin (a protein present in the ganglion cells of our retina) that provides the primary cue for the entrainment of our circadian rhythms.[12] In the presence of daily light that is particularly rich in blue, excitatory signals are sent to the suprachiasmatic nucleus, which prevents the pineal gland from producing melatonin, the hormone that induces to sleep (Eisenstein 2013). Light-dark entrainment of the suprachiasmatic nucleus synchronizes peripheral oscillators throughout the entire body and controls the pace of variation of other physiological responses, such as body temperature, blood pressure, blood levels of hormones, appetite, and gut motility that induces bowel movements. The peripheral oscillators may be phase-shifted by physical activity or by altering meal times, but the light seems to be the most crucial determinant of rhythms driven by the suprachiasmatic nucleus. Of course, for humans and all the other creatures living outside the equatorial zones, the seasonal change in the daily pattern of light/dark exposure has repercussion on the peripheral oscillators. Hence, circannual rhythms are present in sleep, mood, reproduction, diseases, and deaths for humans (Swaab et al. 1996). In other living species, circannual rhythms may be more relevant, because they involve relevant processes such as migration, hibernation, or flowering, to make just a few examples (Gwinner 1986).

Finally, despite the common belief that our mental health and other behaviors, such as the menstrual cycle, are modulated by the phase of the moon, is persistent, there is no convincing evidence that human biology is in any way regulated by the lunar cycle (Foster and Roenneberg 2008).

7.6 AMPLIFICATION AND ADAPTATION IN REGULATORY AND SENSORY SYSTEMS

Two relevant additional properties of the biological sensing and regulatory systems are amplification and adaptation. Amplification is the ability to generate amplified responses to low levels of stimuli. Adaptation is the ability to adapt to constant backgrounds of stimuli (Koshland et al. 1982). Specific signals, such as light of low intensity, a faint sound, the aroma due to a few odor molecules, do not have the energy to generate a behavioral response and must be amplified within the organism. When we talk about amplification of a signal, we may mean two distinct events: (1) magnitude amplification and (2) sensitivity amplification.

7.6.1 MAGNITUDE AMPLIFICATION

Magnitude amplification is the production of some output molecules far higher than the elementary particles of the stimulus; it trusts in chain reactions. For example, in vision, the absorption of one photon can be amplified almost 5,000-fold and becomes perceptible. The first step in human vision is the absorption of one photon by the chromophore of the photoreceptor protein rhodopsin, which is 11-*cis* retinal. The excited state of 11-*cis* retinal isomerizes to all-*trans* retinal (see Figure 7.22) in 200 fs and with a quantum yield of $\Phi = 0.67$. The isomerization is so fast that is vibrationally coherent (Wang et al. 1994).

The isomerization of retinal induces conformational rearrangement of the protein embedding the chromophore. The rhodopsin assumes a signaling state that has a lifetime of about 50 ms. One "activated" rhodopsin has the power of activating about 800 G-proteins. Like "activated" rhodopsin, an activated G-protein has a brief lifetime where it can interact with phosphodiesterase (PDE)

---

[12] Blind people, who do not have cones and rods to orient themselves spatially, are still able to orient temporally if they are provided with melanopsins in the ganglion cells.

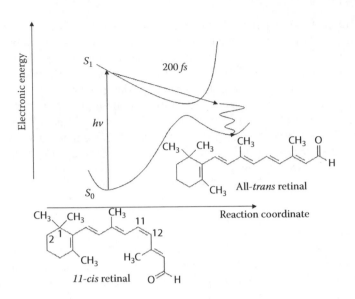

FIGURE 7.22   Scheme of the first step of human vision: the ultrafast photo-isomerization of 11-*cis* retinal to all-*trans* retinal.

and activates it. The latter occurs without any amplification. One activated PDE can catalyze the breakdown of more than one molecule of 3′,5′-cyclic guanosine monophosphate (cGMP) to 5′-guanosine monophosphate (5′-GMP). Usually, six cGMP molecules are converted to 5′-GMP by each activated PDE. Regarding the number of molecules, the absorption of one photon has been amplified about $800 \times 6 = 4800$-fold. Activation of one rhodopsin molecule takes ≈ 4800 cGMP molecules out of circulation. They are not anymore available to maintain the cGMP-gated Na$^+$ channels open, and about 200 Na$^+$ channels close, which is $\approx 2\%$ of the total number of ion channels in a photoreceptor rod cell. Closing 2% of the channels reduces the ionic current, flowing inside the cell, by 2%. The corresponding voltage change across the membrane is ≈ 1 mV hyperpolarization (it means the interior becomes more negative than the exterior). The absorption of just one photon determines an electrical effect felt throughout the entire rod. The photo-induced hyperpolarization of the photosensitive cell reduces the rate of the release of the neurotransmitter glutamate (Oyster 1999).

7.6.2   Sensitivity Amplification

The sensitivity amplification refers to the percentage change in response compared to the percentage change in the stimulus. Three types of mechanisms for sensitivity amplification have been uncovered. One involves a stimulus-protein interaction that is allosteric with a large Hill coefficient. Such interactions are ultrasensitive, with positive cooperativity. A second mechanism for obtaining high sensitivity amplification is to have the same effector at several steps in a pathway. In principle, if an effector S participates in n elementary steps of a multistep pathway, we might expect that the rate of the overall chemical transformation is proportional to $[S]^n$. A third mechanism, shown in Figure 7.23, involves the interconversion of a key enzyme in active (I_a) and inactive (I_i) forms by the cyclic coupling of covalent modification and de-modification reactions. An example of modification and de-modification are the phosphorylation and dephosphorylation reactions, already mentioned in this chapter.

Assuming that the concentration of ATP is maintained at a constant level that is several orders of magnitude higher than the substrate enzymes, the reactions for the interconversion of I_i to I_a and vice versa can be written as follows:

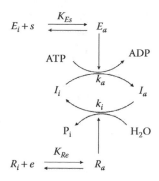

FIGURE 7.23 A cyclic phosphorylation and dephosphorylation of enzyme *I* catalyzed by two distinct enzymes, *E* (a protein kinase) and *R* (a protein phosphatase), respectively; *s* represents a stimulus and *e* is an effector for *R*.

$$E_a + I_i \xrightleftharpoons{K_a} (E_a I_i) \xrightarrow{k_a} E_a + I_a$$

$$R_a + I_a \xrightleftharpoons{K_i} (R_a I_a) \xrightarrow{k_i} R_a + I_i$$

[7.33]

At the stationary state, $k_a[(E_a I_i)] = k_i[(R_a I_a)]$. We make the following assumptions. First, the equilibria in the formation of the enzyme-effector and enzyme-enzyme complexes are reached very fast. Second, the concentrations of the enzyme-enzyme complexes are negligible compared to the concentrations of the active and inactive enzymes. Third, the concentrations of *s* and *e* are at constant levels for any signaling state. It derives that, at the stationary state,

$$k_a K_a [E_a][I_i] = k_i K_i [R_a][I_a]$$

[7.34]

From the mass balances for *E* and *R* and the assumptions listed above, we obtain (for the meaning of the symbols, see also Figure 7.23)

$$[E_a] = \left(\frac{K_{Es}[s][E]_{tot}}{K_{Es}[s]+1} \right) \text{ and } [R_a] = \left(\frac{K_{Re}[e][R]_{tot}}{K_{Re}[e]+1} \right)$$

[7.35]

From the equations [7.34], [7.35], and the assumption that $[I]_{tot} \approx [I_a] + [I_i]$, a steady-state expression for the fraction of the modified interconvertible enzyme is:

$$\frac{[I_a]}{[I]_{tot}} = \frac{k_a K_a [E_a]}{k_i K_i [R_a] + k_a K_a [E_a]}$$

$$= \frac{1}{1 + \frac{k_i K_i}{k_a K_a} \left(\frac{K_{Re}[e][R]_{tot}}{K_{Re}[e]+1} \right) \left(\frac{K_{Es}[s]+1}{K_{Es}[s][E]_{tot}} \right)}$$

[7.36]

The high sensitivity amplification of an interconvertible enzyme derives from the fact that any one of the ten terms in equation [7.36] can be altered by allosteric interaction with one or more allosteric effectors. The curves in Figure 7.24 show the enormous indirect effect of the signal [*s*] on the fractional activation of the interconvertible enzyme. Curves A and B have been obtained maintaining fixed K_{Es} and with two-fold changes of the other parameters, favoring the activation of *I* (the numerical values

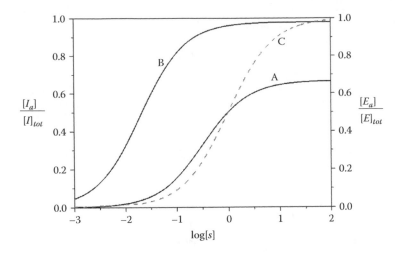

FIGURE 7.24 Ratios of activated interconvertible enzyme I and E as a function of $[s]$ (plotted in logarithmic scale). Curve A and B refer to I and they have been obtained with $K_{Es} = 1$ (in both A and B); $K_{Re} = 1$ (in A) and ½ (in B); $k_a K_a = 1$ (in A) and 2 (in B); $k_i K_i = 1$ (in A) and ½ (in B), $[E]_{tot} = 1$ (in A) and 2 (in B); $[R]_{tot} = 1$ (in A) and ½ (in B), $[e] = 1$ (in A) and ½ (in B). The dashed grey curve C refers to enzyme E.

of the parameters are listed in the caption of Figure 7.24) (Stadtman and Chock 1977). Such changes promote a significant upward shift (from 0.66 to 0.98) of the ratio $[I_a]/[I]_{tot}$, and a downward shift in the concentration of the stimulus s required to produce 50% of the activation of the interconvertible enzyme (log[s] drops from 0.0016 in A to −1.67 in B). It is noteworthy that even in the more unfavorable conditions of curve A, the concentrations of s required to activate indirectly the interconvertible enzyme, when the ratio $[I_a]/[I]_{tot}$ is less than 0.5, are less than those required for comparable activations of the converter enzyme E (compare curve A with the dashed grey curve C).

The results shown in Figure 7.24 demonstrate the high sensitivity amplification offered by a monocyclic cascade system (whose mechanism is depicted in Figure 7.23) consisting of two oppositely directed "converter" enzymes that catalyze the interconversion of a single interconvertible enzyme. It has been demonstrated that bicyclic and multicyclic cascade systems are endowed with considerably greater amplification potential (Chock and Stadtman 1977). The great amplification potential stems from the fact the response of the last interconvertible enzyme in a cascade to a primary stimulus, caused by the allosteric interaction of an effector to the first converter enzyme in the cascade, is a multiplicative function of various parameters in the cascade.

7.6.3 ADAPTATION

A highly sensitive amplification system would cause problems because living beings are almost constantly bombarded by stimuli, which could saturate the sensory system. The organism prevents this by the adaptation that tends to inactivate the sensing apparatus. There are two possible adaptive responses to changes in stimuli. They are (I) absolute and (II) partial adaptation. In absolute adaptation after a transient response, the system adapts absolutely. This behavior happens when a stimulus s (see Figure 7.25a) activates the rate of formation (v_f) of the response regulator X rapidly and the rate of X decomposition (v_d) slowly (see Figure 7.25b). A change in s produces a transient signal that adapts absolutely as shown in Figure 7.25c. When s is unchanging, even though present at a high level, it does not trigger any response. Absolute adaptation occurs in sensory systems that must detect small changes such as in the visual system and bacterial chemotaxis (Koshland 1980). The response regulator could be a small molecule or a combination of small molecules, a protein, a combination of proteins or a protein-ligand complex.

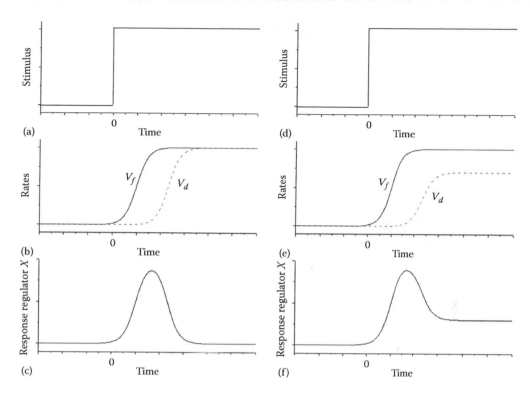

FIGURE 7.25 Behavior of an absolute adaptive system (left column: graphs a, b, and c) and a partial adaptive system (right column: graphs d, e, and f).

In some systems, absolute adaptation may not be needed. For example, hormones provide stimuli from one cell to another. After a short period of stimulation, the hormone level reduces because it is no longer produced in the primary organ and is washed away from the bloodstream. In such case, the partial adaptation can provide a large initial signal that is partially damped to prevent excessive stimulation. Partial adaptation occurs when X inhibits the rate of its decomposition through a slow feedback action (see Figure 7.25d–f).

From Figure 7.25 it is evident that the adaptation action eliminates the ability to detect absolute levels of stimuli, but only in order to increase sensitivity to percentage changes in the background over a wide range of background intensities.

7.7 KEY QUESTIONS

- Why are proteins so important for cells?
- Describe the possible mechanisms of enzyme-substrate interactions.
- Describe the features of allosteric proteins.
- What is the power of the Maximum Entropy Method?
- How does glycolysis originate oscillations?
- Describe the signal transduction model without feedback.
- Describe the signal transduction model with linear feedback
- Describe the signal transduction model with linear feedback and Michaelis-Menten de-activation reaction.
- Describe the signal transduction model with ultrasensitive feedback.
- What is an epigenetic event, and which are the protagonists?
- Which are the favorable conditions to observe oscillations in an epigenetic event?
- Make examples of biological rhythms.

- Explain the mechanism of magnitude amplification.
- Which are the mechanisms for having sensitivity amplification?
- How do we adapt to stimuli?

7.8 KEY WORDS

Proteins; Metabolism; Signal Transduction Pathway; Ultrasensitivity; Hysteresis; Epigenesis; Homeostasis; Ultradian, circadian and infradian rhythms; Amplification in sensing; Adaptation in sensing.

7.9 HINTS FOR FURTHER READING

- Walsh et al. (2018) pinpoint eight compounds as essential ingredients of metabolism within a cell. Seven of them use group transfer chemistry to drive otherwise unfavorable biosynthetic equilibria. They are ATP (for phosphoryl transfers), acetyl-CoA and carbamoyl phosphate (for acyl transfers), S-adenosylmethionine (for methyl transfers), Δ^2-isopentenyl-PP (for prenyl transfers), UDP-glucose (for glucosyl transfers), and NAD(P)H/NAD(P)$^+$ (for electron and ADP-ribosyl transfers). The eighth key metabolite is O_2.
- The possible mechanisms for generating ultra-sensitivity can be learned by reading the series of three papers by Ferrell and Ha (2014a–c).
- If you want to deepen the subject of biochemical oscillators and cellular rhythms, you can read Goldbeter (1996); Goldbeter and Caplan (1976); Ferrell et al. (2011); Glass and Mackey (1988).
- A delightful book, about the action of the human brain on physiological processes, such as hunger, thirst, sex, and sleep, is that by Young (2012).

7.10 EXERCISES

7.1. Compare mechanism A, B, and C of Figure 7.26 and determine which one admits $tr(J) = 0$ as possible solution.

7.2. Consider the simplest model of a signaling event (Figure 7.10). Determine its response curve, i.e., how $(\chi_{R^*})_{ss}$ changes with the ratio $k_1[S]/k_{-1}[P]$.

7.3. Regarding the model of signaling process with linear feedback depicted in Figure 7.12, determine the expression of the maximum for the forward rate. In case the term $k_1[S]$ is negligible, which is the expression of the response curve?

7.4. Build the rate balance plot and the stimulus-response curve for the signaling system with linear feedback and a back reaction that becomes saturated (based on the Michaelis-Menten mechanism). Such system is described by the mechanism of Figure 7.14. Assume that the values of the parameters appearing in equation [7.23] are: $k_f[R]_{tot} = 0.7$ t^{-1}; $(k_{h2}C_P/[R]_{tot}) = 0.16$ t^{-1}; $K_{M2}/[R]_{tot} = 0.029$. Plot the curves of the forward reaction for the following values of the term $k_1[S]$: 0; 0.04; 0.073; 0.093.

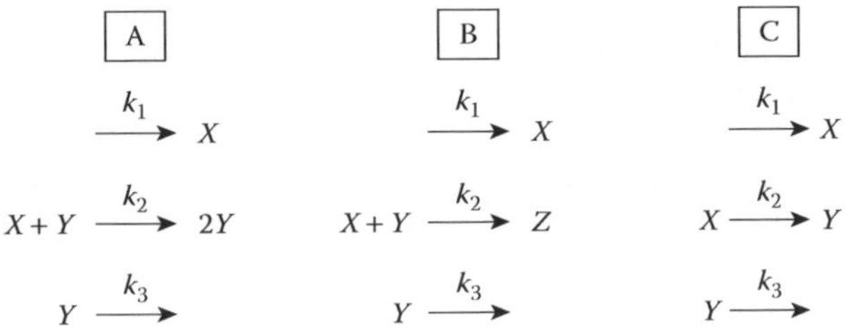

FIGURE 7.26 Three distinct mechanisms.

7.5. Determine the properties of the stationary state for the system of two differential equations [7.27] describing how the [mRNA] and [Y] change over time. Assume, for simplicity, that there is only one *TF*, its association coefficient is $n_i = 1$ and its concentration [*TF*] is fixed.

7.11   SOLUTIONS TO THE EXERCISES

7.1. Regarding mechanism A, the differential equations regarding *X* and *Y* are:

$$\frac{d[X]}{dt} = v_1 - k_2 [X][Y]$$

$$\frac{d[Y]}{dt} = k_2 [X][Y] - k_3 [Y]$$

The terms of the Jacobian are

$$X_x = -k_2 [Y]_{ss}; \; X_y = -k_2 [X]_{ss}; \; Y_x = k_2 [Y]_{ss}; \; Y_y = k_2 [X]_{ss} - k_3.$$

The determinant $\det(J) = k_3 k_2 [X]_{ss} > 0$.

The trace is $tr(J) = k_2 ([X]_{ss} - [Y]_{ss}) - k_3$ can be positive, negative or null.

Regarding mechanism B, the differential equations are

$$\frac{d[X]}{dt} = v_1 - k_2 [X][Y]$$

$$\frac{d[Y]}{dt} = -k_2 [X][Y] - k_3 [Y]$$

Therefore, the terms of the Jacobian are

$$X_x = -k_2 [Y]_{ss}; \; X_y = -k_2 [X]_{ss}; \; Y_x = -k_2 [Y]_{ss}; \; Y_y = -k_2 [X]_{ss} - k_3.$$

The determinant $det(J) = k_3 k_2 [X]_{ss} > 0$.

The trace is $tr(J) = -k_2 ([X]_{ss} + [Y]_{ss}) - k_3$ can be only negative.

Regarding mechanism C, the differential equations are

$$\frac{d[X]}{dt} = v_1 - k_2 [X]$$

$$\frac{d[Y]}{dt} = k_2 [X] - k_3 [Y]$$

The terms of the Jacobian are

$$X_x = -k_2; \; X_y = 0; \; Y_x = k_2; \; Y_y = -k_3.$$

The determinant $det(J) = k_3 k_2 > 0$.

The trace is $tr(J) = -k_2 - k_3$ can be only negative.

Only mechanism A, having an autocatalytic step, can give a null trace of Jacobian and hence a limit cycle as a possible stable solution.

7.2. The response curve describes the dependence of $(\chi_{R^*})_{ss}$ on the ratio $k_1[S]/k_{-1}[P]$. To obtain such a relation, we impose equation [7.19] equal to zero. After a few simple mathematical steps, we obtain

$$\left(\chi_{R^*} \right)_{ss} = \frac{1}{1 + \dfrac{k_{-1}[P]}{k_1[S]}}$$

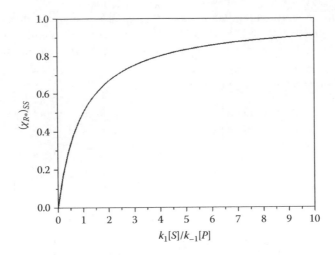

FIGURE 7.27 Response curve for the most straightforward signaling system.

This equation is a hyperbolic function. It is shown in Figure 7.27. Note that when the ratio $k_1[S]/k_{-1}[P]$ is 1, $(\chi_{R^*})_{ss}$ is equal to 0.5, that is 50% of $[R]_{tot}$ is in its activated state. Moreover, $(\chi_{R^*})_{ss}$ goes asymptotically to 1.

7.3. The equation relative to the forward rate (v_{for}) for a signaling system with linear feedback is

$$v_{for} = k_1[S] + \chi_{R^*}\left(k_f[R]_{tot} - k_1[S]\right) - k_f[R]_{tot}\,\chi_{R^*}^2$$

To find its maximum as a function of χ_{R^*}, we calculate the first derivative of v_{for} with respect to χ_{R^*}. We obtain

$$\frac{dv_{for}}{d\chi_{R^*}} = \left(k_f[R]_{tot} - k_1[S]\right) - 2k_f[R]_{tot}\,\chi_{R^*}$$

It is equal to zero when $\chi_{R^*} = \dfrac{\left(k_f[R]_{tot} - k_1[S]\right)}{2k_f[R]_{tot}}$.

In case $k_f[R]_{tot} \gg k_1[S]$, $\chi_{R^*} \approx 0.5$.

When $k_1[S]$ is negligible when compared to the other terms of equation [7.22], the overall rate of R^* formation becomes:

$$\frac{d\chi_{R^*}}{dt} \approx \chi_{R^*}\left(k_f[R]_{tot}\right) - k_f[R]_{tot}\,\chi_{R^*}^2 - k_{-1}[P]\chi_{R^*}$$

In such case, the equation of the response curve is

$$\chi_{R^*} \approx 1 - \frac{k_{-1}[P]}{k_f[R]_{tot}}.$$

7.4. In the case of a signaling system with a linear feedback and a back reaction with saturation (based on a Michaelis-Menten mechanism), a continuous growth of the stimulus from zero value, shifts the system from a stable fixed point to a new stable one in an irreversible

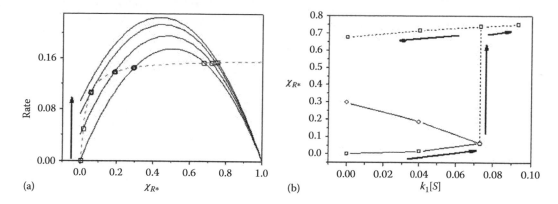

FIGURE 7.28 Rate balance plot (a) and stimulus-response curve (b) for a signaling system with linear feedback and a Michaelis-Menten back reaction. In both (a) and (b), the stable and unstable fixed points are indicated by squares and circles, respectively.

manner (This behavior is also exhibited in the case of a signaling system with sigmoidal positive feedback. Read the next part of paragraph 7.3.2). The rate balance plot and the stimulus-response curve built through the values of the exercise are plotted in Figure 7.28.

7.5. Based on the assumptions of the exercise, the differential equations [7.27] become:

$$\frac{d[\text{mRNA}]}{dt} = k_{tr}\left(\frac{[TF]}{[TF]+K}\right) - k_d[\text{mRNA}] = 0$$

$$\frac{d[Y]}{dt} = k_{tl}[\text{mRNA}] - k_h E_T\left(\frac{[Y]}{K_M+[Y]}\right) = 0$$

The Jacobian for the system is:

$$J = \begin{pmatrix} -k_d & 0 \\ k_{tl} & -\dfrac{k_h E_T K_M}{\left(K_M+[Y]_{ss}\right)^2} \end{pmatrix}$$

The trace and the determinant of J are:

$$tr(J) = -k_d - \frac{k_h E_T K_M}{\left(K_M+[Y]_{ss}\right)^2} < 0$$

$$\det(J) = \frac{k_d k_h E_T K_M}{\left(K_M+[Y]_{ss}\right)^2} > 0$$

The discriminant $\Delta = \left[\left(tr(J)\right)^2 - 4\det(J)\right] = \left(k_d - \frac{k_h E_T K_M}{\left(K_M+[Y]_{ss}\right)^2}\right)^2 > 0$

These conditions correspond to a steady-state solution that is a stable node.

8 The Emergence of Temporal Order in a Chemical Laboratory

Experiment is the interrogation of Nature.

Robert Boyle (1627–1691 AD)

8.1 INTRODUCTION

Oscillations are everywhere in nature: in the Universe, in our Solar System, in our planet, in ecosystems, in the economy, and inside every living being, as we have learned in the previous chapters. It was a great surprise to observe oscillations also in chemical laboratories because chemical oscillations were imagined as a sort of perpetual motion machines in a beaker, in sharp contradiction with the Second Law of Thermodynamics.

8.2 THE DISCOVERY OF OSCILLATING CHEMICAL REACTIONS

The first report of chemical oscillations in a lab was made by Robert Boyle in the late seventeenth century. At that time, Boyle was investigating the luminescence of phosphorus in the gaseous phase, in liquid solutions, and in the solid phase (Harvey 1957). He found with wonder that the oxidation of phosphorous produces flashes of light. The second report didn't occur until the beginning of the nineteenth century when A. T. Fechner (1828) described an electrochemical cell producing an oscillating current. About seventy years later, in the late 1890s, J. Liesegang (1896) discovered periodic precipitation patterns in space and time (first mentioned by F. F. Runge in 1855), and Ostwald (1899) observed that the rate of chromium dissolution in acid periodically increased and decreased. But all these experiments, involving heterogeneous systems, did not undermine the widespread idea that a chemical reaction proceeding towards its equilibrium state cannot show oscillations.

During the early decades of the twentieth century, Lotka (1910, 1920a, 1920b) published a handful of theoretical papers on chemical oscillations. His models inspired ecologists. The most famous one is named as the Lotka-Volterra model, which is used to characterize the predator-prey interaction in an ecosystem as we have seen in Chapter 5.

The report of the first homogeneous isothermal chemical oscillator is ascribed to Bray (1921), and Bray and Liebhafsky (1931), who studied the catalytic decomposition of hydrogen peroxide by the iodic acid-iodine couple. The mixture consisted of H_2O_2, KIO_3, and H_2SO_4 and the reactions were:

$$5H_2O_2 + I_2 = 2HIO_3 + 4H_2O \qquad [8.1]$$

$$5H_2O_2 + 2HIO_3 = 5O_2 + I_2 + 6H_2O \qquad [8.2]$$

where hydrogen peroxide plays as both an oxidizing and reducing agent.

Bray serendipitously observed that the concentration of iodine and the rate of oxygen evolution vary almost periodically. He offered no explanation for the oscillation other than referring to Lotka's theoretical work. He suggested that autocatalysis was involved in the mechanism of its reaction. Bray's results triggered more works devoted to debunking them than to explaining it mechanistically (Nicolis and Portnow 1973). Most chemists tried to prove that the cause of the oscillations was some unknown heterogeneous impurity.

The next experimental observation of an oscillating reaction was made again accidentally by the Russian biophysicist Boris Belousov. During the Second World War, he was working on projects in radiation and chemical warfare protection. By investigating methods for removing toxic agents from the human body, Belousov developed a keen interest in the role of biochemical metabolic processes (Kiprijanov 2016). After the war, part of the Belousov's research included the Krebs cycle. He was looking for an inorganic analog of the citric acid cycle.[1] In 1950, he studied the reaction of citric acid with bromate and ceric ions (Ce^{+4}) in sulfuric acid. He expected to see the monotonic depletion of the yellow ceric ions into the colorless cerous (Ce^{+3}) ions. Instead, he astonishingly observed that the color of the solution repeatedly oscillated with a period of one minute or so between pale yellow and colorless. Belousov also noted that if the same reaction was carried out unstirred in a graduated cylinder, the solution exhibited awesome traveling waves of yellow. Belousov carefully characterized the phenomenon and submitted a manuscript containing the recipe in 1951 to the Soviet periodical *Journal of General Chemistry*. Unfortunately, his paper was quickly rejected because the phenomenon described by Belousov was in contradiction with the Second Law of Thermodynamics. Belousov attempted a second submission to the Russian *Kinetics and Catalysis* in 1955. But his manuscript was rejected again for the same reason—phenomenon like that could not occur. Although Belousov furnished a recipe and photographs of the different stages of the oscillations, the evidence was judged insufficient. Therefore, he decided to work more on that reaction, and he eventually settled for publishing a short abstract in the unrefereed proceedings of a conference on radiation medicine (Belousov 1958). Belousov's recipe circulated among Moscow laboratories, and in 1961 Anatol Zhabotinsky, a graduate student in biophysics at Moscow State University, began to investigate the reaction. Zhabotinsky was pursuing postgraduate research on biorhythms. His supervisor, Simon E. Shnoll, recommended that he investigate Belousov's reaction. Simon E. Shnoll was a professor of biochemistry at Moscow State University and one of the few scientists aware of the Belousov's results at that time (Kiprijanov 2016). Zhabotinsky replaced citric acid with malonic acid and used the redox indicator ferroin to sharpen the color change during the oscillations. In fact, ferroin gave a more drastic red-blue color change than the cerous-ceric pair. He noticed that when, the currently called, Belousov-Zhabotinsky (BZ) reaction is performed in an unstirred thin layer of solution, it spontaneously gives rise to a "target pattern" or spirals of oxidized blue ferritin in an initially homogeneous red ferroin-dominated solution. Zhabotinsky presented some of his results in a conference on "Biological and Biochemical Oscillators" that was held in Prague in 1968. The meeting motivated many chemists of the Eastern bloc to study the reaction, and since the proceedings were published in English (Chance et al. 1973), the news of the oscillating reaction could also spread among the Western chemical community.

Meanwhile, Prigogine and his group in Brussels developed a model, dubbed the Brusselator, which showed homogeneous oscillations and propagating waves like those seen in the BZ system. The Brusselator was relevant because it demonstrated that an open chemical system kept far from equilibrium can exhibit spontaneous self-organization by dissipating energy to the surroundings. Prigogine named as "dissipative structures" all the temporal oscillations and the spatial structures that emerge in very far from equilibrium conditions. When a reaction like the BZ one, is carried out in a closed reactor, it always reaches the equilibrium; it can exhibit only transitory oscillations as it approaches equilibrium.

---

[1] The citric acid cycle, also known as the Krebs cycle, is a collection of key chemical reactions by which all aerobic organisms generate energy through the oxidation of acetate derived from carbohydrates, fats, and proteins. In eukaryotic cells, the citric acid cycle occurs in the matrix of the mitochondria, whereas in prokaryotic cells, it occurs in the cytoplasm.

Stationary oscillations require an open system. In 1977, Nicolis and Prigogine summarized the results of the Brussels group in a book titled *Self-Organization in Nonequilibrium Systems*. Ilya Prigogine was awarded the 1977 Nobel prize in chemistry for his contribution to the nonequilibrium thermodynamics.

Despite the proof offered by the abstract Brusselator model, still missing to legitimize the study of oscillating reactions was a detailed chemical mechanism for a real oscillating reaction, like the BZ one. A crucial step in this direction was achieved by Field, Körös, and Noyes (FKN) who succeeded in explaining the qualitative behavior of the BZ reaction using the grounding principles of chemical kinetics and thermodynamics that rule "normal" non-oscillating chemical reactions. They published their mechanism consisting in twenty or so elementary steps, known nowadays as FKN mechanism, in 1972 (Field et al. 1972). The FKN mechanism was rather large and heavy for computational work. A few years later, Field and Noyes managed to simplify their FKN mechanism; they abstracted a three-variable model, the "Oregonator," that maintained the essential features of the entire BZ reaction (Field and Noyes 1974a).

Finally, in the 1970s, many chemists were convinced that oscillating chemical reactions were possible, genuine, and even very appealing. From the two parent oscillating reactions, the Bray and the BZ ones, many variants were developed, changing some of the ingredients. In 1973, two high school chemistry teachers in San Francisco, Briggs and Rauscher, organized a surprising classroom demonstration to draw the students' attention by combining the malonic acid and a metal ion catalyst (Mn^{+2}), typical of the BZ reaction, with the ingredients of the Bray reaction (iodate and hydrogen peroxide). They showed to their students how this magic mixture could spontaneously change color from amber to deep blue to colorless, and again to amber, repeating several cycles until the solution remains deep blue indefinitely, due to the attainment of the equilibrium.

In the mid-1970s, two groups, one at Brandeis University headed by Epstein and Kustin and the other at the Paul Pascal Research Center in Bordeaux led by Pacault, strove to find out a systematic approach to discovering chemical oscillators. Two members of the Bordeaux group, De Kepper and Boissonade, proposed an abstract model to explain how oscillations may be obtained by perturbing a bistable chemical system in a Continuous-flow Stirred Tank Reactor (CSTR)[2] (Boissonade and De Kepper 1980). Then, De Kepper joined the Brandeis group, and within a few months, the new team developed the first systematically designed oscillator: the arsenite-iodate-chlorite reaction (De Kepper et al. 1981a, 1981b). The approach was then refined and exploited to discover dozens of new chemical oscillators spanning much of the periodic table. At the turn of the twentieth and twenty-first century, the implementation of chemical oscillators has been spurred by the achievements of synthetic biology (Purcell et al. 2010). The fundamental goal of synthetic biology is to understand the principles of biological circuitry from an engineering perspective to synthesize biological circuits both *in vivo* and *in vitro*. Achieving this goal requires a combination of computational and experimental efforts. Much attention has been focusing on the synthesis of gene regulatory networks. So far, the known chemical oscillators may be grouped into five classes: (1) natural biological oscillators within living cells (such as the circadian rhythm we studied in Chapter 7); (2) synthetic oscillators engineered into living organisms; (3) biological oscillators reconstructed *in vitro*; (4) synthetic oscillators involving bio-molecules in cell-free reactions; (5) synthetic oscillators involving chemical compounds but not bio-molecules.

8.3 THE SYSTEMATIC DESIGN OF CHEMICAL OSCILLATORS

The essential requirements for any chemical system to oscillate are:

1. Far-from-equilibrium conditions;
2. Processes of positive and negative feedback that interplay with a time delay or an adequately delayed negative feedback process.

---

[2] Look back to Chapter 3 for a definition of a CSTR.

Feedback means "self-influence." In a chemical reaction, we have feedback when a product affects the rate at which one or more species involved in the same mechanism are produced or consumed. The feedback is positive when the self-influence accelerates the rate of the reaction, whereas it is negative when it slows down the rate.

The positive feedback destabilizes the steady states. The simplest form is the direct autocatalysis. Let us consider the elementary reaction

$$A \xrightarrow{k_r} X \tag{8.3}$$

which is accelerated by its own product X according to the following trimolecular autocatalytic reaction [8.4]:

$$A + 2X \xrightarrow{k_{r'}} 3X \tag{8.4}$$

Of course, k_r' is much greater than k_r. Let us suppose that the overall chemical transformation of A to X is carried out in a CSTR. The differential kinetic equation describing how the concentration of A changes over time will be:

$$\frac{d[A]}{dt} = k_0 \left(C_0 - [A]\right) - k_r [A] - k_r' [A][X]^2 \tag{8.5}$$

where:
   k_0 is the flow rate,
   C_0 is the total analytical concentration of A.

Expressing $[X]$ as $C_0 - [A]$, equation [8.5] becomes

$$\frac{d[A]}{dt} = k_0 \left(C_0 - [A]\right) - [A]\left\{k_r + k_r' \left(C_0 - [A]\right)^2\right\} \tag{8.6}$$

If we divide both terms of equation [8.6] by C_0 and we indicate the ratio ($[A]/C_0$) with a, we obtain

$$\frac{da}{dt} = k_0 \left(1 - a\right) - a\left\{k_r + k_r' C_0^2 \left(1 - a\right)^2\right\} \tag{8.7}$$

At the steady state, $k_0(1-a) = a\{k_r + k_r' C_0^2(1-a)^2\}$. Note that the term on the left represents a straight line, whereas the term on the right is a cubic function. Depending on the values of k_0, k_r, k_r' and C_0, we can have either one or two or three solutions of steady state. When the flow rate is very high, we have only one steady state solution (a_{SS}) characterized by a high value of a (see Figure 8.1a). Such solution refers to a stable steady state because when a is larger than a_{SS}, the rate of A consumption dominates, whereas when $a < a_{SS}$, it is the flow rate to dominate (see the inset in Figure 8.1a). For smaller values of k_0, we have three solutions of steady state. Two of them represent stable steady states, whereas one is unstable as we can easily infer by looking at the plot of Figure 8.1b. With a further reduction of k_0, only a stable steady state, characterized by a low concentration of A, is possible as we can notice by looking at Figure 8.1c. When k_0 goes to zero, also $[A]$ approaches zero because the CSTR becomes similar to a closed system, and the reaction can almost reach the equilibrium, i.e., the complete depletion of A according to our mechanism. If we plot the a_{SS} values as a function of k_0, we obtain a graph that illustrates the important phenomena of bistability and hysteresis (see Figure 8.1d). The grey squares define an arm that is called thermodynamic, or equilibrium branch, because it extends from the equilibrium condition we have at zero flow rate. The black circles define a second arm that is called the flow branch because it is traced starting from

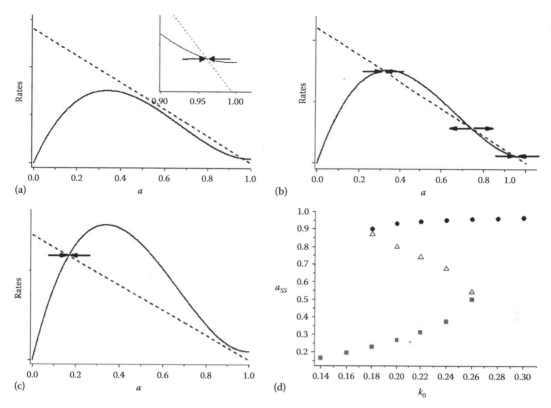

FIGURE 8.1 Steady-state solutions for the chemical system described by equations [8.3], [8.4], working in a CSTR. The cases of high, medium and low flow rates are shown in (a), (b) and (c), respectively. The steady-state solutions (a_{SS}) as a function of k_0 are plotted in (d).

high flow rates. The blank triangles are relative to the solutions of unstable steady states. When the system admits three possible solutions, which state the system resides in depends on from where it comes. The system exhibits hysteresis—it remembers its history.

TRY EXERCISE 8.1

After finding out positive feedback, oscillations may arise by introducing antagonistic negative feedback. The positive feedback pushes the system far away from the initial state, whereas the delayed coupled negative feedback tends to restore the initial state. Therefore, the chain of reactions can start again. Let us assume that species Y exerts a negative feedback effect on the system described by equations [8.3 and 8.4]. The negative feedback effect consists in depleting X. Let us assume that the independent differential equations describing the dynamics of the systems are:

$$\frac{dx}{dt} = f(x, y) = (1-x)(k_r + k_r' C_A^2 x^2) - k_0 x - k_f y$$

$$\frac{dy}{dt} = \frac{1}{\tau} g(x, y)$$

[8.8]

where:
    $x = ([X]/C_A)$, $y = ([Y]/C)$, the term $k_f y$ represents the negative feedback on x;
    τ is a large time constant.

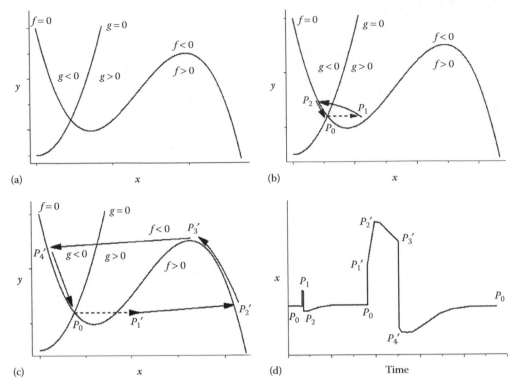

FIGURE 8.2 (a) Phase space for a system of two variables, x and y, originating an excitable state. The continuous arrows in (b) and (c) trace the dynamical path of the system after the perturbation highlighted by the dashed arrows. The temporal evolution of the variable x after the perturbations is plotted in (d).

The presence of a large τ is a way of specifying that x changes much more quickly than y. We get insight on the dynamical behavior of the system if we plot the x and y nullclines (i.e., the curves on which the rate of change of each variable is zero) on their phase space (see Figure 8.2). The x-nullcline is S-shaped because $f(x, y)$ is a cubic equation, and y-nullcline intersects it once. The intersection of the two nullclines represents the steady state of the system. If we know the signs of the derivatives on either side of each nullcline, we have a qualitative picture of the flow in the phase space. The time derivatives of the variables change sign as we cross the nullclines. Two situations are worthwhile to be dealt with and are presented in the next two paragraphs.

8.3.1 EXCITABILITY

The first situation is when the steady-state is on the left or right branches of the S-shaped curve. This steady state is stable because all trajectories in its neighborhood point in toward the intersection of the two nullclines. If the system undergoes a small perturbation, for instance from P_0 to P_1 of Figure 8.2b, the system reacts jumping back to the x-nullcline at P_2 with a small variation in y because the latter one changes much more slowly. Finally, the system moves along the x-nullcline, from P_2 to the original steady state P_0. If another perturbation is applied to the steady-state P_0, so strong to reach point P_1' (see Figure 8.2c), where f is positive, the effect of the perturbation spontaneously grows up to the right-hand branch of the x-nullcline with a small change in the value of y. As soon as the system reaches the x-nullcline in P_2', it starts to move up the nullcline as y increases (in fact we are in a region where $g > 0$), until it attains the maximum in P_3'. From P_3', the system jumps quickly to the left-hand branch of the x-nullcline, and finally it slides along the nullcline back to the

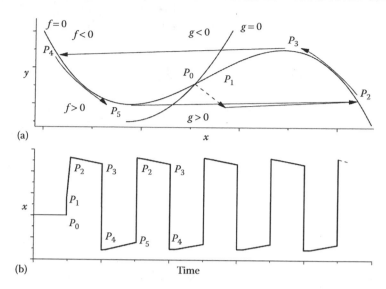

FIGURE 8.3 Representation of the relaxation oscillation in the $x - y$ phase plane in (a); temporal evolution of the x variable in (b).

steady state. If we follow x as a function of time, we record a trace like that shown in Figure 8.2d. The phenomenology depicted in Figure 8.2 is called excitability. A system is excitable if perturbing it in a small amount, it rapidly returns to the steady state, but if it is given a significant perturbation, it makes a large amplitude excursion before returning to the initial state. An excitable system has two essential characteristics: (I) a threshold of excitation and (II) a refractory period. The threshold of excitation is the smallest value of perturbation that causes the system to make a large excursion before recovering the resting state. The refractory period is the time needed to recover the steady state after a perturbation, which is the time that must elapse before the system is ready for receiving another stimulation.

8.3.2 Oscillations

The second important situation is shown in Figure 8.3, and it occurs when the intersection of the two nullclines lies on the middle branch of the S-shaped curve, which is an unstable steady state as inferred from the signs of the first derivatives, f and g. Even a tiny perturbation pushes the system away from the steady state: from P_0 to P_1 and suddenly to P_2 because f changes very quickly. Then, it moves more slowly along the x-nullcline until it reaches point P_3. From there, the system jumps to the other branch of the x-nullcline (point P_4). This branch is in the negative region of $g(x, y)$, so the system moves along the x-nullcline toward P_5 as y slowly decreases. After reaching P_5, the system makes a fast transition back to P_2, and the cycle repeats. It is clear that we have periodic temporal oscillations, as shown in Figure 8.3b, called relaxation oscillations. If we exclude the initial transient stage soon after the perturbation, then the dynamics is the same no matter what the perturbation is.

8.3.3 In Practice

From a practical point of view (Epstein and Pojman 1998), after finding a bistable system, confirmed by the experimental observation of a hysteresis loop, and after choosing Y as a candidate for negative feedback, we must look for two effects to happen. First, the inflow of a small amount of Y to our reactor input stream should shrink the range of bistability measured as a function of some tunable parameters, for instance, the flow rate k_0 (like in Figure 8.4). Second, although the range of bistability should narrow, the states of the two branches should retain their identities. If this occurs,

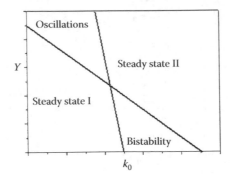

FIGURE 8.4   Cross-shaped phase diagram[3] for a bistable system where species Y is added at different flow rates, k_0.

at a certain point by increasing Y the bistability vanishes and oscillations start (see Figure 8.4). As shown in the cross-shaped phase diagram of Figure 8.4, often the region of oscillation widens as Y increases, although this may depend on the system we are studying. If by adding an even small amount of Y, the states of the two branches begin to approach one another in character and blend into one, it means that the negative feedback is too fast with respect to the reactions that give rise to the bistability. In such case, it is necessary to try different species for the negative feedback.

The first success of the methodology explained in this paragraph dates back to 1980s when De Kepper joined the Epstein's group at Brandeis University. After a careful investigation, the team proved that the chlorite-iodide reaction [8.9] is autocatalytic in iodine and gives rise to a bistable chemical system in a CSTR (Dateo et al. 1982):

$$ClO_2^- + 4I^- + 4H^+ \rightarrow Cl^- + 2I_2 + 2H_2O \qquad [8.9]$$

In the presence of excess chlorite, the iodine is oxidized further to iodate according to the following reaction:

$$5ClO_2^- + 2I_2 + 2H_2O \rightarrow 5Cl^- + 4IO_3^- + 4H^+ \qquad [8.10]$$

The introduction of arsenite generates the necessary negative feedback required to observe oscillations in a quite wide range of conditions. In fact, arsenite reacts with iodate and produces iodide (Epstein and Pojman 1998):

$$IO_3^- + 3H_3AsO_3 \rightarrow I^- + 3H_3AsO_4 \qquad [8.11]$$

It was also found that the chlorite-iodide system can oscillate by itself in a flow reactor, though in a much narrower range of conditions than the full chlorite-iodate-arsenite system (Dateo et al. 1982). In fact, reaction [8.10] and/or [8.12] may provide the necessary feedback to allow it to oscillate.

$$IO_3^- + 5I^- + 6H^+ \rightarrow 3I_2 + 3H_2O \qquad [8.12]$$

---

[3] Boissonade and De Kepper (1980) proposed an abstract model based on two independent variables summarizing the relationships between bistability and limit cycle. They discussed both the linear and nonlinear stability analysis, obtaining a cross-shaped diagram like that shown in Figure 8.4. Epstein and Luo (1991) demonstrated that the same cross-shaped phase diagram may be obtained by replacing the two variable ordinary differential equations of Boissonade and De Kepper, with a single differential delay equation in which the delayed negative feedback of the y variable is mimicked by replacing $y(t)$ in equation [8.8] with $x(t - \tau)$.

8.4 PRIMARY "OSCILLATORS"

The oscillatory chemical reactions have rather complicated mechanisms because they usually involve many species and many steps. Therefore, it usually requires a great experimental and computational effort to determine the complete set of elementary steps that specifies how an oscillatory chemical reaction occurs. Often, to elucidate the main properties of the oscillatory phenomenon, it is satisfactory to focus on the most important processes, overlook the details and propose an abstract model. For defining a taxonomy of chemical oscillators and partition all the synthetic oscillating chemical reactions in families, it is useful to find out a "core" set of reactions that produce the essential oscillatory dynamics. Each minimal set of reactions necessary to observe chemical oscillatory phenomena will be defined as a "primary oscillator." If we focus our attention on homogeneous isothermal reactions, we may sort them into five main families of "primary oscillators." The first four "primary oscillators" will show the antagonism between a positive feedback and a coupled delayed negative feedback restoring the initial state. The fifth "primary oscillator" will rely, only, on properly delayed negative feedback events.

8.4.1 OREGONATOR MODEL: THE "PRIMARY OSCILLATOR" OF COPRODUCT AUTOCONTROL

A family of chemical oscillators is based on the overall reaction of bromination and oxidation of an organic compound by acidic bromate in the presence or not of a catalyst. The catalyst is usually a one-electron redox system of the type $M^{(n+1)+}/M^{n+}$ or $[ML_m]^{(n+1)+}/[ML_m]^{n+}$ where M represents a transition metal and L is a bidentate ligand. Its redox potential is between 1.0 and 1.5 V. When a catalyst is involved in the reaction, the chemical oscillator is referred to as Belousov-Zhabotinsky (BZ) reaction. In the mid-1970s, it was found that even in the absence of a catalyst the reaction between acidic bromate and some organic molecules (like phenol's and aniline's derivatives) may exhibit oscillations (Orbán et al. 1979). In both catalyzed and uncatalyzed oscillations, a relevant role is played by the intermediate bromide ion, Br^- (Ruoff et al. 1988). In fact, the redox state of the system depends crucially on its concentration. It has been demonstrated that when the bromide concentration is higher than a critical value, $[Br^-]_c = (5 \times 10^{-6})[BrO_3^-]_c$, the system is in its reduced state and the main reaction occurring in solution is

$$BrO_3^- + 2Br^- + 3H^+ + 3RH \rightarrow 3RBr + 3H_2O \qquad [8.13]$$

wherein *RH* is the organic substrate. Reaction [8.13] derives from a bunch of four elementary steps:

$$Br^- + BrO_3^- + 2H^+ \rightarrow HOBr + HBrO_2$$
$$Br^- + HBrO_2 + H^+ \rightarrow 2HOBr$$
$$3\left(Br^- + HOBr + H^+ \rightarrow Br_2 + H_2O\right) \qquad [8.14]$$
$$3\left(Br_2 + RH \rightarrow RBr + Br^- + H^+\right)$$

In [8.14], three important intermediates appear: hypobromous acid (HOBr), bromous acid (HBrO₂), and bromine (Br₂). Br₂ brominates the organic substrate. While the elementary steps [8.14] proceed, the concentration of Br^- is progressively consumed. As it becomes smaller than its critical value, $[Br^-]_c$, the entire system shifts to its oxidized state and another set of elementary reactions takes place:

$$2(BrO_3^- + HBrO_2 + H^+ \rightarrow 2BrO_2^\bullet + H_2O)$$

$$4\left(BrO_2^\bullet + M^{n+} + H^+ \rightarrow M^{(n+1)+} + HBrO_2\right)$$

$$\qquad\qquad\qquad\qquad\qquad\qquad [8.15]$$

$$2HBrO_2 \rightarrow HOBr + BrO_3^- + H^+$$

$$HOBr + RH \rightarrow RBr + H_2O$$

In [8.15], we discern the involvement of an autocatalytic production of bromous acid if we sum the first and the second steps. In the case of uncatalyzed brominations, the organic substrate replaces the reduced metal ion. By summing the elementary steps in [8.15], we obtain that the overall transformation is

$$BrO_3^- + 4M^{n+} + RH + 5H^+ \rightarrow RBr + 4M^{(n+1)+} + 3H_2O \qquad [8.16]$$

wherein the catalyst is oxidized. When the concentration of Br^- drops below its critical value, the concentration of $HBrO_2$ increases autocatalytically of about five orders of magnitudes. The processes [8.14] and [8.15] take place under different conditions in the same system, and a solution reacting by [8.14] will of necessity convert itself to one reacting by [8.15]. If we want to have oscillations, there must be a process producing bromide and pushing the system from the set of reaction [8.15] to [8.14]. The oxidized form of the catalyst $M^{(n+1)+}$, produced in [8.15], reacts with the organic species. In the case of malonic acid as organic substrate, the reactions are (Noyes et al. 1972):

$$6M^{(n+1)+} + CH_2(COOH)_2 + 2H_2O \rightarrow 6M^{n+} + HCOOH + 2CO_2 + 6H^+ \qquad [8.17]$$

$$4M^{(n+1)+} + BrCH(COOH)_2 + 2H_2O \rightarrow Br^- + 4M^{n+} + HCOOH + 2CO_2 + 5H^+ \qquad [8.18]$$

As the concentration of bromomalonic acid ($[BrCH(COOH)_2]$) increases, the reaction [8.18] becomes always more relevant. The Br^- produced in [8.18] is consumed by the high concentration of $HBrO_2$ according to the second reaction of [8.14]. However, when the rate of [8.18] becomes sufficiently great, $[HBrO_2]$ is quickly depleted and it drops of many orders of magnitude. This turns off process [8.15], it turns on process [8.14] and completes the cycle, which will start again. Experimental evidence (Varga et al. 1985) suggest that additional bromide may come from intermediate bromo-oxygen species. Moreover, when bromination of the organic substrate is slow or not possible, Br_2 piles up and it may produce bromide by hydrolysis

$$Br_2 + H_2O \rightarrow HOBr + Br^- + H^+ \qquad [8.19]$$

in both uncatalyzed and catalyzed systems.

The basic features of the mechanism proposed by Field, Körös, and Noyes (FKN) are included in the abstract model known as Oregonator advanced by Field and Noyes who were working at the University of Oregon (Field and Noyes 1974a). Such model consists of five steps that involve five chemical species: $X = HBrO_2$, $Y - Br^-$, $Z = 2M^{(n+1)+}$, $A - BrO_3^-$, $P = HOBr$, and $O =$ all oxidizable organic species:

$$\begin{aligned} A + Y &\rightarrow X + P \\ X + Y &\rightarrow 2P \\ A + X &\rightarrow 2X + Z \qquad [8.20] \\ 2X &\rightarrow A + P \\ Z + O &\rightarrow fY \end{aligned}$$

The concentrations of A, P, and O are assumed to be constant as that of H^+, whereas those of X, Y and Z are changeable. The first two steps of [8.20] represent the first two steps of [8.14]. The third step of [8.20] represents the autocatalysis of $HBrO_2$. The fourth is the $HBrO_2$ disproportionation. If the autocatalytic process plays the action of positive feedback, the disproportionation is part of the negative feedback. In fact, disproportionation becomes important when X accumulates, and it works against a further increase of X. However, the disproportionation alone is not strong enough to compete with and inhibit the "explosive" autocatalytic stage. The crucial process of delayed negative feedback is the coproduct autocontrol provided by the fifth and the second process of the Oregonator (Luo and Epstein 1990). As X increases autocatalytically, also Z grows (see the third step of the Oregonator). Then, Z transforms to Y through the fifth step (where f is a stoichiometric coefficient), and Y rapidly consumes X (see the second step). When Y is high and X very low, the first step regenerates X. The transformation of Z to Y provides the essential time delay between the positive and the negative feedbacks to have limit cycles and hence observe oscillations.

TRY EXERCISES 8.2, 8.3, AND 8.4

There are other oscillating reactions whose mechanisms involve coproduct autocontrol as the dominant form of negative feedback. For instance, the reaction involving trypsin, 4-[2-aminoethyl]benzenesulfonyl fluoride, and aminopeptidase M (Semenov et al. 2015); the oxidation of benzaldehyde by air catalyzed by $CoBr_2$ (Colussi et al. 1990); the bromate-iodide reaction in acidic solution (Citri and Epstein 1986), and the H_2O_2–KSCN–$CuSO_4$ reaction that oscillates only above pH 9 (Orbán 1986). The latter, also known as Orbán reaction, was the first example of a homogeneous, liquid phase, halogen-free system that oscillates even under batch conditions. If we mix hydrogen peroxide, potassium thiocyanate, and copper sulfate in alkaline solution, we can observe oscillations in color (between yellow and colorless states), redox potential, and O_2 evolution. If we also add luminol, we can record chemiluminescent oscillations. The original mechanism proposed in 1989 (Luo et al.) was composed of 30 reactions and 26 independent variables. These reactions may be partitioned in three groups (Orbán et al. 2000). The key steps of the first group regard the alkaline decomposition of H_2O_2 catalyzed by Cu(II):

$$H_2O_2 + Cu^{2+} + OH^- \rightarrow HO_2Cu(I) + H_2O$$
$$HO_2Cu(I) + nSCN^- \rightarrow Cu^+\left\{SCN^-\right\}_n + HO_2^\bullet \qquad [8.21]$$

The intermediate copper-peroxide complex ($HO_2Cu(I)$) is yellow and is responsible for the color oscillations. This first group of reactions provides two essential intermediates: $HO_2^\bullet$ and $Cu^+\{SCN^-\}_n$. The key steps of the second group involve the oxidation reactions of KSCN by H_2O_2:

$$2H_2O_2 + SCN^- \rightarrow OS(O)CN^- + 2H_2O$$
$$2OS(O)CN^- \rightarrow OOS(O)CN^- + OSCN^- \qquad [8.22]$$

In [8.22] some new intermediates are produced such as cyanosulfite, $OS(O)CN^-$, and peroxocyanosulfite, $OOS(O)CN^-$ (the oxygen in parentheses is doubled bonded to sulfur). The third group is the pivotal one for oscillations and becomes relevant when the concentrations of $OS(O)CN^-$ and $Cu^+\{SCN^-\}_n$ become high enough:

$$H_2O + OS(O)CN^- + OOS(O)CN^- \rightarrow 2OS(O)CN^\bullet + 2OH^-$$
$$OS(O)CN^\bullet + Cu^+\left\{SCN^-\right\}_n \rightarrow OS(O)CN^- + Cu^{2+} + nSCN^- \qquad [8.23]$$
$$OS(O)CN^- + HO_2^\bullet \rightarrow SO_3^{2-} + HOCN$$

In the mechanism of the Orbán reaction, $OS(O)CN^-$ plays like species X of the Oregonator model [8.20], being generated auto-catalytically through formation and reduction of the $OS(O)CN^\bullet$ radical. Y is $HO_2^\bullet$ that depletes X and is produced by the coproduct $Z = HO_2Cu(I)$. The intermediates SO_3^{2-} and HOCN end up as final products SO_4^{2-}, HCO_3^- and NH_4^+ by further oxidations. The overall transformation is

$$4H_2O_2 + SCN^- \xrightarrow{\quad Cu^{2+} \quad} HSO_4^- + NH_4^+ + HCO_3^- + H_2O \qquad [8.24]$$

The oscillations in the Orbán reaction are not perturbed by addition of luminol (Sattar and Epstein 1990). The release of a photon by a molecule of luminol is a multi-stage process (Figure 8.5).

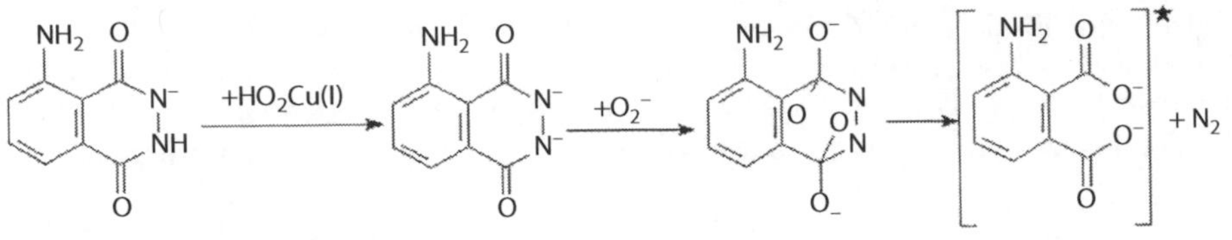

FIGURE 8.5   Generation of the chemiluminescent state of luminol.

First, a conjugated base of luminol is oxidized by copper(I)-peroxide complex. Then superoxide binds to it, and finally it decomposes to give 3-aminophthalate in its excited state and nitrogen. The excited state of 3-aminophthalate relaxes to the ground state by emitting a blue photon.[4] You can experience the Orbán reaction by doing *exercise 8.5*.

8.4.2   THE MODIFIED LOTKA-VOLTERRA OR PREDATOR-PREY "PRIMARY OSCILLATOR"

The Lotka-Volterra model proposed originally in an ecological context (see Chapter 5), suffers from the deficiency that the oscillations are sensitive to initial conditions and perturbations. In fact, the direct coupling between the two variables, prey and predator, H and C, X and Y, and the lack of sufficient time delay give rise to an infinite set of oscillatory solutions. Fluctuations push the system from one solution to another, resulting in irregular behavior in any real system.

By adding a third variable that reacts with one of the two autocatalytic species, we obtain a modified Lotka-Volterra model that generates a considerably stable limit cycle.

$$A + X \xrightarrow{\ k_1\ } 2X$$
$$Y + X \xrightarrow{\ k_2\ } 2Y \qquad\qquad [8.25]$$
$$Z + X \xrightarrow{\ k_3\ } P$$

The third step of [8.25] replaces the first-order consumption of Y in the original Lotka-Volterra model. If we introduce flow terms for X, Y, and Z and assume that the other elements are constant, we can solve the system of three differential equations by numerical integration and find that, for certain contour conditions, the system oscillates as shown in Figure 8.6.

Since the reagent A is maintained constant, the first step of [8.25] is an "explosive" autocatalytic step. The "explosion" is damped by the consumption of X by Y in the second step. This second step is a non-explosive autocatalytic process and terminates when X is consumed. The third variable Z reacts with X and provides the essential time delay between X and Y. Z accumulates sharply as Y consumes X, then decreases until X regains its high value and the cycle repeats again.

TRY EXERCISE 8.6

The oscillations can also emerge when the third species, Z, depletes Y instead of X:

$$A + X \xrightarrow{\ k_1\ } 2X$$
$$Y + X \xrightarrow{\ k_2\ } 2Y \qquad\qquad [8.26]$$
$$Z + Y \xrightarrow{\ k_3\ } P$$

---

[4] The energy of a photon having 430 nm as wavelength is $E = hc/\lambda = (6.63 \times 10^{-34}\,\text{Js})(3 \times 10^{8}\,\text{ms}^{-1}/4.3 \times 10^{-7}\,\text{m}) = 4.6 \times 10^{-19}\,\text{J}$. The energy of one mole of such photons is about 277 KJ. This is a huge amount of energy! Oxidation reactions can release such amount of energy. However, the oxidation reaction must occur very fast and the energy produced must not be dispersed in the surrounding molecules, otherwise the product will not be electronically excited. The concerted process of peroxide rings decomposition has the double property of being highly exothermic and ultrafast. In fact, many chemiluminescent events are based on peroxide rings decomposition.

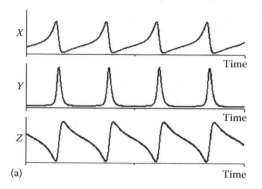
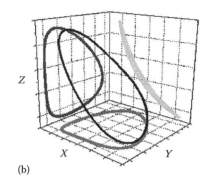

FIGURE 8.6   Simulated oscillations[5] for the modified Lotka-Volterra model described by the mechanism [8.25]. In graph (a), there are the trends of X, Y, and Z versus time. In graph (b), the behavior of the system in its three-dimensional phase space.

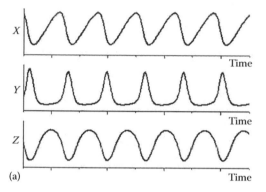
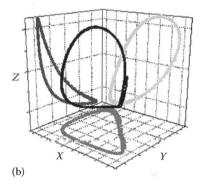

FIGURE 8.7   Simulated oscillations[6] for the modified Lotka-Volterra model described by the scheme of reactions [8.26]. Graph (a) shows the trends of X, Y, and Z versus time. Graph (b) represents the behavior of the system in its three-dimensional phase space.

The essential time delay between the two autocatalytic steps is provided by the third reaction that is a sort of negative feedback of a negative feedback (i.e., the second step). A graphical example of oscillations for the system represented by [8.26] is plotted in Figure 8.7.

Whenever Z is high, Y is low, and Y becomes high only after X has reached its maximum.

A concrete example of the "modified Lotka-Volterra oscillator" is the Briggs-Rauscher reaction. It is an oxidation of malonic acid by a mixture of hydrogen peroxide and iodate catalyzed by manganese ion (Mn^{2+}) in acidic solution. When appropriate amounts of the reactants are mixed in the presence of starch as an indicator, the system repeats more than ten times the sequence colorless, yellow, and blue, in a stirred batch reactor before expiring as a purplish solution with a strong odor of iodine.

---

[5] The results shown in Figure 8.6 can be achieved by using the following values for the parameters of mechanism [8.25]: $k_1A = 0.153$; $k_2 = 0.055$; $k_3 = 0.18$; $k_0 = 0.1$ (representing the flow rate). The concentrations of X, Y and Z constantly injected inside the reactor have been fixed to $X_{in} = 1.8$, $Y_{in} = 0.001$, and $Z_{in} = 3.5$. These values of the parameters and concentrations have also been used by Franck (1985). The initial conditions were $[X_0 \, Y_0 \, Z_0] = [1 \, 1 \, 1]$.

[6] The results depicted in Figure 8.7 have been obtained by using the following values of the parameters characterizing model [8.26]: $k_1A = 0.0055$, $k_2 = 0.00028$, $k_3 = 0.00084$, $k_0 = 0.005$ (representing the flow rate). The concentrations of X, Y and Z constantly injected inside the reactor have been fixed to $X_{in} = 40$, $Y_{in} = 15$, and $Z_{in} = 95$. The initial conditions were $[X_0 \, Y_0 \, Z_0] = [10 \, 10 \, 10]$.

The mechanism of the Briggs-Rauscher reaction consists of many elementary steps (Luo and Epstein 1990) where we distinguish two autocatalytic reactions. The first involves the iodous acid HIO_2:[7]

$$HIO_2 + IO_3^- + H^+ + 2Mn^{2+} \rightarrow 2HIO_2 + 2MnOH^{2+} \qquad [8.27]$$

When the concentration of HIO_2 is high enough, the second autocatalytic process is activated; it exerts a negative feedback on HIO_2 because it consumes iodous acid by producing iodide:

$$I^- + HIO_2 + 2H_2O_2 \rightarrow 2I^- + 2O_2 + H^+ + 2H_2O \qquad [8.28]$$

The crucial time delay or phase shift, between the two autocatalytic reactions is provided by a reaction that consumes iodide in analogy to the third step of [8.26]. It is a negative feedback on the species that exerts the negative feedback on HIO_2:

$$HOI + I^- + H^+ \rightarrow I_2 + H_2O \qquad [8.29]$$

The calculated profiles of $[HIO_2]$, $[I^-]$, and $[HOI]$ reproduce those shown in Figure 8.7, wherein the two autocatalytic processes are well separated in time.

Another concrete example of the modified Lotka-Volterra model is offered by DNA biochemistry (Fujii and Rondelez 2013).

BOX 8.1   A FEW NOTES ABOUT DNA

The building block of a DNA strand is a nucleotide whose structure is shown below.

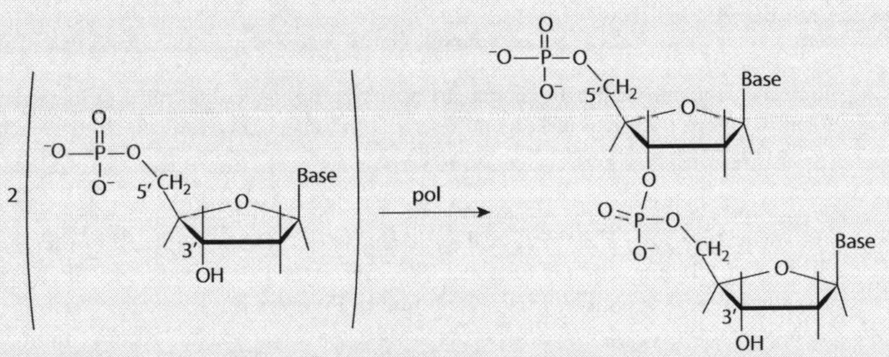

Note that the carbon atom labeled as 5′ is bound to the phosphate, whereas the 3′ carbon atom has a hydroxyl group. DNA is synthesized by the enzyme polymerase (pol) that binds a new nucleotide through the 5′-phosphate group to the 3′-hydroxyl group of the terminal nucleotide of the strand. Two complementary strands of nucleic acids hybridize when they link to each other through hydrogen bonds establishing between them complementary bases: Cytosine (C, a Pyrimidine) with Guanine (G, a Purine), and Thymine (T, a Pyrimidine) with Adenine (A, a Purine). The nicking enzyme, or nicking endonuclease (nick), recognizes specific nucleotide sequences and cleaves only to one of the strands. The exonuclease (exo) is an enzyme that cleaves nucleotides one at a time from the end of a DNA strand by hydrolyzing phosphodiester bonds at either the 3′ or the 5′ end.

---

[7] The iodous acid is produced in the first step: $I^- + IO_3^- + 2H^+ \rightarrow HOI + HIO_2$.

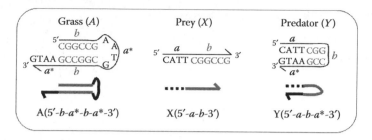

FIGURE 8.8 Schematic structures of the main single-stranded DNA protagonists of the prey-predator process.

The main characters of the biochemical version of the predator-prey interaction are presented in Figure 8.8. The prey X is a 10-base single-stranded DNA whose sequence is labeled as (5'-a-b-3'). The food of the prey (A) is a 20-base single-stranded DNA that is composed of four domains: two four-base domains a* that are the complement of the four-base domain a, and two six-base domains b that are self-complementary. Species A has the sequence 5'-b-a*-b-a*-3'. Finally, the predator Y is a palindromic 14-base single-stranded DNA with sequence 5'-a-b-a*-3'.

The mechanism of this biochemical prey and predator system is depicted in Figure 8.9. The species A serves as a template for the growth of the prey X. Prey proliferation is a multistep process: X hybridizes to the 3' end of A to form the complex $A:X$, which is extended by a polymerase (pol) to yield the double strand $A:2X$. The complex $A:2X$ bears a recognition site for a nicking enzyme (nick), which cuts its top strand into two equal parts, yielding two copies of X upon dehybridization. During predation, X hybridizes over Y, and polymerase extends this adduct to form the double-stranded $Y:Y$. Upon dehybridization, $Y:Y$ yields two copies of Y. The two main characters, X and Y, are degraded by a specific exonuclease.

To ascertain the oscillatory dynamics of such biochemical system, try to solve *exercise 8.7.*

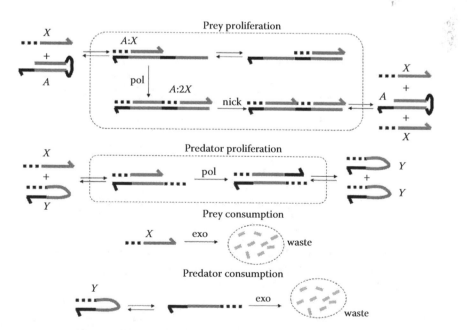

FIGURE 8.9 Mechanism of the biochemical version of the predator-prey process. Harpoon-ended arrows denote DNA strands. The elementary steps are both reversible hybridization/dehybridization reactions and irreversible enzymatic transformations.

8.4.3 THE "FLOW CONTROL PRIMARY OSCILLATOR"

There is a third group of oscillatory reactions wherein the inflow of reagents plays a crucial role, beyond the fundamental function of simple replenishment (Luo and Epstein 1990). Its model, named as "primary oscillator with flow control" consists of three main processes:

$$X + Z \xrightarrow{\ k_1\ } 2X$$
$$X + Y \xrightarrow{\ k_2\ } P_1 \qquad\qquad [8.30]$$
$$A + X \xrightarrow{\ k_3\ } P_2$$

In the first reaction, there is a situation of antagonistic feedback. In fact, it combines a positive feedback action regarding X and a negative feedback action regarding Z. The variable X is, also, cross-coupled with Y independently from Z. Y is consumed first when X grows, as shown in Figure 8.10.

If the flow rate and the inflow concentrations are such that Y recovers before Z, then, a sufficient phase-shift between X and Z can be established to obtain sustained oscillations. If the chosen parameters do not guarantee the proper delay, the first reaction of [8.30] resumes too early, and the cycle is never completed. The third process is a necessary consumption reaction for X to avoid the autocatalytic explosion. Concrete examples of oscillatory reactions complying with the "flow control" model are (Luo and Epstein 1990): (1) the chlorite-iodide reaction where ClO_2^- is reduced to Cl^-, whereas I^- is oxidized to I_2; (2) the minimal bromate reaction, where bromate, bromide, and the catalyst flow in the reactor without the organic substrate; (3) an autocatalytic process that produces thiols from thioesters and diallyl disulfides, inhibited by the inflow of acrylamide (Semenov et al. 2016); (4) the $BrO_3^- - SO_3^{2-} - Fe(CN)_6^{4-}$ reaction that is an example of pH oscillator. In a pH oscillator, the concentration of protons plays a crucial role in the kinetic behavior of redox chemical reactions. The reactants are one oxidant species (like $BrO_3^-, IO_3^-, H_2O_2, IO_4^-$) and either one or two reductants (like $SO_3^{2-}, S^{2-}, Fe(CN)_6^{4-}, S_2O_3^{2-}, NH_2OH$, thiourea, et cetera). A list of some representative

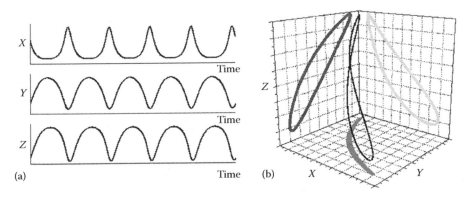

FIGURE 8.10 Simulated oscillations[8] for the model "with flow control" described by the set of reactions [8.30]. Time evolution of X, Y, and Z in graph (a). Dynamical behavior of the system in its three-dimensional phase space in graph (b).

---

[8] The results plotted in Figure 8.10 have been achieved by using the following values of the parameters characterizing the model [8.30]: $k_1 = 0.55$; $k_2 = 11.8$; $k_3 A = 0.396$; $k_0 = 0.1$ (representing the flow rate). The concentrations of X, Y and Z constantly injected inside the reactor have been fixed to $X_{in} = 1.6$, $Y_{in} = 2.35$, and $Z_{in} = 3.04$. Such values have also been used by Franck (1985). The initial conditions have been fixed to $[X_0\ Y_0\ Z_0] = [0.157\ 0.12\ 1.63]$.

TABLE 8.1

A List of Some pH Oscillators Where *Ox* is the Oxidant, Whereas *Red* Indicates the Reductants. The Minimum (pH$_{min}$) and the Maximum (pH$_{max}$) pH Values Are Reported Along with Their Discrepancy (ΔpH)

	pH Oscillator				
Ox	*Red*	pH$_{min}$	pH$_{max}$	ΔpH	**Authors**
BrO_3^-	$SO_3^{2-} + Fe(CN)_6^{4-}$	2.7	6.2	3.5	Edblom et al. (1989)
IO_3^-	NH_2OH	3.0	5.5	2.5	Rábai and Epstein (1990)
IO_3^-	$SO_3^{2-} + Fe(CN)_6^{4-}$	3.0	7.3	4.3	Edblom et al. (1986)
IO_3^-	$SO_3^{2-} + HCSNH_2$ (thiourea)	3.5	7.0	3.5	Rábai et al. (1987)
BrO_3^-	$SO_3^{2-}(Mn^{2+})$	3.5	7.2	3.7	Okazaki et al. (1999)
H_2O_2	$S_2O_4^{2-}$	3.5	9.5	6.0	Kovács and Rábai (2001)
IO_4^-	$S_2O_3^{2-}$	4.0	5.5	1.5	Rábai et al. (1989a)
IO_4^-	NH_2OH	4.0	6.0	2.0	Rábai and Epstein (1989)
H_2O_2	$SO_3^{2-} + HCO_3^-$	4.5	6.5	2.0	Rábai (1997)
BrO_3^-	$NH_2OH + C_6H_5OH$	4.5	7.5	3.0	Orbán and Epstein (1994)
BrO_3^-	$S_2O_3^{2-} + C_6H_5OH$	4.5	7.5	3.0	Orbán and Epstein (1995)
H_2O_2	$SO_3^{2-} + Fe(CN)_6^{4-}$	4.8	7.8	3.0	Rábai et al. (1989b)
H_2O_2	$S_2O_3^{2-} + Cu^{2+}$	4.9	8.5	3.6	Orbán and Epstein (1987)
IO_3^-	$SO_3^{2-} + S_2O_3^{2-}$	5.0	7.0	2.0	Rábai and Beck (1988)
H_2O_2	$SO_3^{2-} + Na_2CO_3$	5.0	7.0	2.0	Frerichs and Thomson (1998)
H_2O_2	$SO_3^{2-} + S_2O_3^{2-}$	5.0	7.0	2.0	Rábai and Hanazaki (1999)
H_2O_2	S^{2-}	6.0	8.0	2.0	Orbán and Epstein (1985)
BrO_3^-	I^-	6.0	8.2	2.2	Orbán and Epstein (1992)
H_2O_2	$SO_3^{2-} + Hemin$	6.5	7.5	1.0	Hauser et al. (2002)

pH oscillators is reported in Table 8.1 (see also Orbán et al. 2015). In such systems, the variation of pH, which can be as large as four pH-units, is the driving force of the oscillations. In fact, if such reactive systems are buffered, the oscillations are suppressed.

Three are the key events in a pH oscillator. First, a fast equilibrium of protonation and deprotonation of a reductant (Red_A):[9]

$$H + Red_A \rightleftarrows HRed_A \qquad [8.31]$$

where:

H corresponds to H^+, and the protonated form of the reductant $HRed_A$, is more reactive than Red_A.

The second relevant event is an autocatalytic generation of H^+ through a redox reaction between the oxidant (Ox) and $HRed_A$:

$$Ox + HRed_A + H \rightarrow Y$$
$$HRed_A + Y \rightarrow 3H \qquad [8.32]$$

---

[9] Note that in the equations [8.31], [8.32], and [8.33], we do not introduce explicitly the electric charges because such equations are not stoichiometric reactions, but schematic descriptions of chemical transformations.

where Y represents one or more intermediates. The third fundamental ingredient to have oscillations is a time-delayed negative feedback process represented by

$$Y \rightarrow P$$
$$Red_B + H \rightarrow Q \qquad\qquad [8.33]$$

In [8.33], Red_B is the possible second reductant, and P and Q are the products.

TRY EXERCISE 8.8

The presence of Red_B is not always necessary. When the rate constant of the $Y \rightarrow P$ process is sufficiently high, we have oscillations even without Red_B. In the two-component system, the reductant can be oxidized at two different states. For instance, in the case of the $H_2O_2 - S^{2-}$ system, the following two transformations are possible after the acid-base equilibrium for sulfide:

$$H_2O_2 + HS^- + H^+ \rightarrow \frac{1}{8}S_8 + 2H_2O \qquad\qquad [8.34]$$

$$4H_2O_2 + HS^- \rightarrow SO_4^{2-} + 4H_2O + H^+ \qquad\qquad [8.35]$$

The partial oxidation of sulfide to sulfur (S_8) requires acid to proceed. Consequently, the pH rises when reaction [8.34] prevails. This step is self-inhibitory. As the concentration of protons falls, the partial oxidation gradually slows down and essentially stops when the pH becomes high enough. When the concentration of H^+ is low, the reductant is wholly oxidized to sulphate. Reaction [8.35] is accompanied by the formation of acid, giving rise to favorable conditions for the acid-consuming reaction [8.34] to revitalize (Rábai et al. 1990) if fresh reactants have flowed into the reactor at the appropriate rate. Before 2011, the pH oscillators were known to produce oscillations only in Continuous-flow Stirred Tank Reactors (CSTRs) or in semi-batch arrangements. In the semi-batch configuration, the Red_A species that is involved in the autocatalytic process is introduced continuously in a mixture of the oxidant and Red_B, producing long-lasting but only transient oscillations. Recently, long lasting pH oscillations have also been achieved in batch-like equipment (Poros et al. 2011). The species Ox (like either BrO_3^- or IO_3^-) and Red_B (like $Fe(CN)_6^{4-}$) have been poured in a beaker, mixed and stirred above a silica gel layer that was loaded with Na_2SO_3. Such layer releases sulfite ions by slow and continuous dissolution into the water solution containing the other reactants. A similar idea has been exploited to achieve pH-oscillations in a closed chemical system containing H_2O_2, HCO_3^- entirely dissolved in water and the poorly soluble solid $CaSO_3$ as a source of sulfite ions (Rábai 2011). There is considerable interest in developing batch pH-oscillators because they are more convenient to use than the equivalent CSTR variants. Batch pH-oscillators have been used to induce periodic changes in the shape and volume of pH-sensitive gels (Labrot et al. 2005), and they will probably be used in the manufacture of "Chemical Robots" (see more information in Chapter 13). For instance, they can be exploited to produce "artificial glands" when combined with pH-sensitive gels. The pH-oscillator coupled with a pH-sensitive sol-gel transition can release drugs periodically inside a diseased organism as if it were a real gland releasing hormones. A pulsating drug delivery device might be particularly appealing in diseases that show strong circadian dependency.

8.4.4 THE COMPOSITE SYSTEM: A CHEMICAL EQUILIBRIUM COUPLED TO A "PRIMARY OSCILLATOR"

For any number of reasons, we might like that a generic species S has a concentration that oscillates. Rather than trying to build an oscillatory reaction directly around the chemistry of S (often a tough riddle), we may seek to couple an equilibrium reaction (ER) involving S with an already known "primary oscillator." First, we select the equilibrium ER that involves S. Then, we must choose a "primary oscillator" that periodically produces and consumes a species C that can shift the position of the equilibrium ER. It is important that C has a rapid effect on ER, while ER has a little or negligible effect on the "primary oscillator." Moreover, the equilibrium of the reaction ER should be swiftly achievable. Finally, we can expect that S oscillates with a period close to that of the unperturbed "primary oscillator" (Kurin-Csörgei et al. 2005). Through this strategy, the oscillations of Ca^{2+} (see Tables 8.2 and 8.5), Al^{3+} (see Table 8.3), F^- (see Table 8.4), Cd^{2+}, Zn^{2+}, Co^{2+}, Ni^{2+} (see Table 8.5) have been achieved.

TABLE 8.2

Example of Chemical Coupling between a pH Oscillator and a Complexation Reaction Involving Ca^{2+} and Ethylenediaminetetraacetic Acid, i.e., EDTA

"Primary Oscillator": pH-Oscillator

$BrO_3^- + SO_3^{2-} + Fe(CN)_6^{4-}$

Equilibrium ER: Complexation Reaction

$$Ca^{2+}_{(aq)} + EDTA^{2-}_{(aq)} \rightleftharpoons CaEDTA_{(aq)}$$

$$-2H^+ \; \big\updownarrow \; +2H^+$$

$$H_2EDTA$$

Source: Kurin-Csörgei, K. et al., *Nature*, 433, 139–142, 2005.

TABLE 8.3

Example of Chemical Coupling between a pH Oscillator and a Precipitation Reaction Involving Al^{3+}

"Primary Oscillator": pH-Oscillator

$BrO_3^- + SO_3^{2-} + Fe(CN)_6^{4-}$

Equilibrium ER: Precipitation Reaction

$$Al^{3+}_{(aq)} + 3OH^-_{(aq)} \rightleftharpoons Al(OH)_{3(s)}$$

$$-3H^+ \; \big\updownarrow \; +3H^+$$

$$6H_2O$$

Source: Kurin-Csörgei, K. et al., *Nature*, 433, 139–142, 2005.

TABLE 8.4

Example of Chemical Coupling between a pH Oscillator and a Complexation Equilibrium Involving Fluoride Anions and Aluminum Ions

"Primary Oscillator": pH-Oscillator

$BrO_3^- + SO_3^{2-} + Fe(CN)_6^{4-}$

Equilibrium ER: Complexation Reaction

$$AlF_4^-{}_{(aq)} \rightleftharpoons 4F^-_{4(aq)} + Al^{3+}_{(aq)} + 3OH^-_{(aq)} \rightleftharpoons Al(OH)_{3(s)}$$

$$-3H^+ \; \big\updownarrow \; +3H^+$$

$$6H_2O$$

Source: Kurin-Csörgei, K. et al., *Nature*, 433, 139–142, 2005.

TABLE 8.5

Example of Chemical Coupling between a pH Oscillator and a Precipitation Equilibrium Involving Divalent Cations (M^{2+})

"Primary Oscillator": pH-Oscillator	Equilibrium ER: Precipitation Reaction
$BrO_3^- + SO_3^{2-}$	$M^{2+}_{(aq)} + SO_3^{2-}_{(aq)} \rightleftharpoons MSO_{3(s)}$
	$-H^+ \big\Vert\, +H^+$
	$HSO_3^-_{(aq)} \longrightarrow SO_4^{2-}$

Source: Horváth, V. et al., *Phys. Chem. Chem. Phys.*, 12, 1248–1252, 2010.

Although the strategy of coupling has made easy to promote the oscillation of cations, also pulses of anions have been constructed. For instance, fluoride oscillations can be generated by coupling a pH oscillator to precipitation and complexation reactions. If we have aluminum and fluoride ions in solution, besides the reactants of a pH oscillator, at high pH, Al^{3+} precipitates as $Al(OH)_3$ and $[F^-]$ is high. When pH decreases, the precipitate dissolves, Al^{3+} is released, and it binds to fluoride anions, determining a drop in free F^-.

The protons do not have the prerogative of playing as the liaison between the "primary oscillator" and the ER reaction. Horváth et al. (2010) have shown that also a reductant like sulfite may play the role of species C and induce the oscillations of divalent cations, like Ca^{2+}, Cd^{2+}, Zn^{2+}, Co^{2+}, and Ni^{2+}. When pH is high, the metal forms a precipitate with sulfite. On the other hand, when $[H^+]$ is high, the precipitate MSO_3 is dissolved, and sulfite is oxidized irreversibly to sulphate by bromate.

8.4.5 "Delayed Negative Feedback Oscillator"

The study and construction of biochemical circuits from simple components has demonstrated (O'Brien et al. 2012) how oscillatory behavior can also arise from just a negative feedback process if it is adequately delayed (see Figure 8.11). In Chapter 7, we learned the basic gene circuits by which proteins are synthesized. In a single gene negative feedback loop, like that depicted in Figure 7.19, where X is mRNA and Y is the repressor protein, we can have at most damped oscillations. Only if we introduce at least one (Z_0) intermediate species that delays the negative action in the feedback loop (see Figure 8.11), we can have sustained oscillations. The situation is even better if there are more delaying intermediate species (Z_i, with $i = 1,...,n$).

This fact has been observed experimentally (Stricker et al. 2008) in a synthetic circuit constructed using Escherichia coli components, where X is the mRNA of the *lac*I gene, Z_0 is the unfolded state

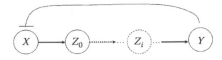

FIGURE 8.11   Circuit topology of a delayed negative feedback loop where Y, derived from one (Z_0) or more intermediates (Z_i with $i = 1,...,n$), inhibits X.

of the protein LacI, Z_i are different aggregated states of LacI, and finally Y is the LacI tetramer repressing the expression of the *lacI* gene.

Essentially, one can model a delayed gene negative feedback oscillator by using the following set of ordinary differential equations:

$$\frac{dX}{dt} = k_{tr}C_{TF}\left(\frac{K^n}{K^n + Y^n}\right) - k_d X \qquad [8.36]$$

$$\frac{dZ_i}{dt} = v_i\left(Z_{i-1}\right) - v_{i+1}\left(Z_i\right) \qquad [8.37]$$

$$\frac{dY}{dt} = v_{n+1}\left(Z_n\right) - \frac{V_{max}Y}{K_M + Y} \qquad [8.38]$$

In [8.36], X represents the concentration of mRNA; $k_{tr}C_{TF}$ is its maximum production rate; K and n are the Hill constant and coefficient, respectively. In [8.37], Z_i with $i = 1,...,n$ represents the i-th intermediate species. Such intermediates may be unfolded states of the proteins, or monomers and different aggregated states of the final species Y, produced at the $v_{n+1}\left(Z_n\right)$ rate. According to equation [8.38], Y degrades by a Michaelis-Menten mechanism, whose V_{max} and K_M are the well-known constants.

The necessary delay to observe oscillations in a negative feedback loop may be achieved by introducing other genes and gene expression processes in between the species X and Y of Figure 8.11. Gene expression can be monitored by the Green Fluorescent Protein. The first reported experimental realization of a synthetic multi-gene oscillator is the well-known "repressilator" (Elowitz and Leibler 2000) that consists of three transcriptional repressor systems. In the network shown in Figure 8.12, the first repressor protein, LacI from Escherichia coli, inhibits the transcription of the second repressor gene, *tet*R from the tetracycline-resistance regulatory gene of transposon Tn*10*, whose protein product, in turn, inhibits the expression of a third gene, *cI* from λ phage. Finally, CI inhibits *lacI* expression, completing the loop. Don't worry if you are not familiar with the main characters of this synthetic biochemical circuit. What is important is to note that with three repressors connected in a negative feedback loop we can observe temporal oscillations.

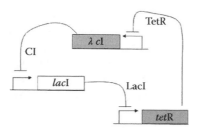

FIGURE 8.12   Scheme of the "repressilator."

The experimental temporal oscillations of the repressilator may be confirmed from a simple model of transcriptional regulation (remember the relevant steps of a process of transcriptional regulation we learned in Chapter 7). Three repressor-proteins, LacI (Z_0), TetR (Z_2) and CI (Y), and their corresponding RNA messengers, $lacI$ (X), $tetR$ (Z_1) and cI (Z_3), participate in transcription, translation and degradation reactions. The kinetics of the system are determined by six coupled first-order differential equations:

$$\frac{dX}{dt} = \alpha_0 + \frac{\alpha}{1 + \frac{Y^n}{K^n}} - k_{dm}X$$

$$\frac{dZ_0}{dt} = k_{dm}X - k_{dp}Z_0$$

$$\frac{dZ_1}{dt} = \alpha_0 + \frac{\alpha}{1 + \frac{Z_0^n}{K^n}} - k_{dm}Z_1$$

[8.39]

$$\frac{dZ_2}{dt} = k_{dm}Z_1 - k_{dp}Z_2$$

$$\frac{dZ_3}{dt} = \alpha_0 + \frac{\alpha}{1 + \frac{Z_2^n}{K^n}} - k_{dm}Z_3$$

$$\frac{dY}{dt} = k_{dm}Z_3 - k_{dp}Y$$

In [8.39], we consider a simplified symmetrical case in which all three repressors are identical except for their DNA-binding specificities. The constants k_{dm}, k_{dp} are relative to the decays of the mRNAs and proteins, respectively. K is the concentration of the repressor needed to repress a promoter half-maximally and n is a Hill coefficient. α_0 is the constant rate of "leaky" initiation, which is the rate of protein production from a given promoter type in the presence of saturating amount of repressor, whereas α is the same rate but in the absence of repressor. For certain conditions, the model shows stable limit cycle oscillations like those shown in Figure 8.13.

The repressilator is the biochemical counterpart of the ring oscillator in electronics. A ring oscillator is a device composed of an odd number of NOT gates. A scheme is shown in Figure 8.14. A NOT gate is a single input device that inverts the logic level of its input signal. In fact, it gives the output logic level 1 when the input is at the logic level 0, and the output logic level 0 when the input is at the logic level 1. In the ring oscillator, the output of the last inverter is fed back into the first. From Figure 8.14 it is evident that the last output of the chain is the logical NOT of the first input. This situation occurs whenever the chain contains an odd number of inverters. The final output of the ring oscillator is asserted a finite amount of time after the first input is asserted; the feedback of this last output to the input produces oscillations.

A ring oscillator requires electric power to operate; a repressilator requires a positive chemical affinity to work. The general rule of thumb is that for electronic/biochemical circuits with a ring architecture, there must be an odd number of inverters/repressors for the entire loop to be a delayed negative feedback circuit and give oscillations. To test this rule *try exercise 8.9.*

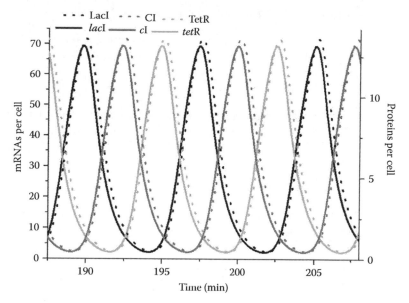

FIGURE 8.13 Temporal oscillations of the number of mRNA and proteins participating in the repressilator.[10]

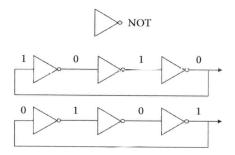

FIGURE 8.14 Scheme of the three-inverter ring oscillator.

---

[10] The oscillations of the repressilator depicted in Figure 8.13 have been obtained by numerical integration of the ODEs [8.39]. In MATLAB, the function file used was

```
function dy = repressilator(t,y)
dy = zeros(6,1);
a0 = 0.1;
a = 100;
n = 2.5;
kdp = 5;
kdm = 1;
K = 1;
dy(1) = a0+[a/(1+((y(6))^n/K^n))]-kdm*y(1);
dy(2) = kdm*y(1)-kdp*y(2);
dy(3) = a0+[a/(1+((y(2))^n/K^n))]-kdm*y(3);
dy(4) = kdm*y(3)-kdp*y(4);
dy(5) = a0+[a/(1+((y(4))^n/K^n))]-kdm*y(5);
dy(6) = kdm*y(5)-kdp*y(6);
```

The script file was

$$[t,y]=ode45('repressilator',[0\ 1000],\ [5\ 5\ 0\ 0\ 0\ 0]);$$

8.5   OVERVIEW AND HINTS FOR FURTHER READING

In this chapter, we have learned a classification of the homogeneous chemical oscillatory reactions, which is based on the types of involved feedback processes. A schematic view of the proposed five "primary oscillators" is shown in Figure 8.15. The schemes report only the main processes involving the essential species X, Y, and Z.

Scheme (a) represents the coproduct autocontrol model where we have autocatalysis for X and a negative feedback action exerted by Y on X, which is delayed because Y is produced by Z.

Scheme (b) describes the "modified predator-prey oscillator." In (b), we have two autocatalytic processes, for X and Y, and a negative feedback action played by Z on either Y or X.

Scheme (c) regards the flow-control model. X is the autocatalytic species that requires the inflow of Z; Y is the inhibitor of X. In a pH oscillator, X is H^+, Z is both Ox and Red_A (see equation [8.32]), and Y is either Red_B (see equation [8.33]) or a second oxidized state of the reductant Red_A.

Scheme (d) depicts a "flow-control oscillator" coupled to an equilibrium reaction through the oscillating concentration of species X. The equilibrium $A \rightleftarrows B$ swings back and forth synchronized with the oscillations of X.

Finally, scheme (e) represents the delayed negative feedback model wherein X is continuously fed from outside.

The list of the five "primary oscillators" presented in this chapter might not be exhaustive. Other ways might exist in which homogeneous chemical oscillations arise. For example, an oscillating chemical reaction may ground on more than one "primary oscillator."

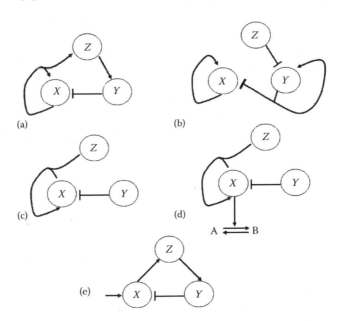

FIGURE 8.15   Schematic representation of the five "primary oscillators." Scheme (a) is for coproduct autocontrol model; (b) for the modified prey-predator model; scheme (c) for the flow-control model; scheme (d) for the "equilibrium-coupled-to-a-flow-control oscillator" model; and (e) for the delayed negative feedback model.

Finally, the classification of oscillating chemical reactions proposed in this book is just one of many others. For instance, Eiswirth et al. (1991) offered another classification based on the stoichiometric network analysis of the chemical mechanisms. Readers who want to know more about such classification of oscillatory reactions can read a review written by Ross and Vlad (1999).

To check if you have a clear picture of the different "primary oscillators" presented in this chapter, *try exercise 8.10.*

Hopefully, the study of the main features of the known chemical oscillators will favor the design of new families of periodic reactions. For this purpose, it is worthwhile reading the review about the mechanisms of autocatalysis by Bissette and Fletcher (2013).

BOX 8.2   THE IMPORTANCE OF MIXING A REACTIVE CHEMICAL SYSTEM

It is really hard to have perfect mixing of a chemical reaction. However, we always need to strive to achieve it. In fact, it guarantees that the concentration of a species is the same everywhere into the reactor. The effect of imperfect mixing in the presence of nonlinear kinetics can be awesome (for a quantitative proof of this statement, try to solve exercise 9.2 in Chapter 9). The effects of imperfect mixing are exacerbated by those reactions that involve more than one phase. If we have liquid and solid phases, the mixing efficiency determines the rate at which fresh material from the liquid phase reaches the surface. On the other hand, if a reaction, occurring in liquid phase, produces a gas, or a gas in the air can participate to the chemical transformation, the rate of mixing exerts a concrete effect on the rate of the chemical reaction. For example, in the case of the BZ reaction, a high rate of stirring tends to increase (I) the rate at which the gaseous bromine generated by the reaction leaves the solution, and (II) the rate at which oxygen can oxidize some of the organic intermediates is taken up by the solution. Since both bromine and oxygen play key roles in the BZ reaction, it is clear that changes in stirring rate affect its dynamical behavior. In fact, there are two sorts of mixing: macro-mixing and micro-mixing. They occur at two different spatial scales and at two different time scales. Macro-mixing guarantees a homogeneous system at the macroscopic level. A solution is homogeneous at the macroscopic level if the average compositions of its macroscopic portions (each having a volume of roughly one cubic centimeter) are identical. The characteristic time needed for a successful macro-mixing can be determined by experiments with colored species. Micro-mixing is the process that guarantees homogeneity at the molecular level. For instance, if we prepare a diluted solution of species X in the solvent L, we may consider the micro-mixing as complete when all the X molecules are surrounded by a shell consisting of at least a layer of L molecules. This situation does not necessarily hold for a solution that is homogeneous at the macroscopic level. Micro-mixing requires much more time than macro-mixing because it occurs through molecular diffusion. Model calculations explain the observation that in a number of systems, very different outputs are obtained if reactants are premixed in a small mixing chamber before entering a CSTR (Puhl and Nicolis 1987). There are even experimental proofs demonstrating that imperfect micro-mixing leads to shifts in bifurcation points and the appearance of aperiodic, transient oscillations in bistable systems (read, e.g., Hu et al. 2010). For more information about stirring and mixing effects, read Chapter 15 of the fascinating book by Epstein and Pojman (1998).

8.6 KEY QUESTIONS

- Why did it take time to accept oscillations in chemistry?
- What does feedback action mean in chemistry?
- Which are the conditions favorable for generating chemical oscillations?
- Which are the features of an excitable system?
- Describe the features of the "primary oscillators."
- Why is so important to mix properly reactive solutions?

8.7 KEY WORDS

Positive and negative feedback; Time delay; Excitability; Oscillations; "Primary Oscillators."

8.8 EXERCISES

8.1. Considering the elementary reactions [8.3], [8.4] and the flow rate $k_0[X]$, find the possible steady-state solutions for X. Build the plot of the rates of the X formation and depletion as functions of $([X]/C_A) = x$, where C_A is the analytical concentration of A flowing inside the CSTR with a flow rate constant k_0. Assuming that $k_r = 1\,t^{-1}$, $k'_r = 100\,M^{-2}t^{-1}$ and $C_A = 1\,M$, find at least three values of k_0 giving (I) a steady-state solution characterized by low $[X]_{ss}$, (II) a steady-state solution characterized by large $[X]_{ss}$, (III) three steady-state solutions.

8.2. The goals of this exercise are to (I) observe the phenomenon of spontaneous temporal self-organization in a chemical laboratory, (II) determine the dependence of the period (T) of the oscillations of the Belousov-Zhabotinsky reaction on the concentration of $KBrO_3$, and (III) become aware of errors that can be made in an experiment carried out in a laboratory. Go to the wet laboratory, wear a white coat, gloves, safety glasses and prepare the following solutions using deionized water as the solvent.

- *Solution A*: H_2SO_4 (sulfuric acid) 0.76 M.
- *Solution B*: $KBrO_3$ (potassium bromate) 0.35 M.
- *Solution C*: $CH_2(COOH)_2$ (malonic acid) 1.2 M.
- *Solution D*: $Ce(SO_4)_2$ (cerium(IV) sulphate) 0.005 M in an aqueous solution containing H_2SO_4 0.76 M.
- *Solution E*: Ferroin (tris(1,10-phenanthroline)iron(III) sulphate) 0.025 M.

Insert a magnetic stir bar into a dry 25 mL glass beaker and place the beaker on a stirrer. Using pipettes and/or automatic pipettor and tips, add the following amounts of stock solutions into the beaker: 2 mL of A, 2 mL of B, 2 mL of C, 2 mL of D, and 200 μL of E. After adding all the reagents and waiting for a few minutes under stirring, transfer 2.5 mL of the reactive solution from the beaker inside a cuvette (with 1 cm as optical path length) having a tiny stir bar. Place the cuvette in a UV-visible spectrophotometer, and maintain it under stirring. If you have a spectrophotometer whose detector is a diode array (that permits simultaneous measurements at multiple wavelengths), fix the spectral range in the interval 200–800 nm, and record how the absorption spectrum of the solution changes over time (by recording one spectrum every 3 seconds). Alternatively, if you have a traditional double beam spectrophotometer, fix the monochromator at one wavelength (I suggest you choose 511 nm), and record how the absorbance of the solution at that specific wavelength changes over time. Monitor the spectral evolution of the system over 40′. By using the trend of absorbance (at a specific wavelength) as a function of time, determine the periods of the oscillations and calculate its average value ($\overline{T_1}$) and the standard deviation of the average ($\sigma_{\overline{T_1}}$). Now, repeat the experiment by changing only the amount of solution B: instead of 2 mL of solution B, add 1.5 mL of it plus 0,5 mL of deionized water (the addition of water is needed to maintain constant the total volume of the solution). Record the spectral evolution of the new BZ reaction and determine the new average period ($\overline{T_2}$) and the new standard deviation of the average period ($\sigma_{\overline{T}_2}$). Finally, repeat the entire procedure after changing only the amount of solution B. Use 1 mL of B + 1 mL of deionized water. Determine the new average period ($\overline{T}_3$) and its uncertainty.

The final goal of this experiment is the determination of the exponent n of the following relation: $T \propto \left[KBrO_3 \right]^n$. Build a log-log plot for its determination.

Warning: for the success of this experiment it is essential to mix the reagents properly. For more information about the importance of stirring solutions, read Box 2 of this chapter.

8.3. The five steps of the Oregonator model [8.20] represent the following five elementary steps of the FKN mechanism (Pellitero et al. 2013):

$$BrO_3^- + Br^- + 2H^+ \xrightarrow{k_1} HBrO_2 + HOBr$$

$$HBrO_2 + Br^- + H^+ \xrightarrow{k_2} 2HOBr$$

$$BrO_3^- + HBrO_2 + H^+ \xrightarrow{k_3} 2HBrO_2 + Ce^{4+}$$

$$2HBrO_2 \xrightarrow{k_4} BrO_3^- + HOBr + H^+$$

$$Ce^{4+} + CH_2\left(COOH\right)_2 \xrightarrow{k_5} f\,Br^-$$

At 293.15 K, the values of the rate constants are:

$$k_1 = 2 \text{ mol}^{-3}\text{dm}^9\text{s}^{-1}$$
$$k_2 = 10^6 \text{ mol}^{-2}\text{dm}^6\text{s}^{-1}$$
$$k_3 = 10 \text{ mol}^{-2}\text{dm}^6\text{s}^{-1}$$
$$k_4 = 2000 \text{ mol}^{-1}\text{dm}^3\text{s}^{-1}$$
$$k_5 = 1 \text{ mol}^{-1}\text{dm}^3\text{s}^{-1}$$

Write the differential equations for the three variables $X = [HBrO_2]$, $Y = [Br^-]$, and $Z = [M^{(n+1)+}]$ and solve them by numerical integration, fixing the value of $[H^+]$ to 0.8 M, of $[CH_2(COOH)_2]$ to 0.02 M, of the coefficient f to 0.6, and assigning to $[BrO_3^-]$ the values of (1) 0.18 M, (2) 0.15 M, (3) 0.12 M, (4) 0.1 M, and (5) 0.08 M, respectively. Determine the relationship between the period of the oscillations and $[BrO_3^-]$, i.e., the exponential n of the relation $T \propto [BrO_3^-]^n$. If you choose MATLAB to solve the differential equations by numerical integration, use stiff solvers, like "ode15s" and "ode23s," because the changes of the concentrations of the intermediates are substantial and show spikes.

8.4. The activation energies (Pellitero et al. 2013) determined at 293.15 K, for the five steps of the Oregonator (see exercise 8.3) are $E_{act,1} = 54$ kJ/mol, $E_{act,2} = 25$ kJ/mol, $E_{act,3} = 60$ kJ/mol, $E_{act,4} = 64$ kJ/mol, and $E_{act,5} = 70$ kJ/mol, respectively. The values of the kinetics constants of the five reactions of the Oregonator are reported in exercise 8.3. Calculate the period (ΔT) of the oscillations at $T_1 = 293.15$ K, $T_2 = 303.15$ K, and $T_3 = 313.15$ K, and for $[BrO_3^-] = 0.18$ M, $[CH_2(COOH)_2] = 0.1$ M, $[H^+] = 0.8$ M, and $f = 0.6$. Finally, determine the activation energy for the Arrhenius-type relationship between ΔT^{-1} and T.

8.5. The goal of this exercise is to observe the spontaneous formation of chemiluminescent oscillations with the Orbán reaction. Go to the laboratory, wear a white coat, gloves, safety glasses and prepare the following solutions using deionized water as the solvent:
  - *Solution A*: H_2O_2 1 M. You may buy H_2O_2 in a drugstore and find that its concentration is expressed in volumes. The concentration in volume is the ratio between the volume of gaseous O_2 (measured in standard conditions, i.e., at 273.15 K and 1 atm)

that evolves for complete decomposition of the H_2O_2 contained in a given volume and the volume of the solution. For example, if we buy a 36-volume solution of H_2O_2, the grams of H_2O_2 we have in 1 L are $x = (36/22.4)68 = 109$ g according to the following decomposition reaction of H_2O_2 considered in standard conditions:

$$2H_2O_{2(aq)} \rightarrow 2H_2O_{(liq)} + O_{2(gas)}$$
$$68 \text{ g} \qquad\qquad 22.4 \text{ L}$$

The concentration of 109 g of H_2O_2 in 1 L of solution can also be expressed as either 10.9% or $[(109\,g/34g\,mol^{-1})/1\,L] = 3.2$ M. For the preparation of H_2O_2 1 M, take 15.6 mL of H_2O_2 3.2 M and add water until you have 50 mL of total volume.
- *Solution B*: KSCN 0.15 M. Dissolve 0.729 g of KSCN (MW = 97.17 g·mol^{-1}) in 50 mL of deionized water.
- *Solution C*: luminol $3.7 \cdot 10^{-3}$ M in NaOH 0.1 N. First, prepare a solution of NaOH 0.1 N by dissolving 2 g of NaOH in 500 mL of deionized water. Then, dissolve 0.0328 g of luminol ($C_8H_7N_3O_2$) in 50 mL of the NaOH 0.1 N solution.
- *Solution D*: $CuSO_4$ $6 \cdot 10^{-4}$ M. Dissolve $9.6 \cdot 10^{-3}$g of $CuSO_4$ in 100 mL of deionized water.

Insert a magnetic stir bar into a small dry beaker and place the beaker on a stirrer. Add the volumes that are listed in Table 8.6 for five distinct experiments.

After stirring the reactive solution for a few minutes, take 3 mL of it and put it in a fluorometric cuvette. The solution within the cuvette must be maintained under stirring.

Record the chemiluminescent oscillations by using a fluorimeter (note that you do not have to switch on the lamp of the fluorimeter to record the chemiluminescent signals). Fix the wavelength of emission at 436 nm. Choose large widths of the slits and do not change them in the five experiments to compare the intensity of the chemiluminescent signals.

Find out the best conditions to observe strong chemiluminescent signals and fast oscillations.

8.6. Write the system of three differential equations for the mechanism [8.25] and solve it numerically for the following values of the parameters: $k_1A = 0.5$; $k_2 = 0.0028$; $k_3 = 0.0084$; $k_0 = 0.05$ (representing the flow rate). The constant concentrations introduced in the open system are $X_{in} = 4$; $Y_{in} = 1$; $Z_{in} = 9$ and the initial conditions are $[X_0, Y_0, Z_0] = [10, 10, 10]$. You can use "ode45" to solve the system of differential equations in MATLAB. Compare the results of this exercise with those depicted in Figure 8.6.

TABLE 8.6
Volumes of the Reactive Solutions A, B, C, D, H_2O_2 3.2 M, and NaOH 0.1 M for Observing the Orbán Reaction in Five Distinct Experiments

	V (sol. A) (mL)	V (H_2O_2 3.2 M) (mL)	V (sol B) (mL)	V (sol C) (mL)	V (NaOH 0.1 M) (mL)	V (sol D) (mL)
Exp. 1	1	0	1	1	0	2
Exp. 2	1	0	1	2	0	1
Exp. 3	1	0	2	1	0	1
Exp. 4	1	0	1	1	1	1
Exp. 5	0	1	1	1	0	2

8.7. It has been demonstrated (Fujii and Rondelez 2013) that the following abstract model can account for the mechanism illustrated in Figure 8.9:

$$X \xrightarrow{v_1} 2X \qquad\qquad v_1 = k_1 \cdot (pol) \cdot A \cdot \frac{X}{1 + b \cdot A \cdot X}$$

$$X + Y \xrightarrow{v_2} 2Y \qquad\qquad v_2 = k_2 \cdot (pol) \cdot X \cdot Y$$

$$X \xrightarrow{v_3} waste \qquad\qquad v_3 = k_3 \cdot (exo) \cdot \frac{X}{1 + \dfrac{Y}{K}}$$

$$Y \xrightarrow{v_4} waste \qquad\qquad v_4 = k_4 \cdot (exo) \cdot \frac{Y}{1 + \dfrac{Y}{K}}$$

Note that prey growth, prey and predator degradations obey Michaelis-Menten kinetics.

Write the system of two differential equations for prey and predator and solve it by numerical integration and the following values for the parameters (Padirac et al. 2013): $k_1 = 3 \times 10^{-3} \, nM^{-2} min^{-1}$; $pol = 1.7$ nM; $A = 140$ nM; $b = 6 \times 10^{-5} \, nM^{-2}$; $k_2 = 4 \times 10^{-3} \, nM^{-2} min^{-1}$; $k_3 = 10^{-2} \, nM^{-1} min^{-1}$; $exo = 25$ nM; $K = 34$ nM; $k_4 = 4 \times 10^{-3} \, nM^{-1} min^{-1}$.

8.8. The fundamental elementary steps of a pH oscillator are listed as follows.

$$H + Red_A \xrightarrow{k_1} HRed_A$$

$$HRed_A \xrightarrow{k_{-1}} H + Red_A$$

$$Ox + HRed_A + H \xrightarrow{k_2} Y$$

$$HRed_A + Y \xrightarrow{k_3} 3H$$

$$Y \xrightarrow{k_4} P$$

$$Red_B + H \xrightarrow{k_5} Q$$

Write the differential equations for the four variables H, Red_A, $HRed_A$, and Y assuming that the concentrations of Ox and Red_B are fixed. Solve numerically the differential equations. In analogy to the real situation, assume that only Red_A and H have nonzero inflow, while all the species are allowed to flow out the reactor. Use the following values for the rate constants and the concentrations of the reagents: $k_1 = (5 \times 10^{10} \, M^{-1}s^{-1})$; $k_{-1} = (3 \times 10^3 \, s^{-1})$; $k_2 = (3.077 \times 10^6 \, M^{-2}s^{-1})$; $k_3 = (10^6 \, M^{-1})$; $k_4 = (11 \, s^{-1})$; $k_5 = (2.5 \, M^{-1}s^{-1})$; k_0 (the inverse of the residence time) $= (10^{-3} \, s^{-1})$; $[Ox] = 6.5 \times 10^{-2}$ M; $[Red_B] = 2.0 \times 10^{-2}$ M; $[Red_A]_0 = 6 \times 10^{-2}$ M; $[H]_0 = 2 \times 10^{-2}$ M. All these values refer to the bromate-sulfite-ferrocyanide reaction (Luo and Epstein 1991).

8.9. Verify that in a biochemical multi-gene circuit similar to the repressilator, if the number of repressors is odd, we can observe oscillations, whereas if the number of repressors is even, the circuit converges to a fixed point. Test this rule of thumb for circuits with 2, 3, 4 and 5 repressors linked in a ring structure. Use the simple model of transcriptional regulation (see equation [8.39]) and the following values for the parameters: $\alpha_0 = 0.1$, $\alpha = 100$, $K = 1$, $n = 2.5$, $k_{dm} = 1$, $k_{dp} = 5$. As initial conditions fix the amount of the mRNA and protein of the first repressor to 10, and the amount of the mRNAs and proteins of all the other repressors to 5.

8.10. An oscillatory dynamic involving a synthetic biochemical circuit containing intercellular communication and affecting the abundance of a population of bacteria has been recently presented (Balagaddé et al. 2005). Many bacteria communicate and coordinate gene expression through quorum sensing. There are bacteria, like Escherichia coli, which produce and release N-acyl-homoserine lactone (*A*) as a chemical signal molecule. As the cell density *N* increases, the concentration of the signal molecule also increases. The signal molecule *A*, in turn, modulates the expression of a killer gene *R* that produces a killer protein *E*. The killer protein reduces the cell density *N*. When the population of the bacteria becomes again low, *A* also becomes low, and the activity of the killer gene becomes negligible. The population enjoys a renewed exponential growth. The delayed negative effect played by *E* on *N* may induce oscillations in the cell density. The ordinary differential equations (ODEs) proposed to describe the dynamics of such system are

$$\frac{dN}{dt} = k_1 N \left(1 - \frac{N}{K_1}\right) - k_d NE - k_f N$$

$$\frac{dA}{dt} = k_A N - \left(d_A + k_f\right) A$$

$$\frac{dR}{dt} = k_R A - d_R R$$

$$\frac{dE}{dt} = k_E R - d_E E$$

In such equations, k_f is the flow rate.

1. Solve the system of ODEs numerically by using the values for the parameters listed in Table 8.7.
2. Transform the differential equations in elementary steps and discuss to which "primary oscillator" may belong such synthetic biochemical circuit.

TABLE 8.7

Values of the Parameters Regarding the Abundance of a Population of Bacteria Influenced by Intercellular Communication

Parameter	Value
k_1	0.7 hr^{-1}
k_f	0.1 hr^{-1}
K_1	3×10^9 mL^{-1}
k_d	0.004 nM^{-1} hr^{-1}
k_A	4×10^{-7} nM mL hr^{-1}
d_A	0.1 hr^{-1}
k_R	1.0 hr^{-1}
d_R	0.7 hr^{-1}
k_E	1.0 hr^{-1}
d_E	0.7 hr^{-1}
$N(t=0)$	1×10^7 mL^{-1}

8.9   SOLUTIONS TO THE EXERCISES

8.1.   The differential equation describing how $[X]$ changes over time is:

$$\frac{d[X]}{dt} = k_r\left(C_A - [X]\right) + k_r'\left(C_A - [X]\right)[X]^2 - k_0[X]$$

If we divide both terms of the equation by C_A and we indicate with x the ratio $\left([X]/C_A\right)$, we obtain:

$$\frac{dx}{dt} = \left(1 - x\right)\left(k_r + k_r' C_A^2 x^2\right) - k_0 x$$

The steady-state solutions are the intersection points between the straight line $k_0 x$ and the cubic curve $-k_r' C_A^2 x^3 + k_r' C_A^2 x^2 - k_r x + k_r$. For the numerical values given in the text of the exercise, i.e., $k_r = 1\, t^{-1}$, $k_r' = 100\, M^{-2} t^{-1}$, $C_A = 1\, M$ (note that the autocatalytic process speeds up the rate of the transformation $A \to X$ of two orders of magnitude), we find that when k_0 is small or large (for instance $15\, t^{-1}$ and $35\, t^{-1}$, respectively), the straight line intersects the cubic curve only at one point. Graphically, we see that such steady states are stable (see plots a and c of Figure 8.16). On the other hand, when k_0 is medium, the straight line intersects the cubic curve at three points. Two of them are stable whereas the third (at the intermediate value of x) is unstable (see plot b of Figure 8.16).

8.2.   For the preparation of the stock solutions, we weight the salts and we dissolve them in specific amounts of water. By performing such operations, we make systematic and perhaps also random errors. The systematic errors depend on the limited accuracy of our balance, our pipettes, and our flasks. For instance, we weight 11.7023 g of $KBrO_3$ by using a technical balance whose inherent uncertainty is ± 0.0001 g. Let us suppose that the execution of the operation is perfect, and we do not introduce any random error. The final value of the $KBrO_3$ mass will be $m = (11.7023 \pm 0.0002)$ g. The next step is to solubilize the salt in deionized water. If we use a flask of volume $V = (200.00 \pm 0.15)$ mL and we avoid any random error, like, for instance, the parallax error (see Appendix D), the concentration of $KBrO_3$ solution will be: $\left[KBrO_3\right]_0 = 0.3503$ mol/L. Its uncertainty is estimated by using

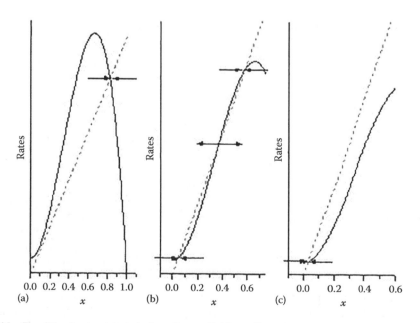

FIGURE 8.16   Possible steady-state solutions for small (a), medium (b), and large (c) value of k_0.

the formula of maximum a priori absolute error (knowing that the $KBrO_3$ molecular weight is $MW = (167.01 \pm 0.01)$ g/mol):

$$\Delta[KBrO_3]_0 = \left|\frac{1}{(MW)(V)}\right|\Delta m + \left|-\frac{m}{(V)(MW)^2}\right|\Delta(MW) + \left|-\frac{m}{(MW)(V)^2}\right|\Delta V$$

$$= 6 \times 10^{-6} + 2.1 \times 10^{-5} + 2.63 \times 10^{-4} = 2.9 \times 10^{-4} \text{ mol/L}$$

The largest contribution to the total uncertainty on $[KBrO_3]_0$ comes from the determination of the volume. Finally, the value of the concentration is $[KBrO_3]_0 = (0.3503 \pm 0.0003)$ mol/L. Similarly, we can estimate the uncertainties in the concentration of the other stock solutions. After preparing the stock solutions, we mix specific amounts of them to start the BZ reaction. In the first run of the experiment, the initial concentration of $KBrO_3$ is:

$$[KBrO_3]_1 = \frac{[KBrO_3]_0 V_1}{V_{fin}} = \frac{(0.3503 \text{ M})(2 \text{ mL})}{(8.2 \text{ mL})} = 0.08544 \text{ mol/L}$$

where:

V_1 is the volume of the $KBrO_3$ stock solution that is taken and mixed with the other reagents,

V_{fin} is the total volume of the reaction mixture.

The uncertainty in $[KBrO_3]_1$ is again estimated by the formula of maximum a priori absolute error:

$$\Delta[KBrO_3]_1 = \left|\frac{V_1}{V_{fin}}\right|\Delta[KBrO_3]_0 + \left|\frac{[KBrO_3]_0}{V_{fin}}\right|\Delta V_1 + \left|-\frac{[KBrO_3]_0 V_1}{V_{fin}^2}\right|\Delta V_{fin}$$

$$= 7.32 \times 10^{-5} + \left|\frac{0.3503}{8.2}\right|0.016 + \left|-\frac{(0.3503)(2)}{(8.2)^2}\right|0.070 = 0.0015\left(\frac{\text{mol}}{\text{L}}\right)$$

Therefore, the concentrations of $KBrO_3$ in the three runs of the reaction are:

$$[KBrO_3]_1 = (0.0854 \pm 0.0015) \text{mol/L}$$

$$[KBrO_3]_2 = (0.064 \pm 0.001) \text{mol/L}$$

$$[KBrO_3]_3 = (0.0427 \pm 0.0007) \text{mol/L}$$

Then, we calculate the logarithm of this quantities. The uncertainty in $\log[KBrO_3]_i$ is determined by the maximum a priori absolute error's formula:

$$\Delta\left(\log[KBrO_3]_i\right) = \left|\frac{\log_{10} e}{[KBrO_3]_i}\right|\Delta[KBrO_3]_i$$

Figure 8.17 reports some spectra of the BZ reaction recorded in the absence (a) and the presence (b) of the indicator ferroin. In (a), it is evident that the solution is uncolored when the chemical system is in its reduced state (i.e., when Ce^{+3} is the dominant form of the cerium ions), and it becomes slightly yellow when Ce^{+4} becomes the dominant form. In the presence of the redox indicator (see plot b), the solution is red when the coordination compound is in its reduced state (with iron in the Fe^{+2} state) because it absorbs all the visible wavelengths except those of the red region. When the solution is in its oxidized state, the complex ferriin (having the iron ion in its Fe^{+3} state) absorbs all the visible wavelengths except those of the blue region.

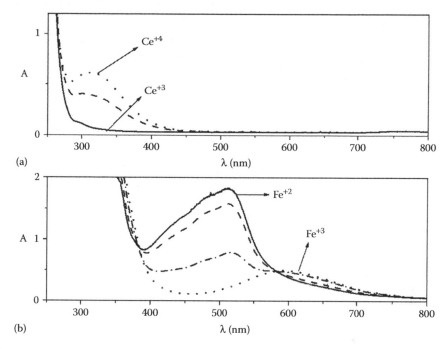

FIGURE 8.17   Spectral evolution of the BZ reaction with the cerium ions as catalysts and in the absence (a) and the presence (b) of ferroin as the redox indicator.

The profiles of the absorbance at 511 nm as a function of the time are plotted in Figure 8.18 for the three different concentrations of $KBrO_3$, labeled as 1 (in a), 2 (in b), and 3 (in c), respectively.

The traces have been recorded for at least 30 minutes. From the kinetics, we determine the periods of the oscillations. Then, we calculate the $\log T$. We will have as many $\log T$ as are the oscillations in each run. The best estimate will be the average, and its uncertainty will be the standard deviation of the mean. Finally, we plot $\log(T_i) \pm \Delta\log(T_i)$ versus $\log([KBrO_3]_i) \pm \Delta\log([KBrO_3]_i)$, and we determine the exponent n by fitting the data with the least-squares method, as shown in Figure 8.19.

It results that $\log(T) = \text{cost} + n\log([KBrO_3]) = (0.32 \pm 0.05) + (-1.61 \pm 0.04)\log([KBrO_3])$ with a correlation coefficient $r = 0.99972$. The lower the $[KBrO_3]$, the longer the period. The exponent is roughly $-(3/2)$. This result is consistent with a mechanism in which bromate plays a role in the rate-limiting steps.

8.3. During the progress of the BZ reaction, changes in the concentrations of protons, bromate, malonic acid, and hypobromous acid are small because the amount of catalyst used is almost two orders of magnitude smaller. Therefore, assuming that $[H^+]$, $[BrO_3^-]$, $[CH_2(COOH)_2]$, and $[HBrO]$ remain unchanged during the reaction, it is possible to write the following system of three nonlinear differential equations:

$$\frac{d[X]}{dt} = k_1[H^+]^2[A][Y] - k_2[H^+][X][Y] + k_3[H^+][A][Z] - 2k_4[X]^2$$

$$\frac{d[Y]}{dt} = -k_1[H^+]^2[A][Y] - k_2[H^+][X][Y] + fk_5[O][Z]$$

$$\frac{d[Z]}{dt} = k_3[H^+][A][Z] - k_5[O][Z]$$

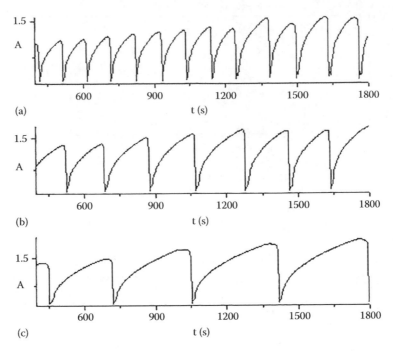

(a)

(b)

(c)

FIGURE 8.18 Trends of A recorded at 511 nm for the three distinct concentrations of $KBrO_3$: (1) 0.0854 M in a; (2) 0.064 M in b; (3) 0.0427 M in c.

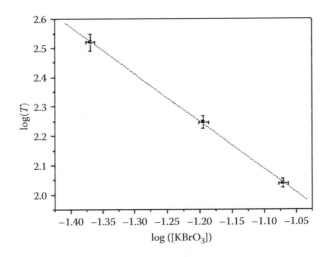

FIGURE 8.19 Linearization of the relation between $\log(T)$ and $\log([KBrO_3])$.

The function file to integrate the system of differential equations in MATLAB should look like this:

```
function dy = Oregonator(t, y)
dy = zeros(3,1);
options=odeset('RelTol',10^-6,'AbsTol',10^-10);
k1 = 1.28;
k2 = 800000;
k3 = 8;
```

```
k4 = 2000;
k5 = 1;
A = 0.12;
O = 0.1;
f = 0.6;
dy(1)=k1*A*y(2)-k2*y(1)*y(2)+k3*A*y(1)-2*k4*y(1)*y(1);
dy(2)=-k1*A*y(2)-k2*y(1)*y(2)+f*k5*O*y(3);
dy(3)=k3*A*y(1)-k5*O*y(3);
```

where $k1$ is $k_1[H^+]^2$, $k2$ is $k_2[H^+]$, $k3$ is $k_3[H^+]$, $y(1)$ is X, $y(2)$ is Y, $y(3)$ is Z. The script file should look like the following:

$$[t, y] = ode15s('Oregonator', [0\ 1200], [0.0001\ 0.0001\ 0.002])$$

wherein [0 1200] is the time window, whereas [0.0001 0.0001 0.002] are the initial conditions for $y(1)$, $y(2)$, and $y(3)$, respectively. The profiles of how the concentrations of X, Y and Z change over time (calculated in the case of $A = 0.18$) are shown in Figure 8.20. It is evident that when the concentration of the activator X is high, then the concentration of inhibitor Y is very low, and vice versa. This behavior is due to the coproduct autocontrol, which is the time needed to store Z that finally regenerates Y.

The three-dimensional plot of Figure 8.21 shows that the chemical system approaches a limit cycle whatever is the initial condition. Moreover, if we look at the projection on the XY plane, we confirm that X and Y exclude reciprocally.

To determine the average value of the period of the oscillations we use the Fourier transform of the trend of [Z] versus time. An example of Fourier Transform is shown in Figure 8.22.

The uncertainty on the frequency is given by the Half Width at Half Maximum of the peak with the largest amplitude (see Figure 8.22). The period is given by $T = 1/v$. The uncertainty of the period is calculated by using the formula of the propagation of the "a priori maximum absolute error":

$$\Delta T = \left| -\frac{1}{v^2} \right| \Delta v$$

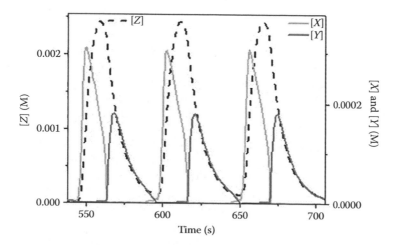

FIGURE 8.20 Trends of [Z], [X] and [Y] over time, when [A] = 0.18 M.

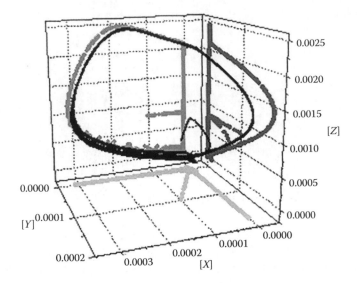

FIGURE 8.21   Three-dimensional phase space of the Oregonator model.

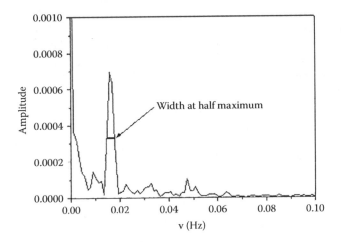

FIGURE 8.22   Fourier transform of the time series [Z] versus time when [A] = 0.18 M.

The numerical integration of the system of three differential equations and the determination of the period of the oscillations after the calculation of the Fourier Transform are repeated for all the conditions proposed in the text of this exercise. Finally, to establish the exponential term n of the relation $T \propto [BrO_3^-]^n$, it is convenient to build a log-log plot, like that shown in Figure 8.23. The straight line determined by the least-squares method has a slope $n = -(0.6 \pm 0.1)$.

8.4.   The function file to integrate the system of three differential equations in MATLAB should look like that presented in exercise 8.3. The values of the kinetic constants at the three temperatures proposed in the text of the exercise are reported in Table 8.8.

They have been calculated through the Arrhenius equation $\left(k(T_i)/k(T_0)\right) = e^{(E_{act}/R(T_0 - T_i))}$. The script file to solve this exercise should look like the following one:

```
[t y]=ode15s('Oregonator',[0 1200], [0.0001 0.0001 0.002]);
```

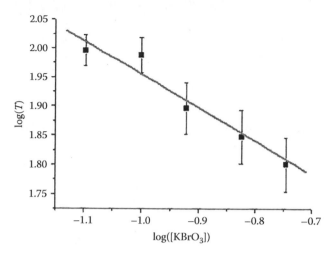

FIGURE 8.23 Linearization of the relation between $\log(T)$ and $\log([KBrO_3])$.

TABLE 8.8

Values of the Kinetic Constants for the Elementary Steps Appearing in the Oregonator Model

T (K)	k_1 (mol⁻³dm⁹s⁻¹)	k_2 (mol⁻²dm⁶s⁻¹)	k_3 (mol⁻²dm⁶s⁻¹)	k_4 (mol⁻¹dm³s⁻¹)	k_5 (mol⁻¹dm³s⁻¹)
293.15	2	1,000,000	10	2000	1
303.15	4.16	1,402,878	22.5	4758	2.58
313.15	8.24	1,925,972	48.2	10709	6.27

wherein [0 1200] is the time window, whereas [0.0001 0.0001 0.002] are the initial conditions for y(1), y(2), and y(3), respectively. Feel free to change slightly the initial conditions.

The calculated average values of the periods are 79 s (at 293.15 K), 30 s (at 303.15 K), and 13 s (at 313.15 K). If we plot $\ln(1/\Delta T)$ versus $1/T$ and we determine the fitting straight line with the least-squares method, we obtain a slope of (-8300 ± 130), which corresponds to $E_{act} = (69000 \pm 1000)\,\mathrm{J/mol} = (69 \pm 1)\,\mathrm{kJ/mol}$.

8.5. The concentrations of the reactants in the proposed five experiments are listed in Table 8.9 (such values have been inspired by the demonstration on chemiluminescence by Prypsztejn et al. [2005]).

TABLE 8.9

Concentrations of the Reagents for the Orbán Reaction in Five Distinct Experiments

	$C(H_2O_2)$ M	V(KSCN) M	V(Luminol) M	V(NaOH) M	V(CuSO₄) M
Exp. 1	0.2	0.03	7.4·10⁻⁴	0.02	2.4·10⁻⁴
Exp. 2	0.2	0.03	1.48·10⁻³	0.04	1.2·10⁻⁴
Exp. 3	0.2	0.06	7.4·10⁻⁴	0.02	1.2·10⁻⁴
Exp. 4	0.2	0.03	7.4·10⁻⁴	0.04	1.2·10⁻⁴
Exp. 5	0.64	0.03	7.4·10⁻⁴	0.02	2.4·10⁻⁴

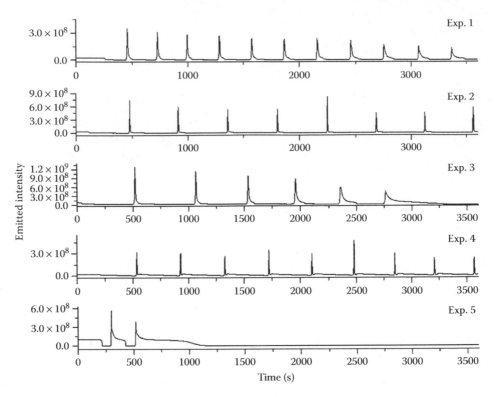

FIGURE 8.24 Chemiluminescent signals recorded in the five experiments whose conditions are listed in Table 8.9.

The time evolutions of the chemiluminescent signals are shown in Figure 8.24.

The most intense peaks have been achieved in "Exp. 3" when the concentration of KSCN is at its maximum value. However, the oscillations die within less than one hour. The shortest periods of the oscillations have been observed in "Exp. 1" when [CuSO$_4$] is at its maximum value. In fact, in "Exp. 1" the average period is 291 seconds. Unfortunately, the intensity of peaks progressively decreases. In "Exp. 2" and "Exp. 4" (when the concentration of NaOH is at its maximum value), the "artificial firefly" lasts longer: it does not stop to give blue chemiluminescent flashes even after one hour. When we use a very high [H$_2$O$_2$] in "Exp. 5," we record only two peaks and, then, "the artificial firefly" dies. According to these results, we may optimize the intensity and the persistence of the chemiluminescent signal by selecting the following two combinations of reagents concentrations.

Combination 1: [H$_2$O$_2$] = 0.32 M, [KSCN] = 0.03 M, [luminol] = 1.1·10^{-3} M, [NaOH] = 0.03 M, [CuSO$_4$] = 2.4·10^{-4} M;

Combination 2: [H$_2$O$_2$] = 0.21 M, [KSCN] = 0.03 M, [luminol] = 1.2·10^{-3} M, [NaOH] = 0.033 M, [CuSO$_4$] = 2.4·10^{-4} M.

For these two conditions, the chemiluminescence becomes visible by sight in a dark room. The recorded chemiluminescent traces are shown in Figure 8.25.

We confirm that a higher [NaOH] guarantees a longer persistence of the oscillations, whereas a larger [H$_2$O$_2$] assures shorter periods of oscillations at the beginning of the reaction.

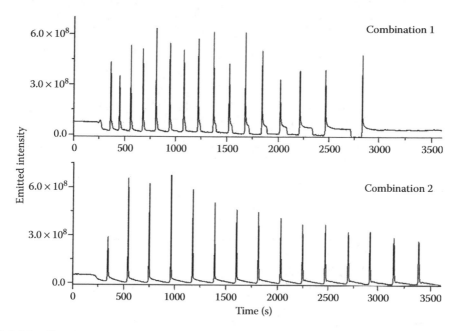

FIGURE 8.25 Chemiluminescent signals recorded for two new conditions of the Orbán reaction (read the text for the details).

8.6. The system of differential equations will be:

$$\frac{dX}{dt} = k_1 \Lambda X - k_2 XY - k_3 XZ + k_0 \left(X_{in} - X \right)$$

$$\frac{dY}{dt} = k_2 XY + k_0 \left(Y_{in} - Y \right)$$

$$\frac{dZ}{dt} = -k_3 XZ + k_0 \left(Z_{in} - Z \right)$$

If we solve the system by numerical integration with the solver "ode45" of MATLAB, the results are those shown in Figure 8.26.

Since the value of k_1 is larger in this example than in the case of Figure 8.6, X reacts more quickly, and the profile of its spikes appears sharper. On the other hand, k_2 and k_3 are much smaller in this example. Therefore, Y changes more smoothly than before.

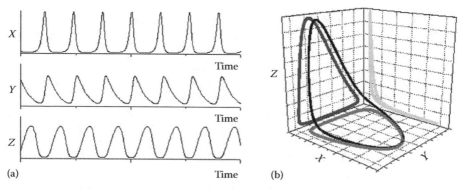

FIGURE 8.26 Simulated oscillations for the modified Lotka-Volterra model. Graph (a) shows the temporal trends of X, Y, and Z. Graph (b) is the representation of the three-dimensional phase space for the system.

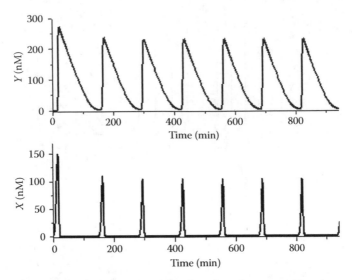

FIGURE 8.27 Temporal trends of the concentrations of predator (upper graph) and prey (lower graph).

8.7. In MATLAB, the function file for the abstract model should look like this:

```
function dy = DNA(t, y)
dy = zeros(2,1);
k1 = 0.003;
pol = 1.7;
A = 140;
b = 0.00006;
k2 = 0.004;
kN = 0.01;
exo = 25;
K = 34;
kp = 0.004;
dy(1) = (k1*pol*A*y(1)/(1+b*A*y(1))) - (k2*pol*y(1)*y(2)) - (kN*exo*y(1)/
(1+y(2)/K))
dy(2) = (k2*pol*y(1)*y(2)) - (kp*exo*y(2)/(1+y(2)/K))
```

wherein $y(1)$ is the prey X, and $y(2)$ is the predator Y.
The script file is:

$$[t, y] = ode15s('DNA', [0\ 1000], [3.0\ 0.1]);$$

The results are plotted in Figure 8.27. There are evident oscillations of the biochemical predator and prey species.

8.8. The function file to integrate the system of differential equations in MATLAB should look like this

```
function dy = pHoscillator(t, y)
dy = zeros(4,1);
k1 = 50000000000;
k11 = 3000;
k2 = 3077000;
k3 = 1000000;
k4 = 11;
k5 = 2.5;
k0 = 0.001;
```

```
X0 = 0.06;
Y0 = 0.02;
A = 0.065;
B = 0.02;
dy(1) = -k1*y(1)*y(2)+k11*y(3)+k0*(X0-y(1));
dy(2) = -k1*y(1)*y(2)+k11*y(3)-k2*A*y(3)*y(2)+3*k3*y(3)*y(4)-
k5*B*y(2)+k0*(Y0-y(2));
dy(3) = k1*y(1)*y(2)-k11*y(3)-k2*A*y(3)*y(2)-k3*y(3)*y(4)-k0*y(3);
dy(4) = k2*A*y(3)*y(2)-k3*y(3)*y(4)-k4*y(4)-k0*y(4);
```

wherein $y(1) = \text{Red}_A$, $y(2) = \text{H}$, $y(3) = \text{HRed}_A$, and $y(4) = \text{Y}$.
The script file should look like the following:

$$[t, y] = \text{ode15s('pHoscillator',[0 1000], [0.065 0.025 0.001 0.001])};$$

The results of the simulation are plotted in Figure 8.28. It is evident that the mechanistic model of the pH oscillator reproduces quite well the experimental oscillations.

8.9. If we use MATLAB to solve this exercise, the function file to integrate the system of ordinary differential equations should look like the following one:

```
function dy = tworepressors(t, y)
dy = zeros(4,1);
a0 = 0.1;
a = 100;
n = 2.5;
kdp = 5;
kdm = 1;
K = 1;
dy(1) = a0+[a/(1+((y(4))^n/K^n))]-kdm*y(1);
dy(2) = kdm*y(1)-kdp*y(2);
dy(3) = a0+[a/(1+((y(2))^n/K^n))]-kdm*y(3);
dy(4) = kdm*y(3)-kdp*y(4);
```

Such file is valid for a circuit having two repressors linked in a ring. The results are shown in Figure 8.29. The system reaches a fixed point where the concentrations of the second mRNA and protein vanish.

When the number of repressors is three, we have the repressilator we studied in paragraph 8.4.5. The results are depicted in Figure 8.30. We do have oscillations and the period is 7 minutes and 24 seconds.

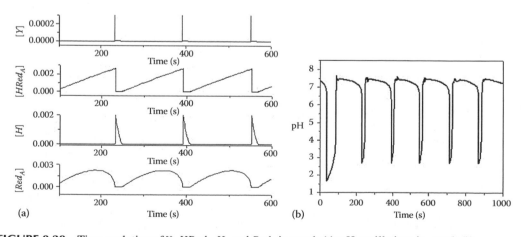

FIGURE 8.28 Time evolution of Y, $HRed_A$, H, and Red_A in graph (a); pH oscillations in graph (b).

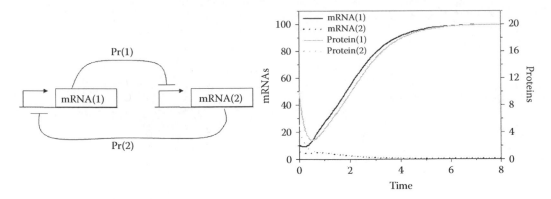

FIGURE 8.29 Scheme of a biochemical circuit with two repressors linked in a ring (on the left) and its dynamical evolution (on the right).

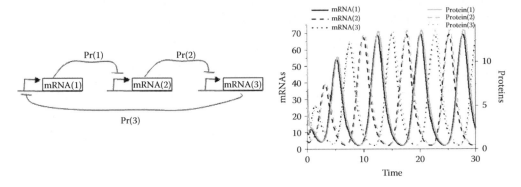

FIGURE 8.30 Scheme of a biochemical circuit with three repressors linked in a ring (on the left) and its dynamical evolution (on the right).

When the network consists of four repressors, the circuit reaches a fixed point (see Figure 8.31). In the fixed point, the concentrations of the components of repressors (2) and (4) vanish, whereas those of repressors (1) and (3) reach not-null steady state values.

When the biochemical circuit consists of five repressors, oscillations emerge, again (see Figure 8.32). The period of the oscillations is 14 minutes, i.e., longer than that found for the repressilator. The larger the number of repressors, the longer the period of the oscillations.

8.10. The solution of the first part requires MATLAB or an analogous software. In case you use MATLAB, the function file to integrate the system of differential equations should look like this:

```
function dy = bacteria(t, y)
dy = zeros(4,1);
k1 = 0.7;
kf = 0.1;
K1 = 3 * 10^9;
kd = 0.004;
kE = 1.0;
dE= 0.7;
kR = 1.0;
dR = 0.7;
kA = 4 * 10^-7;
dA = 0.1;
```

```
dy(1) = k1*y(1)*(1-y(1)/K1)-kd*y(1)*y(4)-kf*y(1);
dy(2) = kA*y(1)-(dA+kf)*y(2);
dy(3) = kR*y(2)-dR*y(3);
dy(4) = kE*y(3)-dE*y(4);
```

The script file is

$$[t, y]=ode45(\text{‘bacteria’},[0\ 500], [1 * 10^7\ 0\ 0\ 0]);$$

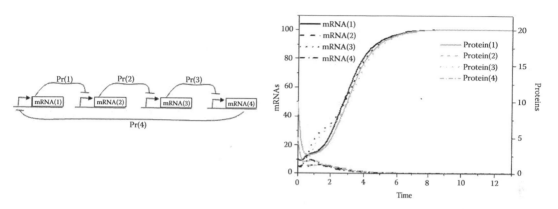

FIGURE 8.31 Scheme of a biochemical circuit with four repressors linked in a ring (on the left) and its dynamical evolution (on the right).

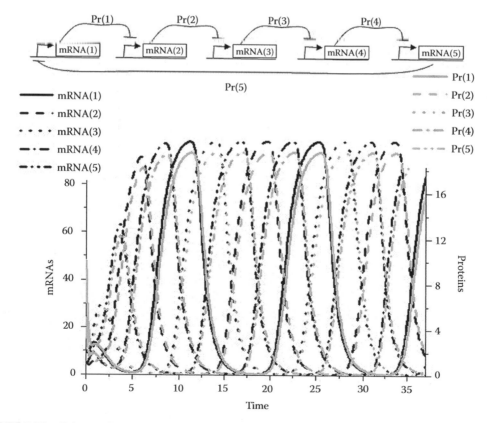

FIGURE 8.32 Scheme of a biochemical circuit with five repressors linked in a ring (on top) and its dynamical evolution (at the bottom).

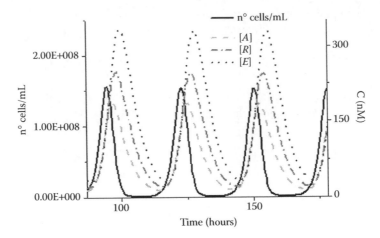

FIGURE 8.33 Oscillations in the number of bacteria and in the concentrations of the molecular signal (A), the killer gene (R) and protein (E).

$$\text{I)} \quad N + F \underset{k_{-1}}{\overset{k_1}{\rightleftharpoons}} 2N$$

$$\text{II)} \quad N \xrightarrow{k_A} A + N$$

$$\text{II)} \quad A \xrightarrow{d_A} P$$

$$\text{III)} \quad A \xrightarrow{k_R} A + R$$

$$\text{III)} \quad R \xrightarrow{d_R} P'$$

$$\text{IV)} \quad R \xrightarrow{k_E} R + E$$

$$\text{IV)} \quad E \xrightarrow{d_E} P''$$

$$\text{I)} \quad N + E \xrightarrow{k_d} D + E$$

FIGURE 8.34 The mechanism of the proliferation of bacteria sending chemical messages that induce inhibition.

The results are plotted in Figure 8.33.

The second part of the exercise requires to assign the synthetic biochemical circuit to one of the "primary oscillators" we studied in this chapter. If we transform the differential equations of the text in elementary steps, we may write the mechanism shown in Figure 8.34.

The numbers I, II, III and IV, appearing in Figure 8.34, refer to the number of the differential equations. The first step is autocatalytic in N. Then, a bunch of steps follows, where the bacteria produce the signal molecule A; A activates the killer gene R, and, finally, R produces the killer enzyme E. E exerts a negative feedback on N. Such negative feedback is delayed by the processes needed to produce E. Therefore, such synthetic biochemical circuit may be embedded in the "coproduct autocontrolled oscillator" that is at the basis of the Oregonator model.

9 The Emergence of Order in Space

The scientist does not study nature because it is useful; he studies it because he delights in it, and he delights in it because it is beautiful.

Henri Poincaré (1854–1912 AD)

9.1 INTRODUCTION

What happens if an autocatalytic process or an oscillating reaction are carried out in a medium that is not under stirring? Well, surprising phenomena and structures appear, like those shown in Figure 9.1. In this chapter, we will discover that systems in out-of-equilibrium conditions may self-organize even in space. Beautiful phenomena and patterns, like chemical waves, and Turing structures and periodic precipitations (Figure 9.1) may emerge even if we start from initial spatially homogeneous conditions. The spontaneous pattern formation occurs not only in a chemistry laboratory, but also everywhere in nature (Ball 2001), in both living and azoic systems.

A spontaneous question arises: how is it possible that chemical structures emerge from a spatially homogeneous system? Such phenomena of self-organization (or dynamic self-assembly) are in sharp contrast with Curie's symmetry principle, which we learned in Chapter 3. In fact, they are examples of spatial "Symmetry Breaking" transitions. The Symmetry Breaking is characteristic of the nonlinear regime.

In this chapter, we will discover that diffusion combined with proper reaction kinetics may contribute to the development of instabilities beyond which the system gives rise to spatial or spatiotemporal ordered patterns. The resulting patterns have a reduced number of symmetry elements if compared with the initial uniform states. They are puzzling because it seems that entropy is dissipated rather than produced. But, they do not violate the Second Law of Thermodynamics. They produce entropy that is discharged in the surrounding environment, like any other "Dissipative Structure" that we have already known.

9.2 THE REACTION-DIFFUSION MODEL

The time evolutions of the many systems we studied in the previous chapters give rise to diversified dynamical landscapes. Such landscapes may be imagined as mountain ranges. Locally, there are many shapes, like those sketched in Figure 9.2. Each one represents a particular dynamical situation. We may stumble across a stable periodic orbit in a groove (like in (a)); an unstable periodic orbit on a brim of a crater (like in (b)); an unstable fixed point on top of a bowl (like in (c)); a stable fixed point at the bottom of a bowl (like in (d)); a saddle point (like in (e)). Such shapes may be mapped by the linear stability analysis applied locally to each fixed point of the dynamical landscape. The eigenvalues and the eigenvectors that we determine tell us how the system responds to local perturbations.

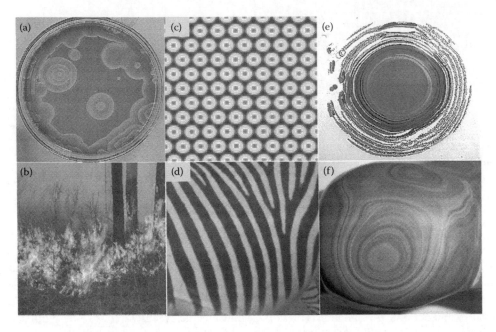

FIGURE 9.1 Examples of chemical waves in (a) and (b), Turing patterns in (c) and (d), and periodic precipitation in (e) and (f) observed in a laboratory (top row) and nature (bottom row).

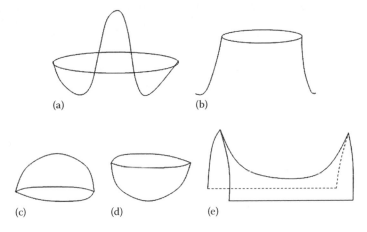

FIGURE 9.2 Possible shapes in a dynamical landscape: a groove in (a), a crater in (b), a bowl in (c) and (d), and a saddle in (e).

If we focus on non-linear chemical reactions enriched by diffusion, we expect to see more complicated, odd and bizarre dynamical landscapes. To unveil them, we use our traditional approach: we linearize the non-linear system around the steady-state solutions, and we determine the eigenvectors and the eigenvalues. The eigenvalues tell us how fast the system rolls along the eigenvectors.

Let us imagine having non-linear chemical reactions taking place inside an open and unstirred system of volume V and area A (see Figure 9.3). Let us indicate with m the concentration of M species within V. The change in the amount of m depends on the flow of M in $\left(\vec{J}_{o \to i} \right)$ and out $\left(\vec{J}_{i \to o} \right)$ of V, and on the rates of all the chemical reactions involving M within V. If we indicate with $\vec{J}_M = \left(\vec{J}_{i \to o} - \vec{J}_{o \to i} \right)$, the net flow through the infinitesimal area $d\vec{A}$, and with $R(M)$ the net rate of chemical transformations involving M within the infinitesimal volume element dV (see Figure 9.3), the balance equation, based on the Conservation Law of Mass, is

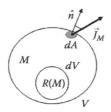

FIGURE 9.3 Terms of the balance equation for M: $\vec{J}_M$ represents the net flow of M out of V in a small area $d\vec{A}$ having $\hat{n}$ as its normal vector; $R(M)$ represents the net rate of M's production within the volume V.

$$\int_V \left(\frac{\partial m}{\partial t} \right) dV = \int_V R(M) dV - \int_A \vec{J}_M \, d\vec{A} \qquad [9.1]$$

According to the divergence theorem (see Box 9.1 of this chapter)

$$\int_A \vec{J}_M \, d\vec{A} = \int_V \left(\vec{\nabla} \vec{J}_M \right) dV \qquad [9.2]$$

BOX 9.1   THE DIVERGENCE THEOREM

For the demonstration of the divergence theorem, let us consider the flow of M out of a tiny volume having the shape of a parallelepiped with dx, dy, and dz as sides lengths (see the following figure, within this box). Let us start to calculate the flow of M through the two sides that are orthogonal to the x-axis. The flow through the $ABCD$ side (having A_{ABCD} as area) is given by:

$$d\Phi_{ABCD} = \vec{J}_M d\vec{A}_{ABCD} = -J_{M,x}\left(x, \bar{y}, \bar{z} \right) dydz \qquad [B1.1]$$

where x represents the position of the $ABCD$ face with respect to the x-axis; $\bar{y}$ and $\bar{z}$ are the average values of the coordinates y and z for the $ABCD$ face. Since the normal vector to the $ABCD$ face has opposite orientation with respect to the x-axis, we need to introduce the minus sign in equation [B1.1].

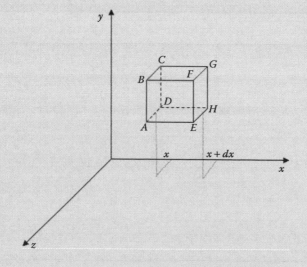

(Continued)

BOX 9.1 (Continued)   THE DIVERGENCE THEOREM

Similarly, the flow of M through the $EFGH$ side, having $x + dx$ as its own coordinate along the x-axis, is

$$d\Phi_{EFGH} = \vec{J}_M d\vec{A}_{EFGH} = +J_{M,x}(x+dx,\bar{y},\bar{z})dydz \qquad [\text{B1.2}]$$

The equation [B1.2] can be transformed into the equation [B1.3] by developing $J_{M,x}(x+dx,\bar{y},\bar{z})$ in Taylor series with $(x,\bar{y},\bar{z})$ as its accumulation point, and truncating the development at its first term, i.e., at its first derivative:

$$d\Phi_{EFGH} = J_{M,x}(x,\bar{y},\bar{z})dydz + \left(\frac{\partial J_{M,x}}{\partial x}\right)_{(x,\bar{y},\bar{z})} dxdydz \qquad [\text{B1.3}]$$

If we sum the contributions $d\Phi_{ABCD}$ and $d\Phi_{EFGH}$, we obtain

$$d\Phi_{ABCD} + d\Phi_{EFGH} = \left(\frac{\partial J_{M,x}}{\partial x}\right)_{(x,\bar{y},\bar{z})} dV \qquad [\text{B1.4}]$$

Analogously, we may determine the contributions of $\vec{J}_M$ through the pairs of faces that are orthogonal to the y- and z-axis, respectively. Finally, the total flux will be

$$d\Phi_{tot} = \left(\frac{\partial J_{M,x}}{\partial x} + \frac{\partial J_{M,y}}{\partial y} + \frac{\partial J_{M,z}}{\partial z}\right)dV \qquad [\text{B1.5}]$$

The term $\left(\frac{\partial J_{M,x}}{\partial x} + \frac{\partial J_{M,y}}{\partial y} + \frac{\partial J_{M,z}}{\partial z}\right)$ is the divergence of $\vec{J}_M$: $div\left(\vec{J}_M\right)$. The flux of M can also be defined by equation [B1.6]:

$$d\Phi_{tot} = \vec{J}_M d\vec{A}_{tot} \qquad [\text{B1.6}]$$

The term $d\vec{A}_{tot}$ represents the sum of the area of the six faces of the parallelepiped. By integration, we conclude that the flow through the macroscopic area A, encompassing the volume V, is

$$\int_A \vec{J}_M d\vec{A} = \int_V div\left(\vec{J}_M\right)dV \qquad [\text{B1.7}]$$

Note that the divergence is a differential operator. When it is applied to the vector $\vec{J}_M$, it gives the scalar function: $div\left(\vec{J}_M\right)$. We can use the nabla $\vec{\nabla}$ operator, which is a vector differential operator $\vec{\nabla} = \hat{i}\frac{\partial}{\partial x} + \hat{j}\frac{\partial}{\partial y} + \hat{k}\frac{\partial}{\partial z}$, to formulate the divergence theorem. With $\vec{\nabla}$, the divergence theorem is reported in equation [9.2]. The divergence of a vector, like $\vec{J}_M$, can be expressed as a scalar product between the nabla operator $\vec{\nabla}$ and the vector itself.

Introducing equation [9.2] into [9.1], we may rewrite the balance equation in the following form:

$$\int_V \left(\frac{\partial m}{\partial t}\right) dV = \int_V R(M) dV - \int_V \left(\vec{\nabla} \vec{J}_M\right) dV \tag{9.3}$$

Since equation [9.3] must be valid for any volume, we can equate the integrands. This step gives us the differential form of the balance equation:

$$\left(\frac{\partial m}{\partial t}\right) = R(M) - \vec{\nabla} \vec{J}_M \tag{9.4}$$

Reminding Fick's law [3.66] ($\vec{J}_M = -D_M \vec{\nabla} m$), and introducing it in [9.4], we obtain

$$\left(\frac{\partial m}{\partial t}\right) = R(M) + D_M \nabla^2 m \tag{9.5}$$

The operator $\nabla^2 = \left(\frac{\partial^2}{\partial x^2} + \frac{\partial^2}{\partial y^2} + \frac{\partial^2}{\partial z^2}\right)$ is the Laplacian, which is the sum of the second derivatives with respect to the spatial variables.

Now, we need to express the term $R(M)$. We imagine having nonlinear chemical processes inside the volume V. For instance, we may have an autocatalytic reaction of the type B + M → 2M, whose rate is kbm (where k is the kinetic constant, and b is the concentration of B). For simplicity, we imagine having just one spatial coordinate, r. Hence, the balance equation [9.5] becomes

$$\left(\frac{\partial m}{\partial t}\right) = kbm + D_M \left(\frac{\partial^2 m}{\partial r^2}\right) \tag{9.6}$$

To find the steady-state solution, we impose $(\partial m / \partial t) = 0$. It derives that

$$kbm = -D_M \left(\frac{\partial^2 m}{\partial r^2}\right) \tag{9.7}$$

The solution of equation [9.7] is a function whose second derivative with respect to r is equal to the same function with the changed sign. A solution is

$$m = m_0 \sin\left(\frac{2\pi}{\lambda} r + \varphi\right) \tag{9.8}$$

In fact, if we introduce [9.8] in [9.7], we obtain

$$\left(\frac{\partial^2 m}{\partial r^2}\right) = -m_0 \left(\frac{2\pi}{\lambda}\right)^2 \sin\left(\frac{2\pi}{\lambda} r + \varphi\right) = -\frac{kbm_0}{D_M} \sin\left(\frac{2\pi}{\lambda} r + \varphi\right) \tag{9.9}$$

The meaning is that a stationary sinusoidal pattern emerges. Its wavelength is

$$\lambda = 2\pi \sqrt{\frac{D_M}{kb}} \tag{9.10}$$

If the mono-dimensional system extends in the range $[-L, +L]$ and the boundary conditions are $\left(\frac{\partial m}{\partial r}\right)_{+L} = \left(\frac{\partial m}{\partial r}\right)_{-L} = 0$. Then, it derives that

$$\left(\frac{\partial m}{\partial r}\right)_{+L} = \left(\frac{\partial m}{\partial r}\right)_{-L} = m_0 \left(\frac{2\pi}{\lambda}\right)\cos\left(\frac{2\pi}{\lambda}(\pm L)+\varphi\right)=0 \qquad [9.11]$$

The latter equation is verified when

$$\left(\frac{2\pi}{\lambda}(\pm L)+\varphi\right)=\pm\frac{n\pi}{2} \qquad \text{with } n=1,3,5,\ldots \qquad [9.12]$$

If
$$\varphi=0, \ \lambda=\frac{4L}{n} \qquad [9.13]$$

TRY EXERCISES 9.1 AND 9.2

9.3   TURING PATTERNS

The Reaction-Diffusion (RD) model we have developed in the previous paragraph is rather trivial. We need more sophisticated models to interpret the emergence of spatially ordered structures, like Turing patterns, chemical waves, and periodic precipitation.

For the interpretation of the Turing patterns (see, e.g., Figure 9.1 c and d), we need a model that includes two species, X and Y, participating in non-linear kinetic laws and diffusing freely in space. Let us indicate the concentrations of the two chemical variables as x and y, respectively. The Partial Differential Equations (PDE) of the type [9.5] become

$$\left(\frac{\partial x}{\partial t}\right)=R(X)+D_X\nabla^2 x$$
$$\left(\frac{\partial y}{\partial t}\right)=R(Y)+D_Y\nabla^2 y \qquad [9.14]$$

In [9.14], $R(X)$, D_X, $R(Y)$, and D_Y represent the net production rate and the diffusion coefficients for X and Y, respectively. We simplify the mathematical treatment of the chemical system if we consider only one spatial dimension that we label as r. Therefore, equation [9.14] transforms into equation [9.15]:

$$\left(\frac{\partial x}{\partial t}\right)=R(X)+D_X\left(\frac{\partial^2 x}{\partial r^2}\right)$$
$$\left(\frac{\partial y}{\partial t}\right)=R(Y)+D_Y\left(\frac{\partial^2 y}{\partial r^2}\right) \qquad [9.15]$$

Let us assume that, initially, the system is in a stable steady-state. The concentrations of X and Y are (x_s, y_s), and the chemical compounds are homogeneously distributed in space. Since we are in a steady-state, we have that $[R(X)]_s = [R(Y)]_s = 0$: the sums of the rates of the transformations involving X and Y, both determined at the steady-state (x_s, y_s), are equal to zero. Moreover, the steady-state is stable. Therefore, $tr(J) < 0$ and $\det(J) > 0$,

$$\text{where } J = \begin{pmatrix} \left[R_x(X) \right]_s & \left[R_y(X) \right]_s \\ \left[R_x(Y) \right]_s & \left[R_y(Y) \right]_s \end{pmatrix}$$

is the Jacobian matrix, whose elements are the partial derivatives of the rates evaluated at the steady-state (x_s, y_s).

Now, let us imagine slightly perturbing the system at one point. The terms δx and δy represent the extents of the perturbations on x and y, respectively. In the absence of diffusion, when the chemical system is under stirring, the perturbations annihilate because the steady-state is stable. What happens in the presence of diffusion?

We expect that diffusion should act to lessen, and eventually annihilate, the inhomogeneities induced by the perturbations, promoting the restoration of the original uniform distribution of the chemical species into space. Does the system behave as we expect? The answer is: "it depends."

Let us have a look at the possible solutions of the partial differential equations linearized about the steady-state (x_s, y_s):

$$\frac{\partial}{\partial t}\begin{pmatrix} \delta x \\ \delta y \end{pmatrix} = \begin{pmatrix} \left[R_x(X) \right]_s & \left[R_y(X) \right]_s \\ \left[R_x(Y) \right]_s & \left[R_y(Y) \right]_s \end{pmatrix}\begin{pmatrix} \delta x \\ \delta y \end{pmatrix} + \begin{pmatrix} D_X \dfrac{\partial^2}{\partial r^2} & 0 \\ 0 & D_Y \dfrac{\partial^2}{\partial r^2} \end{pmatrix}\begin{pmatrix} \delta x \\ \delta y \end{pmatrix} \qquad [9.16]$$

The solution of equation [9.16], for both x and y, is the product of two functions: one depending on the temporal coordinate t and the other on the spatial coordinate r:

$$\delta x = c_1 e^{\lambda t} \cos(Kr)$$
$$\delta y = c_2 e^{\lambda t} \cos(Kr) \qquad [9.17]^1$$

In [9.17], λ is the eigenvalue; K is the wavenumber whose value depends on the contour conditions. Introducing the solutions of equation [9.17] into the system of equation [9.16], and dividing by $e^{\lambda t}\cos(Kr)$, we obtain:

$$\lambda \begin{pmatrix} c_1 \\ c_2 \end{pmatrix} = \begin{pmatrix} \left[R_x(X) \right]_s - D_X K^2 & \left[R_y(X) \right]_s \\ \left[R_x(Y) \right]_s & \left[R_y(Y) \right]_s - D_Y K^2 \end{pmatrix}\begin{pmatrix} c_1 \\ c_2 \end{pmatrix} \qquad [9.18]$$

In [9.18], the Jacobian matrix, J', has additional terms in each of its diagonal elements if compared to J, defined in the absence of diffusion. The trace of the Jacobian is now

$$tr(J') = \left[R_x(X) \right]_s - D_X K^2 + \left[R_y(Y) \right]_s - D_Y K^2 \qquad [9.19]$$

and its determinant is

$$\det(J') = D_X D_Y K^4 - \left(D_X \left[R_y(Y) \right]_s + D_Y \left[R_x(X) \right]_s \right) K^2$$
$$+ \left(\left[R_x(X) \right]_s \left[R_y(Y) \right]_s - \left[R_x(Y) \right]_s \left[R_y(X) \right]_s \right) \qquad [9.20]$$

---

[1] The general solution of the system of partial differential equations [9.16] is $\begin{aligned} \delta x &= c_1 e^{\lambda t} e^{ikr} \\ \delta y &= c_2 e^{\lambda t} e^{ikr} \end{aligned}$.

Since the original trace, regarding the homogeneous system, was negative, the new $tr(J')$ (equation [9.19]) remains negative because the diffusion coefficients are positive. The perturbation will be damped unless $\det(J')$ becomes negative. It is worthwhile noticing that in equation [9.20], defining $\det(J')$, the terms $D_X D_Y K^4$ and $\left([R_x(X)]_s[R_y(Y)]_s - [R_x(Y)]_s[R_y(X)]_s\right)$ are both positive; the only possibility for the new $\det(J')$ to be less than zero is that

$$\left(D_X\left[R_y(Y)\right]_s + D_Y\left[R_x(X)\right]_s\right) > 0 \qquad [9.21]$$

and the discriminant of the equation that determines the roots of the second order equation in K^2 is positive.[2]

We know that the sum $\left([R_x(X)]_s + [R_y(Y)]_s\right) = tr(J)$ is negative. If both terms, $[R_x(X)]_s$ and $[R_y(Y)]_s$, are negative, the relation [9.21] cannot be verified. Therefore, one of the two terms, $[R_x(X)]_s$ and $[R_y(Y)]_s$, must be negative and the other positive. This situation means that one species enhances the rate of its own production, whereas the other decreases its rate of production. Let us suppose that the derivative $[R_x(X)]_s$ is positive. Species X plays as an activator of itself. In other words, X is an autocatalytic species. On the other hand, $[R_y(Y)]_s$ is negative. Hence, Y is an inhibitor of itself. For the sum $\left([R_x(X)]_s + [R_y(Y)]_s\right)$ to be negative, we must have

$$\left|\left[R_x(X)\right]_s\right| < \left|\left[R_y(Y)\right]_s\right| \qquad [9.22]$$

Moreover, $\det(J) = \left([R_x(X)]_s[R_y(Y)]_s - [R_x(Y)]_s[R_y(X)]_s\right)$ is positive if $[R_x(Y)]_s[R_y(X)]_s < 0$. The product $[R_x(Y)]_s[R_y(X)]_s$ is negative if the two derivatives have opposite signs. For example, it might be that $[R_x(Y)]_s > 0$ and $[R_y(X)]_s < 0$, i.e., X is also an activator of Y, and Y is also an inhibitor of X.

We may rearrange equation [9.21] in $D_X[R_y(Y)]_s > -D_Y[R_x(X)]_s$, or in $D_X\left(-[R_y(Y)]_s\right) < D_Y[R_x(X)]_s$. Finally, also considering relation [9.22], we achieve

$$\frac{D_X}{D_Y} < \frac{\left[R_x(X)\right]_s}{\left(-\left[R_y(Y)\right]_s\right)} < 1 \qquad [9.23]$$

This last expression means that the inhibitor, Y, must diffuse more rapidly than the activator, X, to generate instability. If the conditions expressed by the relations $tr(J') < 0$, $\det(J') < 0$, and [9.23] are all verified, along with the subsidiary condition expressed in note 2, the eigenvalues of the solutions [9.17] of the PDE will be one positive (λ_+) and the other negative (λ_-). In other words, the original steady-state that was stable in the absence of diffusion encounters a bifurcation and becomes a saddle point in the presence of diffusion. The solution having a positive eigenvalue, $ce^{\lambda_+ t}\cos(Kr)$, represents the exponential growth of a co-sinusoidal or sinusoidal spatial pattern. An example of what can occur in the saddle-node is shown in Figure 9.4. Graph (1) in Figure 9.4 shows the initial situation, where we have a

---

[2] If we think of equation [9.20] as a quadratic equation in K^2 of the form $a(K^2)^2 + bK^2 + c$, then, to have real roots for the inequality $a(K^2)^2 + bK^2 + c < 0$, the discriminant must be positive (see Figure in this note). In other words, $\Delta = b^2 - 4ac = (D_X[R_y(Y)]_s + D_Y[R_x(X)]_s)^2 - 4D_X D_Y([R_x(X)]_s[R_y(Y)]_s - [R_x(Y)]_s[R_y(X)]_s) > 0$. Reminding that $([R_x(X)]_s[R_y(Y)]_s - [R_x(Y)]_s[R_y(X)]_s) = \det(J) > 0$ and that relation [9.21] must hold, it derives that it is necessary to verify $(D_X[R_y(Y)]_s + D_Y[R_x(X)]_s) > 2[D_X D_Y([R_x(X)]_s[R_y(Y)]_s - [R_x(Y)]_s[R_y(X)]_s)]^{1/2} > 0$.

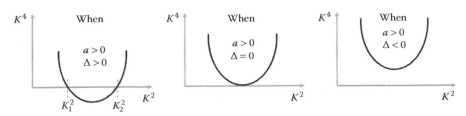

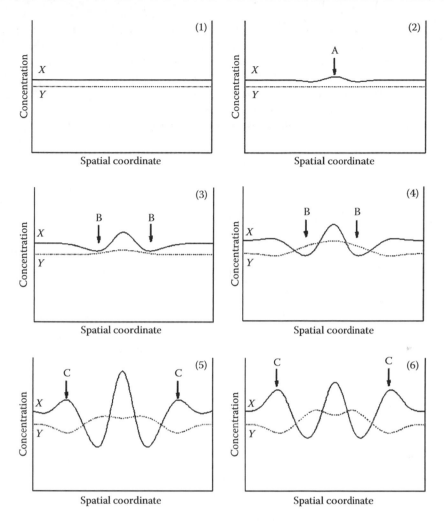

FIGURE 9.4 Schematic and qualitative description of the Turing pattern formation. The graphs from (1) to (6) are six snapshots in the formation of a Turing pattern.

uniform distribution of the two species, X and Y, over the spatial coordinate. If the concentration of the activator X becomes slightly higher in one point (labelled as A in graph 2) by random fluctuation or after a perturbation, then, A can be the trigger point for the emergence of a Turing pattern. Since the activator X self-enhances, its concentration increases in A (see graph 2). This event determines an increase of the inhibitor in the nearby regions labeled as B because Y has a diffusion rate much larger than that of X (graph 3). The presence of Y depresses the activator, resulting in a depletion of its concentration in B (graph 4). At the regions labeled as C, since the inhibitor concentration is getting lower, then X increases and becomes dominant (graph 5). These local situations are enough to generate a chemical structure (graph 6). In the end, the system evolves to a stationary Turing pattern. The wavenumber of the spatial pattern can be found by minimizing $\det\left(J'\right)$ with respect to K (look at the Figure in note 2).

The result (*solve exercise 9.3*) is

$$K = \left[\frac{1}{2}\left(\frac{\left[R_x(X)\right]_s}{D_X} + \frac{\left[R_y(Y)\right]_s}{D_Y}\right)\right]^{1/2} \qquad [9.24]$$

A finite wavenumber exists if $\left(\left([R_x(X)]_s/D_X\right) > \left(-[R_y(Y)]_s/D_Y\right)\right)$. After noticing that the ratio $(D_X/[R_x(X)]_s)^{1/2}$ defines the characteristic decay length of the activator (l_X), whereas the ratio $(D_Y/[R_y(Y)]_s)^{1/2}$ the corresponding decay length of the inhibitor (l_Y), the criterion for the existence of a finite wavenumber is $l_X < l_Y$, which means that the inhibitor has a longer range of the activator. This condition is sometimes referred to as "local self-activation and lateral inhibition" (Meinhardt and Gierer 2000), since the activator tends to be confined to the neighborhood of its initiation site, in contrast to the inhibitor that can spread more rapidly throughout the medium.

TRY EXERCISE 9.4

The possibility to generate patterns, from a homogeneous and stable steady-state by diffusion-driven instability, was inferred for the first time by Alan Turing[3] in 1952. Turing was interested in accounting for the phenomena of morphogenesis in biology. He predicted that the crucial diffusion-driven instability in an initially homogeneous state might be triggered by random disturbances, such as Brownian motion. Since Turing's seminal paper, several Reaction-Diffusion models have been proposed (Maini et al. 1997). Most of them are abstract models of real or imaginary chemical reactions. Some examples are shown in Table 9.1 and *exercise 9.5, 9.6 and 9.7*.

The Reaction-Diffusion models originating Turing patterns can be grouped in two sets depending on the sign of the cross-terms $[R_x(Y)]_s$ and $[R_y(X)]_s$, reminding that the only restriction that must hold is $[R_x(Y)]_s[R_y(X)]_s < 0$. Therefore, we must have $[R_x(Y)]_s > 0$ and $[R_y(X)]_s < 0$, or the opposite. The combinations of the signs for the terms of the Jacobian matrix

$$J = \begin{pmatrix} [R_x(X)]_s & [R_y(X)]_s \\ [R_x(Y)]_s & [R_y(Y)]_s \end{pmatrix} \text{ must be either } \begin{pmatrix} + & - \\ + & - \end{pmatrix} \text{ or } \begin{pmatrix} + & + \\ - & - \end{pmatrix}.$$

In the first case, the self-activator X also plays as an activator of Y, whereas Y inhibits itself, but also X. In the second case, X is an activator of itself but not of Y, and Y is an inhibitor of itself but not of X. The original Turing model along with the Brandeisator, the Gierer-Meinhardt, and the Thomas models are examples of the first case, whereas the Schnackenberg model is an example of the second case (see Table 9.1).

The original Turing model involves only linear differential equations (Turing 1952). The Brandeisator (Lengyel and Epstein 1991) is a model of the first example of Turing pattern obtained in a chemical laboratory (as mentioned in the next paragraph). The Thomas' model (1975) is based on a specific reaction involving the substrates oxygen and uric acid in the presence of the enzyme uricase. The Gierer and Meinhardt (1972) model is a hypothetical but biologically plausible mechanism to account for important types of morphogenesis observed in the development of living beings. Finally, the Schnakenberg model (1979) is based on a hypothetical mechanism involving a cubic autocatalytic process.

TRY EXERCISE 9.8

---

[3] Alan Turing (1912–1954 AD) was a British mathematician who is widely considered to be the father of theoretical computer science and artificial intelligence. He provided a formalization of the concepts of algorithm and computation with the Turing machine, which is a model of a general-purpose computer (read Chapter 10 for more information). He also devised a test to measure the intelligence of a machine.

During the Second World War, Turing devised techniques for breaking secret codes of the German army. Winston Churchill said that Turing made the single most significant contribution to Allied victory in the war against Nazi Germany. In fact, Turing's pivotal role in cracking intercepted coded messages enabled the Allies to defeat the Nazis in several crucial battles. After the Second World War, Turing became interested in mathematical biology, and he wrote his seminal paper on the chemical basis of morphogenesis (Turing, 1952).

TABLE 9.1

Examples of Reaction-Diffusion (RD) Models Wherein x is the Activator, and y is the Inhibitor

Name	Equations	Schematic Model
Original Turing model (Turing 1952)	$$\frac{\partial x}{\partial t} = a_1 x - b_1 y + c_1 - d_1 x + D_X \nabla^2 x$$ $$\frac{\partial y}{\partial t} = a_2 x + b_2 y + c_2 - d_2 y + D_Y \nabla^2 y$$	Activator / Inhibitor
Brandeisator (Lengyel and Epstein 1991)	$$\frac{\partial x}{\partial t} = k_1' - k_2' x - \frac{4 k_3' x y}{\alpha + x^2} + D_X \nabla^2 x$$ $$\frac{\partial y}{\partial t} = k_2' x - \frac{k_3' x y}{\alpha + x^2} + D_Y \nabla^2 y$$	Activator / Inhibitor
Gierer-Meinhardt (Gierer and Meinhardt 1972)	$$\frac{\partial x}{\partial t} = k_1 - k_2 x + \frac{k_3 x^2}{y} + D_X \nabla^2 x$$ $$\frac{\partial y}{\partial t} = k_3 x^2 - k_4 y + k_5 + D_Y \nabla^2 y$$	Activator / Inhibitor
Thomas (1975)	$$\frac{\partial x}{\partial t} = k_1 - k_2 x - \frac{k_5 x y}{k_6 + k_7 x + k_8 x^2} + D_X \nabla^2 x$$ $$\frac{\partial y}{\partial t} = k_3 - k_4 y - \frac{k_5 x y}{k_6 + k_7 x + k_8 x^2} + D_Y \nabla^2 y$$	Activator / Inhibitor
Schnakenberg (1979)	$$\frac{\partial x}{\partial t} = k_1 a - k_2 x + k_3 x^2 y + D_X \nabla^2 x$$ $$\frac{\partial y}{\partial t} = k_4 b - k_3 x^2 y + D_Y \nabla^2 y$$	Activator / Inhibitor

9.4 TURING PATTERNS IN A CHEMICAL LABORATORY

The first example of a Turing structure in a chemical laboratory occurred in 1990, nearly forty years after Turing's seminal paper, thanks to De Kepper and his colleagues who were working with the chlorite-iodide-malonic acid (CIMA) reaction in an open unstirred gel reactor, in Bordeaux. The mechanism of the CIMA reaction is quite complicated. After a relatively brief initial period, chlorine dioxide and iodine build up within the gel and play the role of reactants whose concentrations vary relatively slowly compared with those of ClO_2^- and I^-. This observation allows us to state that it is indeed the chlorine-dioxide-iodine-malonic acid (CDIMA) reaction that governs the formation of Turing patterns. Fortunately, the CDIMA and CIMA reactions may be described by a model that consists of only three stoichiometric processes and their empirical laws (see equations, from [9.25] up to [9.30]). Such model holds because there is no significant interaction among the intermediates of the three main stoichiometric processes and because none of the intermediates builds up to high concentrations (Lengyel and Epstein 1991).

$$CH_2(COOH)_2 + I_2 \rightarrow ICH(COOH)_2 + I^- + H^+ \qquad [9.25]$$

$$v_1 = \frac{k_{1a}\left[CH_2(COOH)_2\right]\left[I_2\right]}{k_{1b} + \left[I_2\right]} \qquad [9.26]$$

$$ClO_2 + I^- \rightarrow ClO_2^- + \frac{1}{2}I_2 \qquad [9.27]$$

$$v_2 = k_2\left[ClO_2\right]\left[I^-\right] \qquad [9.28]$$

$$ClO_2^- + 4I^- + 4H^+ \rightarrow Cl^- + 2I_2 + 2H_2O \qquad [9.29]$$

$$v_3 = k_{3a}\left[ClO_2^-\right]\left[I^-\right]\left[H^+\right] + k_{3b}\frac{\left[ClO_2^-\right]\left[I_2\right]\left[I^-\right]}{\alpha + \left[I^-\right]^2} \qquad [9.30]$$

The malonic acid serves only to generate iodide via reaction [9.25]. The reaction [9.27] between chlorine dioxide and iodide produces iodine. The reaction [9.29] is the key process: it is autocatalytic in iodine and inhibited by iodide (see equation [9.30]). The differential equations [9.26], [9.28] and [9.30] contain six variables. They are $[CH_2(COOH)_2]$, $[I_2]$, $[ClO_2]$, $[I^-]$, $[H^+]$, and $[ClO_2^-]$. The system of three differential equations ([9.26], [9.28] and [9.30]) have been integrated numerically, and the computational results have reproduced quite well the experimental results and the possibility of having oscillations (Lengyel et al. 1990). The numerical analysis has also revealed that during each cycle of oscillation, $[CH_2(COOH)_2]$, $[I_2]$, $[ClO_2]$, and $[H^+]$ change very little, while $[ClO_2^-]$ and $[I^-]$ vary by several orders of magnitude. This evidence suggests that the model can be reduced to a two-variable system by treating the four slightly varying concentrations as constants. The simplified model looks like the following set of schematic reactions, wherein $X = [I^-]$, $Y = [ClO_2^-]$ and $A = I_2$.

$$A \xrightarrow{k_1'} X$$
$$X \xrightarrow{k_2'} Y \qquad [9.31]$$
$$4X + Y \xrightarrow{k_3'} P$$

The new rate laws are:

$$v_1' = k_1' = \frac{k_{1a}\left[CH_2(COOH)_2\right]_0\left[I_2\right]_0}{k_{1b} + \left[I_2\right]_0}$$
$$v_2' = k_2'\left[X\right] = k_2\left[ClO_2\right]_0\left[X\right] \qquad [9.32]$$
$$v_3' = k_3'\frac{\left[X\right]\left[Y\right]}{\alpha + \left[X\right]^2} = k_{3b}\left[I_2\right]_0\frac{\left[X\right]\left[Y\right]}{\alpha + \left[X\right]^2}$$

In the definition of v'_3 in [9.32], the first term of [9.30] has been neglected, since it is much smaller than the second term under the conditions of interest. From the mechanism [9.31] and equation [9.32], it derives that the system of partial differential equations describing the spatiotemporal development for $[X]$ and $[Y]$ is

$$\frac{\partial[X]}{\partial t} = k'_1 - k'_2[X] - 4k'_3\frac{[X][Y]}{\alpha + [X]^2} + D_X\frac{\partial^2[X]}{\partial r^2}$$

$$\frac{\partial[Y]}{\partial t} = k'_2[X] - k'_3\frac{[X][Y]}{\alpha + [X]^2} + D_Y\frac{\partial^2[Y]}{\partial r^2}$$

[9.33]

where r is the spatial coordinate. In [9.33], we have four variables (two are independent, i.e., t and r, and two dependent, i.e., $[X]$ and $[Y]$) and six constants. We may simplify further the system of differential equations [9.33] if we nondimensionalize it.[4] The final result is

$$\frac{\partial X^*}{\partial \tau} = \gamma\left(a - X^* - 4\frac{X^*Y^*}{1 + X^{*2}}\right) + \frac{\partial^2 X^*}{\partial \rho^2}$$

$$\frac{\partial Y^*}{\partial \tau} = \gamma\left(bX^* - \frac{bX^*Y^*}{1 + X^{*2}}\right) + d\frac{\partial^2 Y^*}{\partial \rho^2}$$

[9.34]

In [9.34], $X^* = (1/\sqrt{\alpha})[X]$, $Y^* = (k'_3/k'_2\alpha)[Y]$, $\tau = (D_X/L^2)t$, $\rho = (1/L)r$ are the dimensionless variables and $d = (D_Y/D_X)$, $\gamma = (k'_2L^2/D_X)$, $a = (k'_1/k'_2\sqrt{\alpha})$, $b = (k'_3/k'_2\sqrt{\alpha})$ are the new four parameters. To observe Turing structures, the homogeneous system of ODEs

$$\frac{dX^*}{d\tau}\left(a - X^* - 4\frac{X^*Y^*}{1 + X^{*2}}\right) = f(X^*,Y^*) = 0$$

$$\frac{dY^*}{d\tau} = \left(bX^* - \frac{bX^*Y^*}{1 + X^{*2}}\right) = g(X^*,Y^*) = 0$$

[9.35]

must have a steady-state $\left(X^*_s = \left(\frac{a}{5}\right), Y^*_s = \left(1 + \frac{a^2}{25}\right)\right)$ that is stable to homogeneous perturbation. In other words, the steady-state solution (X^*_s, Y^*_s) must have $tr(J) < 0$, $\det(J) > 0$, and one species must play as the activator whereas the other as the inhibitor.

TRY EXERCISE 9.9

---

[4] The first step of the procedure of nondimensionalization requires the definition of dimensionless variables. For the system of differential equations [9.33], the dimensionless variables are $X^* = X_C[X], Y^* = Y_C[Y], \tau = t_Ct, \rho = r_Cr$. Introducing these new variables in [9.33], we obtain:

$$\frac{\partial X^*}{\partial \tau} = k'_1\frac{X_C}{t_C} - k'_2\frac{X^*}{t_C} - 4k'_3\frac{X^*Y^*}{t_C Y_C}\frac{1}{\left(\alpha + \left(\frac{X^*}{X_C}\right)\right)^2} + \frac{D_X r_C^2}{t_C}\frac{\partial^2 X^*}{\partial \rho^2}$$

$$\frac{\partial Y^*}{\partial \tau} = k'_2\frac{Y_C}{X_C t_C}X^* - k'_3\frac{1}{X_C t_C}\frac{X^*Y^*}{\left(\alpha + \left(\frac{X^*}{X_C}\right)\right)^2} + \frac{D_Y r_C^2}{t_C}\frac{\partial^2 Y^*}{\partial \rho^2}$$

The next step is to choose the nondimensionalizing constants. If L is the extension of the spatial coordinate, $r_C = \left(\frac{1}{L}\right)$. Introducing the definition of r_C, we derive that $t_C = \left(\frac{D_X}{L^2}\right)$. If we fix $X_C = \left(\frac{1}{\sqrt{\alpha}}\right), Y_C = \left(\frac{k'_3}{k'_2\alpha}\right)$, and $d = \left(\frac{D_Y}{D_X}\right)$, $\gamma = \left(k'_2L^2/D_X\right), a = \left(k'_1/k'_2\sqrt{\alpha}\right), b = \left(k'_3/k'_2\sqrt{\alpha}\right)$, the differential equations can be written into dimensionless form [9.34].

Since $\left(\partial g/\partial Y^*\right) < 0$ and $\left(\partial f/\partial Y^*\right) < 0$, the species $Y^* = ClO_2^-$ is the inhibitor of itself and X^*. The species $X^* = I^-$ plays as the self-activator (or auto-catalyst) if $a > 5\sqrt{5/3}$. The steady state becomes a saddle point in the presence of the diffusion, if the determinant of the new Jacobian J' (containing the extra terms of diffusion in its diagonal elements) is negative, i.e., if

$$\det\left(J'\right) = \det \begin{pmatrix} \left(\dfrac{\partial f}{\partial X^*}\right)_s - K^2 & \left(\dfrac{\partial f}{\partial Y^*}\right)_s \\[2ex] \left(\dfrac{\partial g}{\partial X^*}\right)_s & \left(\dfrac{\partial g}{\partial Y^*}\right)_s - K^2 d \end{pmatrix} < 0 \qquad [9.36]$$

The determinant of J' is a negative real number if the relations [9.37] and [9.38] are true.

$$\left(\frac{\partial g}{\partial Y^*}\right)_s + d\left(\frac{\partial f}{\partial X^*}\right)_s > 0 \qquad [9.37]$$

$$\left(\frac{\partial g}{\partial Y^*}\right)_s + d\left(\frac{\partial f}{\partial X^*}\right)_s > 2\sqrt{\left[d\left[\left(\frac{\partial f}{\partial X^*}\right)_s\left(\frac{\partial g}{\partial Y^*}\right)_s - \left(\frac{\partial f}{\partial Y^*}\right)_s\left(\frac{\partial g}{\partial X^*}\right)_s\right]\right]} > 0 \qquad [9.38]$$

If we use plausible values of the kinetic constants for the CDIMA reaction (Lengyel and Epstein 1991), we find that the ratio $d = \left(D_Y/D_X\right)$ must be at least 10 for the relation [9.38] to hold. If you want to verify this result, try to solve *exercise 9.10*.

In aqueous solution, substantially all small molecules and ions have diffusion coefficients that lie within a factor of two of $2\times10^{-5}\,cm^2s^{-1}$. Therefore, it seems impossible that the diffusion coefficient of ClO_2^- can be ten times greater than that of I^-. What did it make possible to get around this problem? Here, serendipity entered. De Kepper and colleagues (Castets et al. 1990) were working with the CIMA reaction in an unstirred continuous flow gel reactor. In such a reactor, the two broad faces of a cylindrical slab of gel, 2 or 3 mm thick and made of either polyacrylamide or agar (just to make a pair of examples), are in contact with two solutions of different compositions: one containing iodine and the other chlorine dioxide and malonic acid. The reactants diffuse into the gel, and they encounter in a region near the middle of the gel, where the pattern, less than 1 mm thick, can emerge. The continuous flow of fresh reactants maintains the system as open, and the gel prevents convective motion. De Kepper and colleagues employed starch as an indicator to increase the color contrast between the oxidized and reduced states of the reaction. They introduced starch into the acrylamide monomer solution before polymerization to the gel because the bulky and heavy starch molecules cannot diffuse into the gel from the reservoirs. The starch (St) forms a blue complex with tri-iodide (I_3^-):

$$St + I_2 + I^- \rightarrow \left(StI_3^-\right) \qquad [9.39]$$

The complex (StI_3^-) is practically immobile. We may imagine the starch molecules being dispersed randomly throughout the gel. When an iodide encounters a "trap" made of starch and iodine, it remains blocked until the complex breaks up. The net result is that the effective diffusion rate of iodide decreases noticeably. If the concentration of starch is high enough, it provides the way of slowing down the diffusion rate of the activator iodide with respect to the inhibitor chlorite, which, on the other hand, maintains the typical diffusion rate it has in aqueous solution. It was the fortuitous choice of starch as the indicator that produced the first experimental Turing structure (for a quantitative treatment of the starch's effect read the paper by Lengyel and Epstein (1992)). The patterns, observed for the CIMA and CDIMA reactions performed in a gel reactor, appear to occur in a single plane parallel to the faces of the gel at which the reactor is fed. The patterns are essentially bi-dimensional. Why? The formation of Turing patterns requires that concentrations of the species involved in the reaction lie within ranges that allow the four inequalities (I) $tr(J) < 0$, (II) $\det(J) > 0$, (III) [9.37] and (IV) [9.38] to be satisfied. In a gel reactor, the values of the concentrations of all the species are position-dependent, ranging from their input feed values on one side of the gel where they enter to essentially zero on the other side. Evidently, the conditions for generating Turing structures can be satisfied only within a thin slice of the gel, if at all.

Recently, Epstein and his collaborators (Bánsági et al. 2011) at Brandeis University have obtained the first examples of three-dimensional Turing structures in a chemical laboratory. Such wonderful three-dimensional Turing patterns are based on the Belousov-Zhabotinsky (BZ) reaction carried out in an oil-water mixture in the presence of the amphiphilic surfactant sodium bis(2-ethylhexyl) sulfosuccinate, known as Aerosol OT (AOT). In an AOT-water-oil system when the ratio $R_{w/o} = ([water]/[oil])$ is low, a reverse microemulsion forms, in which droplets of water, surrounded by a monolayer of surfactant molecules, are dispersed in oil. Such droplets have nanometric dimensions (their radius depends on the ratio $R_{w/s} = ([H_2O]/[AOT])$) and behave like nanoreactors because the reactants of the BZ reactions are polar and they partition within the aqueous droplets. By playing with the ratio $R_{w/o} = [water]/[oil]$, it is possible to have either isolated nanodroplets wandering through the oil phase by Brownian motion (when $R_{w/o}$ is low), or droplets that coalesce into water channels (when $R_{w/o}$ is high). The formation of Turing patterns is favored by a low value of $R_{w/o}$ and when two distinct conditions of mass transport are satisfied. First, when the BZ reaction begins, apolar intermediates, notably the inhibitor bromine, are produced within the nanodroplets and diffuse through the oil phase. Second, the polar species, including the activator $HBrO_2$, diffuse together with the entire water droplet. The isolated droplets move around much slower than single molecules, and when they collide, they mix their contents through a fission-fusion mechanism. The average time between collisions is about 1 millisecond (ms), which is much shorter than the period of the oscillations. Hence, the medium can be treated as macroscopically continuous.

Since the movement of apolar single Br_2 molecules, which play as an inhibitor, occurs at rates much faster than that of the nanodroplets containing the activator $HBrO_2$, Turing patterns can emerge. If the microemulsions are sandwiched between a pair of glass plates (Vanag and Epstein 2001a) separated by an 80-mm-thick Teflon gasket, the Turing structures are bi-dimensional. On the other hand, if the microemulsions are placed in a cylindrical quartz capillary with an inner diameter (0.3–0.6 mm) that exceeds the wavelength of the patterns, the Turing patterns are three-dimensional. These experiments are remarkable because they show that Turing patterns, persisting for one hour or more, can also be obtained in closed systems. They are transient because they are not sustained like those in an open system. Previously, a few cases of transient Turing patterns in closed systems have been found. For instance, the CDIMA reaction performed in the presence of starch at 4°C in a Petri dish (Lengyel et al. 1993) giving rise to Turing patterns of mixed spots and stripes or network-like structures that remain stationary for 10–30 minutes. Another example is the polyacrylamide-methylene blue-sulfide-oxygen reaction carried out in a Petri dish and originating a variety of spatial patterns such as hexagons and stripes (Watzl and Münster 1995).

TRY EXERCISES 9.11, 9.12 AND 9.13

9.5 TURING PATTERNS IN NATURE

The brilliant idea contained in the paper titled "The chemical basis of morphogenesis" written by Turing in 1952 did not emerge into the spotlight until two decades later. In 1972, two developmental biologists, Hans Meinhardt and Alfred Gierer at the Max Planck Institute for Virus Research in Tübingen (Germany), proposed a theory of biological pattern formation that paralleled that described by Turing. In Turing's, Gierer's, and Meinhardt's model, the spontaneous formation of patterns occurs when two morphogens or generic ingredients interact non-linearly. One must be an autocatalyst.

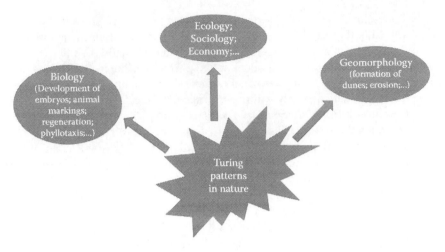

FIGURE 9.5   Sketch that shows the broad applicability of the Reaction-Diffusion model.

The other must be a self-inhibitor. The autocatalyst is also an activator of the self-inhibitor. On the other hand, the self-inhibitor inhibits the autocatalyst (see Table 9.1).[5] Crucially, the two species must have different rates of diffusion, the inhibitor being faster. Such model, named as Reaction-Diffusion (RD) model, does not need to be limited to discrete molecules as interacting elements, and diffusion is not the only mode of transmission. In fact, Turing's RD model has been having a profound impact on vast range of disciplines, such as physiology, ecology, botany, chemistry, as well as geomorphology, social sciences and economy (see Figure 9.5). Turing's RD model is mathematically easy and effective in extracting the nature of many phenomena in Complex Systems, although it omits many details of them. As Turing said, his original model is a simplification and an idealization of Complex Systems. Therefore, it could be falsified, in agreement with the view of the epistemologist Popper.[6]

9.5.1   Biology: The Development of Embryos

The Really Big Question that Turing raised in his paper "The chemical basis of morphogenesis" is this: "How is it possible that fertilized eggs give rise to such complex forms as are the living beings?" A fertilized egg, named zygote (from the ancient Greek word ζυγωτός that means "joined"), has spherical symmetry, but, in the end, it gives rise to an animal with well-defined axes.[7]

---

[5] We must remind also the Schnackenberg's model that involves different relations between the self-activator and the self-inhibitor.

[6] Karl Popper (Vienna, 1902–London, 1994) is considered one of the greatest philosophers of science of the twentieth century. According to Popper, scientists are "problem-solvers;" the growth of human knowledge proceeds from our problems and from our attempts to solve them. These efforts involve the formulation of theories that can never be proven, but they can be falsified, meaning that they can and should be scrutinized by decisive experiments. According to Popper, the advance of scientific knowledge is an evolutionary process similar to the biological evolution. To respond to a given problem, some tentative theories are proposed. These theories are, then, checked for error elimination. The error elimination procedure performs a similar function for science that natural selection plays for biological evolution. The theories that survive the process of refutation are not truer, but they fit better than the others to the data available. Just as the biological fitness of a species does not assure unlimited survival, neither does rigorous testing protect a scientific theory from refutation in the future. According to the Popper's view, the evolution of theories reflects a certain type of progress towards greater and greater problems in a process very much akin to the interplay between genetic modifications and natural selection.

[7] There are two main axes in almost all animals that are called bilateria. There is the antero-posterior axis that defines the "head" and the "tail" ends. But there is also the dorso-ventral axis that is at right angle to the former. For instance, human faces are ventral, whereas the back of human heads are dorsal.

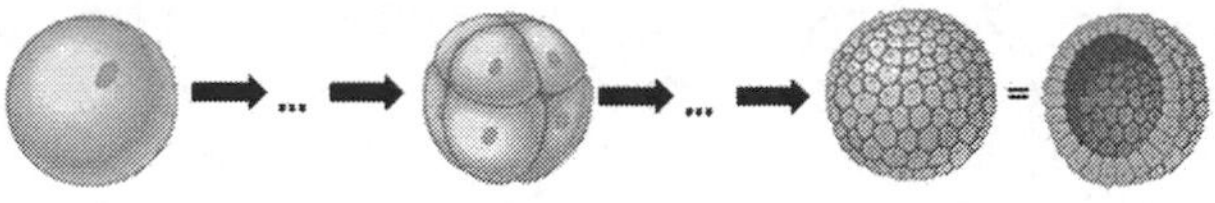

Zygote   Eight-cell body   Blastula

FIGURE 9.6   Sketch that represents the development of a zygote into a blastula.

How does it happen? Development begins when the fertilized egg divides into two cells. Then, there is a second cleavage at right angle with respect to the first, and a third cleavage again at right angle producing eight cells. After many more cleavages, the embryo becomes a hollow ball, called blastula, with the cells arranged as a spherical sheet (see Figure 9.6). At this stage, there is no indication of the asymmetric animal into which the blastula develops. Then, the gastrulation starts. The cells of the blastula rearrange so that the front and back, the top and the bottom of an animal become evident, and the basic body plan is laid down. It is only after the gastrulation that the form of the animal begins to emerge. This is the reason why Wolpert (2008) stated that gastrulation is the truly important event in our life.

During gastrulation, movement and folding of cell sheets form the basis of the early development of many structures as diverse as the heart, lungs, and brain. Small changes in how fast and how far the cell contraction spreads have profound effects on the forms. How do cells know where and when to specialize, change shape, or move? It is evident that there exists positional information (Wolpert 2011) and a generative program somewhere within the zygote. The spatial distribution of specific chemicals encodes the positional information. The instructions for molding the embryo are served into the DNA that works as if it were a memory of a modern electronic computer. A memory of a computer having the Von Neumann architecture stores both data and instructions (remember what we learned in Chapter 2). The same can be said of the DNA of the nucleus. All the cells have the same DNA and the same sequence of genes. Within the DNA, there are the instructions for making all the proteins in the cell and the program that controls their synthesis.[8] How do the cells differentiate and specialize? Animals are made up of different types of cells, such as nerve cells, muscle cells, blood cells, germ cells, skin cells, and so on. Humans have about 350 different types of cells (Wolpert 2008), while lower animals have less. The function of a cell depends on the proteins it contains. Proteins perform either structural or catalytic functions (as we learned in Chapter 7). There are proteins that are common to most cells and are needed to carry out basic functions, such as the production of energy or the synthesis of key molecules. But there are also proteins that are present only in certain types of cells. For example, albumin is peculiar to liver cells; hemoglobin is only within red blood cells; insulin belongs to pancreas cells, keratin is expressed in skin cells, the contractive actin and myosin are synthesized within muscle cells, et cetera. The proteins that are made within each cell depend on the cells receiving positional information, which is the mutual communication between the nucleus and the cytoplasm and the extracellular signals. A useful picture to describe cell diversification is that of a ball on top of a mountain. The ball can slide along different downhill pathways ending on distinct valleys. The ball represents an undifferentiated cell that can transform into a specialized one, depending on the epigenetic landscape (read Box 9.2 of this chapter). In the case of humans, there are so much as 350 possible valleys! Which factors rule the selection of a particular branch? The development of a zygote into an organism is truly an astonishing phenomenon that is still under investigation (Wolpert 2008). Scientists have unveiled only few scenes but not the entire film. We are aware that different mechanisms are involved into

---

[8] The mathematician Gregory Chaitin (2012), in his book *Proving Darwin. Making Biology Mathematical*, wherein he looks for a mathematical demonstration of Darwin's theory of evolution, asserts that the DNA of living beings is a software. It is a particular software, because it evolves, relentlessly.

the embryonic development. One of these is the Turing's Reaction-Diffusion model. For example, the proteins Nodal and Lefty appear to work as an activator-inhibitor pair during the induction of the mesoderm that is one of the three primary layers of germ cells[9] in the very early embryo of bilaterian animals (Nakamura et al. 2006). Turing's RD models have been also proposed for the limb development when digits emerge from the undifferentiated limb bud (Raspopovic et al. 2014), and for embryonic feather branching in birds (Harris et al. 2005). It is reasonable to expect that the Turing's RD model will be used to interpret other events in embryonic development, in the next future. But it is not the only mechanism to produce shapes and structures in an embryo. For example, it has been found that another important morphological mechanism is that based on gradients of chemicals. Thomas Hunt Morgan (1866–1945 AD), an American evolutionary biologist, geneticist, embryologist, who, for his discoveries concerning the role played by the chromosome in heredity, was awarded the Nobel Prize in 1933, clearly proposed how gradients could control patterning. The combination of genetic and embryological studies allowed for the identification of the regulatory genes that control patterning in the early embryo of the fruit-fly *Drosophila*. One of the most important is the gene *bicoid*, which is involved in patterning the anterior end of embryo. The polarity of the embryo, i.e., which end will become the head, and which end the tail, is defined into the egg. A special chemical composition of the cytoplasm is located at the future anterior end. In this special portion of cytoplasm there is the message for synthesizing the protein coded by the *bicoid* gene. When the egg is laid by the mother fly, the *bicoid* protein begins to be synthesized at the anterior end and diffuses along the egg setting up a concentration gradient. The largest value of the *bicoid* protein is at the front end. This gradient controls the position of the boundary between the head and the thorax and also activates other genes involved in patterning the posterior end of the embryo (see Figure 9.7). If there is not the right gradient, the *bicoid* proteins are not synthesized at the anterior end and the embryo develops into a larva lacking both head and

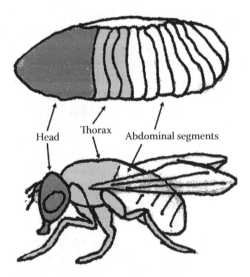

FIGURE 9.7 Structure of the Drosophila (at the bottom) and its embryo (on top).

---

[9] Animals with bilateral symmetry (also called plane symmetry), having one symmetry plane that divides an organism into roughly two mirror-image halves with respect to the external appearance only, produce three primary layers of germ cells within their embryos. The ectoderm is the external one from which the epidermis and the nervous system will develop. The endoderm is the internal one; it will give rise to the digestive system. The mesoderm is the intermediate layer; it will give rise to the muscular system, the heart, blood and other internal organs.

thorax. The resulting abnormal egg can be rescued by injecting the chemicals of a normal anterior cytoplasm into the anterior region of the egg. After restoring the right gradient, normal development takes place. If, on the other hand, the chemicals of the normal anterior cytoplasm are injected into the middle of the egg, the head develops in the middle of the embryo.

TRY EXERCISE 9.14

Also, the gradient mechanism is reasonably involved in the migration of cells. In fact, there is no doubt that some cells exhibit chemotaxis, meaning, cells receive signals from the surrounding tissues that direct them along the appropriate developmental pathway. During gastrulation, it is the difference in cell adhesion that guides the cells. In fact, change in cellular adhesiveness is another essential mechanism in the developmental program. The adhesiveness of a cell depends on specific proteins that are embedded in the cell's surface membrane and have one portion that sticks out and binds to a similar or complementary molecule on adjacent cells. The crucial point is that cells express different Cell Adhesion Molecules (CAMs) at various stages in the development. CAMs play a pivotal role in the spatial arrangements of cells. The spatial organization of the different types of cells is essential for the formation of organs. Consider our arms and legs. Our arms and legs contain the same types of cells, such as muscle, tendon, skin, bone, and so on, yet they are different. The explanation lies in how these different types of cells arrange spatially. The differentiation of cells occurs through either environmental signals or unequal distribution of some special cytoplasmic factors at cell division.

Finally, even chemical oscillations can play a role in the development of an embryo and the formation of specific patterns. For example, in vertebrates, shortly after gastrulation, the brain can be seen forming at the anterior end of the embryo. Behind the brain, there are the somites that are blocks of tissue that will develop the vertebrae and the muscles of the back. The somites look like two lines of paving stones (see Figure 9.8). Somitogenesis is an example of a dynamic embryonic process that relies on precise spatial and temporal control of gene expression. In fact, somitogenesis involves the oscillation of the expression of certain genes, ruled by an internal clock (Goldbeter and Pourquié 2008). The oscillations of the clock are converted into separated blocks of tissue. One can think of each cell as having a clock and those behind them a clock that is set ticking a little later. When such cells synchronize, they form a somite (Baker et al. 2006; Tsiairis and Aulehla 2016).

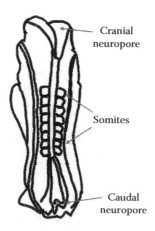

FIGURE 9.8 Schematic structure of a human embryo at roughly the 29th day. The top of the cranial neuropore corresponds to the terminal lamina of the adult brain and the posterior neuropore (or caudal neuropore) to the terminal filum at the end of the spinal cord.

BOX 9.2   EPIGENESIS

Epigenesis is the study of environmental influences acting upon and modifying the individual genetic program (Wessel 2009). The theory of epigenesis was first proposed in the fourth century BC by Aristotle in *On the Generation of Animals*. Aristotle, a truly Philo-physicist (remember Chapter 1), asked himself if an organism exists from the beginning, or if it develops in a process similar to the knitting of a fisherman's net. In this way, Aristotle sparked the controversy between two possible answers: the preformationist and the epigenetic theories. Of course, Aristotle privileged the knitting metaphor—the epigenetic theory. However, the debate prolonged many centuries, and it was insoluble without awareness of the existence of cells, DNA and genes. Only in the latter half of the twentieth century, more information about the cell and its constituents were gathered, and it was recognized that embryos develop by epigenesis. Nowadays, we are aware that embryogenesis provides the most striking example of epigenetics at work, although we still need to know the details of how a ball of identical pluripotent cells differentiates into many cell types and organs in response to a network of physical and chemical environmental signals (Cañestro et al. 2007). After sequencing the genome of different species, it is evident there are extensive similarities in DNA among diverse organisms. A natural question arises: Where do the differences come from? The theory of Evo-Devo, which is the nickname for Evolutionary Developmental Biology, proposes that morphological diversity among species is, for the most part, not due to differences in functional genes, which encode proteins for cellular maintenance and building, but in genetic switches that are used to turn genes on and off (Carroll 2005). These switches are sequences of DNA that do not code for any protein and are part of what was called "junk DNA." They are activated by proteins encoded by regulatory genes. Humans share many functional genes with other creatures, but according to the Evo-Devo theory, the diversity of organisms is due primarily to evolutionary modifications of switches and the genetic regulatory network. The master genes constrain the morphology that organisms can assume, and the notion that every trait can vary indefinitely is not valid according to Evo-Devo.

9.5.2   BIOLOGY: REGENERATION OF TISSUES

Regeneration of biological tissues is a phenomenon that is closely related to embryonic development. Animals, such as newts and hydra, show remarkable powers of regeneration when parts of their body are removed. Not many animals can survive decapitation; hydra not only survives but regenerates a new head. Hydra is a small glove-shaped animal with tentacles for catching prey at one end and a sticky foot at the other. One can split a hydra into two parts, and within a few days, two new complete hydrae will form (Figure 9.9). Regeneration requires changing the state of the cells and remodeling the tissues because it occurs even when all cell multiplication is prevented. Growth happens only when the animal starts feeding again.

A formal explanation for the regeneration of the hydra entails the possession by cells of positional information. If the cells know where they are in the animal, they can work appropriately (Wolpert 2008). The cells know where they are if they are in the head or the foot because the head produces an inhibitor that diffuses down the body and prevents any of the other tissues from making a head. When the head is removed, the concentration of the inhibitor falls, and a new head can now be made. Once the head has regenerated the gradients are reestablished. A similar mechanism, but of course reversed in space, operates at the food end.

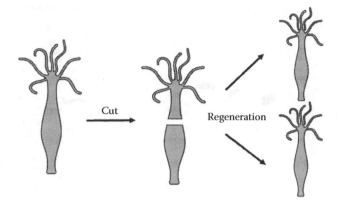

FIGURE 9.9   Regeneration of hydra.

9.5.3   BIOLOGY: PHYLLOTAXIS

In his 1952 paper, Turing proposed that his theory might be useful to explain also phyllotaxis, which is the regular arrangement of leaves on plant stems. Turing's idea was that an activator-inhibitor system of hormones acting at the growing tip of a plant defines the spots that grow into buds on the cylindrical stem. Leaves and flowers are formed from the shoot apical meristem,[10] triggered by the plant hormone auxin. Recently, it has been shown (Reinhardt et al. 2003) that auxin is transported through plant tissues by specific carrier proteins. Existing leaf primordia act as sinks, redistributing auxin and creating its heterogeneous distribution in the meristem. Auxin accumulation occurs only at certain minimal distances from existing primordia, defining the position of future primordia and inhibiting the formation of any new buds nearby. This model for phyllotaxis accounts for its reiterative nature, as well as its regularity and stability.

9.5.4   BIOLOGY: ANIMAL MARKINGS

Many animals have fascinating color patterns on their skin. There is not a real consensus on what many of these markings are for. In some species, the markings play as camouflage. The camouflage can be cryptic or disruptive. It is cryptic when the patterns blend in and match their surroundings, so the animal can hide from predators. An example is the baby tapir who is born with a reddish-brown pattern of stripes and spots helping to hide it in the dappled forest light. The camouflage is disruptive when it works by breaking up the outlines of the animal, like in the leopard. In other species, the color patterns may play other important roles in kinds of behavior such as shoaling, mate choice, and antagonistic displays.

In the 1980s, Meinhardt (1982) and the mathematical biologist Murray (2003) showed independently that Turing's theory offers a plausible explanation for a wide range of animal pigment patterns, from mammals like zebras, leopards, to fishes and seashells (Figure 9.10). The basic idea is that morphogens turn on or off genetic pathways that stimulate the production of pigments.

---

[10] In plants, the meristem is the tissue containing undifferentiated cells (meristematic cells), found in zones of the plant where growth can take place. The shoot apical meristem gives rise to organs like the leaves and flowers, while the root apical meristem contains the meristematic cells for the root growth. The meristematic cells of plants can be compared to animal stem cells, which have an analogous behavior and function. The term meristem derives from the Greek word μερίζειν (merizein), meaning to divide, in recognition of its inherent function.

FIGURE 9.10 Examples of pigment patterns in animals.

In mammal skins, the pigment is melanin, and the cells that contain melanin are called melano-cytes, which are found in the basal, or innermost, layer of the epidermis. There are essentially only two kinds of melanin: eumelanin (from the Greek $\varepsilon\upsilon$-$\mu\acute{\varepsilon}\lambda\alpha\varsigma$ that means good-black), which results in black or brown, and pheomelanin (from $\varphi\acute{\alpha}\varepsilon o\varsigma$ that means bright), which results in yellow or reddish. Both types of melanin are made of cross-linked polymers based on the amino acid tyrosine; they differ in the overall chemical composition and structure. It is believed that whether or not melanocytes produce melanin depends on the presence or the absence of chemical activators and inhibitors. Turing's RD model has been demonstrated effective in reproducing the broken ring markings characteristic of leopards and jaguars (Liu et al. 2006) and the patterns of lady beetles (Liaw et al. 2001).

An essential characteristic of the stationary pattern made by the RD mechanism is its robustness against perturbations. As far as the patterning process works, the resulting structure can autono-mously rearrange in a manner that is very specific to the RD mechanism. When an animal with a stripe pattern grows, the spacing of the stripes becomes wider. Therefore, the growth of the body tends to change the spacing of the pattern continuously. Since an RD mechanism makes the pattern, some rearrangement occurs to keep the original spacing. Kondo and Asai (1995) observed in vivo precisely these changes in tropical angelfish *Pomacanthus*. If the stripes of the growing fish exhibit branch points (Figure 9.11), their division occurs by the horizontal movement of the branch points, whereas if there are no branch points, new stripes emerge between the old stripes.

Time

FIGURE 9.11 The growth of the angelfish *Pomacanthus* is accompanied by horizontal movement of branch points and division of stripes. (Reprinted from *Semin. Cell Dev. Biol.*, 20, Kondo, S. and Shirota, H., Theoretical analysis of mechanisms that generate the pigmentation pattern of animals, 82–89, Copyright 2009, with permission from Elsevier.)

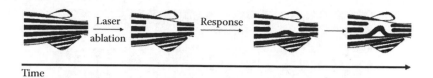

Time

FIGURE 9.12 Four snapshots of the experiment performed by Yamaguchi et al. (2007) and regarding the laser ablation of melanophores belonging to two dorsal stripes of zebrafish. (Reprinted from *Semin. Cell Dev. Biol.*, 20, Kondo, S. and Shirota, H., Theoretical analysis of mechanisms that generate the pigmentation pattern of animals, 82–89, Copyright 2009, with permission from Elsevier.)

Yamaguchi et al. (2007) ablated by laser the melanophores in two dorsal stripes of zebrafish and continuously eliminated new melanophores developing in the area (Figure 9.12). This operation made a broad region lacking melanophores, and induced a dynamic dorsal of the remaining ventral melanophore stripe to fill the space. The rearrangement began by forming a bell-shaped pattern to fill the vacant space. This dynamic response of the pattern confirms its intrinsic RD mechanism.

9.5.5 ECOLOGY, SOCIOLOGY, AND ECONOMY

Social animals and plants might use the local self-activation and long-range inhibition Turing-type mechanism to produce a wide variety of ecological structures. A proven example is the cemetery formation in Messor Sancta ant colonies (Theraulaz et al. 2002). Messor Sancta ants collect the bodies of expired colony members and arrange them in piles. The ants constantly pick up and redistribute the bodies, producing a kind of "diffusion" of corpses. Some piles form, and after a certain time, several clusters are generated. Over time, some clusters grow and others disappear, leading finally to a steady state with a stable number of clusters. Because ants are more likely to drop a body on a pile as the pile gets larger and larger, there is a positive feedback action controlling their growth. There is also long-range inhibition because the region surrounding a big pile gets free of bodies and it reduces the probability that a new cluster forms nearby. The example of cemetery formation in Messor Sancta ant colonies spurs researchers to look for Turing's morphogenesis in other collective behavioral patterns, such as network formation, nest construction in termites, and grouping even in higher organisms.

For example, it has been demonstrated that a Reaction-Diffusion model can explain the phenomenon of crime hotspots (Short et al. 2010) in neighborhoods with anomalously high crime rates. Potential crime targets, such as homes, automobiles, or people, are continuously distributed in the model, and motivated offenders (activators) search for suitable targets. In the absence of an inhibiting agency, such as a security measure or a police force, the offenders commit their crimes. When the known positive feedback action, in which crime induces more crime, holds, a Turing structure of hotspots originates. Such a pattern of hotspots is tough to be eradicated. In fact, focused inhibition induces merely the Turing pattern of hotspots to move or mutate in shape, but not to disappear.

Turing's patterns are important also in political economy. When science tries to rationalize human agglomerations at different spatial scales, like the North-South dualism, the emergence of cities, or the emergence of commercial districts within cities (Xepapadeas 2010), the theory of Turing's patterns is valuable. Moreover, Turing's model is useful when we deal with the economic management of ecosystems, like fishery management, spatial pollution or water pricing. In fact, the theory may suggest the formulation of regulatory policies with spatial features.

9.5.6 GEOMORPHOLOGY

The generation of structure from initially structure-less systems is not restricted to living beings. An example is the formation of sand dunes. One would expect that the wind distributes the sand evenly

FIGURE 9.13   Examples of sand ripples.

in the desert. But the opposite occurs. Since sand transport is highly sensitive to wind speed, local changes in wind velocity alter the pattern of sand deposition. A local minor accumulation of sand may be triggered by tiny stochastic variations or local surface variation that affects airspeed. Sand dunes arise from sand deposition behind a minor wind shelter (Figure 9.13). This deposition accelerates the further accumulation of sand. In fact, the growth of the dune increases the wind-shelter effect and enhances sand deposition even further. In doing so, the dune acts as a sink, removing sand from the wind and suppressing the formation of other dunes nearby. Similarly, erosion of rocks by water proceeds faster at some slight depression. In fact, more water collects in the incipient valleys, causing that the erosion proceeds there more rapidly. Sand or water that accumulates at a particular location cannot contribute to the process at another position. In fact, next to a river or a lake, there are no other rivers or lakes (Meinhardt 2009).

9.5.7   The Next Development of Turing's Theory: The Mechanochemical Patterning

In his seminal paper on the chemical basis of morphogenesis, Turing declared the importance of mechanics in morphogenesis. Although he was aware of the relevant role of both the chemical and mechanical part of morphogenesis, he neglected the mechanical contribution for computational reasons. There were no digital computers available, yet. But, today it is indeed clear that active mechanical processes, like transport along cytoskeletal filaments (for more information about the cytoskeleton, read Box 9.3 of this chapter), cytoplasmic flow and endocytosis, play essential roles in patterning at the cell and tissue levels along with reactions and diffusion events.

The diffusion grounds on the Brownian motion. The Brownian motion is fast over short distances, but slow over long ones because it is not unidirectional but random. The average distance traveled by a morphogen having D as its own diffusion coefficient, and after a time interval Δt is $\lambda \approx \sqrt{D(\Delta t)}$.[11] If the morphogenetic species participates in chemical reactions that deplete it with a rate constant k_d, its lifetime is $\tau = \left(\frac{1}{k_d}\right)$, and the distance it covers will be $\lambda \approx \sqrt{D\tau} = \sqrt{D/k_d}$. In the case of Turing's mechanism, the two morphogens having D_X and D_Y as diffusion coefficients, and τ_X and τ_Y as lifetimes, originate a pattern that has a characteristic wavelength λ given by

$$\frac{1}{\lambda^2} = \frac{1}{\lambda_X^2} - \frac{1}{\lambda_Y^2} \approx \frac{1}{D_X \tau_X} - \frac{1}{D_Y \tau_Y}$$

[9.40]

---

[11] Remind exercises 3.10 and 3.11.

BOX 9.3   CYTOSKELETON

The cytoskeleton is a network of protein fibers that permeate the entire cell. It gives the cell structure as well as motility. In metaphorical terms, the network of protein fibers plays as the bone structure of the cell but also as the highway system of the cell. There are three types of fibers within the cytoskeleton: (I) microfilaments, (II) intermediate filaments, (III) microtubules. (I) The microfilaments are the thinnest of the protein fibers; their diameter ranges from 6 to 7 nm. They are composed entirely of one type of linear protein called actin. One important use of microfilaments is in muscle contraction. During muscle contraction, the protein myosin bound to actin "walks" along the filament creating the contractile motion. Actin filaments are also involved in the cell movement and division. (II) The intermediate filaments are rope-like and are composed of different types of proteins; they are thicker than microfilaments. In fact, their diameter is about 10 nm. Just like microfilaments, intermediate filaments give the cell tensile strength and provide structural stability. All the cells have intermediate filaments, but the protein subunits of these structures vary. Some intermediate filaments are associated with specific cell types. For example, neurofilaments are found specifically in neurons, desmin filaments are found specifically in muscle cells, and keratins are found specifically in epithelial cells. (III) Microtubules are the largest of the three types of filaments. Their diameter is about 25 nm. A microtubule is a hollow, rigid tube made of a protein called tubulin. Microtubules are ever-changing, with reactions continually adding and subtracting tubulin dimers at both ends of the filament. The rates of change, at either end, are not balanced; one end grows more rapidly and is called the plus end, whereas the other end is known as the minus end. In cells, the minus ends of microtubules are anchored in structures called microtubule-organizing centers (MTOCs). The primary MTOC in a cell is called the centrosome, and it is usually located adjacent to the nucleus. Microtubules are involved in the separation of chromosomes during cell division, in the transport within the cell and the formation of specialized structures called cilia and flagella (the latter is true only in eukaryotic cells).

If the morphogenetic process, which we are focusing on, is slow enough that morphogens cover the required distances by diffusion, then the Brownian motion is fast enough to account for the spatial patterns that can be observed within cells and tissues. However, when the development of patterns is too fast than the diffusion, other processes are necessary for pattern formation. One is the phenomenon of traveling chemical waves that will be described in paragraphs 9.6 and 9.7. Two others are mechanical processes within cells and tissues that act faster over long distances than Brownian motion, and they are presented right here. They are two types of advection that require chemical energy resources (Horward et al. 2011).

The first is transport driven by motor proteins (read Box 9.4 of this chapter to know more about motor proteins). Typically, a motor protein proceeds with a speed (v) of 1 μm s^{-1}. A useful way to compare the relative importance of advection and diffusion over the distance L is to calculate the Péclet number (Pe). The Péclet number is the ratio of the diffusion time driven by thermal energy to the advection time driven by motor proteins.

$$Pe = \frac{\tau_{dif}}{\tau_{adv}} = \frac{vL}{D}$$
[9.41]

When $Pe > 1$, the advection is faster than diffusion, whereas when $Pe < 1$ the opposite is true. The Péclet number is less than 1 when we consider short distances. For example, if the diffusion coefficient of a protein is 5 μm^2 s^{-1}, Pe is less than 1 as long as $L < 5$ μm. On the other hand,

BOX 9.4   MOTOR PROTEINS

The myriad of chemical reactions that occur within a cell cannot be described simply as the result of a complex series of parallel and consecutive chemical transformations brought about by diffusion and random collisions of chemical species embedded in different cellular compartments. Cells are polar structures, and their internal composition and structure are not homogeneous neither isotropic. Moreover, most of the essential cellular functions, like the maintenance of a transmembrane electric potential, cell division, and translocation of organelles, all require directional movement and vectorial transport of chemical species. To overcome the randomizing effect of Brownian motion and to carry out directional processes, cells have macromolecules that behave like tiny machines. These are the so-called molecular motors: the energy to drive their motion ultimately comes from chemical reactions taking place in the catalytic site(s) of the motors.

Molecular motors dominate the randomness of single molecular events and generate unidirectional processes in the cell. Cells have hundreds of different types of molecular motors (Bustamante et al. 2001). Some of them operate cyclically: the steps of the mechanical cycle are coupled to the steps of a chemical cycle that fuels the molecular motor. The chemical steps typically involve the two most common energy repositories, the nucleotides ATP or GTP (in myosin, kinesin, and helicases) or translocation of an ion across an electrochemical gradient (in F_0 ATPase and the bacterial flagellar motor). A parameter that measures the mechanochemical coupling is the ratio χ:

$$\chi = \frac{\langle v \rangle}{L \langle v_r \rangle} \qquad [B4.1]$$

In [B4.1], $\langle v \rangle$ is the mean velocity of the motor, L its step size and $\langle v_r \rangle$ the mean reaction rate. If the mechanical motion and the chemical reaction are tightly coupled, $\chi \sim 1$ because the product $L \langle v_r \rangle$ will equal $\langle v \rangle$. On the other hand, if many chemical steps do not lead to mechanical motion, χ will be less than 1. Other molecular motors are "one-shot": they unleash previously stored elastic potential energy and then disassemble (like in spasmoneme and actin polymerization).

A molecular motor moving on a molecular route and interacting with all the surrounding molecules has many degrees of freedom. Many of them are in thermal equilibrium with the microenvironment and define the so-called *bath variables*. *Bath variables* are not involved directly in the directional movement of the motor, but they affect its motion as fluctuating stochastic forces, as sources of friction, and as entropic contributions to the thermodynamics of molecular movement. The rest of the variables that are involved in the unidirectional motion of the motor are called *system variables*. Since a molecular motor must be fueled by a chemical reaction, at least one of the *system variables* measures the progress of the chemical reaction and is called *reaction coordinate*. All the other *system variables* are *mechanical variables*. The simplest case is when the motor has only one *reaction coordinate* (x_1) and one *mechanical variable* (x_2). If it were a macroscopic motor, its motion would be determined only by the shape of the energy potential surface. But we are considering microscopic molecular motors, and they suffer the random thermal forces. The motion of a molecular motor is stochastic and only statistically biased by the potential. In fact, the inertial forces due to the potential are negligible compared to friction forces. The molecular motors always operate at low Reynold's number. The Reynold's number is the ratio of inertial (F_{in}) to viscous forces (F_v) on a body moving in a fluid:

$$Re = \frac{F_{in}}{F_v} = \frac{dvl}{\eta} \qquad [B4.2]$$

(Continued)

BOX 9.4 (Continued)   MOTOR PROTEINS

In equation [B4.2], d is the density of the fluid, v is the velocity of the motor having length l and η is the viscosity of the fluid. Macroscopic motors usually operate at high Reynold's numbers, where inertial forces dominate viscous forces. For example, a car 4 m long, moving in the air (with $d = 1.225$ Kg/m^3 and $\eta = 1.81 \times 10^{-5}$ Pa*s) at the velocity of 50 Km/h, has $Re = 3.8 \times 10^6$. On the other hand, all microscopic motors, either unicellular organisms like protists (they use flagella or cilia to move), or organelles or molecular motors within cells, move in an environment characterized by a low Reynold's number. For example, a protein 10 nm long moving at a velocity of 1 μm/s within the water (having $d = 1000$ Kg/m^3 and $\eta = 0.001$ Pa*s) has $Re = 1 \times 10^{-8}$! It is as if our car moved through tar; it is like swimming through a very viscous liquid. Locomotion at low Re requires the continuous expenditure of work. If work stops, the motion stops immediately. Locomotion at low Re drags along a large added mass of fluid by viscosity, and it is more energetically demanding than motion at high Re.

For molecular motors, the random forces are rapid compared with the time scale of the movement of the motor. Therefore, the random forces loose correlation between steps of the walk and the motion is Markovian. It is "memoryless" motion: its next step depends on its present state but not on the motion's full history. Therefore, the movement of the molecular motor over the potential energy surface in the mechanochemical space is a uniform motion described by the Langevin equations:

$$-\gamma_1 \frac{dx_1}{dt} - \frac{\partial V\left(x_1, x_2\right)}{\partial x_1} + f_1(t) + F_{B1}(t) = 0 \qquad [B4.3]$$

$$-\gamma_2 \frac{dx_2}{dt} - \frac{\partial V\left(x_1, x_2\right)}{\partial x_2} + f_2(t) + F_{B2}(t) = 0 \qquad [B4.4]$$

where the subscripts 1 and 2 refer to the reaction and mechanical coordinates, respectively. $V\left(x_1, x_2\right)$ is the energy potential of the motor; $f_1(t)$ and $f_2(t)$ represent the external mechanical forces acting on the motor and exerted by a laser trap, an atomic force microscope, or the forces due to the load that the motor is driving. The other two terms of equations [B4.3] and [B4.4] derive from the surrounding fluid, with γ_1 and γ_2 being the friction coefficients, and $F_{B1}(t)$ and $F_{B2}(t)$ being the Brownian random bath forces. Because of the random forces, individual trajectories obtained for solving these equations are meaningless; only statistical distributions over many trajectories are useful. The final solution of the Langevin equations will be a probability distribution $w\left(x_1, x_2, t\right)$. Since the total probability is constant, it will obey a continuity equation without a source term:

$$\frac{\partial w}{\partial t} + \nabla J = \frac{\partial w}{\partial t} + \sum_{i=1}^{2} \frac{\partial J_i}{\partial x_i} = 0 \qquad [B4.5]$$

where J represents the probability current (with $[L]/[t]$ as dimensions). In our case of a biased diffusion process, J is the sum of two contributions:

$$J_i = -\left(\frac{kT}{\gamma_i}\right)\frac{\partial w}{\partial x_i} + \left(\frac{f_i}{\gamma_i}\right)w \qquad [B4.6]$$

(Continued)

BOX 9.4 (Continued)   MOTOR PROTEINS

The first term represents the diffusion, whereas the second is the advection driven by the forces f_i including that due to the potential and external forces, but excluding the stochastic forces, which are included in the first term.

Inserting equation [B4.6] into [B4.5], we obtain the Smoluchowski equation:

$$\frac{\partial w}{\partial t} + \nabla J = \frac{\partial w}{\partial t} + \left[\sum_{i=1}^{2} \left(-\frac{k_B T}{\gamma_i} \frac{\partial^2}{\partial x_i^2} + \frac{1}{\gamma_i} \frac{\partial}{\partial x_i} f_i \right) \right] w = 0 \qquad [B4.7]$$

Molecular motors convert chemical energy into mechanical motion by two general mechanisms: the power stroke and the Brownian ratchet. In the power stroke mechanism, the chemical reaction and the mechanical movement are tightly coupled. For example, the chemical step could be the binding of a substrate to the site of the protein, and the movement, the corresponding conformational change of the macromolecule. In the Brownian ratchet mechanism, the chemical process is not coupled to the mechanical movement. The motor is driven by thermal fluctuations, but the chemical reaction rectifies or selects the forward fluctuations. The path of a Brownian ratchet drawn in a free-energy surface having one mechanical coordinate and one chemical coordinate will be zigzag. It will consist of pairs of transitions at right angles: one nearly parallel to the chemical axis and one nearly parallel to the mechanical axis. The chemical step is irreversible with its ΔG large and negative. The free energy is expended to favor forward motion, rather than doing mechanical work on the protein directly. A Brownian ratchet is efficient when each fuel molecule burned results in exactly one step forward.

Pe is larger than 1, when $L > 5$ μm. It means that the fastest way to cross an entire eukaryotic cell that has a diameter of 10 μm is through advection. In fact, it will take 10 s, whereas the Brownian motion would take the double: $(\lambda^2/D) = 20$ s. When the advection moves the motor proteins in a direction that is the opposite with respect to the diffusion, a new length scale emerges that is $\lambda = D/v$.

The second type of advection, which is faster than diffusion, is movement due to mechanical stress. Forces exerted by motor proteins locally on the cytoskeleton create an active stress gradient and lead to a velocity gradient. If there is friction with the surroundings, then the velocity gradient will have the characteristic length $\lambda = \sqrt{\eta / \gamma}$ where η is the viscosity and γ is the friction coefficient. Such velocity gradient leads to rapid movement of material, at the speed of sound, called cytoplasmic streaming (Quinlan 2016). Morphogens embedded in this material are moved along with it, like objects transported by a flowing river. Within a cell, such bulk-material transport can be driven by flows of cytoplasm. But bulk-material flow also occurs at the tissue scale, resulting from changes in cell position, orientation, and shape.

The mechanochemical patterning is responsible for cell polarity (Goehring and Grill 2013). Cell polarity is the asymmetric organization of a cell. Cell polarity is crucial for certain cellular functions, such as cell migration, directional cell growth, and asymmetric cell division. It is also relevant in the formation of tissues. For example, epithelial cells are examples of polarized cells that feature distinct apical and basal plasma membrane domains (see Figure 9.14). The apical-basal polarity drives the opposing surfaces of the cell to acquire distinct functions and chemical components.

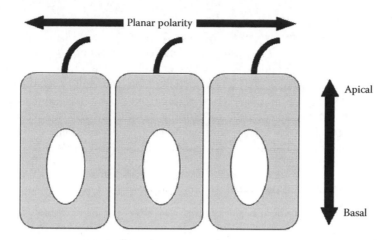

FIGURE 9.14 Polarity of epithelial cells.

There is also planar cell polarity that aligns cells and cellular structures such as hairs and bristles within the epithelial plane. Cell polarity is so fundamental that it is ubiquitous among living beings: it is present not only in animals and plants, but also in fungi, prokaryotes, protozoa, and even archaebacteria.

9.6 CHEMICAL WAVES

Spontaneously, we are inclined to think of diffusion as a process that tends to homogenously spread chemicals in space. However, in this chapter we are discovering that this is not always the case. For example, in paragraph 9.2 we have learned that a periodic structure emerges from a homogeneous system when diffusion is coupled to an autocatalytic reaction. In paragraph 9.3, we have discovered how diffusion combined with activator-inhibitor dynamics can give rise to stationary patterns having several shapes. In this paragraph, we are going to learn that, when diffusion works in the presence of an excitable system or an oscillatory reaction carried out in an unstirred reactor, amazing phenomena, called chemical waves, emerge (Figure 9.15). The most common type of chemical wave is the single propagating front that is a relatively sharp boundary between reacted and unreacted

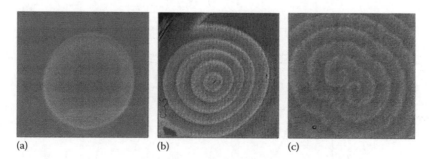

(a)   (b)   (c)

FIGURE 9.15 Examples of chemical waves generated in a chemistry laboratory by using the BZ reaction: a propagating wave front in (a); a target pattern in (b), and a spiral pattern in (c).

chemicals and that can be circular like that shown in picture (a) of Figure 9.15. In some systems, the medium restores its original state, relatively quickly, after the passage of the front. Such phenomenon is referred to as a pulse. Even more astonishing are the target patterns in which pulses are emitted periodically from the same point, named as "pacemaker" or "leading center" (see picture (b) in Figure 9.15). If an expanding pulse is broken at a point, it begins to curl up at the broken ends and produces oppositely rotating spiral patterns (see picture (c) in Figure 9.15). The images of chemical waves shown in Figure 9.15 have been collected in a chemistry laboratory by exploiting the BZ reaction. But in nature there are many other examples of chemical waves. Some of them are described in paragraph 9.7.

9.6.1 Propagator-Controller Model

An easy mathematical description of how chemical waves form is the Propagator-Controller model (Fife 1984). Let us assume that our chemical system consists of two variables, x and y, which react according to the kinetic laws $f(x, y)$ and $g(x, y)$, and diffuse along the spatial coordinate r. The system of partial differential equations is:

$$\frac{\partial x}{\partial t} = \frac{1}{\eta} f(x, y) + D_x \left(\frac{\partial^2 x}{\partial r^2} \right) \tag{9.42}$$

$$\frac{\partial y}{\partial t} = g(x, y) + D_y \left(\frac{\partial^2 y}{\partial r^2} \right) \tag{9.43}$$

In equations [9.42] and [9.43], D_x and D_y are the diffusion coefficients of x and y; the parameter η that is assumed to be much smaller than one is introduced to highlight the difference in the rates of x and y transformations: the variable x changes much more quickly than y. Moreover, we assume that the kinetic laws f and g avoid that the concentrations of the variables grow to infinity. In particular, the nullcline $f(x, y) = 0$ has the S shape we found in paragraph 8.3 (Figure 9.16). Finally, the time scale of the chemical transformations is much shorter than that associated with the diffusion processes. Therefore, the system of two partial differential equations [9.42 and 9.43] can be reduced nearly everywhere in time and space to the following equations:

$$f(x, y) = 0 \tag{9.44}$$

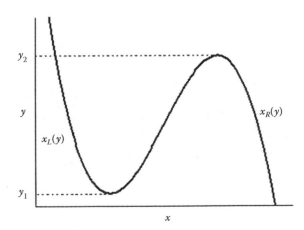

FIGURE 9.16   The S-shape of the $f(x, y) = 0$ nullcline.

$$\frac{dy}{dt} = g(x(y), y) \qquad\qquad [9.45]$$

In [9.45], the variable x, which changes rapidly, assumes instantaneous values that depend on that of y.

As we learned in chapter 8, the points on the left-hand and right-hand branches of the $f(x, y) = 0$ nullcline (labeled as $x_L(y)$ and $x_R(y)$ respectively, in Figure 9.16) represent stable solutions, whereas the points on the middle branch represent unstable solutions.

9.6.1.1 Phase Waves

Let us suppose that the second nullcline $g(x, y) = 0$ intersects the middle branch. This condition means the chemical system is in oscillatory regime (remember paragraph 8.3). Since the system is not under stirring, the variable x may be discontinuous in r. There will be regions where x assumes a value of the right-hand branch and regions where it assumes a value of the left-hand branch. The area that separates the two regions is called boundary layer. For $\eta \rightarrow 0$, the boundary layer is a point in a one-dimensional medium, whereas it is a very thin slice for a bi-dimensional medium. In the boundary layer, x assumes a value belonging to the middle branch whereas y assumes a value between y_1 and y_2. The layer will move with a velocity that depends on the "Controller" y and in its movement the "Propagator" variable x jumps from x_L to x_R or vice versa. If we fix our attention to a specific point in space, the frequency of the appearance of the boundary layer is equal to the frequency of oscillations in that location. Such kind of chemical wave is called phase wave. Nothing moves as a phase wave propagates. In other words, the propagation of a phase wave does not rely upon the physical interchange of materials between adjacent points. In fact, in the case of the BZ reaction it has been demonstrated (Kopell and Howard 1973) that phase waves, run in a cylinder, pass through a barrier impenetrable to diffusion.[12] If we define the phase φ of an oscillation as the ratio between the time delay (t) and the period (τ) of an oscillation (see the upper plot of Figure 9.17), then it derives that the velocity of a phase wave depends inversely on the phase gradient, $\Delta\varphi$, between adjacent points in space. For example, in Figure 9.17, it is shown the situation where adjacent points along the spatial coordinate r have a phase difference $\Delta\varphi = 0.1$. At time t_0, in the five points, labeled as r_1, r_2, r_3, r_4, r_5, φ is different from both 0 and 1. Hence, in all five points, $[X]$ is low. An instant later, at t_1, $\varphi = 1$ in r_1, and just in r_1, $[X]$ becomes large. At time t_2, the condition $\varphi = 1$ is verified in the adjacent point r_2, and the maximum of $[X]$ is now in r_2. In r_1, φ has become 1.1, and $[X]$ has dropped, again. At time t_3, the condition $\varphi = 1$ is verified only in r_3. And so on. In this way, the wave front propagates as shown in Figure 9.17.

9.6.1.2 Trigger Waves

When the phase gradient is large, the phase wave moves slowly. In this case, the diffusion of reagents is not anymore negligible, and the front can propagate faster than a phase wave (Reusser and Field 1979) because another type of chemical wave propagates, called trigger wave. The velocity of a trigger wave is controlled by the rates of chemical reactions and diffusion; it cannot go through an impenetrable barrier as a phase wave does. Trigger waves do not require necessarily an oscillatory medium to appear. They can be originated even in an excitable medium. We learned the definition of excitability in the previous chapter (see paragraph 8.3.1). Our system is excitable when the nullcline $g(x, y) = 0$ intersects the first nullcline, $f(x, y) = 0$, in any point of either the left-handed x_L or the right-handed x_R branch (in Figure 9.18, $g(x, y) = 0$ intersects $f(x, y) = 0$ on the left-hand branch). Before any perturbation, the system stays indefinitely in its stable steady state that is the intersection point, labeled as 0, in Figure 9.18. If the system feels a sufficiently strong perturbation that pushes it

---

[12] It is possible to observe phase waves between physically isolated oscillators as described in Ross et al. (1988). Such phase waves are also called kinematic waves. They do not involve mass transfer from one oscillator to another.

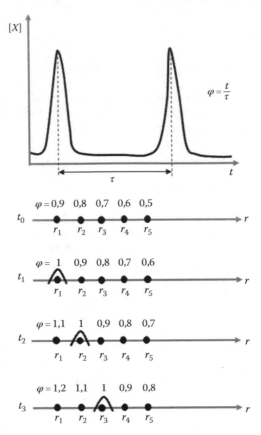

FIGURE 9.17 The upper plot illustrates the definition of phase φ for an oscillating reaction. In the lower part, the sketch, which must be read from t_0 up to t_3, helps to understand how a phase wave propagates along the spatial coordinate r.

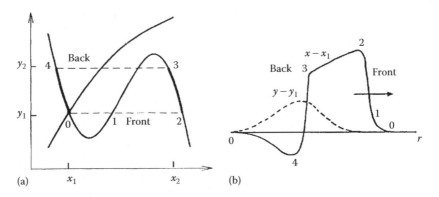

FIGURE 9.18 Propagating pulse in an excitable chemical system. Graph (a) shows the variations of the Propagator x and the Controller y in their phase plane, with the front and the back of the pulse depicted as dashed traces. Graph (b) shows the same pulse along the spatial coordinate r.

up to reach point 1, then it initiates a large excursion especially in the x variable that follows the cycle 0-1-2-3-4-0 in graph a of Figure 9.18. A pulse of a trigger wave propagates (see graph b of the same figure). The leading edge of the wave is at 0-2, whereas its trailing edge is at 3-4.

The perturbations that promote trigger waves can be induced by the experimenter or be spontaneous. A chemical system can be pushed deliberately beyond its threshold of excitability in several ways.

For example, chemically, by sinking a silver wire in a solution containing the BZ reagents in a non-oscillatory regime, or by adding drops of acid into a pH-sensitive bistable reaction. Otherwise, it is possible to induce a trigger wave thermally, by dipping a hot wire in the solution. In one case, it has been demonstrated that waves can be electrochemically initiated in an unstirred thin film of a solution containing iodate and arsenous acid (Hanna et al. 1982). Waves in photosensitive systems can be initiated by irradiation at the appropriate wavelength and intensity. A remarkable example in literature is offered by the light-sensitive version of the BZ reaction that involves the ruthenium bipyridyl complex $[Ru(bpy)_3]^{2+}$ as catalyst (Kuhnert et al. 1989). The spontaneous generation of trigger waves may be due to concentration fluctuations at a particular point in the solution or to the presence of an artificial pacemaker point such as a dust particle. The point where there is the dust particle may be in oscillatory regime, while around it the system is excitable. The pacemaker point is the source of phase waves that ultimately turn in trigger waves when they propagate in the excitable bulk medium.

9.6.2 Shapes of Chemical Waves

Chemical waves can assume different shapes depending on the geometry of the medium hosting them.

9.6.2.1 Mono- and Bi-Dimensional Waves

One-dimensional waves are formed in a narrow medium like a test tube, and they may consist of a single pulse or a train of fronts (Tyson and Keener 1988). Two-dimensional waves occur more frequently in nature and can be easily observed in the laboratory by using thin layers (1 mm deep) of a solution (with proper chemicals) held in a Petri dish. In a two-dimensional medium, a wave originating in a point produces a circular front when the wave propagates at the same velocity in all directions. When the system gives rise to repeated waves, we observe a pattern of concentric waves, called target pattern (like that shown in picture b of Figure 9.15). When the fronts of two waves belonging to two distinct target patterns collide, they annihilate and lead to cusp-like structures in the vicinity of the area of collision. When an expanding circular wave front is disrupted, spiral waves are formed. In the case of the BZ reaction, the disruption of a circular wave front can be performed by pushing a pipette through the solution or with a gentle blast of air from a pipette onto the surface of the reacting solution (Ross et al. 1988). The wave curls in and forms a pair of counter-rotating spirals as shown in picture c of Figures 9.15 and 9.19.

TRY EXERCISE 9.15

It is possible to obtain multi-armed spirals if a drop of KCl (the chloride anion interferes with the oxidation-reduction chemistry of the BZ reaction) is added to the center of rotation of a forming spiral wave (Agladze and Krinsky 1982). More recently, it has been shown that when the BZ reaction is carried out inside the nanoreactors of a water-in-oil microemulsion with sodium bis(2-ethylhexyl) sulfosuccinate (AOT) as the surfactant, antispiral waves are formed (Vanag and Epstein 2001b). Antispiral waves are spirals traveling from the periphery toward their centers. In the same AOT-microemulsions, packets of waves traveling coherently either toward or away from their centers of curvature have been observed (Vanag and Epstein 2002). They range from nearly plane waves to target-like patterns. Under certain conditions (Vanag and Epstein 2003), packet, target, and spiral

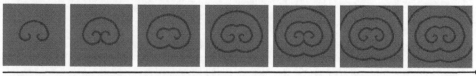

Time

FIGURE 9.19 Simulation of the propagation of a spiral wave.

waves may break into small segments. These short segments, or dashes, propagate coherently in the direction perpendicular to the breaks.[13]

9.6.2.2 Three-Dimensional Waves

When we increase the thickness of the chemical system, we expect to observe three-dimensional chemical waves. Experimentally, it is difficult to detect them because there are many disturbing phenomena, like convection and bubble formation. However, some results have been achieved in sealed vessels or using gels that hinder convective motions. In three-dimensions, we can observe either spherical waves or scroll waves. Spherical and scroll waves are 3D extensions of target patterns and spiral waves, respectively. A scroll wave is made of a two-dimensional surface that rotates around a one-dimensional filament or axis (Figure 9.20) that terminates at the boundary of the three-dimensional spatial domain. The filament may assume a vertical orientation, or it may curve or twist. Sometimes, it joins itself into a ring within the spatial domain. Several types of scroll waves have been observed experimentally and numerically. A list and more-in-depth analysis can be found in the papers by Tyson and Keener (1988), and Winfree and Strogatz (1984).

9.6.2.3 Effect of Curvature

In two and three dimensions the wave fronts may be curved. Any curvature is quantitatively characterized by specifying its radius of curvature that is the radius of the circle best fitting the front. If the radius is r, the curvature is $c = (1/r)$. By convention, the curvature c is taken with a positive sign when the curved front propagates towards the center of the circle; on the other hand, c is negative when the curved front propagates far from the center. It has been proved theoretically (Tyson and Keener 1988) and experimentally (Foerster et al. 1988) that the velocity of a curved wave front depends on its curvature, according to the eikonal equation

$$v_c = v_p + c * D \qquad\qquad [9.46]$$

In [9.46], v_p is the velocity of the plane wave ruled by the concentration of the controller species (i.e., y in equations [9.42 and 9.43]) at the front; D is the diffusion coefficient of the propagator (i.e., species x in equations [9.42 and 9.43]) and c is the curvature. It is clear that for a curved front

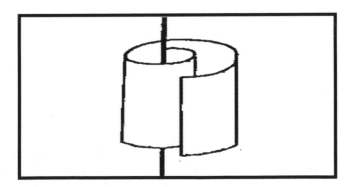

FIGURE 9.20 Sketch that depicts the two essential elements of a scroll wave embedded in a medium represented by the rectangle: (I) a surface that rotates around (II) a filament that is the vertical segment.

---

[13] Wave segments can also be achieved by generating waves in an excitable medium and propagating in a sub-excitable medium. A medium is sub-excitable when the threshold of excitation is sufficiently large that a chemical wave cannot be maintained. The sub-excitable regime has been easily obtained by using a light of the proper intensity and the photosensitive version of BZ reaction where the photocatalyst ruthenium(II)-bipyridyl produces the inhibitor bromide (Kádár et al. 1998).

propagating far from the center of its curvature, the velocity of the curved wave is smaller than that of a planar wave front considered at the same chemical composition. When the curved wave propagates toward the center of the circle, the opposite is true.

TRY EXERCISE 9.16

Note that in the case of three-dimensional waves, the eikonal equation becomes

$$v_c = v_p + (c_1 + c_2) * D \qquad\qquad [9.47]$$

where c_1 and c_2 are the principal curvatures of the wave front surface (Tyson and Keener 1988).

9.7 "CHEMICAL" WAVES IN BIOLOGY

The phenomenon of traveling chemical waves is widespread in biology. The reason is that it is an effective means of transmitting chemical information. In fact, it can be much faster than pure diffusion. For example, imagine that a protein has the task of transporting a message through the axon of a neuron. A neuron is a cell specialized in receiving, integrating, processing and transmitting information. The axon of a neuron is a cable-like structure that carries signals (see Figure 9.21). Let us consider an axon that is 1 cm long.[14] The protein diffuses through the axon with $D \approx 5$ $\mu m^2/s$.[15] The time the protein spends to cross the axon only by diffusion is enormously long: almost eight months! It is evident that neurons must exploit other strategies to transfer information. In fact, information crosses an axon of 1 cm in just a few milliseconds! How is it possible? Such fast transfer is made possible by the involvement of electrochemical waves. What are electrochemical waves?

9.7.1 WAVES IN A NEURON

Any neuron within our brain receives chemical signals through the dendrites (Figure 9.21). Such information is transduced in transmembrane electrochemical potential in the soma and, finally,

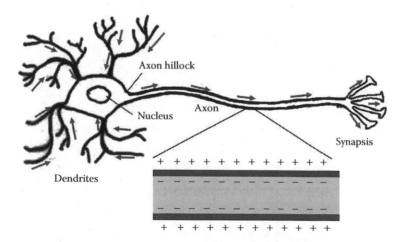

FIGURE 9.21   Schematic structure of a neuron. The grey arrows indicate the flow of information from the dendrites up to the synapse through the axon.

---

[14] There are neurons whose axon can be 1.5 m long! For example, those connecting our toes with the base of the spine.

[15] Note that in pure water D would be ≈ 100 $\mu m^2/s$, but the crowded environment of a cell reduces the velocity of diffusion significantly, especially for a macromolecule. Read Box 9.4 of this chapter.

integrated into the axon hillock. Neurons, like the other cells in our body, are electrically polarized: they have an electric potential gradient between the exterior and the interior of the cellular membrane. This gradient of electric potential is made possible by the presence of proteins within the membrane; they work as ionic pumps and channels.

The information that is integrated into the axon hillock can trigger an electrochemical wave that crosses the entire axon quickly without weakening and reaches the bottom part of the neuron where there are the synapses. At the synapses, the electrochemical information is transduced back in chemical signals, and neurotransmitters are released to the dendrites of other neurons.

Every excitable portion of the neural membrane has two critical values of electric potential: one is the resting potential and the other is the threshold potential. The resting potential represents the value that the membrane maintains as long as it is not perturbed; usually, in the axon hillock, it is −70 mV. The inputs, which a neuron receives, induce either a hyperpolarization or depolarization of the membrane. In other words, the inputs either decrease or increase its electric potential. When a sufficiently strong excitatory input depolarizes the membrane, and its potential overcomes the threshold value (that is ≈ −55 mV in the axon hillock), the neuron triggers an action potential (Figure 9.22). The electric potential of the axon hillock soars quickly, in roughly 1 ms or less, up to reach a positive value, as large as +50 mV. Then, it drops to values that are more negative than the resting potential. Finally, it recovers its initial state, and the neuron is ready to discharge another action potential.

The discharge of an action potential involves voltage-gated ion channels and pumps that are transmembrane proteins shown schematically in Figure 9.23. We start from the resting state (a) with the membrane that is hyperpolarized. If the neuron receives an excitatory input, the membrane potential raises because the sodium channels open (see step b in Figure 9.23). The increase of the membrane potential induced by the income of the first sodium ions exerts a positive feedback action. In fact, it causes a further opening of the Na^+ channels and a stronger depolarization. Overall, this autocatalytic action determines a sharp jump of the potential membrane (see part c of Figure 9.23). Then, within a fraction of one-thousandth of a second, the sodium channels switch to an inactivated state due to the highly positive membrane potential (see part d of Figure 9.23). In the inactivated state, the sodium channels are closed and also refractory to reopening. At the same time, potassium channels open. They allow K^+ ions to leave the neuron. Such transfer repolarizes the membrane. Finally, the ion pumps restore the initial transmembrane gradients of K^+ and Na^+ ions by exploiting energy supplied by the cell. After all, the portion of the axon where all the described events have occurred is ready to transmit another action potential (see part e of Figure 9.23). Meanwhile, the transmitted action potential propagates only forwards. The refractory states of the sodium channels prevent the depolarization from spreading backward along the axon. The action potentials propagate vectorially towards the synapsis, as genuinely electrochemical waves.

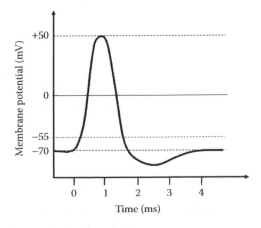

FIGURE 9.22 Schematic profile of an action potential.

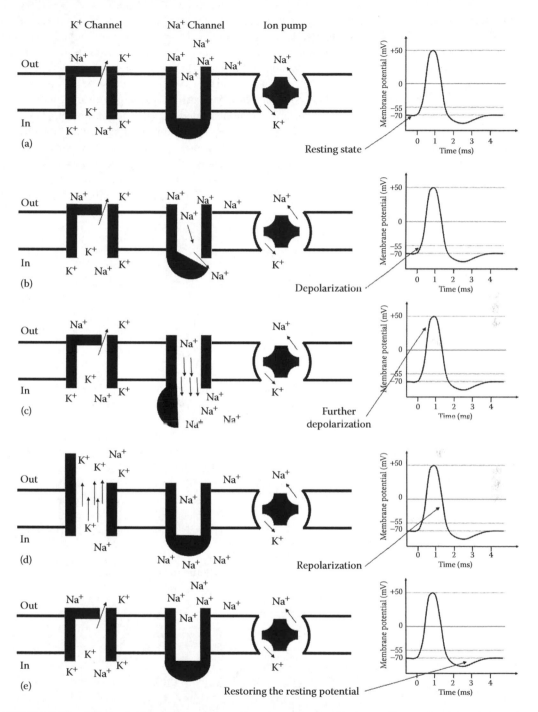

FIGURE 9.23   Schematic description of five relevant steps in the discharge of an action potential: resting state (a); depolarization (b); further depolarization (c); repolarization (d), and restoring the resting potential (e).

When the action potential reaches the synapse, voltage-gated Ca^{+2} channels open and calcium ions enter the axon terminals. Ca^{+2} entry causes neurotransmitter-containing synaptic vesicles to move to the plasma membrane, fuse with it and release their content according to the process of exocytosis. The neurotransmitter diffuses and binds to ligand-gated ion channels of the dendrites of other neurons, exerting either an inhibitory or an excitatory action.

Our understanding of how voltage-gated ion channels give rise to propagating action potentials is primarily due to the physiologists Alan Hodgkin and Andrew Huxley. They were the first to record the potential difference across the membrane of a squid giant axon by inserting a fine capillary electrode in it. It was the year 1939. The Second World War began only a few weeks after their first experiments were published in *Nature* (Hodgkin and Huxley 1939). Therefore, they were forced to quit their research. After the war, Hodgkin and Huxley could restart their fruitful collaboration that culminated with the publication of their mathematical model of the action potential in 1952 (Hodgkin and Huxley 1952). The electrical behavior of the axon membrane may be described by the electrical circuit represented in Figure 9.24. Current flows through the membrane either by charging the membrane capacity $\left(I_C = C_M \left(dE/dt\right)\right)$ or by movement of ions through resistances that are in parallel with the capacity. The ionic current consists of three contributions: sodium (I_{Na}), potassium (I_K) and leakage (I_L) current, the latter being due to other ions, such as chloride anions. Each term of the ionic current is given by Ohm's law:

$$I_{Na} = \frac{1}{R_{Na}}\left(E - E_{Na}\right) \qquad [9.48]$$

$$I_K = \frac{1}{R_K}\left(E - E_K\right) \qquad [9.49]$$

$$I_L = \frac{1}{R_L}\left(E - E_L\right) \qquad [9.50]$$

In the equations [9.48 through 9.50], E is membrane potential, whereas E_{Na}, E_K, and E_L are the equilibrium potentials for sodium, potassium, and leakage ions, respectively. The terms $(1/R_{Na})$, $(1/R_K)$ and $(1/R_L)$ are the sodium, potassium and leakage ions conductances.

The sodium and potassium conductances are time- and membrane potential-dependent. In fact, depolarization causes a temporary rise in sodium conductance, whereas repolarization determines an increase in potassium conductance but also a decrease in sodium conductance. If we refer all the potentials to the absolute value of the membrane resting potential E_r, and we fix $V = E - E_r$, $V_{Na} = E_{Na} - E_r$, $V_K = E_K - E_r$, $V_L = E_L - E_r$, the equation that gives the total membrane current I is:

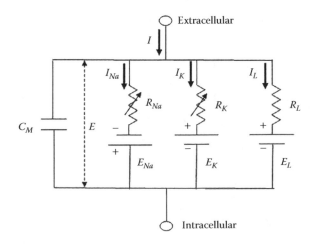

FIGURE 9.24 Electrical circuit proposed by Hodgkin and Huxley as a model to explain the formation of action potentials in axons.

$$I = C_M \frac{dV}{dt} + \frac{1}{R_{Na}}(V - V_{Na}) + \frac{1}{R_K}(V - V_K) + \frac{1}{R_L}(V - V_L) \qquad [9.51]$$

After defining the best non-linear analytic functions for $(1/R_{Na})$ and $(1/R_K)$ by fitting experimental data of potassium and sodium conductances, Hodgkin and Huxley succeeded in reproducing the shape of an action potential. For the description of how an action potential propagates through an axon, we must consider the concatenation of circuits like that of Figure 9.24. If we indicate with r the spatial coordinate that represents the principal axis of the cylindrical axon, perpendicular to the membrane, the current flowing along r is

$$I = \frac{1}{2\pi(R_{out} + R_{in})} \frac{\partial^2 V}{\partial r^2} \qquad [9.52]$$

where R_{out} and R_{in} are the external and internal resistances of the axon.[16] If we consider an axon surrounded by a large amount of conducting fluid, R_{out} is negligible compared with the resistance of the axoplasm (that is the cytoplasm within the axon of a neuron), i.e., R_{in}. If we indicate with a the radius of the axoplasm and with ρ_{in} its resistivity, the final equation describing the propagation of an action potential is

$$\frac{a}{2\rho_{in}} \frac{\partial^2 V}{\partial r^2} = C_M \frac{\partial V}{\partial t} + \frac{1}{R_{Na}}(V - V_{Na}) + \frac{1}{R_K}(V - V_K) + \frac{1}{R_L}(V - V_L) \qquad [9.53]$$

Equation [9.53] can be rearranged in the following form:

$$\frac{\partial V}{\partial t} = \frac{a}{2\rho_{in}C_M} \frac{\partial^2 V}{\partial r^2} + f(V) \qquad [9.54]$$

9.7.2 THE FISHER-KOLMOGOROV EQUATION

Equation [9.54] is formally equivalent to the equation that describes a traveling wave. If you remember equations [9.42 and 9.43], it is evident that the traveling waves of a chemical concentration x that diffuses and reacts, are expressed by

---

[16] It is important to notice that, so far, we have learned and modeled how electrochemical signals propagate within an axon of small diameter. In large diameter axons, the propagation of electrochemical signals is assisted by the myelin sheath. The myelin sheath is made of cells wrapping the axon with concentric layers of their cell membranes. Such layers are made of lipids. Therefore, they are not conductive. In a myelinated neuron, the portions of the axon not covered are called nodes of Ranvier (see Figure in this note). The layer of insulation generated by the myelin coat avoids current leakage from the axon by blocking the movement of ions through the axon membrane. As a result, action potentials propagate faster in a myelinated axon rather than an unmyelinated one.

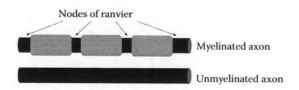

Nodes of ranvier

Myelinated axon

Unmyelinated axon

$$\frac{\partial x}{\partial t} = D\frac{\partial^2 x}{\partial r^2} + f(x)$$ [9.55]

where D is the diffusion coefficient, and $f(x)$ represents the kinetic law. The simplest case of a kinetic law that generates propagating chemical waves involves an autocatalytic reaction of the type:

$$X + S \xrightarrow{k_a} 2X$$ [9.56]

The net reaction is $S \to X$. S is the reacting substrate and X is the autocatalytic species. If we have S distributed homogeneously in our system, the addition of a small amount of X at one point can trigger the autocatalytic reaction, like a burning fire. Then, the autocatalysis coupled with diffusion makes the rest, and chemical waves spread throughout the system. If the initial concentration of S is s_0 and that of X is negligible, the mass balance allows us to write $s_0 = s + x$. Therefore, equation [9.55] becomes

$$\frac{\partial x}{\partial t} = D\frac{\partial^2 x}{\partial r^2} + k_a x(s_0 - x)$$ [9.57]

where k_a is the kinetic constant of the autocatalytic reaction. We may rewrite equation [9.57] in terms of the dimensionless variable $\xi = x/s_0$ and obtain

$$\frac{\partial \xi}{\partial t} = D\frac{\partial^2 \xi}{\partial r^2} + k_a \xi(1 - \xi)$$ [9.58]

Equation [9.58] was proposed by the British statistician and mathematical biologist Ronald Fisher (1937) for describing the spatial spread of an advantageous mutant gene through a population of interbreeding organisms in a linear habitat, such as a shoreline. He wrote a partial differential equation like [9.58], where ξ represents the frequency of the advantageous mutant gene X at the expense of the allelomorph S occupying the same locus;[17] k_a represents the rate of mutation and D is the coefficient of diffusion of the advantageous gene. Fisher showed that waves of stationary shape advance through the population with speeds

$$v \geq 2\sqrt{k_a D}$$ [9.59]

The same result was achieved by the Russian mathematician Andrey Kolmogorov and his co-workers, exactly in the same year (Kolmogorov et al. 1937). This is the reason why equation [9.58] is now known as Fisher-Kolmogorov equation.[18] It is surprising that the result [9.59] was anticipated

---

[17] In genetics, allelomorphs (or shortly, alleles) are alternative forms of a gene that can occupy the same locus on a particular chromosome and control the same character.

[18] A detailed mathematical treatment of the Fisher-Kolmogorov equation that does not have an analytical solution can be found in chapter 13 of the book "Mathematical Biology I" by Murray (2002), and in the feature article by Scott and Showalter (1992) published in the Journal of Physical Chemistry. In the latter, the authors deal not only with a quadratic but also a cubic autocatalytic process. A representation of propagation of chemical waves according to the Fisher-Kolmogorov equation is shown in the Figure of this note.

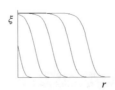

TABLE 9.2

Examples of Chemical and Electrochemical Waves Described by Equation [9.59]. The Data have been Extracted from the References Indicated in the Table

Chemical/Electrochemical Waves	k_a	D	v
Squid giant axon (Cross and Hohenberg 1993)	$3 \times 10^3\,s^{-1}$	$3.4 \times 10^2\,cm^2/s$	$10^3\,cm/s$
Spreading Depression (Dahlem and Müller 2004)	$8 \times 10^{-2}\,s^{-1}$	$8 \times 10^{-5}\,cm^2/s$	$5 \times 10^{-3}\,cm/s$
Mechano-chemical waves in heart (Cross and Hohenberg 1993)	$3 \times 10^2\,s^{-1}$	$0.6\,cm^2/s$	$13\,cm/s$
Calcium waves in fertilized mammal eggs (Whitaker 2006; Donahue and Abercrombie 1987)	$1.2\,s^{-1}$	$5.3 \times 10^{-6}\,cm^2/s$	$5 \times 10^{-3}\,cm/s$
cAMP waves in aggregation stage of *Dictyostelium discoideum* (Cross and Hohenberg 1993)	$10^{-2}\,s^{-1}$	$4 \times 10^{-6}\,cm^2/s$	$2 \times 10^{-4}\,cm/s$
Muskrat's biological invasion in Europe (Murray 2002)	$0.15\,year^{-1}$	$15\,km^2/year$	$3\,km/year$
BZ reaction	$2.25\,s^{-1}$	$2 \times 10^{-5}\,cm^{-2}\,s^{-1}$	$0.014\,cm/s$

intuitively, thirty-one years before, by the German chemist Robert Luther (1906) who was a pioneer in the study of chemical waves.[19]

In Table 9.2, there are examples of chemical waves whose speed of propagation is calculated by using equation [9.59].

9.7.3 Waves in Our Brain

It is clear that the brain is an electrochemical organ. Its electrical activity is studied either noninvasively by electrodes placed along the scalp, which record the voltage over time and originate the electroencephalograms (EEGs), or invasively by intracranial electrodes that collect electrocorticograms (ECoGs). Both techniques have facilitated investigating the patterns of brain activity. It has been observed that spatially and temporally organized activity among distributed populations of cells often takes the form of synchronous rhythms. These collective electrical oscillations are called "brain waves" (Gray 1994). The brain waves are sorted into five categories, the δ-, θ-, α-, β-, and γ- bands, depending on their characteristic frequencies and purposes. The α- (9–14 Hz), β- (14–40 Hz), and γ- (40–100 Hz) frequency band oscillations all contribute to the neuronal underpinnings of attention, working memory and consciousness (Palva and Palva 2007). When the brain is aroused and actively engaged in mental activities, it generates β waves. When our brain is involved in particular complex tasks, it generates γ waves. On the other hand, when we have completed a task, and we sit down to rest, we are often in a α state. Also, when we take time out to reflect or meditate, we are usually in a α state. If we begin to dream and mentally relax, we are in a θ state (4–8 Hz). We can maintain this state even when we perform repetitive tasks, and we are mentally disengaged from them. The final brainwave state is δ (1.5–4 Hz). Deep dreamless sleep takes us down to the lowest frequencies of the δ state. It is a well-known fact that humans dream in about 90-minute cycles (Hartmann 1968). When the delta (δ) brainwave frequencies increase into the frequency of

---

[19] The original paper by Luther has been translated by Arnold et al. (1987) and commented by Showalter and Tyson (1987).

theta (θ) brainwaves, active dreaming takes place. Typically, when this occurs there is Rapid Eye Movement (called REM sleep). The problem that the brain has to solve during sleep is how to integrate memories of experiences that happened during the day with old memories. Scientists know that waves of electrical activity, referred to as spindles, help to consolidate and integrate memories during sleep. Spindles are active in the cerebral cortex in the time between dream sleep and deep sleep. The spindle is a wave that begins in a portion of the cortex close to the ear, spirals through the cortex towards the top of the back of the head and, then, on the forehead area before circling back. It propagates with a speed of 2–5 m/s. The spindles strengthen connections among brain cells in distant parts of the brain. For example, these circular waves of electrical activity may strengthen the links between cells of the cortex that store memories of the sound with those storing memories of the sight, et cetera (Muller et al. 2016).

Electrochemical waves are also involved in pathophysiological events, such as Spreading Depression that is at the basis of a migraine (Dahlem and Müller 2004). A wave of Spreading Depression propagates when the ionic gradients across the plasma membrane of neuronal cells break down, leading to a massive efflux of potassium ions into the extracellular space and influx of sodium, calcium, and chloride ions into the cells. During this phenomenon, single neurons remain electrically silent, until the proper ionic gradients across the membrane are reestablished. A wave of Spreading Depression propagates within the different cell layers of the brain with a speed of 3 mm/min, which is 200,000 times slower than the speed of an action potential propagating within a neuron (Table 9.2).

9.7.4 Waves in Our Heart

Another beautiful example of the important role played by chemical waves in nature is in our heart. Our heart (see the left part of Figure 9.25) is a muscle that works continuously and pumps blood through the blood vessels of the circulatory system. Blood carries oxygen and nutrients and assists the removal of metabolic wastes, such as carbon dioxide. The heart's beating originates tiny electrical changes on the skin that are monitored by using electrodes placed on the patient's body. The signal that doctors record is called an electrocardiogram (see the right part of Figure 9.25). Each beat of our heart originates locally, in the cells of the "sinus node" (located in the heart's right atrium), which are in oscillatory regime, synchronized, and with a period of roughly 0.8 seconds. Such oscillations arise from self-sustained periodic changes of the intracellular calcium concentration, and, therefore, can be regarded as originated by calcium oscillators (Nitsan et al. 2016). From the sinus node, an electrical signal spreads across the cells of our heart's left and right atria. In fact, outside the sinus node, the cells are excitable, and they transmit the outgoing waves produced periodically in the sinus-node. The waves cause the atria to contract. The right atrium has previously collected deoxygenated blood from veins, whereas the left atrium has previously collected oxygenated blood via the pulmonary veins. The contraction of the atria, marked by the P wave in the electrocardiogram (cf. Figure 9.25), pumps the blood through the open valves from the atria into both ventricles. The signal arrives at the Atrioventricular node near the ventricles (Figure 9.25). Here it is slowed to allow our heart's right and left ventricles to fill with blood. On the electrocardiogram, this step is represented by the horizontal segment between the P and Q (Figure 9.25). Then, the signal is released and moves downwards, dividing into left and right bundle branches (this step corresponds to the Q wave in the electrocardiogram). As the signal spreads across the cells of the ventricle walls, both ventricles contract, although not at the same time. In fact, the left ventricle contracts slightly before than the right ventricle. On the electrocardiogram, the R wave marks the contraction of the left ventricle, whereas the S wave marks the contraction of the right ventricle. The contraction of the left ventricle pushes oxygenated blood through the aortic valve to the rest of the body. The contraction of the right ventricle pushes blood to our lungs for saturating it with oxygen. As the signal passes, the walls of the ventricles relax and await the next wave. Such step is represented by the T wave on the electrocardiogram.

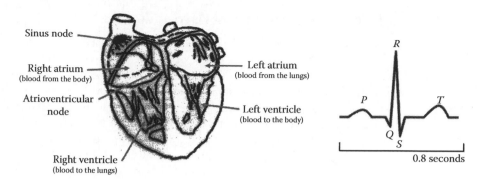

FIGURE 9.25   Sketch of the human heart (on the left) and the characteristic profile of an electrocardiogram for a healthy human heart (on the right).

The entire cycle, from P to T, repeats over and over. As it is shown in Table 9.2, the waves propagate at a speed of 13 cm/s, which is two orders of magnitude smaller than the propagation speed of the electrochemical waves in a squid giant axon.

A recent study (Nitsan et al. 2016) has demonstrated that the communication among the cardiac cells is not based only on electrochemical signals. In fact, there is also a mechanical communication that is essential for converting electrical pacing into synchronized beating. As mechanical coupling depends on the elastic properties of tissue, disruption of the normal mechanical environment can impair this interaction. For example, if there exists even a small region with disordered features, heterogeneities or defects, such as an unexcitable tissue, the excitation waves may be perturbed in passing through that region, and they may break apart (Glass 2001). When an excitation wave breaks, it leaves two free ends. These ends tend to curl up into spirals. Spiral waves have a larger speed (remember the eikonal equation [9.46]). Eventually, the heart tissue will oscillate at a higher frequency. This effect is thought to cause a heart disorder known as "paroxysmal tachycardia," when the frequency of the heartbeat increases by a factor of ten (Cross and Hohenberg 1993). Another cardiac pathology, fibrillation, is thought to involve an inhomogeneous excitable medium in which an excitation target wave breaks up to form many spirals. The net result is the presence of many different structures vibrating asynchronously in a chaotic way.

9.7.5   CALCIUM WAVES

Chemical waves of Ca^{+2} are relevant not only for the control of rapid and frequently repeated responses such as heartbeat, neurotransmitter release, and muscle contraction. They are apparently ubiquitous. They are present in somatic cells and sex cells. Calcium waves are known to trigger transformations of the cell cortex (that is the layer of proteins on the inner face of the plasma membrane of the cell) and cytoplasm, as well as to stimulate many enzymatic and metabolic processes (Whitaker 2006). For example, the activation of eggs by sperm is accompanied by a significant transient in intracellular calcium concentration. The calcium wave initiates at the point of the sperm entry and crosses the egg as a tsunami-like wave at a speed of about 5–50 μm/s (Table 9.2 reports the values relative to mammal eggs of 100 μm crossed by calcium waves in ~2 s) up to reach the antipode of the egg. The large calcium signal triggers a reorganization of the entire egg cortex and cytoplasm (Sardet et al. 1998). There is evidence that calcium signals are important in all three relevant stages of embryogenesis: in embryonic axis formation (anterior-posterior, dorsoventral and left-right axes), in coordinated cell migrations forming tissues (like in gastrulation forming gut and in neurulation generating the spinal cord), and in organogenesis (once the overall body plan is laid out, local differentiation gives rise to organs) (Whitaker 2006).

9.7.6 cAMP Waves: The Case of Dictyostelium Discoideum

The chemical waves of another compound, 3′,5′-cyclic adenosine monophosphate (cAMP), are responsible for aggregation and morphogenesis of a simple organism, the slime mold *Dictyostelium discoideum* (*Dd*) (Dormann et al. 1998). Slime molds live as single amoebae in the upper layers of the soil and leaf litter in the eastern Northern Hemisphere and eastern Asia. They feed on bacteria and divide. Starvation, caused by scarce bacteria and a crowded environment, triggers a survival strategy: the single amoebae aggregate to form a mass of 10^4–10^5 cells (see Figure 9.26). In the aggregated state or mound, the cells cooperate, specialize, and start to differentiate into a number of different cell types. The mound elongates, falls over and forms a cylindrically shaped slug (after ~15 h). The slug has a distinct polarity with a tip at the anterior end that guides its movement. Food finding is enhanced in the slug that acts as an interacting assembly of sensors and effectors, gathering and analyzing more information about the world than could a single amoeba. These sensory integration systems transduce physical and chemical signals into social cues, which amplify or attenuate group responses. The slug has a photo- and thermo-tactic power and can migrate to the surface of the soil. Finally, on the soil, the slug transforms into a fruiting body (up to 4 mm high) consisting of a stalk supporting a spore mass. The spores that are dormant cells, under suitable conditions, germinate into new amoebae and the cycle, requiring 24 h at room temperature, can start again.

The developmental program of *Dd* is controlled by the interplay between the shape and dynamics of cAMP waves and the cAMP chemotactic cell movement of the cells (Dormann and Weijer 2001). In the aggregation stage, it has been demonstrated that the chemotactic power of an amoeba grounds on a transmembrane cAMP receptor that can assume two states: an active state promoting the intracellular synthesis of cAMP and working at low extracellular [cAMP], and a desensitized state working when the extracellular [cAMP] is very high and inhibiting a further production of cAMP (Halloy et al. 1998). When stimulated with cAMP, *Dd* cells, suffering starvation, respond by synthesizing and secreting more cAMP, which results in non-dissipating cAMP waves. Such waves, propagating at a speed of ~10^2–10^3 µm/min (cf. Table 9.2), guide the aggregation of the individual amoebae (Gregor et al. 2010). It has been shown (Dormann and Weijer 2001) that cAMP waves are also involved in the slug's movement. The waves are initiated periodically in the anterior part of the slug tip and propagate backward at a speed of ~30 µm/min. They reflect the coordinated periodic

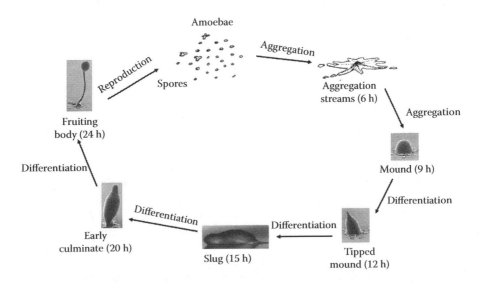

FIGURE 9.26 *Dictyostelium discoideum* (*Dd*) life cycle.

movement behavior of the cells in the slug. Moreover, also orange light stimulates the slug's tip to release cAMP, which may play a role in its photo-tactic response (Miura and Siegert 2000). In its final metamorphosis, the cells at the slug's tip migrate down through the center of the aggregate and initiate stalk formation. They die and transform in sturdy cellulose walls to hold up the spherical ball full of spores (Figure 9.26). The final fruiting body consists of about 20% stalk cell and 80% spore cells. This social stage is remarkable because it involves altruism: the stalk cells die to support the spore cells. It is vulnerable to cheaters. Cheating is a social action that takes place in the context of cooperative acts that the cheaters somehow violate. In *Dd*, the expected social contract is that the frequency of each genetically distinct clone among the spores will be the same as it was in the original mixture of aggregated cells. The same should be true in the stalk tissue. If this is not the case, the dominant clone has cheated the minority clone by getting more than its fair share into spores. Therefore, it is evident that this property makes *Dd* a great model even for social evolution. However, it is worth stating that it is not possible to attribute any conscious awareness to cheating in *Dd* that lacks any nervous system. In humans, it is different because cheating is value-based and assumes a certain awareness of the moral grounds of actions (Strassmann and Queller 2011).

9.7.7 SPREADING OF SPECIES, EPIDEMICS AND ... FADS

Biological invasions, such as the spreading of insects, animals, seeds, and diseases into new territories, and favored by intercontinental journeys, economic relationships and markets, and even the circulation of ideas and fears among human communities fed by telecommunication technology, all of those phenomena can be modelled by the theory of chemical waves and the Fisher-Kolmogorov equation. Such theory is reasonable as long as the scale of the individual's movement is small compared with the size of the observations. There are several cases of monitored biological invasions.

A classic example is the spread of muskrat (*Ondatra zibethica*) in Central Europe in the first half of the twentieth century (Skellam 1951). The muskrat was brought to Europe from North America for fur-breeding. In 1905, a few muskrats escaped from a farm near Prague. This small group started multiplying, and in a few decades, the muskrats spread over the whole continental Europe with a rate of ~3 km/year (Table 9.2).

Another example is the invasion of the gypsy moth (*Lymantria dispar*) in North America. This insect is thought to have been brought from France to a place near Boston by an amateur entomologist. When it escaped around 1870, it spread at an estimated speed of ~10 km/year. By 1990, the whole Northwest of the United States became heavily infested with resulting damages to the American agriculture.

Many more examples of biological invasions can be found in the book by Shigesada and Kawasaki (1997), and in that by Petrovskii and Li (2006).

TRY EXERCISES 9.17 AND 9.18

9.8 LIESEGANG PATTERNS

Other exciting phenomena wherein chemical reactions coupled with diffusion originate stunning patterns are the periodic precipitations discovered by the German chemist Raphael Eduard Liesegang (1896). Some pictures of this impressive phenomenon are shown in Figure 9.27.

In the usual experimental setup, two strong electrolytes are needed. One is dissolved in a gel matrix (the so-called inner electrolyte), whereas the other is poured onto the gel (outer electrolyte). The concentration of the outer electrolyte is much larger than that of the inner one. The ions of the outer electrolyte diffuse into the gel and react with the inner ions. An exchange reaction occurs and two new salts form within the gel. One of the new two salts precipitates and forms a beautiful pattern. The features of the pattern depend on the materials involved (included the gel) and the geometry of the system. If the precipitation occurs in a test tube, we observe a sequence of solid bands (Figure 9.27); if it occurs in a Petri dish, a pattern of concentric rings emerges (see picture e of Figure 9.1).

FIGURE 9.27 Periodic precipitation of Ag_2CrO_4 in gelatin, within a test tube.

The shapes of the patterns can be described quantitatively by specifying the formation times (t_n) of the bands, their positions (x_n, measured from the interphase separating the solution and the gel), and their width (w_n). The values of the variables t_n, x_n, and w_n, can be predicted by using three phenomenological rules (Rácz 1999; Antal et al. 1998). The first is the *time law*:

$$x_n \sim \sqrt{t_n} \qquad [9.60]$$

The time law is a consequence of the diffusive motion of the reaction front within the gel (in fact, $x \sim \sqrt{Dt}$). The positions of the bands form a geometric series according to the *spacing law*:

$$x_n \sim Q(1+p)^n \qquad [9.61]$$

where $p > 0$ (typically included between 0.05 and 0.4) is the spacing coefficient, and Q represents the amplitude of the spacing. The *spacing law* can also be written in the following form known as Jablczynski law:

$$\frac{x_n}{x_{n-1}} \sim 1 + p \qquad [9.62]$$

The spacing coefficient is not a universal quantity. For example, it depends on the concentrations of the outer $([Ou]_0)$ and inner $([In]_0)$ electrolytes according to the *Matalon-Packter law*:

$$p = F\left([In]_0\right) + \frac{G\left([In]_0\right)}{[Ou]_0} \qquad [9.63]$$

where $F\left([In]_0\right)$ and $G\left([In]_0\right)$ are decreasing functions of their argument $[In]_0$. Finally, there is the *width law* stating that the width w_n of the n-th band is an increasing function of n:

$$w_n \sim x_n^\alpha \qquad [9.64]$$

with $\alpha > 0$. Since the uncertainty in the measurements of w_n is usually rather large, the width of the bands has been ignored many times in the quantitative determination of the Liesegang patterns.

Although the Liesegang structures have been known for more than 100 years and despite an intensive theoretical work devoted to them, there remain some experimental observations that have not been explained.[20] Moreover, there are still many theories to interpret periodic precipitations (Henisch 1988). The theories that have been proposed differ in the details treating the nucleation and the kinetics of the precipitate growth (Antal et al. 1998; Henisch 1988). The simplest theory is based on the *supersaturation of ion-product*: when the local product of the concentrations of the outer electrolyte (A) and the inner electrolyte (B), [A][B], reaches a critical value, nucleation of the precipitate starts. The nucleated particles grow and deplete A and B in their surroundings. The local value of [A][B] drops and no new nucleation takes place. The reaction zone moves far from the precipitation zone, and the critical value of the product [A][B] is reached again. The repetition of these processes leads to the formation of the bands. In the theory of *nucleation and droplet growth*, A and B react to produce a new species C, which also diffuses in the gel. When the local concentration of C reaches some threshold value, nucleation occurs and the nucleated particles, D, act as aggregation seeds. When the concentration of C around D is larger than a critical value, the species C binds to D, and the droplet grows in dimension. This theory considers two threshold values: one for nucleation and the other for droplet growth. A third theory considers the formation of the bands as an *induced sol-coagulation process*. The electrolyte A and B bind to form a sol C. The sol C coagulates if [C] exceeds a supersaturation threshold and if the local concentration of the outer electrolyte is above a threshold. The band formation is a consequence of the coagulation of the sol and the motion of the front defined by the critical concentration of the outer electrolyte. A fourth theory interprets the phenomenon of periodic precipitation as the *propagation of an autocatalytic precipitation reaction*. Such theory involves three chemical species and two chemical steps. The species are the outer (A) and inner (B) electrolytes that diffuse within the gel and produce nuclei S through an energetically unfavorable nucleation step:[21]

$$A + B \rightarrow S \qquad\qquad [9.65]$$

The nuclei S favor further associations between A and B:

$$A + B + (n-1)S \rightarrow nS \qquad\qquad [9.66]$$

The transformation in equation [9.66] is called deposition step, and it represents the energetically favorable autocatalytic growth of the nuclei S that gives rise to a precipitate (Lebedeva 2004). When a deposition occurs, it automatically inhibits further precipitation in its neighborhood because it depletes the concentration of A, B, and S. This model based on autocatalysis is corroborated by the features the periodic precipitation patterns share with the transient pattern observed in exercise 9.1, generated by the BZ reaction. After all, it is now accepted that a growing crystal is an example of an excitable medium (Cartwright et al. 2012; Tinsley et al. 2013). It can give rise to target or spiral patterns. The growth of spirals may be responsible for the helical structures we sometimes detect in Liesegang structures.

TRY EXERCISE 9.19

---

[20] Examples of observations that are tough to be explained are listed here. (I) Patterns form only under certain conditions. (II) Some patterns are chiral and show helical structures. (III) A few substances can give rise to periodic precipitations where the interspacing of successive rings decreases with their ordinal numbers. These periodic precipitations are called revert Liesegang patterning. (IV) The coexistence of the so-called primary and secondary patterns wherein two types of patterns are superimposed with different frequencies (Tóth et al. 2016). Many conundrums that we may find in the literature are due to the use of non-standardized gels, reagents, and methods (Henisch, 1988).

[21] The nucleation step is energetically unfavorable because when the nuclei are still small, molecules are relegated on the energetically unpleasant superficial positions. When the nuclei become larger, the ratio between the number of superficial molecules over those within the volume decreases and the growth of the nuclei becomes energetically favorable.

9.9 LIESEGANG PHENOMENA IN NATURE

The wonderful phenomenon of periodic precipitations is around us in nature.

9.9.1 IN GEOLOGY

The earth has provided and still provides chemical environments that mimic those employed by Liesegang to obtain periodic precipitations. The Liesegang rings are obtained in closed systems, in the presence of a gel that prevents convective motions of the fluid and allows only the diffusion of the reactants, one of which is placed on the gel.[22] Therefore, the optimal geological environment for observing Liesegang rings should be a homogeneous gelatinous substrate with pronounced boundaries and in physical contact with a chemical reagent. Many patterns resembling Liesegang rings are found in quartz in its many natural forms. Beautiful examples are the agates (an agate is shown in Figure 9.28). Some geologists justify the presence of distinct layers in agates, by assuming that all quartz on earth was at one time a silica hydrogel. This hypothesis was reinforced over one hundred years ago when a vein of gelatinous silica was found in the course of deep excavations through the Alps for the Simplon Tunnel, which is a railway tunnel that connects Italy and Switzerland (Henisch 1988). However, there are some geologists who believe that the formation of rings, like those in agates, derive from the deposition of layers of silica filling voids in volcanic vesicles or other cavities. Each agate forms its own pattern based on the original cavity shape. The layers form in distinct stages and may fill a cavity completely or partially. When the filling is not complete, a hollow void can host crystalline quartz growth, and the agate becomes the outer lining of a geode (in the agate shown in Figure 9.28, there is a small geode indicated by the black arrow).

Although natural specimens of silica hydrogel are rarely encountered, the same cannot be said for quicksand that is a hydrogel made of fine sand, clay and salt water. When quicksand dehydrates, it forms sandstone. Liesegang rings are frequently encountered in sandstone. They appear in concretions that are discrete blocks having an ovoid or spherical shape formed by the precipitation of mineral cement within the spaces between the sediment grains. One example is offered by the Sydney sandstone that is composed of pure silica and small amounts of siderite ($FeCO_3$), bound with a clay matrix; it shows beautiful concentric yellow-brown rings.

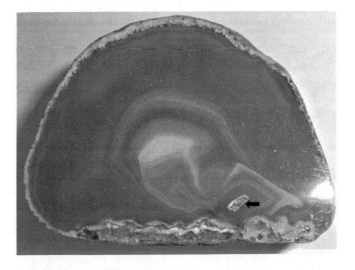

FIGURE 9.28 An example of agate with a small geode.

---

[22] It is worthwhile noticing that the Turing Reaction-Diffusion patterns require an open system to be maintained over time. On the other hand, periodic precipitations do not need an open system as long as they have been formed.

The Liesegang phenomenon is sometimes invoked also to justify the crystal inclusions when a material is trapped within a mineral. In the laboratory, it has been demonstrated that large crystals can form after dissolving the microcrystalline material in, say, an acid and allowing its solution to diffuse into a gel medium having a pH at which the solubility of the material is much lower (Henisch 1988). Most geologists do not interpret inclusions in minerals as examples of Liesegang phenomena. In fact, they comply with the "Father of Modern Geology," James Hutton, who formulated a law that bears his name and states that "fragments included in a host rock are older than the host rock itself." However, Hutton's Law is unlikely to be universally valid. For example, in the case of gold crystals in quartz. Many gold deposits in quartz could originally have been deposited in natural silica gel because it is reasonable to suppose that high temperatures and an ample supply of reducing agents must have been present during eras of mineralization (Stong 1962).

9.9.2 In Biology

The formation of patterns in biology similar to the Liesegang rings has been observed in bacterial growth. When bacteria spread more or less circularly from a point, in search of food contained in a gelatinous medium such as agar, they consume their nutrient within a ring of a certain thickness. Within that ring, bacteria could not flourish due to the competition for food. Therefore, they move further and form another ring (Henisch 1988). In the end, we observe a pattern of concentric rings that look like a Liesegang ring structure.

The periodic precipitation phenomenon is also important in the context of human health. Cystic and inflammatory lesions may give rise to concentric non-cellular lamellar structures, occurring as a consequence of the accumulation of insoluble products in a colloidal matrix. The shape and size of such Liesegang rings may vary significantly (for example in Tuur et al. 1987, the Liesegang rings measured from 7 up to 800 μm). Some pathologists may mistake Liesegang rings for eggs and larvae of parasites or tumoral lesions, but the use of polarized light microscopy combined with the right stains can be really useful for a correct diagnosis (Pegas et al., 2010).

9.10 A FINAL NOTE: THE REACTION-DIFFUSION STRUCTURES IN ART AND TECHNOLOGY

Despite many efforts, there is no a universal theory of pattern formation for system out-of-equilibrium. This situation is in sharp contrast to what happens at equilibrium. For predicting the structural properties of a system at equilibrium, we have the universal principle of minimization of its free energy. Such principle is a local version of the Second Law of Thermodynamics because it is focused on the system rather than on the entire Universe. Based on the principle of minimization of system's free energy, it is possible to build phase diagrams. A phase diagram allows us to predict the most stable phase for a material, after specifying the contour conditions. An analogous diagram, which we might call "morphology diagram," for the prediction of pattern formation out-of-equilibrium, does not exist, because we lack universal principles of dynamic evolution (Ben-Jacob and Levine 2001). Nevertheless, the unpredictable phenomena of spontaneous pattern formation have unthinkable aesthetic and technological possibilities that spur us to explore more deeply.

9.10.1 Reaction-Diffusion Processes as Art

Originally, painters tried to represent the surrounding world in realistic manners with their works of art. Correspondingly, the artists used professional skills and specific techniques to reproduce even the most minute details on canvas. The advent of photography, in the first half of the nineteenth century, contributed to change the role of painters. A picture can reproduce reality with extreme accuracy. Therefore, a painter should not make efforts to describe the surrounding world with fidelity, but rather he should just interpret it and communicate his impressions. And in fact, new artistic movements were born, shifting the content of the paintings from the figurative to the abstract. Correspondingly, techniques evolved, tracing new artistic trends. They moved from the Impressionism of Claude Monet (for instance, the oil in canvas titled "Impression, soleil levant," painted in 1872), where the painting starts to be made of relatively small, thin, yet visible brush strokes, to the Pointillism of Georges Seurat and Paul Signac in 1886 in which small, distinct dots or patches of color are applied in patterns to form an image, to the unique style of Jackson Pollack's drip paintings (created in the 1930s), where the artist only marginally controls the structure of his composition.

Reaction-Diffusion processes performed by using micro-patterned hydrogel stamps to deliver a solution of one or more reactants into a film of dry gels doped with chemicals that react with those delivered from the stamp can be regarded as a new micro-scale painting technique (Grzybowski 2009). It is possible to control the initial conditions, but the images form on their own. Since fluctuations are always present and they may be emphasized in far-from-equilibrium conditions, each RD process creates a unique pattern. In this respect, the RD phenomena are a form of artistic expression that has two authors: the person who selects the reagents, prepare the hydrogel stamp and the dry gel, and nature that lays down the colors and creates the ultimate structures.

9.10.2 REACTION-DIFFUSION PROCESSES IN TECHNOLOGY

The Reaction-Diffusion processes are also useful for the technological development in two fields, at least. First, in the field of material science. In fact, RD processes originate macroscopic patterns that have ordered features spanning many spatial scales up to the micro- or even the nano-level (Grzybowski 2009). What is remarkable is the possibility of generating structures having dimensions significantly smaller than those characterizing the initial distribution of reagents. Moreover, the tiny features of the structures may be finely shaped by changing macroscopic initial conditions, such as the concentrations of the reagents, their gradients, and the identity of the hosting medium.

Another appealing technological application of the RD phenomena is in the development of systems of artificial intelligence (Adamatzky and De Lacy Costello 2012; Gentili 2013). The artificial RD systems, described in this chapter, are smart because they imitate the computational strategies of their biological counterparts tightly. The next generation of computing machines might be just Reaction-Diffusion Computers. An RD computer would be a spatially extended chemical system, which would process information using interacting growing patterns, excitable, sub-excitable and diffusive waves. In RD processors, the elementary computing elements are micro-volumes, working in parallel and communicating with the closest neighbors through diffusion. Information is encoded as the concentration of chemicals, and the computation is performed via the interaction of wave fronts. The RD processors could be implemented in either geometrically constrained architectures or free space or encapsulated in elastic membrane or hybrid designs. On the other hand, stationary patterns could work as memory elements of the futuristic Chemical Computers wherein processors and memory will not be physically separated like in our current electronic computers that are based on the Von Neumann architecture.

Presumably, in the next future, we will witness a great contribution of the RD processes to the development of materials science and the unconventional information technology.

9.11 KEY QUESTIONS

- What happens if an autocatalytic reaction is carried out in an unstirred container?
- Describe the two-variables model for the formation of Turing patterns.
- Which are the conditions required to observe Turing patterns?
- What was the fortuitous event that allowed discovering Turing patterns in a chemical laboratory? Which was the trick to achieve three-dimensional Turing patterns?
- Repeat the procedure for nondimensionalizing a system of differential equations.
- Which are the principal processes responsible for the embryological development?

- Make examples of the application of the Turing's Reaction-Diffusion model.
- Which are the roles of cytoskeleton within a cell?
- Which are the transport processes that can compete with diffusion in pattern formation within a cell?
- Describe how a molecular motor works.
- What distinguishes a chemical wave from a physical wave?
- Present the Fisher-Kolmogorov equation.
- Present some examples of chemical waves in biology.
- What is a Liesegang pattern and how does it form?
- Which are possible applications of Reaction-Diffusion patterns?

9.12   KEY WORDS

Balance equation; Turing's Reaction-Diffusion model; Positional Information; Molecular motor; Advection; Cell polarity; Trigger waves; Phase waves; Periodic precipitations.

9.13   HINTS FOR FURTHER READING

- For deepening the subject of self-organization and evolution in biological systems, I suggest reading the books *The Origins of Order* by Kauffmann (1993), *On Growth and Form* by Thompson (1917), which allows contemplating the forms of life, *Life's Ratchet. How Molecular Machines Extract Order from Chaos* by Hoffmann (2012), and the review by Lander (2011).
- More information about chemical waves can be found in the book *Chemical Waves and Pattern* edited by Kapral and Showalter (1995), and in the book *Chemical Oscillations, Waves and Turbulence* by Kuramoto (2003). In biology, there are many examples of chemical waves as shown in the review by Deneke and Di Talia (2018), and in that by Volpert and Petrovskii (2009). The chemical processes responsible for intracellular calcium waves are described in the review by Berridge et al. (2000). Strogatz (2004) dedicates two chapters of his intriguing book, titled *Sync*, to the theme of brain waves.

9.14   EXERCISES

9.1. If a reaction with an autocatalytic step is left unstirred, exciting phenomena can be observed. So, go to your wet laboratory, wear a white coat, gloves, and safety glasses. Prepare the following solutions using deionized water as the solvent.
   - Solution A: H_2SO_4 (sulfuric acid) 0.76 M.
   - Solution B: $KBrO_3$ (potassium bromate) 0.35 M.
   - Solution C: $CH_2(COOH)_2$ (malonic acid) 1.2 M.
   - Solution D: $Ce(SO_4)_2$ (cerium(IV) sulphate) 0.005 M in an aqueous solution containing H_2SO_4 0.76 M.
   - Solution E: Ferroin (tris(1,10-phenanthroline)iron(III) sulphate) 0.025 M.

   In a test tube having a diameter of about 1.6 cm, introduce the following amounts of the stock solutions: 2 mL of A, 2 mL of B, 2 mL of C, and 2 mL of D. Stir the final solution with a glass rod, efficiently and quickly. Then, drop 200 μL of E, vigorously. Do not shake the solution, but fix the test tube to a laboratory stand through a clamp. Then, wait and stare at the solution.
   a.   What do you see?
   b.   Try to determine the wavelength of the periodic spatial structure you observe.
   c.   Apply equation [9.10] to determine the rate constant of the autocatalytic step responsible for the pattern you see. Remember that, according to the

FKN model, the autocatalytic step in the mechanism of the BZ reaction is $HBrO_2 + BrO_3^- + 3H^+ + 2Ce^{+3} \rightarrow 2Ce^{+4} + 2HBrO_2 + H_2O$. In equation [9.10], b represents $[BrO_3^-]$, D_M is the diffusion coefficient of $HBrO_2$, which is equal to $2 \times 10^{-5}\,cm^2s^{-1}$.

d.   Repeat the experiment by changing only the amount of $KBrO_3$. Instead of 2 mL, add 200 µL of solution B plus 1.8 mL of water in the test tube. What happens to the wavelength of the periodic spatial structure?

9.2.   In Figure 9.29, there are three spatial distributions of the concentration u in a mono-dimensional "box" of dimension 1.

The first case (trace labeled as 1) refers to a situation of uniform distribution of u along x: in fact, $u = 1$ everywhere along x in the interval [0, 1]. Such spatial distribution of u comes from a perfect mixing. On the other hand, trace 2 refers to the case of a linear increase of the concentration of u from 0 at $x = 0$, up to $u = 2$ at $x = 1$. Finally, trace 3 refers to a step function: $u = 0$ in the first half of the x box, whereas it becomes equal to 2 in the second half. In any case, the average value of u within the range [0, 1], is 1.

Calculate the average rates of hypothetical reactions that are governed by the following kinetic laws: (a) $rate \propto u$; (b) $rate \propto u^2$; (c) $rate \propto u^3$; (d) $rate \propto e^{(u-1)}$; (e) $rate \propto 10^{(u-1)}$. Makes final considerations about the importance of mixing, especially when the kinetic law is non-linear.

9.3.   Find the minimum of $det(J')$, defined in equation [9.20], with respect to K.

9.4.   Try to find an analogy of the "local self-activation and lateral inhibition" that gives rise to Turing structures in nature and/or in our daily life.

9.5.   The conditions required to have diffusion-driven instability are:

I. $tr(J) = \left[R_x(X) \right]_s + \left[R_y(Y) \right]_s < 0$

II. $det(J) = \left(\left[R_x(X) \right]_s \left[R_y(Y) \right]_s - \left[R_x(Y) \right]_s \left[R_y(X) \right]_s \right) > 0$

III. $\left(D_X \left[R_y(Y) \right]_s + D_Y \left[R_x(X) \right]_s \right) > 0$

IV. $\left(D_X \left[R_y(Y) \right]_s + D_Y \left[R_x(X) \right]_s \right) > 2 \left[D_X D_Y \left(\left[R_x(X) \right]_s \left[R_y(Y) \right]_s - \left[R_x(Y) \right]_s \left[R_y(X) \right]_s \right) \right]^{1/2}$

Apply these conditions to the following set of Reaction-Diffusion equations for the stationary state $(x_s, y_s) = (0,0)$:

$$\frac{\partial x}{\partial t} = \alpha x \left(1 - r_1 y^2 \right) + y \left(1 - r_2 x \right) + \left(D\delta \right) \nabla^2 x$$

$$\frac{\partial y}{\partial t} = \beta y \left(1 + \frac{\alpha r_1}{\beta} xy \right) + x \left(\gamma + r_2 y \right) + \delta \nabla^2 y$$

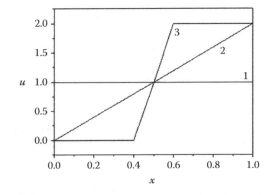

FIGURE 9.29   Three trends of the concentration u versus the x-coordinate.

Such model has been obtained by expanding the functions $R(X)$ and $R(Y)$ in a Taylor series about the homogeneous stationary state (x_s, y_s), neglecting terms of order higher than cubic (Barrio et al. 1999).

Moreover, determine the critical wavenumber of the Turing structure as a function of the diffusion-reaction parameters, using equation [9.24]. If the length of the domain is L, how many waves appear in the pattern?

9.6. Consider the reaction-diffusion equations of the exercise 9.5. The parameters have the following values: $D = 0.516$; $\delta = 0.0021$; $\alpha = 0.899$; $\beta = -0.91$; $r_1 = 3.5$; $r_2 = 0$; $\gamma = -\alpha = -0.899$. What do you expect if the system is in a mono-dimensional space?

Let us assume that the spatial domain is L = [−1, 1] and that we have zero-flux boundary conditions. Note that the Laplacian operator for diffusion must be calculated by a finite difference method that approximates the derivatives in each direction. Consult any textbook regarding the numerical solutions of partial differential equations. For instance, the textbook by Morton and Mayers (2005).

9.7. Consider the following reaction-diffusion equations:

$$\frac{\partial u}{\partial t} = \alpha u\left(1 - r_1 v^2\right) + v\left(1 - r_2 u\right) + \left(D\delta\right)\left(\frac{\partial^2 u}{\partial x^2} + \frac{\partial^2 u}{\partial y^2}\right)$$

$$\frac{\partial v}{\partial t} = \beta v\left(1 + \frac{\alpha r_1}{\beta} uv\right) + u\left(\gamma + r_2 v\right) + \delta\left(\frac{\partial^2 v}{\partial x^2} + \frac{\partial^2 v}{\partial y^2}\right)$$

These equations are formally identical to those presented in exercises 9.5 and 9.6, but now we use different symbols for the variables for a matter of clarity.

Solve the system of partial differential equations on a square domain with a side included between [−1, +1], over time and in three different situations that differ only in the values of the r_1 and r_2 parameters. Use the Euler's method (see Appendix A).

The parameters r_1 and r_2 affect the interaction of the activator and inhibitor. First case: $r_1 = 3.5$ and $r_2 = 0$.

Second case: $r_1 = 0.02$ and $r_2 = 0.2$.

Third case: $r_1 = 3.5$ and $r_2 = 0.2$.

The values of the other parameters are always the same in the three cases. They are: $D = 0.516$, $\delta = 0.0021$, $\alpha = 0.899$, $\beta = -0.91$, $\gamma = -\alpha$.

Note that $D = (D_U / D_V)$ is the ratio between the diffusion coefficients of the activator and inhibitor, respectively. D must be less than 1 since the diffusion of the inhibitor must be greater than that of the activator. The value of δ acts to scale the diffusion compared to the chemical reactions. In every case, assume to have zero-flux boundary conditions.

9.8. On the webpage of Prof. Shigeru Kondo's research group (http://www.fbs.osaka-u.ac.jp/labs/skondo/) it is possible to download a Reaction-Diffusion simulator. An easy guide to the simulator can be found on the Supporting Material of the paper published by Kondo and Miura in 2010. The simulator solves the following system of partial linear differential equations of the Turing model:

$$\frac{\partial u}{\partial t} = a_u u - b_u v + c_u - d_u u + D_u \nabla^2 u$$

$$\frac{\partial v}{\partial t} = a_v u + b_v v + c_v - d_v v + D_v \nabla^2 v$$

It is possible to play with the values of the parameters and see how they affect the final pattern. Try the combinations of the parameters values proposed by the authors of the simulator.

In the end, try your combination of the parameters values, after checking that the four conditions required to generate a Turing structure hold.

9.9. Determine the steady-state solution for the CIMA reaction in the homogeneous case (equation [9.35]). Define the values of the a and b parameters that guarantee a stable steady state.

9.10. In the paper by Lengyel and Epstein (1991) there are the values of the kinetic constants for the simplified kinetic model of the CDIMA reaction [9.31 and 9.32]. They are: $k_{1a} = 7.5 \times 10^{-3} \text{s}^{-1}$, $k_{1b} = 5 \times 10^{-5}\text{M}$, $k_2 = 6 \times 10^3 \text{M}^{-1}\text{s}^{-1}$, $k_{3b} = 2.65 \times 10^{-3}\text{s}^{-1}$, $\alpha = 1 \times 10^{-14}\text{M}^2$. If the initial concentrations of the reactants are $\left[\text{CH}_2(\text{COOH})_2\right]_0 = 10^{-3}\text{M}$, $\left[\text{I}_2\right]_0 = 10^{-3}\text{M}$, $\left[\text{ClO}_2\right]_0 = 6 \times 10^{-4}\text{M}$, and they are maintained constant, determine how large should be the ratio $d = \left(D_Y/D_X\right)$ to observe the formation of Turing patterns.

9.11. In the Schnackenberg model, the reaction steps are:

$$A \xrightarrow{k_1} X$$

$$X \xrightarrow{k_2} P$$

$$2X + Y \xrightarrow{k_3} 3X$$

$$B \xrightarrow{k_4} Y$$

Assuming that the concentrations of A and B are kept constant by continuous matter supply, write the two partial differential equations (PDEs) for $[X]$ and $[Y]$ imaging to have a non-stirred bi-dimensional squared reactor, whose sides have lengths $r_1 = r_2 = L$. Nondimensionalize the system of PDEs.

9.12. Find the steady-state solution of the Schnackenberg model without considering diffusion. Determine the relations among the parameters needed to have the possibility of observing the formation of Turing patterns in the presence of diffusion.

9.13. The nondimensionalized partial differential equations of the Schnackenberg model are

$$\left(\frac{\partial u}{\partial \tau}\right) = \gamma\left\{a - u + u^2 v\right\} + \left(\frac{\partial^2 u}{\partial x^2}\right) + \left(\frac{\partial^2 u}{\partial y^2}\right)$$

$$\left(\frac{\partial v}{\partial \tau}\right) = \gamma\left\{b - u^2 v\right\} + d\left(\frac{\partial^2 v}{\partial x^2}\right) + d\left(\frac{\partial^2 v}{\partial y^2}\right)$$

Solve them over time, on a square domain with a side included between [0, +1], and in four conditions that differ only in the values of a and b.

a. $a = 0.07$ and $b = 1.61$;

b. $a = 0.14$ and $b = 1.34$;

c. $a = 0.02$ and $b = 1.77$;

d. $a = 0.1$ and $b = 1.35$

The values of d and γ are fixed to 20 and 10000, respectively. To integrate the partial differential equations, use the finite difference method with a spatial step size $\Delta x = \Delta y = 0.01$. For more details on the computational method read the paper by Dufiet and Boissonade (1992).

9.14. The formation of the chemical gradients within a fertilized egg is a relevant process in the embryological development. In fact, the inhomogeneous distribution of a morphogen can induce the partition of an egg into distinct regions. Let us imagine that a morphogen, for instance, protein P, is synthesized continuously in the anterior part of the egg (Figure 9.30). Protein P diffuses into the rest of the egg, but it has a limited lifetime $\tau = (1/k_d)$. The kinetic constant k_d refers to a degradation process of P.

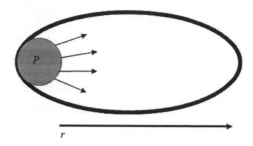

FIGURE 9.30 Sketch of a cell wherein protein P is synthesized in anterior cellular portion and diffuses towards the posterior cellular portion.

Which is the spatial distribution of the concentration of P along the spatial coordinate r, i.e., $[P] = f(r)$?

9.15. This exercise has three main goals. First, we want to observe live chemical waves. Second, we determine the dependence of the rate of chemical waves on the $[H^+][BrO_3^-]$ product. Finally, we define the relationship between the period T of the oscillations and $[H^+][BrO_3^-]$ product. Go to the wet laboratory, wear a white coat, gloves and safety glasses, and prepare the following solutions using deionized water as the solvent:

- Solution A: $KBrO_3$ 0.6 M in H_2SO_4 0.6 M.
- Solution B: $CH_2(COOH)_2$ 0.48 M
- Solution C: KBr 0.97 M
- Solution D: Ferroin (tris(1,10-phenanthroline) iron(III) sulphate) 0.025 M.

Into a small Erlenmeyer flask, introduce 7 mL of A, 3.5 mL of B and 1 mL of C. Close the container with a stopper and stir the solution with a magnetic stirrer. Bromate oxidizes bromide to bromine, and the solution looks brown. The brown color slowly disappears because bromine reacts with malonic acid to form bromomalonic acid. When the solution becomes transparent, add 0.5 mL of D and stir. Use a pipet to transfer 2.5 mL of the mixture into a cuvette having a tiny stir bar and pour the rest into a Petri dish (about 10 cm in diameter) to cover its surface uniformly. Place the cuvette in a UV-Visible spectrophotometer and maintain its solution under stirring. Record how the absorbance at 620 nm changes over time (every three seconds). Measure the period of the oscillations. Place the Petri dish on a sheet of millimeter graph paper and leave it quiescent for a while. Then, chemical waves will appear; measure the change in the radial distance from the center of a target pattern as the function of time. Also, measure the wavelength, which is the distance between consecutive fronts. Swirl the Petri dish gently to remove bubbles of CO_2 and after leaving it quiescent for a while, observe the appearance of new patterns. Measure the speed of the waves again. Does it remain constant? Repeat the experiment using the following combinations of the solutions:

- 6 mL of A, 1 mL of deionized water, 3.5 mL of B, 1 mL of C, and finally 0.5 mL of D;
- 5 mL of A, 2 mL of deionized water, 3.5 mL of B, 1 mL of C, and finally 0.5 mL of D.

Determine how the speed of the waves and the period of the oscillations depend on the product $[H^+][BrO_3^-]$. In other words, determine the exponents, n' and n'', appearing in the following two relations:

$$v \propto \left(\left[H^+ \right] \left[BrO_3^- \right] \right)^{n'}$$

$$T \propto \left(\left[H^+ \right] \left[BrO_3^- \right] \right)^{n''}$$

According to Field and Noyes (1974b), the wave velocity $v \geq (4kD[H^+][BrO_3^-])^{1/2}$, where $D = (2 \times 10^{-5} \mathrm{cm}^2 \mathrm{s}^{-1})$ is the diffusion coefficient of $HBrO_2$ and k is the kinetic constant of the autocatalytic reaction, i.e.,

$$\frac{d[HBrO_2]}{dt} = k[H^+][BrO_3^-][HBrO_2]$$

Assuming that $v = (4kD[H^+][BrO_3^-])^{1/2}$, determine k and compare your estimate with the expected value of $20 \ \mathrm{M}^{-2}\mathrm{s}^{-1}$ (Pojman et al. 1994).

9.16. Use the eikonal equation [9.46] to estimate the size of a spiral wave, knowing that the velocity of a front wave is $v_p = 0.0120 \ \mathrm{cm/s}$ and $D = 2 \times 10^{-5} \mathrm{cm}^2 \mathrm{s}^{-1}$.

9.17. A fire that spreads in a forest can be assumed to be an example of chemical waves. Why? Is it a trigger or a phase wave? Which is the autocatalyst? How long is the refractory period?

9.18. Compare the properties of chemical waves with those of physical waves, like light and sound. Which features are peculiar to chemical waves?

9.19. In this exercise that requires roughly three weeks to be completed, you will observe the phenomenon of periodic precipitation for Ag_2CrO_4 in gelatin. The Liesegang patterns take many days to form because ions must diffuse over long distances. Go to your wet laboratory, wear a white coat, gloves, and safety glasses. Dissolve 0.2 g of $K_2Cr_2O_7$ in 200 mL of distilled or deionized water, within a beaker. The solution appears yellow. Then, add 8 g of gelatin and dissolve it by stirring and heating up to 50°C. Pour the hot solution into two or more Petri dishes generating thin layers of gels, and within test tubes filling them up to 3 cm from the top.

Cover the Petri dishes with glass plates and the test tubes with parafilm. Let all of them stand for, at least, 24 h at room temperature. Meanwhile, prepare two solutions of $AgNO_3$ at different concentrations. One by dissolving 1 g of $AgNO_3$ in 10 mL of distilled or deionized water; the other by dissolving 0.5 g of $AgNO_3$ in 10 mL of distilled or deionized water. After one day at ~25°C, the gel is formed.

Place a few drops of silver nitrate solution on the surface of the gels formed in Petri dishes and cover them with glass plates to slow down evaporation.

Place 5 mL of the silver nitrate solutions on top of the gels contained in large test tubes and 2 mL on top of small test tubes. Cover the test tubes, again, with parafilm. Let the silver nitrate to diffuse within the gels.

Monitor the number of rings as a function of time, their spacing and the speed of spreading of the most advanced ring.

Which is the effect of the $AgNO_3$ concentration on the properties of the patterns? Are the patterns easily reproducible? Does the size of the test tube affect the features of the patterns?

Compare the Liesegang patterns with the transient structures obtained with the BZ reaction in exercise 9.1 and 9.15. Which are the similarities and differences? Do we confirm the plausible presence of an autocatalytic process in the Liesegang phenomenon? Verify the *time law* and apply equation [9.10] to determine the diffusion coefficient of silver ions within the gel and the kinetic constant of the autocatalytic process of nuclei association.

9.15   SOLUTIONS TO THE EXERCISES

9.1. (a) Soon after the addition of component E, the solution wherein there is the redox indicator becomes blue. Quickly, it turns to red. Then, blue stripes originate spontaneously at the bottom of the colored part of the solution (see Figure 9.31 that is in black and white).

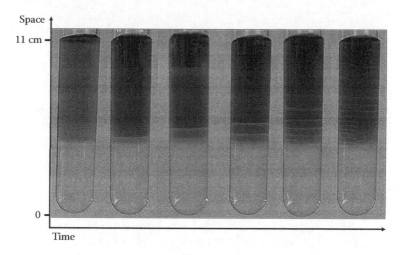

FIGURE 9.31   The spontaneous emergence of a spatial structure inside a test tube wherein the BZ reaction takes place in unstirred conditions.

The stripes move slowly upward. After more than 30', the colored portion of the solution becomes covered by thin blue stripes over a red background (see Figure 9.31).

(b) The wavelength of the spatially periodic structure can be determined by using a graph paper. The wavelength of the pattern corresponds to the distance between two adjacent stripes. It is evident that the wavelength of the wave train is not constant, and it increases moving towards the top of the solution. Figure 9.32 shows a schematic quantitative representation of the pattern that can be obtained.

If we consider the following first five stripes, the average value of the wavelength is $\lambda_a = 0.17$ cm.

(c) By using equation [9.10], we can determine the kinetic constant k of the autocatalytic process, responsible for the periodic structure. We obtain

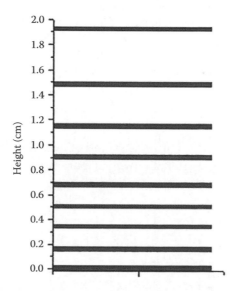

FIGURE 9.32   Schematic structure of the pattern shown in Figure 9.31.

$$k = \frac{4\pi^2}{b}\left(\frac{D_M}{\lambda_a^2}\right) = \frac{4\pi^2}{\dfrac{(0.35\,M)(0.002\,L)}{(0.0082\,L)}}\left[\frac{2\times10^{-5}\,cm^2s^{-1}}{(0.17\,cm)^2}\right] = 0.32\,M^{-1}s^{-1}$$

(d) If we consider the formula [9.10], we expect that by reducing b, the wavelength of the periodic structure increases. This relation is what we see by performing the experiment. Since the concentration of $KBrO_3$ is ten times smaller, the speed of the formation of the periodic structure is definitely much smaller than before. It takes a lot of time to observe the final structure. The final pattern has a wavelength λ_d that is more than three times longer according to the ratio $\frac{\lambda_d}{\lambda_a} = \sqrt{\frac{(b)_a}{(b)_d}} = \sqrt{10}$.

9.2. The results of the average rates are listed in Table 9.3.

When we have perfect mixing and uniform distribution of u (it is the case represented by trace 1), the rate is independent of x. On the other hand, when the mixing is imperfect, and we have a gradient, the average value of the rate maintains equal to 1 only in the case of a linear kinetic law. From the other values of the average rate, we also notice that the higher the non-linearity of the kinetic law, the greater the difference between the uniform and non-uniform distribution of u along x. This result means that non-linear reactions amplify the effect of imperfect mixing.

9.3. If we think of equation [9.20] as a quadratic equation in K^2 of the form $a(K^2)^2 + bK^2 + c$, we calculate its derivative with respect to K^2 and we impose it equal to zero, to find its minimum.

$$\frac{d\left(det\left(J'\right)\right)}{d\left(K^2\right)} = 2a\left(K^2\right) + b = 0$$

It derives that

$K^2 = -(b/2a)$, i.e., $K^2 = \left(D_X\left[R_y\left(Y\right)\right]_s + D_Y\left[R_x\left(X\right)\right]_s\right)/2D_X D_Y$. The latter coincides with equation [9.24].

9.4. Two intuitive examples of how a "local self-activation and lateral inhibition" can give rise to Turing structures have been proposed by Murray (1988, 2003). In both cases, the scenario is that of dry forest, and the activator is a fire that tends to spread throughout the forest. The inhibitor may be a team of firefighters (Murray 1988) or a large group of grasshoppers (Murray 2003). The firefighters equipped with helicopters can disperse a fire-retardant spray, whereas the grasshoppers can generate a lot of moisture by sweating when they get warm. If the fire moves too quickly, then the outcome will be a uniform charred area.

TABLE 9.3

Average Rates for Different Kinetic Laws and the Three Situations of u versus x Shown in Figure 9.29

Kinetic Law	Trace 1	Trace 2	Trace 3
u	1	1	1
u^2	1	1.4	2
u^3	1	2.2	4
$e^{(u-1)}$	1	1.2	1.54
$10^{(u-1)}$	1	2.4	5.05

However, when the firefighters travel faster, or the grasshoppers sweat profusely and generate enough moisture, they prevent the fire spreading everywhere. Eventually, a stable pattern of charred black and uncharred green regions will be established.

Other examples of "local self-activation and lateral inhibition" may be found in societies, ecosystems, et cetera.

9.5. The Jacobian matrix is $J = \begin{pmatrix} \left[R_x(X)\right]_s & \left[R_y(X)\right]_s \\ \left[R_x(Y)\right]_s & \left[R_y(Y)\right]_s \end{pmatrix} = \begin{pmatrix} \alpha & 1 \\ \gamma & \beta \end{pmatrix}$.

The four conditions for having a Turing pattern are:

I. $\alpha + \beta < 0$;

II. $\alpha\beta - \gamma > 0$;

III. $\delta D\beta + \alpha\delta > 0$;

IV. $\delta D\beta + \alpha\delta > 2\sqrt{D\delta^2(\alpha\beta - \gamma)}$

The critical wavenumber is $K = \left[\frac{1}{2}\left(\frac{\alpha}{D\delta} + \frac{\beta}{\delta}\right)\right]^{1/2}$.

Reminding that the wavelength is $\lambda = (2\pi/K)$, the number of waves in the domain of length L is given by the ratio $(L/\lambda) = (LK/2\pi)$.

9.6. Based on the parameters listed in the text of the exercise, the differential equations become

$$\frac{\partial x}{\partial t} = \alpha x\left(1 - r_1 y^2\right) + y + (D\delta)\left(\frac{\partial^2 x}{\partial r^2}\right)$$

$$\frac{\partial y}{\partial t} = \beta y\left(1 + \frac{\alpha r_1}{\beta} xy\right) + xy + \delta\left(\frac{\partial^2 y}{\partial r^2}\right)$$

All the four conditions for having a Turing pattern, considering a diffusion-driven instability from the stationary state $(x_s, y_s) = (0,0)$, are verified. In fact,

I. $\alpha + \beta = 0.899 - 0.91 < 0$;

II. $\alpha\beta - \gamma = (0.899)(-0.91) - (-0.899) > 0$;

III. $\delta D\beta + \alpha\delta = (0.0021)(0.516)(-0.91) + (0.899)(0.0021) > 0$;

IV. $\delta D\beta + \alpha\delta \cong 9 \times 10^{-4} > 2\sqrt{D\delta^2(\alpha\beta - \gamma)} \cong 8.6 \times 10^{-4}$

The critical wavenumber is $K = \left[(1/2)((\alpha/D\delta) + (\beta/\delta))\right]^{1/2} \cong 14 \text{ cm}^{-1}$. The wavelength is $\lambda = (2\pi/K) \cong 0.449 \text{ cm}$. Since the spatial domain is $L = [-1, +1]$, its total length is 2. It follows that the number of waves in the domain is $(2/\lambda) \approx 4.5$. This can be confirmed by solving numerically our system of partial differential equations, using the following MATLAB file (adapted from Schneider 2012):

```
%Grid size
Tf=3000;
a=-1;                      % Lower boundary
b=1;                       % Upper boundary
M=100;                     % M is the number of spaces between points a and b.
dr=(b-a)/M;                % dr is delta r
r=linspace(a,b,M+1);       % M+1 equally spaced r vectors including a and b.
%Time stepping
dt=0.04;                   % dt is delta t the time step
N=Tf/dt;                   % N is the number of time steps
```

```
%Constant Values
D=0.516;                  % D is the Diffusion coefficient Du/Dv
delta=0.0021;             % sizes the domain for particular wavelengths
alpha=0.899;              % a is alpha, a coefficient in f and g
beta=-0.91;               % b is beta, another coefficient in f and g
r1=3.5;                   % r1 is the cubic term
r2=0;                     % r2 is the quadratic term
gamma=-alpha;             % g is for gamma
%pre-allocation
xnp1=zeros(M+3,1);
ynp1=zeros(M+3,1);
%Initial Conditions
xn=-0.5+rand(M+3,1);       %Begin with a random point between [-0.5,0.5]
yn=-0.5+rand(M+3,1);
for n=1:N
xn(1)=xn(3);              %Boundary conditions on left flux is zero
xn(M+3)=xn(M+1);          %Boundary conditions on right
yn(1)=yn(3);
yn(M+3)=yn(M+1);
for j=2:M+2
%Source function for x and y
srcx=alpha*xn(j)*(1-r1*yn(j)^2)+yn(j)*(1-r2*xn(j));
srcy=beta*yn(j)*(1+(alpha*r1/beta)*xn(j)*yn(j))+xn(j)*(gamma+r2*yn(j));
Lapx=(xn(j-1)-2*xn(j)+xn(j+1))/dr^2; %Laplacian x
Lapy=(yn(j-1)-2*yn(j)+yn(j+1))/dr^2; %Laplacian y
xnp1(j)=xn(j)+dt*(D*delta*Lapx+srcx);
ynp1(j)=yn(j)+dt*(delta*Lapy+srcy);
end
xn=xnp1;
yn=ynp1;
% Graphing
if mod(n,25000)==0
%subplot(2,1,2)
plot(r, xn(2:M+2),r, yn(2:M+2))
axis([-1,1,-1,1]);
fprintf('Time t = %f\n',n*dt);
input('Hit enter to continue:')
end

end
```

In such MATLAB file, we apply the Euler's method (see Appendix A) to a semi-discretized Reaction-Diffusion system. Moreover, the second derivative is expressed as finite differences:

$$\frac{d^2x}{dr^2} = Lapx = \frac{x(r-\Delta r)-2x(r)+x(r+\Delta r)}{(\Delta r)^2} = (xn(j-1)-2*xn(j)+xn(j+1))/dr^2$$

The initial conditions are fixed as a randomly generated set of values between [−0.5, +0.5].

The result of the calculation is shown in Figure 9.33.

We notice that the x and y species are spatially distributed in anti-phase conditions: in the point where x is at its maximum, y is at its minimum and vice versa. This result verifies that in the places where the activator prevails, the inhibitor succumbs and vice versa. Moreover, we confirm that the number of waves in the spatial domain is about 4.5.

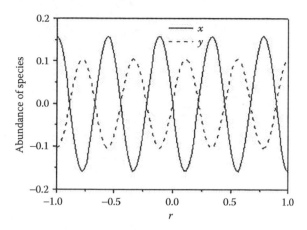

FIGURE 9.33 Turing pattern involving x and y in one-dimensional space of coordinate r.

9.7. If you have MATLAB at your disposal, an example of .m file that can be used to find the solution of the exercise is the following one (adapted from Schneider 2012):

```
%Grid size
Tf=100000;
a=-1;                       % Lower boundary
b=1;                        % Upper boundary
M=100;                      % M is the number of spaces between points a
                            and b.
dx=0.04;                    %(b-a)/M; % dx is delta x
dy=0.04;                    %(b-a)/M;
x=linspace(a, b,M+1);       % M+1 equally spaced x vectors including a and b.
y=linspace(a, b,M+1);
%Time stepping
dt=0.08;                    %100*(dx^2)/2; % dt is delta t the time step
N=Tf/dt;                    % N is the number of time steps in the interval
                            [0,1]

%Constant Values
D=0.516;                    % D is the Diffusion coefficient Du/Dv
delta=0.0021;               % sizes the domain for particular wavelengths
alpha=0.899;                % a is alpha, a coefficient in f and g (-a is
                            gamma)
beta=-0.91;                 % b is beta, another coefficient in f and g
r1=3.5;                     % r1 is the cubic term
r2=0;                       % r2 is the quadratic term
gamma=-alpha;               % g is for gamma
%pre-allocation
unp1=zeros(M+3,M+3);
vnp1=zeros(M+3,M+3);
%Initial Conditions
un=-0.5+rand(M+3,M+3);      %Begin with a random point between [-0.5,0.5]
vn=-0.5+rand(M+3,M+3);
    for n=1:N
    for i=2:M+2
un(i,1)=un(i,3);            %Boundary conditions on left flux is zero
un(i,M+3)=un(i,M+1);        %Boundary conditions on right
vn(i,1)=vn(i,3);
vn(i,M+3)=vn(i,M+1);
end
```

```
for j=2:M+2
un(1,j)=un(3,j);          %Boundary conditions on left
un(M+3,j)=un(M+1,j);      %Boundary conditions on right
vn(1,j)=vn(3,j);
vn(M+3,j)=vn(M+1,j);
end

for i=2:M+2
for j=2:M+2
   %Source function for u and v
   srcu=alpha*un(i, j)*(1-r1*vn(i, j)^2)+vn(i, j)*(1-r2*un(i, j));
   srcv=beta*vn(i, j)*(1+(alpha*r1/beta)*un(i, j)*vn(i, j))
   +un(i, j)*(gamma+r2*vn(i, j));
   uxx=(un(i-1,j)-2*un(i, j)+un(i+1,j))/dx^2; %Laplacian u
   vxx=(vn(i-1,j)-2*vn(i, j)+vn(i+1,j))/dx^2; %Laplacian v
   uyy=(un(i, j-1)-2*un(i, j)+un(i, j+1))/dy^2; %Laplacian u
   vyy=(vn(i, j-1)-2*vn(i, j)+vn(i, j+1))/dy^2; %Laplacian v
   Lapu=uxx+uyy;
   Lapv=vxx+vyy;

      unp1(i, j)=un(i, j)+dt*(D*delta*Lapu+srcu);
      vnp1(i, j)=vn(i, j)+dt*(delta*Lapv+srcv);
end
end
   un=unp1;
   vn=vnp1;

% Graphing
if mod(n,6250)==0

   %subplot(2,1,2)
   hdl = surf(x,y,un(2:M+2,2:M+2));
   set(hdl,'edgecolor','none');
   axis([-1, 1,-1,1]);
   %caxis([-10,15]);
   view(2);
   colorbar;
   fprintf('Time t = %f\n',n*dt);
   ch = input('Hit enter to continue:','s');
   if (strcmp(ch,'k') == 1)
     keyboard;
   end
end

end
```

The patterns achieved when $r_1 = 3.5$, $r_2 = 0$, and there are not quadratic terms in the PDEs, are shown in Figure 9.34. They are made of stripes.

The patterns obtained when $r_1 = 0.02$, $r_2 = 0.2$, and the contribution of the cubic terms in the PDEs is pretty small, are shown in Figure 9.35. They are made of spots.

The patterns obtained when $r_1 = 3.5$, $r_2 = 0.2$ and both the cubic and the quadratic terms are relevant, are shown in Figure 9.36. They contain both stripes and spots.

In synthesis, when there is only the contribution of the cubic terms, we observe stripes; when the quadratic term is dominant, we have spots. Finally, when both the quadratic and the cubic terms contribute appreciably, we observe patterns with spots and stripes (Barrio et al. 1999).

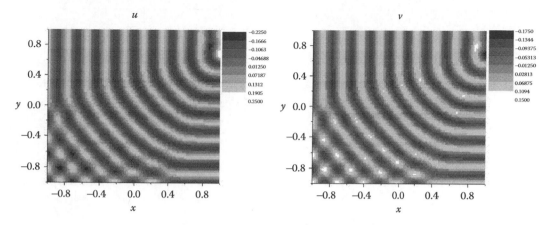

FIGURE 9.34 Profiles of u (on the left) and v (on the right) obtained when $r_1 = 3.5$, $r_2 = 0$, and within the square box having x and y as spatial coordinates.

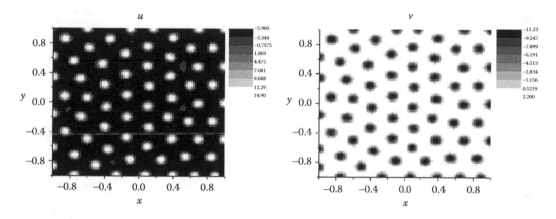

FIGURE 9.35 Profiles of u (on the left) and v (on the right) obtained when $r_1 = 0.02$, $r_2 = 0.2$, and within the square box having x and y as spatial coordinates.

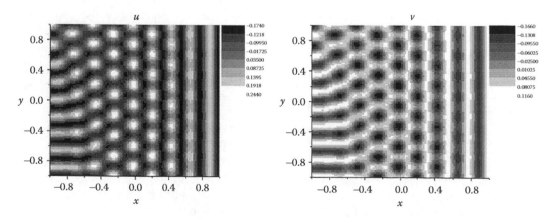

FIGURE 9.36 Profiles of u (on the left) and v (on the right) obtained when $r_1 = 3.5$, $r_2 = 0.2$, and within the square box having x and y as spatial coordinates.

TABLE 9.4

Six Combinations of the Parameters of the Turing Model and Pictures of the Patterns They Generate. In the Pictures, the Activator is Gray, Whereas the Inhibitor is Black

Name	a_u	b_u	c_u	d_u	a_v	b_v	c_v	d_v	D_u	D_v	Pattern
Spot	0.08	−0.08	0.005	0.03	0.1	0	−0.15	0.06	0.02	0.5	
Net	0.08	−0.08	0.2	0.03	0.1	0	−0.15	0.06	0.02	0.5	
Stripe2	0.08	−0.08	0.04	0.03	0.1	0	−0.15	0.08	0.02	0.5	
Stripe3	0.08	−0.08	0	0.03	0.1	0	−0.15	0.08	0.02	0.5	
Dapple	0.15	−0.08	0.025	0.03	0.15	0	−0.15	0.06	0.02	0.5	
Spot2	0.1	−0.08	0.025	0.03	0.15	0	−0.1	0.06	0.02	0.5	

9.8. The combinations of the parameters available in the RD simulator by Kondo and his team are listed in Table 9.4. In the last column of Table 9.4, there are the pictures of the patterns generated by numerical solution of the partial differential equations.

A new combination of parameters values is the following one:
$a_u = 0.08$, $b_u = −0.08$, $c_u = 0.2$, $d_u = 0.03$, $D_u = 0.02$, $a_v = 0.1$, $b_v = 0$, $c_v = −0.15$, $d_v = 0.07$, $D_v = 0.5$. This combination differs from that named as "Net" in Table 9.4 only in d_v. The new combination verifies the four conditions required to have a Turing pattern. In fact,

I. $tr(J) = a_u - d_u + b_v - d_v = 0.08 - 0.03 + 0 - 0.07 < 0$

II. $\det(J) = (a_u - d_u)(b_v - d_v) - b_u a_v = (0.05)(-0.07) - (-0.08)(0.1) = 0.0045 > 0$

III. $D_u(b_v - d_v) + D_v(a_u - d_u) = (0.02)(-0.07) + (0.5)(0.05) = 0.0236 > 0$

IV. $D_u(b_v - d_v) + D_v(a_u - d_u) = 0.0236 > 2[D_u D_v \det(J)]^{1/2} = 2[(0.02)(0.5)(0.0045)]^{1/2} = 0.013$

If we start from a randomized condition, the final stationary pattern is shown in Figure 9.37 as graph a. If we start from configuration a and we reduce the values of the diffusion coefficient to $D_u = 0.002$ and $D_v = 0.05$, we obtain the pattern shown in b of Figure 9.37. On the other hand, if we use the values $D_u = 0.002$ and $D_v = 0.05$ from an initial randomized distribution of u and v into space, we obtain the pattern labeled as c in Figure 9.37.

If we take the pattern obtained by the combination named as "Stripe2" in Table 9.4 as the initial condition and we reduce the values of the diffusion coefficients to $D_u = 0.002$ and $D_v = 0.05$, we achieve pattern D of Figure 9.38. If we do the same from the output of the combination named as "Stripe3," we achieve pattern E of Figure 9.38.

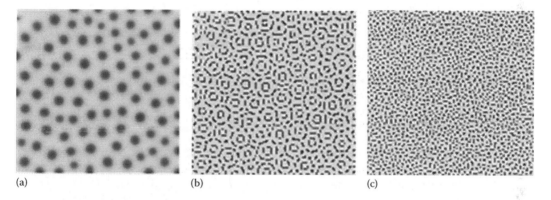

(a)   (b)   (c)

FIGURE 9.37   Three Turing patterns obtained for different values of the parameters and distinct initial conditions. Read the text for details.

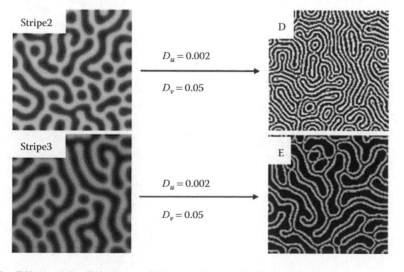

FIGURE 9.38   Effects of the diffusion coefficients values on the Turing patterns.

These modifications demonstrate the plasticity of the Turing patterns. Such plasticity has been experimentally demonstrated by Watanabe and Kondo (2012) in transgenic zebrafish.

9.9. The steady state solution for the homogeneous case can be determined by fixing the two differential equations, devoid of the diffusion terms, equal to zero:

$$\frac{dX^*}{d\tau} = \left(a - X^* - 4\frac{X^*Y^*}{1+X^{*2}} \right) = f\left(X^*, Y^* \right) = 0$$

$$\frac{dY^*}{d\tau} = \left(bX^* - \frac{bX^*Y^*}{1+X^{*2}} \right) = g\left(X^*, Y^* \right) = 0$$

The steady state solution is when $X_s^* = (a/5)$, $Y_s^* = (1+(a^2/25))$
This steady state is stable if $tr(J) < 0$, $\det(J) > 0$. The Jacobian is

$$J = \begin{pmatrix} \left(\frac{\partial f}{\partial X^*}\right)_s & \left(\frac{\partial f}{\partial Y^*}\right)_s \\ \left(\frac{\partial g}{\partial X^*}\right)_s & \left(\frac{\partial g}{\partial Y^*}\right)_s \end{pmatrix} = \begin{pmatrix} \left[-1 - 4\frac{\left(1-\frac{a^2}{25}\right)}{\left(1+\frac{a^2}{25}\right)} \right] & \left[-\frac{20a}{(25+a^2)} \right] \\ \left[b - b\frac{\left(1-\frac{a^2}{25}\right)}{\left(1+\frac{a^2}{25}\right)} \right] & \left[-\frac{5ba}{(25+a^2)} \right] \end{pmatrix} = \begin{pmatrix} \frac{3a^2 - 5(25)}{(25+a^2)} & -\frac{20}{(25+a^2)} \\ \frac{2a^2 b}{(25+a^2)} & -\frac{5ba}{(25+a^2)} \end{pmatrix}$$

The trace of the Jacobian J is negative if $b > (3/5)a - (25/a)$; the determinant of J is positive if $a > 5$.

9.10. The values of the kinetic constants and the concentrations presented in the text of the exercise allow us calculating the rate constants that appear in [9.32]. We find that $k_1' = 7.14 \times 10^{-6}$ s^{-1}M, $k_2' = 3.6$ s^{-1}, $k_3' = 2.65 \times 10^{-6}$ s^{-1}M. It derives that the parameters $a = \left(k_1' / k_2' \sqrt{\alpha} \right)$ and $b = \left(k_3' / k_2' \sqrt{\alpha} \right)$ are equal to 19.8 and 7.36, respectively. Note that a and b are dimensionless. Now, we determine the values of the terms of the Jacobian J.

$$J = \begin{pmatrix} 2.68 & -0.96 \\ 14.14 & -1.75 \end{pmatrix}$$

Note that two of the four terms of the Jacobian are positive: they regard the activator I^-. The other two terms are negative, and they regard the inhibitor ClO_2^-. $\det(J) = 8.88 > 0$. The determinant J' becomes negative if the relationships [9.37] and [9.38] are both verified. The first relation imposes that $d > 0.65$. The second one imposes that $d > 6$.

9.11. The PDEs are:

$$\left(\frac{\partial [X]}{\partial t} \right) = k_1' - k_2[X] + k_3[X]^2[Y] + D_X\left(\frac{\partial^2 [X]}{\partial r_1^2} \right) + D_X\left(\frac{\partial^2 [X]}{\partial r_2^2} \right)$$

$$\left(\frac{\partial [Y]}{\partial t} \right) = k_4' - k_3[X]^2[Y] + D_Y\left(\frac{\partial^2 [Y]}{\partial r_1^2} \right) + D_Y\left(\frac{\partial^2 [Y]}{\partial r_2^2} \right)$$

where $k_1' = k_1[A]$ and $k_4' = k_4[B]$. Note that in the original PDEs we have five variables and six parameters. To obtain the nondimensionalization of the PDEs, we define the following dimensionless variables: $u = X_C[X]$, $v = Y_C[Y]$, $\tau = t_C t$, $x = x_C r_1$, $y = y_C r_2$.

Introducing these new variables in the PDEs, we get:

$$\left(\frac{\partial u}{\partial \tau}\right) = k_1'\left(\frac{X_C}{t_C}\right) - k_2\left(\frac{u}{t_C}\right) + k_3\left(\frac{u^2}{X_C t_C}\right)\left(\frac{v}{Y_C}\right) + \frac{D_X x_C^2}{t_C}\left(\frac{\partial^2 u}{\partial x^2}\right) + \frac{D_X y_C^2}{t_C}\left(\frac{\partial^2 u}{\partial y^2}\right)$$

$$\left(\frac{\partial v}{\partial \tau}\right) = k_4'\left(\frac{Y_C}{t_C}\right) - k_3\left(\frac{u}{X_C}\right)^2\left(\frac{v}{t_C}\right) + \frac{D_Y x_C^2}{t_C}\left(\frac{\partial^2 v}{\partial x^2}\right) + \frac{D_Y y_C^2}{t_C}\left(\frac{\partial^2 v}{\partial y^2}\right)$$

The next step requires the definition of nondimensionalizing constants. Since L is the length of the sides of the square, it derives that $x_C = y_C = (1/L)$. Introducing these definitions of x_C and y_C in the PDEs, it is spontaneous to fix $t_C = (D_X/L^2)$. If we also impose $d = (D_Y/D_X)$, we rewrite the PDEs as

$$\left(\frac{\partial u}{\partial \tau}\right) = \frac{k_2}{t_C}\left\{\frac{k_1'}{k_2}X_C - u + \frac{k_3}{k_2}\left(\frac{u^2}{X_C}\right)\left(\frac{v}{Y_C}\right)\right\} + \left(\frac{\partial^2 u}{\partial x^2}\right) + \left(\frac{\partial^2 u}{\partial y^2}\right)$$

$$\left(\frac{\partial v}{\partial \tau}\right) = \frac{k_2}{t_C}\left\{\frac{k_4'}{k_2}Y_C - \frac{k_3}{k_2}\left(\frac{u}{X_C}\right)^2 v\right\} + d\left(\frac{\partial^2 v}{\partial x^2}\right) + d\left(\frac{\partial^2 v}{\partial y^2}\right)$$

If we fix $X_C = (k_3/k_2)^{1/2}$ and $Y_C = (k_3/k_2)^{1/2}$, we have:

$$\left(\frac{\partial u}{\partial \tau}\right) = \frac{k_2}{t_C}\left\{\frac{k_1'}{k_2}\left(\frac{k_3}{k_2}\right)^{1/2} - u + u^2 v\right\} + \left(\frac{\partial^2 u}{\partial x^2}\right) + \left(\frac{\partial^2 u}{\partial y^2}\right)$$

$$\left(\frac{\partial v}{\partial \tau}\right) = \frac{k_2}{t_C}\left\{\frac{k_4'}{k_2}\left(\frac{k_3}{k_2}\right)^{1/2} - u^2 v\right\} + d\left(\frac{\partial^2 v}{\partial x^2}\right) + d\left(\frac{\partial^2 v}{\partial y^2}\right)$$

Finally, if we fix $\gamma = (k_2/t_C)$, $a = \left((k_1'/k_2)(k_3/k_2)\right)^{1/2}$, and $b = (k_4'/k_2)(k_3/k_2)^{1/2}$, the PDEs assume their final nondimensionalized form:

$$\left(\frac{\partial u}{\partial \tau}\right) = \gamma\left\{a - u + u^2 v\right\} + \left(\frac{\partial^2 u}{\partial x^2}\right) + \left(\frac{\partial^2 u}{\partial y^2}\right)$$

$$\left(\frac{\partial v}{\partial \tau}\right) = \gamma\left\{b - u^2 v\right\} + d\left(\frac{\partial^2 v}{\partial x^2}\right) + d\left(\frac{\partial^2 v}{\partial y^2}\right)$$

wherein we still have five variables, but now only four parameters.

9.12. The nondimensionalized ordinary differential equations of the Schnackenberg model are:

$$\left(\frac{\partial u}{\partial \tau}\right) = \gamma\left\{a - u + u^2 v\right\}$$

$$\left(\frac{\partial v}{\partial \tau}\right) = \gamma\left\{b - u^2 v\right\}$$

If we put the ODEs equal to zero, we find that the steady-state solution is

$$u_s = a + b, \; v_s = \frac{b}{(a+b)^2} \text{ with } b > 0 \text{ and } a + b > 0.$$

The Jacobian for the Schnackenberg model is:

$$J = \begin{pmatrix} 2u_s v_s - 1 & u_s^2 \\ -2u_s v_s & -u_s^2 \end{pmatrix} = \begin{pmatrix} \dfrac{(b-a)}{(a+b)} & (a+b)^2 \\ \dfrac{-2b}{(a+b)} & -(a+b)^2 \end{pmatrix}$$

Since $\left[R_u(U)\right]_s$ and $\left[R_v(V)\right]_s$ must have opposite signs, we must have $b > a$. The conditions required to have diffusion-driven instability are:

I. $tr(J) = \dfrac{(b-a)}{(a+b)} - (a+b)^2 < 0$, i.e., $(b-a) < (a+b)^3$;

II. $\det(J) = (a+b)^2 > 0$;

III. $\left[R_v(V)\right]_s + d\left[R_u(U)\right]_s > 0$, i.e., $d(b-a) > (a+b)^3$;

IV. $\left\{\left[R_v(V)\right]_s + d\left[R_u(U)\right]_s\right\}^2 > (4d)\det(J)$, i.e., $\left[d(b-a) > (a+b)^3\right]^2 > 4d(a+b)^2$

The relations (I)–(IV) define a domain in (a, b, d) parameter space that is favorable to observe Turing structures.

9.13. I report an example of .m file that is useful to solve the exercise by MATLAB as follows.

```
function []=RDzeroflux(N,a,b,d,gamma)
% defining the mesh
N = 100;
h = 1/N; % step size in x and y
x = h*(0:N); % x coordinates of grid
y = h*(0:N); % y coordinates of grid
[xx, yy] = meshgrid(x, y); % 2D x and y mesh coordinates
dt =.01*h^2; % time step - usually small
% parameter values
a =.1;
b = 1.35;
d = 20;
gamma = 10000;
% Initial data at t=0:
us = (a+b); % u steady state
vs = b/us^2; % v steady state
u = (a+b)*ones(size(xx)); % u steady state
v = (b./u.^2); % v steady state
u = u +.1*randn(size(xx)); % add small perturbations about the steady
state
v = v +.1*randn(size(xx)); % add small perturbations about the steady
state
% Time-stepping:
t = 0;
tmax = 10;
nsteps = round(tmax/dt); % number of time steps
for n = 1:nsteps % main time-stepping loop
t = t+dt;
uE = u(:,[2:N+1 N]);
```

```
uW = u(:,[2 1:N]);
uN = u([2 1:N],:);
uS = u([2:N+1 N],:);
vE = v(:,[2:N+1 N]);
vW = v(:,[2 1:N]);
vN = v([2 1:N],:);
vS = v([2:N+1 N],:);
% finite difference formula
u2v = u.^2.*v;
u = u + gamma*dt*(a-u+u2v)+dt*(uE+uW+uN+uS-4*u)/h^2;
v = v + gamma*dt*(b-u2v)+d*dt*(vE+vW+vN+vS-4*v)/h^2;
end;
% plot solution:
colormap('grey')
surf(x, y,u)
title(['u at t = 'num2str(t)],'fontsize',16)
zlim([0 4])
shg
end
```

The results are shown in Figure 9.39. The first situation originates pattern a with parallel stripes. The second situation gives rise to pattern b with diagonal parallel stripes. The third and fourth conditions originate the hexagonal patterns c and d, respectively. The activator is abundant in the white regions of the graphs, whereas the inhibitor is abundant in the black regions of the same graphs.

9.14. The differential equation describing how [P] changes over time and space is:

$$\frac{\partial[P]}{\partial t} = -k_d[P] + D_P \frac{\partial^2[P]}{\partial r^2}$$

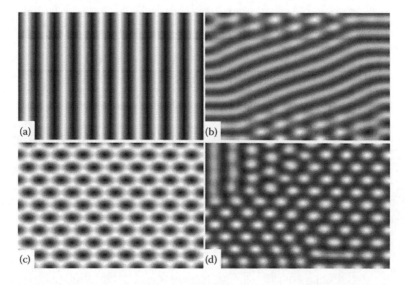

FIGURE 9.39 Turing patterns obtained from the Schnackenberg's model in the in the four conditions (a, b, c, and d) listed in the text of the exercise.

At the stationary state

$$\frac{k_d}{D_P}[P] = \frac{\partial^2 [P]}{\partial r^2}$$

The solution will be a function of the type $[P] = e^{-br}$, where $b = \sqrt{k_d/D_P}$. The analytical solution suggests us that $[P]$ decays exponentially. The spatial distribution of the morphogen depends on the ratio between the kinetic constant k_d, representing the reaction(s) depleting P, and the diffusion coefficient D_P.

9.15. For the preparation of the stock solutions, we weight the salts, and we dissolve them in deionized water. By performing such operations, we make systematic errors, and we might also make random errors. The unavoidable systematic errors are due to the use of the balance, the pipettes, and the flasks. For instance, for the preparation of the $KBrO_3$ solution in sulfuric acid and water, we weight (10.0200 ± 0.0002) g of $KBrO_3$; we dissolve them in (20.00 ± 0.03) mL of a previously prepared stock solution of H_2SO_4 (2.988 ± 0.036) M and, then, we add deionized water until we have (100.0 ± 0.1) mL. If we avoid introducing any random error in our operations, the final concentrations of potassium bromate and sulfuric acid in the stock solution are $[H_2SO_4] = (0.598 \pm 0.009)$ M and $[KBrO_3] = (0.6000 \pm 0.0006)$ M, respectively. The uncertainties have been estimated through the formula of maximum a priori absolute error (read Appendix D). The uncertainty of $[H_2SO_4]$ is larger than that of $[KBrO_3]$ because the sulfuric acid solution was prepared through two dilution steps.

The exercise requires to mix the reagents in three distinct conditions, maintaining constant the total volume of the reactive solution. $V_{tot} = (12.0 \pm 0.1)$ mL. Table 9.5 reports the volumes of the solution A taken in the three distinct experiments. Moreover, it reports the calculated final concentrations of $KBrO_3$ ($[BrO_3^-]_f$), H_2SO_4 ($[H_2SO_4]_f$), H^+ ($[H^+]_f$), and the product $[H^+]_f[BrO_3^-]_f$. The values of $[H^+]_f$ have been calculated by considering a complete first dissociation of H_2SO_4, and, for the second dissociation, the equilibrium constant $K_{a2} = 1 \times 10^{-2}$ M.

Figure 9.40 shows an example of the chemical waves that can be observed by performing this exercise. Chemical waves do not behave like physical waves. In fact, when one wave front runs into a wall of the Petri dish or when two wave fronts collide, we do not see the phenomenon of reflection. A chemical wave can come back only if new reagents are supplied in the path it traced. Chemical waves do not pass through each other when they collide, but instead, they destroy each other because reagents have been exhausted. The time required to replenish the system of new reagents is named as the refractory period. Only after elapsing the refractory period, a new chemical wave can pass through, again.

For each one of the three $[H^+]_f[BrO_3^-]_f$ values tested, we have measured several wave velocities. Then, we have calculated their average and the standard deviation of the average

---

TABLE 9.5

Volumes of the Solution A and the Final Concentrations of BrO_3^-, H_2SO_4, H^+ Used in the Three Experiments

Volume of A (mL)	$[BrO_3^-]_f$ (M)	$[H_2SO_4]_f$ (M)	$[H^+]_f$ (M)	$[H^+]_f[BrO_3^-]_f$ (M^2)
(7.0 ± 0.1)	(0.350 ± 0.008)	(0.349 ± 0.013)	(0.358 ± 0.019)	(0.125 ± 0.009)
(6.0 ± 0.1)	(0.300 ± 0.008)	(0.299 ± 0.012)	(0.308 ± 0.018)	(0.0924 ± 0.008)
(5.0 ± 0.1)	(0.250 ± 0.007)	(0.249 ± 0.011)	(0.258 ± 0.016)	(0.0645 ± 0.006)

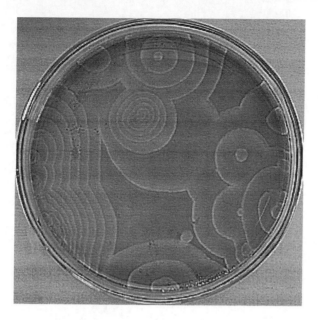

FIGURE 9.40 A picture showing chemical waves.

(see Table 9.6). We have done the same for the determination of the period of the oscillations (see Table 9.6).

After plotting $\log(v)$ and $\log(T)$ versus $\log([H^+]_f[BrO_3^-]_f)$, we determine the exponents n' and n'' as the slopes of straight lines determined by the least squares method (see Figure 9.41).

Note that the period of the oscillations shortens whereas the velocity of the chemical waves grows up by increasing the product $[H^+]_f[BrO_3^-]_f$. The linear relationship between $\log(T)$ and $\log([H^+]_f[BrO_3^-]_f)$ is:

$$\log(T) = c' + n'\log\left(\left[H^+\right]_f\left[BrO_3^-\right]_f\right) = (0.4\pm0.2) - (1.37\pm0.23)\log\left(\left[H^+\right]_f\left[BrO_3^-\right]_f\right).$$

The correlation coefficient is $r = 0.986$.

The linear relationship between $\log(v)$ and $\log([H^+]_f[BrO_3^-]_f)$ is:
$$\log(v) = c'' + n''\log([H^+]_f[BrO_3^-]_f) = (-0.8\pm0.2) + (1.22\pm0.18)\log([H^+]_f[BrO_3^-]_f),\text{ with}$$
$r = 0.989$.

The estimated $n' = -(1.37\pm0.23)$ includes the value of the slope we found for the relationship $\log(T) = \text{cost} + n\log([KBrO_3]) = (0.32\pm0.05) + (-1.61\pm0.04)\log([KBrO_3])$ in exercise 8.2. We may assume that n' corresponds to $(-3/2)$ and n'' to $(+1)$, within

TABLE 9.6

Velocities and Periods Measured as a Function of the $[H^+]_f[BrO_3^-]_f$ Product

v (cm/s)	$\log(v)$	T (s)	$\log(T)$	$\log([H^+]_f[BrO_3^-]_f)$
(0.0120 ± 0.0003)	(-1.92 ± 0.01)	(44.2 ± 0.6)	(1.646 ± 0.005)	(-0.903 ± 0.03)
(0.00755 ± 0.00006)	(-2.122 ± 0.003)	(76 ± 2)	(1.88 ± 0.01)	(-1.03 ± 0.04)
(0.00533 ± 0.00006)	(-2.273 ± 0.005)	(111 ± 1)	(2.045 ± 0.005)	(-1.19 ± 0.04)

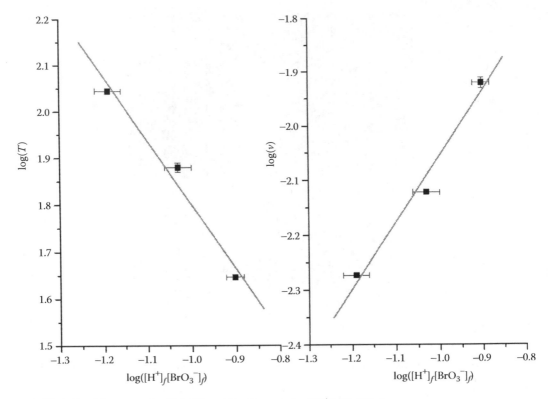

FIGURE 9.41   Linear trends of $\log(T)$ and $\log(v)$ versus $\log([\mathrm{H}^+]_f[\mathrm{BrO}_3^-]_f)$.

the uncertainties of our determinations. If we remind the definition of the velocity of a wave as the ratio between its wavelength and its period $\left(v = \left(\lambda/T\right)\right)$, we find that $\lambda = 10^{c'+c''}([\mathrm{H}^+]_f[\mathrm{BrO}_3^-]_f)^{n'+n''}$. Therefore, $n' + n'' \approx -(1/2)$, as expected.

According to Field and Noyes (1974b), $v = \left(4kD[\mathrm{H}^+][\mathrm{BrO}_3^-]\right)^{1/2}$. If we plot the data of velocity vs. the products $([\mathrm{H}^+]_f[\mathrm{BrO}_3^-]_f)^{1/2}$, we find that the slope of the straight line (determined by the least squares method) is $(4kD)^{1/2} = (0.048 \pm 0.009)\mathrm{cmM}^{-1}\mathrm{s}^{-1}$. From the latter result, it is possible to estimate the value of the kinetic constant: $k = (28 \pm 11)\mathrm{M}^{-2}\mathrm{s}^{-1}$. The value $k = 20\ \mathrm{M}^{-2}\mathrm{s}^{-1}$ falls within the range determined in this exercise.

More accurate results can be achieved by eliminating convection in our experiments. The Petri dish allows evaporation, causing temperature gradients between the top and the bottom part of the solution and hence convection. There are two strategies to avoid convection. A strategy is to minimize the thickness of the solution layer, maintaining wet the entire surface of the Petri dish. The other strategy is to add a surfactant to the solution. The risk of having convection can be eliminated if the BZ reaction is carried out within a gel.

9.16. If we consider an expanding circle of a spiral wave, we have $v_c = 0$ when $v_p = -c * D$. It derives that $r = \left(D/v_p\right) = 0.0017\,\mathrm{cm} \approx 20\,\mu\mathrm{m}$.

9.17. A fire spreading in a forest can be described as it were a chemical wave. Any forest is an excitable medium. Therefore, the fire is an example of a trigger wave. It is peculiar because the autocatalyst of the oxidation reactions responsible for the fire is the high local temperature. The higher the temperature, the faster the oxidation reactions. Usually, the refractory period for a fire is pretty long, because wherever it goes through, it leaves only ashes. Therefore, new vegetation needs to be born before a new fire can pass, again.

9.18. Two properties are peculiar to chemical waves. One is the refractory period. The other is the possibility of not having attenuation. If the fuel of the propagating chemical reaction is distributed uniformly in the space, the chemical wave does not attenuate.

9.19. The precipitation of the red Ag_2CrO_4 within the gels contained in cylinders and test tubes starts within one hour or two after the addition of the silver nitrate solution. The precipitation and the formation of the rings' pattern continue to develop over a period that lasts even more than twenty days. In Figure 9.42, there is a sequence of pictures taken at different delay times. Picture A has been taken soon after the addition of the outer electrolyte solution over the gel. The other images refer to the same sample after 1 day (B), 4 days (C), 7 days (D), 13 days (E), and 27 days (F). It is evident that the number of rings increases with time. Their formation starts at the interface between the solution of $AgNO_3$ and the gel. Then, day by day, always new rings develop towards the bottom part of the cylinder. The direction of their appearance is opposite to that observed for the BZ reaction in exercise 9.1. In fact, in the case of the BZ reaction, the rings form from the bottom part of the solution, and they move towards the top.

Sometimes, the nucleation process is made evident by the formation of opaque yellow layers of Ag_2CrO_4 nanoparticles within the gel (see Figure 9.43).

For the verification of the time law $x_n \sim \sqrt{t_n}$, we need to determine the distance L swept by the front of the precipitation reaction at various delay time t_n. By plotting L versus $\sqrt{t_n}$ (Figure 9.44), we obtain a straight line whose slope may be assumed to be $\sqrt{D}$ (with D being the diffusion coefficient of silver nitrate).

The slope of the plot is $s = (0.518 \pm 0.014)$cm/h$^{0.5}$. It derives that $D = (0.27 \pm 0.01)$cm^2/h $= (7.5 \pm 0.3) \times 10^{-5}$cm^2/s. The diffusion coefficient of the silver nitrate in the gel is of the same order of magnitude of its D in water.

The trends of the number of rings as a function of the time (expressed in hours) for the two concentrations of $AgNO_3$ are shown in Figure 9.45.

A way for estimating the average wavelength of the pattern (λ) is to divide the distance between the interphase layer and the farthest ring (d) by the total number of rings (N) (Table 9.7).

The wavelengths of the Liesegang patterns are of the same order of magnitude of the wavelengths for the pattern generated by the BZ in exercise 9.1.

According to equation [9.10], the lower the concentration, the longer the wavelength. The ratio between the two estimated wavelengths is

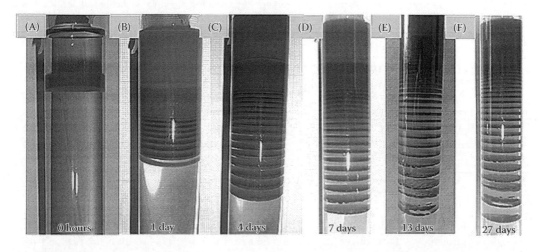

FIGURE 9.42 Snapshots that show the formation of a Liesegang pattern within a test tube soon after the addition of the outer electrolyte (A), and after 1 day (B), 4 days (C), 7 days (D), 13 days (E), 27 days (F).

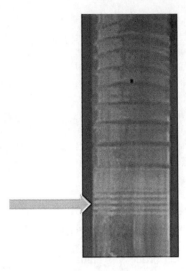

FIGURE 9.43   Temporary formation of thin layers of Ag_2CrO_4 nanoparticles, indicated by the arrow.

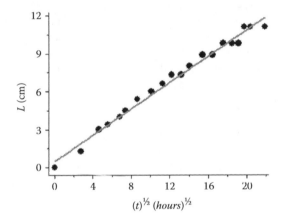

FIGURE 9.44   Trend of the distance L traveled by silver nitrate as a function of $t^{1/2}$ where t is the time elapsed.

$$R = \frac{\lambda_{dil}}{\lambda_{conc}} = \frac{(0.60 \pm 0.01)}{(0.48 \pm 0.01)} = (1.25 \pm 0.05)$$

R is close to the expected value of $\sqrt{b_{conc} / b_{dil}} = \sqrt{2} \approx 1.4$.

If we use equation [9.10], it is possible to determine the kinetic constant value of the deposition step [9.66] that is autocatalytic for the nuclei S:

$$k = \frac{4\pi^2 D}{b\lambda^2} = (0.011 \pm 0.001) M^{-1} s^{-1}$$

The value of k for the deposition step [9.66] is an order of magnitude smaller than the value of the autocatalytic step for the BZ reaction (see exercise 9.1). This result is the reason why the formation of the final pattern in the case of the periodic precipitations takes more time than in the case of the BZ reaction.

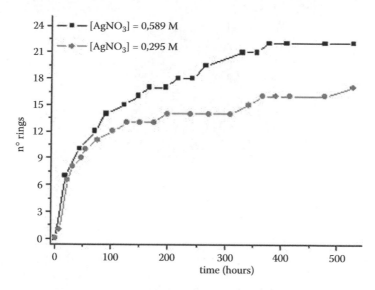

FIGURE 9.45 Trends of the number of rings appearing in the Liesegang structures as a function of the time.

TABLE 9.7
Wavelengths (λ) of the Liesegang Patterns for the Two Outer Electrolyte Solutions.
The Length of the Pattern is d, and the Number of Rings is N

Outer Electrolyte Solution	d (cm)	N	λ (cm)
(0.589 ± 0.001) M	(10.5 ± 0.2)	22	(0.48 ± 0.01)
(0.2945 ± 0.0005) M	(10.2 ± 0.2)	17	(0.60 ± 0.01)

The wavelength of the Liesegang rings does not depend appreciably on the diameter of the cylinder or test tube. Usually, the rings are horizontal and parallel to each other (see the pattern on the right of Figure 9.46). Sometimes amazing spiral structures appear unpredictably (see the pattern on the left of Figure 9.46). Probably, macroscopic curved gel interphase could force spiral growth of the crystals.

If we look at the Liesegang patterns in two dimensions, we notice that they show some features of the chemical waves (see Figure 9.47).

FIGURE 9.46 Two Liesegang patterns with spiral rings (on the left) and parallel rings (on the right).

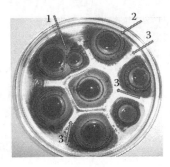

FIGURE 9.47 Liesegang patterns in two dimensions. Phenomena of annihilation in 1 and 2. In 3, the gel is white because silver ions have sucked the chromate ions.

In fact, the Liesegang rings annihilate when they collide with each other or when they encounter the wall of the Petri dish (see the points highlighted by the arrows labeled as 1 and 2, respectively). It is worthwhile noticing that some regions of the gel included between two or more patterns are white and not yellow (see, for example, the points highlighted by the arrow labeled as 3), presumably because the chromate anions have been sucked by silver ions to form the precipitate Ag_2CrO_4. This experimental evidence confirms that when deposition occurs, it automatically inhibits further precipitation in its neighborhood.

In this exercise, all the uncertainties have been estimated by the propagation formula of the a priori maximum absolute error.

10 The Emergence of Chaos in Time

Prediction is very difficult, especially if it's about the future.

Niels Bohr (1885–1962 AD)

10.1 INTRODUCTION

This book describes some of the exciting phenomena we observe when we explore the properties of out-of-equilibrium systems. So far, we have discovered that, although the Second Principle of Thermodynamics still holds, systems that are out-of-equilibrium can self-organize in both time and space. In the linear regime, the transformations usually comply with the Curie's symmetry principle. In other words, the causes of the phenomena have the same or a smaller number of symmetry elements compared with the effects. In the nonlinear regime, we can witness remarkable events of symmetry breakage. In other words, the results have fewer symmetry elements than the causes.

But the surprises are not finished here. In fact, the nonlinear regime retains another fantastic phenomenon: Chaos! In this chapter, we are going to discover that the dynamics of very-far-from equilibrium systems, working in the nonlinear regime, may be aperiodic and very sensitive to the initial conditions. The dynamics can be so sensitive to the initial conditions that they become unpredictable in the long term! When does it happen and why? The answer is in the next paragraphs. I introduce chaos by using a standard and very cheap scientific facility—a pendulum. Every student in his/her first years of study learns the dynamic of a pendulum because it is simple and gives rise to entirely periodic motion. But this is true if we idealize the behavior of the pendulum by considering its dynamic only in the linear regime. In paragraph 10.2, we explore the behavior of a pendulum in the nonlinear regime, and we experience unexpected scenarios. Such scenarios are astonishing because they demonstrate that starting from apparently identical initial conditions, the dynamics of two double-pendula can diverge pretty fast and in a remarkable way. Chaos and Order are two faces of the same coin: The coin of the nonlinearity. Two enjoyable examples show how very-far-from-equilibrium systems can self-organize or exhibit chaotic dynamics, depending on the contour conditions. The first example regards a famous model of population growth, the logistic map. The second case concerns convection. Convection is everywhere and around us. It is in the stars, but also in our planet, beneath the terrestrial crust, and in our atmosphere. Moreover, the convection can also be studied in a laboratory, for instance, by implementing a "hydrodynamic photochemical oscillator."

After becoming aware of what chaos is, we will find out that in the nonlinear regime, a general thermodynamic rule is the Maximization of Entropy Production. Moreover, we will study how to deal with time series and distinguish chaotic from stochastic dynamics. Finally, we will discover that nowadays chaotic dynamics are not just curious scientific phenomena to study or nuisances to avoid when we expect periodicity. Chaotic systems are rich sources of dynamical behaviors that can be mastered and exploited in many fields. For instance, in Information Technology, wherein alluring scenarios of chaos-communication and computation are outlining.

10.2 NONLINEARITY AND CHAOS: THE CASE OF THE DOUBLE PENDULUM

The boundary between linear and nonlinear systems delimits what is understandable and predictable from what is elusive and difficult to be predicted in the long term. It is quite easy to find exact solutions of linear differential equations. Therefore, the dynamics of linear systems can be predicted. On the other hand, the evolution of nonlinear systems is hardly ever predictable, especially in the long term.

TRY EXERCISE 10.1

Linear systems can exhibit four possible dynamics. They can show an exponential growth whenever the differential equations are of the type [10.1].

$$\frac{dy}{dt} = ky \tag{10.1}$$

If the differential equation looks like [10.2],

$$\frac{dy}{dt} = -ky \tag{10.2}$$

the linear system will decay exponentially.

If, on the other hand, the differential equation looks like [10.3]

$$\frac{d^2y}{dt^2} = -by \tag{10.3}$$

the system will oscillate periodically. Finally, a linear system can show a behavior that is a combination of the three cases described by equations [10.1] through [10.3], respectively.

The nonlinear systems can exhibit much more strange dynamics. Let us consider the simple case of a pendulum (see Figure 10.1). It is a popular nonlinear system. To make things easy, we assume that our pendulum is simply a point mass m, attached to a massless and rigid rod of length L, which oscillates due to the gravitational force mg (where g is the gravitational acceleration). A pendulum with these features is an idealization. In fact, in reality, the string has a mass; the mass m bound at the bottom of the string has a volume; the air exerts resistance to the movement of the pendulum; the pivot generates friction, and usually, the rod is a flexible string. However, even the simplified

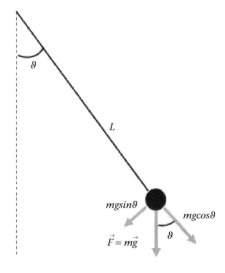

FIGURE 10.1 A sketch of a single pendulum where m is the mass, L is the length of the string, g is the gravitational acceleration, and ϑ is the angular displacement.

model is enough to explain the difference between linear and nonlinear dynamics. The moment τ of the gravitational force that works on the point mass m and has $I = mL^2$ as its moment of inertia is equal to:

$$\tau = I\frac{d^2\vartheta}{dt^2} = -mgL(sin\vartheta) \qquad [10.4]$$

Equation [10.4] can be rearranged in

$$\frac{d^2\vartheta}{dt^2} = -\frac{g(sin\vartheta)}{L} \qquad [10.5]$$

Equation [10.5] is nonlinear. However, for small values of the angle ϑ, the function $sin\,\vartheta$ can be approximated with ϑ, and equation [10.5] becomes linear. The solution is the harmonic oscillator:

$$\vartheta = Acos(\omega t + \varphi) \qquad [10.6]$$

The period of the oscillations depends on L, and the amplitudes of the oscillations depend on its total energy.

When the angle ϑ is large, the function $sin\,\vartheta$ cannot be approximated with ϑ, and we must use the nonlinear differential equation [10.5] to describe the dynamics of the system. For large angular displacements, the energy of the pendulum is high. The pendulum can exhibit strange evolutions. For example, it could rotate 360°; or when it has slightly smaller energy, it could describe a large rotation, but before ending the cycle, the pendulum could collapse because the string loses its tension and the mass falls perpendicularly. Usually, it is difficult, if not impossible, to predict the dynamics of the pendulum in the nonlinear regime. In fact, it is in the nonlinear regime that we experience the extreme sensitivity to the initial conditions, and chaotic dynamics can emerge.

A concrete example of a chaotic system that can be easily found in the physics departments, but it can also be built at home, is the double pendulum (Shinbrot et al. 1992). A double pendulum consists of two simple double pendula, which are bound to one another as shown in Figure 10.2. The first point mass m_1 is suspended from a fixed point (that is also the origin of the reference system) by a rigid weightless rod of length L_1 and the second point mass m_2 is suspended from m_1 by another weightless rod of length L_2 (see Figure 10.2). The top and center pivots are assumed frictionless,

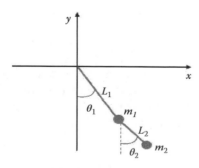

FIGURE 10.2 Structure and parameters characterizing a double pendulum. The parameters defining the upper pendulum are labeled by the subscript 1, whereas those of the lower pendulum by subscript 2. The origin of the reference system has been placed at the pivot point of the upper pendulum.

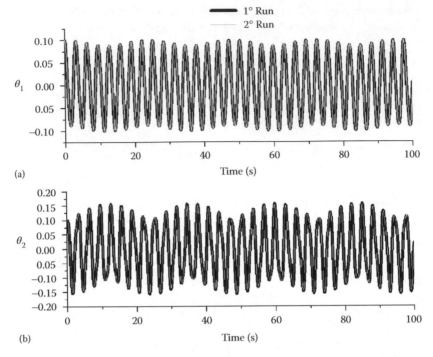

(a)

(b)

FIGURE 10.3 Trends of θ_1 (a) and θ_2 (b) versus time calculated for a double pendulum with $L_1 = 2$, $m_1 = 2$, $L_2 = 1$, $m_2 = 1$, and for two distinct initial conditions: $\theta_1(t = 0) = 0.100$ radians, $\theta_2(t = 0) = 0.100$ radians for the 1° run shown as tick black traces, and $\theta_1(t = 0) = 0.100$ radians, $\theta_2(t = 0) = 0.105$ radians for the 2° run shown as thin grey traces. In the simulation (performed by using the MATLAB ode45 solver) all of the pendulums begin from rest.

and the pendula rotate under the action of gravity in the absence of air. The double pendulum's total energy, which is the sum of its kinetic and potential energies, is conserved. The double pendulum is a Hamiltonian system that exhibits chaotic behavior.

The sensitivity to the initial conditions can be appreciated when we consider motions of large amplitudes. In Figure 10.3, the time evolution of the angular displacements, θ_1 and θ_2, are reported for two runs characterized by small initial values. In the first run of the simulation, $\theta_1(t = 0) = 0.100$ radians (that corresponds to 5.7°) and $\theta_2(t = 0) = 0.100$ radians. In the second run, $\theta_1(t = 0) = 0.100$ radians and $\theta_2(t = 0) = 0.105$ (i.e., 6°) radians. Note that the two runs differ only for 0.005 radians (i.e., 0.5°) in θ_2. The resulting time evolutions are perfectly overlapped for both θ_1 and θ_2, and the motions of the pendulum remain periodic.

When we compare the evolutions of the double pendulum for two other very similar initial conditions, but starting from large θ_1 and θ_2 values (like in the 3° and 4° runs), the result is entirely different (see Figure 10.4). After a short temporal interval (roughly 7 seconds and half), the two trajectories diverge significantly from one another although the two dynamics derive from initial conditions that differ for only 0.5° in θ_2 (like between the 1° and 2° runs of Figure 10.3). It is simply astonishing. The motions are aperiodic and unpredictable.

TRY EXERCISES 10.2 AND 10.3

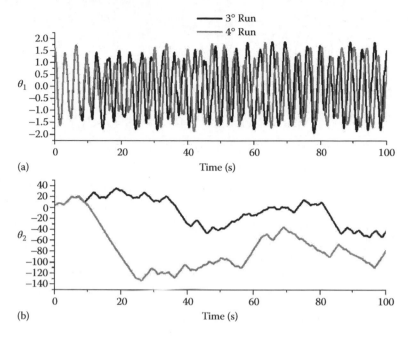

FIGURE 10.4 Trends of θ_1 (a) and θ_2 (b) versus time calculated for a double pendulum with $L_1 = 2$, $m_1 = 2$, $L_2 = 1$, $m_2 = 1$, and for two distinct initial conditions: $\theta_1(t = 0) = 1.570$ radians, $\theta_2(t = 0) = 4.710$ radians for the 3° run shown as black traces, and $\theta_1(t = 0) = 1.570$ radians, $\theta_2(t = 0) = 4.715$ radians for the 4° run shown as grey traces. In the simulation (performed by using the MATLAB ode45 solver) all of the pendulums begin from rest.

10.3 NONLINEARITY AND CHAOS: THE CASE OF THE POPULATION GROWTH AND THE LOGISTIC MAP

A simple model to describe the growth of a population of living organisms (X) in the presence of limited food resources (F) is inherently autocatalytic:

$$X + F \underset{k_{-1}}{\overset{k_1}{\rightleftarrows}} 2X \qquad [10.7]$$

The following differential equation describes the variation of the population over time:

$$\frac{d[X]}{dt} = k_1 [X][F] - k_{-1}[X]^2 \qquad [10.8]$$

If the food supply is fixed (i.e., $[F]$ is a constant), the X population reaches a stationary state given by

$$[X]_{ss} = \frac{k_1[F]}{k_{-1}} \qquad [10.9]$$

This stationary state is stable (as we found in exercise 3.13). Biologists call the ratio $\left(k_1[F]/k_{-1}\right)$ as the carrying capacity K, and equation [10.8] is usually presented in the following form:

$$\frac{dN}{dt} = rN\left(1 - \frac{N}{K}\right) \qquad [10.10]$$

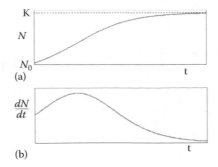

(a)

(b)

FIGURE 10.5 Profile of the logistic function (a), and rate of population growth (b) over time.

where N is the number of organisms and $r = k_1 \left[F \right]$ is the growth rate. Equation [10.10] is called the logistic equation, and it was proposed for the first time in 1838 by Verhulst[1] to describe the growth of human populations. By integrating equation [10.10] (*try exercise 10.4*) between time 0 when the population is N_0 and time t when the population is N_t, we achieve the logistic function:

$$N_t = \frac{K}{1 + ((K - N_0)/N_0)e^{-rt}}$$
[10.11]

A plot of the logistic function is shown in Figure 10.5a. It has a sigmoid shape: the population starts from $N_0 > 0$ (of course if $N_0 = 0$, the population cannot grow at all) and goes to the carrying capacity K. The growth rate of the population initially increases (see Figure 10.5b), reaches a maximum and then begins to decrease when the population approaches the carrying capacity K.

There exists a discrete-time analogue of the logistic equation for the population growth. It is called the logistic map and it is

$$x_{n+1} = rx_n \left(1 - x_n \right)$$
[10.12]

In equation [10.12], x_n is a dimensionless measure of the population in the n-th generation. It is given by the ratio between the actual number of organisms over the carrying capacity K, i.e., $x_n = (N_n/K)$. x_{n+1} is the dimensionless measure of the population in the next generation, and r represents the growth rate.[2] A plot of the logistic map is depicted in Figure 10.6. The logistic map has a maximum at $x_n = (1/2)$, and its value is $(r/4)$. Therefore, if we assume that r lies between 0 and 4, x_{n+1} will stay within the [0–1] range. When x_{n+1} becomes 0, it means that the population extinguishes. When x_{n+1} becomes 1, the population is at its carrying capacity K. Can we predict the dynamical evolution of a population if we know the initial value x_0? In the long term, the population should converge to the steady state values, called also fixed points, where $x_{n+1} = x_n = x_{ss}$. In other words, at the steady state, $x_{ss} = rx_{ss}(1 - x_{ss})$ and the population will stay at x_{ss} for all the future iterations if the external conditions do not change and the system is not perturbed. The fixed points are $x_{ss} = 0$ and $x_{ss} = (1 - 1/r)$. The former is valid for any value of r, whereas the latter gives allowable population values only when $r \geq 1$.

What kind of steady states are they? What is their stability? To answer this question, we can apply a method that is similar to the linear stability analysis we learned in the case of ordinary differential

---

[1] Pierre-François Verhulst (Brussels, 1804–1849) was a mathematician who proposed the logistic equation to describe the population growth after he had read *An Essay on the Principle of Population* by Thomas Malthus. In his book, Malthus pointed out that human population tends to increase geometrically (wherein successive changes in a population differ by a constant ratio), while the food supply grows only arithmetically (wherein successive changes differ by a constant amount).

[2] Equation [10.12] derives from equation [10.10], by dividing both its terms by K. Thus, we obtain $(1/K)(dN/dt) = r((N/K) - (N^2/K^2))$ and hence $x_{n+1} = rx_n(1 - x_n)$.

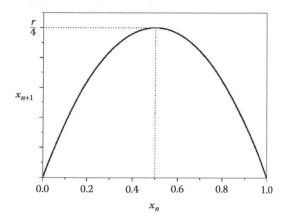

FIGURE 10.6 Profile of the logistic map.

equations. We assume to perturb the population slightly from x_{ss} to $x_{ss} + \xi_0$, and we monitor whether the population is drawn back to x_{ss} or it goes far away from it. In other words, we monitor if ξ_0 goes to zero or it becomes larger and larger after successive iterations. If we indicate with $f(x)$ the logistic map, we can write

$$x_{n+1} = x_{ss} + \xi_{n+1} = f(x_{ss}) + f'(x_{ss})\xi_n + \ldots \qquad [10.13]$$

In the development of the Taylor's series, we neglect the terms including the second and higher order derivatives. By the definition of fixed points, equation [10.13] reduces to

$$\xi_{n+1} = f'(x_{ss})\xi_n \qquad [10.14]$$

The iteration of equation [10.14] gives

$$\xi_1 = f'(x_{ss})\xi_0$$
$$\xi_2 = f'(x_{ss})\xi_1 = \left[f'(x_{ss})\right]^2 \xi_0$$
$$\ldots \qquad [10.15]$$
$$\xi_n = f'(x_{ss})\xi_{n-1} = \left[f'(x_{ss})\right]^n \xi_0$$

If $\left|f'(x_{ss})\right| < 1$, i.e., the first derivative of $f(x)$, calculated at the fixed point, lies between -1 and $+1$, then x_{ss} is stable. On the other hand, if $\left|f'(x_{ss})\right| > 1$, the fixed point is unstable. The first derivative of the logistic map calculated at x_{ss} is

$$f'(x_{ss}) = r - 2rx_{ss} \qquad [10.16]$$

Therefore, the fixed point $x_{ss} = 0$ is stable when $r < 1$.

For the other fixed point, it results that $f'(x_{ss}) = r - 2r(1-1/r) = 2 - r$. Therefore, it will be stable when $\left|2 - r\right| < 1$. i.e., when $1 < r < 3$.

In synthesis, when r is smaller than 1, the population extinguishes. Conversely, when $1 < r < 3$, the population approaches the steady state $x_{ss} = (1 - 1/r)$ value.

What about the marginal case $\left|f'(x_{ss})\right| = 1$?

When $r = 1$, there is just $x_{ss} = 0$ as a fixed point. The definition of its stability requires the determination of the second and higher order terms in equation [10.13]. Since $f''(x_{ss}) = -2r$, it derives that $x_{ss} = 0$ is a stable fixed point.

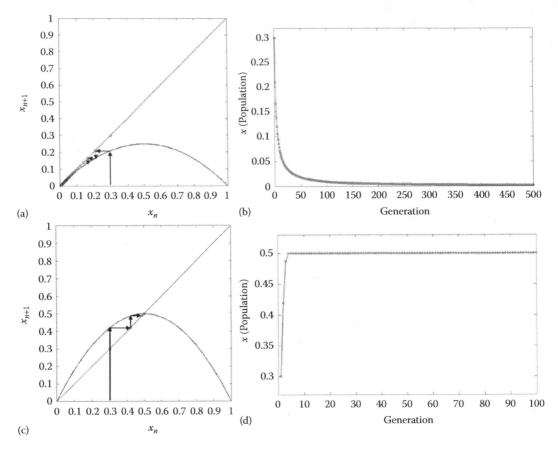

(a)   (b)   (c)   (d)

FIGURE 10.7 Evolution of a population determined by the cobweb analysis and by successive iterations when $r = 1$ in (a) and (b), and when $r = 2$ in (c) and (d), respectively.

All these considerations regarding the dynamic of the population and the stability of the fixed points can be confirmed through two alternative methods. The first method is by iterating equation [10.12] on a calculator starting from an initial value x_0. The second method is by the so-called cobweb construction, which is a graphical approach. Two examples of cobweb construction are illustrated in Figure 10.7a and c. We must plot the logistic map along with the straight line $x_{n+1} = x_n$. The points where the logistic function crosses the straight line, i.e., $x_{n+1} = x_n = f(x_n)$, represent the steady states. The evolution of the system starting from x_0 is traced as it follows. We find x_0 on the abscissa, and we trace a vertical straight line until we reach the logistic map. The ordinate of the intersection point between the vertical line and the logistic map represents x_1. At this stage, we move horizontally to the diagonal line, so that the abscissa becomes x_1. From here, we move vertically to the graph of $f(x)$ again, and we find x_2. If we now move horizontally to the line $x_{n+1} = x_n$, we find the abscissa of x_2. The repetition of these two operations, which involve moving vertically up to intersect the graph of the logistic map and moving horizontally to the diagonal, generates the sequence of x_n values.

How does the population evolve when $r \geq 3$?

When $r = 3$, the abundance of the population oscillates, alternating between a large population ($x_k = 0.69$, after 50 iterations) in one generation and a smaller population ($x_{k+1} = 0.64$) in the next one (see Figure 10.8a and b). This type of oscillation, in which the population repeats every two generations, is called "cycle of period 2." At the larger r, when it becomes equal to 3.45, the population approaches an asymptotic solution in which it repeats every four generations, meaning the cycle has "period 4" (see Figure 10.8c and d).

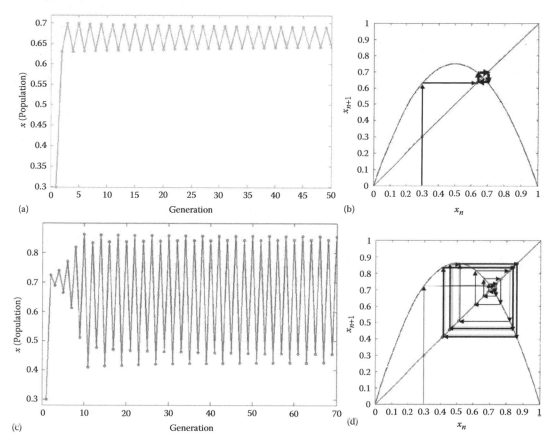

FIGURE 10.8 Evolution of a population described by the logistic map with $r = 3$ (a) and $r = 3.45$ (c). Cobweb analysis in the case of $r = 3$ (b) and $r = 3.45$ (d).

A further increase of the growth rate to $r = 3.54$ yields a cycle of period 8. If we continue to increase r in small increments, we find further period doublings to 16, 32, 64, ... cycles. When $r \approx 3.57$, the values of x_n never repeat. The sequence never settles down to either a fixed point or a periodic orbit: we experience the Chaos!

Chaos emerges not only for $r = 3.57$ but also for the other larger values of r, as shown in the bifurcation diagram of Figure 10.9a, where the asymptotic values of x_n are plotted as a function of r. When we have a chaotic solution, the graph appears covered by a cloud of infinite dots. If we look at Figure 10.9b carefully, we may notice that there are narrow white patches interspersed between the chaotic clouds of dots. These narrow white patches represent "periodic windows." The bifurcation diagram reveals an unexpected mixture of Order and Chaos! It is incredible that zooming in on different portions of the bifurcation diagram, for example in the range of $3.847 \leq r \leq 3.857$, copies of the overall diagram reappear (see Figure 10.9c). This property has a profound relevance that we will understand in the next chapter.

When we are in a chaotic regime, we discover that the system is extremely sensitive to the initial conditions and its dynamic is aperiodic. Starting from two values of populations that differ for a tiny amount, we notice that, after few iterations, the abundances of the populations diverge reciprocally.

TRY EXERCISE 10.5

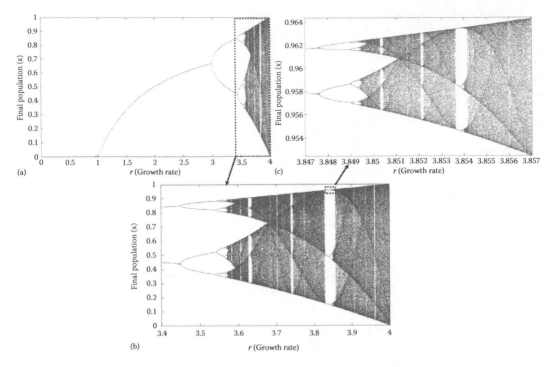

FIGURE 10.9 Bifurcation diagram for the logistic map. In (a) r assumes values between 0 and 4; in (b) r assumes values between 3.4 and 4, and in (c) there is a zoom in on the range $3.847 \leq r \leq 3.857$.

10.4   THE UNIVERSALITY OF CHAOS

Many of the properties of the population dynamics anticipated by the logistic map has been confirmed by the experimental study of the vole *Clethrionomys rufocanus* of Hokkaido, the northernmost island of Japan (May 1998). The populations of voles exhibited either steady abundance, or regular cycles, or unpredictable fluctuations driven by "internal" mechanisms and external environmental noise.

Plots, such as those depicted in Figure 10.9, show successions of bifurcations culminating in chaotic regimes and are not prerogatives of the logistic map. Any map whose graph is smooth, concave down, and with a single maximum is parabola-shaped (in mathematical terms it is defined "unimodal") and follows a period-doubling route to chaos as the logistic map does.

TRY EXERCISE 10.6

In the 1970s, the mathematician Mitchell Feigenbaum discovered that all the bifurcation diagrams of unimodal maps are not only qualitative similar, but they also exhibit a universal quantitative feature (Feigenbaum 1980). In fact, if we indicate with r_k the critical value of r at which the unimodal map bifurcates into a cycle of period 2^k, it results that

$$\lim_{k \to \infty} \left(\frac{r_k - r_{k-1}}{r_{k+1} - r_k} \right) = \delta = 4.669201609\ldots \qquad [10.17]$$

Equation [10.17] means that the distance between period doublings decreases as the control parameter r increases.

TRY EXERCISE 10.7

Its limit for $k \to \infty$ is a universal constant (δ) as it is $\pi = 3.1415\ldots$ for the circles. The universality of δ, now called the Feigenbaum's constant, has been demonstrated mathematically[3] and experimentally in many laboratories studying fluid dynamics, electronic circuits, lasers and chemical reactions (Cvitanovic 1989). One proof has been obtained by analyzing convection that is the subject of the next three paragraphs.

10.5 CONVECTION

Convection is one of the three possible mechanisms of heat transfer (the other two are conduction and irradiation). Convection may occur when a fluid is heated from below, and the thermal gradient, ΔT, between the top and the bottom parts of the fluid is quite large, whereby conduction is not the most efficient mechanism of heat transfer.[4] The heating is so intense that it expands the bottom layer of the fluid. The bottom layer becomes less dense than the upper layers. This situation is unstable within the terrestrial gravitational field. The bottom layer tends to rise, whereas the colder and denser top layers tend to sink. The buoyancy force triggers convection. The bottom fluid cannot advance as a whole since there is no place for the fluid above it to go. In fact, a pattern made of roll-shaped cells, having a long tubular shape, emerge (see Figure 10.10). The fundamental unit of the pattern consists of two counterrotating cylindrical rolls; one clockwise (labeled as C in Figure 10.10) and the other anti-clockwise (labeled as AC). Warm fluid rises along one edge of a roll, traverses the upper surface and loses its heat. Then, it plunges to the bottom layer along the opposite side of the roll. There, at the bottom, the temperature of the fluid rises again.

A systematic study of the convective mechanism was undertaken since the beginning of the twentieth century (Velarde and Normand 1980). At that time, the most influential experiments were performed by the French physicist Henri Bénard, and a remarkable theoretical investigation was carried out by the English physicist John William Strutt, Lord Rayleigh. The theory proposed by Rayleigh refers to a thin layer of a fluid confined between two flat, rigid, horizontal plates, heated from below and cooled from above.[5] The temperature difference between the bottom and the top parts of the fluid is maintained constant. Gravity is the only force acting within the fluid. The thermal gradient affects the density of the fluid significantly. In particular, the density declines as the temperature rises. Now, let us consider a small parcel of the fluid at the bottom, having volume V and density d_0. Due to the random fluctuations, the parcel of the fluid may move infinitesimally upwards. If this event occurs,

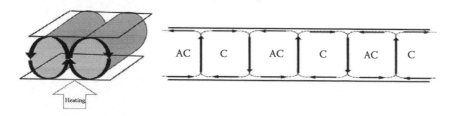

FIGURE 10.10 Scheme of two adjacent roll-shaped cells in the convection of a fluid heated from the below (on the left). On the right, frontal view of an array of adjacent cells rotating clock-wise (C) and anti-clockwise (AC).

---

[3] Feigenbaum demonstrated the universality of δ by using the concept of renormalization borrowed from statistical physics. If you are interested to his proof, read Feigenbaum (1979).

[4] This paragraph presents Natural Convection. Natural Convection is different from Forced Convection. In Forced Convection, the fluid of motion is promoted by an external force generated, for instance, by a pump, a fan or other devices.

[5] The model system studied by Lord Rayleigh did not coincide tightly with the features of the system investigated by Bénard. Read Box 1 of this chapter to know more about the theory explaining the experimental results obtained by Bénard.

the parcel is now surrounded by a cooler and denser (d) portion of the fluid (with $d > d_0$). As a result, the parcel feels the buoyancy force (F_b) due to the gravity force (being g its acceleration):

$$\vec{F}_b = -(d - d_0)\vec{g}V \qquad [10.18]$$

This buoyancy force pushes the parcel farther upwards. In contrast to $\vec{F}_b$, there is the drag force ($\vec{F}_V$) dependent on the fluid shear viscosity (η):

$$\vec{F}_V = -\eta\vec{v}L \qquad [10.19]$$

In [10.19], $\vec{v}$ represents the velocity of the parcel and L is its radius. When the drag force is larger or equal to the buoyancy force, there can be no motion. On the other hand, when the buoyancy force is stronger than the drag one, the ordered convective motions can start. However, there exists a second phenomenon that opposes the appearance of convection. And this is the heat conduction. The molecules in the warm parcel, in contact with a cooler surrounding, dissipate their high kinetic energy by impact with the slower encompassing molecules. Heat flows spontaneously from the displaced warm parcel. The time needed for the parcel to reach thermal equilibrium is inversely proportional to the thermal diffusivity of the fluid. If this time is comparable to the time required for the parcel to move a distance such as its diameter, the convective flow cannot be formed.

Of course, as there is the chance that a warm parcel of fluid moves upwards by fluctuations and enters a colder region, so there exists the chance that a cool parcel moves downwards and finds a less dense portion of the fluid. The buoyancy force will push the cool parcel farther downwards. But the drag force will contrast its movement. Moreover, heat will flow by conduction from the surrounding environment into the cooler parcel, and it will tend to annihilate the thermal gradient (see Figure 10.11). The fluid self-organizes, and the convective rolls emerge when the buoyancy forces overcome dissipative processes due to the viscosity and heat conduction.

There exists a dimensionless parameter that allows predicting if an extended layer of fluid, having height h and under a vertical thermal gradient (ΔT), self-organizes producing convective rolls. This parameter is the Rayleigh number (Ra), which is a ratio between two counteracting forces. The first is the buoyancy force [10.18] that, expressed as a function of the isobaric thermal expansion coefficient ($\beta = (1/V)(\partial V/\partial T) = -(1/d)(\partial d/\partial T)$) and per unit of volume, becomes

$$\frac{F_b}{V} = \beta g d_0 \Delta T \qquad [10.20]$$

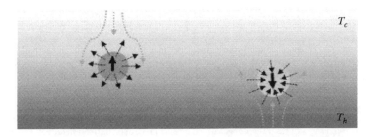

FIGURE 10.11   Forces acting on two parcels of fluid. One warm parcel moves upwards, and one cool parcel moves downwards. The formation of convective rolls depends on the action of the buoyancy force (continuous black arrow) against the drag force (dotted grey arrows) and the heat conduction (dashed black arrows). The dimensions of the two parcels are exaggerated with respect to the linear dimensions of the fluid.

The second is the dissipative force (F_d) estimated per unit of volume and generated by the joined action of the drag force and heat conduction:

$$\frac{F_d}{V} = \frac{\alpha \mu d_0}{h^3}$$ [10.21]

In [10.21], $\alpha = \left(k/C_P d_0\right)$ is the thermal diffusivity that depends on the thermal conductivity (k), the specific heat capacity at constant pressure (C_P), and the average density of the fluid (d_0); $\mu = \left(\eta/d_0\right)$ is the kinematic viscosity. It derives that

$$Ra = \frac{\beta g \Delta T h^3}{\mu \alpha}$$ [10.22][6]

The onset of convection is for $Ra_c \sim 1708$, independent of the fluid under consideration, and for two flat rigid boundaries at the top and bottom layer surfaces (Chandrasekhar 1961). As long as Ra is slightly above Ra_c, the cylindrical rolls of the convective pattern remain straight, and their motion is periodic. At any fixed point in space, the temperature is constant. If we increase the thermal gradient, another instability sets in when Ra reaches a second critical value. A wave starts to propagate back and forth along the longest axis of the cylindrical rolls, causing the temperature to oscillate at each point. Further heating determines the emergence of other instabilities. In fact, waves with other frequencies start to propagate back and forth along each roll, and temperature oscillates with more than one frequency in each point of the fluid. When large Ra values are reached, the local temperatures start to change chaotically, with an infinite number of frequencies. Finally, for even larger Ra values, the motion of the fluid turns from a laminar to turbulent one (see Figure 10.12).

The first elegant experiments demonstrating the transition of natural convection from periodic to chaotic states were performed by the French physicist Albert Libchaber in the 1970s and 1980s

---

[6] The Rayleigh number can also be presented as the product of two other dimensionless parameters: the Prandtl number (Pr), and Grashof number (Gr).

$$Ra = \left(Pr\right)\left(Gr\right)$$

The Prandtl number is the ratio between the time scale for the diffusion of heat (h^2/α) and the time scale for the diffusion of momentum (h^2/μ):

$$Pr = \frac{\mu}{\alpha}$$

When Pr is small, it means that heat diffuses more quickly than momentum. When Pr is large, the opposite is true. The Grashof number (Gr) is the ratio between the buoyancy force [10.18] and the drag force [10.19]:

$$Gr = \frac{\left(d - d_0\right) g V}{\eta v L}$$

If we equate the drag and momentum forces $\eta v L = d_0 v^2 L^2$, we obtain that $v = \left(\eta/L d_0\right)$. Introducing this definition of velocity v and the definition of thermal expansion coefficient β in the equation of Grashof number, we obtain

$$Gr = \frac{g \beta \Delta T h^3}{\eta^2}$$

Of course, when Gr is small, the drag force dominates over the buoyancy force, and the motion of the fluid is laminar. On the other hand, when Gr is large, it is the buoyancy force to overwhelm the drag force, and the motion of the fluid is turbulent. In the case of Forced Convection, the Reynold number (see Box 4 of Chapter 9) is analogous to the Grashof number.

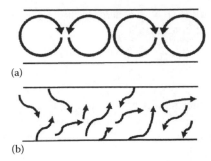

(a)

(b)

FIGURE 10.12 Examples of laminar (a) and turbulent (b) convection. The arrows represent the velocity vectors.

when he was working at the École Normale Supérieure (EMS) of Paris. At first, he used liquid helium at 3 K placed in a very tiny cell, having dimensions of the order of mm, surrounded by a chamber at high vacuum (Libchaber and Maurer 1978; Maurer and Libchaber 1979). The low temperature is guaranteed by thermal determinations of very high resolution and high accuracy. All the thermal measures were carried out by exploiting microbolometers engraved in the cell.

TRY EXERCISE 10.8

Then, Libchaber used liquid mercury that assured the better stability of the convective rolls in the presence of a stationary magnetic field (Libchaber et al. 1982). In both cases, by a constant increase of the thermal gradient, Libchaber and his coworkers measured the value of Ra at the period-doubling bifurcations. At Ra_c, the temperature at each point was constant over time. At $Ra_0 = 2Ra_c$, the temperature started to oscillate. For a farther increase of the thermal gradient, there was another bifurcation, and T began to oscillate with two frequencies. For larger ΔT, the frequencies became four, then, eight, sixteen, … and finally infinite. The trend became chaotic. By using the critical Ra values estimated at each bifurcation, Libchaber and coworkers could confirm the universality of the Feigenbaum number [10.17]. Although convection is not directly related with the unimodal maps, it shows the same feature of them as far as the evolution to chaotic conditions is concerned.

TRY EXERCISE 10.9

BOX 10.1   THE MARANGONI-BÉNARD CONVECTION

The system investigated experimentally by Bénard had one feature different from the model system studied by Rayleigh. The difference was that the layer of the fluid was not confined between two rigid boundaries but instead was open to the air at its upper surface. This change was relevant for the shape of the pattern of the convective flow. In the thin layer of the fluid heated from below and maintained open to the air above, Bénard observed the formation of a polygonal tessellation of the fluid surface. When the pattern was fully developed, it appeared as an almost perfect array of hexagons, as it were a honeycomb. The center of each hexagon is a region of warm upwelling fluid that reaches the surface; it is dragged to the perimeter of the polygon, where it sinks because cooled down. There was also a slight depression of the upper free surface of the fluid at each cell center. When the fluid is a thin layer and is maintained open to the air, the surface tension of the fluid

(Continued)

BOX 10.1 (Continued)　THE MARANGONI-BÉNARD CONVECTION

is the primary factor responsible for the convection pattern and not the buoyancy. Surface tension (γ) is the force per unit of length whose effect is to minimize the surface area of fluid. It is responsible for the spherical shapes of liquid drops. Surface tension plays like the propulsive force in the Bénard convection because it varies with temperature: in particular, it decreases when the fluid is warmed. Therefore, when the temperature of the surface of the liquid is not uniform, local gradients of surface tension are generated. The gradients give rise to flows on the surface. These flows are communicated to the bulk of the fluid as a result of its shear viscosity (see Figure in this B10.1).

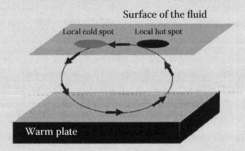

FIGURE B10.1　Convection induced by the local surface tension gradients. The surface tension in local cold spots is larger than in local hot spots.

The mechanism for the onset of convection is similar to that proposed by Rayleigh when the buoyancy is the driving force. However, in this case, the driving force is the surface tension. The surface tension force (F_{st}) per unit volume of fluid is

$$\frac{F_{st}}{V} = \left(\frac{d\gamma}{dT}\right)\frac{\Delta T}{h^2} = \gamma_T\left(\frac{\Delta T}{h^2}\right) \qquad \text{[B1.1]}$$

being h the thickness of the layer, and ΔT the thermal gradient between the bottom and the top of the fluid. Friction and thermal diffusion oppose the action of the surface tension. Both effects combine to yield a dissipative force (F_d) per unit of volume that is equal to $F_d/V = \alpha\mu d_0/h^3$ (remember equation [10.21]). The ratio between $\left(F_{st}/V\right)$ and $\left(F_d/V\right)$ is the Marangoni number (named after the Italian physicist lived between the nineteenth and the twentieth centuries):

$$Ma = \frac{\gamma_T h\Delta T}{\alpha\mu d_0} \qquad \text{[B1.2]}$$

The Marangoni number is the control parameter for the convection driven by the surface tension. When Ma is above a critical value Ma_c, which depends on the boundary conditions, but it is usually ~80 (Bragard and Velarde 1997), the surface tension dominates and overcomes the dissipative effects of viscous drag and thermal diffusion.

TRY EXERCISE 10.10

10.6   THE ENTROPY PRODUCTION IN THE NONLINEAR REGIME: THE CASE OF CONVECTION

In the previous paragraph, we learned that when a fluid is heated from below but the thermal gradient ΔT is small, heat is transferred through conduction. At larger ΔT, above a critical value ΔT_c, convection becomes the fastest mechanism of heat transfer. There exists a dimensionless parameter that measures the ratio between the rates of heat transfer by convection and conduction. This parameter is the Nusselt number (Nu):

$$Nu = \frac{convection\ heat\ transfer\ rate}{conduction\ heat\ transfer\ rate} \qquad [10.23]$$

When the Rayleigh number is less than its first critical value, Ra_c, $Nu < 1$, and conduction is more effective in transferring heat. When $Ra \sim Ra_c$, $Nu \sim 1$. When $Ra > Ra_c$, $Nu > 1$ and convection is more efficient.

In Chapter 3, we knew that when conduction is the most effective mechanism of heat transfer because ΔT is small and we are not very far-from-equilibrium, the system evolves spontaneously to a stable stationary state where the Entropy Production P^* is minimized. What happens when ΔT is large, and convection becomes the fastest mechanism of heat transfer? In other words, how does P^* change when we are very far-from-equilibrium?

In the case of heat conduction in a system of volume V and surface area A, the Entropy Production is

$$P^* = \frac{d_i S}{dt} = \int_V J_q \nabla\left(\frac{1}{T}\right) dV = \int_V L_q \left[\nabla\left(\frac{1}{T}\right)\right]^2 dV \qquad [10.24]$$

because we are in linear regime and $J_q = L_q \nabla(1/T) = -k\nabla T$.

How does the Entropy Production P^* change over time? We consider the time derivative of equation [10.24] to answer this question. Assuming L_q constant over the temperature range of the system, we obtain:

$$\frac{dP^*}{dt} - 2\int_V L_q \nabla\left(\frac{1}{T}\right)\frac{\partial}{\partial t}\left[\nabla\left(\frac{1}{T}\right)\right] dV - 2\int_V J_q \nabla\left[\frac{\partial}{\partial t}\left(\frac{1}{T}\right)\right] dV \qquad [10.25]$$

Equation [10.25] transforms in [10.26], after integration by parts:[7]

$$\frac{dP^*}{dt} = 2\int_A \left[\frac{\partial}{\partial t}\left(\frac{1}{T}\right)\right] J_q\, \hat{n}\, dA - 2\int_V \left[\frac{\partial}{\partial t}\left(\frac{1}{T}\right)\right] \nabla J_q dV \qquad [10.26]$$

In [10.26], $\hat{n}$ is the unit vector perpendicular to the surface A and pointed outwards. The first integral of equation [10.26] vanishes when the thermal gradient is maintained constant at the boundary of

---

[7] Integration by parts is a method to solve integrals when the integrand is a product of two functions, for example, $x \cdot y$. The method derives from the formula for the derivative of a product: $d(x \cdot y) = ydx + xdy$. After integrating both terms of the equation, we obtain $\int d(x \cdot y) = x \cdot y = \int ydx + \int xdy$. Suppose we need to calculate either the indefinite integral $\int ydx$ or the definite integral $\int_{t_1}^{t_2} ydx$. The solution will be $\int ydx = x \cdot y - \int xdy$ and $\int_{t_1}^{t_2} ydx = (x \cdot y)_{t_2} - (x \cdot y)_{t_1} - \int_{t_1}^{t_2} xdy$, respectively. This method of integration by parts makes sense when the solution of the integral $\int xdy$ is known.

the system. Using the Fourier's law and assuming the thermal conductivity k constant, the second integral of [10.26] becomes:

$$\frac{dP^*}{dt} = 2\int_V k\nabla^2 T\left[\frac{\partial}{\partial t}\left(\frac{1}{T}\right)\right]dV = -2\int_V \frac{k\nabla^2 T}{T^2}\left(\frac{\partial T}{\partial t}\right)dV \qquad [10.27]$$

The continuity (or balance) equation for the internal energy of the fluid (Anderson Jr. 2009) considering (I) heat advection, (II) heat conduction, (III) viscous heating and (IV) cooling by volume expansion is

$$\frac{\partial}{\partial t}\left(\rho u\right) = -\rho\vec{v}\nabla u + k\nabla^2 T + \tau\nabla\vec{v} - P\nabla\vec{v} \qquad [10.28]$$

In [10.28], ρ is the density, $\vec{v}$ the velocity, u the internal energy per unit of mass, τ the viscous stress, and P the pressure. If we express $u = c_v T$, where c_v is the specific heat at constant volume, and we assume that c_v is constant in the fluid, we get (Ozawa and Shimokawa 2014)

$$\rho c_v \frac{\partial T}{\partial t} = -\rho\vec{v}c_v\nabla\left(T\right) - P\nabla\vec{v} + \tau\nabla\vec{v} + k\nabla^2 T \qquad [10.29]$$

It is now evident that the temporal change of T depends on the cooling induced by the heat advection and volume expansion, and on the heating due to the viscous force and conduction. The contribution of heat conduction, $k\nabla^2 T$, can be expressed as a function of the other terms of equation [10.29]:

$$k\nabla^2 T = \rho c_v\frac{\partial T}{\partial t} + \rho\vec{v}c_v\nabla\left(T\right) + P\nabla\vec{v} - \tau\nabla\vec{v} \qquad [10.30]$$

We can insert this definition of $k\nabla^2 T$ into [10.27] and achieve

$$\frac{dP^*}{dt} = 2\int_V \left[\rho c_v\frac{\partial T}{\partial t} + \rho\vec{v}c_v\nabla\left(T\right) + P\nabla\vec{v} - \tau\nabla\vec{v}\right]\left[\frac{\partial}{\partial t}\left(\frac{1}{T}\right)\right]dV \qquad [10.31]$$

In the absence of convection, $\vec{v} = 0$, equation [10.31] reduces to

$$\frac{dP^*}{dt} = -2\int_V \frac{\rho c_v}{T^2}\left(\frac{\partial T}{\partial t}\right)^2 dV \leq 0 \qquad [10.32]$$

Equation [10.32] tells us that P^* tends to decrease until it reaches a minimum in its final steady state, when $\left(\partial T/\partial t\right) = 0$. This is the theorem of Minimum Entropy Production that we have learned in Chapter 3.

What happens in the presence of convection?

$$\frac{dP^*}{dt} = -2\int_V \frac{k\nabla^2 T}{T^2}\left[\frac{\partial T}{\partial t}\right]dV = -2\int_V \frac{k\nabla^2 T}{T^2}\left[-\vec{v}\nabla\left(T\right) - \frac{P}{\rho c_v}\nabla\vec{v} + \frac{\tau}{\rho c_v}\nabla\vec{v} + \frac{k}{\rho c_v}\nabla^2 T\right]dV \qquad [10.33]$$

If we assume that the cooling due to volume expansion and the viscous heating are negligible with respect to terms describing the conduction and heat advection, equation [10.33] reduces to [10.34]:

$$\frac{dP^*}{dt} = -2\int_V \frac{k\nabla^2 T}{T^2}\left[-\vec{v}\nabla\left(T\right) + \frac{k}{\rho c_v}\nabla^2 T\right]dV \qquad [10.34]$$

When the thermal gradient ΔT is so large that convection is the most effective mechanism of heat transfer, the Nusselt number is larger than 1, and $\bar{v}\nabla(T) \geq (k/\rho c_v)\nabla^2 T \geq 0$ or $\bar{v}\nabla(T) \leq (k/\rho c_v)\nabla^2 T \leq 0$. In both cases, it derives that

$$\left(\frac{dP^*}{dt}\right) \geq 0 \qquad\qquad [10.35]$$

In the nonlinear regime, the Entropy Production always grows until it maximizes after fixing the contour conditions. Typically, in very far-from-equilibrium conditions, there are many available states. According to the most recent investigations, it seems there exists a selection criterion that is useful to predict the evolution of such systems. And this is the Maximum Entropy Production principle (MaxEP) that has been proved in some circumstances, such as fluid turbulence, crystal growth morphology, biological evolution (Dewar and Maritan 2014). The out-of-equilibrium system evolves spontaneously to the state that maximizes Entropy Production (Martyushev and Seleznev 2006). The theoretical basis of MaxEP is a subject of open debate. One hypothesis is that MaxEP has a statistical explanation (Dewar 2009): when a system is forced very far-from-equilibrium, the MaxEP condition is selected simply because it is by far the most probable one. The MaxEP state can be microscopically obtained in an overwhelmingly greater number of ways than any other non-equilibrium state. The MaxEP principle needs further confirmations and studies to become a general evolutive criterion. However, it is worthwhile stressing that the discovery of general principles in nonlinear regime is a tough task. In fact, in the nonlinear regime, the events are often unique and unreproducible. They are irreversibly and make history. Hence, history is not a prerogative of humans but is also made by inanimate physical and chemical systems.

10.7 THE "BUTTERFLY EFFECT"

10.7.1 THE COMPLEXITY OF CONVECTION IN THE TERRESTRIAL ATMOSPHERE

As anticipated in paragraph 10.1, the phenomenon of convection is widespread in nature. For example, it is the mechanism of heat transfer in the stars, like our sun. In our planet, convection is the motive force for the slow migration of the continents and the vast oceans currents. Also, the global circulation of the atmosphere depends on convective flows. The UV-Visible radiation emitted by the sun and reaching the biosphere is absorbed mainly by the terrestrial crust, and it is re-emitted as infrared. Such thermal radiation having small frequencies is absorbed by the lowest layers of the atmosphere due to the presence of water, carbon dioxide, and other gaseous compounds. A vertical thermal gradient originates, and small-scale convective flows are induced. Other more significant convective flows are generated by the temperature gradients existing between the warm tropics and the cold poles. An accurate theoretical analysis of these natural convective phenomena is not straightforward, at all. The Rayleigh's model studied in paragraph 10.5 assumes that only the density of the fluid changes with temperature; the other physical properties are considered to be constant. Of course, this is an approximation that in real fluids is quite coarse. In reality, also viscosity and thermal diffusivity change with temperature. Moreover, fluids are compressible, especially if they are gaseous, and pressure is a significant variable, which in turn affects density and other properties. The variables are quite tangled. For example, viscous drag dissipates kinetic energy in the form of heat, and so it raises the temperature. The local heating determines a reduction in viscosity. A theory that explicitly considers all the relations and the positive and negative feedback actions among the variables becomes too cumbersome. Therefore, we must seek a compromise between the natural complexity and the complexity of our theories.

10.7.2 THE LORENZ'S MODEL

In the 1960s, an idealized dissipative hydrodynamical model for the convective flows in the atmosphere, formulated for pursuing weather forecasts, consisted in the following system of three nonlinear differential equations, developed by Edward Lorenz (1963), who started from the Navier-Stokes equations:[8]

$$\frac{dX}{dt} = \sigma(Y - X) \qquad [10.36]$$

$$\frac{dY}{dt} = rX - Y - XZ \qquad [10.37]$$

$$\frac{dZ}{dt} = XY - bZ \qquad [10.38]$$

In these equations, there are only two nonlinear terms, which are the products XZ in [10.37] and XY in [10.38]. The variable X is proportional to the velocity of the convective motion, and Y is proportional to the temperature difference between the ascending and descending currents. The variable Z is proportional to the deviation of the vertical temperature profile from linearity (the linearity is verified before convection starts, when the heat is transferred only by conduction) and a positive value indicates that the strongest gradients are close to the boundaries. The parameter σ is the Prandtl number,[6] $r = (Ra)/(Ra_c)$, and b is related to the aspect ratio of the rolls. The steady state solutions are $(X_s, Y_s, Z_s) = (0, 0, 0)$ (labelled as S_0), $\left(\sqrt{b(r-1)}, \sqrt{b(r-1)}, (r-1)\right)$ (labelled as S_+), $\left(-\sqrt{b(r-1)}, -\sqrt{b(r-1)}, (r-1)\right)$ (labelled as S_-) when $r > 1$, and just $S_0 = (0, 0, 0)$ when $r < 1$. The solution S_0 represents the absence of convection, whereas the other two solutions, S_+ and S_-, differing just in the signs of X_s and Y_s, represent left- and right-turning convection rolls. The stability of the solutions can be investigated by considering the evolution of small perturbations and applying the linear stability analysis we learned in Chapter 3. If we indicate with $x(t) = (X(t) - X_s)$, $y(t) = (Y(t) - Y_s)$ and $z(t) = (Z(t) - Z_s)$ the distance of the variables from their steady-state values, they evolve in time according to the following linearized equations:

$$\begin{pmatrix} \dot{x}(t) \\ \dot{y}(t) \\ \dot{z}(t) \end{pmatrix} = \begin{pmatrix} -\sigma & \sigma & 0 \\ r - Z_s & -1 & -X_s \\ Y_s & X_s & -b \end{pmatrix} \begin{pmatrix} x(t) \\ y(t) \\ z(t) \end{pmatrix} \qquad [10.39]$$

For the steady-state solution S_0, the characteristic equation of [10.39] is

$$(b + \lambda)\left[\lambda^2 + \lambda(1 + \sigma) + \sigma(1 - r)\right] = 0 \qquad [10.40]$$

Equation [10.40] has three real negative roots when $0 < r < 1$, and two negative and one positive when $r > 1$. This result confirms that when $0 < r < 1$, the solution S_0 is stable and the fluid transfers heat only by conduction and not by convection. On the other hand, when $r > 1$, convection starts and S_0 is not any more stable.

For the other two steady-state solutions, i.e., S_+ and S_-, the characteristic equation becomes

$$\lambda^3 + \lambda^2(\sigma + b + 1) + \lambda b(r + \sigma) + 2\sigma b(r - 1) = 0 \qquad [10.41]$$

---

[8] The Navier-Stokes equations are the fundamental equations that describe how fluids behave, and they are based on the momentum's conservation.

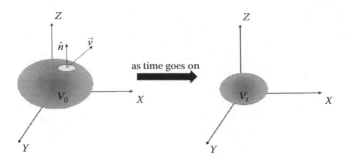

FIGURE 10.13 Time evolution for the volume V_0 of the Lorenz's phase space. V_0 shrinks to V_t after the time interval $(t - t_0)$.

It has been demonstrated (Marsden and McCracken 1976) that when $1 < r < r_c = \sigma(\sigma + b + 3)/(\sigma - b - 1)$ equation [10.41] has one real negative and two complex conjugate roots. S_+ and S_- are stable fixed points and they are surrounded by unstable limit cycles. This behavior means that $r = r_c$ is a subcritical Hopf bifurcation (see Chapter 4). For $r > r_c$, the fixed points are unstable, and there are no attractors in the neighborhood. Nevertheless, the dynamical trajectories cannot diverge to infinity. In fact, the Lorenz system is dissipative: volumes in the three-dimensional phase space contract as we demonstrate, right now.

Let us consider a portion of the Lorenz's phase space having volume V_0 and surface A_0 at time t_0 (see Figure 10.13). We may conceive all the points laying on the surface A_0 as initial conditions. What happens to the volume if they evolve in time? Let dA be an infinitesimal portion of the surface A_0, and $\hat{n}$ be its outward normal. If $\vec{v} = \hat{i}(dX/dt) + \hat{j}(dY/dt) + \hat{k}(dZ/dt)$ is the punctual velocity, the product $\vec{v} * \hat{n}$ represents the projection of the velocity onto the outward normal vector. The infinitesimal volume swept in the infinitesimal time interval dt is $dV = (\vec{v}\,\hat{n})(dt)(dA)$. Hence, the total volume swept in the time interval by all the points of the initial surface A_0 is:

$$\frac{dV}{dt} = \int_A \vec{v}\,\hat{n}\,dA \tag{10.42}$$

Applying the divergence theorem, equation [10.42] transforms in

$$\frac{dV}{dt} = \int_V \nabla \vec{v}\,dV \tag{10.43}$$

The divergence of $\vec{v}$ is

$$\nabla \vec{v} = \frac{\partial}{\partial X}\left(\frac{dX}{dt}\right) + \frac{\partial}{\partial Y}\left(\frac{dY}{dt}\right) + \frac{\partial}{\partial Z}\left(\frac{dZ}{dt}\right) = -\sigma - 1 - b < 0 \tag{10.44}$$

Therefore, [10.43] becomes

$$\frac{dV}{dt} = -(\sigma + 1 + b)V \tag{10.45}$$

After separating the variables and integrating from time t_0 up to time t, we obtain

$$V(t) = V_0 e^{-(\sigma + 1 + b)(t - t_0)} \tag{10.46}$$

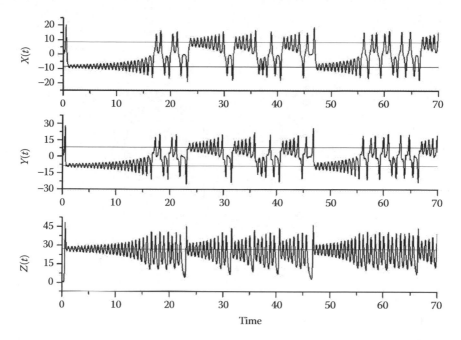

FIGURE 10.14   Time evolutions for X, Y and Z when $\sigma = 10$, $b = 8/3$ and $r = 28$. The horizontal straight lines that are present in each plot represent the solutions S_+ and S_-.

Equation [10.46] tells us that the initial volume of the phase space, V_0, shrinks (as shown in Figure 10.13) with an exponential velocity. This result does not mean that each volume contracts to a point, but it could end up to a limit cycle or something else.

So, now, the question is what happens when we are beyond the critical point r_c? To find an answer, we turn to the numerical integration for solving the system of differential equations [10.36, 10.37, and 10.38]. For the constants, we choose the numerical values also used by Edward Lorenz in (Lorenz 1963): $\sigma = 10$, $b = 8/3$ and $r = 28$ since $r_c \approx 24.74$. As initial condition, we fix the vector $(X(0), Y(0), Z(0)) = (0.1, 0.1, 0.1)$, which is a state close to the absence of convection (represented by the steady state solution S_0). By using the MATLAB solver "ode45", we solve the three differential equations [10.34] numerically and we obtain the time evolutions of the three variables that are depicted in Figure 10.14.

The instability of the initial conditions is evident. All the three variables jump abruptly to large values. Then, they approach solution S_- immediately. They wiggle around S_- with amplified oscillations. After reaching a critical point (at about $t = 17$), X and Y grow abruptly from negative to positive values, and they start to wiggle around the second steady state solution: S_+. But this occurs only for a while. In fact, thereafter, X and Y change sign again. These changes in sign occur at seemingly irregular intervals. The variables X and Y move back to S_- and forth to S_+ in an aperiodic way. A plot of the trajectory of the system in its three-dimensional phase space is shown in Figure 10.15.

The trajectory appears to describe a surprising shape. It looks like a pair of butterfly wings. If we solve the differential equations numerically for more extended periods of time, we discover that the Lorenz's system remains entrapped in this "strange attractor." This attractor is really strange because, at a first sight, it looks like a pair of surfaces that merge into one in the central part of the form. But, this is an illusion due to the volume contraction of the trajectory. In reality, by zooming in, we find that the trajectory crosses an infinite number of surfaces. The orbits move on and never intersect each other nor exactly repeat themselves. The attractor is a set of points with zero volume for $t \to \infty$ (remember equation [10.46]) but an infinite surface area because the object we see in Figure 10.15 is just an approximation of the real attractor. How strange is this attractor? Actually, the attractor is an example of a fractal having dimension 2.05. It is something more than a surface

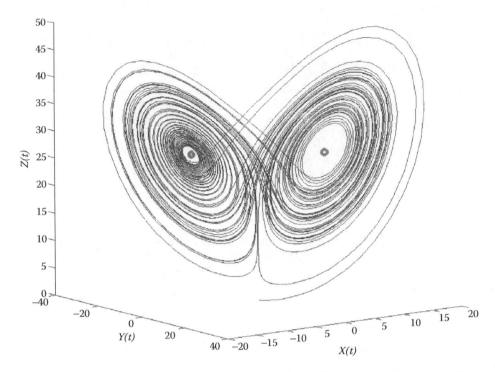

FIGURE 10.15 Projection of the Lorenz trajectory for $\sigma = 10$, $b = 8/3$ and $r = 28$ in the three-dimensional $X(t)$, $Y(t)$, $Z(t)$ space. The two gray points represent the steady state S_- (point on the left) and S_+ (point on the right) solutions.

(having dimension 2) and less than a solid (having dimension 3). We will learn more about fractals in Chapter 11. For the moment, let us remain focused on our simple model for climate because it hides another great surprise.

10.7.3 THE SENSITIVITY TO THE INITIAL CONDITIONS

In 1961, Edward Lorenz was a mathematician and meteorologist at the Massachusetts Institute of Technology. He was testing the prediction capabilities of his model for weather. One day, he decided to prolong a simulation for a longer interval. Instead of calculating the entire run, he decided to restart it from somewhere in the middle. He entered data from his printouts and launched the calculations. In the end, he found something unexpected. The output of the second run matched the data achieved in the first run only at the beginning. After a few steps, the two runs diverged exceptionally, similarly to what is shown in Figure 10.16.

At first, Lorenz thought that something had gone awry in his computer (it was a Royal McBee, a crude and slow computer compared with current standards). After many trials, he realized that there was no malfunction in his computer. The origin of the discrepancy between the two predictions was another one. His printouts showed numbers with only three digits, whereas the computer was making calculations by using six digits. When Lorenz restarted the second run, he entered rounded-off data, assuming that the difference was inconsequential. But, this was not the case. His nonlinear model for weather forecasts showed a strong sensitivity to the initial conditions. More precisely, if $\|\delta_0\|$ is the tiny Euclidean distance between two initial conditions in three-dimensional phase space, their distance over time $\|\delta(t)\|$ grows exponentially:

$$\|\delta(t)\| = \|\delta_0\|e^{\lambda t} \qquad [10.47]$$

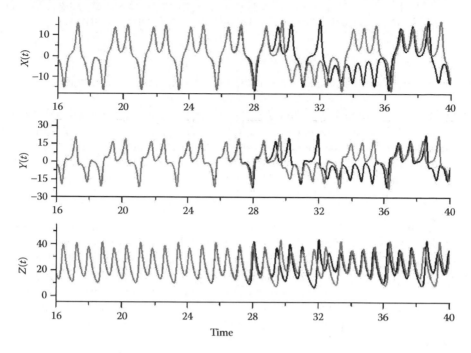

FIGURE 10.16   Trends of $X(t)$ (on top), $Y(t)$ (in the middle) and $Z(t)$ (at the bottom) for two different initial conditions whose distance is only $\delta_0 = 0.0001$.

In [10.47], λ is a positive constant named as the Lyapunov exponent. If we plot $\ln\left(\left\|\delta\left(t\right)\right\|\right)$ versus time t, we find a wiggling trend that at first grows and then levels off. The growth part can be fitted by a straight line (see Figure 10.17). The slope of the straight line is equal to 0.9 and represents the value of the Lyapunov exponent λ. The exponential divergence does not continue indefinitely in time, but it stops when $\left\|\delta\left(t\right)\right\|$ is comparable to the size of the attractor. This behavior explains why $\ln\left(\left\|\delta\left(t\right)\right\|\right)$ shows a value of leveling or saturation.

TRY EXERCISE 10.11

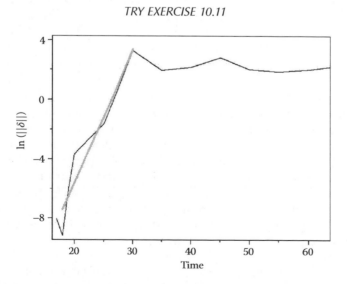

FIGURE 10.17   Trend of the natural logarithm of $\delta(t)$ versus time (thin black trace). The first part has been fitted by a straight line (thick gray trace) whose slope is $\lambda \sim 0.9$.

Roving the phase space of the Lorenz's system is an enjoyable leisure activity.

TRY EXERCISES 10.12 AND 10.13

If you like to glean more information about the Lorenz equations, refer to the book by Sparrow (1982).

The practical implication of equation [10.47] is that the long-term prediction in the Lorenz's system is impossible. In fact, the initial conditions can always be known with a limited degree of accuracy (read Appendix D). The finiteness of our knowledge limits our predictive power. Let us make a numerical example. Suppose we determine the initial conditions with an accuracy $\|\delta_0\| = 10^{-3}$. Our predictions are valuable as long as $\|\delta(t)\| = 10^{-1}$. In other words, we can trust in them for a period of the order of $t \sim (1/\lambda) ln(10^{-1}/\|\delta_0\|)$. Now, we make every effort to improve our facilities and collect more accurate initial conditions. Suppose our accuracy increases thousand times and $\|\delta_0'\|$ is pushed to the limit of 10^{-6}. Our predictions last longer: $t' \sim (1/\lambda) ln(10^{-1}/\|\delta_0'\|)$. Although a thousand-fold improvement in the accuracy of our tools, used to determine the initial conditions (imagine the efforts and expenses needed to succeed in this goal), has been achieved, our predictions become valuable only for $t'/t = 5/2 = 2.5$ more on the timescale.

The sensitivity on initial conditions of chaotic dynamics is also referred to as the "butterfly effect." This metaphor derives from a talk titled "Predictability: Does the Flap of a Butterfly's wings in Brazil Set Off a Tornado in Texas?", presented by Lorenz at the 139th meeting of the American Association for the Advancement of Science in 1972. The provocative question raised in his title was intended to suggest that the flap of the delicate and vulnerable butterfly's wings might create tiny changes in the atmosphere, which may ultimately modify the path of a violent and destructive tornado. In a nonlinear system, like the Lorenz's model or the real climate, a small difference in the initial conditions can rise and originate enormous macroscopic consequences. The flap of a butterfly's wings in Brazil is not the cause of the tornado, but it is part of the uncountable initial conditions that may lead or not lead to a tornado in Texas. In a chaotic system, a small input might have an enormous impact.

10.7.4 THE HYDRODYNAMIC PHOTOCHEMICAL OSCILLATOR

It is possible to simulate the butterfly effect in action in a wet laboratory by performing exercise 10.14. This exercise calls for using a "Hydrodynamic Photochemical Oscillator" (Gentili et al. 2014). When a vertical column of a solution containing a photo-excitable species is irradiated steadily at the bottom by UV or visible radiation, absorbed by either the solvent or the solute or both, the local heating generates a vertical thermal gradient that triggers convective motions of the fluid. As soon as the hydrodynamic motions of the solvent begin, the solute plays the role of a probe of the convective motions. Hence, the optical signal emitted or transmitted by the solute oscillates periodically, or aperiodically, depending on the thermal gradient generated between the bottom and the top parts of the solution. The solute can be a luminescent species, like 9,10-dimethylanthracene (DMA, see Figure 10.18a and c). Otherwise, it could be a photochromic species, like a spirooxazine (as SpO shown in Figure 10.18b). SpO is a direct, thermally reversible photochromic compound. In other words, it is a molecular switch. Under UV, the uncolored spirooxazine transforms into a colored merocyanine (MC in Figure 10.18b and d). The reaction responsible for the color appearance is a photo-electrocyclization, which is a ring-opening process where one σ bond (the spiro C-O bond) is converted into one π bond (the carbonyl group C=O). The molecular transformation is reversible. After discontinuing the irradiation, the color bleaches spontaneously, in a few tens of seconds, at room temperature.

The convective motions in the "Hydrodynamic Photochemical Oscillators" occur in a vertical column and not in an almost-infinite, horizontal fluid layer as in the case of the Rayleigh- or

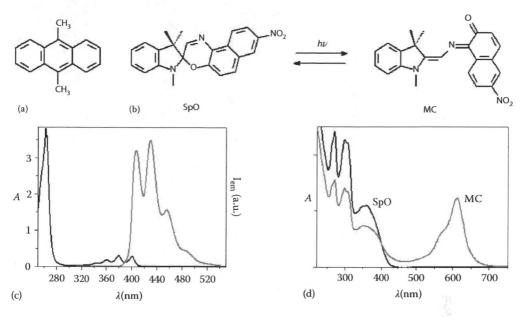

FIGURE 10.18 Structures of 9,10-dimethylanthracene, DMA, (a) and the photochromic spirooxazine (1,3-dihydro-1,3,3-trimethyl-8′-nitro-spiro[2*H*-indole-2,3′-[3*H*]naphth[2,1-*b*][1,4]oxazine]) SpO that transforms reversibly into the merocyanine (MC) through a photo-electrocyclization (b). In graph (c) there are the absorption (black trace) and emission (gray trace) spectra of DMA. In graph (d) the absorption spectra of SpO (black trace) and MC (gray trace) are shown.

Marangoni-Bénard convections studied in paragraph 10.5. In this new geometry, the Rayleigh number assumes the following form (Olson and Rosenberger 1979):

$$Ra = \frac{\beta g \Delta T l^4}{\mu \alpha h}$$
[10.48]

where:

h is the height of the vertical column

l is the length of one side of the square base

The onset of convection is for $Ra_{c,1} \sim 200$, independent of the fluid under consideration. As long as Ra maintains below a second critical value that is $Ra_{c,2} \sim 2400$, there are oscillatory convective motions with a finite number of frequencies. When Ra overcomes $Ra_{c,2}$, aperiodic oscillations are detectable.

The patterns of the convective rolls can be easily observed by using a photochromic compound. For example, in Figure 10.19, you can see an example of the convection planform with three rolls (traced by the black curved arrows). One roll is at the bottom; the other two are above it, and one moves clockwise and the other counterclockwise. The snapshot of the convective motions in Figure 10.19 has been obtained for an estimated vertical thermal gradient of 0.3 K (Gentili et al. 2014). The merocyanine, produced by UV irradiation (see the optical fiber conveying the UV radiation on the left side of the cuvette, at the bottom), is spread at a speed of about 2 mm/s throughout the solution by hydrodynamic waves.

TRY EXERCISE 10.14

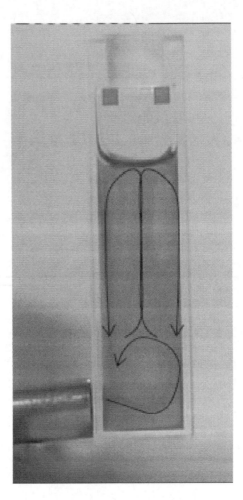

FIGURE 10.19 Convection planform with rolls in a cuvette containing a methanol solution of the photochromic SpO. The solution is UV irradiated at the bottom, on the left side, through an optical fiber. MC colors the solution in blue. The curved thin black arrows trace the convective rolls.

10.8 APERIODIC TIME SERIES

One important goal of science is to predict the future of natural phenomena. If we know the laws that govern a phenomenon, we can make predictions by solving the equations that describe its dynamics. If we study phenomena that are multivariate and nonlinear, the knowledge of their governing laws does not guarantee the formulation of meaningful predictions in the long term, but only shortly. An example is the flow of a fluid described by the Navier-Stokes equations, based on the momentum's conservation.

There are cases wherein the principles governing the evolution of the systems are not rationalized, yet, and their formulation is a tough task because the systems involve a vast number of agents. This situation is widespread indeed if we look at Complex Systems. Examples are the trends stock prices in the economic markets. They depend on macroeconomic forces and a massive number of deals due to agents that act almost independently of each other.

In both situations, the case of both deterministic chaotic system and non-deterministic natural Complex System, it is still possible to make predictions by learning from the past. We can learn from the past, if we collect time series data and we process them, properly. In the next two paragraphs, we are going to learn how to deal with aperiodic time series, to answer two central questions. First question: are the aperiodic time series we record, chaotic or stochastic? Which are the exclusive features of chaotic time series? Second question: is it possible to predict aperiodic time series?

It should be emphasized that, sometimes, our determinations are affected by significant casual errors and environmental noise. The consequent apparent fluctuations may mask a genuinely deterministic chaotic dynamic.

10.8.1 How Do We Recognize Chaotic Time Series?

Nonlinear systems whose dynamics are aperiodic and extremely sensitive to the initial conditions give rise to chaotic time series. When we project the chaotic time series in their phase space, we see that the distance between initial close points diverges exponentially in time. Moreover, the trajectory of any initial point traces a strange attractor that has a fractal structure, and typically a non-integer dimension.

Therefore, it is clear that to recognize a chaotic time series, at first, we need to build its phase space. Even if we do not know the state variables that define the phase space, it is possible to construct the phase space by using the Takens' time-delay embedding theorem (Takens 1981) and a recorded time series with a sufficiently large number of data. We collect a time series $A*$. It could be the absorbance of a "Hydrodynamic Photochemical Oscillator," the electroencephalogram signals of a patient with epileptic seizures, the abundance of a species in an ecosystem, the temperature of the earth over the centuries, the price of a commodity in the stock market, or the number of visit of a website, or any other time series. Whatever is the time series $A*$, it consists of N observations collected with sampling interval Δt, $A^* = \{A_i, i = 1, \dots, N\}$. The time series is transformed into a matrix whose elements are a time-lagged version of the original data:

$$A = \begin{pmatrix} A_1 & A_{1+\tau} & \cdots & A_{1+(m-1)\tau} \\ A_2 & A_{2+\tau} & \cdots & A_{2+(m-1)\tau} \\ \vdots & \vdots & \vdots & \vdots \\ A_{N-(m-1)\tau} & A_{N-(m-2)\tau} & \cdots & A_N \end{pmatrix} = \begin{pmatrix} \bar{A}_1 \\ \bar{A}_2 \\ \vdots \\ \bar{A}_{N-(m-1)\tau} \end{pmatrix} \qquad [10.49]$$

Each row vector of the matrix [10.49] is a single point in the phase space. The integer time delay τ is a dimensionless parameter defined as the actual time delay, τ_R, divided by the sampling interval, Δt. The parameter m is the embedding dimension of the phase space. The number of observation, N, must be much larger than τ and m.

10.8.1.1 Time Delay τ

First, we need to determine the time delay τ. Its calculation must be carried out carefully. If we choose a τ that is too small, then, adjacent components A_i and $A_{i+\tau}$ are too highly correlated for them to serve as independent coordinates. If τ is too large, then, neighboring components are too uncorrelated. We can use two methods for the determination of τ: one based on the calculation of the autocorrelation and the other that grounds on the calculation of the mutual information. The autocorrelation of the matrix A is determined by using the following algorithm:

$$c(\tau) = \frac{1}{N} \sum_{i=1}^{N} A_i A_{i+\tau} \qquad [10.50]$$

A good choice for the time delay is the smallest τ value for which it results (Lai and Ye, 2003)

$$\frac{c(\tau)}{c(0)} \leq \frac{1}{e} \qquad [10.51]$$

The mutual information for a partition of real numbers in a certain number of boxes is given by:

$$I(\tau) = \sum_{ij} p_{ij}(\tau) ln \frac{p_{ij}(\tau)}{p_i p_j}$$ [10.52]

where:

p_i is the probability of finding a time series value in the i-th interval,

p_j in the j-th interval,

$p_{ij}(\tau)$ is the joint probability that an observation falls in the i-th interval and at the observation time τ later falls into the j-th.

The calculation is repeated for different values of τ. The best choice for the time delay is the τ value for which the mutual information has the first marked minimum (Fraser and Swinney 1986).

10.8.1.2 Embedding Dimension m

For the determination of the embedding dimension m, there is the false nearest neighbor method (Kennel et al. 1992). For each point $\overline{A}_i$ in the m-dimensional phase space, the method looks for its nearest neighbor $\overline{A}_j$. It calculates the square of the Euclidean distance:

$$R_{m,i}^2 = \sum_{k=0}^{m-1} \left(A_{i+k\tau} - A_{j+k\tau}\right)^2$$ [10.53]

Then, it goes from dimension m to $m + 1$. A new coordinate is added to each vector $\overline{A}_i$ and $\overline{A}_j$, and the distance becomes

$$R_{m+1,i}^2 = R_{m,i}^2 + \left(A_{i+m\tau} - A_{j+m\tau}\right)^2$$ [10.54]

Any neighbor for which

$$\left(\frac{R_{m+1,i}^2 - R_{m,i}^2}{R_{m,i}^2}\right)^{1/2} = \frac{\left|A_{i+m\tau} - A_{j+m\tau}\right|}{R_{m,j}} > R_t$$ [10.55]

where R_t is a given heuristic threshold, is designated as false. When the number of false nearest neighbors is zero, m is the suitable embedding dimension. The value of R_t is arbitrary. Cao (1997) proposed another algorithm to avoid the arbitrary choice of R_t. Cao's algorithm is:

$$E(m) = \frac{1}{N - m\tau} \sum_{i=1}^{N-m\tau} \frac{R_{m+1,i}}{R_{m,i}}$$ [10.56]

$E(m)$ depends on m and τ. In [10.56], we must consider only the distances of the nearest neighbors. For the investigation of $E(m)$ variation when the embedding dimension increases of one unit, Cao defined $E1(m) = E(m+1)/E(m)$. $E1(m)$ stops changing when m is greater than a value m_0 if the time series comes from an attractor. Then, $m_0 + 1$ is the best estimate of the minimum embedding dimension. For a stochastic time series, $E1(m)$ never attains a saturation value as m increases. Sometimes it is difficult to resolve whether $E1(m)$ is slowly increasing or has stopped changing if m is large. There is another quantity to circumvent this difficulty. It is $E2(m)$. First, we calculate

$$E^*(m) = \frac{1}{N - m\tau} \sum_{i=1}^{N-m\tau} R_{m,i}$$ [10.57]

Then, we define $E2(m) = E^*(m+1)/E^*(m)$. For stochastic data, since the future values are independent of the past values, $E2(m)$ will be equal to 1 for any m. On the other hand, for deterministic time series, $E2(m)$ is certainly related to m. As a result, it cannot be a constant for all m; there must exist some m's such that $E2(m) \neq 1$. $E2(m)$ is useful to distinguish deterministic from random time series.

10.8.1.3   Lyapunov Exponents

After building the phase space of a time series, an essential indicator of the possible presence of deterministic chaos is the exponential divergence of initially nearby trajectories. Such divergence can be probed by determining the Lyapunov exponents. The Rosenstein's (Rosenstein et al. 1993) and Kantz's (1994) methods allow calculating the largest Lyapunov exponent.

The Rosenstein's method locates the nearest neighbor of each point in the phase space. The nearest neighbor of $\bar{A}_j$ is the point $\bar{A}_{\hat{j}}$, which minimizes the distance

$$d_j(0) = min \left\| \bar{A}_j - \bar{A}_{\hat{j}} \right\| \qquad [10.58]$$

where $\| \ \|$ denotes the Euclidean norm. The j-th pair of nearest neighbor diverges approximately at a rate proportional to the largest Lyapunov exponent (λ). In fact, the distance between the two points, after a time $i\Delta t$ (where Δt is the sampling period of the time series) is

$$d_j(i) \sim d_j(0) e^{\lambda(i\Delta t)} \qquad [10.59]$$

In logarithmic form, equation [10.59] becomes

$$ln d_j(i) = ln d_j(0) + \lambda(i\Delta t) \qquad [10.60]$$

The linear relation [10.60] can be defined for all the points in the phase space. Finally, the largest Lyapunov exponent is calculated using a least-squares fit to the "average" line defined by

$$y(i) = \frac{1}{\Delta t} \left\langle ln d_j(i) \right\rangle \qquad [10.61]$$

Kantz's method consists in computing

$$D(\Delta t) = \frac{1}{N} \sum_{j=1}^{N} ln \left(\frac{1}{\|U_j\|} \sum_{i \in U_j} dist(T_j, T_i; \Delta t) \right) \qquad [10.62]$$

where N is the length of time series. The term $dist(T_j, T_i; \Delta t)$ is the distance between a reference trajectory T_j and a neighboring trajectory T_i after the relative time, Δt, determined as

$$dist(T_j, T_i; \Delta t) = \left\| \bar{A}_j - \bar{A}_i \right\| \qquad [10.63]$$

where $\bar{A}_i$ is a neighbor of $\bar{A}_j$ after Δt. U_j is the ε-neighborhood of $\bar{A}_j$, i.e., the set of all the delay vectors of the series having a distance less than or equal to ε with respect to $\bar{A}_j$. The quantity $D(\Delta t)$ is calculated for different sizes of ε (ranging from a minimum given by (data-interval)/1000 to a maximum given by (data-interval)/100) and for different embedding dimensions, m. The slope of D versus Δt is an estimation of the maximal Lyapunov exponent.

10.8.1.4   Kolmogorov-Sinai Entropy

In a phase space of dimension m, a hypersphere of initial conditions, having hypervolume V, evolves into a hyper-ellipsoid whose principal axes change at rates given by the Lyapunov exponents. In any chaotic conservative system, the so-called chaotic Hamiltonian systems, the sum of the positive Lyapunov exponents is counterbalanced by the sum of the negative Lyapunov exponents. In fact, according to the Liouville's theorem, the total hypervolume V of a Hamiltonian system remains constant during its evolution in time. In the case of the dissipative chaos, when there is degradation of energy due to friction, viscosity, or other processes, the total hypervolume V shrinks. The hypervolume V contracts and the system remains confined in a strange attractor. The sum of the negative Lyapunov exponents outweighs the sum of the positive ones. This does not mean that the hypervolume V contracts lengths in all directions. Intuitively, some directions stretch, provided that some others contract so much that the final hypervolume is smaller than the initial one. The exponential separation of close trajectories, which is a feature of any chaotic dynamic, takes place along the directions characterized by positive Lyapunov exponents (Eckmann and Ruelle 1985).

The sum of all the positive Lyapunov exponents equals the Kolmogorov-Sinai entropy (S_K):

$$S_K = \sum_{\lambda_i > 0} \lambda_i \qquad [10.64]$$

The Kolmogorov-Sinai entropy is related to the Information Entropy (Frigg 2004). It has been formulated to describe the uncertainty we have in predicting a chaotic dynamic in its phase space. A dynamical system in the phase space of total hypervolume M is like an information source. This is evident if we partition M in a certain number of rigid, non-intersecting boxes: $\beta = \{b_i | i = 1, \ldots, k\}$, covering the entire phase space. The evolution of a differentiable dynamical system is described by m differential equations of the first order of the type

$$\frac{d\vec{A}(t)}{dt} = f\left(\vec{A}(t)\right) \qquad [10.65]$$

in the case of continuous time, or by a map

$$\vec{A}(n+1) = f\left(\vec{A}(n)\right) \qquad [10.66]$$

in the case of discrete time.

If we start from an initial condition $\vec{A}(0)$, the system traces a trajectory in its evolution. In other words, the system generates a string that is the list of boxes it visits.[9] For example, if we consider a partition consisting in two boxes, $\beta = \{b_1, b_2\}$, then the dynamical system generates a string of the same sort as the one we obtain from a source that sends only binary digits. As it occurs when we receive an unknown message, also in the case of a chaotic dynamic, we are not acquainted with

---

[9] The definition of S_K grounds on the idea that the volume in the phase space can be interpreted as a probability. The phase space is partitioned in a certain number of rigid, non-intersecting boxes: $\beta = \{b_i | i = 1, \ldots, k\}$. Together, they cover the entire phase space of hypervolume M. It is worthwhile noticing that if $\beta = \{b_i | i = 1, \ldots, k\}$ is a partition of M, then $f\beta = \{fb_i | i = 1, \ldots, k\}$ is as well, where f is the function [10.65] or [10.66] describing the evolution of the system. All we know at each iteration (for discrete systems) or after each fixed time step (for a continuous system) is in which box the trajectory ($\vec{A}$) lies. The measurements that $\vec{A}$ lies in box b_0 at time t_0, and in box b_1 at time t_1, tell us that $\vec{A}$, in fact, lies in the region $b_0 \cap f^{-1}(b_1)$, i.e., in the intersection of b_0 with the pre-image of b_1. The preimage $f^{-1}(b_i)$ of each b_i is all points that will be mapped into b_i after a single iteration or time step. The intersection of the two sets b_i and $f^{-1}(b_i)$ gives a finer partition β_1 of the attractor: $\beta_1 = \{b_i \cap f^{-1}(b_j)\}$ and the entropy of this partition is $S(\beta_1) = -\sum_{ij}(p_{ij}) log(p_{ij})$, where p_{ij} is the probability of finding a point over the box $b_i \cap f^{-1}(b_j)$. The Kolmogorov-Sinai entropy is the supremum, taken over all choices of initial partition β_0, of the limit over an infinite number of iteration or time steps ($N \to \infty$): $S_K = \sup(\lim_{N \to \infty}(1/N) S(\beta_N))$, with $\beta_N = \{b_0 \cap f^{-1}b_1 \cap f^{-2}b_2 \ldots \cap f^{-N}b_N\}$. A positive value of S_K is proof of chaos.

what box the system will be in the next. Due to the restrictions on the accuracy of the definition of the initial conditions, we may not know accurately what the system's initial state is, and this uncertainty is even amplified by the chaotic dynamic as time goes on. Therefore, we gain information when we learn that the system is in one box, let us say b_1, rather than into the other, b_2. And the information amounts to one bit. The replacement of the two-boxes-partition with one having more boxes is the analogue to the transition from an information source with two symbols to one with more symbols. If a dynamic system has a positive S_K means that whatever the past history of the system, we are not able to predict with certainty in what box of the partition the system's state will line next. Moreover, the magnitude of S_K is a measure of how high our failure to predict the future will be; the greater the Lyapunov exponents, the larger the Kolmogorov-Sinai entropy, the more uncertain we are about the future.

10.8.1.5 Correlation Dimension

Dissipative systems with chaotic dynamics generate attractors that are strange. Strange attractors have fractal structures and typically non-integral dimensions. A way of determining the strangeness of an attractor consists in measuring the correlation dimension D_C by the Grassberger-Procaccia (1983) algorithm. The Grassberger-Procaccia algorithm requires calculation of the correlation sum:

$$C(m,\varepsilon) = \frac{1}{N'^2} \sum_{i \neq j}^{N'} \Theta\left(\varepsilon - \left\| \bar{A}_i - \bar{A}_j \right\| \right)$$
[10.67]

where:
  ε is the size of the cell (whose minimum is given by (data interval)/1000 and whose maximum is the (data interval)),
  $N' = N - \tau(m-1)$,
  Θ is the Heaviside step function, which is 1 when $\left\| \bar{A}_i - \bar{A}_j \right\| \leq \varepsilon$, whereas it is 0 when $\left\| \bar{A}_i - \bar{A}_j \right\| > \varepsilon$.

Then, the correlation sum is used to determine the correlation dimension through equation [10.68]:

$$C(m,\varepsilon) \propto \varepsilon^{D_C}$$
[10.68]

In practice, the plot of log(C) versus log(ε) is fitted by a straight line determined by the least-squares method. Its slope is the correlation dimension D_C. The procedure is repeated for different value of m. If the dynamic is chaotic, D_C converges to a finite value that, often, is not an integer. If the dynamic is stochastic, D_C does not converge and does not show a saturating value even at high m values (Osborne and Provenzale 1989).

10.8.1.6 Permutation Entropy

Another parameter that is useful for discriminating between chaotic and stochastic data is the permutation entropy (Bandt and Pompe 2002). It represents the Shannon entropy of the permutation patterns of the elements of the time series. In the phase space of embedding dimension m, we account for all $m!$ possible permutations of all the vectors $\bar{A}_i = \{A(t_i), A(t_{i+\tau}), ..., A(t_{i+(m-1)\tau})\}$, for $i = 1,...,N-(m-1)\tau$. Every pattern among the $m!$ possible permutations is labelled as π_j (with $j = 1,..., m!$). We determine the abundance $q(\pi_j)$ of each pattern π_j and calculate ts frequency $v(\pi_j) = q(\pi_j)/(N-(m-1)\tau)$. The permutation entropy (S_P) is the Shannon entropy associated with the distribution of the permutation patterns:

$$S_P = -\sum_{j=1}^{m!} v(\pi_i) log_2 v(\pi_i)$$
[10.69]

The range of variability for S_P is 0 (for monotonically increasing or decreasing data) and $log_2(m!)$ (for completely stochastic data). For this reason, we can normalize S_P by diving it by $log_2(m!)$. The normalized permutation entropy is:

$$S_P = -\frac{\sum_{j=1}^{m!} v(\pi_i) log_2 v(\pi_i)}{log_2(m!)}$$

[10.70]

It ranges between 0 and 1.

10.8.1.7   Surrogate Data

The surrogate data methods are useful and reliable statistical tests to explore the presence of chaos in a time series (Theiler et al. 1992). These methods have two ingredients: a null hypothesis tested against observations and a discriminating statistic. The null hypothesis is a potential explanation that we want to check is inadequate for interpreting the data we have collected. A discriminating statistic is a number that quantifies some features of the time series. If the statistical number is different for the observed data with respect to the surrogate data generated under a specific null hypothesis, then the null hypothesis can be rejected. Examples of null hypothesis are the following: a time series that is (I) linearly filtered noise; (II) linear transformation of linearly filtered noise; (III) monotonic nonlinear transformation of linearly filtered noise. Different methods for generating surrogate data have been proposed so far. For example, the Iterative Amplitude Adjusted Fourier Transformed (IAAFT) surrogate data method (Schreiber and Schmitz 1996) considers a nonlinear rescaling of a Gaussian linear stochastic process as the null hypothesis. On the other hand, the cycle surrogate data method (Small and Judd 1998) is based on the null hypothesis that each cycle in aperiodic dynamics is independent of its adjacent cycles. Usually, one begins with simple assumptions and progresses to more sophisticated models if the collected time series is inconsistent with the surrogate data.

10.8.1.8   Short-Term Predictability and Long-Term Unpredictability

A feature of a chaotic time series is its unpredictability in the long term. It has been demonstrated (Sugihara and May 1990) that the accuracy of a nonlinear forecasting method falls off when it tries to predict chaotic time series, and the prediction-time interval is increased. On the other hand, the nonlinear forecasting method is roughly independent of the prediction time interval when it tries to predict time series that are uncorrelated noise.[10] For the discernment of a chaotic from a white noise time series, we first need to build its phase space by exploiting the Takens' theorem. Then, we plot the time series in its phase space. For each point $\bar{A}_i$ of the trajectory, it is possible to find its neighbors, which are $\bar{A}_j$ (with $j = 1, ..., k$). For the prediction of the value $\bar{A}_{i+\tau\Delta t}$, which lies $\tau\Delta t$ ahead in its phase space (see Figure 10.20), the nonlinear predictor exploits the corresponding $\bar{A}_{j+\tau\Delta t}$ (with $j = 1, ..., k$) and the following algorithm:

$$\bar{A}_{i+\tau\Delta t} = \sum_{j=1}^{k} W(\bar{A}_j, \bar{A}_i) * \bar{A}_{j+\tau\Delta t}$$

[10.71]

---

[10] Noise is a signal generated by a stochastic process. Stochastic phenomena produce time series having power spectral density (i.e., power per unit of frequency) functions that follow a power law of the form

$$f = \frac{L(v)}{|v|^b}$$

where v is the frequency, b is a real number included in the interval $[-2, 2]$, and $L(v)$ is a positive slowly varying or constant function of v. When $b = 0$, we have white noise or uncorrelated noise. When $b \neq 0$, we have colored noise that has short-term autocorrelation (Kasdin 1995).

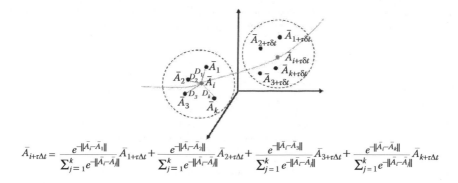

$$\bar{A}_{i+\tau\Delta t} = \frac{e^{-\|\bar{A}_i - \bar{A}_1\|}}{\sum_{j=1}^{k} e^{-\|\bar{A}_i - \bar{A}_j\|}} \bar{A}_{1+\tau\Delta t} + \frac{e^{-\|\bar{A}_i - \bar{A}_2\|}}{\sum_{j=1}^{k} e^{-\|\bar{A}_i - \bar{A}_j\|}} \bar{A}_{2+\tau\Delta t} + \frac{e^{-\|\bar{A}_i - \bar{A}_3\|}}{\sum_{j=1}^{k} e^{-\|\bar{A}_i - \bar{A}_j\|}} \bar{A}_{3+\tau\Delta t} + \frac{e^{-\|\bar{A}_i - \bar{A}_k\|}}{\sum_{j=1}^{k} e^{-\|\bar{A}_i - \bar{A}_j\|}} \bar{A}_{k+\tau\Delta t}$$

FIGURE 10.20 Sketch and formula of the nonlinear local predictor.

In [10.71], $W\left(\bar{A}_j, \bar{A}_i\right)$ is a nonlinear function that depends on the Euclidean distances $d_j = \left\|\bar{A}_i - \bar{A}_j\right\|$ (with $j = 1, ..., k$) as indicated in [10.72]

$$W\left(\bar{A}_j, \bar{A}_i\right) = \frac{e^{-\left(\|\bar{A}_i - A_j\|\right)}}{\sum_{j=1}^{k} e^{-\left(\|\bar{A}_i - A_j\|\right)}} \qquad [10.72]$$

For the quantitative comparison of the predictions with the real data, we calculate the correlation coefficient C:

$$C = \frac{N_{ts} \sum_i (A_{i,\text{exp}} A_{i,\text{pred}}) - \sum_i (A_{i,\text{exp}}) \sum_i (A_{i,\text{pred}})}{\left[\left(N_{ts} \sum_i (A_{i,\text{exp}}^2) - \left(\sum_i A_{i,\text{exp}}\right)^2\right)\left(N_{ts} \sum_i (A_{i,\text{pred}}^2) - \left(\sum_i A_{i,\text{pred}}\right)^2\right)\right]^{1/2}} \qquad [10.73]$$

In [10.73], N_{ts} is the number of testing data, $A_{i,\text{exp}}$ is the i-th experimental value, and $A_{i,\text{pred}}$ is the i-th predicted value. The correlation coefficient always lies between +1 and −1. When C is close to 1, there is a high degree of correlation. If C is negative, data $A_{i,\text{exp}}$ and $A_{i,\text{pred}}$ are anti-correlated. When C is nearly zero, the predicted and the experimental data are independent, and the predictions are not reliable. The procedure is repeated for increasing value of the prediction time $\tau\Delta t$. A decreasing trend of C versus $\tau\Delta t$ is a proof that the aperiodic time series is not uncorrelated noise. On the other hand, if C is independent of $\tau\Delta t$, the time series is white noise. Predictions of uncorrelated noise have a fixed amount of error, regardless of how far, or close, into the future one tries to project.[11]

TRY EXERCISE 10.15

10.8.2 Prediction of the Chaotic Time Series

The beginning of "modern" time series prediction is set at 1927, when Yule invented the autoregressive technique in order to predict the annual number of sunspots. His model predicted the next value as a weighted sum of the previous observations of the series (Yule 1927). In the half-century

---

[11] A deterministically chaotic time series may be distinguished from colored noise when the correlation coefficient obtained by the nonlinear local predictor is significantly better than the corresponding C obtained by the best-fitting autoregressive linear predictor (Sugihara and May 1990) where the predicted value $\bar{A}_{i+\tau\Delta t}$ is a linear function of previous values. For instance, $\bar{A}_{i+\tau\Delta t} = b\bar{A}_i + \varepsilon_{i+\tau\Delta t}$, which is a first-order autoregressive model (read also next paragraph).

following Yule, the reigning paradigm remained that of linear models driven by noise (Gershenfeld and Weigend 1993). An autoregressive model of order N looks like equation [10.74]:

$$x_t = \sum_{i=1}^{N} a_i x_{t-i} + \epsilon_t \qquad [10.74]$$

In [10.74], $a_i \left(i = 1, \ldots, N\right)$ are the parameters of the model and ϵ_t is white noise. Linear time series models have two particularly desirable features; they can be understood in depth, and they are straightforward to implement. However, they have a relevant drawback; they may be entirely inappropriate for even moderately complicated systems. Two crucial developments in aperiodic time series prediction occurred around 1980. The first development was the state-space reconstruction by the time-delay embedding; the second was the research line of machine learning, typified by the nonlinear artificial neural networks, which can adaptively explore a large space of potential models. Both developments were enabled by the availability of powerful computers that allowed much longer time series to be recorded and more complex algorithms to be used. Since the 1980s, several models have been proposed to understand and predict aperiodic time series. Such investigations interest many disciplines, such as meteorology, medicine, economy, engineering, astrophysics, geology, chemistry, and many others. In 1991, Doyne Farmer, head of the Complex Systems Group at the Los Alamos National Laboratory, quit his job and cofounded, along with his longtime friend and fellow physicist, Norman Packard, a firm called Prediction Company. The mission of their new firm was to develop fully automated trading systems, based on predictive models of markets. In the same year, the Santa Fe Institute organized a competition, the Santa Fe Time Series Prediction and Analysis Competition, to compare different prediction methods (Gershenfeld and Weigend 1993). Six time-series data sets were proposed: fluctuations of a far-infrared laser (data set A); physiological data from a patient with sleep apnea (data set B); currency exchange rate data (data set C); a numerically generated series (data set D); astrophysical data from a variable star (data set E); and Bach's final fugue (data set F). The main benchmark was data set A consisting of 1000 points and with 100 points in the future to be predicted by the competitors. The winner was E. A. Wan, who used a finite impulse response neural networks for autoregressive time series prediction. In 1998, there was the K.U. Leuven Competition within an international workshop titled "Advanced Black-Box Techniques for Nonlinear Modeling: Theory and Applications" (Suykens and Vandewalle, 1998). The benchmark was a time series with 2000 data sets generated from a computer simulation of Chua's electronic circuit (read Box 10.2 of this Chapter for more information about the Chua circuit). The task was to predict the next 200 points of the time series. The winner was J. McNames who used the nearest trajectory method, which incorporated local modeling and cross-validation techniques. More time series prediction competitions have been organized in the twenty-first century, several international symposia on forecasting have been arranged and research groups focused on time series prediction, have been born. Usually, the best predictions have been guaranteed by artificial neural network methods (see, e.g., Gentili et al. 2015 where an artificial neural network is compared with Fuzzy logic and the nonlinear local predictor).

10.8.2.1 Artificial Neural Networks

Artificial neural networks (ANNs) are algorithmization architectures simulating the behavior of real nerve cells networks in the central nervous system (Fausett 1994; Hassoun 1995). They are well suited to predict chaotic and stochastic time series, but also solve problems that are complex, ill-defined, highly nonlinear, and to recognize variable patterns (also read Chapters 12 and 13). There are infinite ways of organizing a neural network, although there are just four ingredients needed to build one of them.

The first ingredient is its architecture or connection patterns. Based on the architecture, ANNs are grouped into two categories: (a) feed-forward networks, where graphs have no loops, and (b) feedback (or recurrent) networks, where loops occur because of feedback connections. In feed-forward networks, neurons are organized into layers that have unidirectional connections between them.

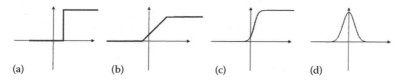

FIGURE 10.21   Examples of activation functions: (a) threshold, (b) piecewise linear, (c) logistic, (d) Gaussian functions.

The second ingredient to generate an ANN is the ensemble of its activation functions that transform the inputs of a node into output values. A mathematical neuron computes a weighted (w_i) sum of n inputs, x_i $(i = 1, 2, ..., n)$, and generates an output through an activation function $f(.)$

$$y = f\left(\sum_{i=1}^{n} w_i x_i \right)$$ [10.75]

Examples of $f(.)$ are the threshold, piecewise linear, logistic, and Gaussian functions shown in Figure 10.21.

The third ingredient is the cost function that estimates if the output is acceptable. The most frequently used cost function is the squared error $E(t)$

$$E(t) = \left\| out(t) - target(t) \right\|^2$$ [10.76]

where:

$out(t)$ is the output at time t calculated from the input recorded at time t,
$target(t)$ is the desired output at t.

The fourth and last ingredient is the training algorithm, also known as the learning rule, which modifies the parameters w_i to minimize the chosen cost function. There are three types of learning paradigms: (1) supervised, (2) unsupervised, (3) hybrid learning. In supervised learning, the network is provided with a correct output. Weights are determined to allow the network to produce answers as close as possible to the known correct answers. Unsupervised learning does not require a correct answer associated with each input pattern in the training data set. It explores the underlying structure of the data, or correlations between patterns in the data, and organizes patterns into classes from these correlations. Finally, hybrid learning combines supervised and unsupervised learning. Part of the weights is usually determined through supervised learning, while the others are obtained through unsupervised learning. There are many learning rules. The most frequently used in time series prediction are the error-correction rules within the supervised learning paradigm. They iteratively update the weights w_i by taking a small step (parametrized by the learning rate η) in the direction that decreases the squared error [10.76] the most. The updated weight $\tilde{w}_i$ will be obtained from the previous weight w_i through the algorithm [10.77]:

$$\tilde{w}_i = w_i - \eta \frac{\partial E(t)}{\partial w_i} = w_i + 2\eta x_i \left(out - target \right)$$ [10.77]

When in the network there are hidden layers, the error-correction rule transforms in back-propagation one (Rumelhart et al. 1993). In time series prediction, the available data are often divided into two sets: training and testing sets. The data of the training set are used during the learning stage when the weights of the network are optimized. The data of the testing set are used to check whether the network is suitable to make reliable predictions in the future or not.

BOX 10.2   CHUA'S ELECTRONIC CIRCUIT

The Chua's circuit is an electronic circuit that gives rise to a chaotic dynamic. It was invented in 1983, by Leon Chua, an American electrical engineer and computer scientist working at the University of California, in Berkley, since 1971. Chua wanted to design a concrete laboratory circuit that could be formally described by a system of differential equations close to that formulated by Lorenz (remember equations [10.36–10.38]). The final goal was that of demonstrating that chaos is a physical phenomenon and not an artifact of computer simulations and round-off errors in computations. Chua's circuit is shown in the following figure B10.2.

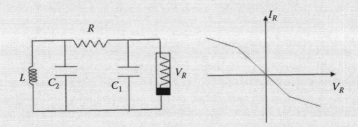

FIGURE B10.2   Chua circuit (on the left) and the function relating the current and voltage of the Chua diode (on the right).

It consists of five elements. Four of them are traditional linear passive components, which are one inductance ($L > 0$), two capacitors and one positive resistance. Interconnection of passive elements gives rise to trivial dynamics, with all element voltages and currents decaying to zero over time. The most straightforward circuit that gives rise to oscillatory or chaotic dynamics requires, at least, one locally active nonlinear element, such as the Chua's diode V_R. Chua's diode must be characterized by a nonlinear current (I_R) versus voltage (V_R) function, such as that depicted in the right part of the earlier figure. There exist several generalized versions of the Chua's Circuit. The Chua's Circuit has been used as a physical source of pseudo-random signals, and in numerous experiments on synchronization studies, and simulations of brain dynamics. Arrays of Chua's circuits have been used to generate 2-dimensional spiral waves, 3-dimensional scroll waves, and stationary patterns, such as Turing patterns (Chua 1998).

10.9   MASTERING CHAOS

The deterministic optimism, flourished at the beginning of the nineteenth century and promoted by Pierre-Simon Laplace, was shattered by two events. First, the formulation of the "Uncertainty Principle" by Heisenberg in 1927 and regarding the microscopic world. Second, the discovery of chaotic dynamics in the macroscopic world. In the second half of the nineteenth century, it was Henri Poincaré the first who realized the existence of nonlinear macroscopic dynamics extremely sensitive to the initial conditions. Poincaré was studying the three-body problem interacting through the gravitational force, and he discovered the sensitivity to the initial conditions when he tried to solve the case wherein the third body had a tiny mass (M_3) compared with the other two ($M_3 \ll M_2 < M_1$). After many years of research, we now know that chaotic systems are common in nature. Sometimes, we do not see the catastrophic effects of chaos only because those effects manifest after a long time. For example, the Solar System is intrinsically chaotic, but "astronomically" stable. In fact, for collisions between planets, such as Mercury and the Sun or Mercury and Venus, we should wait for at least 10^9 years (Cencini et al. 2009).

In the previous paragraphs, we have learned that extreme sensitivity to tiny perturbations characterizes chaotic systems. This feature, labeled as "butterfly effect," has been considered troublesome, for a long time. Since nobody can predict precisely how chaotic systems evolve over long periods, everybody has dealt with them in just one way: avoiding chaotic dynamics as long as it has been possible.

The first who realized that the "butterfly effect" may be fruitful in practical situations was John von Neumann in 1950 (Shinbrot et al. 1993). If it is true that the climate is a chaotic system, then, von Neumann thought, small, carefully chosen atmospheric disturbances could lead to the desired large-scale change in weather. We still struggle to find out what might be those tiny perturbations that would reduce the average temperature of our planet, which has been heating up in the last decades (Salawitch et al. 2017). Nevertheless, the von Neumann's idea is in principle sensible. In fact, the "butterfly effect" permits the exploitations of tiny perturbations to master chaotic systems. This possibility is tough to pursue in case of multi-dimensional chaotic systems, like the climate, but it is much easier in the case of low-dimensional chaotic systems. For example, in 1985, NASA scientists succeeded in sending their spacecraft ISEE-3/ICE more than 80 million km across the Solar System, achieving the first scientific cometary encounter, by using only small amounts of residual hydrazine fuel. This feat was made possible by the "butterfly effect" on the dynamic of the three-body system, composed of the Earth, Moon and the spacecraft. It would not have been possible in a nonchaotic system (Farquhar et al. 1985).

Why is it so appealing mastering chaotic dynamics? Because chaotic systems have a wealth of dynamical solutions. In fact, the skeleton of a chaotic attractor is a collection of an infinite number of periodic orbits, each one being unstable. Every unstable periodic orbit gives a specific performance. The overall chaotic orbit has a performance that is a weighted average of the performances attained by the single periodic orbits. There are periodic orbits that give better performances than the weighted average. Therefore, if the goal is to optimize the performance of the system, it is useful to select the unstable high performance periodic orbits.

How is it possible to succeed in this challenge? It seems like taming a crazy horse! The dynamics in the chaotic attractor is ergodic, which means that during its time evolution, the system visits the neighborhood of every point of the unstable periodic orbits embedded within the chaotic attractor. Therefore, there are two main strategies for mastering a chaotic dynamic: one based on feedback and the other on non-feedback methods (Boccaletti et al. 2000). The first approach includes those means that select the perturbation based upon a knowledge of the state of the system. The dynamic of the system is observed for suitable learning time, and when the ergodic chaotic orbit comes close to the desired unstable periodic orbit, a small "kick" is given to place the system on or very close to the desired periodic orbit. The small "kick" could be a tiny perturbation to either a control parameter of the system or a system state variable accessible to the operator. Even if the "kick" is effective, the system, quite easily, will move far apart from the desired unstable periodic orbit due to the noise and the intrinsic instability of the periodic orbit. Therefore, other small "kicks" will be needed to be reapplied to reposition the system orbit closer to the desired periodic orbit. By following this procedure, continuously, the system can be kept close to the desired periodic orbit indefinitely. The second strategy for mastering chaotic dynamics consists in perturbing periodically or stochastically the system, even without knowing the actual dynamical state, producing drastic changes and leading eventually to the stabilization of some periodic behavior. The limit of the second strategy is that it is not goal-oriented, and the operator cannot decide the final periodic orbit.

What is impressive is that thanks to chaos, it is possible to produce an infinite number of dynamical behaviors using the same system, with the only help of adequately chosen tiny perturbations. This action is not the case for a non-chaotic system; typically, small disturbances can only change their dynamics slightly. We need disturbances of the same order of magnitude of the values that the dynamical variables assume in the unperturbed evolution to steer a non-chaotic system in the direction we desire.

10.9.1 APPLICATIONS

The idea of mastering chaos has stimulated applications in widely diverse fields of study; for example, in mechanics (by controlling the chaotic vibrations of a magneto-elastic ribbon [Ditto et al. 1990]), in fluid mechanics (by controlling chaotic convection [Singer et al. 1991]), in electronics (by controlling a chaotic diode resonator circuit [Hunt 1991]), in nonlinear optics (by controlling the chaotic output of a laser [Roy et al. 1992]), in chemistry (by controlling the Belousov-Zhabotinsky [Petrov et al. 1993] or the combustion [Davies et al. 2000] or the peroxidase-oxidase [Lekebusch et al. 1995] reactions when they run in chaotic regime), in physiology (by controlling chaos in heart beating [Garfinkel et al. 1992], and maintaining the chaotic state in the neuronal activity of hippocampal slices that become periodic in the case of epileptic seizures [Schiff et al. 1994]).

Two applications have attracted considerable attention in the scientific community over the past few years; namely the control of chaotic dynamics for communicating and for computing.

10.9.1.1 Communication by Chaotic Dynamics

When we studied the Kolmogorov-Sinai entropy, we learned that any chaotic system could be viewed as an information source that produces signals. In fact, a chaotic dynamic is a point that follows a "strange" trajectory in its phase space. If the phase space is partitioned in many pieces, each labeled by a different symbol, the chaotic system becomes a symbol source, because a symbolic orbit is obtained by writing down the sequence of symbols corresponding to the successive partition elements visited by the point in its orbit. Since the chaotic dynamic is a continuous-time waveform source, it can also be transformed into a digital signal source. Slaving the output of a chaotic oscillator allows for the transmission of the desired message, namely a sequence of desired "high" and "low" values encoded with binary symbols "1" and "0", respectively (Boccaletti et al. 2000).

There is another approach for exploiting chaos in communication, and it is based on chaos synchronization (Boccaletti et al. 2002). Two chaotic systems starting from different initial conditions evolve in an unsynchronized manner. The feeding of the right bias from one system to another can push the two systems into a synchronized state. The two chaotic systems remain in step with each other during the transmission of a signal masked by the chaotic contribution. When the receiver synchronizes to the transmitter, the message is decoded by subtraction between the signal sent by the transmitter and its copy generated at the receiver. Finally, the true message is decoded (Boccaletti et al. 2000).

10.9.1.2 Computing by Chaotic Dynamics

In principle, any chaotic system can mimic a Turing machine. A Turing machine (Turing 1936) is a computing device consisting of a programmable read/write head with a paper tape passing through it. It has inspired the design of electronic computers that are currently based on the Von Neumann's architecture (Burks et al. 1963). Such architecture consists of four components (remember what we have learned in Chapter 2): a processor, making the computation; a memory, storing data and instructions; an information exchanger, allowing the flow of information in and out of the computer; an information bus, connecting the other three elements. The tape of the Turing machine works as both the memory and the information exchanger. In fact, it is divided into squares, each square bearing a single symbol (0 or 1), and it is of unlimited length. It serves both as a vehicle for input and output, and as working memory for storing the intermediate results of a computation. The read/write head is programmable, and it works as both the processor and the bus bearing information from the head to the tape and vice versa. Although the head can perform a limited number of operations, like reading and writing on the square of the tape under the head, moving the head and halting, the Turing machine is capable of computing everything, if appropriately programmed.

A chaotic system can mimic a Turing machine. The time series that it generates acts as the tape, which acts as information exchanger and memory, whereas the strange attractor that the chaotic system traces when it evolves in its phase space plays as the head-processor of the peculiar chaos-based Turing machine. In fact, it has been demonstrated that nonlinear deterministic systems can emulate all the fundamental binary logic functions[12] through a threshold-based morphing mechanism (Ditto et al. 2010). The morphing mechanism grounds in three steps. For a chaotic system, whose state is represented by the variable x, the first step is to consider the input. If x' represents input 0, then, in the second step, the temporal update of the state of the system, $f(x')$ (where $f(x)$ is the function describing the time evolution of the system) is the output. Finally, in the third step, after fixing a threshold value, x_{th}, the output $f(x')$ is transformed in binary values: if $f(x') \leq x_{th}$, the output is 0, whereas if $f(x') > x_{th}$, the output is 1. For encoding input 1, a specific increment Δx will be added to x', and $f(x' + \Delta x)$ will be its output. The most promising property of chaos computing is the ability to reconfigure the chaotic dynamic to any logic gate by exploiting the techniques developed for controlling chaos.

So far, chaos-based computing has been implemented, almost exclusively, by conventional Complementary Metal-Oxide Semiconductor (CMOS)-based Very Large Scale Integrated (VLSI) circuits. Since nonlinear systems that exhibit chaotic dynamics are abundant in nature, we may expect that several other physical and chemical systems can implement chaos-based computing; for example, the "Hydrodynamic Photochemical Oscillators" we have studied in this chapter (Hayashi et al. 2016; Gentili et al. 2017b).

10.10   KEY QUESTIONS

- When can we observe chaotic dynamics?
- Which are the possible evolutions of a system described by linear differential equations?
- When does a double pendulum originate chaotic dynamics?
- What is the equation that describes the dynamics of a population as a function of the food available?
- Describe the possible solutions of the logistic map and the features of its bifurcation diagram.
- When do convective rolls emerge?
- What is the difference between the Rayleigh and Marangoni numbers?
- How does Entropy Production change in the case of convection?
- Why is convection so important?
- Describe the features of the Lorenz's model.
- What is a "Hydrodynamic Photochemical Oscillator?"
- How do we recognize chaotic time series?
- Can we predict chaotic time series?
- Is it possible to master chaotic dynamics?
- Which are possible applications of mastering chaos?

10.11   KEY WORDS

Linear and nonlinear equations; Double pendulum; Logistic map; Bifurcation diagram; Rayleigh and Marangoni numbers; MaxEP; Butterfly effect; Lyapunov exponent; Strange attractors; Transient Chaos; Hamiltonian Chaos; Dissipative Chaos; Artificial Neural Networks.

---

[12] Any Boolean circuit can be built by an adequate connection of NOR and NAND logic gates. This implies that universal computing is feasible if the NOR and NAND gates can be implemented.

10.12   HINTS FOR FURTHER READING

- An enjoyable and introductory reading about chaos is the book by Gleick (1987). An easy and pleasant introduction to Chaos has been written by Feldman (2012), who has also proposed interesting and basic courses about dynamical systems and chaos in the Complexity Explorer website.
- The subject of the analysis of aperiodic time series applied to physics, engineering, biology, and medicine can be studied in deep with the book by Kantz and Schreiber (2003), the book edited by Kantz et al. (1998), which focused on physiological data, and that by Abarbanel (1996). Another text that presents the role of chaos in physiological rhythms is that by Glass and Mackey (1988). A book dealing with the role of Chaos in the brain is that edited by Lehnertz et al. (2000).
- Schöll and Schuster (2008) have edited a handbook for whoever is interested in chaos control.
- The idea that nonlinear dynamics can be "message sources" has been developed also by Crutchfield and Young (1989), and Crutchfield (1994).

10.13   EXERCISES

10.1. Include all the following equations in two sets: the set of linear equations and the set of nonlinear equations. $(a) \sin x = y$; $(b) (dy/dt) + y + lny = 0$; $(c) 2(dy/dt) + 3y = 0$; $(d) 5(d^2y/dt^2) - (dy/dt) + 2y = 0$; $(e) 5(d^2y/dt^2)^2 - (dy/dt) + 2y = 0$; $(f) x^2 + x + 2 = y$; $(g) 2x + 3 = y$; $(h) (dy/dt) - xy + 2x = 0$; $(i) y = lnx - x$.

10.2. Via numerical integration (using, e.g., the MATLAB ode45 solver), determine the evolution of the double pendulum for $\theta_1(t = 0) = -0.100$, $\theta_2(t = 0) = 0.520$, and $\theta_1'(t = 0) = \omega_1 = 0$, $\theta_2'(t = 0) = \omega_2 = 0$. The motion (Shinbrot et al. 1992) is expressed by the following first order equations:

$$\theta_1' = \omega_1$$

$$\theta_2' = \omega_2$$

$$\omega_1' = \frac{g\left(sin\theta_2 \cos(\Delta\theta) - \mu sin\theta_1\right) - \left(L_2\theta_2'^2 + L_1\theta_1'^2 \cos(\Delta\theta)\right)\sin(\Delta\theta)}{L_1\left(\mu - cos^2(\Delta\theta)\right)}$$

$$\omega_2' = \frac{g\mu\left(sin\theta_1 \cos(\Delta\theta) - sin\theta_2\right) + \left(\mu L_1\theta_1'^2 + L_2\theta_2'^2 \cos(\Delta\theta)\right)\sin(\Delta\theta)}{L_2\left(\mu - cos^2(\Delta\theta)\right)}$$

where $\Delta\theta = \theta_1 - \theta_2$ and $\mu = (m_2 + m_1)/m_2$.

Compare the solution with those shown in Figure 10.3.

10.3. Via numerical integration (using, e.g., the MATLAB ode45 solver), determine the evolution of the double pendulum for $\theta_1(t = 0) = 1.57$, $\theta_2(t = 0) = 1.57$ (case (a) in Figure 10.22), and $\theta_1'(t = 0) = 0$, $\theta_2'(t = 0) = 0$. For the equations of motion see exercise 10.2. Compare the solution of this exercise with the behavior obtained in the 3° run of Figure 10.4 (its initial conditions are shown in graph (b) of Figure 10.22).

10.4. Integrate the logistic equation [10.10].

10.5. Use the logistic map and determine the evolution of two populations that start from two slightly different initial conditions: $x_0' = 0.30000$ and $x_0'' = 0.30001$, respectively. Use two different values of growth rate: (a) $r = 3$, and (b) $r = 4$. What differences do you ascertain in the two cases? Calculate the abundance of the populations for, at least, 50 iterations. You can use either a spreadsheet or MATLAB or any other software for the calculations.

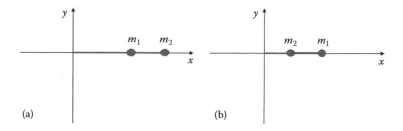

FIGURE 10.22   (a and b) Two different initial conditions for the double pendulum.

10.6. Another example of a unimodal map is $x_{n+1} = rsin(\pi x_n)$, with $0 \leq r \leq 1$ and $0 \leq x \leq 1$. Verify that it is a unimodal map and generate the bifurcation diagram (e.g., by using MATLAB). Compare this latter bifurcation diagram with that built for the logistic map.

10.7. Apply equation [10.17], defining the Feigenbaum's constant, to estimate the value of r at which the logistic map becomes chaotic (r_∞). For the calculation, use the coordinates of the following two bifurcation points: a cycle with period 16 is at $r_{k-1} = 3.564407$, and the subsequent bifurcation point, originating a cycle with period 32, is at $r_k = 3.568750$. Calculate $r_{k+1} \approx r_\infty$.

10.8. Libchaber investigated the transition of the convective motion from the laminar to the turbulent regime by using liquid helium at 3 K. By knowing that the isobaric thermal expansion coefficient is $\beta = 6.15 \times 10^{-2} \, K^{-1}$ (Lide 2005), dynamic viscosity $\mu = 2.7 \times 10^{-4} \, cm^2 s^{-1}$, thermal diffusivity $\alpha = 4.3 \times 10^{-4} \, cm^2 s^{-1}$ (Libchaber and Maurer 1978), $g = 9.8 \, m^2 s^{-1}$, and reminding that the critical Ra value for observing spontaneous formation of cylindrical convective rolls is 1708, calculate the thermal gradient needed when the height of the cell is 0.1 cm and when it is 1 cm long.

10.9. In the experiments described by Libchaber et al. (1982), the authors increased constantly the thermal gradient between the two plates of the cell. They wanted to confirm the phenomenon of the period doubling cascade as a route to chaos as predicted for the unimodal maps by Feigenbaum. Using equation [10.17], and knowing that the critical values of the ratio Ra_k/Ra_c for $k = 2, 3$ and 4 were 3.4850, 3.6183 and 3.6486, respectively, calculate the value of the parameter δ. Estimate the final uncertainty by using the propagation formula for the "maximum a priori absolute error" (see Appendix D) knowing that the uncertainty in each ratio (Ra_k/Ra_c) is ±0.0005.

10.10. This exercise consists in observing "in vivo" the Bénard cells described in Box 10.1 of this chapter. Go to your wet laboratory, wear a white coat, gloves, safety glasses, and recover the following equipment: a hot plate, a small metallic pan, aluminum powder (or bronze powder) and silicone oil. There are many types of silicone oils, which are all polysiloxanes, thus, polymers having a chain with alternating silicon and oxygen atoms, and with organic side chains. They cover a broad range of viscosities. It is convenient to use a silicone oil with a dynamic viscosity of at least 0.5 cm²/s (Van Hook and Schatz 1997). Mix the aluminum powder and the oil. Pour the mixture into the metallic pan to form a thin layer of fluid (around 1 mm thick). Heat the pan gently, and observe the pattern formation (it could be enough to heat the layer between 50°C and 100°C). You do not need very high temperatures to see Bénard cells. It is important to avoid burning your oil. After observing a pattern, destroy it by shaking the mixture. Then leave the system alone and look at the recovery of the pattern. What does it mean? What is the shape of the Bénard cells? Does the shape of the cells depend on the thermal gradient?

10.11. Solve the system of three differential equations [10.36–10.38] of Lorenz's model numerically after fixing $\sigma = 10$, $b = 8/3$, $r = 28$. Start the calculations by using two slightly different initial conditions that are both close to the steady state solution S_+: $[X(0); Y(0); Z(0)] =$ [13.824; 15.368; 32.474] and $[X(0)'; Y(0)'; Z(0)'] = [13.824; 15.369; 32.474]$. Perform the calculation in the time interval [0–30]. Calculate (I) how the Euclidean distance $\|\delta(t)\|$ between the two trajectories changes over time, and (II) the Lyapunov exponent. Repeat the calculations using another pair of initial conditions that are close to the steady solution S_0: $[X(0)''; Y(0)''; Z(0)''] = [0; 1.000; 0]$ and $[X(0)'''; Y(0)'''; Z(0)'''] = [0; 1.001; 0]$. What is the difference between the two situations?

10.12. Solve the system of nonlinear differential equations [10.36–10.38] of the Lorenz's model numerically after fixing $\sigma = 10$, $b = 8/3$ and $r = 20$. Start the calculations by using two initial conditions that differ slightly: [14.0 12.0 10.0] and [14.0 12.1 10.0]. What do you observe?

10.13. Solve the system of nonlinear differential equations [10.36–10.38] of the Lorenz's model numerically after fixing $\sigma = 10$, $b = 8/3$ and $r = 350$. Start the calculations by using initial conditions that differ appreciably. What do you observe?

10.14. In this exercise, you make an experience of the "butterfly effect" in action. Go to your wet laboratory, wear a white coat, gloves and safety glasses. If you have a spectrofluorimeter in your laboratory, you can use 9,10-dimethylanthracene (DMA) as a probe of the convective motions and follow the list A of instructions. On the other, if you have a UV-visible spectrophotometer, but also a UV lamp for irradiation, you can use a photochromic compound and follow the list B of instructions. The photochromic species could be the spirooxazine depicted in Figure 10.18 or any other compound as long as it colors at room temperature under the photon flux of your lamp. The two possible experimental set-ups are sketched in Figure 10.23, wherein I_{EM} is the light emitted by DMA, whereas I_{probe} is the intensity of the probe ray emitted by the lamp of the spectrophotometer in case we use a photochromic compound.

List A of instruction.

A_1–Prepare at least 10 mL of a 9,10-dimethylanthracene (DMA) solution in chloroform having a concentration of about 4×10^{-5} M.

A_2–Pour 3 or slightly more milliliters of the solution into a fluorometric cuvette having 1 cm as path length. Measure the height of the solution (it should be ≥ 3 cm).

A_3–Record the UV-visible absorption spectrum of the solution.

A_4–Collect its emission spectrum by exciting at 260 nm (select the widths of the slits and the time integration to avoid saturation of the detector).

A_5–Maintaining the cuvette inside the sample holder of the fluorimeter, uncap it, and wait for a time interval of 10′–15′. Use a timer to determine the exact period waited

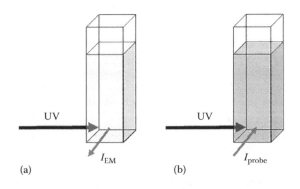

(a)   (b)

FIGURE 10.23 Experimental set-up when we use a fluorescent probe in (a) or a photochromic probe in (b).

before executing the next step. Meanwhile, monitor the environmental conditions in your room, by measuring temperature, pressure, and humidity.

A_6–Run an experiment wherein you collect how the intensity emitted by DMA, at a specific wavelength, changes over time (for at least 2000 s) under continuous irradiation at 260 nm. Store data every 1 s after fixing the integration time to 1 s.

A_7–At the end of the experiment, cap the cuvette and measure the final height of the solution. Did the solution evaporate? How much? Estimate the power (Pw) absorbed by the solution through the following formula: $Pw = \Delta H_{vap} \cdot \rho \cdot \Delta V / (MW) \cdot \Delta t$, where ΔH_{vap} is the heat of vaporization (31.4 kJ/mol), ρ the density (1.489 g/cm³), and MW the molecular weight (119.38 g/mol) of chloroform. ΔV is the amount of volume that evaporated, and Δt is the time when the cuvette was maintained uncapped.

A_8–Collect the absorption spectrum of the solution. What do you observe? Did it change respect to the spectrum recorded at point A_3?

A_9–Repeat the instructions from A_2 up to A_8 by using a fresh solution of DMA in chloroform. Do you observe perfect reproducibility, especially for the fluorescence signal recorded versus time?

A_{10}–Based on the profile of the fluorescence signal versus time, and the number of frequencies determined by calculating its Fourier Transform (see Appendix C), try to estimate the vertical thermal gradient by using the formula of the Rayleigh number [10.48]. For chloroform, $\beta = 1.107 \times 10^{-3}\,\mathrm{K^{-1}}$, $\mu = 4 \times 10^{-3}\,\mathrm{cm^2\,s^{-1}}$, $\alpha = 7.31 \times 10^{-4}\,\mathrm{cm^2\,s^{-1}}$.

List B of instructions.

B_1–Prepare at least 10 mL of the photochromic solution in acetone as the solvent, and at a concentration of about $9 \times 10^{-5}\,\mathrm{M}$.

B_2–Pour 3 or slightly more milliliters of the solution into a fluorometric cuvette having 1 cm as path length. Measure the height of the solution (it should be ≥ 3 cm).

B_3–Maintaining the cuvette inside the sample holder of the UV-visible spectrophotometer, uncap it and wait for a time interval of 10'–15'. Use a timer to determine the exact time waited before executing the next step. Meanwhile, turn on the lamp for irradiation and monitor the environmental conditions in your room by measuring temperature, pressure, and humidity.

B_4–The UV irradiation must be performed at 90° respect to the spectrophotometric rays. You can convey the UV radiation to the bottom part of the fluorometric cell through an optical fiber, as it is shown in Figure 10.19. If your spectrophotometer has a diode array as the detector, you can record the complete UV-visible spectra during irradiation at every single shot. Otherwise, if you have a traditional double- or single-beam spectrophotometer, with a photomultiplier as the detector, you should know the wavelength that corresponds to the maximum for the band into the visible, and then tune your monochromator to that wavelength. Run an experiment and record how the UV-visible spectrum or the absorbance at the selected wavelength evolves under constant irradiation in the UV. Store data every 3 s, and for at least one hour.

B_5–At the end of the experiment, cap the cuvette and measure the final height of the solution. Did the solution evaporate? How much? Estimate the power (Pw) absorbed by the solution through the following formula: $Pw = \Delta H_{vap} \cdot \rho \cdot \Delta V / (MW) \cdot \Delta t$, where ΔH_{vap} is the heat of vaporization (32.1 kJ/mol), ρ the density (0.7845 g/cm³), and MW the molecular weight (58.08 g/mol) of acetone. ΔV is the amount of volume that evaporated, and Δt is the time interval wherein the cuvette was maintained uncapped.

B_6–Repeat the instructions from B_2 up to B_5 by using a fresh solution of the photochromic compound in acetone. Do you observe perfect reproducibility, especially for the signal of absorbance vs. time?

B_7–Based on the profile of the absorbance vs. time and the number of frequencies determined by calculating its Fourier spectrum (see Appendix C), try to estimate the vertical thermal gradient by using the formula of the Rayleigh number [10.48]. For acetone, $\beta = 1.43 \times 10^{-3}\,K^{-1}$, $\mu = 4 \times 10^{-3}\,cm^2\,s^{-1}$, $\alpha = 9.48 \times 10^{-4}\,cm^2\,s^{-1}$.

10.15. The nonlinear time series analysis is being successfully applied in a variety of disciplines. For instance, in medicine. In fact, clinical diagnoses depend on the ability to record and analyze physiological signals, such as electrocardiograms (ECG), electroencephalograms (EEG), heart rate recordings, concentrations of hormones, et cetera. Typically, these signals are generated by processes that are nonlinear and nonstationary. An analysis of their features promises to be of clinical value, distinguishing states of normal or pathological functioning, and forecasting worsening conditions. The National Center for Research Resources of the National Institutes of Health has created a Research Resource for Complex Physiologic Signals (Goldberger et al. 2000). The resource consists of three elements. One is PhysioBank that is an archive of physiological signals. The second is PhysioToolkit that is a library of open-source software for physiological signal processing and analysis. The third is PhysioNet that is an online forum for the dissemination and exchange of biomedical signals and open-source software. In PhysioBank, you can find many time series, and you can play by analyzing their nonlinear features. For example, you can download a dataset of heart rate and chest volume recorded from a patient in the sleep laboratory of the Beth Israel Hospital in Boston (MA, USA) (Rigney et al. 1993; Ichimaru and Moody 1999). For its analysis, you can freely download the TISEAN software (Hegger et al. 1999). As a first attempt, you may try to determine the phase space of the time series "b1.txt" (https://physionet.org/physiobank/database/santa-fe/) by calculating the Mutual Information and/or the Autocorrelation, and the False Nearest Neighbors. After estimating the time delay (τ) and the embedding dimension (m), try to calculate the largest Lyapunov exponent by using the Kantz's method.

10.14   SOLUTIONS TO THE EXERCISES

10.1. A linear equation is an expression of the type: $y = a + b * x$, where x and y are the independent and dependent variables, respectively. The equation (g) is an example. When we plot a linear equation, it looks like a straight line. An equation is nonlinear when the independent variable does not appear simply at the first power. Nonlinear equations contain power functions, product functions and/or transcendental functions. Examples of nonlinear equations are (a), (f), and (i). A differential equation is linear when the dependent variables and their derivatives appear only to the first power, and there are not products of dependent variables. Examples are the equations (c) and (d). In equation (d) there is a second derivative of the dependent variable, but it is to the first power. On the other hand, the equation (e) is nonlinear because the second derivative is elevated to the power of 2. Other examples of nonlinear differential equations are (b) and (h).

10.2. The trends of the angles ϑ_1 and ϑ_2 over the first 20 seconds are shown in Figure 10.24.

The dynamics are periodic for both angles, in agreement with what we found in the first and second run plotted in Figure 10.3. A difference is that now the oscillations are anti-phase, whereas they were in phase condition in Figure 10.3. The trends for the spatial coordinates (x_1, y_1) and (x_2, y_2) are plotted as follows. They have been calculated by using trigonometry and the estimated values of θ_1 and θ_2 (Figure 10.25).

$$x_1 = L_1 \sin\theta_1$$

$$y_1 = -L_1 \cos\theta_1$$

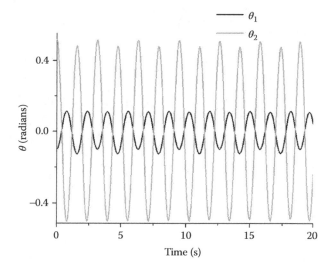

FIGURE 10.24   Trends of θ_1 and θ_2 over time.

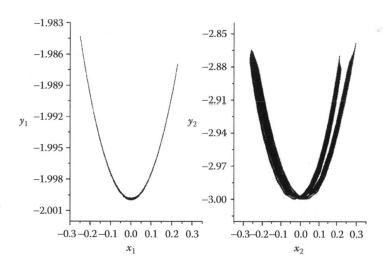

FIGURE 10.25   Profiles of the Cartesian coordinates for the masses of the double pendulum.

$$x_2 = x_1 + L_2 \sin \theta_2$$

$$y_2 = y_1 - L_2 \cos \theta_2$$

10.3. Since the initial values of θ_1 and θ_2 are large, the motion of the double pendulum is highly nonlinear. In fact, it is aperiodic. The trends for the spatial coordinates (x_1, y_1) and (x_2, y_2) are plotted in Figure 10.26.

It is interesting to compare the motion for $\theta_1(t = 0) = 1.57$ (90°), $\theta_2(t = 0) = 1.57$ (90°), with that for $\theta_1(t = 0) = 1.57$ (90°), $\theta_2(t = 0) = 4.71$ (270°). In the two conditions, the double pendulum has the same potential energy. Although the total energy of the Hamiltonian system is equal, the motion is significantly different, especially for mass 2 of the double pendulum (see Figure 10.27).

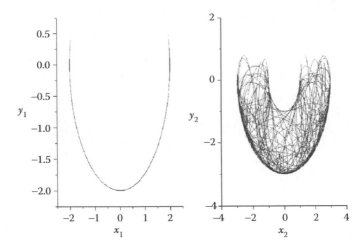

FIGURE 10.26   Profiles of the Cartesian coordinates for the two masses of the double pendulum.

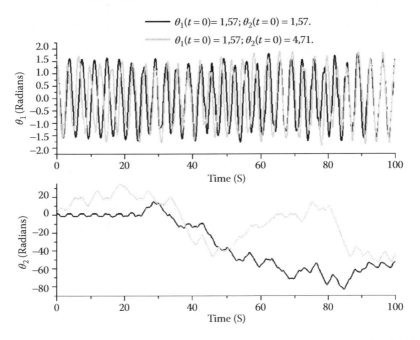

FIGURE 10.27   Trends of θ_1 and θ_2 over time for two initial conditions that correspond to the identical total energy of the double pendulum.

10.4. For the integration of the differential equation [10.10], $(dN/dt) = rN(1 - (N/K))$, first of all, we separate the variables:

$$\frac{KdN}{N(K-N)} = rdt$$

Then, we set

$$\frac{KdN}{N(K-N)} = \frac{A}{N}dN + \frac{B}{(K-N)}dN = \frac{\left[A(K-N) + BN\right]dN}{N(K-N)} = rdt$$

When $N = 0, AK = K$, i.e., $A = 1$. When $N = K, BK = K$, i.e., $B = 1$ Therefore,

$$\int_{N_0}^{N_t}\left(\frac{1}{N}\right)dN + \int_{N_0}^{N_t}\left(\frac{1}{K-N}\right)dN = \int_0^t rdt$$

Finally, we achieve the logistic function [10.11].

10.5. From the data reported into Table 10.1 and depicted in Figure 10.28, it is evident that when $r = 3$, the evolution of the population is not sensitive to the initial conditions. The two sequences oscillate between two values, and the discrepancy between the two series maintains equal to the initial value of 0.00001. An entirely different behavior emerges in the case of $r = 4$. The evolution is extremely sensitive to the initial conditions. The two sequences stay reasonably close together for the first ten iterations. Then, they diverge, and there is no way to tell that the sequences ever started off a mere 10^{-5} away from each other. When $r = 3$, the population evolves towards an ordered state, whereas when $r = 4$, the evolution is chaotic because it is aperiodic and really sensitive to the initial conditions.

TABLE 10.1

Values of the Populations According to the Logistic Map Calculated for 50 Iterations and Fixing the Parameter r Equal to 3 and 4, Respectively

	$r = 3$		$r = 4$	
0	0.30000	0.30001	0.30000	0.30001
1	0.63000	0.63001	0.84000	0.84002
2	0.6993	0.69929	0.5376	0.53756
3	0.63084	0.63085	0.99434	0.99436
4	0.69864	0.69864	0.02249	0.02244
5	0.63162	0.63163	0.08794	0.08775
6	0.69803	0.69802	0.32084	0.32019
7	0.63236	0.63237	0.87161	0.87068
8	0.69745	0.69744	0.44762	0.4504
9	0.63305	0.63305	0.98902	0.99016
10	0.6969	0.69689	0.04342	0.03898
11	0.6337	0.6337	0.16615	0.14985
12	0.69638	0.69637	0.55416	0.50958
13	0.63431	0.63432	0.98826	0.99963
14	0.69588	0.69588	0.04639	0.00147
15	0.63489	0.6349	0.17695	0.00586
16	0.69542	0.69541	0.58256	0.02331
17	0.63544	0.63545	0.97273	0.09108
18	0.69497	0.69496	0.1061	0.33115
19	0.63596	0.63597	0.37936	0.88595
20	0.69454	0.69454	0.94178	0.40416
21	0.63646	0.63647	0.21931	0.96326
22	0.69414	0.69413	0.68484	0.14158
23	0.63693	0.63694	0.86333	0.48614
24	0.69375	0.69374	0.47196	0.99923
25	0.63738	0.63739	0.99685	0.00307
26	0.69338	0.69337	0.01254	0.01225
27	0.63782	0.63782	0.04955	0.04842

(Continued)

TABLE 10.1 (*Continued*)
Values of the Populations According to the Logistic Map
Calculated for 50 Iterations and Fixing the Parameter *r*
Equal to 3 and 4, Respectively

	r = 3		*r* = 4	
28	0.69302	0.69302	0.18837	0.18428
29	0.63823	0.63824	0.61155	0.60129
30	0.69268	0.69267	0.95023	0.95896
31	0.63863	0.63863	0.18918	0.15742
32	0.69235	0.69234	0.61356	0.53055
33	0.63901	0.63901	0.94841	0.99627
34	0.69203	0.69203	0.1957	0.01488
35	0.63937	0.63938	0.62961	0.05863
36	0.69173	0.69172	0.93281	0.22078
37	0.63972	0.63973	0.25071	0.68815
38	0.69143	0.69143	0.75142	0.8584
39	0.64006	0.64006	0.74715	0.48621
40	0.69115	0.69115	0.75566	0.99924
41	0.64039	0.64039	0.73855	0.00304
42	0.69088	0.69087	0.77237	0.01213
43	0.6407	0.6407	0.70325	0.04793
44	0.69061	0.69061	0.83475	0.18254
45	0.641	0.64101	0.55176	0.59688
46	0.69035	0.69035	0.98928	0.96246
47	0.6413	0.6413	0.04241	0.14454
48	0.69011	0.6901	0.16243	0.49459
49	0.64158	0.64158	0.5442	0.99988
50	0.68987	0.68986	0.99219	4.69E-4

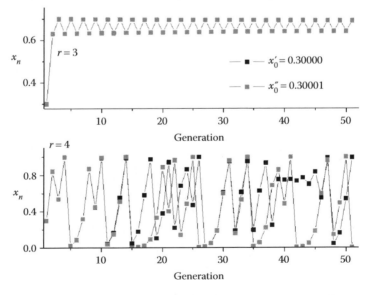

FIGURE 10.28 Evolutions of two populations for two different initial conditions: $x_0' = 0.30000$ (dark points) and $x_0'' = 0.30001$ (gray points), and for $r = 3$ (upper graph) and $r = 4$ (lower graph).

If you want to calculate the iterations by using MATLAB, you can use the following script:

```
Logistic map.m

numsteps=50;
x(1)=0.30000;
r=3.0;

for i=1:numsteps
     x(i+1)=r*x(i)*(1-x(i));
end;

plot(1:numsteps, x, 'bo-');
```

10.6. The unimodal map $x_{n+1} = rsin(\pi x_n)$ has a shape that is pretty similar to that of the logistic map (see Figure 10.29). It is smooth, concave down, and its maximum is equal to $x_{n+1} = r$ at $x_n = 0.5$ (in fact, its first derivative $y' = r\pi cos(\pi x_n)$ becomes null when $x_n = 1/2$).

The bifurcation diagram can be built by using the following script of MATLAB:

```
bifurcation.m

numr=1000;
startr=0.00;
endr=1.00;

r=linspace(startr, endr, numr);
skipnum=500;
num=500;

for j=1:length(r)
   x=0.1;

     for i=1:skipnum
         x=r(j) * sin(pi*x);
     end;
```

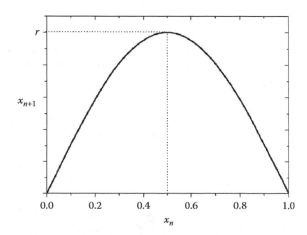

FIGURE 10.29 Profile of the unimodal map $x_{n+1} = rsin(\pi x_n)$.

```
        for i=1:num
           x=r(j) * sin(pi*x);
           results(j, i)=x;
        end;
    end;

    plot(r, results, "b.", "MarkerSize", 0.5);
    xlabel("r (Growth Rate)")
    ylabel("Final x Values (Population)")
    title ("Bifurcation Diagram for Unimodal Map")

    return;
```

The unimodal map $x_{n+1} = rsin(\pi x_n)$ exhibits a bifurcation diagram (built by discarding the first 500 iterations) that is pretty similar to that obtained in the case of the logistic map (compare Figure 10.30 with Figure 10.9). Both unimodal maps undergo period-doubling evolutions up to reach a chaotic regime. The chaotic regimes are interspersed with periodic windows.

10.7. We rewrite equation [10.17] in the following simplified form:

$$\delta \approx \frac{r_k - r_{k-1}}{r_\infty - r_k}$$

After rearranging it, we achieve

$$r_\infty \approx r_k + \left(\frac{r_k - r_{k-1}}{\delta}\right)$$

Introducing the numerical values of r_k, r_{k-1} and δ, the result is $r_\infty = 3.569680$. A better estimate can be obtained using values of r for larger k.

10.8. By rearranging equation [10.22] in the following form

$$\Delta T = \frac{(Ra_c)\mu\alpha}{\beta g h^3}$$

and introducing the values of the parameters listed in the text of the exercise, we find that when $h = 0.1$ cm, $\Delta T = 0.003$ K, whereas when $h = 1$ cm, $\Delta T = 3 \times 10^{-6}$ K. Libchaber used a cell with $h = 0.1$ cm because its microbolometers could detect thermal gradients of the order of one thousandth of degree K. He did not use a cell with $h = 1$ cm, because he could not measure the thermal gradients with an accuracy of one millionth of degree K.

10.9. From the list of the critical ratio values, which are Ra_4/Ra_c, Ra_3/Ra_c, and Ra_2/Ra_c, we can estimate the parameter δ, through the equation as follows:

$$\delta = \frac{(Ra_3/Ra_c) - (Ra_2/Ra_c)}{(Ra_4/Ra_c) - (Ra_3/Ra_c)} = \frac{3.6183 - 3.4850}{3.6486 - 3.6183} = 4.4$$

By applying the formula of the error propagation, we find that the uncertainty in δ is 0.2.

Therefore, $\delta = 4.4 \pm 0.2$ in reasonable agreement with the theoretical result of 4.669... Note that it is not an easy task that of determining the appearance of a bifurcation in the presence of experimental noise. Usually, it is not possible to measure more than about five period-doublings (Cvitanovic 1989). Nevertheless, the agreement between experiments and theory is remarkable.

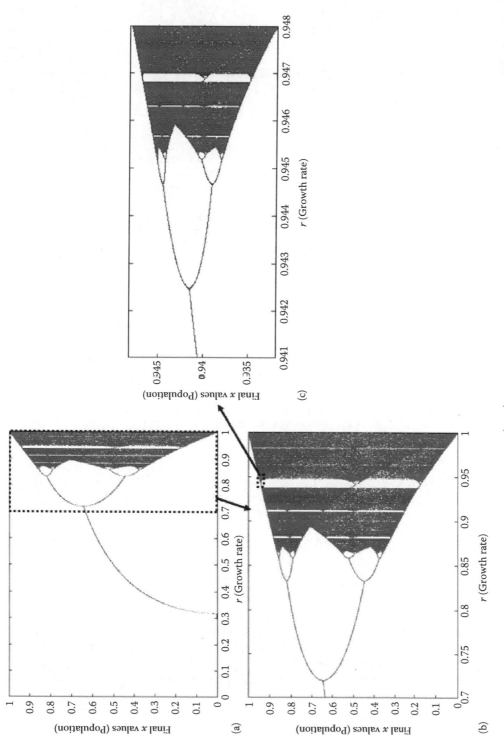

FIGURE 10.30 Bifurcation diagram of the unimodal map $x_{n+1} = r\sin(\pi x_n)$. In (a), r assumes values between 0 and 1; in (b) r is included between 0.7 and 1; in (c) r is included between 0.941 and 0.948.

FIGURE 10.31   An example of Bénard convection with a silicone oil plus aluminum powder.

10.10. The Bénard cells are examples of what Ilya Prigogine called as dissipative structures. A system maintained very far-from-equilibrium self-organizes, producing entropy that is dissipated in the environment. When the Marangoni number (Ma) is $Ma_c < Ma \leq 2Ma_c$, the cells show a hexagonal shape (Lappa 2010). The circulation of the fluid in the hexagonal cells is upwards in the center and downwards along the edge. Optical investigations have demonstrated that the surface of the fluid is depressed over the center of the cells of a few micrometers. For Ma values larger than $2Ma_c$, the cells become squared. The transition from hexagons to squares is mediated by the occurrence of pentagonal cells. The squared cells are more efficient in heat transfer than hexagonal cells, and their formation is accompanied by an increase in the Nusselt number (see equation [10.23]). The patterns that you observe also depend on the boundary conditions, i.e., the material of the plate and its shape. An example of Bénard convection is shown in Figure 10.31 obtained with a silicone oil having 100 mPa*s as viscosity at room temperature.

10.11. For the numerical solution of the system of nonlinear differential equations [10.36 through 10.38], you can use the ode45 solver of MATLAB. An example of script is the following one:

```
function dy = lorenz(t, y)
dy = zeros(3,1);
si = 10;
r = 28;
b = 8/3;
dy(1) = si*(y(2) - y(1));
dy(2) = -y(1)*y(3) + r*y(1) - y(2);
dy(3) = y(1)*y(2) - b*y(3);
```

The time evolutions of variable X for the two pairs of initial conditions are shown in Figure 10.32. The data graphed in plot (a) refer to the first two initial conditions: $[X(0); Y(0); Z(0)] = [13.824; 15.368; 32.474]$ and $[X(0)'; Y(0)'; Z(0)'] = [13.824; 15.369; 32.474]$. The data graphed in plot (b) refer to the other pair of initial conditions that are close to S_0. It is evident that when the initial conditions are closer to S_0, the two trajectories take more time to diverge.

By using the data of plot (a), we can calculate how the Euclidean distance $\|\delta\|$ between the two initially very close trajectories increases over time. The output is depicted in Figure 10.33. We confirm what we see in Figure 10.17. The $\ln(\|\delta\|)$ first grows linearly, until it saturates. The slope of the straight line fitting the growth part (see the gray straight line in Figure 10.33) is $\lambda \sim 0.9$.

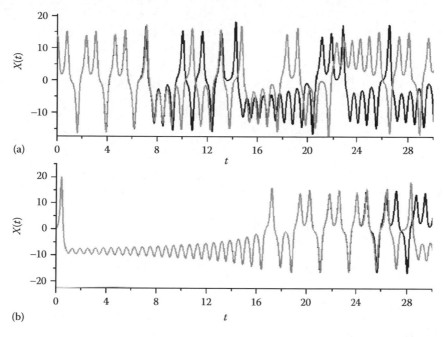

(a)

(b)

FIGURE 10.32 Trends of $X(t)$ for two pairs of initial conditions: for the pair close to S_+ in (a), and for the pair close to S_0 in (b).

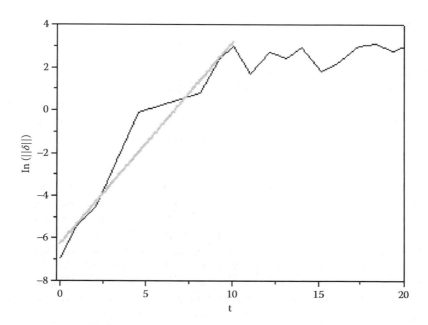

FIGURE 10.33 Trend of $\ln(\|\delta\|)$ versus time (thin black trace). The first part of this trend has been fitted by a straight line (thick gray trace).

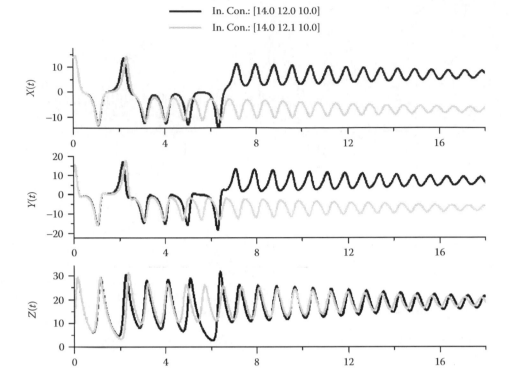

FIGURE 10.34   Trends of $X(t)$ (on top), $Y(t)$ (in the middle), and $Z(t)$ (at the bottom) for the two initial conditions indicated into the legend.

10.12. The solutions of the calculations obtained for the two distinct initial conditions proposed in the text of the exercise are shown in the Figures 10.34 and 10.35.

In the first stages, both trajectories seem to describe a strange attractor. But, in the long term, we discover that both of them spiral down toward one of the two stable solutions, which are S_+ and S_-. What is surprising is that although the dynamic is not chaotic because it is not aperiodic in the long term, it shows sensitive dependence to the initial conditions. In fact, although we start the calculations from two points that are only $\|\delta_0\| = 0.1$ far apart in the phase space, the system evolves towards two very distant solutions: S_+ and S_-. Therefore, the behavior of the system is unpredictable, at least for certain initial conditions, and it is referred to as transient chaos or metastable chaos or pre-turbulence (Strogatz 1994). In the cases of transient chaos, the final states are simple: they are periodic or stationary states. Nevertheless, the presence of a transient chaotic stage can make the entire process hard to be predicted in its outcome. This situation is familiar in everyday life. It suffices to think about gambling like throwing a dice or a coin. The outcome depends sensitively on the initial orientation and velocity if the latter is sufficiently large. A system exhibits transient chaos when a non-attracting chaotic saddle is present in its phase space. The system stays near the chaotic saddle for a finite amount of time exhibiting chaotic behavior, before exiting that region and approaching its final state asymptotically.

10.13. In paragraph 10.7, we learned that when $r > r_c$ we can observe chaotic dynamics. In this exercise, we calculate the evolution of the Lorenz's system for a considerable value of

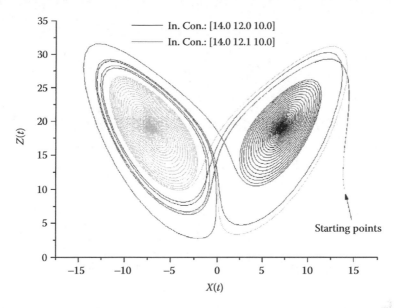

FIGURE 10.35 Trajectories traced in the $Z(t)$–$X(t)$ space, starting from the two initial conditions indicated into the legend.

parameter r ($r = 350$). Surprisingly, we find that the dynamic is periodic. The period of the oscillations does not depend on the initial conditions we choose (Figures 10.36 and 10.37).

Usually, when $r > r_c$, one finds chaotic dynamics but there are also small windows of periodic behavior interspersed (Strogatz 1994). This phenomenology resembles that observed in the case of the logistic map presented in paragraph 10.3.

10.14. First, I report the results achieved by following the instructions of list A. In Figure 10.38, the absorption spectra of DMA, before (trace 0), and after irradiation (traces 1 and 2) are

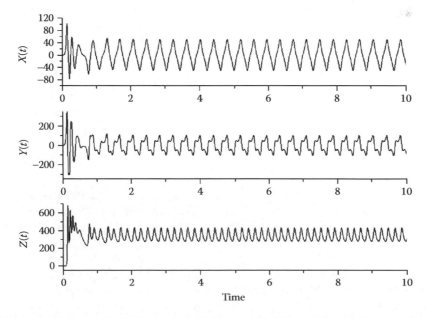

FIGURE 10.36 Periodic trends of $X(t)$ (on top), $Y(t)$ (in the middle), and $Z(t)$ (at the bottom) obtained for $\sigma = 10$, $b = 8/3$, $r = 350$, and $[X(0), Y(0), Z(0)] = [0, 1.00, 0]$ as initial condition.

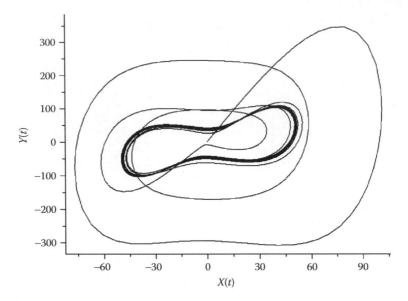

FIGURE 10.37 Trajectory traced in the $Y(t)$–$X(t)$ space when $\sigma = 10$, $b = 8/3$, $r = 350$, and $[X(0), Y(0),$ $Z(0)] = [0, 1.00, 0]$ is the initial condition.

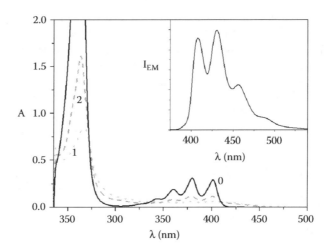

FIGURE 10.38 Absorption spectra of DMA before (continuous black trace, labeled as 0) and after two distinct irradiations (dotted light gray and dashed gray traces, labeled as 1 and 2, respectively). The fluorescence spectrum of DMA is shown in the inset.

reported. The spectra 1 and 2 have been recorded after two distinct irradiation experiments. They are significantly different from spectrum 0. This evidence suggests that UV irradiation at 260 nm triggers reactions that decompose DMA. In fact, it is known that when anthracene derivatives, substituted in positions 9 and/or 10, are excited in their higher energy optically active transition (i.e., below 300 nm), they participate in photodissociation reactions. In other words, they release their substituents as radical groups (Favaro et al. 2007). Moreover, the radiation at 260 nm is absorbed not only by DMA but also the solvent $CHCl_3$. Upon irradiation at 260 nm, chloroform produces chlorine radicals. Then, chlorine radicals induce chain reactions involving oxygen and promoting a more or less noticeable oxidative degradation of DMA (Laplante and Pottier 1982).

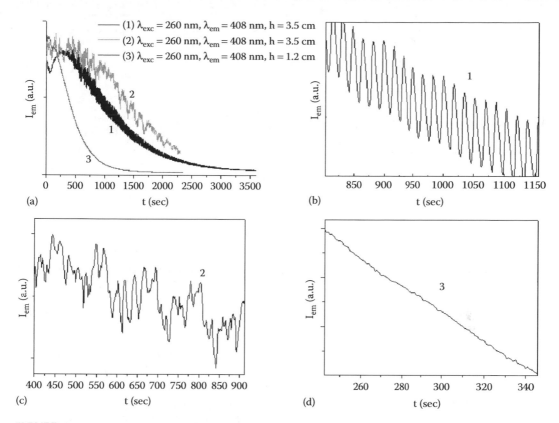

FIGURE 10.39 Time evolutions of the DMA emission (recorded at $\lambda_{em} = 408$ nm) upon steady irradiation at 260 nm recorded in three experiments labeled as 1, 2 and 3, respectively. The conditions of the three experiments are indicated in the legend of graph (a), wherein an overall image of the three kinetics is shown. Enlargements of 1 in (b), 2 in (c) and 3 in (d).

Examples of DMA fluorescence signals that can be recorded upon steady irradiation at 260 nm are shown in Figure 10.39.

Graph (a) of Figure 10.39 shows the time evolutions of DMA fluorescence collected at 408 nm in three distinct experiments, numbered consecutively 1, 2 and 3. Experiments 1 and 2 are two replicas. In fact, the UV irradiation was carried out only at the bottom of 3.5 mL. On the other hand, in case 3, all the 1.2 mL of the solution were under UV. In all the three experiments, the luminescence intensity decreased because DMA was degraded. If we zoom in on every kinetics (see the other three graphs of the same figure), we appreciate their differences. Only in traces 1 and 2, we see luminescence oscillations. In 3, they are absent, because there was no convection. The oscillations are just due to the convective motions of the solution, and they do not depend on the photo-induced reactions. In fact, the oscillations recorded in 1 and 2 can also be observed in the light scattered by the solution, and they disappear under stirring (Laplante et al. 1984). Although experiments 1 and 2 were performed in the same conditions, they originated two significantly different results. This behavior is the "butterfly effect" in action. We quantify the difference between the two traces by calculating their Fourier spectra (see Figure 10.40).

The Fourier spectrum of trace 1 consists of just one fundamental frequency that is $v_0 = 0.0586$ s^{-1} and corresponds to a period of 17 s. The other peaks at 0.1172 s^{-1} (i.e., $2v_0$) and 0.1758 s^{-1} (i.e., $3v_0$) are the second and third harmonics, respectively. The Fourier spectrum of trace 2 is more complicated than that of trace 1. It has the largest

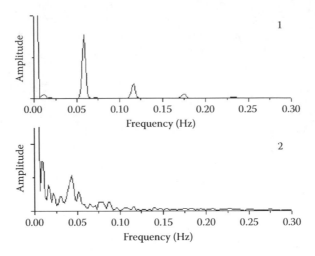

FIGURE 10.40   Fourier spectra of the fluorescence kinetics 1 (graph on top) and 2 (graph at the bottom) shown in Figure 10.39.

amplitude at the frequency of 0.0078 s⁻¹ that corresponds to a period of 128 s. There is also another important frequency of 0.043 s⁻¹ that corresponds to a period of 23 s. Based on the kinetic traces and their Fourier spectra, we infer that the Rayleigh numbers of experiments 1 and 2 were both included between $Ra_{c,1}$ and $Ra_{c,2}$, but Ra for 2 was slightly larger than Ra for 1. Therefore, from the definition of Ra (equation [10.48]), it derives the that thermal gradients in the experiments 1 and 2, i.e., $(\Delta T)_1$ and $(\Delta T)_2$, were $0.0019\,K < (\Delta T)_1 < (\Delta T)_2 < 0.0227\,K$. Finally, the height of the solution decreased 0.2 cm in 2200 s. Therefore, the power absorbed by the solution was about 0.036 J s⁻¹.

It is possible to experience the "butterfly effect" also by following the instructions of list B. The spectral evolution for the photochromic SpO under irradiation at 360 nm is reported in Figure 10.41. Initially, the solution in uncolored. Upon UV, a band centered at 612 nm forms and the solution becomes colored. The absorbance of the spectral band into the visible region grows until we reach the photo-stationary state. Then, it oscillates due to the convective motion of the solvent.

Figure 10.42 reports two spectral trends recorded at 612 nm (graphs a and c), and in two independent experiments (labeled as 1 and 2, respectively).

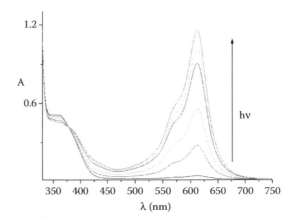

FIGURE 10.41   Spectral evolution of the photochromic SpO upon UV irradiation.

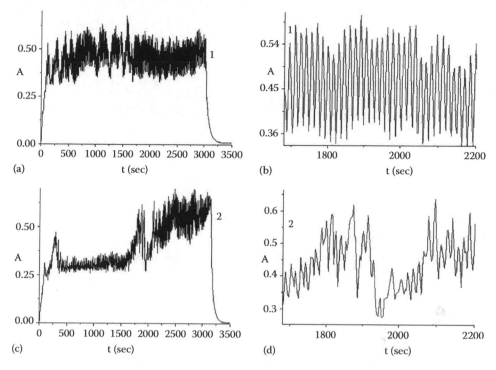

FIGURE 10.42 Spectral profiles recorded at 612 nm and generated by the "Hydrodynamic Photochemical Oscillator" based on the photochromic SpO in two distinct experiments, labeled as 1 (graph a) and 2 (graph c). Graphs b and d show two enlargements of the kinetics 1 and 2, respectively.

Although the experimental conditions were very similar (see the data in Table 10.2, where the subscripts i and f refer to the initial and final states, respectively), the two profiles of A versus time are significantly different. It is worthwhile noticing that the oscillations shown in Figure 10.42 are recordable also at wavelengths where both SpO and MC do not absorb, for example at 800 nm, confirming that are due to the hydrodynamic motions of the solutions.

The discrepancy between the two kinetics (1 and 2) is further proof of how convection is sensitive to the initial conditions. In both cases, we have oscillations in the signals of A versus time, but they are not the same. The Fourier Transforms (see Figure 10.43) demonstrate that in both cases we have a discrete number of frequencies. However, in experiment 1,

TABLE 10.2

Contour Conditions Relative to the Experiments 1 and 2 Shown in Figure 10.42.

Conditions	Exp. 1	Exp. 2
h_i	3.1	3.1
h_f	3.0	3.0
T_i	25.3°C	25.0°C
T_f	25.4°C	25.2°C
P_i	967.0 hPa	967.7 hPa
P_f	966.2 hPa	967.0 hPa
(Humidity)$_i$	43%	42.6%
(Humidity)$_f$	41.4%	43.7%

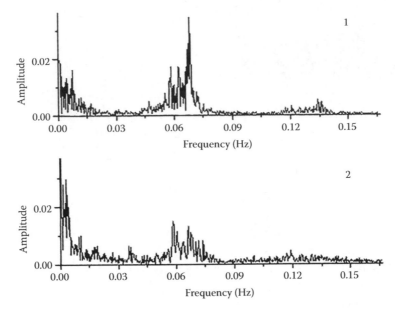

FIGURE 10.43 Fourier spectra for the spectrophotometric kinetics 1 (graph on top) and 2 (graph at the bottom), respectively.

there is a dominant frequency of 0.0678 s^{-1}, whereas, in experiment 2, more than one frequency has similar amplitudes. For instance, the frequencies of 0.0584 s^{-1} and 0.0663 s^{-1}.

It is reasonable to assume that in both cases $Ra_{c,1} < (Ra)_{1or2} < Ra_{c,2}$. By using the values of the physical parameters for acetone, we deduce that the vertical thermal gradients in the two runs are in the range $0.0017\,K < (\Delta T)_1 \leq (\Delta T)_2 < 0.0201\,K$. In both experiments, the volume that evaporated was 0.1 cm^3 after 3500 s. The power absorbed by the solutions was 0.012 J s^{-1}. Figure 10.44 shows the hydrodynamic waves spreading MC throughout the solution.

10.15. The data "b1.txt" are plotted in Figure 10.45. The upper graph shows the heart rate, whereas the lower graph reports the trend of chest volume over time. Both the time series refer to a patient exhibiting sleep apnea, namely a sleep disorder characterized by pauses in breathing or periods of shallow breathing. The anomalous breathing periods can last for a few seconds to several minutes.

Before running a routine of TISEAN, you can be acquainted with the necessary inputs and the right syntax, by typing >*name_of_routine -h*.

In Figure 10.46, you can see the outputs of the calculations. The graphs in the left column of the Figure refer to the heart rate time series, whereas the charts in the right column

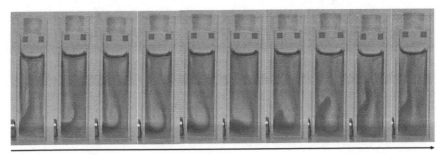

Time

FIGURE 10.44 Sequence of snapshots showing hydrodynamic convective waves spreading MC throughout the solution.

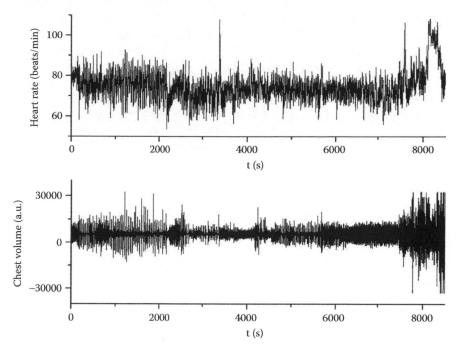

FIGURE 10.45 Time series of heart rate (top graph) and chest volume (lower graph) for a patient showing sleep apnea.

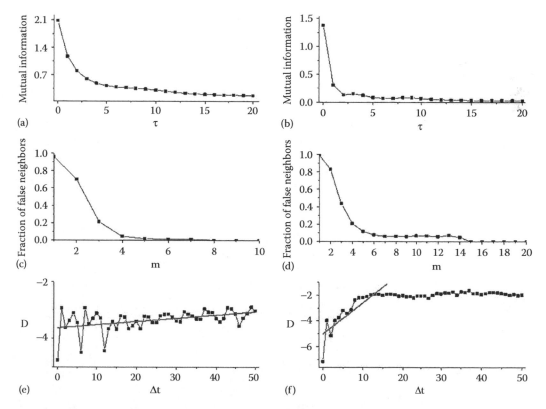

FIGURE 10.46 Determination of the time delays (a and b), embedding dimensions (c and d), and Lyapunov exponents (e and f) for the heart rate and chest volume time series, respectively.

refer to the chest volume time series. The plots of the first row regard the determination of the time delay (τ) through the calculation of mutual information (routine *mutual*). It results that the first minimum in (a) is for $\tau = 6$, whereas in (b$\|\delta(t)\|$) is for $\tau = 2$. The plots of the second row regard the calculations of the fraction of the false nearest neighbors (routine *false_nearest*). The fraction is 0 for $m = 8$ in (c) and $m = 15$ in (d). Therefore, by using the Takens' theorem, it results that the phase space for the heart rate time series has $\tau = 6$ as time delay and $m = 8$ as embedding dimension, whereas that for the chest volume time series has $\tau = 2$ and $m = 15$. The largest Lyapunov exponent calculated by the Kantz's method (routine *lyap_k*) for heart rate is $\lambda = 0.011$ (graph (e)), whereas that for the chest volume is $\lambda = 0.033$ (graph (f)), indicating the possibility that both time series are chaotic.

11 Chaos in Space
The Fractals

A fractal is a way of seeing infinity

Benoit Mandelbrot (1924–2010 AD)

11.1 INTRODUCTION

In Chapter 10, we have learned that chaotic dynamics can originate "strange" structures. An example is the bifurcation diagram of the logistic map (see Figure 10.9). You probably remember that it describes the evolution of the relative abundance x_n of a population as a function of its growth rate r. If we zoom in on specific regions of the bifurcation diagram, where "white patches" are interspersed with clouds of infinite points, copies of the overall diagram reappear at different spatial scales. The bifurcation diagram is self-similar whatever is the magnification scale.[1]

Even the Lorenz model for weather forecasts originated a "strange" attractor in the chaotic regime (see Figure 10.15). At first sight, the Lorenz attractor looks like a butterfly: it consists of a pair of surfaces, as if it were a pair of wings, merging into the central part—the body of the graceful flying insect. If we look at the details of the attractor, we find, just as Lorenz realized, every chaotic trajectory crosses an infinite number of surfaces and not just two. To gain more insight into the geometrical structure of the Lorenz's strange attractor, it is useful to exploit the strategy contrived by Poincaré while he was studying the three body problem mentioned in Chapter 10. Poincaré discovered that the complexity of the analysis of a swirling trajectory can be reduced by focusing on its intersections with a fixed surface. Such a surface, now known as Poincaré section, transforms a continuous N-dimensional trajectory into an $(N-1)$-dimensional map if the surface is transverse to the trajectory (see Figure 11.1). The resulting $(N-1)$-dimensional map is called Poincaré map.

In particular, the Poincaré map of the Lorenz attractor for $z = 27$ looks like bundles of sequences of points (Viswanath 2004). Every bundle of points is self-similar because it maintains its regular spacing at different levels of magnification (in principle, *ad infinitum*).

In 1976, the theoretical astronomer, Michel Hénon, after being acquainted with the Lorenz attractor, formulated a two-dimensional map as a simple model having the essential properties of the Lorenz system (Hénon 1976). The Hénon map is given by

$$x_{n+1} = y_n - ax_n^2 + 1 \qquad [11.1]$$

$$y_{n+1} = bx_n \qquad [11.2]$$

where a and b are two adjustable parameters. When $a = 1.4$ and $b = 0.3$, the Hénon map gives rise to a strange attractor, looking like a boomerang and shown in Figure 11.2.

Successive enlargements of the exterior border of the boomerang-like shape reveal the fine structure of the attractor. In the first picture, the border consists of three lines. By zooming into the

---

[1] The only limit that exists is imposed by the degree of accuracy of our calculations. In other words, the limit depends on the number of significant figures we used in building the bifurcation diagram of the logistic map.

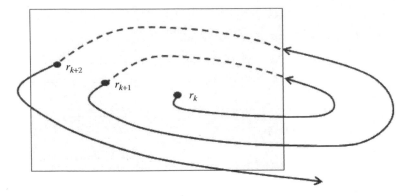

FIGURE 11.1 Representation of a Poincaré section that transforms a continuous three-dimensional trajectory into a bi-dimensional map.

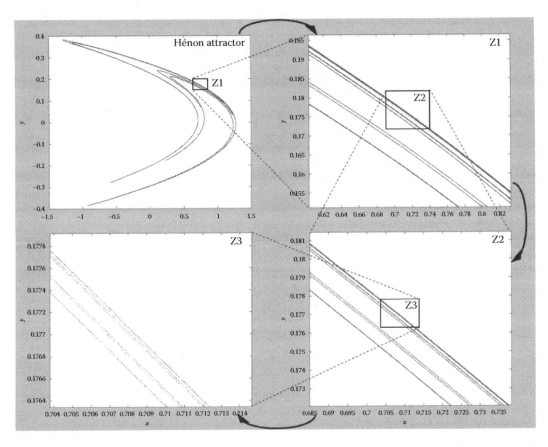

FIGURE 11.2 A series of three enlargements (Z1, Z2, and Z3) into a portion of the exterior border of the Hénon attractor.

rectangle Z1 of Figure 11.2, we see that there are six lines: a single curve, then two closely spaced curves above it, and then three more. An enlargement in Z2 reveals that the three lines are actually six curves; grouped one, two, three, precisely as in Z1. A further expansion of the last three lines unveils the structure also observed in Z1; i.e., one, two and three lines. And so on. Self-similarity can continue indefinitely. The limit is imposed only by the number of decimal figures we use in our calculations.

TRY EXERCISE 11.1

Self-similarity can be encountered in every strange attractor. In fact, the strange attractors are examples of FRACTALS.

11.2   WHAT IS A FRACTAL?

A fractal is a complex geometric shape. It maintains its sophisticated form, or the main features of its complex structure, under different levels of magnification, i.e., under changes of spatial scale. This peculiarity of fractals is defined as either "scale invariance" or "scale symmetry," or more often, "self-similarity." Several fractals can be built by following simple rules. A fractal that often appears in the strange attractors generated by chaotic dissipative dynamics is the Cantor set. For example, it is the fractal that can be detected in the Lorenz attractor. The set is named after the nineteenth-century mathematician Georg Cantor. However, it was created by Henry Smith, who was a nineteenth-century geometry professor at the Oxford University.[2] To build a Cantor set, we start with a bounded interval C_0. We remove an open interval from inside C_0. C_0 is divided into two subintervals, each containing more than one point. Iteratively, we remove an open interval from each remaining interval. In this way, the length of the remaining intervals at each step shrinks to zero as the construction of the set proceeds. After an infinite number of iterations, we obtain the Cantor set. The most often mentioned Cantor set is that obtained by removing the middle third of every interval (see Figure 11.3). In the end, it consists of an *infinite* number of infinitesimal pieces, separated by gaps of various lengths, and having a *negligible* length.[3] Of course, it is impossible to print a fractal on a piece of paper.

TRY EXERCISE 11.2

Another well-known and perfectly self-similar fractal is the Koch curve, contrived by the Swedish mathematician Helge von Koch. For its construction, we start with a horizontal segment (K_0). Then, we delete the middle third of K_0, and we replace it with the other two sides of an equilateral triangle, and we obtain K_1 (see Figure 11.4). In the subsequent stages, we repeat recursively the same rule:

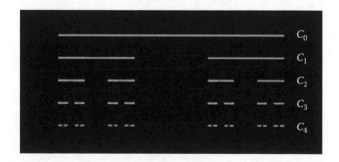

FIGURE 11.3   Scheme showing the first four steps for constructing the Cantor set by removing the middle third of every interval. It has been obtained by using the "Examples of Fractals" model, Complexity Explorer project, http://complexityexplorer.org.

---

[2] The first self-similar fractals were invented earlier than the discovery of strange attractors. In the beginning, they were merely "pathological phenomena" interesting only mathematicians.

[3] Cross sections of strange attractors are often Cantor set devoid of exact self-similarity. However, they are topological Cantor sets. A topological Cantor set C has two properties. First, C contains only single points and no intervals. Second, every point in C has a neighbor arbitrarily close by. A famous example of topological Cantor set is the logistic map (Strogatz 1994).

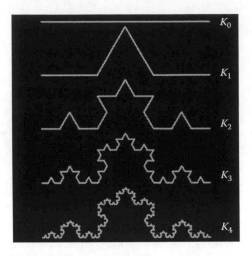

FIGURE 11.4 The first four steps in the construction of the Koch curve. It has been obtained by using the "Examples of Fractals" model, Complexity Explorer project, http://complexityexplorer.org.

K_n is achieved by replacing the middle third of each line segment of K_{n-1} by the other two sides of an equilateral triangle. In the end, after an infinite number of iterations, we obtain K_∞ that is the Koch curve. It consists of an *infinite* number of segments, and it has an *infinite* length, although it is confined to a finite region of the plane.

TRY EXERCISE 11.3

It is evident that the Cantor set and the Koch curve have extraordinary properties. The Cantor set consists of an infinite number of segments, but its length approaches zero. Moreover, it is not continuous. The Koch curve has an infinite length, and it is not differentiable. They are "monsters"! Try to imagine what happens if you build the Koch curve on the three sides of an equilateral triangle (see Figure 11.5). We obtain the so-called Koch snowflake. It is a "geometrical monster" because it has an infinite perimeter that encompasses a finite area.

The tools of Analysis cannot deal with fractals. In fact, they are not smooth curves; sometimes, they are neither continuous curves. Hence, they are not differentiable everywhere. Isaac Newton and Gottfried Leibniz, when inventing Calculus, in the middle of the seventeenth century, proposed the fundamental idea of approximating any smooth function by tiny little segments of straight lines (see Figure 11.6). This approach is reasonable. It suffices to zoom into any of such smooth curves, to understand why. However, the intuition of Newton and Leibniz becomes useless in the case of fractals. In fact, a fractal is a geometric figure that does not become simpler when analyzed into smaller and smaller parts. Straight lines cannot describe a fractal on each of its points.

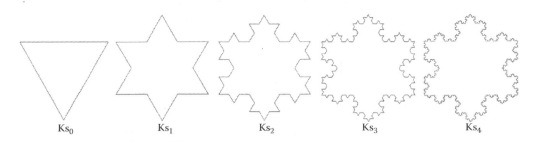

FIGURE 11.5 The first four steps for constructing the Koch snowflake.

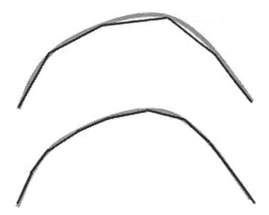

FIGURE 11.6 The smooth gray function is approximated by black segments. The shorter the segment, the better the approximation.

11.3 FRACTAL DIMENSION

What is the dimension of the Cantor set that is a set of points? And what is the dimension of the Koch curve that has an infinite length? If we think about Euclidean geometrical objects, it is easy the determination of their dimensions. For instance, segments are one-dimensional. But even any smooth curve is one-dimensional because every point on it is determined by one number, which is the arc length from some fixed reference point on the curve. The dimension can be defined as the minimum number of coordinates needed to describe every point in the object. Therefore, planes and smooth surfaces are two-dimensional, solids are three-dimensional. In fact, what happens if we bisect a one-dimensional object, such as a segment? We obtain two equal half-sized segments. What happens if we trisect the original segment? We obtain three equivalent copies (see the top left part of Figure 11.7). Now, let us consider a square. What happens if we bisect both sides of the square? We obtain four copies of the original square. What happens if we trisect each side? We obtain nine copies (see the bottom left part of Figure 11.7). Now let us consider a cube. What happens if we bisect each side of the cube? We achieve eight copies of the original cube. What if we trisect each side? We obtain 27 copies (see the right part of Figure 11.7). From these observations, we infer that the number of copies N depends on the scale factor r and the dimension d (known as the Hausdorff dimension), according to the following equation:

$$N = r^d \qquad [11.3]$$

In the case of a line, $d = 1$; in the case of a square, $d = 2$, and in the case of a cube, $d = 3$. In fact, $2^1 = 2, 3^1 = 3, 2^2 = 4, 3^2 = 9, 2^3 = 8, 3^3 = 27$.

We can, now, apply equation [11.3] for determining the dimension of the Cantor set represented in Figure 11.3. The scale factor is $r = 3$ because we trisected the initial segment into three equivalent parts. Then, we obtained two copies, $N = 2$, because we removed the middle third. Therefore, the dimension of the Cantor set is:

$$d = \frac{\log(N)}{\log(r)} = \frac{\log(2)}{\log(3)} \approx 0.63 \qquad [11.4]$$

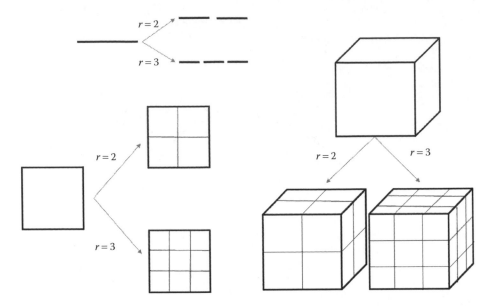

FIGURE 11.7 What happens to a segment (top left), a square (bottom left) and a cube (on the right) when they are bisected ($r = 2$) and trisected ($r = 3$).

Puzzling! The dimension is not an integer! And it is between 0 and 1. In fact, if we take a moment to reflect, we realize that $d \approx 0.63$ is reasonable because the Cantor set is not a simple point (with $d = 0$), but neither a line (with $d = 1$). It is in between a point and a line.

And what about the Koch curve? Let us apply equation [11.3], again. For the Koch curve, the scale factor is 3, because we trisect the initial segment. The number of copies is 4 because the middle part is replaced by two equivalent segments. Therefore:

$$d = \frac{\log(N)}{\log(r)} = \frac{\log(4)}{\log(3)} \approx 1.26 \qquad [11.5]$$

Amazing! The dimension is not an integer, again. It is larger than 1 because the Koch curve has an infinite length. But it is smaller than 2 because it does not have an area.

Let us consider another famous fractal: the Sierpinski gasket. To construct it, we start with an equilateral triangle (S_0 in Figure 11.8). We bisect each side, and we link each midpoint through a segment. We obtain another equilateral triangle that is inverted with respect to the first. We remove it, and we obtain S_1. We repeat this step for the remaining three triangles, and we have S_2. And so on. After repeating an infinite number of times that procedure, we achieve the Sierpinski gasket. The scale factor is 2 because we bisect the sides of the triangles. The number of copies is 3. Hence, the dimension is

$$d = \frac{\log(N)}{\log(r)} = \frac{\log(3)}{\log(2)} \approx 1.58 \qquad [11.6]$$

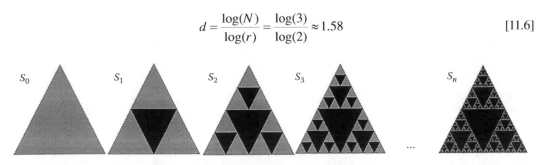

FIGURE 11.8 The Sierpinski gasket, and the way it is constructed.

Again, the dimension is between 1 and 2, as in the case of the Koch curve. However, the Sierpinski gasket has a dimension larger than that of the Koch curve because it is closer to a two-dimensional object, as a triangle is.

TRY EXERCISES 11.4 THROUGH 11.7

11.4   FRACTALS THAT ARE NOT PERFECTLY SELF-SIMILAR

It is possible to create fractals also by calculating a simple non-linear equation iteratively. The foundations of this methodology were laid by Gaston Maurice Julia,[4] at the beginning of the twentieth century. He worked on the iteration of polynomials and rational functions. For example, he studied the following quadratic equation:

$$f(z) = z^2 + c \qquad [11.7]$$

In [11.7], c is a constant and $z = a + ib$ is a complex number. Starting from an initial point z_0, Julia calculated if $f(z)$ diverged to infinity or approached zero after many iterations: $f(z_n) = f(z_{n-1}) = \ldots = f(z_1) = f(z_0)$. He repeated these iterations for a heap of points in the complex plane, and he found that some of them tended to a limiting position, whereas others never settled down, but diverged. However, Julia could not visualize the shapes of the two kinds of sets because, at his time, there were no computers available. In the 1970s, the Polish-born, French-educated mathematician Benoit Mandelbrot (1924–2010), working at IBM's research center in New York (USA) and having the most powerful computers at his disposal, revived Julia's work. He developed the first not perfectly self-similar fractal, the famous "Mandelbrot set." He solved equation [11.7] iteratively, always starting from the seed $z_0 = 0$, and for different complex c values. If, after n iterations, z_n, the so-called orbit of z_0, did not tend to infinity, the seed $(0, c)$ belonged to the Mandelbrot set. Otherwise, it did not. Thus, the Mandelbrot set is a record of the fate of the orbit of $z_0 = 0$ under iteration of $z^2 + c$: the c numbers are represented graphically and colored in different ways depending on the fate of the orbit of 0. An example is shown in Figure 11.9. All the c values in the black region cause z to stay finite. All the c values outside the black region cause z to go to infinity. The shades of gray are proportional to the speed with which the value of z goes to infinity. The speed increases by going from light gray to gray up to dark gray. In Figure 11.9, there are different magnifications of the initial set. It is evident that the Mandelbrot set is self-similar. However, the self-similarity is not perfect, like in the case of the fractals we studied in the previous paragraphs.

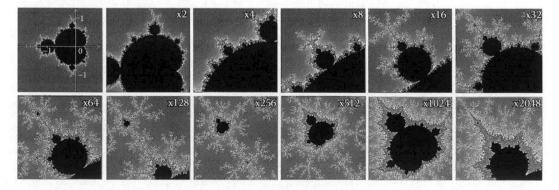

FIGURE 11.9   Increasing magnification of the Mandelbrot set built by using "Easy Fractal Generator" available at the website http://www.easyfractalgenerator.com/Home.aspx.

---

[4] Gaston Maurice Julia (born in 1893, Sidi Bel Abbès, Algeria, and died in 1978, Paris, France) was one of the main inventors of iteration theory.

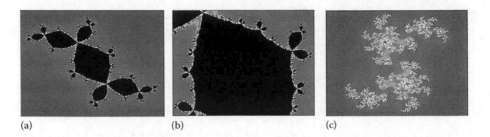

(a)   (b)   (c)

FIGURE 11.10   Julia set when $c = -0.12 + 0.75i$ in (a) and a magnification of it in (b); Julia set when $c = 0.4 - 0.3i$ in (c).

TRY EXERCISE 11.8

There is another famous fractal that is based on the iteration of the quadratic equation [11.7]: it is the Julia set. For the construction of the Julia set, it is necessary to fix the value of the constant c and calculate the fate of all possible seeds in the plane of the complex numbers. Those seeds that do not escape to infinity, but get closer to the origin of the axes, form the black filled Julia set. Of course, there are as many Julia set as are the possible values of c.

TRY EXERCISE 11.9

Two types of Julia sets are depicted in Figure 11.10. The one in picture (a) is an example of a connected Julia set because the black parts are all connected. It is called the "Douady's rabbit," after the French mathematician Adrien Douady. It looks like a fractal rabbit because it has a main body with two ears attached. By zooming in, we see pairs of ears along the borderlines, at every scale (a magnification is reported in [b]). In (c), there is an example of a "cloud" Julia set because it consists of infinitely many pieces, each of which is a single point. These points pile up and form a structure that looks like a Cantor set.[5]

TRY EXERCISE 11.10

11.5   THE FRACTAL-LIKE STRUCTURES IN NATURE

Many natural forms, in both the animate and the inanimate worlds, look like fractals. Examples are ferns, Romanesco broccoli, flames, clouds (see Figure 11.11), but also coastlines, leaf veins, blood vessels in lungs, rivers, mountain ranges, plant roots, neural networks, and many more.

Even the crowns of trees are beautiful examples of fractal-like structures. In fact, if we zoom in one of them, we obtain images that maintain the structural complexity of the overall crown (see Figure 11.12).

The first who had the idea that many natural forms have fractal structures was Benoit Mandelbrot (Mandelbrot 1982). Mandelbrot coined the name "fractal" from the Latin word *fractus* that means "broken" or "fractured." This term was chosen in explicit cognizance of the fact that the irregularities found in fractal sets are often strikingly reminiscent of the fracture surfaces in metals (Mandelbrot et al. 1984). Mandelbrot was attracted by the beauty and irregularity of fractal-like natural forms and was frustrated by the inability of usual differential geometries[6] to describe them properly. Mandelbrot was wondering himself: "Why is geometry often described as 'cold' and 'dry'?" Mandelbrot (1982) replied, "One reason lies in its inability to describe the shape of a cloud,

---

[5] The Julia set is of the "connected-type" (said also "filled-type") when c belongs to the Mandelbrot set. If c escape from the Mandelbrot set, the corresponding Julia set is of the "Cantor-type" or "cloud-type."

[6] The non-fractal geometries are the Euclidean (built on the plane), the spherical (relative to surfaces with positive curvatures) and the hyperbolic (relative to surfaces with negative curvatures) geometries.

FIGURE 11.11   Pictures of natural fractal-like structures: ferns, broccoli, flames, and clouds.

FIGURE 11.12   Four successive magnifications (Z1–Z4) of the original image (0) of a tree.

a mountain, a coastline or a tree. Clouds are not spheres, mountains are not cones, coastlines are not circles and bark is not smooth, nor does lightning travel in a straight line."

To be honest, the natural forms are fractal-like and not perfect fractals (Shenker 1994). An infinite number of iterations generates the true fractals. Genuine fractals are analogous to irrational numbers, which are infinite sequences of numbers. There are irrational numbers that can be expressed just by one symbol. For example, $\sqrt{2}$, or π, or e. But, there are much more irrational numbers that cannot be compressed. Similarly, some fractals can be described by a short algorithm; for instance, the Cantor set or the Koch curve. But, there are also fractals that are not exactly self-similar. In other words, their self-similarity is approximate or just statistical. Natural forms that look like fractals have always a self-similarity that is not exact. Moreover, their self-similarity holds on only for a finite number of spatial scales, not ad infinitum. In fact, whatever is the object, either a cloud, or a tree, or else, after several magnifications, the object loses its original identity, and what we see are molecules or subatomic particles. On the other hand, if we look at the same object but, in a scale, quite larger than its dimensions, we lose its details and roughness, and we can describe it by using the tools of the traditional geometries and analysis.

TRY EXERCISE 11.11

11.6 THE DIMENSIONS OF FRACTALS THAT ARE NOT PERFECTLY SELF-SIMILAR

There are a lot of different methods for determining the dimension of fractals that are not perfectly self-similar (Falconer 1990). One method used commonly is the "Box Counting." Let us imagine having the picture of a fractal or fractal-like structure on a plane. We cover it with a grid of squares or boxes having size l smaller than the dimensions of the fractal. The number of boxes required to cover the object is proportional to its dimension. For example, the number of squares of side l, $N(l)$, needed to cover the black tick curve of Figure 11.13, being of total length L ($>l$), is

$$N(l) \propto \frac{L}{l} \qquad [11.8]$$

On the other hand, the squares necessary to cover the grey spot of Figure 11.13, having area Ar, is

$$N(l) \propto \frac{Ar}{l^2} \qquad [11.9]$$

Similarly, for a fractal or fractal-like object having dimension D, the number of boxes, required to overlay it, is $N(l) \propto 1/l^D$. The Box Counting method requires that we repeat the calculation of $N(l)$, many times, by changing the box size, l. Then, we plot the $\log(N(l))$ versus $\log(1/l)$. We fit the data by a straight line, determined by the least-squares method. The slope of the straight line gives us D, because

$$D = \lim_{l \to 0} \frac{\log(N(l))}{\log\left(\frac{1}{l}\right)} \qquad [11.10][7]$$

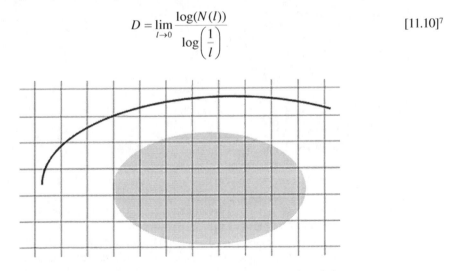

FIGURE 11.13   The black curve and the grey spot covered by a grid of squares.

---

[7] The dimension D, determined by the Box Counting method for a fractal-like structure that we find in nature, is named as its Hausdorff dimension.

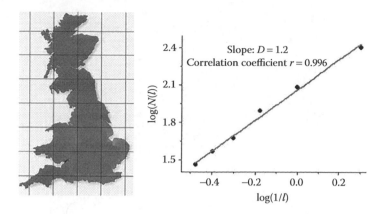

FIGURE 11.14   Determination of the dimension D of the Britain coast by the Box Counting method.[8]

The jagged seacoast shapes are examples of statistical self-similar fractals because every portion can be considered a reduced-scale image of the whole, from a statistical point of view. The contoured path of a coastline fills more space than a line, but less space than a plane, and therefore its fractal dimension will be included between 1 and 2. Highly corrugated coastlines fill more space and have fractal dimensions close to 2. Smooth coastlines have dimensions closer to 1. It is possible to measure its dimension by using the Box Counting method. Before the advent of computers, cartographers have encountered many problems in measuring the length of the coasts and borders. For example, the determination of the length of Britain's coast was really tough because it is particularly complex. The coastline measured on a large-scale map was approximately half the length of the coastline measured on a detailed map. The closer cartographers looked, the more detailed and longer the coastline became. If we apply the Box Counting method, we find that the dimension of Britain's coast is $D = (1.2 \pm 0.1)$ (see Figure 11.14).

The value of the fractal dimension D has some positive correlation with the irregularity or jaggedness of a frontier. In fact, the smoothest coast on the atlas is that of South Africa, and its Box Counting Dimension has resulted in being $D = 1.02$ (Mandelbrot 1967). On the other hand, the land frontier between Spain and Portugal, which is quite irregular, has $D = 1.14$. The rougher the border, the larger the D, because it is not a simple line and it has space-filling characteristics in the plane.

The Box Counting method can be applied not only to complex structures projected in two dimensions but also to objects observed in volumetric space. Instead of using squares, we must use cubes that form lattices. The possible values of the fractal dimension are $0 \leq D \leq 1$ for the fractal dusts; $1 \leq D \leq 2$ for the fractal lines; $2 \leq D \leq 3$ for the fractal surfaces, and $3 \leq D \leq 4$ for the fractal volumes.

TRY EXERCISE 11.12

11.7   A METHOD FOR GENERATING FRACTAL-LIKE STRUCTURES IN THE LAB

An easy way for generating fractal-like structures in a laboratory is to use the Hele-Shaw cell (Vicsek 1988). The Hele-Shaw cell is a pair of transparent, rigid plates separated by a small gap, typically 0.5 mm high,[9] with a hole in the center of the top plate (see Figure 11.15).

---

[8] Mandelbrot (1967) calculated a dimension $D = 1.25$ for the coast of Great Britain. The data in Figure 11.14 refers to a Box Counting method applied to a Great Britain printed on A4 paper, and by using six distinct grids with the sides of the squares being 3, 2.5, 2, 1.5, 1, and 0.5 cm long, respectively (see also exercise 11.12). A reasonable estimate of the uncertainty is 10%. In fact, if we apply the same procedure to the grey speckle of Figure 11.13, we find that its Box Counting Dimension is $D = (1.8 \pm 0.2)$.

[9] Henry Selby Hele-Shaw (1854–1941) was an English engineer, who studied the flow of water around a ship's hull by using two parallel flat plates separated by an infinitesimally small gap. You can build your own Hele-Shaw cell by using 30×30 cm plates made of glass or clear plastic, like plexiglass.

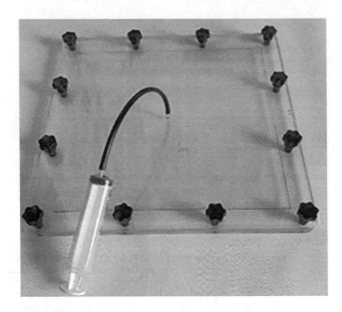

FIGURE 11.15   Picture of a Hele-Shaw cell.

A high-viscosity fluid, such as glycerin or a 4% aqueous solution of polyvinyl alcohol (PVA), fills the space between the plates. For generating patterns that are visible easily, a colored aqueous solution must be injected by a plastic syringe connected to the hole through a rubber tubing.[10]

The aqueous solution displaces the viscous fluid generating beautiful structures characterized by fingers of different shapes and sizes. The phenomenon of viscous fingering has been studied for a long time because it is relevant in engineering. For instance, the petroleum industry has been trying consistently to find ways of inhibiting viscous fingering because it limits oil recovery in a porous media. Viscous fingering is also important for the study of pattern formation in non-equilibrium conditions when the interplay between microscopic interfacial dynamics and external macroscopic forces plays a crucial role (Ben-Jacob and Garik 1990). Figure 11.16 shows just a few examples of the uncountable morphologies that can be obtained in the Hele-Shaw cell.

We can obtain either dense-branching morphologies (that consist of fingers of well-defined width confined within a roughly circular envelope) like in (a) and (c) or dendritic morphologies like in (b), (d), and (e). The splitting at the tip distinguishes the fingers from dendrites. The edge of the low-viscosity fluid moves forward into the high-viscosity fluid because the pressure just behind the interface (P_{in}) is higher than that in the viscous fluid (P_{out}) just in front of it. The speed at which the interface advances depends on how steep this pressure gradient is ($P_{in} - P_{out}$). When the interface randomly develops bulges, the pressure gradients become steeper, rising to their highest values at the tips. In fact, the excess pressure on the tip of a bulge is, according to the Laplace equation, $\Delta P = 2\gamma / r$, where γ is the surface tension and r is the curvature ray (see the sketch in Figure 11.17).

Therefore, a self-amplifying effect induces a faster growth of the bulges with respect to the other parts of the interface. This positive feedback is counterbalanced by the surface tension that tends to reduce the surface extension. When the viscous fluid has a limited solubility into the advancing fluid, the negative effect exerted by the surface tension is relevant, and the pattern that sprouts, has a dense-branching morphology, like in (a) and (c) of Figure 11.16. On the other hand, when the two fluids are perfectly miscible, like water and an aqueous polymer solution (for instance, 4% PVA made

---

[10] The rubber tubing has an internal diameter of 2 mm.

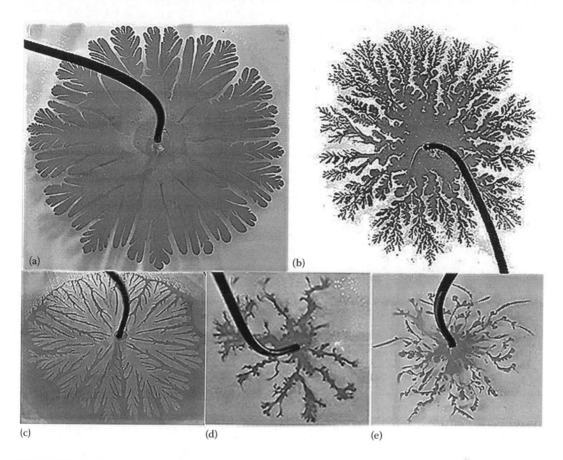

FIGURE 11.16 Examples of dense-branching morphologies (a) and (c), obtained by using pure glycerin plus a red-dye aqueous solution in (a), and red-colored glycerin plus air in (c). Examples of dendrites in (b), (d), and (e) obtained by using 4% PVA reticulated by sodium borate plus a red-dye aqueous solution.

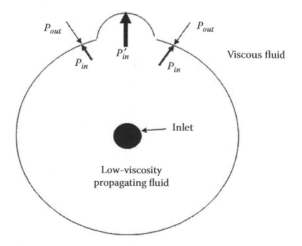

FIGURE 11.17 A circular interface between the viscous fluid and the low-viscosity propagating fluid having a bulge generated by fluctuation.

viscous by addition of a small amount of sodium borate[11]), the surface tension at the interphase is reduced at its minimum. Therefore, the bulges that sprout by fluctuations self-amplify easily. The emerging patterns look like dendrites (like in b, d, and e of Figure 11.16).

TRY EXERCISE 11.13

11.8   DENDRITIC FRACTALS

Dendritic structures[12] are spread throughout the animate and the inanimate worlds. The tree branches, the plant roots, the branched projections of neurons, the alveoli in the lungs, the cell colonies (for instance, cells of Bacillus subtilis, as shown by Matsushita et al. [1993]), the mineral dendrites, the frost that forms on automobile windshields, dust, soot, and the dielectric breakdown are just a few examples of dendritic shapes that we may find around us.[13] Dendritic structures (see Figure 11.18) are statistically self-similar objects and have scale-invariant properties. For example, the density correlation function $\delta(r)$, defined as the average density of units making up the object at distance r from a point on the structure, is a measure of the average environment of a particle. $\delta(r)$ is a scale-invariant function because it is a power law in r (Sander 1986):

$$\delta(r) = kr^{-(d_E - d)} \qquad [11.11]$$

In [11.11], k is a constant, d_E is the Euclidean dimension of the space, and d is the fractal dimension. $\delta(r)$ of dendrites is a decreasing function of r: the average density decreases as the fractal becomes larger. For the determination of how the total mass M of the fractal scales with its radius of gyration R,[14] we simply multiply equation [11.11], when $r = R$, by the volume R^{d_E} and we obtain:

$$M(R) = KR^{d - d_E} R^{d_E} = KR^d \qquad [11.12]$$

where K is another constant. For an ordinary curve, $d = 1$ because by doubling the length, the mass doubles, too. For a disk, $d = 2$, and for a solid $d = 3$. For a dendrite projected on a plane, d is not an integer; it has been measured to be $d \approx 1.7$.

Dendrites can be grouped in two sets, based on the process of their formation. There are dendritic structures that form by aggregation having "out-to-in" nature: particles come from outside and join the growing cluster. But, there are also dendritic structures that grow from an "in-to-out" dynamics, like those generated by viscous fingering or dielectric breakdown. Among the "out-to-in" dendrites, many are generated by isotropic aggregation. Isotropic aggregation is the process of suspended particles that move randomly and stick to one another to form connected structures. It is different from the anisotropic aggregation that is a more constrained aggregation where particles connect only in favored lattice directions. Basically, anisotropic aggregation is crystallization and is responsible, for instance, of the dendritic six-branched structures of frost on

---

[11] When $Na_2B_4O_7$ is dissolved in water, it produces boric acid (H_3BO_3) and borate ions ($B(OH)_4^-$). If a small amount of 4% sodium borate solution is added to the 4% PVA, the borate ions bind to hydroxyl groups of PVA, promoting cross-linking among the macromolecules and increasing the viscosity of the polymer. The degree of cross-linking depends on the amount of borate that is added (see the Figure in this note). The resulting viscous fluid is "Non-Newtonian" because its viscosity grows by increasing pressure.

[12] The word dendrite derives from the Greek δένδρον that means "tree."

[13] The ways the lichen spreads across rock, the malignant cells grow in tumors, the tea or coffee stains wet napkins, give rise to forms with ruffled, frond-like boundaries that radiate from a central focus. These boundaries are fractal (Ball 2009).

[14] The radius of gyration is defined by $R^2 = (1/M)\sum_i (r_i - r_c)^2$, where $r_c = (1/M)\sum_i r_i$ is the cluster's center of mass.

FIGURE 11.18   Example of dendrite built by using the Diffusion Limited Aggregation model implemented in the software NetLogo. (From Wilensky, U., and Rand, W., *Introduction to Agent-Based Modeling: Modeling Natural, Social and Engineered Complex Systems with NetLogo*, MIT Press, Cambridge, MA, 2015.)

the windshield. The isotropic "out-to-in" aggregation is well described by the diffusion-limited aggregation (DLA) model (Witten and Sander 1981, 1983). The initial condition is an immobile particle (or "seed") at the center of a disk. Then, at the boundary of the disk, a particle is released, and it wanders by Brownian motion. Eventually, it hits the seed, and it sticks to it. Then, another particle is released at the boundary, and it starts its random walk that stops when it hits the two-particles cluster and sticks to it. This process repeats, again and … again. The possibility that the particles rearrange after sticking, to find a more energetically favorable position, is excluded. The dendrite in Figure 11.18 has been obtained by this model implemented in NetLogo (Wilensky and Rand 2015). It looks wispy and open because holes are formed and not filled up. In fact, a random walker cannot wander down one of the channels in the cluster without getting stuck on the sides.

The growth of "in-to-out" dendritic shapes are triggered by fluctuations and are molded by two forces: one negative and another positive. Fluctuations (due to the random motions of atoms and molecules) on a propagating flat front may produce small bumps. The surface tension tends to suppress bumps smaller than a certain limit and smooths the interface. On the other hand, a positive feedback process self-amplifies the small bumps producing larger bulges, and hence growing fingers. In the case of viscous fingering, the positive feedback action is played by the stronger pressure gradient that is present on top of a tip, as we learned in the previous paragraph (see Figure 11.17). In the case of dielectric breakdown, which is the passage of a spark through an insulating material, the electric field around the tips of a branching discharge is stronger than elsewhere. In the case of a solidification process occurring in an undercooled liquid and initiated at a seed crystal, the positive action is the temperature gradient that is steeper around the bulge than elsewhere; the latent heat of solidification is shed around more rapidly at the tip of growing branches than on flat interfaces (Langer 1980). When the positive self-amplifying action dominates, a dendritic shape is formed, whereas when the negative force is overwhelming, a dense-branching morphology is produced.

TRY EXERCISE 11.14

11.9 MULTIFRACTALS

Look at the dendrites shown in Figures 11.16 and 11.18. You may notice that the density of points on the periphery of the dendrites is much lower than that in the cores of them. A fractal structure, like that of a dendrite, has a scale symmetry that is different from place to place. If we characterize a dendrite by determining just one dimension, we give an approximate description of its features. As far as such a complex fractal structure is concerned, it is more appropriate to define a distribution function that describes how the dimension varies in it. Such structure is called multifractal. A convenient way for quantifying the local variations in scale symmetry of a multifractal is the determination of its pointwise fractal dimensions. In the fractal, F, we select a point x_i and we consider a ball of radius r centered in x_i. We count the number of points that are inside the ball, $N_i(r)$, for $r \to 0$. We, then, define $\alpha = \lim_{r \to 0}\left(lnN_i(r)/\ln(r)\right)$. We repeat the determination for many other points of the structure. Let S_α the subset of F consisting of all the points with a pointwise dimension α. If α is a typical scaling factor in F, S_α will be a large subset; on the other hand, if α is unusual, S_α will be small. Each set S_α is itself a fractal, so it is reasonable to measure its size by its fractal dimension, $f(\alpha)$. $f(\alpha)$ is called the multifractal spectrum of F (Halsey et al. 1986). Therefore, a multifractal is an interwoven set of fractals of different dimensions α having $f(\alpha)$ as relative weights. The function $f(\alpha)$ is bell-shaped as shown in Figure 11.19. The maximum value of $f(\alpha)$ is the dimension calculated by the box counting method.

Multifractal structures arise in many nonlinear events and not just in DLA, viscous fingering and dielectric breakdown. Multifractals may be encountered in both fully-developed turbulences (Paladin and Vulpiani 1987) and fluid flows within random porous media (Stanley and Meakin 1988) and fractures in disordered solids (Coniglio et al. 1989) and chaotic attractors (Paladin and Vulpiani 1987). The concept of multifractality is also used to describe and interpret complex images and time series.

11.9.1 ANALYSIS OF THE COMPLEX IMAGES

Images that are discontinuous and fragmented are challenging to interpret. Such images may be analyzed by using the multifractals theory. A grayscale image can be described by the distribution of the intensity: bright areas have high densities, whereas dark areas have low densities. The local dimension α describes the pointwise regularity of the objects by considering the distribution of intensity, and it must be calculated for each point of the image. Then, the image can be partitioned into subsets consisting of points where the local dimension is α. For a two-dimensional image, points with $\alpha \approx 2$ are points where the intensity is regular. Points with $\alpha \neq 2$ lie in regions where there are singularities. If α is much larger or much smaller than 2, the region is characterized by discontinuities of the intensity. The value of $f(\alpha)$ gives the global information of the image. Points in a homogeneous region are abundant; therefore, they have a significant value of $f(\alpha)$. Edge points

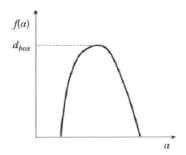

FIGURE 11.19   Shape of a multifractal spectrum $f(\alpha)$.

have smaller $f(\alpha)$ values because they are rare. If many "edge" points, having $\alpha \neq 2$, are present, we are in the presence of a homogeneous textured image.

Multifractal analysis has revealed itself as a powerful tool in the analysis of complex medical images for distinguishing healthy and pathological tissues (Lopes and Betrouni 2009); in geology for exploring geochemical data that have rugged and singular spatial variability (Zuo and Wang 2016). Multifractals may also be useful for predicting earthquakes (Harte 2001). The intuitive motivation is that earthquakes derive from fractal fractures of the terrestrial crust. Major fractures occur along major faults, the most dramatic being the tectonic plate boundaries. Within major fault systems, there are smaller faults that branch off and form a self-similar hierarchy of networks. These networks are often characterized by multifractal dimensions.

11.9.2 ANALYSIS OF THE COMPLEX TIME SERIES

Most of the time series that we encounter in the economy, biology, geology, and engineering (Tang et al. 2015) display many singularities, which appear as step-like or cusp-like features. Such singularities may be treated as fractals having different dimensions α and weights $f(\alpha)$. For instance, it has been demonstrated that heart rate fluctuations of healthy individuals are multifractal, whereas congestive heart failure, which is a life-threatening condition, leads to a loss of multifractality (Ivanov et al. 1999). In economy, the classical financial theories conceive stock prices as moving according to a random walk. Information digested by the market is the motor of the random movement of the prices. Since traditional finance theory assumes that the arrival of news is random, and no one knows if it will be good or bad news, then, the prices move randomly. However, because economies of the industrialized countries tend to grow, on average, economists model the randomness of news as having a slight bias toward good news. This model is referred to as the theory of random walk with drift. This theory cannot account for the sudden substantial price changes and spikes that are frequent in specific periods. According to Mandelbrot and Hundson (2004), markets are like roiling seas. Multifractals can describe their turbulence. Multifractal simulations of the market activity provide estimates of the probability of what the market might do and allow one to prepare for inevitable changes.

11.10 DIFFUSION IN FRACTALS

The diffusion exhibits a peculiar behavior when it occurs in fractals rather than in Euclidean spaces. In Euclidean spaces, the root-mean-square distance $\langle R^2 \rangle^{1/2}$ traveled by a random walker is proportional to $(Dt)^{1/2}$, where D is the diffusion coefficient (remind exercises 3.10 and 3.11). In a fractal, a random walker travels shorter distances after the same time interval because the path is winding, having many looped and branched routes. Therefore, the root mean square distance is

$$\langle R^2 \rangle^{\frac{1}{2}} \propto (Dt)^{\frac{1}{d_w}} \qquad \qquad [11.13]$$

wherein $d_w > 2$ (Havlin and Ben-Avraham 1987). Equation [11.13] means that the Brownian motion on a fractal substrate, forcing the walker into many detours, is delayed: $(Dt)^{1/d_w} < (Dt)^{1/2}$.

The probability of finding a random walker at position r in Euclidean space d_E, departing from $r = 0$, and after time t is given by the Gaussian distribution (Rothschild 1998)

$$p(r,t) = \frac{1}{(4\pi Dt)^{\frac{d_E}{2}}} e^{-\frac{r^2}{4Dt}} \qquad \qquad [11.14]$$

The Gaussian distribution [11.14] has zero mean and standard deviation $\sigma = \sqrt{2Dt}$. The probability of returning to the starting point $p(0,t)$ is proportional to $(4\pi Dt)^{-d_E/2}$. The smaller $p(0,t)$ and the wider the random walker's excursion. Hence, the reciprocal $[p(0,t)]^{-1}$ is a measure of the spatial excursion taken by the random walker. If the random walk occurs in a fractal having the internal dimension d_{fr}, the spatial excursion will be proportional to $(t)^{d_{fr}/2}$. But such excursion can be expressed as the root mean square distance traveled in the fractal having dimension d:

$$[p(0,t)]^{-1} \propto (t)^{\frac{d_{fr}}{2}} \propto \left\langle R^2 \right\rangle^{\frac{d}{2}} \propto (t)^{\frac{d}{d_w}} \qquad [11.15]$$

By comparing the time exponents in equation [11.15], we obtain the relationship between the exponents d, d_w, and d_{fr}:

$$d_{fr} = \frac{2d}{d_w} \qquad [11.16]$$

d_w is called the fractal dimension of the path of the anomalous random walk (Havlin and Ben-Avraham 1987), whereas d_{fr} is called fracton. d_{fr} is also called spectral dimension of the fractal having dimension d (Alexander and Orbach 1982) because the time-dependent probability of a random walker to return to a certain, previously visited site (i.e., $[p(0,t)]^{-1} \propto (t)^{d_{fr}/2}$), has a frequency-dependent counterpart, which is the density of vibrational states of the low-frequency phonons

$$\rho(v) \sim v^{d_{fr}-1} \qquad [11.17]$$

The dynamics of acoustic phonons, which are longitudinal displacement modes, conceptually equivalent to translational particle diffusion, are probed by vibrational spectroscopy using wavelengths of the order of mm (i.e., wavenumbers of a few cm^{-1}). If we use too long or too short wavelengths, we cannot detect the details of the connectivity of a fractal. It is like trying to characterize a fractal surface of an adsorbent material by exploiting molecules whose linear sizes are not comparable to the irregularities that we want to probe.

Dimension d is related to the morphology of a fractal, whereas d_{fr} and d_w are related to its topology. The fracton includes both the static attributes of the medium, characterized by the fractal dimension d and the dynamic aspects of the motion of the random walker, characterized by d_w. For the whole class of random fractals (such as the dendritic DLA) embedded in Euclidean two- or three-dimensions, $d_w > 2$. Hence, based on equation [11.16], it is $d_{fr} < d < d_E$, and d_{fr} is always $\approx 4/3$ (Alexander and Orbach 1982).

The diffusion in a linear polymer corresponds to a self-avoiding random walk, and it has $d_{fr} = 1$, $d_w = 2d$, no matter what its fractal dimensionality is. For a branched polymer, $d_{fr} \neq 1$. When the density of bridges is high enough, d_{fr} equals d because $d_w = 2$, independently of d (Helman et al. 1984).

TRY EXERCISE 11.15

11.11   CHEMICAL REACTIONS ON FRACTALS AND FRACTAL-LIKE KINETICS IN CELLS

Usually, we assume that the chemical reactions occur in dilute solutions that are spatially homogeneous. However, when diffusion-limited reactions occur on fractals or dimensionally restricted architectures,[15] the elementary chemical kinetics are quite different from those valid in traditional systems. In fact, in unconventional media, it has been found that, both phenomenologically and theoretically, the rate law of either a homodimeric ($A + A \rightarrow P$) or heterodimeric ($A + B \rightarrow P$)

---

[15] Examples of dimensionally restricted architectures are those encountered within a cell where chemical transformations may be confined to two-dimensional membranes or one-dimensional channels.

diffusion-limited reaction exhibits a characteristic reduction of the rate constant with time (Kopelman 1988):

$$k = k_0 t^{-h} \qquad [11.18]$$

In [11.18], k is the instantaneous rate coefficient,[16] $t \geq 1$, and $0 \leq h < 1$. For a homodimeric reaction of the type

$$A + A \rightarrow P \qquad [11.19]$$

the differential equation describing its anomalous evolution looks like

$$-\frac{d[A]}{dt} = k_0 t^{-h} [A]^2 \qquad [11.20]$$

After separating the variables and integrating between $(t = 0; [A] = [A]_0)$ and $(t; [A])$, we obtain

$$\frac{1}{[A]_0} - \frac{1}{[A]} = \frac{k_0}{(h-1)} t^{(1-h)} \qquad [11.21]$$

When the homodimeric reaction occurs in a homogeneous medium or systems made homogenous by effective stirring, $h = 0$, the equation [11.21] becomes the traditional integrated kinetic law with k as a constant.

Equation [11.20], having the rate coefficient dependent on time, is equivalent to a time-invariant rate law with an increased kinetic order, n. In fact, from [11.21], if we assume that $[A]_0^{-1}$ is negligible respect to $[A]^{-1}$, we achieve

$$t \approx \left(\frac{1-h}{k_0[A]} \right)^{\frac{1}{1-h}} \qquad [11.22]$$

The insertion of [11.22] into [11.20] yields

$$-\frac{d[A]}{dt} = k_0 \left(\frac{k_0}{1-h} \right)^{\frac{h}{1-h}} [A]^{2+\frac{h}{1-h}} \qquad [11.23]$$

It is evident that the reaction order $n = 2 + (h/(1-h))$ with respect to A exceeds 2. As the reaction becomes increasingly diffusion-limited or dimensionally restricted, h increases, the rate constant decreases more quickly with time (see equation [11.20]), and the kinetic order in the time-invariant rate law [11.23] grows well beyond the molecularity of the reaction (Savageau 1995). When the homodimeric reaction [11.19] takes place in a random fractal, embedded in a Euclidean two- or three-dimensional space,

$$h = 1 - \frac{d_{fr}}{2} \qquad [11.24]$$

Introducing [11.24] in equation [11.23], we obtain

$$-\frac{d[A]}{dt} \propto [A]^{1+\frac{2}{d_{fr}}} \qquad [11.25]$$

---

[16] The term "coefficient" rather than "constant" must be used when k depends on time.

Since $d_{fr} \approx 4/3$ for random fractals in two or three dimensions, the reaction order respect to A results in $n \approx 2.5$ (Kopelman 1988). The reaction [11.19] may also proceed on catalytic islands distributed on non-catalytic supports. Whether such "catalytic dust" is strictly fractal or is just made of monodisperse islands, the result is anomalous fractal-like reaction kinetics. In particular, for fractal dust, with $0 < d_{fr} < 1$, the reaction order with respect to A might assume large values: $3 < n < \infty$.

TRY EXERCISE 11.16

The relevant role that a fractal structure may have on the chemical kinetics is also proved in the nucleus of a eukaryotic cell. In the nucleus, the chromatin is a long polymer (each human cell contains about two meters of DNA) that fills the available volume with a compact polymorphic structure on which the transcription, replication, recombination, and repair of the genome occur. The three-dimensional architecture of chromatin is crucial to the functioning of a cell. Chromatin is a hierarchical structure spanning four spatial scales: from few nanometers to tens of micrometers. In fact, it is constituted by DNA (≈ 2 nm tick); nucleosomes (i.e., segments of DNA, each one wrapping eight histone proteins and forming a complex ≈ 10 nm long); chromatins that are fibers ≈ 30 nm high, generated by the nucleosomes that fold up; higher order chromatin loops and coils ≈ 300 nm in length, and, finally, ≈ 1.4 μm chromosome territories contained in the nucleus ≈ 20 μm large. Chromatin is actually a fractal (Bancaud et al. 2012). Some areas of the nucleus contain heterochromatin made of DNA packed tightly around histones. Other nuclear areas contain euchromatin that is DNA loosely packed. Usually, genes in euchromatin are active, whereas those in heterochromatin are inactive. The different activity is related to the different fractal structure (Bancaud et al. 2009). In fact, euchromatin seems to have a higher fractal dimension ($d \approx 2.6$), exposing a broader and rougher surface to the proteins scanning for their target sequences. Heterochromatin has a smaller fractal dimension ($d \approx 2.2$) because it is flatter and smoother and has a less extended surface. The more accessible DNA surfaces in euchromatin can be scanned more efficiently by nuclear factors favoring active transcription than in the heterochromatin. These conclusions have been drawn by collecting the time evolution of luminescent signals emitted by fluorescent proteins and markers (Bancaud et al. 2009). The decays of light-emitted intensities ($I(t)$) coming from fractal structures can be fitted by the stretched exponential (or Kohlrausch or KWW) function:

$$I(t) = I_0 e^{-\left(\frac{t}{\tau}\right)^{\beta}} \qquad\qquad [11.26]$$

where $0 < \beta \leq 1$, and τ is a parameter having, of course, the dimension of time (Berberan-Santos et al. 2005). The stretched exponential function [11.26] is the integrated form of the first order differential equation

$$\frac{dN}{dt} = -k(t)N \qquad\qquad [11.27]$$

where N is the concentration of the excited luminophores, and $k(t)$ is the time-dependent rate coefficient. When k is time-independent, we obtain the traditional exponential decay function [11.26] with $\beta = 1$. The stretched exponential function is used to describe the luminescence decay collected for any inhomogeneous, disordered medium in alternative to the maximum entropy method presented in Appendix B. Moreover, it is suitable to fit the transient signal originated by Resonance Energy

Transfer (RET) phenomenon based on the electrostatic interaction between dipoles and/or multipoles. In the case of RET, $\beta = d/s$, where d is the fractal or Euclidean dimension and s represents a parameter whose value depends on the RET mechanism. In fact, s is equal to 6, 8, or 10 for dipole-dipole, dipole-quadrupole, and quadrupole-quadrupole interactions, respectively.

11.12 POWER LAWS OR STRETCHED EXPONENTIAL FUNCTIONS?

If we look back at the equations appearing in this chapter, we might notice that many of them are power laws. For example, the equation describing the relationship between the mass of a dendritic fractal and its radius of gyration (equation [11.12]), or the relationship between the kinetic constant of a dimeric elementary reaction and time (equation [11.18]). Actually, power laws are widespread in science (West 2017): they may be encountered in many disciplines. In biology, the basal metabolic rate of an organism (R_E), the rate at which its cells convert nutrients to energy, scales with the organism's body mass (M) according to the following empirical power law (Schmidt-Nielsen 1984):

$$R_E \propto M^{\frac{3}{4}}$$ [11.28]

Moreover, the shape of a growing biological organism is ruled by the empirical power law

$$y \propto x^b$$ [11.29]

where x is the size of one part of the organism, whereas y is the size of another part, and b is a constant (Huxley 1932). The power law [11.29] describes the allometric morphogenesis, which is the development of form or pattern by differential growth among the component parts. It has been confirmed many times in biological context, but it is not confined only to organisms (Savageau 1979). Examples of other phenomena that obey power laws are the intensity of solar flares, the ranking of cities by size, the distribution of income within an economic system, the relative growth of staff within industrial firms, the fluctuations in stock prices, and the social differentiation and division of labor in primitive societies. Power laws recur even in geology. The Gutenberg-Richter law (Gutenberg and Richter 1956), describing the link between earthquake magnitude (Ma)[17] and its frequency (f) is

$$f = 10^{a-b(Ma)}$$ [11.30]

where a and b are two constants. Omori's law (Utsu et al. 1995) that predicts the frequency of aftershocks (ν) is a power law of time (t):

$$\nu = \frac{k}{(c+t)^p}$$ [11.31]

with c and p being constants.

---

[17] The magnitude in the Richter scale is the base-10 logarithmic scale of the ratio (R) of the amplitude of the seismic waves to an arbitrary minor amplitude, as recorded on a standardized seismograph at a standard distance. Since $Ma = \log_{10}(R)$, equation [11.30] is a power law: $f = 10^a(R^{-b})$.

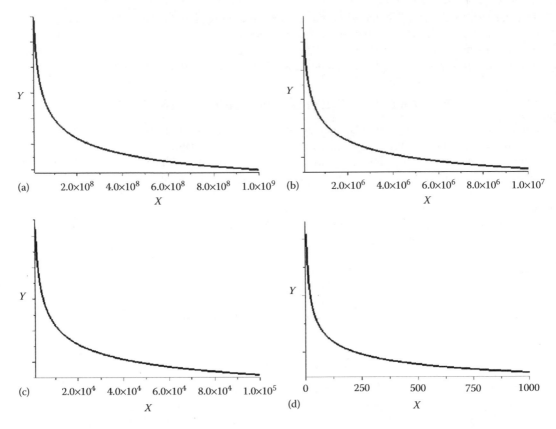

FIGURE 11.20 Self-similarity of a power law. The graphs (a)–(d) zoom in on the power law.

The recurrence of power laws is reminiscent of fractal phenomena. A power law is itself self-similar (Mitchell 2009) like any fractal. Figure 11.20 demonstrates that the power law is scale-invariant. It has the same shape, whatever is the scale of the plot.

An effective strategy to recognize power laws is to build log-log plots of experimental data. In fact, a true power law appears as a straight line in a double logarithmic axis plot (see Figure 11.21). Most of the natural, social, and economic phenomena, cited earlier, when analyzed in a sufficiently wide range of possible values of their variables, may exhibit a limited linear regime followed by a significant curvature. Such trend is well-fitted by a stretched exponential function of the type

$$y = Ae^{\left[-\left(\frac{x}{x_0}\right)^c\right]}$$
[11.32]

where $0 < c < 1$ and x_0 are constant parameters (Figure 11.21). When $c = 1$, the stretched exponential function transforms in a pure exponential function.

The fact that most of the natural, social and economic distributions display a curved trend in a double logarithmic axis plot, decaying faster and leading to thinner tails than a power law, has been interpreted in terms of finite-size effects (Laherrère and Sornette 1998). The natural shapes and phenomena look like fractals, but they are not genuine fractals in the mathematical sense because this would mean a self-similarity ad infinitum.

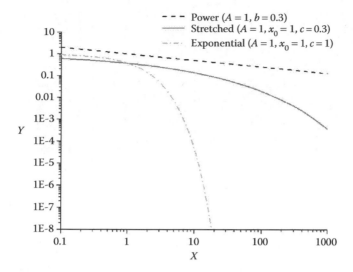

FIGURE 11.21 Representation of a power law with its exponent $b - 0.3$ (dashed line); a stretched exponential with $c = 0.3$, $x_0 = 1$(continuous line), and an exponential function (dashed-dotted line).

11.13 WHY DOES CHAOS GENERATE FRACTALS?

At the beginning of this chapter, we learned that chaotic systems manufacture fractals. For this reason, fractals are also named as "Chaos in Space." How is it possible that the "Chaos in Time" gives rise to fractals? Chaos appears only in the nonlinear regime. Whereas the linearity means "stretching," the nonlinearity implies "stretching" and "folding." In the nonlinear regime, the continual repetition of the two actions, "stretching" and "folding," generates fractals. Let us consider a chaotic system in its phase space. When we study its time evolution, we start from a restricted portion of the phase space, which includes all the possible initial conditions that are indistinguishable (due to the unavoidable uncertainty in defining the initial conditions). This portion of the phase space could be a sphere or a cube or any other simple volume.[18] If we let the system evolve, each point traces its trajectory. The portion of the phase space embedding all the initial points changes its shape during the evolution of the dynamics. The countless repetition of "stretching" and "folding" actions transforms the original shape into a fractal. And the metamorphosis is complete at the infinite time. We can make this idea appetizing and easy to remind if we think about a pastry chef making a croissant. The chef maneuvers the dough patiently. The dough represents the portion of the phase space embedding all the possible indistinguishable initial conditions. The chef stretches the dough with a rolling-pin. Then, he puts a layer of butter on the dough and folds it. He stretches it again, puts another layer of butter and folds it again. These actions are repeated many times. Ideally, ad infinitum. Finally, after cooking the dough, the chef gets a croissant. And a croissant is a fractal-like structure. It is evident if we cut the croissant into two halves, and we observe them in the direction parallel to the table (see Figure 11.22). Both halves of the croissant show an almost infinite number of layers.

This culinary similarity is also suitable to explain the sensitivity to the initial conditions of the chaotic dynamics. In fact, if we consider two close points in the initial spherical dough (representing two distinct initial conditions), then, just after the first stretching, they are pulled apart. The folding can approach them. But, when the chef stretches the dough again, they even get farther apart. In one of the many folding operations, the two points may end up in different layers. Since then, the two points have two completely different evolutions, likewise the two trajectories in their phase space.

---

[18] When the dimensions of the phase space are larger than three, the portion of the phase space including the initial conditions will be either a hypersphere or a hypercube or any other simple hypervolume.

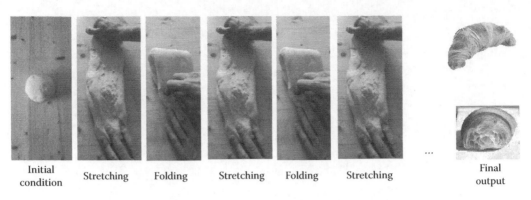

| Initial condition | Stretching | Folding | Stretching | Folding | Stretching | ... | Final output |

FIGURE 11.22   A fractal-like croissant is obtained by stretching and folding the dough, many times.

11.14 CHAOS, FRACTALS, AND ENTROPY

We know very well that any "dissipative" system abides by the Second Law of Thermodynamics and increases entropy because, in the course of its evolution, it degrades potential and/or kinetic energy into heat. We also know that any "conservative" non-chaotic system does not produce entropy because the sum of its kinetic and potential energies remains constant when it evolves. Non-chaotic Hamiltonian systems are theaters of reversible transformations. But, what about the entropy change in the evolution of a chaotic "conservative" system (Baranger 2000)? As we already know very well, the initial condition of the system can only be defined with a finite degree of accuracy. All the equally likely initial points are confined within a portion of the phase space whose volume is $V_i = \sum_{k=1}^{N_i} n_k$, where N_i is the total number of points contained in V_i. The initial uncertainty[19] we have about the state of the system is

$$H_i = -\sum_{k=1}^{N_i} p_k \log_2 p_k = \log_2(V_i) \qquad [11.33]$$

If we now let the system evolve, the shape of the volume containing all the initial points changes drastically. In fact, each point inside V_i follows its own trajectory. Points that are nearby at the beginning, then, diverge away from each other in an exponential manner. Due to the constant action of "stretching" and "folding" events, tendrils appear and knots of all sorts form. The structure becomes finer and finer until we get a fractal (see Figure 11.23).

In the Hamiltonian mechanics, an important theorem holds. It is the Liouville's theorem stating that the initial phase space volume, defined by the sum of all possible initial conditions, does not change. In other words, the final volume of the phase space, V_f, including the trajectories of all the

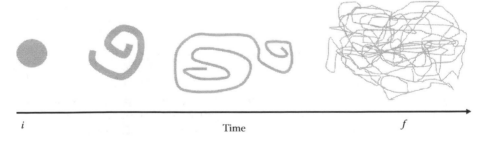

| i | Time | f |

FIGURE 11.23   Metamorphosis of a spherical portion of the phase space for a chaotic Hamiltonian system.

[19] Remember the Theory of Information formulated by Claude Shannon and presented in Chapter 2.

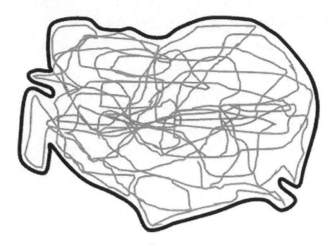

FIGURE 11.24 The smooth black bag circumscribes the fractal generated by the chaotic Hamiltonian system.

initial points, contained in V_i, is equal to the initial volume: $V_f = V_i$. This condition would mean that the uncertainty at the end of the transformation, $H_f = -\sum_{k=1}^{N_f} p_k \log_2 p_k = \log_2(V_f)$, would be equal to H_i. However, the final shape is a fractal, full of holes, knots, and tendrils. It is a complex structure (see Figure 11.23)! For the determination of its volume, we are forced to circumscribe it within a larger, smooth bag, like in Figure 11.24.

The outcome is that the final volume $V_f' > V_i$. Therefore, the entropy variation given by $\Delta S = H_f - H_i$ is positive. The chaotic dynamics gives rise to a fractal, which is "Chaos in Space." The fractal-genesis brings about an increase of our ignorance and hence entropy.

11.15 KEY QUESTIONS

- What is the common feature of the strange attractors?
- What is a fractal?
- How do we determine the dimension of a self-similar fractal?
- How do we obtain the Mandelbrot and the Julia sets?
- Can we describe the "roughness" of many natural complex shapes through the fractal geometry? Are there genuine fractals in nature?
- How can you determine the dimension of a fractal-like structure?
- Describe the factors that contribute to the formation of patterns in a Hele-Shaw cell.
- Describe the properties of dendritic fractals.
- What is a multifractal?
- What characterizes the diffusion in a fractal?
- What does mean fractal kinetics?
- What is the difference between power laws and stretched exponential functions?
- Why does chaotic dynamics generate fractals?
- Can a Hamiltonian system produce entropy?

11.16 KEY WORDS

Strange attractors; Self-similarity; Dimension of an object; Dendrites; Multifractals; Fractons; Power Law; Stretched Exponential Function.

11.17   HINTS FOR FURTHER READING

Three other introductory books regarding fractals are that by Feldman (2012) and the two volumes by Peitgen et al. (1992a, 1992b).

11.18   EXERCISES

11.1. Prove that the Hénon attractor is self-similar if you expand the exterior border of the boomerang-like structure after computing at least one million successive iterates of equations [11.1 and 11.2], starting from the origin (i.e., $[x_0, y_0] = [0, 0]$).

11.2. The first steps needed to build a Cantor set by eliminating the middle third of every interval are depicted in Figure 11.3. Assume that the length of C_0 is 1 in arbitrary units (*a.u.*). Which is the length of the two segments of C_1? What is the length of each line segment in C_2? And C_2 total length? Try to answer the same questions for C_3. Then, determine the general expression defining the number of line segments, the length of each line segment, and the total length at the n-th iteration. After an infinite number of iterations, which is the number of line segments, and which is the full length?

11.3. The first steps needed to build a Koch curve are shown in Figure 11.4. Assume that the length of K_0 is 1 in arbitrary units (*a.u.*). Which is the total length of K_1? What is the length of each line segment in K_2? And K_2 entire length? Try to answer the same questions for K_3. Then, determine the general expression defining the number of line segments, the length of each line segment, and the total length at the n-th iteration. After an infinite number of iterations, which is the number of line segments and which is the full length?

11.4. Consider the Sierpinski gasket shown in Figure 11.8. Find the answers to the following questions in iterations 1, 2, 3, and $n \rightarrow \infty$:
1. What are the number of triangles?
2. If A is the total area of the triangle at S_0, which is the area of each triangle and the full area of the structure?
3. If L is the length of one side in S_0, which is the perimeter of each triangle and the total perimeter of the figure?

11.5. A Cantor set different from that depicted in Figure 11.3 can be obtained by dividing the initial segment into five equal pieces, delete the second and the fourth subintervals, and, then, repeat this process, indefinitely. Determine the dimension of this fractal, called the even-fifths Cantor set since the even fifths are removed at each iteration.

11.6. If a square is scaled up by a factor of 3, what happens to its area? If the Sierpinski gasket is scaled up by a factor of 3, what happens to its size?

11.7. Construct the Lévy C curve fractal by using the following instructions: start from a segment (LC_0). Use the segment as the hypotenuse of an isosceles triangle with angles of 45°, 90°, and 45°. Then, remove the original segment, and you obtain LC_1. At the second stage, repeat the procedure for the two segments, conceiving both of them as the hypotenuses of two equivalent isosceles triangles. Repeat the steps many times. How does the fractal look like? Which is its dimension?

11.8. Solve iteratively equations [11.7] by starting by the following seeds: $(z_0, c) = (0, +i)$; $(0, -i)$; $(0, 2 - i)$; $(0, 0.01 - 0.5i)$. Calculate how the distance r of $z = a + ib$ from the origin of the graph, changes after each iteration by using the formula $r = \sqrt{a^2 + b^2}$.

11.9. Calculate the fate of the seeds $z_0 = 0, 0.5, 1$ for $c = 0$ and 1.

11.10. Do the seeds $(-0.12 + 0.75i)$ and $(0.4 - 0.3i)$ belong to the Mandelbrot set? In other words, do their distances decrease after some iterations of equation [11.7]?

11.11. The universe is another example of a natural fractal. Which fractal set does the universe look like?

11.12. Print the map of Norway on an A4 paper and determine the dimension of its border by using the Box Counting method. For the application of the Box Counting method, prepare four grids, based on unit-squares having sides 2.5, 2, 1.5, 1 cm long, respectively, and print them on transparencies.

11.13. Build your Hele-Shaw cell. By using both pure glycerin and 4% PVA cross-linked by addition of sodium borate (for example, add 0.5 mL of 4% $Na_2B_4O_7$ to 25 mL of 4% PVA) as viscous fluids, try to achieve dense-branching morphologies and dendrites. Film the experiments and extract images of the best patterns you obtained. Print them on A4 papers and determine their dimensions by using the Box Counting method. For this purpose, prepare four grids, based on unit-squares having sides 2.5, 2, 1.5, 1 cm long, respectively, and print them on transparencies.

11.14. If a spherical silver nanoparticle has a mass M and a radius r, how much do its mass increase by doubling its radius? If we have synthesized a fractal silver nanoparticle having a radius of gyration R, mass M, and dimension $d = 3.5$, how much do its mass increase by doubling its radius of gyration?

11.15. The fractal dimension of a protein can be determined by calculating the mean value of the number of alpha carbon atoms that lie within a sphere of radius R centered at an arbitrary alpha carbon. According to this procedure, it results that myoglobin has $d \approx 1.67$ (Helman et al. 1984). The Raman electron-spin-relaxation measurements indicate that the protein vibrates like a compact structure (MacDonald and Jan 1986). What does this evidence mean for d_{fr} and d_w?

11.16. Imagine having a fractal-like homodimeric reaction described by the differential equation [11.20]. We carry out this chemical transformation in two reactors, maintained at the same thermodynamic conditions of temperature, pressure, and so forth. The only difference is the initial concentration of A. In the two reactors, we reach the same instantaneous concentration $[A]_i$ at two distinct delay times: in the first, at time $t = 15$ s, whereas, in the second, at time $t = 5$ s. Is the reaction proceeding with the same rate in the two reactors, both having the same $[A] = [A]_i$? If $h = 0.33$, how much is the instantaneous rate coefficient?

11.19 SOLUTIONS TO THE EXERCISES

11.1. If you use MATLAB to solve this exercise and you plot the output disregarding the first 50 points, the script could be like that reported as follows.

```
function Henon_map(a, b)
N=1000000;
x=zeros(1,N);
y=zeros(1,N);
x(1)=0;
y(1)=0;
for i=1:N
  x(i+1)=1+y(i)-a*(x(i))^2;
  y(i+1)=b*x(i);
end
axis([-1,2,-1,1])
plot(x(50:N),y(50:N),'.','MarkerSize',2);
fsize=15;
set (gca,'xtick',[-1:1:1],'FontSize',fsize)
set (gca,'ytick',[-1:1:2],'FontSize',fsize)
xlabel('\itx','FontSize',fsize)
ylabel('\ity','FontSize',fsize)
end
```

The command line will be *Henon_map(1.4 0.3)*, and the picture you obtain looks like Figure 11.2.

11.2. If the length of C_0 is 1 *a.u.*, the length of two segments of C_1 is $(1/3+1/3=2/3)a.u.$ In C_2, we have four segments. The length of each segment is $(1/9)$ *a.u.*, and the total length is, of course, $(4/9)$ *a.u.* In C_3, we have eight segments. The length of each segment is $(1/27)$ *a.u.*, and the total length is $(8/27)$ *a.u.* After n iterations, the number of segments is 2^n, and the total length is $(2/3)^n$. For $n \to \infty$, the number of segments goes to infinite, and the full length goes to zero.

11.3. If the length of K_0 is 1 *a.u.*, the total length of K_1 is $(4 \times 1/3)a.u.$ In K_2, we have 16 segments, and the length of each one of them is $(1/9)$ *a.u.* In K_3, we have $4 \times 16 = 64$ segments; the length of each of them is $(1/27)$ *a.u.* The total length will be $(64/27)$ *a.u.* After n iterations, the number of segments will be 4^n, and the full length will be $(4/3)^n$. For $n \to \infty$, the number of segments and the total length go to infinite.

11.4. (1) After one iteration, the number of triangles is 3; after two iterations, it is $3^2 = 9$; after three iterations, it is $3^3 = 27$, and after n iterations, it is 3^n. In the end, the number of triangles becomes infinite.

(2) If A is the initial full area of the triangle, after one iteration, the area of each triangle is $(1/4)A$, and since the number of triangles is 3, the full area is $(3/4)A$. Analogously, after two iterations, the area of each triangle is $(1/4)^2A$. Knowing the number of triangles, it derives that the total area is $(3/4)^2A$. After three iterations, the area of the single triangle is $(1/4)^3A$. The total area will be $(3/4)^3A$. After n iterations, the entire area will be $(3/4)^nA$, i.e., it shrinks to zero for $n \to \infty$.

(3) If L is the length of one side of the first triangle, the total perimeter is $3L$. The length of one side after one iteration is $(\frac{1}{2})L$, and the perimeter of just one triangle will be $(3/2)L$. The total perimeter will be $3(3/2)L$. After the second iteration, the length of one side of a triangle is $(1/4)L$. The perimeter of one triangle is $(3/4)L$, and the total perimeter will be $3^2(3/4)L$. After the third iteration, the length of one side is $(1/8)L$; the perimeter of one triangle is $(3/8)L$, and the total perimeter will be $3^3(3/8)L$. Now, we can generalize and state that, after n iterations, the length of one side is $(1/2)^nL$, the perimeter of one triangle is $(3/2^n)L$, and the total perimeter will be $3^n(3/2^n)L$. In other words, the perimeter becomes longer and longer.

In the end, the Sierpinski gasket is made of an infinite number of triangles having a negligible area but an infinite perimeter.

11.5. For the determination of the dimension of the even-fifths Cantor set, we apply equation [11.3]. The scale factor is $r = 5$. The number of copies is $N = 3$. Therefore, the dimension is $d = \log(3)/\log(5) \approx 0.68$.

11.6. If a square is scaled up by a factor of 3, its area increases by $3^2 = 9$ because the square is bi-dimensional. On the other hand, if the Sierpinski gasket is scaled up by a factor of 3, its size increases by $3^{1.58} \approx 5.67$.

11.7. The structure of the fractal after the first five iterations is shown in Figure 11.25.

After 11 iterations, it looks like a highly-ornamented letter C (see Figure 11.25).

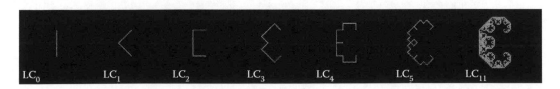

LC$_0$ LC$_1$ LC$_2$ LC$_3$ LC$_4$ LC$_5$ LC$_{11}$

FIGURE 11.25 The first five iterations, and the output of the eleventh one (LC$_{11}$) for the generation of the Lévy C curve fractal. The pictures have been obtained by using "Examples of Fractals" model, Complexity Explorer project, http://complexityexplorer.org.

If L is the length of the original segment, the scale factor is $(\sqrt{2})L$. The number of copies is 2. Therefore, the dimension is $d = \log(2)/\log(\sqrt{2}) = 2$. Surprise: d is an integer number, which corresponds to the value of bi-dimensional objects. After n iterations, the structure consists of 2^n segments, whose length is $(1/\sqrt{2})^n L$. The total length grows at each stage: after n iterations, it is $(2/\sqrt{2})^n L$.

11.8. The results of the iterations for the different seeds are reported in the following tables. The results listed in Tables 11.1 and 11.2 demonstrate that starting from either $(0,+i)$ or $(0,-i)$, the distance r oscillates between two values.

In the third case, starting from $(0,2-i)$, we see that the distance escapes to infinity (Table 11.3).

TABLE 11.1

Iterative Solutions of Equations [11.7], Starting from $(z_0, c)=(0,+i)$

Step (n)	z_n	r
0	0	0
1	$+i$	1
2	$-1+i$	$\sqrt{2}$
3	$-i$	1
4	$-1+i$	$\sqrt{2}$
5	$-i$	1

TABLE 11.2

Iterative Solutions of Equations [11.7], Starting from $(z_0, c)=(0,-i)$

Step (n)	z_n	r
0	0	0
1	$-i$	1
2	$-1-i$	$\sqrt{2}$
3	$+i$	1
4	$-1-i$	$\sqrt{2}$
2	$+i$	1

TABLE 11.3

Iterative Solutions of Equations [11.7], Starting from $(z_0, c) = (0,2-i)$

Step (n)	z_n	r
1	$2-i$	$\sqrt{5}$
2	$5-5i$	$\sqrt{50}$
3	$2-51i$	$\sqrt{2605}$
4	$-2595-205i$	2603

Finally, in the fourth case (see Table 11.4), it never escapes to infinity.

11.9. The outputs of the iterations are reported in the following tables (Tables 11.5 and 11.10). When $c = 0$, the seeds do not escape. On the other hand, when $c = 1$, all the three seeds escape. The Julia sets for $c = 0$ and $c = 1$ are shown in Figure 11.26, obtained through the Julia set generator at the link http://www.shodor.org/interactivate/activities/JuliaSets/.

Note that for $c = 1$, there are no black areas, but only scattered black dots. The black color denotes points that do not escape, but they approach the origin. In other words, they are "prisoners." The shades of gray colors represent the points that escape. Different shades of gray are associated with different speeds of escape. For $c = 0$, the Julia set is a circle of radius 1 containing "prisoners."

TABLE 11.4
Iterative Solutions of Equations [11.7],
Starting from $(z_0, c) = (0, 0.01 - 0.5i)$

Step (n)	z_n	r
1	$0.01 - 0.5i$	0.5
2	$-0.2399 - 0.51i$	0.56
3	$-0.19255 - 0.255i$	0.3195
4	$-0.018 - 0.4017i$	0.402
5	$-0.151 - 0.4855i$	0.508
6	$-0.203 - 0.3534i$	0.408

TABLE 11.5
Iterative Solutions of Equations [11.7],
Starting from $(z_0, c) = (0, 0)$

Step (n)	z_n
0	0
1	0
2	0

TABLE 11.6
Iterative Solutions of Equations [11.7],
Starting from $(z_0, c) = (0, 1)$

Step (n)	z_n
0	0
1	1
2	2
3	5
4	26

TABLE 11.7
Iterative Solutions of Equations [11.7],
Starting from $(z_0, c) = (0.5, 0)$

Step (n)	z_n
0	0.5
1	0.25
2	0.0625
3	0.0039

TABLE 11.8
Iterative Solutions of Equations [11.7],
Starting from $(z_0, c) = (0.5, 1)$

Step (n)	z_n
0	0.5
1	1.25
2	2.56
3	7.566
4	58.25

TABLE 11.9
Iterative Solutions of Equations [11.7],
Starting from $(z_0, c) = (1, 0)$

Step (n)	z_n
0	1
1	1
2	1
3	1
4	1

TABLE 11.10
Iterative Solutions of Equations [11.7],
Starting from $(z_0, c) = (1, 1)$

Step (n)	z_n
0	1
1	2
2	5
3	26
4	677

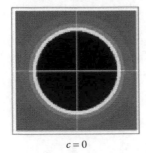
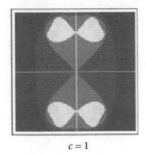

$c = 0$ $c = 1$

FIGURE 11.26   Julia sets for $c = 0$ (on the left) and $c = 1$ (on the right).

11.10. The outputs of the first iterations for the two seeds proposed in this exercise are reported in Tables 11.11 and 11.12. Both seeds belong to the Mandelbrot set. For the first seed ($c = -0.12 + 0.75i$), the distance from the origin oscillates periodically. For the second seed ($c = 0.4 - 0.3i$), the distance maintains small, but it changes irregularly.

By using the two seeds as c values of two distinct Julia sets, we know (remember Figure 11.10) that the first gives rise to a "connected-type" and the second to a "Cantor-type" Julia set, respectively (Tables 11.11 and 11.12).

11.11. The universe observed from the Earth looks like a Cantor-set. The stars are the points of this "natural" Cantor set.

11.12. Norway is famous for its fjords that make its coast particularly irregular (see Figure 11.27). If we apply the Box Counting method, we find that the dimension of Norway coast is $D = (1.42 \pm 0.14)$. Feder in his book (1988) reports $D \approx 1.52$. Since the Norway coast is more jagged than that of Britain, its fractal dimension is larger.

11.13. By using glycerin as the viscous fluid, we obtain dense-branching morphologies (see patterns a and c in Figure 11.16) because the surface tension at the interface between the viscous and non-viscous fluids is quite strong. By using aqueous solutions of cross-linked PVA as the viscous fluid, the surface tension at the interface is reduced, and it is more likely to obtain dendrites. The dimensions of the fractal-like patterns shown in

TABLE 11.11

Iterative Solutions of Equations [11.7], Starting from $(z_0, c) = (0, -0.12 + 0.75i)$

Step (n)	z_n	r
1	$-0.12 + 0.75i$	0.76
2	$-0.668 + 0.57i$	0.878
3	$0.00146 - 0.0116i$	0.0117
4	$-0.12 + 0.75i$	0.76
5	$-0.668 + 0.57i$	0.878
6	$0.00156 - 0.0113i$	0.0114

TABLE 11.12

Iterative Solutions of Equations [11.7], Starting from $(z_0, c) = (0, +0.4 - 0.3i)$

Step (n)	z_n	r
1	$0.4 - 0.3i$	0.5
2	$0.47 - 0.54i$	0.716
3	$0.329 - 0.808i$	0.873
4	$-0.144 - 0.832i$	0.84
5	$-0.271 - 0.06i$	0.28
6	$0.4698 - 0.2675i$	0.54
7	$0.55 - 0.55i$	0.78
8	$0.4 - 0.9i$	0.98
9	$-0.26 - 1.02i$	1.05
10	$-0.575 + 0.23i$	0.62

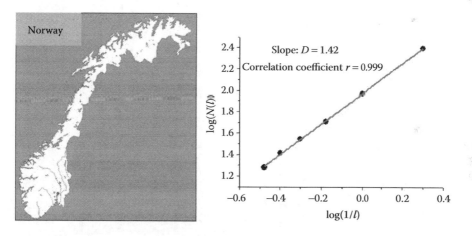

FIGURE 11.27   Determination of the dimension D of Norway coast by the Box Counting method.

Figure 11.16, determined by the Box Counting method, are reported in Figure 11.28. It is evident that the dense-branching morphologies have $d = 1.72$ as the dimension. The well-developed dendrite in (b) has $d = 1.70$. On the other hand, the other two dendrites in (d) and (e) has smaller dimensions: $d = 1.40$ and 1.53, respectively.

11.14. The mass of the spherical silver nanoparticle having density d increases eight times because $M_i = d\left(\frac{4}{3}\pi r_i^3\right)$, $r_f = 2r_i$, and hence $M_f = 8M_i$. For the fractal nanoparticle, the growth of the mass is larger. In fact, $M_f = 2^{3.5} M_i \approx 11.3M_i$.

11.15. Since myoglobin vibrates as a compact structure, $d \approx d_{fr}$. Thus, according to equation [11.16], $d_w \approx 2$.

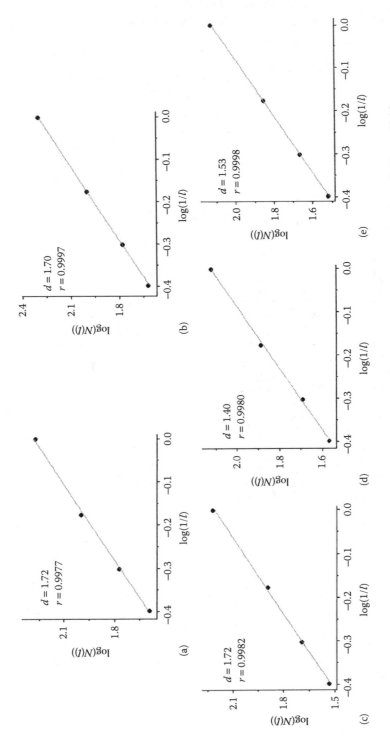

FIGURE 11.28 Determination by the Box Counting method of the dimensions of the five fractal-like shapes (a–e) shown in Figure 11.16.

11.16. Although the species A has the same instantaneous concentration in the two reactors, the rates of reaction are different. In fact, $k = k_0/(15)^{0.33}$ in the first reactor, whereas $k = k_0/(5)^{0.33}$ into the second one. At the time of the observation, the reaction proceeds faster into the second reactor. How is it possible? The two reactors have a different history. The longer time elapsed in the first reactor has "degraded" the rate coefficient k. If they were catalytic reactions, we would say that the catalyst got "poisoned." But there is no "poison." Only when we deal with homogenous distribution of the reactants in Euclidean spaces, we should expect that two or more systems, with identical conditions, will have equal reactions rates.

12 Complex Systems

The whole is something besides the parts.

Aristotle (384–322 BC)

As long as a branch of science offers an abundance of problems,
so long it is alive; a lack of problems foreshadows extinction or
the cessation of independent development.

David Hilbert (1862–1943)

12.1 THE NATURAL COMPLEXITY CHALLENGES

A major goal of science is that of solving problems. In fact, the scientific knowledge has allowed humanity to reach outstanding achievements in the technological development and economic growth, especially in the last three hundred years (Beinhocker 2007), or so. Despite many efforts, there is still a list of compelling challenges that has to be won. First, we are unable to predict catastrophic events on our planet, like earthquakes or volcanic eruptions, which, unfortunately, may cause many victims and damages every time they occur. Second, we are still striving to protect our environment and ecosystems from some risks, such as the climate change and the shrinkage of biodiversity. Third, we need to find innovative solutions to guarantee a worldwide sustainable economic growth, primarily by focusing on the energy issue. Fourth, we struggle to ensure stability in our societies. Finally, there are still incurable diseases that must be defeated, such as cancer, Alzheimer, Parkinson, diabetes, HIV, to cite just a few.

Why do we find so many difficulties in winning such challenges? Because, whenever we face them, we deal with Complex Systems, such as the geology and the climate of our planet; the ecosystems; the living beings; the human brain; the immune system; the human societies, and the economy.

We encounter substantial limitations in describing Complex Systems exhaustively when we start from the fundamental physical laws. Why?

12.2 THE COMPUTATIONAL COMPLEXITY OF THE NATURAL COMPLEX SYSTEMS

The solution of the equations that describe Natural Complex Systems, based on the knowledge of their ultimate constituents, such as atoms, remains a task that sounds unreasonable. In fact, it is entirely beyond our reach. The difficulty stems from the exponential growth of the computational

$$51$$
$$\times \quad 13$$
$$\overline{153}$$
$$+ \quad 51$$
$$\overline{663}$$

FIGURE 12.1   An example of pencil-and-paper algorithm for solving a multiplication.

cost with the number of interacting particles (Whitfield et al. 2013).[1] The computational cost, also called computational complexity of an algorithm, is a measure of the number of steps the algorithm requires to solve a problem, expressed as a function of the size of the problem. In the field of the Computational Complexity, a problem is a computational task coupled to a question (Goldreich 2008). An example is the Schrödinger equation that wants to answer the question: "Which is the total energy of a chemical system consisting of many interplaying particles?" Another example is the famous Traveling Salesman Problem (TSP). It asks: "Given a graph with nodes (i.e., a map with cities), edges and costs associated with the edges (i.e., connections and their costs), which is the least-cost closed tour containing every node (i.e., every city) just once?" An algorithm for solving a problem is a finite set of instructions or steps that guarantee to find the correct solution to any instance of that problem. If we imagine using a Turing machine, the steps of an algorithm can be the addition, subtraction, multiplication, finite-precision division, and the comparison of two numbers. An example of an algorithm is the pencil-and-paper procedure of multiplication learned at school. Let us recall it with an instance shown in Figure 12.1.

The pencil-and-paper method applied to the example reported in Figure 12.1 requires a total of four multiplications and three additions to achieve the final output. It is easy to infer that the total number of steps depends on the size of the numbers. For having an idea of the running time of an algorithm solving a particular problem, it is useful to determine the function that relates the number of computational steps to the size of the problem. Each elementary step is performed at the same speed, or, in other words, each operation requires the same amount of time that depends on the intrinsic speed of our computational machine. The size of a problem is expressed in terms of bits, N, required to encode an instance of that problem. For example, in the case of TSP, the running time of an algorithm is expressed as a function of the number of nodes, the number of edges plus the number of bits required to encode the numerical information of the edge costs. A problem is easy if

---

[1] The equation of conventional nonrelativistic quantum mechanics, which in principle can describe every Complex System we mentioned in paragraph 12.1, is:

$$i\hbar \frac{\partial}{\partial t}\big|\Psi\big\rangle = H\big|\Psi\big\rangle$$

where the Hamiltonian operator is:

$$\mathcal{H} = -\sum_{j=1}^{N_e} \frac{\hbar^2}{2m}\nabla_j^2 - \sum_{k=1}^{N_n} \frac{\hbar^2}{2M_k}\nabla_k^2 - \sum_{j=1}^{N_e}\sum_{k=1}^{N_n} \frac{Z_k e^2}{\big|\vec{r}_j - \vec{R}_k\big|} + \sum_{j<l}^{N_e} \frac{e^2}{\big|\vec{r}_j - \vec{r}_l\big|} + \sum_{k<m}^{N_n} \frac{Z_k Z_m e^2}{\big|\vec{R}_k - \vec{R}_m\big|}$$

In the definition of $\mathcal{H}$ (wherein intranuclear interactions and gravity are not included), e and m are the electron charge and mass, $\vec{r}_j$ is the position vector of the jth electron, N_e their total number; $Z_k e$, M_k, and $\vec{R}_k$ are the charge, the mass and the position of the kth nucleus, and $\hbar$ is the Planck's constant divided by 2π. The Schrödinger equation can be solved accurately only for the very simple hydrogen atom involving two particles (the nucleus and one electron). Any problem with three or more particles cannot be solved accurately. This statement also holds for macroscopic bodies interacting through the gravitational force. We need to apply an approximate method, such as the perturbation approach, to achieve a reasonable solution (Kwapień and Drożdż, 2012).

it is polynomial (*P*), for instance if the number of computational steps increases polynomially with the size of the problem. This idea is usually expressed by the Big-O notation. For two functions, *f*(*x*) and *g*(*x*) of a non-negative parameter *x*, we say that $f(x) = O(g(x))$, if there exists a constant $c \geq 0$ such that, for $x \to \infty$, $f(x) \leq cg(x)$. In terms of Big-O notation, a problem, *f*(*N*), is tractable if

$$f(N) = O(N^k) \tag{12.1}$$

wherein *k* is a constant, which could be 1, 2, etc. The amount *N* never appears in the exponent. If $k = 1$, it means that the number of computational steps grows linearly with the size of the problem; if $k = 2$, quadratically, and so on. The class of the polynomial problems, *P*, is the class of the recognition problems. An example of a recognition problem is: "track the telephone number of a person in a phone book." If *N* is the total number of elements in the book, the computational steps are, at most, equal to *N*.

However, there also exists exponential problems for which the number of computational steps grows exponentially with the size of the problem itself:

$$f(N) = O(N^N) \text{ or } f(N) = O(2^N) \; f(N) = O(N!) \tag{12.2}$$

These problems are intractable when *N* is large. Examples are the Schrödinger equation for which $f(N) = O(2^N)$ holds (Bolotin 2014), and the TSP for which $f(N) = O(N!) \approx O((N/e)^N)$. It suffices to make a few calculations to understand why exponential problems cannot be solved accurately and in a reasonable time, even if we have the best supercomputers in the world at our disposal. Let us consider the Schrödinger equation. For ten interacting particles, the maximum number of computational steps needed to determine the energy of the system is $2^{10} = 1024$; if $N = 20$, the n° of steps is $2^{20} \approx 1 \times 10^6$. According to the TOP500 list,[2] updated in November 2017, the fastest supercomputer in the world is the Chinese Sunway TaihuLight that reaches the astonishing computational rate of 93 PFlop/s.[3] With TaihuLight at our disposal, we need just ten femtoseconds and ≈ 10 picoseconds to solve the Schrödinger equation for a system with 10 and 20 particles, respectively. But, if our system consists of 500 particles, the number of computational steps becomes so huge, $2^{500} \approx 3.3 \times 10^{150}$, even TaihuLight would require an unreasonable amount of time to find the exact solution: $\approx 1 \times 10^{126}$ years. This amount of time is much, much longer than the age of the Universe, which has been estimated to be 14×10^9 years.

TRY EXERCISE 12.1

We must abandon the idea of finding the exact solutions of large exponential problems if the only possible algorithm is that of brute force.[4] This statement holds even if we parallelize the computation (Aaronson 2005). In fact, the benefit of making the calculations with, let us say, 10^{20} processors working in parallel, can never be appreciable for enormous exponential problems.[5] Therefore, we are obliged to transform the original exponential problems into problems of recognizing acceptable solutions, generated non-deterministically and in a reasonable time. In other words, the exponential problems are changed in Non-Deterministic Polynomial problems or NP-problems. For example, let us imagine facing a TSP. Given a graph G with a certain number of nodes, edges, and costs, we get

---

[2] The TOP500 project (website: https://www.top500.org/) ranks the 500 most powerful non-distributed computer systems in the world. This project started in 1993 and publishes an updated list of the best supercomputers twice a year: in June and November.

[3] 1 (PFlops/s)= 1×10^{15} floating-point operations per second. P is for Peta that corresponds to 1×10^{15}.

[4] The brute force solution of an exponential problem, such as the TSP, involves an exhaustive search that consists in measuring the length of all the permutations of edges and, then, seeing which is the cheapest.

[5] If a processor requires an amount of time equal to $[10^{20} \Delta t]$ for solving a problem, 10^{20} processors, working in parallel, would need a period equal to $[\Delta t]$.

satisfied as soon as we find a complete tour requiring a total cost that is less than or equal to some predefined maximum amount F. Possible solutions might be generated by heuristic algorithms; for instance, by guessing. Then, the problem becomes polynomial because it reduces to recognize if the solutions, generated non-deterministically, verify the imposed conditions or not.

In synthesis, the P-class contains computational problems that are solved efficiently, whereas the NP-class includes problems whose solutions are effectively verifiable. The NP-class encompasses many scientific problems. Following, is a summary list.

- In the scientific experiments regarding the characterization of the time evolution of systems, we gather data in time, which are snapshots of the states of the systems. For the comprehension of the laws behind the systems' behavior, we must reconstruct the underlying dynamical equations from the snapshots. Such a task, named "system identification," is, in general, an intractable computational problem (Cubitt et al. 2012), especially for the Natural Complex Systems.
- In 1925, the German physicists Ernst Ising proposed a model for the phase transitions. Ising's model can be described in graph-theoretic terms (Cipra 2000), in which vertices represent atoms in a lattice and edges represent bonds between adjacent atoms. Each vertex i can be in one of two states, usually indicated as $\sigma_i = \pm 1$. Each edge has an assigned coupling constant, J_{ij}, where i and j are the two vertices. When neighboring vertices i and j are in states σ_i and σ_j, respectively, the interaction between them contributes an amount $-J_{ij}\sigma_i\sigma_j$ to the total energy E_{tot} of the system. The total energy is given by $E_{tot} = -\sum_{i,j} J_{ij}\sigma_i\sigma_j$ where the sum is extended to all the pairs of neighbors—all the edges of the graph. If two vertices, i and j, are in the same state and the J_{ij} is positive, then they decrease the total energy. If all the coupling constants are positive, the overall system has an evident lowest-energy configuration, where all the atoms are in the same state, either $+1$ or -1. But if the coupling constants are a mixture of positive and negative numbers, as it happens in the case of spin glasses, finding the ground state (i.e., the state with the lowest energy) for a three-dimensional lattice is an NP-complete problem. In fact, with N vertices, each one having two possible states, finding the ground state can be done by brute force, computing all the 2^N possible combinations.
- The protein folding problem consisting in predicting its three-dimensional structure from a string giving the protein's amino acid sequence is also *NP*-complete (Lathrop 1994; Berger and Leighton 1998).
- Many other computational tasks of practical interest, in planning, scheduling, machine-learning, financial-forecasting, the design of computers' hardware with an optimal arrangement of transistors on a silicon chip, and computational biology belong to the class of NP-complete problems (Monasson et al. 1999).

NP-complete problems, cited earlier (whose other examples are the TSP, the Schrödinger equation, and the satisfiability problems[6]), form a special subset of the *NP*-class set. In fact, it embodies the secret of computational intractability since a polynomial time algorithm for one of them would immediately imply the tractability of all the problems in NP. This result would mean $NP = P$ (see Figure 12.2). Then, suddenly, Complexity would melt like snow under the sun, and we would become able to understand nature as never we have done, so far, in the history of humanity.

---

[6] The satisfiability (SAT) problem consists of a logical propositional formula with N variables, and it requires to find a value (true or false) for each variable that makes the formula true. For example, the formula "x AND NOT y" is satisfiable because the values x = TRUE and y = FALSE makes "x AND NOT y" = TRUE. In contrast, "y AND NOT y" is not satisfiable. When the formula consists of a conjunction of clauses, and each clause is a disjunction of k variables, any of which can be negated, the problems is called k-SAT. For $k > 2$, the k-SAT is NP-complete.

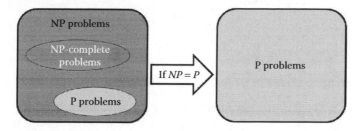

FIGURE 12.2   The three main classes of computational problems (on the left) that would become just one if it was demonstrated that *NP* = *P*.

Aware of the relevance of the Computational Complexity theory, in 2000, the Clay Mathematics Institute in Cambridge, Massachusetts, has named "*P* versus *NP*" as one of its "Millennium" problems, and it offers one million dollars to anyone who provides verified proof that either *NP* = *P* or *NP* ≠ *P*. If the relation *NP* = *P* were demonstrated to be true, then, as the same Gödel said in a letter to Von Neumann, in 1956, that discovery would have "consequences of the greatest magnitude" (Sipser 1992). Our life would not be the same (Fortnow 2009). Everything would be much more efficient. The transportation schedules of all forms of transit would be optimized, allowing people and goods to move quicker and cheaper. Manufacturers and businessmen would improve their production processes and increase profits. It would become much easier to find new effective treatments for incurable diseases, make weather forecasts, predict catastrophic events, and the trends of stock markets. Humanity would have new powerful algorithms and tools to tackle all the Complexity Challenges.

The discovery that *NP* = *P* would bring benefit (Fortnow 2009) to another challenge we encounter in dealing with Natural Complex Systems. It is the challenge of formulating universally valid and effective algorithms for recognizing variable patterns. Variable patterns are objects (both animate and inanimate) or events whose recognition is made difficult by their multiple features (since they contain a significant amount of information), variability, and extreme sensitivity on the context. Examples are human faces, voices, and fingerprints (biometrics), handwritten cursive words and numbers, patterns in medical diagnosis, patterns in apparently uncorrelated scientific data (data mining), and chaotic and stochastic time series. Human beings are good at recognizing some variable patterns, such as faces and voices. However, we need to formulate algorithms for identifying every type of pattern, whatever is their context. The steps of pattern recognition are: (I) data acquisition by instruments; (II) selection of the features (any pattern is represented by a k-dimensional vector if k is the number of selected features); (III) application of an algorithm for the classification step. In the literature (Bishop 2006), there are different examples of algorithms that have been proposed, such as statistical, structural, or based on artificial neural networks (Jain et al. 2000). All of them still suffer in universality and effectiveness.

12.3   IF IT WERE *NP* = *P*, WOULD BE THE COMPLEXITY CHALLENGES SURELY WON?

In the previous paragraph, we knew the computational hurdles science encounters whenever it tries to describe Complex Systems. Even if, one day, someone demonstrated that *NP* = *P*, one relevant limit in the scientific knowledge would remain. It is the limit in determining the initial conditions for every Complex System. The famous Uncertainty Principle formulated, at first, by Heisenberg (1927), and then extended by others (Erhart et al. 2012; Furuta 2012), declares the impossibility of determining, accurately and simultaneously, position and momentum of a microscopic particle. The Uncertainty Principle places limits to the deterministic dream of describing the dynamics of the entire Universe starting from the description of its tiny particles. We might get satisfied by describing Complex Systems from the macroscopic point of view, neglecting their ultimate atomic constituents. However, the intrinsic unpredictability of chaotic dynamics, in the long term and at both the microscopic and macroscopic level, would remain valid. In fact, any chaotic dynamics is extremely sensitive to the initial conditions. And we are fully aware that the initial conditions of both microscopic and macroscopic systems cannot be determined with an infinite degree of accuracy because any experimental measurement is affected by unavoidable uncertainties (see Appendix D).

I think that we are now completely aware of the sharp limits we encounter in describing Natural Complex Systems by using the fundamental physical laws. Therefore, we need to develop models to interpret Natural Complexity. The first requirement for formulating a scientific model is to know, in depth, the features of the systems we want to describe. Therefore, in the next pages of this chapter, I present the characteristics of Complex Systems.

12.4 THE FEATURES OF COMPLEX SYSTEMS

The Universe is the most extensive Complex System. Within our Universe, all the galaxies are Complex Systems. Within each galaxy, all the stars and planets are Complex Systems, as well. If we assume that life is present only on the Earth (so far, we do not have proofs in contrast to this hypothesis), then, we can plainly state that the Earth is the most complex planet in the Universe. What makes our planet complex is its climate, its geology, and, particularly, its biosphere. The biosphere is the global ecological system that consists of all the living beings, their relationships, and their interactions with the geosphere, hydrosphere, and atmosphere. Every ecosystem included in the biosphere is complex. Even every living being, either pluricellular or even unicellular, is complex. Among the living creatures, we, humans, are the most complex, mainly due to our brains, which confer us the power of computing with numbers and words. Of course, even human societies and the economy are other examples of Complex Systems.

Which are the common features of so diverse Complex Systems?

12.4.1 NETWORKS

A Complex System consists of many elements, often diverse, if not unique. For example, the components of an ecosystem are plainly diverse, and those belonging to a human community are evidently unique. The elements of a Complex System are also highly interconnected. Complex Systems are networks. They are intertwined systems. In fact, the etymology of the adjective "Complex" derives from the Latin verb *cum-plectere* that means "to intertwine together." It is different from the etymology of "complicated." "Complicated" derives from the Latin verb *cum-plicare* that means "to fold together." Anything that is "complicated," is "folded" and can be "unfolded" (see Figure 12.3). On the other hand, anything that is "complex," is interwoven and cannot be "unfolded." Instead, it needs to be untangled.

A network is described as a collection of nodes (also called vertices) connected by links (or edges) (Newman 2010). The nodes represent the elements of the network, and the links are the connections between them (see Figure 12.4). For instance, if we consider our brain, a node is a neuron, whereas a link is either a connection between the synapse of a neuron and the dendrite of another or a gap junction.[7] In an ecosystem, the nodes could be the single species, and the links could be the ecological relationships between them. In economy, the nodes are the different individual agents that can be firms, banks, or countries, and links are their mutual interactions, be it trade, ownership, credit-debt relationships, or research and development alliances.

In undirected networks (see Figure 12.4a), the links between nodes do not have an assigned direction. In directed networks, the interplay between any two nodes has a well-defined direction, indicated by an arrow (see Figure 12.4b).

---

[7] A gap junction is an aggregate of intercellular channels that permits transfer of ions and small molecules.

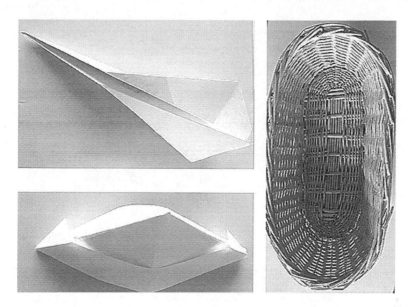

FIGURE 12.3   Examples of "complicated" or "folded" objects, on the left, and a "complex" or "intertwined" basket, on the right.

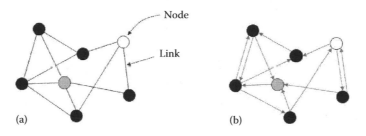

(a)   (b)

FIGURE 12.4   Simple undirected (a) and directed (b) networks.

An essential property of a node within a network is its connectivity or degree, d, which is the number of links the node has to other nodes. For example, in (a), the grey node has degree $d = 4$. In a directed network, we can define an incoming degree, d_{in}, which denotes the number of links that point to a node ($d_{in} = 3$ for the grey node in network b), and an outgoing degree, d_{out}, which indicates the number of links that starts from it ($d_{out} = 1$ for the grey node). It is useful to define the degree distribution, $P(d)$, of a network. $P(d)$ is obtained by counting the number of nodes with $d = 1,2,\ldots,m$ and dividing each value by the total number of nodes, N. The degree distribution of the small network, depicted in Figure 12.4a, is reported in Figure 12.5. An undirected network with L links is characterized by an average degree $\bar{d} = 2L/N$. High-degree nodes are called hubs.

Another relevant feature of a network is the distance between nodes, which is measured by counting the number of links we need to pass through to go from one node to another. Usually, there are many alternative paths between two nodes. Among them, the most interesting is the shortest path, sp, which is the path with the smallest number of links between the selected nodes. For example, the shortest path between the white and grey nodes of the network represented in Figure 12.4a, is 2. In the directed network Figure 12.4b, the shortest path from the white node to the grey node, sp_{WG}, is 2 and is different from the shortest path from the grey to the white, sp_{GW}, which is equal to 3. The overall navigability of a network is described by specifying its mean path length, $\overline{sp}$, which is the average over the shortest paths between all pairs of nodes.

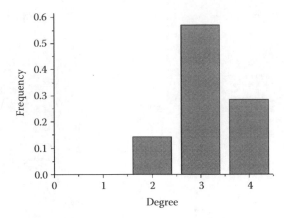

FIGURE 12.5 Degree distribution of the small network depicted in Figure 12.4(a).

A final attribute of a network is its clustering. The clustering coefficient C of a node represents the fraction of pairs of neighbors that are connected to one another. If a node has d neighbors, there are $d(d-1)/2$ pairs of neighbors. C is the fraction of the $d(d-1)/2$ pairs that are linked. For example, for the grey node of the network in Figure 12.4a, $C = 2/6$, whereas for the white node, $C = 0$. The average clustering coefficient, $\bar{C}$, of a network quantifies the tendency of the nodes of that network to form clusters or groups.

TRY EXERCISE 12.2

After an analysis of the structures and properties of many real networks, scientists have proposed six main models. The first is the model of *regular networks* (or the so-called *regular lattice structures*). An example is shown in Figure 12.6, and it is labeled as a. In a, we have 18 nodes. Each node is linked to its four nearest neighbors. The average clustering coefficient is 0.5, and the average path length is 2.5. In general, in a regular network, all the nodes have the same degree and $P(d)$, plotted as a histogram, is just a single vertical segment. The mean path length, $\overline{sp}$, is long and the clustering is high.

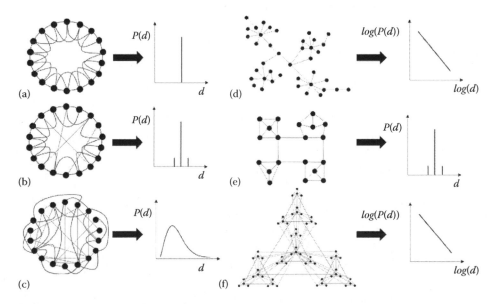

FIGURE 12.6 Six models of networks and their characteristic degree distributions. (a–f) represent regular, small-world, random, scale-free, modular, and hierarchical networks, respectively.

If we substitute a few of the nearest-neighbor links of a into long-distance links, chosen randomly, we obtain the second model, a *small-world network*, like that labeled as b in Figure 12.6. In a small-world network (Watts and Strogatz 1998) $\overline{sp}$, becomes short because the long-range edges are "short-cuts" connecting vertices that would otherwise be much farther apart. Nonetheless, the nodes of the network maintain high clustering coefficients and $P(d)$ is still a discrete distribution. When the links of a regular network are completely randomized, we obtain the third model, a *random network* like that labeled as c in Figure 12.6. In a random network, the degree distribution is a Poisson function (Erdös and Rényi 1960) peaked at $\overline{d}$ and decaying exponentially for $d \gg \overline{d}$. This trend implies that most of the nodes have approximately the same number of links, close to the average $\overline{d}$. Since the tail of $P(d)$ decreases exponentially, nodes with degree larger than $\overline{d}$ are extremely rare. The mean path length is short, like in a small-world network, but the clustering coefficient is low and independent of a node's degree, so $C(d)$ appears as a horizontal line when plotted as a function of d (Barabási and Oltvai 2004).

TRY EXERCISE 12.3

The fourth model is the *scale-free network* (see d in Figure 12.6) that is characterized by a power-law degree distribution. The probability that a node has degree d is $P(d) = d^{-\gamma}$, where γ is a positive constant, usually included between 2 and 3 (Barabási and Oltvai 2004). The power-law distribution entails that most of the nodes have low degree, whereas there are just a few hubs. The scale-free network has a fractal-like structure and the features that are typical of a small-world network, i.e., short mean path length and high average clustering coefficient.

The fifth model is the *modular network* (see e in Figure 12.6), where one can easily identify groups of nodes that are highly interconnected with each other (the so-called modules) but have only a few or no links to nodes outside of the group to which they belong to. A modular network is characterized by a discrete degree distribution because nodes within single modules have robust connectivity, whereas those bridging modules have sparse connectivity. The average cluster coefficient is significant.

TRY EXERCISE 12.4

The sixth model is the *hierarchical network* with clusters of nodes that combine in an iterative manner (Ravasz et al. 2002). If we look carefully at the graphical example labeled as f in Figure 12.6, we notice nodes that appear in the center of the network have the highest degree and the lowest clustering coefficient, while the nodes at the periphery of the network have low degrees and the most significant clustering coefficients. In between are nodes with a moderate degree and moderate clustering. It results that both $P(d)$ and $C(d)$ are power-law functions.

TRY EXERCISE 12.5

Most real networks are difficult to deal with, for different reasons (Strogatz 2001). First, their structures can be intricate tangles, and they may continuously evolve. Second, their connections can have different strengths and effects, and they may change, as well. Third, and most often, nodes are diverse, and their internal state can vary in time in complicated, non-linear, ways. An example of an incredibly dynamic network is the immune system that protects our bodies from infections and intruders (antigens) (read Box 12.1 in this chapter). Despite all these difficulties, the study of many biological, technological and social networks has demonstrated the ubiquity of small-world properties. Networks as diverse as the cell, the brain, the food web, the internet, the World Wide Web,[8]

---

[8] The Internet is the network of computers and routers with the wires and cables, which physically connect them, as edges. The World Wide Web is the network of web pages where the edges are the hyperlinks (URL's) that point from one page to another.

collaboration of movie actors, collaboration of scientists, the connections of bank, international trade relationships, the words in human language, and so on, all exhibit short mean path lengths and high average clustering coefficients with respect to hypothetical random networks having the same number of nodes (Albert and Barabási 2002; Schweitzer et al. 2009). Small-world features guarantee an enhanced signal-propagation speed, computational power, spread of fads and infectious diseases, and synchronizability (Watts and Strogatz 1998). Among the investigated real networks, most of them exhibit a degree distribution that follows a power law. This evidence suggests that many Complex Systems are scale-free networks, containing hubs. Why are scale-free networks so common? There are two main reasons. First, most real networks grow continuously, and new nodes join the system over extended periods of time. Second, nodes prefer to connect to nodes that already have many links, according to a process known as the preferential attachment (Barabási and Albert 1999). A new node, joining a network, has a higher probability to connect with a node having, already, many links. This probability Π is, in fact, given by the ratio:

$$\Pi = \frac{d_i}{\sum_{j=1}^{N} d_j} \qquad [12.3]$$

where:
d_i and d_j are the degrees of node i-th and j-th, respectively,
N is the total number of nodes.

The node that has the largest degree increases further its connectivity and becomes a hub. "The rich get richer." Scale-free networks have two key features: robustness to random failure and vulnerability to hubs' attack (Albert et al. 2000). The robustness of the scale-free networks is rooted in their inhomogeneous connectivity. In fact, the power-law distribution guarantees that the majority of nodes have only a few links and that there are just a few hubs. The probability that random failure involves nodes with low connectivity is much higher. Therefore, the failure of nodes scarcely connected does not affect the overall topology and the average length of the shortest paths. However, the removal of a few key hubs splinters the network into small isolated clusters of nodes.

Finally, it has been discovered that real networks that are scale-free and devoid of topological constraints, such as limitations on the link length, have their clustering coefficients that are power-law functions of the degree d: $C(d) \sim d^{-1}$ (Ravasz and Barabási 2003). This evidence means that many Complex Systems can be modelled as hierarchical networks. The best example is a living cell. The structure and functions of a living cell depend on the complex web of the interactions between the cell's numerous constituents. A cell grounds on a network of networks, including metabolic, signaling, transcription-regulatory, and protein-protein interaction modules. None of these modules is independent. A major challenge of contemporary biology is to describe quantitatively the network of networks that rules the behavior of a cell. A recent study reported the successful construction of a whole-cell computational model for the bacterium *Mycoplasma genitalium* (Karr et al. 2012) in which the function of all its 525 genes has been simulated. Marcus Covert and his colleagues at Stanford University have broken the *Mycoplasma* cell's overall functions down into 28 separate functional modules describing different sub-processes, like DNA replication, RNA transcription, protein folding, metabolism. Then, they modeled each module separately based on the information available in literature. They assumed that the modules are approximately independent on short timescales (less than 1 s). Simulations were carried out by running through a loop in which the modules were run independently at each time step but depended on the values of the variables (playing like connecting nodes) determined by the other modules at the previous time step. The simulated bacterium had many expected properties, in its metabolic fluxes, global distribution of protein and RNA, metabolite amounts, and gene function but also provided insights into many previously unobserved cellular behaviors, including in vivo rates of protein-DNA association, and an inverse relationship between the duration of DNA initiation and that of replication. This successful research suggests an approach to model more complex cells and even multicellular systems, tissues, ecosystems, and economy. It could become a foundational platform for interpreting the behaviors of all kinds of Complex Systems, including climate and geology of our planet. In the latter cases, nodes and links would be molecules and intermolecular forces, respectively. For instance, convection starts when strong macroscopic thermal gradients transform a regular dynamic network, which is the liquid described at the molecular level,[9] into a modular network; modules are micrometer-sized clusters of molecules that move either upwards or downwards depending on their mutual densities.

BOX 12.1   THE IMMUNE SYSTEM

The immune system of a human adult contains approximately 1 trillion T-cells (the T indicates that they are lymphocytes developed by the thymus) and 1 trillion B-cells (the B means that they are lymphocytes developed by the bone marrow), located in the lymphoid organs and the blood. Moreover, there are approximately 10 billion antigen-presenting cells (that take up antigens and present them to T-cells) located in the lymphoid organs (Bianconi et al. 2013). This collection of cells works together in an efficient way without any central control. For maximizing the chances of encountering antigens, lymphocytes continually circulate between the blood and specific lymphoid tissues. A given lymphocyte spends an average of 30 minutes per day in the blood and recirculates about 50 times per day between the blood and lymphoid tissues. If lymphocytes encounter an antigen trapped by the antigen-presenting cells of the lymphoid organs, lymphocytes with receptors specific to that antigen stop their migration and settle to mount an immune response locally. As these lymphocytes accumulate in the affected lymphoid tissue, the tissue often becomes enlarged. The immune system network involves many spatial scales: the molecular scale, through the dynamic interactions occurring within cells; the cellular scale, through crosstalk among the many distinct immune cell types; the tissue scale and, finally, the organism scale. The system-level network analyses of the immune system allow a more profound understanding. However, we are still far from applying network-based approaches to the construction of comprehensive predictive models of the immune response to diverse perturbations (Subramanian et al. 2015). One reason is the adaptive power of the immune system. There is an astronomical number of possible antigens, which is much bigger than the number of receptors in lymphocytes. When a new kind of antigen enters our body, there will be at least one receptor that binds to the intruder, though very weakly. The activated lymphocyte migrates to a lymph node, where it divides rapidly, producing many daughter cells with mutations that alter the receptor shapes. These new daughter cells are tested against the new antigen. The cells that do not bind die after a short time. Those that bind more strongly are unleashed into the bloodstream, where they encounter the antigens and bind to them more tightly than did their mother lymphocyte. It is a process of Darwinian Natural Selection. The outcome is an emerging arsenal of antibodies and killer T-cells that evolves through mutation, selection, and replication of the fittest ones to attack the new antigen. Further investigation and advances in mathematical and computational tools are needed for a deep comprehension of the immune system.

---

[9] A crystal is a neat example of a regular static network. On the other hand, a liquid can be modeled as a regular dynamic network because the single particles are much free to move and change their positions. However, the overall system may be still described as a regular network. A rarefied gas looks more like a random dynamic network rather than a regular network.

12.4.2 OUT-OF-EQUILIBRIUM SYSTEMS

Complex Systems are networks that are out-of-equilibrium. They can be either open or closed or isolated (for instance, our Universe is postulated to be an isolated thermodynamic system). They are maintained out-of-equilibrium by external and/or internal gradients of intensive variables. For example, Earth is out-of-equilibrium due to three principal contributions. The first is the gravitational field generated by the sun and the moon. The second is the thermal energy released by the processes of nuclear fissions involving unstable radionuclides that are beneath the terrestrial crust. The last contribution is the electromagnetic radiation and the wind of particles and gamma rays that come from the sun. The sun is maintained out-of-equilibrium by the nuclear fusion reactions occurring in its inner core. Through the so-called proton-proton chain reactions, hydrogen converts to helium, and vast amounts of thermal energy are unleashed.[10] The thermal energy produced within the core of our star migrates by irradiation and convection towards the external surface, the so-called photosphere. The photosphere of our sun has an average temperature of 5,777 K (NASA website) and emits thermal radiation whose frequencies belong to the UV, visible and near-IR regions of the electromagnetic spectrum (see Figure 12.7). The solar thermal radiation is the primary power source for our planet (Kleidon 2010). For this reason, it is essential to know its thermodynamic properties.

12.4.2.1 The Thermodynamics of Thermal Radiation

The electromagnetic, thermal radiation, also called heat radiation, is due to the thermal motion of charged particles that are at the foundation of any material. All the bodies emit thermal radiation: our sun, our planet, ourselves, this book, and so on. The spectrum of thermal radiation depends only on the temperature. It was Max Planck (1914), at the beginning of the twentieth century, who derived the expression for the energy density of thermal radiation, as a function of the temperature (T) and frequency (ν), after introducing the revolutionary quantum hypothesis:

$$u(\nu,T) = \frac{8\pi h \nu^3}{c^3} \frac{1}{\left(e^{h\nu/k_B T} - 1\right)} \qquad [12.4]$$

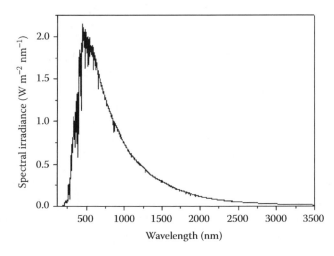

FIGURE 12.7 The spectral irradiance of the sun outside of the atmosphere, i.e., at Air Mass 0. (data extracted from the website http://www.pveducation.org/).

---

[10] A nuclear reaction releases an amount of energy that is of the order of MeV. On the other hand, in a chemical reaction, only the electrons of the outer electronic spheres of atoms and molecules participate, and the amounts of energy that are involved are of the order of eV.

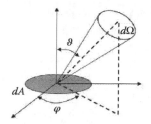

FIGURE 12.8 Definition of the angles ϑ and φ used to obtain the general equation [12.7].

In equation [12.4], $h = 6.626 \times 0^{-34}$ Js is Planck's constant, $c = 2.998 \times 10^{8}$ ms^{-1} is the speed of light, and $k_B = 1.381 \times 10^{-23}$ JK^{-1} is Boltzmann's constant.

Let us consider a cavity with walls at a uniform and constant temperature. The elementary particles, constituting the walls, move, being powered by the thermal energy. They generate electromagnetic waves that fill the cavity. Energy is transferred from the walls to the electromagnetic field. On the other hand, the electromagnetic waves hit the walls and are absorbed by the particles. The system reaches the equilibrium when the energy transferred from the walls to the electromagnetic field is equal to that transferred from the field to the walls. To know the properties of thermal radiation within the cavity, we make a tiny hole in a wall of the cavity. dA is its infinitesimal area, which is so small that does not perturb the equilibrium between the cavity and the radiation. The thermal radiation, which would hit the area dA, escapes from the cavity. The energy emitted per unit of time, in the frequency range v and $v + dv$, within the solid angle $d\Omega$, which defines an angle ϑ with the normal to the surface (see Figure 12.8), is

$$U(v,T)(dv) = u(v,T)(dv)(dA\cos\vartheta)c\frac{d\Omega}{4\pi} \qquad [12.5]$$

The intensity of radiation, in the frequency range v and $v + dv$, is

$$I(v,T)(dv) = u(v,T)(dv)(\cos\vartheta)c\frac{d\Omega}{4\pi} \qquad [12.6]$$

Since $d\Omega = \sin(\vartheta)d\vartheta d\varphi$, by integrating ϑ between 0 and $\pi/2$, and φ between 0 and 2π, we obtain[11]

$$I(v,T)(dv) = \frac{u(v,T)(dv)c}{4} \qquad [12.7]$$

The emissivity of a body k, $E_k(T,v)dv$, is defined as the radiation intensity emitted by that body that is at temperature T and in the frequency range v and $v + dv$. The absorptivity, $a_k(v,T)dv$, is defined as the fraction of the incident intensity $I(v)dv$ that is absorbed by the same body at temperature T and in the frequency range v and $v + dv$. By considering the principle of energy conservation, Kirchhoff (1860) formulated the law that brings his name (Kirchhoff's law):

$$E_k(T,v) = a_k(v,T)I(v) \qquad [12.8]$$

The absorptivity is a dimensionless absorption coefficient, and it assumes values between 0 and 1: $0 \le a_k(v,T) \le 1$. When the body k is in equilibrium with its thermal radiation, $a_k(v,T)$ is equal to 1, k

---

[11] The integrals are $\int_0^{\pi/2} \sin\vartheta\cos\vartheta d\vartheta = \left(-\cos^2\vartheta/2\right)\Big|_0^{\pi/2} = 1/2$ and $\int_0^{2\pi} d\varphi = 2\pi$.

is called blackbody. For a blackbody, $E_k(T,v)$ equals $I(v)$. This is a general rule and does not depend on the chemical composition of the body. It is worthwhile noticing that if $a_k(v,T)=0$, then $E_k(T,v)=0$. This result means that a body transparent to a frequency v is not able to thermally emit that particular frequency. If we transform the energy density of the thermal radiation (defined by the Planck's formula [12.4]) in intensity by using equation [12.7], and we integrate over the entire range of frequencies, we obtain the total emissive power of a blackbody:

$$E_k(T) = \frac{2\pi h}{c^2} \int_0^\infty \frac{v^3 dv}{e^{hv/kT}-1} = \frac{2\pi^5 k^4}{15c^2h^3}T^4 = \xi T^4 \qquad [12.9]$$

Equation [12.9] is known as the Stephan-Boltzmann law. The constant is $\xi = 5.67\times10^{-8}\ \mathrm{W/m^2K^4}$.

From the application of the Second Law of Thermodynamics, it is possible to infer some properties of the spectral composition of the thermal radiation. First, it must be isotropic and uniform, independent of position, within the cavity. If this were not the case, it would be possible to transfer energy from one place to another of the cavity. But the energy transfer is impossible because the system is in thermal equilibrium. Second, it must be independent of the chemical composition of the cavity walls and the shape of the cavity. If this were not the case, it would be possible to transfer energy between two cavities having the same temperature, but differing just in the chemical composition and/or shape.

Thermal radiation can be conceived as a gas of photons. Is it an ideal or a real gas? What kind of gas is it? Which are its properties? The thermal radiation, contained in a cavity, exerts a pressure on the walls of the cavity. To define this pressure, let $N(v,T)$ be the number of photons of frequency v contained in a cubic cavity of volume V and at temperature T. The momentum of each photon is

$$p = \frac{hv}{c} \qquad [12.10]$$

When a photon hits a wall of the cavity, it transfers a momentum $2p$ to the wall, every time it is reflected.[12] Since the photons move randomly, at every instant of time, 1/6 of the photons will be moving in the direction of a wall. The number of photons with frequency v, which will collide a wall, per second, will be $N(v,T)c/6V$. The total momentum transferred to the wall per unit of time and per unit of area corresponds to the pressure exerted by the photons of frequency v:

$$P(v,T) = \frac{N(v,T)c}{6V}\frac{2hv}{c} = \frac{N(v,T)hv}{3V} \qquad [12.11]$$

Since $N(v,T)hv/V$ represents the energy density of photons with frequency v, $P(v,T) = u(v,T)/3$. If we consider all the photons, whose frequencies are in the range $[0,+\infty]$, the total pressure will be

$$P(T) = \int_0^\infty P(v,T)dv = \int_0^\infty \frac{u(v,T)}{3}dv = \frac{u_{tot}(T)}{3} \qquad [12.12]$$

The total energy density $u_{tot}(T)$ depends only on T. The total internal energy of the gas of photons will be

$$U = Vu_{tot}(T) \qquad [12.13]$$

---

[12] The situation is analogous to that of a swimmer when it hits the wall of a swimming pool and turns his direction of motion.

This last equation demonstrates that the gas of photons is not like an ideal gas because its internal energy depends not only on T but also V. In fact, if we expand the volume of the cavity, the internal energy of the gas of photons increases. But, if $dU > 0$ because $dV > 0$, the thermal radiation must absorb heat. In fact, $dU = dq - PdV$. The consequent entropy change will be

$$dS = \frac{dU + PdV}{T} \qquad [12.14]$$

Since U is function of T and V, then

$$dU = \left(\frac{\partial U}{\partial T}\right)_V dT + \left(\frac{\partial U}{\partial V}\right)_T dV \qquad [12.15]$$

If we introduce equation [12.15] into [12.14], we obtain

$$dS = \frac{1}{T}\left(\frac{\partial U}{\partial T}\right)_V dT + \frac{1}{T}\left[\left(\frac{\partial U}{\partial V}\right)_T + P\right]dV \qquad [12.16]$$

Equation [12.16] means that $(\partial S/\partial T)_V = (1/T)(\partial U/\partial T)_V$ and $(\partial S/\partial V)_T = (1/T)\left[(\partial U/\partial V)_T + P\right]$. Knowing that $(\partial^2 S/\partial V\,\partial T) = (\partial^2 S/\partial T\,\partial V)$, it follows that

$$\left(\frac{\partial}{\partial V}\left[\frac{1}{T}\left(\frac{\partial U}{\partial T}\right)_V\right]\right)_T = \left(\frac{\partial}{\partial T}\left[\frac{1}{T}\left[\left(\frac{\partial U}{\partial V}\right)_T + P\right]\right]\right)_V \qquad [12.17]$$

Equation [12.17] can be rewritten as

$$\frac{1}{T}\left(\frac{\partial^2 U}{\partial V \partial T}\right) = -\frac{1}{T^2}\left(\frac{\partial U}{\partial V}\right)_T + \frac{1}{T}\left(\frac{\partial^2 U}{\partial T \partial V}\right) + \left(\frac{\partial}{\partial T}\left(\frac{P}{T}\right)\right)_V \qquad [12.18]$$

It derives that

$$\left(\frac{\partial U}{\partial V}\right)_T = T^2\left(\frac{\partial}{\partial T}\left(\frac{P}{T}\right)\right)_V \qquad [12.19]$$

known as the Helmholtz equation (Kondepudi and Prigogine 1998). By using equations [12.12] and [12.13], equation [12.19] transforms in

$$u_{tot}(T) = T^2\left(\left(-\frac{u_{tot}(T)}{3T^2}\right) + \frac{1}{3T}\left(\frac{\partial u_{tot}}{\partial T}\right)_V\right) \qquad [12.20]$$

The latter can be rearranged in

$$4u_{tot}(T) = T\left(\frac{\partial u_{tot}}{\partial T}\right)_V \qquad [12.21]$$

If we separate the two variables appearing in [12.21], and we integrate, we obtain the well-known Stephan-Boltzmann law:

$$u_{tot}(T) = \beta T^4 \qquad [12.22]$$

wherein $\beta = \xi 4/c = 7.56 \times 10^{-16}\,\text{Jm}^{-3}\text{K}^{-4}$ is a constant. If we introduce equation [12.22] into [12.12], we obtain the equation of state for thermal radiation:

$$P = \frac{\beta T^4}{3} \qquad\qquad [12.23]$$

By introducing the definition of $U = V\beta T^4$ and equation [12.23] into [12.16], we obtain that $(\partial S/\partial V)_T = (4/3)\beta T^3$ and $(\partial S/\partial T)_V = 4\beta T^2$. After separating the variables and integrating, assuming that $S = 0$ when $T = 0$ and $V = 0$, the definition of entropy is

$$S = \frac{4}{3}\beta V T^3 \qquad\qquad [12.24]$$

The other thermodynamic functions are enthalpy H

$$H = U + PV = V\beta T^4 + V\frac{\beta T^4}{3} = \frac{4}{3}V\beta T^4 \qquad\qquad [12.25]$$

Helmholtz work function A

$$A = U - TS = V\beta T^4 - \frac{4}{3}\beta V T^4 = -\frac{1}{3}V\beta T^4 \qquad\qquad [12.26]$$

and the Gibbs free energy G

$$G = H - TS = \frac{4}{3}V\beta T^4 - \frac{4}{3}V\beta T^4 = 0 \qquad\qquad [12.27]$$

We know that ΔA is proportional to the maximum work that a system can perform, and ΔG to the maximum work, excluded that of the type "pressure and volume" (Atkins and de Paula 2010). From equations [12.26] and [12.27], it is evident that, at equilibrium, thermal radiation can perform only work of expansion (i.e., of the type "pressure and volume"). In fact, its chemical potential is $\mu = 0$ (equation [12.27]).

TRY EXERCISES 12.6, 12.7, 12.8, 12.9

12.4.2.2   The Fate of the Solar Thermal Radiation and the Climate Change

The sun, with a surface area of $\approx 6.09 \times 10^{18}\,\text{m}^2$ (NASA website), emits $\approx 3.84 \times 10^{26}$ Joule per second in the surrounding space. Its activity is not steady but oscillates with a period of about 11 years. The exact length of the period can vary. It has been as short as 8 years and as long as 14, and it is associated with the formation and depletion of black sunspots on its surface. More sunspots determine an increased solar activity in the level of solar radiation and ejection of solar material. Each cycle varies dramatically in intensity (NASA website). At the origin of the cycles, there is a flipping phenomenon of the magnetic north and south poles of the sun.

The amount and fate of solar thermal radiation that reaches the earth regulates the climate of our planet. The most recent estimate of the average solar irradiance that reaches the outer surface of the terrestrial atmosphere in one year is $(340.2 \pm 0.1)\,\text{Wm}^{-2}$ (Kopp and Lean 2011). According to the 5th Assessment Report of the Intergovernmental Panel on Climate Change (IPCC 2013), when solar radiation crosses terrestrial atmosphere it is partly (22.4%) reflected and scattered (mainly by clouds and aerosols[13]) back out into space. Moreover, it is absorbed (23.2%) by the components of

---

[13] Aerosols are colloids made of solid particles or liquid droplets dispersed in air.

the atmosphere. 54.4% of the solar irradiance reaches the terrestrial surface. An average of 7.1% is reflected back to space. The total solar irradiance reflected back to space (22.4% + 7.1% = 29.5%) by the earth as a whole is called albedo. The remaining 47.3%(= 54.4%–7.1%) is absorbed. Since the solar irradiance is not uniformly distributed in space and time, both altitudinal and latitudinal and longitudinal thermal gradients are generated. These thermal gradients give rise to the convective motions, winds and oceanic currents. The heat absorbed by water also triggers the water cycle, whereas the radiation is absorbed particularly by the land becomes a source of infrared thermal radiation. Such IR radiation is partly reabsorbed by some components of our atmosphere, such as H_2O, CO_2, CH_4, N_2O, and so on, causing a greenhouse effect.

TRY EXERCISE 12.10

The average temperature of Earth depends on three principal factors:

1. The energy coming from the sun;
2. The chemical composition of the atmosphere;
3. The albedo.

When human activities or volcanic eruptions release greenhouse gases into the atmosphere, they exert a positive radiative force, which increases the temperature. On the other hand, when aerosols are unleashed, a significant amount of solar radiation is reflected back to space, and the average terrestrial temperature decreases (the aerosols exert a negative radiative forcing). According to an ongoing temperature analysis conducted by scientists at NASA's Goddard Institute for Space Studies (GISTEMP Team 2017), the average global temperature on Earth has increased by about 0.8°C since 1880. Two-thirds of the warming has occurred since 1975, at a rate of roughly 0.15°C–0.20°C per decade. There are two opposite theses about the causes of global warming. The most trusted thesis, proposed by the Intergovernmental Panel on Climate Change, supports the hypothesis that global warming is due to the increase of the level of CO_2 in the atmosphere, produced by human activities. In other words, the global warming is anthropogenic. The second thesis (Idso and Singer 2009; Idso et al. 2013), sustained mainly by the Nongovernmental International Panel on Climate Change, states that global warming is more likely to be attributable to natural causes than human activities. When our planet warms due to an increase, for example, in solar activity, there is a positive feedback effect in action. In fact, the consequent melting of the permafrost[14] and the greater evaporation of the oceans determine a more massive average content of CO_2 in the atmosphere. Where is the truth? We know that climate is an example of Complex System. Recently, new models, called Earth system models (ESMs), have been proposed, wherein the climate is studied in terms of the myriad of interrelated physical, chemical, biological, and socioeconomic processes. These models show that biosphere not only responds to climate change, but also directly influences the direction and magnitude of climate change (Bonan and Doney 2018). The dynamics of climate is nonlinear and extremely sensitive to the contour conditions. Therefore, it is not possible to decide the exact and unequivocal contribution of all the potential factors to global warming. More powerful computational resources would be useful for better predictions (see Chapter 13).

12.4.2.3 Solar Radiation and Life on Earth

Solar radiation plays a fundamental role in life on Earth. In fact, it is a source of both energy and information for the living beings. What distinguishes a living being from inanimate matter is that the former is purposeful (or teleonomic) because it has the functions of surviving and reproducing.[15] For accomplishing their tasks, all known living beings are alike because, irrespective of

---

[14] Permafrost is permanently frozen soil, sediment, or rock.

[15] The functions of a human being go beyond survival and reproduction. They extend to love for all the other humans, all the creatures, and, according to the religious wisdom, our Creator.

form, complexity, time or place, they must capture, transform, store and use energy, and collect, transduce, process, store and send information. According to the theory of autopoiesis (from the Greek "*αυτός*" that means "self" and "*ποίησις*" that means "creation, production"), a living being is able to self-maintain and self-reproduce (Maturana and Varela 1980). The information needed to accomplish their purposes is contained in the DNA and in the surrounding environment; the energy required to live is furnished by the metabolic network.[16] Before the discovery of the double helix structure of DNA by Watson and Crick (1953), Schrödinger (1944) reasoned that the genetic material must be an "aperiodic crystal." In fact, heredity requires structural stability and scarce chemical reactivity. Hence, the genetic material must be crystalline. However, standard crystals are based on the periodic repetition of simple units and thus store small amounts of information. Therefore, Schrödinger thought that the genetic material must be aperiodic in order to encode something as complex as a cell. Recently, after sequencing the genome of different living beings, it has been discovered that the number of genes is not enough to specify the entire structure and organization of an organism (Davies 2012). For example, the human genome consists of about 22,000–25,000 genes. Such a small number of genes is not enough to specify the structure of a brain that contains $\approx 10^{12}$ neurons and $\approx 10^{16}$ synaptic connections. Therefore, the genome is not some sort of blueprint for an organism. In fact, a blueprint has a one-to-one correspondence between the symbolic representation and the actual object. Instead, a genome is an algorithm (Davies 2012) for building an organism, and a significant amount of information contained in a cell or an organism derives from its surrounding environment. Genes constrain the development of a living being, but they are not alone in determining it. For example, the wiring of a brain develops and changes over years in response to the external stimuli. Every living being adapts its inner processes to the features of the environment. The cellular activity can be grouped into three types of modules (sometimes indicated as networks if a cell is described as a network of networks): signaling, metabolic and genetic modules (see Figure 12.9). These modules do not work in isolation, but they are interconnected, although each module maintains its own identity. The discretization of the cell in modules is possible because each module is chemically isolated (Hartwell et al. 1999). The chemical isolation comes from spatial localization or chemical specificity. The connections among modules are guaranteed by components that belong to different modules. A cell reacts in parallel to many concurrent inputs, and its behavior is not just a function of the values of its inputs but also their variety, the order in which they arrive, their timing,

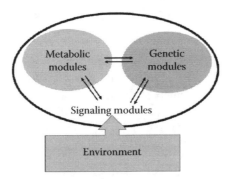

FIGURE 12.9   Discretization of the processes taking place in a cell.

---

[16] Viruses are not alive because they cannot self-sustain. Viruses contain genetic information in the form of DNA or RNA, they are open systems, reproduce, and are subjected to genetic mutations, like any other living being. However, viruses require the metabolic machinery of a living host to produce the energy needed for their replication.

and so forth. A living being, both unicellular and multicellular, is a reactive system maintained constantly out-of-equilibrium. When this condition vanishes, we have death.[17]

12.4.2.4 Solar Radiation as an Energy Source for Life on Earth

All the terrestrial ecosystems are, directly or indirectly, sustained by the energy coming from our closest star: the sun. In fact, solar radiation drives the photosynthesis of carbohydrates in photo-autotrophs organisms, such as plants, algae, and phytoplankton that "eat" light.

$$nH_2O + nCO_2 \xrightarrow{hv} nCH_2O + nO_2 \qquad [12.28]$$

Process [12.28] is a photo-induced redox reaction. The carbohydrates that are the products in process [12.28] become food for both the autotrophs and the heterotrophs. Molecular oxygen is the by-product of the photosynthesis. The primordial atmosphere of our planet was devoid of oxygen because the first photosynthetic organisms, probably living close to hydrothermal vents, were using H_2S and not H_2O as reducing reagent of CO_2. The dominant photosynthetic process was

$$6CO_2 + 12H_2S + hv \rightarrow C_6H_{12}O_6 + 12S + 6H_2O \qquad [12.29]$$

producing S rather than O_2 as a by-product. The appearance of oxygenic photosynthesis dates back to 3.5 billion of years ago (Nisbet and Sleep 2001).[18] It determined a radical change in the evolution of life on Earth. In fact, when O_2 became a relevant component of the terrestrial atmosphere, it protected the biosphere from the most harmful UV radiations by producing ozone and allowed life to colonize the lands.[19] Moreover, it opened a new chemical avenue for the exploitation of the chemical energy of carbohydrates. It made possible the breakdown of carbohydrates not only by fermentation but also combustion, gaining more energy.

The oxygenic photosynthesis consists of two distinct phases (Blankenship 2002): the light-driven and the dark phase, respectively. In the light-driven phase, the first step is the absorption of light by specialized complexes that contain pigments and proteins. The absorption of light occurs on a timescale of femtoseconds (10^{-15} s), and it is efficient due to the large absorption coefficients of pigments (chlorophylls, carotenoids, and, in some bacteria, phycobilins), as well as to the high density of pigments in the photosynthetic complexes (Figure 12.10).

If the light is absorbed by the antenna system consisting of an aggregate of pigments, an ultrafast energy transfer process from the antenna to the chlorophyll present in the photosynthetic reaction center follows. Due to the proximity of the antenna and reaction center pigments, the energy transfer occurs coherently through the excitonic mechanism (Romero et al. 2017). Therefore, the first step of the light-driven phase of photosynthesis is the formation of the excited state of the chlorophyll (Chl^*) present in the reaction center, by indirect or direct absorption. The photosynthetic reaction center is a multi-subunit membrane protein complex that functions as a remarkable photochemical device. In fact, Chl^* participates in an ultrafast unidirectional electron transfer process within the protein, which requires a few picoseconds and allows to separate two opposite charges: a positive

---

[17] In complex living beings, such as humans, the dichotomy between life and death can be not so evident because the different organs can cease to work not all at the same time, but progressively. This evidence raises ethical issues (Aguilar, 2009).

[18] The formation of the Earth dates back to 4.5×10^9 years (4.5 Gyr) ago. The geological evidence demonstrates that life has been present for at least 3.5 Gyr, and it is probable that life began before 3.8 Gyr.

[19] At about 100–150 Km from the land, O_2 is photo-dissociated by the solar radiations having the higher energies: $O_2 + hv\,(\lambda \leq 240\text{ nm}) \rightarrow 2O$. The oxygen atoms couple to form O_2, again, or they can bind to O_2 molecules and give O_3. The concentration of ozone reaches a maximum at a height of 20–30 Km over the equator. Ozone exerts a screening action from solar radiation with wavelengths included between 250 and 350 nm. These radiations could be harmful for the DNA of living beings.

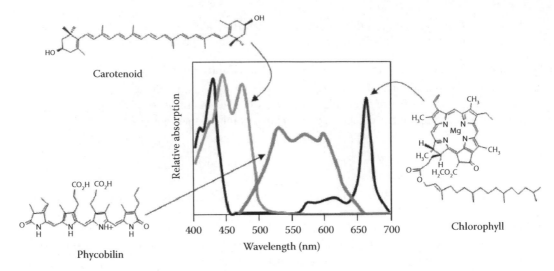

FIGURE 12.10 Spectral profiles of chlorophyll, carotenoid, and phycobilin pigments. A chlorophyll molecule absorbs the violet and red (and it is responsible for the green color of leaves), a carotenoid absorbs the violet and blue (it is responsible for the yellow color of fallen leaves in autumn), whereas a phycobilin absorbs the green and yellow.

charge in Chl and a negative charge in another acceptor molecule (that can be a quinone). The quantum efficiency of the charge-separation process is usually very high (close to 1), due to the well-organized supramolecular structure of the reaction center that avoids recombination of the charges having opposite signs. For the estimation of the maximum work that can be achieved from Chl*, it is necessary to determine the Helmholtz free energy of the pigment: $\Delta A = (\Delta U - T\Delta S)$. ΔU is given by the energy of the absorbed photon ($h\nu$) minus that squandered in vibronic relaxation processes (E_{sq}) from the Franck-Condon state[20] to the potential energy minimum of the electron donor state of Chl* (see Figure 12.11).

$$\Delta U = h\nu - E_{sq} \qquad [12.30]$$

The entropy difference between the ground and the excited state of Chl may be assumed to be small and negligible in the photoreaction center (thanks to the rigid structure of Chl). It is not necessary to consider the contribution of mixing entropy (also called configurational entropy). In fact, each photoreaction center is a separate thermodynamic system, and the free energy available is independent of the relative positions of centers that are excited with respect to those in the ground state (Jennings et al. 2014). Because of the short lifetime of the excited state, there is virtually no spreading out of energy. Therefore, $T\Delta S \approx 0$ and the maximum work that can be achieved from Chl* is

$$\Delta A \approx (h\nu - E_{sq}) \qquad [12.31]$$

In the absorption process (Chl + $h\nu \rightarrow$ Chl*), $\Delta H \approx \Delta U$ because $\Delta(PV) \approx 0$. Therefore, $\Delta G \approx \Delta A$. It is possible to have work different from that of the type pressure-volume because the photosynthetic organism has a different temperature with respect to the sun.

---

[20] The Franck-Condon state is the electronically excited state obtained immediately after the absorption of light (occurring in 10^{-15} s) and having the same molecular conformation of the original ground state. In fact, in 10^{-15}s, the nuclei do not have enough time to change their reciprocal position.

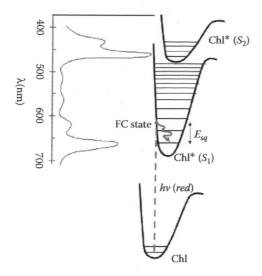

FIGURE 12.11 Schematic representation of the formation of the Franck-Condon (FC) state of Chl, and its relaxation after absorption of a red photon. In the primary electron transfer process occurring inside the reaction center, the lowest excited state of Chl is converted into a charge-transfer state, characterized by the localization of the electron and hole on adjacent molecules (Romero, E. et al., *Nature*, 543, 355–365, 2017.).

The efficient transformation of solar energy into electrochemical energy, taking place in the chlorophyll-based harvesting system, triggers redox chemical reactions that produce O_2 (due to the oxidation of water) and two other essential species, endowed with high chemical energy content. They are nicotinamide adenine dinucleotide phosphate (NADPH) and adenosine triphosphate (ATP). ATP is the well-known energy vector of cells, whereas NADPH is a reducing agent. Both species are involved in the dark phase of the overall photosynthetic process where reduction of CO_2 produces carbohydrates.

There are other circumstances wherein solar radiation is stored as electrochemical energy by living beings. Some examples are reported in Table 12.1

Alternatively, solar radiation is exploited as heat. For instance, cold-blooded living beings (such as the reptiles) bask in the morning sun to raise their internal body temperature and stimulate their metabolism. In thermodynamic terms, the sun is a heat reservoir at $T_s = 5,777$ K that transfers thermal radiation to a cold body, the living being at T_c. According to the Carnot efficiency, the maximum Helmholtz free energy gained by the cold body absorbing a photon of frequency v is:

$$\Delta A = hv \left(1 - \frac{T_c}{T_s} \right)$$

[12.32]

12.4.2.4.1 Human Activities

Of course, solar radiation is an inexhaustible energy source also useful for the human activities. According to the International Energy Outlook 2016 (EIA 2016), the total world consumption of marketed energy should expand to 1.84×10^{14} KWh in 2020. The sun releases an amount of energy that is ten thousand larger than that required by human activities, every year. However, solar energy has one principal drawback if we think about its exploitation. That is its inhomogeneous distribution in time and space. In fact, solar radiation is discontinuous in time, and it is spread over a broad area.

TABLE 12.1

Examples of Exploitation of Solar Radiation as Electrochemical Energy by Living Beings

Living Beings	Photo-induced Processes	Absorption Features	Purposes
Halobacteria living in water saturated with salts	In the protein Bacterio-rhodopsin, the chromophore all-*trans* retinal photo-isomerizes to 13-cis retinal	Green-yellow (max at 568 nm)	Bacterio-rhodopsin is a transmembrane proton-pump: it generates a trans-membrane H^+ gradient that triggers the synthesis of ATP
Halobacteria living in water saturated with salts	In the protein Halo-rhodopsin, the all-*trans* retinal photo-isomerizes to 13-cis retinal	Green-yellow (max at 578 nm)	Halo-rhodopsin is a transmembrane anion-pump: it controls the cellular concentration of salts.
All living beings	Photo-sensitization is, usually, an oxidation reaction of a substrate induced by a sensitizer that absorbs solar radiation	It depends on the absorption spectrum of the sensitizer. Usually, it absorbs visible light	It can be a harmful oxidation reaction for the living being, or it can be a strategy of defense against undesirable guests.
All living beings	Formation of dimers of pyrimidine bases in DNA	UVB (320–280 nm)	Harmful effect by UVB on DNA
Humans	Photo-induced production of the protective pigment melanin	UVA (400–320 nm)	Melanin screens cells from noxious effect of solar radiation (especially UVB).
Humans	7-dehydrocholesterol is converted in previtamin D_3	UVB	Previtamin D_3 thermally isomerizes to Vitamin D_3 that is important for skeletal health
Many living beings but not humans	Photolyase destroys dimers of pyrimidine bases by a photo-induced electron transfer involving a flavin.	Violet and blue	Photolyase repairs DNA that has been damaged by the formation of dimers of pyrimidine bases

Source: Björn, L.O. (ed.), *Photobiology. The Science of Light and Life.* 3rd ed., Springer, New York, 2015.

Nevertheless, it can be exploited as a source of both electrochemical energy and heat. Examples of techniques that use solar radiation as a source of electrochemical energy are:

- Photovoltaics when solar radiation is converted to electricity.
- Solar fuels when solar radiation is converted into chemical energy. An example is natural photosynthesis. It is possible to use the biomass as fuel. Alternatively, it is possible to exploit secondary fuels that are produced by biomass in their metabolism. Examples of secondary metabolites are ethanol, methanol, methane, and hydrogen. Such fuels are renewable, and by burning them, we release oxidized carbon (mainly, CO_2), which is already within the biosphere. When we burn petrol, coal, and natural gas, we indirectly exploit solar energy because fossil fuels are fossilized solar radiation. However, they are exhaustible because their formation required millions of years. Moreover, by burning fossil fuels, we release carbon (mainly, like CO_2), which was sequestered from the biosphere.

Examples of techniques to exploit solar radiation as a source of heat are:

- Heating environments or fluids.
- Heating thermocouples to produce electrical energy by the Seebeck effect.
- Hydroelectric energy grounds on the kinetic energy of water, whose cycle is triggered by solar thermal energy.

- Wind energy grounds on the kinetic energy of air, whose winds are generated by the asymmetric heating of the atmosphere due to the sun.
- By using solar concentrators (lenses and/or mirrors), it is possible to reach high temperatures (thousands of degrees Kelvin), induce thermal reactions, and store solar energy into chemicals

It is worthwhile striving to improve the techniques that exploit the renewable solar energy to respond to the ever-growing energy demand.

12.4.2.5   Solar Radiation as Information Source for Life on Earth

Biological systems collect, process, store, and send information to accomplish their functions of surviving and reproducing. Information exists only in the presence of life forms (or devices, such as robots, built by intelligent beings). In fact, the interactions between inanimate objects are driven by force-fields. On the other hand, the interactions between biological systems are information-based (Roederer 2003).[21] Aware of this fundamental difference, the sharp question about the origin of life on Earth can be re-formulated as it follows: "How was it possible that from an inanimate world, devoid of agents able to process information, matter self-organized in forms able to make decisions?" Life's emergence appears as a phase transition, as a sudden change in how chemistry can process and use information and free energy (Cronin and Walker 2016). If we consider a living being, there are three distinct types of natural (not man-made) information systems. First, the Biomolecular Information Systems (BISs); second, the Immune Information System (IIS) and, third, the Neural Information System (NIS). All unicellular organisms, plants, and fungi, devoid of a nervous system, are driven by BISs[22] and protected by IISs. On the other hand, every animal (except for sponges) relies on BISs at the cellular level, on IISs for protection against pathogens, and on NISs to make decisions as a whole.[23] In a BIS (see Figure 12.12), the intelligence relies on three elements. First, the multiple sensory proteins that collect information. Second, the signaling network, based on proteins, which transduce, amplify, and process the received data. Third, the epigenetic events, such as the activation and/or inhibition of the expression of specific genes in DNA, which are the actuators. The output of computation might have repercussions in the metabolic network of the cell. In an IIS, receptors of B-cells and T-cells bind to specific antigens (Ishida 2004). Intra- and inter-lymphocytes have cooperative signaling events to induce the expression of antibodies and killer T-cells that are the actuators to combat the infection. In a NIS (see Figure 12.12), there are sensory cells that collect and transduce physical and chemical messages; their content is sent to the brain where the information is processed, and decisions are made. Such decisions are signals sent to the effectors, i.e., muscles and glands.

All living beings, except those living permanently under the dark,[24] exploit solar light for their spatial and temporal orientation. For example, plants, which are sessile organisms, have specialized modules to sense their environmental light and adjust their form, orientation, metabolism, and flowering. The phenomena of photomorphogenesis, phototropism, and photoperiodism optimize plants' performances under their local conditions (see Figure 12.13).

---

[21] Humans exploit the states of the inanimate matter to encode information, and specific machines (i.e., computers) to make computations.

[22] A multicellular organism devoid of NIS exploits molecular signals to connect distinct cells.

[23] Animals, micro-organisms, and plants that live in societies give rise to "Social Information Systems" (SISs). If the society has a hierarchical structure, we can distinguish the master, which takes the decisions, and the slaves which execute the orders. If the society does not have a hierarchical structure, the SIS grounds on the so-called "Swarm Intelligence" (read paragraph 12.4.3).

[24] In the deep sea, there are bioluminescent species that emit light useful for their spatial orientation.

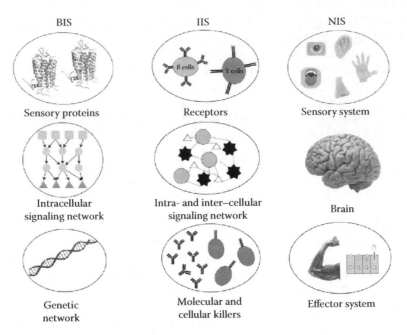

FIGURE 12.12 Schematic structures of a Biomolecular Information System (BIS), an Immune Information System (IIS), and a Neural Information System (NIS).

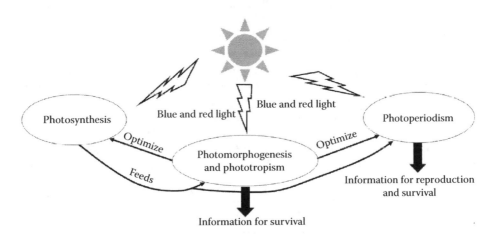

FIGURE 12.13 The role of solar radiation for the fundamental activities of plants.

12.4.2.5.1 Photomorphogenesis

Photomorphogenesis refers to plant development. It is ruled by the photoreceptive protein Phytochrome that absorbs either red (600–700 nm) or far-red (700–800 nm) light (Björn 2015). The chromophore of Phytochrome is an open-chain tetrapyrrole that can switch between two configurations within its apoprotein: one that absorbs the red (R state) and the other the far-red (FR state). When the environmental light is wealthy of red, Phytochrome is switched from R to its FR state:

$$R \xrightarrow{\text{red}} FR \qquad [12.33]$$

After process [12.33], a cascade of molecular signaling events occurs (Björn 2015) and a plant continues to grow steadily and quietly or, if it is still in the stage of a seed, it germinates. On the other hand, when the environmental light is wealthy of far-red, Phytochrome switches from FR to its R state:

$$FR \xrightarrow{\ far-red\ } R \qquad\qquad [12.34]$$

After process [12.34], another cascade of molecular signaling events occurs and, finally, a plant accelerates its growth or, if it is in the stage of a seed, it does not germinate. These opposite responses are linked to the clue a plant receives by absorbing either red or far-red light. Since the red is also absorbed by chlorophyll, when the environmental light is wealthy in red, the plant infers that other plants do not surround it, and it can germinate and grow without fearing competition. When the environmental light is poor in red and wealthy in far-red, which is scattered and not absorbed by other plants, it means there are many competitors, nearby. Therefore, if it is in the stage of a seed, it does not germinate because the surrounding is highly competitive; if it is already a developed plant, it accelerates its growth to become taller than its competitors and, hence, increase the survival probability by "eating" more sun.

12.4.2.5.2 *Phototropism and Photoperiodism*

Phototropism (Björn 2015) is the stimulation of an ensemble of processes that, ultimately, optimize the photosynthetic efficiency of plants. Examples are:

- The movement of chloroplasts, which are the organelles devoted to the photosynthesis;
- Leaf orientation;
- The opening of stomata pores in the leaf epidermis, which regulates gaseous exchange and, in particular, CO_2 uptake;
- Heliotropism that is the power of tracking the movement of the sun.

All these processes are triggered by the blue light that is absorbed by the photoreceptive protein, Phototropin, having a flavin mononucleotide as its chromophore. It is a molecular switch, like phytochrome. The inactive form (D) absorbs the blue and transforms into the active form (L):

$$D \xrightarrow{\ blue\ } L \qquad\qquad [12.35]$$

L does not absorb the blue but lets it be absorbed by chlorophyll, which exploits it for photosynthesis. At the same time, the L state, after a cascade of molecular signaling events, promotes the processes that optimize the photosynthesis. Soon after the sunset, under the dark and at room temperature, L transforms into D, spontaneously, and all the processes, optimizing photosynthesis, are switched off.

$$L \xrightarrow{\ dark\ } D \qquad\qquad [12.36]$$

In this way, the plant rests, and Phototropin becomes ready to cover its role of detecting blue light, the day after, at sunrise. Actually, plants measure the length of the day by another photoreceptor protein, Cryptochrome, which is a flavoprotein sensitive to the blue. The acquisition of a circadian rhythm is essential to rule both flowering time, a necessary trait for the optimization of pollination and seed production, and dormancy, which is the temporary cessation of growth for the protection of growing tissues in winter.

Of course, during their lifetimes, plants must deal not only with the light conditions but also with variable temperatures, nutrient and water availability, as well as toxins and symbiotic, antagonistic and commensal biota. Abiotic and biotic signals interact with the BIS of a plant. Due to the variety of possible inputs, several sensory units operate in parallel and are "weighted" appropriately to regulate the "actuators." The actuators are genes involved in cell division, cell expansion, and those that affect the positioning, growth, and differentiation of primordia[25] (Scheres and van der Putten 2017).

12.4.2.5.3 Unicellular Organisms

Unicellular organisms, such as bacteria, algae, and protozoa trust in BISs like plants do. The main difference is that micro-organisms can move thanks to their motor apparatus, cilia, or flagella. Photoreceptor proteins allow micro-organisms to probe quality and quantity of the light present in the environment. The light stimuli, after transduction and amplification, induce a modification of the movement patterns and guide the cells into environmental niches where the illumination conditions are the best for growth, survival, and/or development (Lenci 2008).

12.4.2.5.4 Animals

Animals that have NISs at their disposal can form images of their surroundings. This power derives from the highly organized architecture of their visual sensory systems. If we focus just on the human visual sensory system, we can state that its complex architecture allows us to see the shape, color, and movement of objects, and recognize variable patterns. How is this possible? First of all, we have organs specialized for vision: The eyes (Oyster 1999). The structure of an eye is similar to that of a camera (see Figure 12.14): the cornea and crystalline lens play as the objective; the pupil and iris work as the diaphragm; the retina as the photosensitive film or CCD.

On the retina, there are four types of photoreceptor cells: one kind of rod (with 120 million of replicas) and three types of cones (6 million altogether). The rods are abundant on the periphery of the retina and work in the presence of scattered light (scotopic vision). The cones are concentrated in the center of the retina (called fovea) and work in the presence of daylight (photopic vision). The rod and the three types of cones differ in the spectral position of their lowest energy electronic absorption bands. These bands are all generated by seven transmembrane α-helices proteins having 11-*cis* retinal as the chromophore. The first elementary step in human color vision is always the photo-isomerization of the 11-*cis* retinal (see Chapter 7). The different spectral positions of the bands for the rod and the three cones are due to a distinct aminoacidic composition of the pocket embedding the retinal chromophore. The three types of cones are labeled as "Blue," "Green," and "Red," based on the spectral position of their absorption bands (see Figure 12.15). They allow us to distinguish colors. How is it possible? The multiple information of a light stimulus, which is its modality (M) or spectral composition, intensity (I_M), spatial distribution ($I_M(x, y, z)$), and time evolution ($I_M(x, y, z, t)$), is

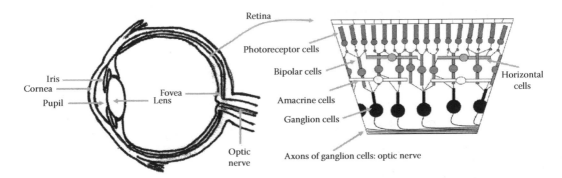

FIGURE 12.14 Schematic structure of a human eye (on the left) and retina (on the right).

---

[25] In plants, leaf primordia are a group of cells that form new leaves near the top of the shoot. Flower primordia are the little buds that form at the end of stems, from which a flower develops.

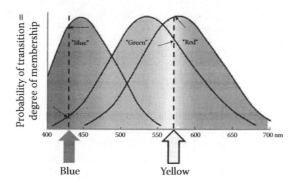

FIGURE 12.15 Absorption spectra of the Blue, Green, and Red photoreceptor proteins, and representation of the degrees of membership of blue and yellow lights to the three Molecular Fuzzy sets.

encoded hierarchically. To rationalize the mechanism of encoding, we can invoke the theory of Fuzzy sets (Gentili 2014a). A Fuzzy set, proposed by the electrical engineer Lotfi Zadeh (1965), is different from a Boolean set because it breaks the Law of Excluded-Middle. In fact, an item may belong to a Fuzzy set and its complement, at the same time, and with the same or different degree of membership. The degree of membership (μ) of an item to a Fuzzy set can be any real number included between 0 and 1:

$$0 \le \mu \le 1 \qquad [12.37]$$

Let us make an example by considering the absorption bands of the three photoreceptor proteins we have in the fovea. The absorption bands play like three Molecular Fuzzy sets. In fact, when, for example, a blue light (having a wavelength of 425 nm) hits our fovea, it is absorbed by both the Blue and the Green proteins, but not by the Red one. This behavior means that the blue ray belongs to both the Blue and the Green absorption bands, but with different degrees: $\mu(Blue) > \mu(Green) > \mu(Red) = 0$. On the other hand, a yellow light (having a wavelength of 570 nm) is absorbed by both the Green and the Red proteins. The yellow light belongs to both the Green and the Red Molecular Fuzzy sets, but not to the Blue one. Moreover, it belongs more to the Red protein rather than to the Green one: $\mu(Red) > \mu(Green) > \mu(Blue) = 0$. These examples show that the information regarding the modality is encoded as degrees of membership of the light stimulus to the Molecular Fuzzy sets. In other words, the modality is encoded as Fuzzy Information at the Molecular Level ($\bar{\mu}_{ML}$).

The intensity of a light stimulus determines the number of retinal molecules that photo-isomerize within a specific cone and per unit of time. Therefore, the information regarding the intensity is encoded as degrees of membership of the light stimulus to the cones, which work as Cellular Fuzzy sets. The intensity is encoded as Fuzzy Information at the Cellular Level ($\bar{\mu}_{CL}$). The information regarding the spatial distribution $I_M(x, y, z)$ is encoded as degrees of membership of the light stimulus to the array of Cellular Fuzzy sets, covering the retina. In other words, the spatial distribution of the light stimulus is encoded as Fuzzy information at the Tissue Level ($\bar{\mu}_{TL}$). The overall information of a light stimulus is a matrix of data. The shape of the matrix reproduces the distribution of cones on the fovea (assumed to lay on the x, y plane). Each term of the matrix, $\bar{\mu}_{x_i, y_i}(\lambda)$, is given by:

$$\bar{\mu}_{x_i, y_i}(\lambda) = \left((\bar{\mu}_{ML}) \times (\bar{\mu}_{CL}) \right)_{x_i, y_i} = \left(\frac{\Phi_{C_l} I_{0,\lambda} \left(1 - 10^{-(\varepsilon_{C_l})(C_{C_l})L} \right)}{\left(n_{c \to t}^{MAX} \right) / \Delta t} \right)_{x_i, y_i} \qquad [12.38]$$

In [12.38], $I_{0,\lambda}$ is the intensity of the incident light having wavelength λ. Φ_{C_l} is the quantum yield for the photo-isomerization of retinal within cone C_l. ε_{C_l} and C_{cl} are the absorption coefficient and concentration of the photo-receptor protein contained in cone C_l, respectively. L is the length of the photosensitive part of the same cone, and $\left(n_{c \to t}^{MAX} \right)/\Delta t$ is the maximum number of photo-isomerization that can occur per unit time within C_l.

The information of the light stimulus collected by the array of photoreceptor cells is not relayed to the brain as such, but it is processed by an ensemble of neurons that is present in the retina. There are bipolar and horizontal cells connected directly to rods and cones. Then, there are ganglion cells joined to bipolar cells, and amacrine cells allowing the interconnection among bipolar and ganglion cells that are far apart. The axons of ganglion cells form the optic nerve that relays the optical signals to the visual cortex of the brain (see Figure 12.14). Each bipolar cell is connected directly to a number of photoreceptor cells located roughly opposite it. This number ranges from one photoreceptor at the center of the fovea to thousands in the periphery of the retina. In addition to these direct vertical connections, each bipolar cell receives some of its afferences from horizontal cells. The horizontal cells guarantee the connection of a bipolar cell to a set of more distant photoreceptors. All the photoreceptors connected to a bipolar cell define its receptive field that has a circular shape. Since each cone and rod may be connected to more than one bipolar cell, the receptive fields of bipolar cells are Fuzzy sets. A bipolar cell integrates the signals coming from different photoreceptor cells. Its final output depends on the activation pattern of the photosensitive cells belonging to its receptive field. Light shining on the center of a bipolar cell's receptive field and light shining on the surround produce opposite changes in the cell's membrane potential. The purpose of Bipolar Fuzzy sets is to improve the contrast and definition of the visual stimulus. A further increase of contrast is achieved by the ganglion and amacrine cells. Each ganglion cell is connected directly to a number of bipolar cells located roughly opposite it. In addition, each ganglion cell receives some of its afferences from amacrine cells. The amacrine cells guarantee the connection among distant bipolar and ganglion cells. All the photoreceptor cells indirectly connected to a ganglion cell define its receptive field that has a circular shape. The receptive fields of ganglion cells are Fuzzy sets that include more photosensitive cells than the Bipolar Fuzzy sets. In fact, each eye has about 126 million of photosensitive cells and just 1 million of ganglion cells. It is clear that one photoreceptor cell influences the activity of hundreds or thousands of Ganglion Fuzzy sets. The center-surround structure of the receptive fields of bipolar cells is transmitted to the ganglion cells. The accentuation of contrasts by the center-surround receptive fields of the bipolar cells is thereby preserved and passed on to the ganglion cells. The presence of overlapping receptive fields (like overlapping Fuzzy sets) allows processing the information of a light stimulus in parallel and increasing the acuity by highlighting the contrasts in space and time. For example, in the retinal ganglion cells, three channels convey the information about colors from the eye to the visual cortex (Gegenfurtner 2003). In one channel, the signals coming from the Red cones are summed (OR operator) to those coming from the Green cones to compute the intensity of the light stimulus. In a second channel, the signals from the Red and Green cones are subtracted (NOT operator) to compute the red-green component of a stimulus. Finally, in a third channel, the sum of the Red and Green cone signals is subtracted from the Blue cone signals to compute the blue-yellow variation in a stimulus. The signals in these three channels are transmitted in distinct pathways. Whereas a bipolar cell encodes the visual information in the value of a graded potential, which is analog and not discrete, a ganglion cell encodes the information in the analog value of the frequency of firing action potentials. The information encoded through action potentials is relayed more safely over long distances and can reach the visual cortex (located in the back of the head). The risk of losing information by noise is minimized.

The visual cortex is partitioned into different areas: V1, V2, V3, V4, and V5. Each area is divided into compartments. Recent studies (Johnson et al. 2001) reveals that the function of the compartments is not unique, but Fuzzy. For example, the analysis of luminance and color is not separated, but there is a continuum of cells, varying from cells that respond only to luminance, to a few cells that do not respond to luminance at all. Other investigations (Friedmann et al. 2003) have revealed that in areas V1 and V2, orientation and color selectivity are not binary measures; they are Fuzzy. In fact, cells vary continuously in their degree of tuning, and it is possible to assign a membership function for orientation and another for color perception to each cell. The extraction of information in the visual cortex is carried out with the same mechanism observed in the neurons of the retina, i.e., by granulation of neurons in Fuzzy sets that integrate and abstract information.

Based on this description, it might seem that human vision is deterministic, reproducible, objective, and universal. But this is not the case because human vision, like any other sensory perception, depends on the physiological state of the perceiver, his/her past experiences, and each sensory system is unique and not universal. Moreover, every human brain must deal with the uncertainty in the perception. Under uncertainty, an efficient way of performing tasks is to represent knowledge with probability distributions and acquire new knowledge by following the rules of probabilistic inference (Van Horn 2003; De Finetti et al. 1992). This consideration leads to the idea that the human brain performs probabilistic reasoning, and human perception can be described as a subjective process of Bayesian Probabilistic Inference (Ma et al. 2008; Mach 1984). The perception of a stimulus S_M by a collection of cortical cells X_M will be, then, given by the posterior probability $p(S_M | X_M)$:

$$p\left(S_M \mid X_M\right) = \frac{p\left(X_M \mid S_M\right) p\left(S_M\right)}{p\left(X_M\right)} \tag{12.39}$$

In equation [12.39], $p\left(S_M\right)$ is the prior probability, $p\left(X_M \mid S_M\right)$ is the likelihood, and $p\left(X_M\right)$ is the plausibility. The plausibility is only a normalization factor. In agreement with the theory of Bayesian probabilistic inference generalized in Fuzzy context (Coletti and Scozzafava 2004), the likelihood may be identified with the deterministic Fuzzy information described earlier. The prior probability comes from the knowledge of the regularities of the stimuli and represents the influence of the brain on human perception. Human perception is a trade-off between the likelihood and the prior probability (Kersten et al. 2004). The likelihood represents the deterministic and objective part of the human perception. The prior probability represents its subjective contribution. In fact, it is reasonable to assume that all the possible patterns of activity of cortical neurons of a specific area are granulated in Fuzzy sets, whose number and shape depend on the context. Moreover, such Fuzzy sets are labeled by distinct adjectives within our brain. The noisier and ambiguous are the features of a stimulus, the more prior probability driven will be the perception, and the less reproducible and universal will be the sensation.

Sometimes, we receive multimodal stimuli that interact with more than one sensory system. Each activated sensory system produces its own mono-sensory Fuzzy information. Physiological and behavioral experiments have shown that the brain integrates the mono-sensory perceptions (Ernst and Banks 2002) to generate the final sensation. There are brain areas, such as the Superior Colliculus (Stein and Meredith 1993), which contain distinct mono-sensory, as well as multisensory neurons. Neurophysiological data have revealed influences among unimodal and multimodal brain areas, as well (Driver and Spense 2000). Multisensory processing pieces signals of different modality if stimuli fall on the same or adjacent receptive fields (according to the "spatial rule") and within close temporal proximity (according to the "temporal rule"). Finally, multisensory processing forms a total picture that differs from the sum of its unimodal contributions; a phenomenon called multisensory enhancement in neuroscience (Stein and Meredith 1993; Deneve and Pouget 2004), or colligation in the Information theory (Kåhre 2002). The principle of inverse effectiveness states that the multisensory enhancement is stronger for weaker stimuli. Since sensory modalities are not equally reliable, and their reliability can change with context, multisensory integration involves statistical issues, and it is often assumed to be a Bayesian probabilistic inference (Pouget et al. 2002).

From this paragraph, we understand that human vision is extraordinarily complex. Its complexity is magnified by the uniqueness of each sensory system, the dependence of the sensory action on the physiological state of the perceiver, and his/her past experiences. Probably, the human power of recognizing variable patterns derives from the granulations of neurons and their patterns of activity in Fuzzy sets.

12.4.3 EMERGENT PROPERTIES

The Complexity of a natural system can be estimated by the degree of difficulty in predicting its properties when the features of the system's parts are given. In fact, any Complex System is a network that exhibits one or more collective properties that are said emergent because they come to light, as a whole.[26]

Complexity (C) derives from a combination of three features: Multiplicity (M), Interconnection (Ic) and Integration (Ig) (Lehn 2013):

$$C \propto M©Ic©Ig \qquad\qquad [12.40]$$

Many and often diverse nodes (Multiplicity), which are strongly interconnected (Interconnection), exhibit emergent properties because they integrate their features (Integration). This statement is valid especially when the systems are in out-of-equilibrium conditions. The symbol © expresses a peculiar combination of the three parameters: M, Ic, and Ig. Complexity and emergent properties can be encountered along a hierarchy of levels: for instance, at the molecular, supramolecular, and cellular levels. But also passing from cells to tissues, from tissues to organisms, and from organisms to societies and ecosystems. At each level, novel properties emerge that do not exist at lower levels.

In the history of philosophy and science, different taxonomies of the emergent properties have been proposed (Corning 2002; Clayton and Davies 2006; Bar-Yam 2004). Here, I offer a new taxonomy based on the structures of the networks.

1. *Regular* and *random networks* show emergent properties in both equilibrium and out-of-equilibrium conditions.[27] At equilibrium, the emergent properties are not affected significantly by the removal or the addition of a few nodes. Examples are the phases of matter (such as solids, liquids, and gases). Other cases are the pressure and temperature of a macroscopic phase.

   In out-of-equilibrium conditions, examples of emergent properties of regular and random networks are all those phenomena of self-organization, in time and space, which we have studied in the previous chapters of this book. For instance, the predator and prey dynamics, the optimal price of a good in a free market,[28] oscillating reactions, chemical waves, Turing structures, periodic precipitations, convection in fluids, et cetera. These kinds of emergent properties are sensitive to the removal or addition of a few nodes. In fact, large out-of-equilibrium systems can self-organize into highly interactive, critical states where minor perturbations may lead to events, called avalanches, spanning all sizes, up to that of the entire system. This feature is described by the Sandpile model (Bak and Paczuski 1995). Let us consider a pile of sand on a table, where sand is added slowly. When the pile becomes steep, it reaches a stationary state. In fact, the amount of sand

---

[26] The word "emergence" comes from the Latin verb *emergere* that is composed by the preposition *ex* that means "out" and the verb *mergere* that means "to dip, to immerse, to sink."

[27] Machines, software, and sentences in natural language can be described as instances of regular networks exhibiting emergent properties as a whole. Their emergent properties depend on the delicate structure of the network and are strongly affected by the removal of just one or more nodes. The nodes, with known properties, work together predictably, and the whole does or means (if we refer to sentences) what is designed to do or mean.

[28] Adam Smith used the term "invisible hand" to describe the emergent property of trade and market exchange, which spontaneously channel self-interest toward socially desirable ends.

added is balanced, on average, by the amount of sand leaving the system along the edges of the table. This stationary state is critical. In fact, a single grain of sand might cause an avalanche involving the entire pile. Natural Complex Systems exhibit a kind of emergent property called Self-Organized Criticality (SOC): they have periods of stasis interrupted by intermittent bursts of activity. Since Complex Systems are noisy, the occurrence of a burst of action cannot be predicted, and it is hard to be reproduced. At most, what is predictable and reproducible is the statistical distribution of these avalanches, which usually follows a power-law. Thus, if an experiment is repeated, with slightly different random noise, the resulting outcome is entirely different.

Other examples of emergent properties of regular and random networks are the fish schooling and birds flocking, on the one hand, and phase transitions, on the other. A school of fish or a flock of birds exhibits an emergent collective and decentralized intelligent behavior that is called swarm intelligence (Reynolds 1987; Bonabeau et al. 1999). It has been demonstrated by simulation that the collective behavior derives from the interactions among the individuals that follow a few simple rules based on local information. For instance, the behavior of a flock of birds can be reproduced by assigning three simple rules to every agent. The first is the alignment rule: a bird looks at its neighbor birds and assumes a velocity (regarding module, direction and versus) that is close to the mean speed of its nearest group mates. The second is the cohere rule: after alignment, a bird takes a small step in the direction that the center of mass of birds takes. The third is the separate rule: a bird should always avoid any collision. This kind of behavior reminds that of matter at any phase transition, called criticality among statistical physicists (Christensen and Moloney 2005). The phase transition occurs when an external agent finely tunes specific external parameters to particular values. At a phase transition, the many microscopic constituents of material give rise to macroscopic phenomena that can be understood by considering the forces exerting among the single particles. Whereas the actions of living beings, in a flock of birds or school of fish, are information-based, those of particles in a material are force-driven.

TRY EXERCISE 12.11

2. Social insect colonies have features of *Scale-free networks*. The connections among nestmates are nonrandomly distributed for most colony functions (Fewell 2003). A few key individuals disseminate information to many more nestmates than do others; they play like hubs. The most obvious hub is the queen: she does not centrally control all the colony functions but, in honeybees, she secretes a pheromone that represses reproduction in workers and maintains colony cohesion. Essential hubs are also present within worker task groups: they are the scouts or dancers. Such vital individuals communicate most of the information about resource location and availability and maintain the cohesion of the group that goes out to forage. The removal of hubs can disrupt the system severely, whereas the loss of any of the vast majority of workers would have little effect.

3. In *modular networks*, each module can have an emergent property, and their cooperative action gives rise to one or more synergistic effects. Examples are the symbiotic relationships that can be encountered in ecology. For instance, more than 1.2 billion years ago, a cyanobacterium took up residence within a eukaryotic host (Gould 2012). This event gave rise to algae that contain a photosynthetic organelle (plastid), which is remnant of the cyanobacterium and mitochondria, which are organelles derived from the integration of other prokaryotes early in eukaryotic evolution. The endosymbionts optimize both the respiration and the photosynthesis by synergy. "The whole is something over and above its parts, and not just the sum of them all..." as alleged, more than 2000 years ago, by Aristotle in its philosophical treatise titled Metaphysics. In the presence of synergistic effects, $2 + 2$ does not make 4, but more.

4. In a *hierarchical network*, each level has an emergent property. Examples of hierarchical networks are the living beings. Life is the emergent property of the network as a whole (Goldenfeld and Woese 2011). A living being's isolated molecular constituents, such as phospholipids, water, salts, DNA, RNA, proteins, and so on, can never show life. Only, if we consider all the constituents organized in the dynamic hierarchical structure of a cell, we can observe the fantastic phenomenon of life. In a hierarchical network, we have both upward and downward causation. Upward causation is when the features of lower levels rule the emergent properties of the higher levels. Downward causation is the opposite. The properties of higher levels influence those of the lower levels. I report a few examples (Noble 2006). A mother and her environment transmit to the genes of her embryo adverse or favorable influences. The heart of an athlete shows different gene expression patterns from those of a sedentary person. Hormones released by endocrine glands and circulating in the blood system can influence events inside the cells. The act of sexual reproduction ends in fertilizing an egg cell. And so on. It is the highly dynamic, heterogeneous, organized, fractal-like, structure of chromatin (see Chapter 11) that marks the intersection of upward and downward causation (Davies 2012). In fact, its structure and behavior are influenced by both the genes it contains and the macroscopic forces acting on it from the rest of the cell and the cell's environment. The possibility of having both upward and downward causation gives living beings the power of influencing their environment, but also of adapting to it. Living beings and their societies are Complex Adaptive Systems (Miller and Page 2007).

12.5   KEY QUESTIONS

- Make examples of Complex Systems.
- Which are the Natural Complexity Challenges?
- Present the challenges in the field of Computational Complexity.
- What is the link between Natural Complexity and Computational Complexity?
- What is the fundamental and unavoidable limit we will always encounter in the description of Complex Systems even if it were demonstrated that $NP = P$?
- Which are the essential properties of Complex Systems?
- Which are the parameters that characterize the structure of a network?
- Present the features of the model networks.
- Which are the most common network models in nature and why?
- Which are the essential factors maintaining the Earth out-of-equilibrium?
- Describe the thermodynamic properties of thermal radiation.
- Which are the factors affecting the climate of the Earth?
- What is life?
- How does life exploit solar radiation as a source of energy?
- How does life exploit solar radiation as a source of information?
- Where does the complexity of human vision come from?
- What is an emergent property and how does it originate?
- How many kinds of emergent properties do you know?

12.6   KEY WORDS

Natural Complex Systems; Computational Complexity; P, NP, and NP-complete Problems; Regular Networks; Small-World Networks; Random Networks; Scale-free Networks; Modular Networks; Hierarchical Networks; Thermal radiation; Biomolecular Information System; Neural Information System; Immune Information System; Social Information System; Fuzzy set; Bayesian probability; Multiplicity; Interconnection; Integration; Emergent properties.

12.7 HINTS FOR FURTHER READING

- The subject of Computational Complexity can be studied more deeply by reading the book by Garey and Johnson (1979) and that by Papadimitriou (1994). The description of some of the principal NP problems can also be found in (Lewis and Papadimitriou, 1978).
- For those who want to deepen their understanding of networks, I suggest the book by Caldarelli and Catanzaro (2012) and the recent book by Barabási (2016), which is freely available at the link http://barabasi.com/networksciencebook/. Recent exciting investigations on the phenomena of synchronization in large networks are presented in a special issue of the journal *Chaos*, edited by Duane et al. (2017).
- Whoever is interested particularly about Complexity in ecology, I suggest the paper by Montoya et al. (2006) and that by Sugihara et al. (2012).
- The books by Haken (2006) and Walker et al. (2017) are relevant for whom wants to learn more about the role of information in living beings and Complex Systems.
- A program for large network analysis is available at http://mrvar.fdv.uni-lj.si/pajek/
- A list of databases for biochemical networks is available in (Subramanian et al. 2015).

12.8 EXERCISES

12.1. Compare the time required to solve three polynomial problems of the type $O(N^1)$, $O(N^2)$, $O(N^3)$, and two exponential problems of the type $O(2^N)$ and $O(N^N)$, for $N = 10, 20, 100$, and 500. Image using the Chinese supercomputer TaihuLight, whose computational rate is 93 PFlop/s.

12.2. For the network shown in Figure 12.16, determine the degree distribution, the average degree, the average clustering coefficient, and the shortest path between nodes A and D.

12.3. At https://ccl.northwestern.edu/netlogo/index.shtml webpage, download NetLogo software, free of charge (Wilensky 1999). In Models Library, open the NetLogo "Small Worlds" model (Wilensky 2005) or the "SmallWorldNetworks" model (available at Complexity Explorer project, http://complexityexplorer.org). Build a regular network and calculate its mean path length $\overline{sp}$ and its average clustering coefficient $\overline{C}$. Then, randomize the links of the network by increasing the rewiring probability p monotonically. Determine how the ratios $\overline{sp}(p)/\overline{sp}(0)$ and $\overline{C}(p)/\overline{C}(0)$ changes for $p = 0.01; 0.05; 0.1; 0.15; \ldots; 1$.

12.4. Determine the degree distribution and the average clustering coefficient of the modular network shown in Figure 12.17.

12.5. Figure 12.18 shows an example of a hierarchical network and its iterative construction (Ravasz and Barabási 2003). The starting subgraph is a fully connected cluster of five nodes with links also between diagonal nodes. The 1° iteration, required to build the network, consists in creating four identical replicas and connecting the peripheral nodes of

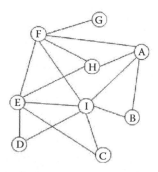

FIGURE 12.16 Structure of an undirected network.

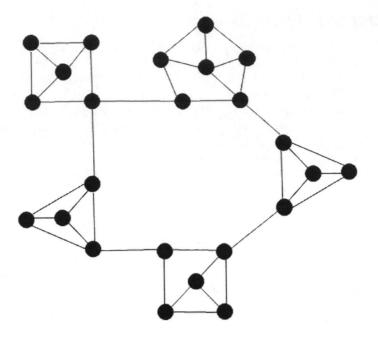

FIGURE 12.17   Example of modular network.

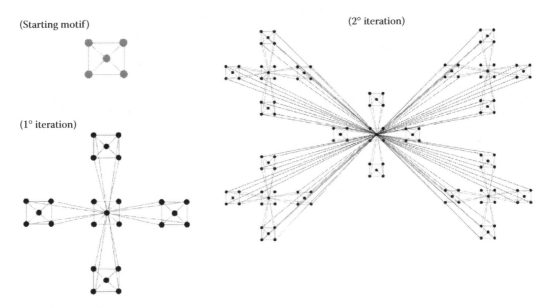

FIGURE 12.18   Construction of a hierarchical network.

each new cluster to the central node of the original subgraph. The output is a network with 25 nodes. In the 2° iteration, four copies of the output of the 1° iteration are created and, again, the peripheral nodes are connected to the central node of the original motif, obtaining a network with 125 nodes. Determine the clustering coefficient of the nodes at the center of (I) the numerous five-node motifs, (II) the 25-node modules, and (III) the 125-node network.

12.6. Calculate the intensity of the thermal radiation emitted by bodies at 5900, 5500, 5000, and 4500 K, in the range $[10 \div 10010]$ nm and Wm^{-2}nm^{-1}. Define the function relating the wavelength that corresponds to the maximum intensity emitted at the various temperatures versus the reciprocal of temperature. Moreover, confirm the Stephan-Boltzmann law.

12.7. Calculate the spectral emissivity of our body that has an average temperature of 36°C. In which spectral region does our body emit?

12.8. Knowing that the solar irradiance at the Earth's atmosphere is 1366 W/m^2, how large is the pressure of the solar radiation exerted on the terrestrial atmosphere, considering a perpendicular incidence?

12.9. Try to explain how the Crookes radiometer works (see Figure 12.19). A set of vanes, black on one side, white or silver on the other, are enclosed within a sealed glass bulb, evacuated at about 1 Pa. The vanes can rotate on a low-friction spindle. When exposed to light or heat source, the dark sides rotate away from the source. On the other hand, if a cold body (for instance, a block of ice) is placed nearby, the vanes rotate in the opposite direction, i.e., the black sides turn towards the cold body. This radiometer was invented by the chemist Sir William Crookes, in 1873, after noticing that sunlight shining on his balance was disturbing his weighing within a partially evacuated chamber. The explanation of the working mechanism of the Crookes radiometer was a source of scientific controversy for over half a century.

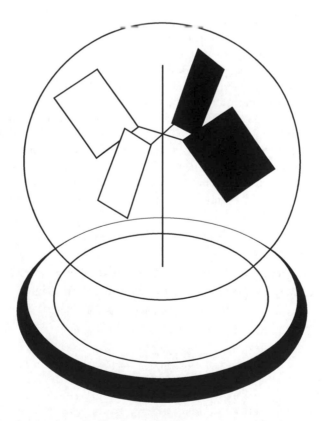

FIGURE 12.19 Schematic structure of a Crookes radiometer.

12.10. Knowing that the average temperature of the Earth is 288 K, how much is its total radiant power, and which is the wavelength of the peak power?

12.11. Open NetLogo software (Wilensky, 1999). Go to the Models Library; select Biology and open the "Flocking" Model (Wilensky, 1998). *I Part*: Play with the model by changing the vision slider. Run simulations with a vision of 1.0 patch, then 3.0 patches, 5.0 patches, 7.0 patches, each time letting the simulation run for at least 300 ticks (Click on "Go" again to stop, then "Setup" before starting a new run). As you increase the vision range, does the group flocking behavior become stronger (converging more quickly on larger flocks) or weaker (converging more slowly, with smaller flocks)?

II Part: Set the vision slider back to 3.0 patches. Play with the model by changing the minimum-separation slider. Run simulations after setting the minimum-separation slider to 1.00, 2.00, and then 4.00. Which of these settings gives you the most "flock-like" behavior after about 300 ticks?

12.9 SOLUTIONS TO THE EXERCISES

12.1. The amount of time required to solve the three polynomial problems and the two exponential problems are listed in Table 12.2. The calculations have been performed for N equal to 10, 20, 100, and 500, and assuming to have the Chinese supercomputer TaihuLight for the computations.

It is amazing how much the time soars abruptly when we deal with exponential problems, and we increase N. It is evident that when the size of the instance is large, any exponential problem cannot be solved accurately and in a reasonable time.

12.2. In Table 12.3, there is a list of the degree and clustering coefficient for each node of the network.

The degree distribution is shown in Figure 12.20.

The average degree is $\bar{d} = 3$; the average clustering coefficient is $\bar{C} = 0.62$. The shortest path between node A and D is equal to 2 links.

12.3. First of all, we build a regular network with 200 nodes. The degree is $d = 4$ for each node, $\overline{sp}(0) = 25.25$ and $C(0) = 0.5$ (obtained by the "SmallWorldNetworks" model). By assigning monotonically growing values to the rewiring probability p, we gain the trends of the ratios $\overline{sp}(p)/\overline{sp}(0)$ and $\bar{C}(p)/\bar{C}(0)$ that are reported in Figure 12.21.

It is noteworthy that the ratio $\overline{sp}(p)/\overline{sp}(0)$ drops immediately to ≈ 0.2 when p is just 0.01. By rewiring only 1% of the total number of links, we obtain the onset of the small-world phenomenon. For $p > 0.2$, $\overline{sp}(p)/\overline{sp}(0)$ remains almost constant to the value of ≈ 0.15, even when we have a completely random network. On the other hand, the ratio $\bar{C}(p)/\bar{C}(0)$ decreases less abruptly, but when $p > 0.4$, it goes to zero, confirming that in a random network the clustering is very low.

TABLE 12.2

Running Times of Polynomial and Exponential Problems Calculated by the Chinese Supercomputer TaihuLight

Running time	Size of the Problem (N)			
	10	**20**	**100**	**500**
O(N)	1×10^{-16} s	2×10^{-16} s	1×10^{-15} s	5.4×10^{-15} s
O(N²)	1×10^{-15} s	4.3×10^{-15} s	1×10^{-13} s	2.7×10^{-12} s
O(N³)	1×10^{-14} s	8.6×10^{-14} s	1×10^{-11} s	1.3×10^{-9} s
O(2ᴺ)	1×10^{-14} s	1×10^{-11} s	4.3×10^5 years	1×10^{126} years
O(Nᴺ)	1×10^{-7} s	36 years	3.4×10^{175} years	1×10^{1325} years

---

TABLE 12.3

Degree (d) and Clustering Coefficient (C) of the Nodes Belonging to the Network of Exercise 12.2

Node	d	C
A	4	3/6
B	2	1
C	2	1
D	2	1
E	5	4/10
F	5	4/10
G	1	0
H	3	2/3

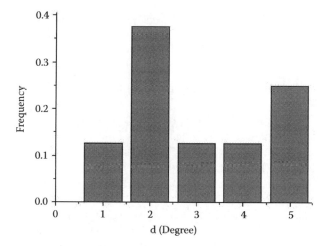

FIGURE 12.20 Degree distribution for the network of this exercise.

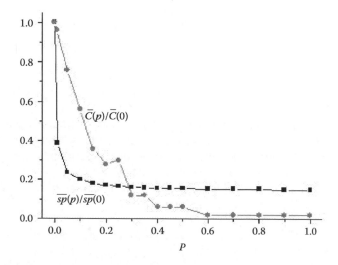

FIGURE 12.21 Trends of $\overline{sp}(p)/\overline{sp}(0)$ (black squares) and $\overline{C}(p)/\overline{C}(0)$ (gray circles).

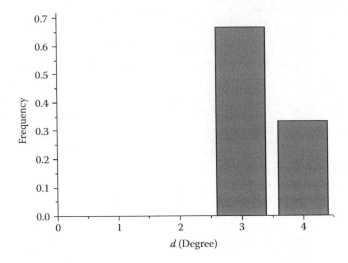

FIGURE 12.22   Degree distribution for the modular network of the exercise 12.4.

12.4. The degree distribution of the modular network is shown in Figure 12.22. The average clustering coefficient is ≈ 0.51.

12.5. The clustering coefficient of the node at the center of the five-node motifs is 1 because its degree is 4 and its neighbors are all mutually linked: $C = (2*6/4*3)$. The node at the center of a 25-node module has $d = 20$ and $C = (2*30/20*19) = 3/19$. The node at the center of the 125-node network has $d = 84$ and $C = (2*126/84*83) = 3/83$. These data confirm that in a hierarchical network, the clustering coefficient is a power law of the degree of connectivity, d.

12.6. The spectral emissivity, as a function of the wavelength λ, can be obtained from the Planck's formula $u(v,T)dv = \left(8\pi h v^3/c^3\right)\left(1/\left(e^{hv/k_BT} - 1\right)\right)dv$. First, we substitute v with c/λ and dv with $\left(c/\lambda^2\right)d\lambda$. Second, we multiply the energy density by $c/4$ (according to equation [12.7]). The final equation looks like

$$E(T,\lambda) = \frac{2\pi hc^2}{\lambda^5} \frac{1}{\left(e^{\frac{hc}{k_BT\lambda}} - 1\right)}$$

By using SI units, we obtain a spectral emissivity in Wm^{-3}. Dividing by 10^9, we achieve the units required by the exercise, which are Wm^{-2}nm^{-1} (see Figure 12.23).

From Figure 12.23, we observe that the hotter the body, the more intense the emissivity of the thermal radiation. If we calculate the integrals of these functions, and we plot the $\log\left(Area\,of\,E(T,\lambda)\right)$ versus $\log(T)$, we can fit the data by a straight line with slope equal to 4, confirming the Stephan-Boltzmann law (see Figure 12.24).

If we plot the wavelength that corresponds to the maximum of $E(T,\lambda)$ versus $1/T$, we obtain a linear trend, like that shown in Figure 12.25.

The relation $\lambda(nm) = A + (2,9\times 10^6/T(K))$ is known as Wien's law. The hotter the body, the shorter the wavelength that corresponds to the maximum spectral emissivity.

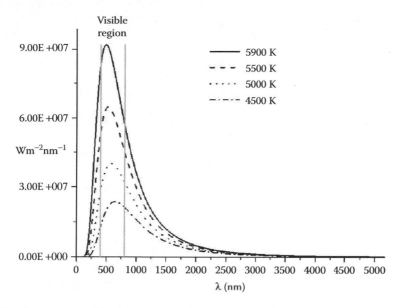

FIGURE 12.23 Spectral emissivity of the black body at four distinct temperatures.

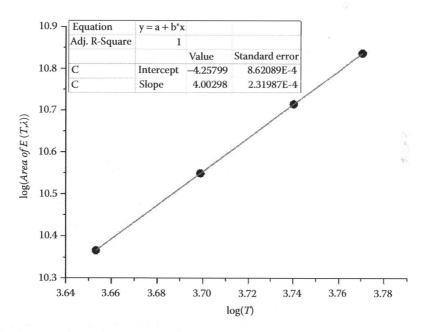

FIGURE 12.24 Linear relation between $\log(Area\ of\ E(T, \lambda))$ versus $\log(T)$.

12.7. The spectral emissivity of our body at 309 K, calculated by using the formula presented in the previous exercise, is plotted in Figure 12.26. It is included between 2.5 and 50 μm, i.e., within the IR region. Snakes that have IR sensors can perceive the thermal IR radiation that we emit.

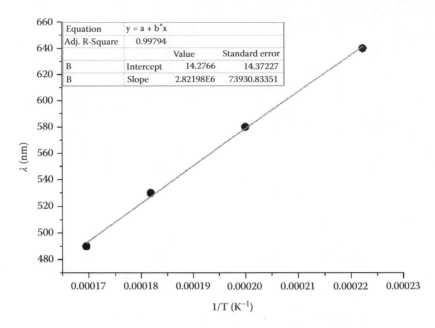

FIGURE 12.25 Linear relation between the wavelength (λ) that corresponds to the maximum of $E(T, \lambda)$ and $1/T$.

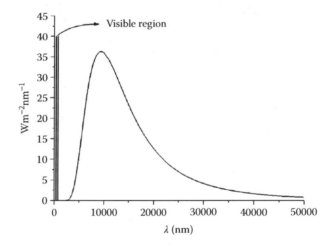

FIGURE 12.26 Spectral emissivity of the human body.

12.8. By using equation [12.7], we transform the intensity 1366 W/m² in energy density:

$$u_{tot} = \frac{4I}{c} = \frac{4(1366)}{3 \times 10^8} = 1.82 \times 10^{-5} \, \text{Wm}^{-3}$$

Then, the pressure is evaluated by using equation [12.12].

$$P = \frac{u_{tot}}{3} = \frac{1.82 \times 10^{-5}}{3} = 6.1 \times 10^{-6} \, Pa \approx 6 \times 10^{-11} \, \text{atm}$$

It is evident that the solar pressure is very tiny.

12.9. The currently accepted explanation of the phenomenon involved in the Crookes radiometer was first proposed by the English physicist Osborne Reynolds, in 1879. It considers the flow of gas around the edges of the vanes from the warmer black sides to the cooler white sides. The slight tangential viscous friction causes the vanes to move in the direction from black to white. When a cold body is placed near the radiometer, the motion of the gas molecules is reversed, and the rotation occurs in the opposition direction. It is not the radiation pressure the responsible force of the movement. In fact, if it were, the vanes would rotate in the opposite direction, since the photons are absorbed by the black surfaces but reflected from the white (or silver) sides, thus transferring twice the momentum. The rotation of vanes is controlled by the radiation pressure only when a high vacuum is operated within the sealed glass bulb.

12.10. For the calculation of the total emissive power of the Earth, we can use the Stephan-Boltzmann law, equation [12.9]:

$$E_k(T) = \xi T^4 = \left(5.67 \times 10^{-8} \frac{\text{W}}{\text{m}^2\text{K}^4}\right)(288\,\text{K})^4 = 390 \frac{\text{W}}{\text{m}^2}$$

To determine the wavelength of the peak power, we can use the Wien's law obtained in the exercise 12.6.

$$\lambda(\text{nm}) = 14.3 + \frac{2{,}9 \times 10^6}{T(K)} = 10084\,\text{nm}$$

The emissive power of the Earth is peaked in the IR, at the wavenumber of 992 cm^{-1}.

12.11. *Part I*: By increasing the vision range from 1.0 patch to 7.0 patches, the flocking birds converge more quickly on more massive flocks.

Part II: The most "flock-like" behavior is achieved by fixing the minimum-separation to 1.0.

13 How to Untangle Complex Systems?

Everything should be made as simple as possible, but not simpler.

Albert Einstein (1879–1955 AD)

Every revolution was first a thought in one man's mind.

Ralph Waldo Emerson (1803–1882 AD)

I think it's fair to say that personal computers have become the most empowering tool we've ever created. They're tools of communication, they're tools of creativity, and they can be shaped by their user.

Bill Gates (1955–)

13.1 INTRODUCTION

In Chapter 12, we learned which are the Complexity Challenges and the reason we find so many difficulties in winning them. Now, the question is: How can we tackle the Complexity Challenges effectively? Whenever we deal with Complex Systems, we need to collect, handle, process Big Data, i.e., massive data sets (Marx 2013). Therefore, for trying to win the Complexity Challenges, we need to contrive smart methods to cope with the vast volume and the fast stream of data we collect, their variety (they come in many types of formats, such as numeric, text documents, video, audio, etc.), and variability and complexity (because they are often interlinked). If we want to extract insights from Big Data, it is essential to speed up our computational rate and find out new ways to process data, i.e., new algorithms. There are two main strategies to succeed. One consists of improving current electronic computers, and the other is the interdisciplinary research line of Natural Computing.

13.2 IMPROVING ELECTRONIC COMPUTERS

Since their first appearance in the 1950s, electronic computers have become ever faster and ever more reliable. Despite this constant progress, current computers have the same architecture as the first ones. A structure that was worked out by John von Neumann (read Chapter 2 for the details). The principal elements of the so-called von Neumann computer are an active central processing unit (CPU) and a passive memory. The memory stores two fundamental types of information. First is instructions, which the CPU fetches in sequence and then tells the CPU what to do—whether it be to make an operation, compare bits or fetch other data from memory. Second is the data. The data are manipulated according to the instructions of the program. Information is encoded as binary digits through electrical signals, and transistors are the basic switches of the CPU. A transistor is a semiconductor device. There are different types of transistors (Amos and James 2000). A Field-Effect-Transistor (FET), broadly used in digital and analog electronics to implement integrated circuits, is made of two regions negatively-charged, called source and drain, separated by a region positively-charged, called

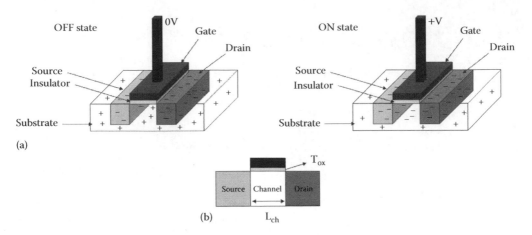

FIGURE 13.1 The schematic structure of a MOSFET in its OFF and ON states (a), and its frontal view (b), wherein L_{ch} is the length of the channel, whereas T_{ox} is the width of the insulator.

substrate (see Figure 13.1a). The substrate is surmounted by a thin layer of insulator (for instance, SiO_2 or HfO_2. In this case the transistor is named as MOSFET wherein MOS stands for Metal Oxide Semiconductor) and an electrode, called gate, which allows applying a voltage to the substrate. A rule of thumb (Cavin et al. 2012) suggests that the width of the insulator, T_{ox}, must be $\approx 1/30$ of the channel length, L_{ch} (see Figure 13.1b). When no voltage is applied to the gate (0V), the positively-charged substrate plays as a barrier that does not allow electrons to flow from the source to the drain. If a positive voltage (+V) is applied to the gate, the positive charges of the substrate are pushed away, and a negatively charged communication channel is established between the source and the drain. It is evident that a FET acts as a switch that is ON when the gate voltage is applied, whereas it is OFF when the gate voltage is not applied. Transistors are used to construct the fundamental elements of integrated circuits and CPU, which are physical devices that perform logical operations and are called logic gates. Devices realized with logic gates can process either combinational or sequential logic. A combinational logic circuit is characterized by the following behavior: in the absence of any external force, it is at equilibrium in its S_0 state. When an external force F_0 is applied, it switches to a new state S_1 and remains in that state as long as the force is present. Once the force is removed, it goes back to S_0. On the other hand, a sequential logic circuit is characterized by the following behavior: if it is in the state S_0, it can be switched to the state S_1 by applying an external force $F_{0\rightarrow1}$. Once it is in the state S_1, it remains in this state even when the force is removed. The transition from S_1 to S_0 is promoted by applying a new force, $F_{1\rightarrow0}$. Once the logic circuit is in the state S_0, it remains in this state even when the force $F_{1\rightarrow0}$ is removed. In contrast to the combinational logic circuit, the sequential one remembers its state even after removal of the force.

Since 1965, the pace of improvement of electronic computers has been described by the law formulated by Gordon Moore (co-founder of Intel) stating that the number of transistors on a chip doubles every year. Ten years later, Moore edited his law and alleged that the number of transistors on a chip actually doubles every two years (Moore 1995). By increasing the number of transistors per chip, the number of computational steps that can be performed at the same cost grows. In fact, by miniaturizing a transistor, the voltage needed to power the transistor scales downward, too (Dennard et al. 1999). In integrated circuit manufacture, billions of transistors and wires are packed in several square centimeters of silicon, with meager defect rates, through precision optics and photochemical processes (Markov 2014). Nowadays, silicon wafers are etched to the limit of 14 nm, and the use of X-ray lasers promises to overcome the limit of 10 nm (Markov 2014). The miniaturization of transistors is limited by their tiniest feature, the width of the gate dielectric, which has reached the size of several atoms. This situation creates some problems. First, a few missing atoms can alter transistor performance. Second, no transistor is equal to the others. Third, electric current tends to

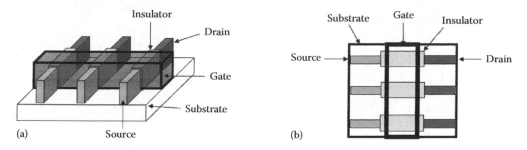

FIGURE 13.2 Lateral view (a) and top-view (b) of a tri-gate FinFET.

leak through thin narrow dielectrics, due to quantum-mechanical tunneling events, making unreliable the states ON and OFF of the transistors. Electronic engineers are striving to go around these problems. For example, they have devised tri-gate FinFET transistors (see Figure 13.2). A FinFET is a redesigned transistor having wider dielectric layers that surround a fin shape. Moreover, the presence of three gates allows checking the flow of current on three sides rather than just one.

One indicator of the CPU performance is the maximum number of binary transitions per unit of time (β):

$$\beta = Mv \qquad [13.1]$$

where M is the number of switches working with the clock frequency v of the microprocessor (Cavin et al. 2012). The computational power of a CPU, measured in the number of instructions per second, is directly proportional to β. Therefore, it is evident that for larger computational power, it is important to increase not only M but also v. Researchers have found that a silicon CPU can work at most at a frequency of 4 gigahertz without melting from excessive heat production. To overcome this hurdle, it is necessary to introduce either an effective cooling system or multi-core CPUs. A multi-core CPU is a single computing element with two or more processors, called "cores," which work in parallel. The speed-up (Sp) of the calculations is described by the Amdahl's law (Amdahl 1967):

$$Sp = \frac{1}{(1-P) + \dfrac{P}{N}} \qquad [13.2]$$

In [13.2], N is the number of cores, P is the fraction of a software code that can be parallelized, and $(1 - P)$ is the portion that remains serial. The speed-up grows by increasing N and P, as shown graphically in Figure 13.3. The challenge becomes the parallelization of the computational problems. One can spend many years getting 95% of code to be parallel, and never achieve a speed-up larger than twenty times, no matter how many processors are involved (see Figure 13.3). Therefore, it is important to continue to increase the number of transistors per chip (M in equation [13.1]) by miniaturization. However, soon or later, Moore's law will stop to hold because transistors will be made of a few atoms.

The twilight of Moore's law does not mean the end of the progress (Waldrop 2016). In fact, chip producers are investing billions of dollars to contrive computing technologies that can go beyond Moore's law. One strategy consists of substituting silicon with other computing materials. For example, graphene: a 2-D hexagonal grid of carbon atoms. Electrons in graphene can reach relativistic speeds and graphene transistors, which can be manufactured using traditional equipment and procedures, work faster than the best silicon chips (Schwierz 2010). However, the main disadvantage is that graphene does not have a band gap like silicon. Therefore, it is not suitable to process binary but rather analog logic, or continuous logic (Bourzac 2012). When sheets of graphene roll in void cylinders, we obtain carbon nanotubes that have small band gaps and become useful for processing digital logic. However, carbon nanotubes are delicate structures. Even a tiny variation

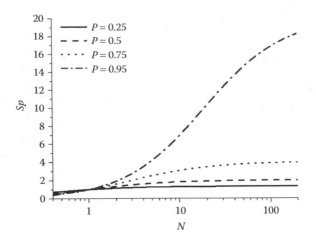

FIGURE 13.3   Graphical representation of the Amdahl's law as a function of N and for different values of P.

in their diameter and chirality (depending on the rolling angle) can induce the disappearance of the band gap. Moreover, engineers must learn how to cast billions of nanotubes in ordered circuits using the technology available for silicon (Cavin et al. 2012).

It is worthwhile noticing that without a fast and high capacity memory, a speedy CPU is useless. For this reason, researchers are trying to revolutionize the hierarchical structure of memory (see Figure 13.4). At the bottom, there is the non-volatile flash memory that grounds on transistors and has high capacity but low speed. Then, there is the moderately fast Dynamic Random-Access Memory (DRAM) where bits are stored in capacitors. Since the capacitors slowly discharge, the information is volatile and eventually fades unless the capacitor charge is refreshed periodically. On top, there is the Static Random-Access Memory (SRAM) that uses bi-stable flip-flop circuits to store the bits that are the most frequently accessed instructions and data (for more details read Shiva 2008).

The hierarchical structure of the memory can be revolutionized by introducing cells of memristors. A memristor (Chua 1971) is the fourth fundamental circuit element along with the resistor, capacitor, and inductor. A memristor—short for memory resistor—is an electronic component whose resistance is not constant but depends on how much electric charge has flowed in what direction through it in the past. In other words, the memristor remembers its history. When the electric power supply is turned off, the memristor remembers its most recent resistance until it is turned on again (Yang et al. 2013). Cells of memristors can be exploited to devise a RAM that is not anymore volatile. To achieve performances similar to the SRAM, it is necessary that memristors cells are close to CPU. This requirement is not easy to be accomplished with the available technology. However, photons may be used to connect the memory based on memristors with the traditional SRAM working closely to CPU.

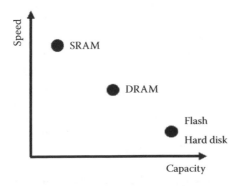

FIGURE 13.4   The features of the three types of memory in an electronic computer.

Another strategy to reduce data-movement demands and data access bottlenecks between logic and memory is to extend the electronic devices into a third, or vertical, dimension, by stacking together microprocessors and memory chips. 3-D architecture could reduce energy use to less than one-thousandth that of standard 2-D chips (Sabry Aly et al. 2015).

The constant improvement of electronic computers along with cloud computing (Waldrop 2016), which is the practice of using a worldwide network of remote servers (computer or software) hosted on the Internet to store, manage, and process data, are bringing benefits to the study of Complex Systems. However, this is not the only path that must be followed to succeed in our goal of winning the Complexity Challenges. We must also trust in the promising route being tracked by the interdisciplinary research line of Natural Computing.

13.3 NATURAL COMPUTING

Researchers working on Natural Computing (Rozenberg et al. 2012) draw inspirations from nature to propose:

- New algorithms;
- New materials and architectures to compute and store information;
- New methodologies and models to interpret Natural Complexity.

Natural Computing is based on the rationale that every natural transformation is a kind of computation because information can be encoded through the states of natural systems (Landauer 1991). The processes that occur in nature can be grouped into two sets: those that involve living beings and are information-driven, and those that involve only inanimate matter and are driven by force fields. We can take inspiration from both kinds of natural transformations, as shown in the next paragraphs. A useful methodology to succeed is that proposed by the cognitive scientists Gallistel and King (2010), and the neuroscientist Marr (2010). Such methodology requires an analysis of natural phenomena at three levels (as indicated in Figure 13.5). The first is the analysis at the "computational level"

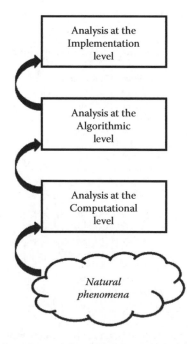

FIGURE 13.5 A useful methodology for studying complex natural phenomena.

that consists in determining the inputs, outputs, and the computations that the system can perform. The second is the analysis at the "algorithmic level" that consists in formulating algorithms that might carry out the computations defined in the previous level of analysis. Finally, the third is the analysis at the "implementation level" that consists in looking for mechanisms that can implement the algorithms.

13.3.1 Computing Inspired by Natural Information Systems

There are four main categories of natural information systems (see Chapter 12): Biomolecular Information Systems (BIS, such as a cell), Immune Information Systems (IIS), Neural Information Systems (NIS), and Social Information Systems (SIS). Their structures, working mechanisms, and peculiar phenomena, such as evolution and learning, constitute wealthy sources of inspiration for new algorithms, new computing materials and architectures, and new models to try to understand Complexity (see Figure 13.6).

In the following section, I report a brief description of some results achieved by imitation of the Natural Information Systems.

13.3.1.1 Artificial Life, Systems Chemistry, Systems Biology, and Synthetic Biology

The research line called "Artificial life" refers to the theoretical studies and the experimental implementation of artificial systems that mimic living beings. Its birth is normally attributed to Chris Langton, who used the term as a title for the "interdisciplinary workshop on the synthesis and simulation of living systems," organized in September 1987, in Los Alamos, New Mexico (Langton 1990). Although life has been studied for a long time, its fundamental principles are still hidden (Schwille 2017). In fact, there is not a universally accepted definition of life, and we do not know if the project of producing artificial life is feasible yet. This state of affairs means that the first level of the analysis of life, the computational level (refer to Figure 13.5), is incomplete. We can have an idea of the complexity of life if we compare a BIS with a von Neumann computer. A peculiarity of BIS is the lack of crisp boundaries between memory, processor, input and output components (Brent and Bruck 2006), and between software and hardware. The hardware of BIS is made of macromolecules, such as DNA, RNA, proteins, and simple molecules and ions. But DNA

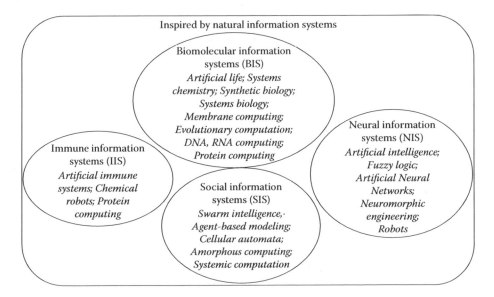

FIGURE 13.6 Some of the algorithms, materials, and architectures for computing, and models for interpreting Complex Systems, proposed by drawing inspiration from the Natural Information Systems.

is also the software (Ji 1999). The DNA code uses the four nucleotides (adenine, cytosine, guanine, thymine) as letters; the genes as words; the conformations of chromatin as sentences. As we learned in Chapter 12, a BIS is more comparable to a Neural Information System. In fact, in a cell there are sensory proteins that collect information; such information is processed within the signaling network; the output of the computation triggers the genetic network; the latter has the power of affecting both the metabolic and the signaling network. Systems Biology investigates the characteristic networks of a cell and their interactions (Klipp et al. 2016). The best way to check our degree of comprehension of life is to develop Systems Chemistry (Ashkenasy et al. 2017) and try to implement a so-called protocell (Luisi 2006). A protocell is a system that should mimic the first unicellular organism supposed to be the ancestor of all forms of life on Earth, called LUCA that is the acronym of Last Universal Common Ancestor (Weiss et al. 2016). This system should have the properties we consider fundamental for life. The first is teleonomy, which is the quality of having the purposes of surviving and reproducing. The second is the ability to use matter and energy to encode information and exploit it to accomplish its objectives. The third is a boundary that separates these information-based processes from the environment. The fourth is a metabolic network to self-sustain the system. The fifth is adaptability to an ever-changing surrounding. The sixth is the possibility of reproduction, hence birth and death. So far, no one has succeeded to obtain life from scratch. Perhaps, the easiest way for implementing artificial biological systems is through "Synthetic Biology." Synthetic biologists take parts of natural biological systems ("bio-bricks") and use them to build artificial biological systems. Their activity is essential not only for understanding the phenomenon of life, but also for biomedical application, and production of chemicals and fuels (Stephanopoulos 2012).

13.3.1.2 Membrane Computing

A cell is a really crowded environment. Nevertheless, sophisticated and reliable computations are carried out within it, continuously. The strategical secret is the presence of many compartments, delimited by membranes, organized hierarchically, and in selective communication. A generic membrane system, called P-system in honor of its inventor (Paun G. 2002), is a schematic nested hierarchical structure of compartments, like that shown in Figure 13.7. Each membrane-enveloped region contains objects and transformation rules that modify these objects, specify whether they will be transferred outside or stay inside the region, dissolve membranes, and create new membranes. The rules in each region of a P-system work in a parallel manner. A computation consists in repeatedly applying rules to an initial configuration of the P system until no rule can be applied anymore, in which case the objects in a priori specified regions are considered the output of the computation.

Membrane computing finds application in biological modeling, linguistics, computer graphics, economics (modeling transformations in economy is more difficult than in biochemistry because the reactions occur not only through encounter of the reagents but also by psychological factors), cryptography, and computer science to devise sorting, ranking algorithms, and innovative procedures to find approximate solutions of some NP-problems (Nishida 2004).

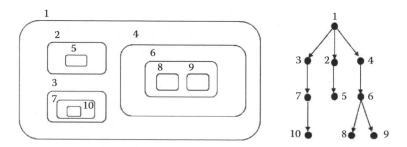

FIGURE 13.7 An example of a cell-like structure (on the left) and its tree representation (on the right).

13.3.1.3 DNA and RNA Computing

The basic idea of DNA computing is to exploit (1) the bases of DNA strands to encode information and (2) molecular biology tools to make computations. It was Leonard Adleman (1994) who pioneered the field when he solved a small instance of the Hamiltonian Path Problem (commonly known as the Travelling Salesman Problem) solely by manipulating DNA strands in a test tube. The Hamiltonian Path is an example of NP-complete problem (read Chapter 12): given a graph with a certain number of nodes, i.e., given a map with a certain number of cities, and given a certain number of interconnections between nodes, find the path that starts at the start node and ends at the end node and passes through each remaining node exactly once. So far, no one has proposed an algorithm that gives the exact solution in a short time for maps with many nodes (whose number is indicated by N). There are only algorithms that give approximate solutions, achievable in reasonable time. An algorithm of this kind requires the generation of random paths through the graph. Then, for each randomly generated path, it is necessary to follow the instructions reported in the flowchart of Figure 13.8.

Adleman (1994) chose a Hamiltonian Path Problem with seven cities and fourteen flights shown in Figure 13.9 (You may try to solve this problem as an exercise. Usually, it takes about one minute).

For understanding the main idea and procedure contrived by Adleman, it is enough (Adleman 1998) to consider a map that contains four cities (Rome, Berlin, London, and Madrid) linked by six flights, as shown in Figure 13.10. The task is to determine a Hamiltonian Path that starts from Rome and ends in London. Each city is named through a sequence of bases, defined randomly. In the example of Figure 13.9, Rome is encoded through a sequence of eight bases, as ACTTGCAG; the first half of the sequence is conceived as the first name of the city, whereas the second half is the last name. So, Rome's last name is GCAG. London is encoded as CCGAGCAA, and its first name is CCGA. Of course, each city has its complementary DNA-name, which is the Watson-Crick complement of the strand used as the name.

The next step is to encode the direct flight numbers with specific DNA sequences. The idea of Adleman was to concatenate the last name of the city of departure with the first name of the city of destination. Therefore, the flight from Rome to London is GCAGCCGA. Then, Adleman took a pinch (roughly 10^{14} molecules) of each complementary DNA city name and a pinch of

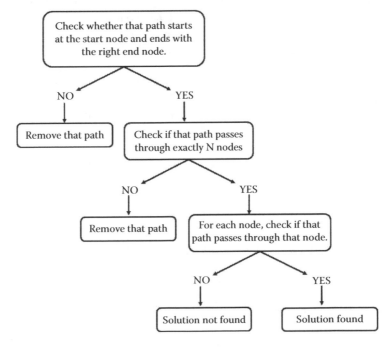

FIGURE 13.8 Flowchart containing the instructions of the algorithm suitable to solve the Hamiltonian Path Problem.

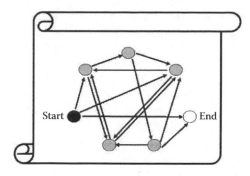

FIGURE 13.9 The Hamiltonian Path Problem solved by Adleman (1994) with DNA strands.

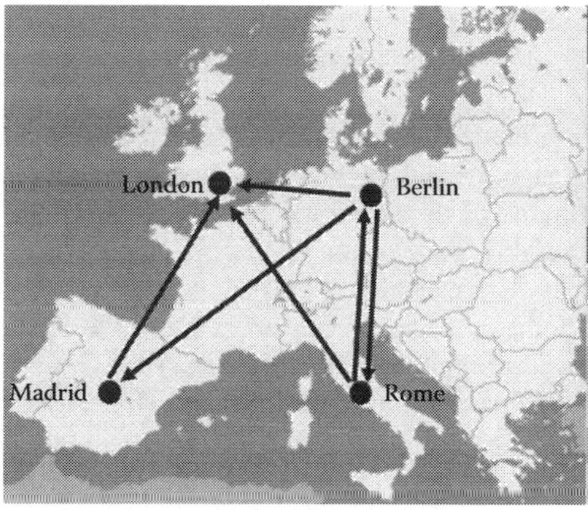

City	DNA Name	Complement
ROME	ACTTGCAG	TGAACGTC
BERLIN	TCGGACTG	AGCCTGAC
MADRID	GGCTATGT	CCGATACA
LONDON	CCGAGCAA	GGCTCGTT
Flight		DNA Flight number
ROME–BERLIN		GCAGTCGG
ROME–LONDON		GCAGCCGA
BERLIN–MADRID		ACTGGGCT
BERLIN–LONDON		ACTGCCGA
BERLIN–ROME		ACTGACTT
MADRID–LONDON		ATGTCCGA

FIGURE 13.10 A trivial Hamiltonian Path Problem solvable by using strands of DNA and the DNA-hybridization reaction.

each DNA flight number and put them together in a test tube. To trigger the computation, he added water, DNA ligase (it binds strands of DNA), salt, and a few other ingredients to mimic a cell environment. Within about one second, the solution of the Hamiltonian Path Problem was inside the test tube. How was it possible? In the solution, a countless number of collisions among the many strands occur every instant. It is possible that the strand representing the Rome-to-Berlin flight number (GCAGTCGG) and that representing the complementary name of Berlin (AGCCTGAC) meet and stick together through hydrogen bonds, because the last four bases of the sequence encoding the flight number is the complement of the first four bases of the sequence encoding the complement of Berlin's name (see Figure 13.11). If the resulting complex

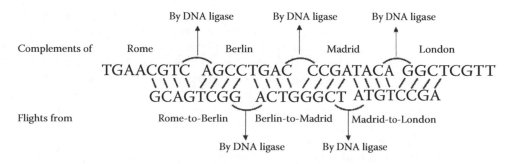

FIGURE 13.11 Solution of the Hamiltonian Path Problem illustrated in Figure 13.10.

encounters the strand representing the Berlin-to-Madrid flight number (ACTGGGCT), they can bind together because the last four bases of the Berlin's complement are complementary with the first four bases of the strand representing the Berlin-to-Madrid flight number. The DNA ligase, present in the solution, concatenates the strands of flight numbers. In this manner, complexes grow in length. Of course, at equilibrium, the solution contains many combinations of the original 8-bases strands introduced in the test tube. Such combinations represent complete or incomplete paths, generated randomly as required by the algorithm mentioned at the beginning of this paragraph. Since Adleman used a large number of DNA strands for each flight and complementary city name, and since the Hamiltonian Path Problem contained just a few cities, he was certain that at least one of the molecules formed in the test tube encoded the Hamiltonian Path: GCAGTCGGACTGGGCTATGTCCGA. The DNA strands found quickly the solution because the simultaneous interactions of hundreds of trillions of strands represented a clear manifestation of parallel molecular processing.

To verify if the chemical system in the test tube found the right solution of the Travelling Salesman Problem, Adleman followed the flowchart illustrated in Figure 13.8 by exploiting molecular biology tools. To discard complexes that did not both begin with the start city and terminate with the end city, Adleman relied on the polymerase chain reaction (PCR). He introduced many copies of two short pieces of DNA as primers to signal DNA polymerase to start its Watson-Crick replication. The primers used were the last name of the start city (GCAG, for Rome) and the complement of the first name of the end city (GGCT, for London). The first primer induced DNA polymerase to copy complements of sequences that had the right start city, and the second primer initiated the duplication of strands that had the correct end city. PCR proceeds through thermocycling. Very hot conditions induce the denaturation of double-stranded DNA. Less warm conditions are favorable for DNA polymerase to duplicate the individual pieces. The result was that strands with both the right start and end compositions were duplicated at an exponential rate. On the other hand, strands that had the right start city but not the right end city, or vice versa, were duplicated much more slowly. DNA sequences that had neither the right start nor the right end were not duplicated at all. In this manner, the first stage of the flowchart in Figure 13.8 was accomplished. Then, by using the gel electrophoresis technique that separates DNA molecules having different length through an electric field, Adleman identified those strands that had the right length (consisting of 24 bases for the example of Figure 13.10). This operation completed the second step of the flowchart. To check if the remaining sequences passed through all the intermediary cities, Adleman used a procedure known as affinity separation. Many copies of the sequence that is the complementary name of Berlin were attached to microscopic iron beads. These tagged beads were suspended in the test tube containing the remaining strands, under conditions favorable for the hybridization reaction. Only those strands that contained the Berlin's sequence annealed to the segments bound to the beads. Then, through a magnet placed in contact with the wall of the test tube, Adleman attracted the beads and could discard the liquid containing strands that did not have the desired city's name. Then, he added new solvent and removed the magnet to resuspend the beads. By heating, he induced the strands to detach from the beads. Next, he reapplied the magnet to attract the beads to the wall of the test tube. The liquid, which now contained the desired DNA strands, was recovered for further screening. The process was repeated for the other intermediary cities, such as Madrid as far as the example of Figure 13.10 is concerned. After finishing the affinity separations, the final step of the flowchart of Figure 13.8 was over, and Adleman was sure that the DNA strands left in the test tube should be those encoding the solution of the Hamiltonian Path Problem. This result was confirmed by an additional PCR, followed by another gel-electrophoresis separation. Although the DNA found the solution just in one second, the entire procedure of separation and recognition of the right solution required seven days in the laboratory. Therefore, the Adleman's idea is brilliant, but the procedure is impractical for solving more massive Hamiltonian Path Problems (Linial and Linial 1995). In fact, the mass of DNA needed for the computation would become too big, and the exhaustive search for the right solution would be like to find a small needle in an enormous haystack. Moreover, the search may be hindered by the unavoidable artifacts that PCR can introduce in the amplification stages.

Probably, DNA computing is more promising *in vivo*, inside living cells. The dream is to implement a "chemical robot" or what has also been called a "Doctor in a cell" by Ehud Shapiro from the Weizmann Institute. The "Doctor in a cell" consists of an autonomous molecular computing device that, after analyzing the surrounding cellular microenvironment, it decides if to release or not a specific drug (Benenson et al. 2004). For example, an oligonucleotide AND gate has been engineered to respond to specific microRNA inputs, which are indicators of cancer, by a fluorescent output. The DNA AND gate has been implanted in alive mammalian cells (Hemphill and Deiters 2013). In other examples, chemical robots have been implemented by exploiting the molecular self-assembly of DNA that can originate non-arbitrary two- and three-dimensional shapes at the nanometer scale, called "scaffolded DNA origami" (Rothemund 2006). For example, nanoscale robots, based on DNA origami, are capable of dynamically interacting with each other in the cells of a living cockroach (Amir et al. 2014). The interactions generate logic outputs, which are relayed to switch molecular payloads on or off. The payload can be a molecule such as a short DNA sequence, an antibody, or an enzyme. If the payload can activate or inactivate another robot, this will create a circuit inside a living cell.

RNA is another promising material for computing within living beings. In fact, RNA is involved in many regulatory networks. Moreover, RNA is structurally more versatile and more thermodynamically stable than DNA. It displays a structural flexibility and functional diversity similar to that of proteins, including enzymatic activities. The discovery of diverse RNA functions and the methods to produce fluorogenic RNA in a cell suggest that small RNAs can work cooperatively, synergistically or antagonistically to carry out computational logic circuits. With all these advantages, RNA computation is promising, but it is still in its infancy (Qiu et al. 2013).

DNA material has another important application: it could be used as a memory of data, maybe of the overwhelming Big Data. Life on Earth teaches us that DNA is an alluring material for information storage. In fact, a whole genome that fits into a microscopic cell contains all the instructions required by a living being to survive. Moreover, if the genome is held in dry and dark conditions, it can remain almost intact hundreds, thousands, and even millions of years (Pääbo et al. 2004). So, DNA is ultracompact and long-lasting storage material. Scientists have been storing digital data in DNA since 2012. That was when George Church and his colleagues encoded a 53,426-word book using less than a picogram of DNA (Church et al. 2012). For encoding the digital file, researchers have divided it into tiny blocks of data and converted these data into DNA's four-letter alphabet. They encoded one bit per base: A or C for zero, G or T for one. Each DNA fragment was replicated thousands of times. Since errors in synthesis and sequencing are rarely coincident, each molecular copy corrects errors in the other copies. Moreover, each DNA fragment contained a digital "barcode" that recorded its location in the original file. Reading the data required a DNA sequencer and a computer to reassemble all of the fragments in order and convert them back into digital format. Density, stability, and energy efficiency are all potential advantages of DNA storage (Extance 2016), while costs and times for writing and reading are currently impractical. So, the DNA storage approach is not promising if data are needed instantly, but it would be better suited for archival applications.

TRY EXERCISE 13.1

13.3.1.4 Evolutionary Computing

Biological evolution is the change in life forms that occurred over the last four billion years, and that continues to happen on Earth. Darwin (1859) supposed that the changes were gradual, but the fossil record also suggests that long periods of no change in the morphology of organisms have been punctuated by relatively short periods of significant change, with the disappearance of some species and the emergence of new ones (Eldredge and Gould 1972). The evolution of biological forms

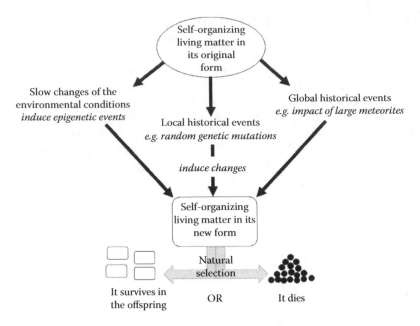

FIGURE 13.12   The main factors contributing to the evolution of biological forms.

is both gradual and sudden, and it depends on more than one factor (see Figure 13.12). First, every living being has the power of self-organizing, of creating order within itself, with the tendency of becoming always more complex (Kauffmann 1993). The evolution towards new forms may be induced by the slow changes of the environmental conditions that trigger epigenetic events: some genetic switches are turned entirely or partly off, and some others are completely or partially turned on. However, the evolution may also be induced by either global historical events (such as the impact of large meteorites on Earth) that provoke significant changes in the environment, or local historical events such as "random" genetic mutations that are sporadic and impossible to predict. The new biological forms are tested continuously on their power of fitting to their ecosystem through natural selection: those that are successful in surviving and reproducing continue to exist through its off-spring. Sexual reproduction favors the variability because the offspring have a genetic code that is the mixture of those of the parents.

The idea of imitating the basic features that govern the natural evolutionary processes in computer science came onto the scene by the early 1960s. Such research is known as Evolutionary Computation. There are different mainstreams, but the most known is Genetic Algorithm (GA) proposed by Holland (1975). The input to a GA consists of a population of candidate programs. A candidate program is written as strings of bits, numbers or symbols to accomplish a task. The programs can be generated randomly. Each program is tested on how well it works on the desired task by determining a fitness value. The programs that have the highest fitness values are selected to be parents of the next generation. The offspring is generated by merging parts of the parents. The new programs are then tested in their fitness, and the entire procedure is repeated over and over again until we get satisfied with the achieved fitness values. Evolutionary Computation and Genetic Algorithms have been used to solve problems in many scientific and engineering areas, as well as in art and architecture (Beasley 2000).

13.3.1.5   Artificial Immune Systems

Our immune system has an innate and an adaptive component (read Box 13.1 in Chapter 12 for more details), which has a diverse set of cells, molecules, and organs that work in concert to protect our body. Our immune system has some attributes that are worthwhile mimicking. First, the pattern recognition ability of the receptors. A healthy immune system is very effective in distinguishing harmful substances from the body's own components and directing its actions accordingly. Second, the immune subsystems, especially the adaptive one, exhibit the ability to learn and remember previously encountered pathogens. When a new pathogen appears, the Immune System learns how to defeat it (the primary response). The next time that pathogen is encountered, a faster and often more aggressive response is elicited (the secondary response). Finally, the immune system is spread throughout the body and acts collectively without any central controller.

Artificial immune systems are designed by following a procedure (de Castro and Timmis 2002) that requires the definition of the (1) representation, (2) affinity measure, and (3) immune algorithm. The first step is the selection of an abstract representation of the components of the system, considering the application domain. Any immune response requires the shape recognition of an antigen by the cell receptors. For modeling this shape recognition process, the concept of shape-space has been introduced (Perelson and Oster 1979). The shape-space approach assumes that the properties of the receptors and antigens can be represented by a data structure. Most often, the data structure is an attribute string, which can be a real-valued vector, a binary string, et cetera (de Castro 2007). The affinity of a receptor towards either an antigen or an element of the organism is measured by calculating the distance between the two corresponding attribute strings in the shape-space. Finally, an immune-inspired algorithm is selected. Various types of immune-inspired algorithms exist. The choice must be taken considering the problem that has to be solved. For example, clonal selection-based algorithms attempt to capture mechanisms of the antigen-driven proliferation of B-cells, which induce an improvement in their binding abilities. Clonal selection algorithms capture the properties of learning, memory, adaptation, and pattern recognition (Timmis et al. 2008). Other examples are the immune network algorithms (Jerne 1974) that view the immune system as a regulated network of molecules and cells that recognize each other. Antibodies and cells are nodes of the network, and the training algorithm determines the growth or the prune of edges between the nodes, based on their mutual affinities. The immune network acts in a self-organizing manner and generates memory effects. Immune network algorithms have been used in clustering, data visualization, control, and optimization domains. In general, the theory of artificial immune systems finds broad applicability in machine-learning, pattern recognition, security of information systems, and data analysis (de Castro 2007).

13.3.1.6 Cellular Automata

A cellular automaton is a mathematical model of a dynamical system that is discrete in both space and time. In fact, a d-dimensional cellular automaton is a d-dimensional grid of cells, each of which can take on a value from a finite set of integers (most often, just two values, 1 or 0, on or off). The value of each cell at time step $t + 1$ is a function of the values of a small local neighborhood of cells at time t. The cells update their state simultaneously according to a given local rule. The idea of cellular automata came to John von Neumann, back in the 1940s, based on a suggestion made by his colleague, the mathematician and nuclear physicist Stanislaw Ulam (both of them participated to the Manhattan Project during the Second World War). von Neumann was trying to implement the self-reproduction phenomenon, which is peculiar to the living matter, in machines, and he thought to succeed by inventing cellular automata. In fact, cellular automata are examples of mathematical models constructed from many simple identical components that, together, are capable of complex behavior. This instance has been demonstrated by the physicist Stephan Wolfram (2002) by using the simplest possible form that is one-dimensional, two-state cellular automaton. It consists of a line of sites, with each site carrying a value (a_i) 0 or 1, white or black, dead or alive. The value a_i of the site at each position i is updated in discrete time steps according to an identical deterministic rule that is a function of the values a_k of the neighbors:

$$a_i^{(t+1)} = f\left(a_{i-1}^{(t)}, a_{i+1}^{(t)}\right)$$

[13.3]

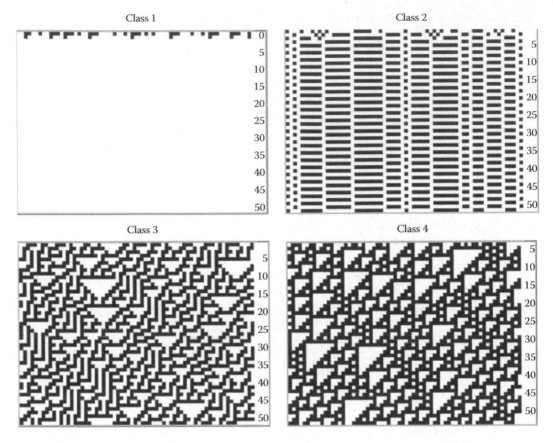

FIGURE 13.13 Examples of the four classes of spatiotemporal patterns generated by one-dimensional cellular automata following deterministic rules. These patterns have been achieved by using "ElementaryCAs" model, Complexity Explorer project, http://complexityexplorer.org.

Different cellular automaton rules yield very different patterns. Different initial states with a particular cellular automaton rule yield patterns that differ in the details but are similar in form and statistical features. All the possible patterns can be grouped into four classes (see Figure 13.13).

- *Class 1*: Independently of the initial configurations, all cells in the lattice switch to the off state and stay that way, indefinitely. The final pattern is spatially homogeneous.
- *Class 2*: The final patterns are uniform or periodic, depending on the initial configuration.
- *Class 3*: Aperiodic and chaotic behavior.
- *Class 4*: Localized structures, moving around and interacting with each other in very complicated ways.

Another famous cellular automaton is that invented by the mathematician John Conway (Gardner 1970) and called the "Game of Life." It is an infinite two-dimensional grid of square cells, each of which is in one of two possible states: off or on, 0 or 1, white or black, dead or alive, unpopulated or populated. Every cell has eight neighbors: four adjacent orthogonally and four adjacent diagonally (see Figure 13.14). When we start from a certain configuration of the grid, its evolution over discrete time steps is determined by a set of "genetic laws" listed as follows.

- *Birth*: a dead cell, surrounded by three live cells, becomes alive at the next time step.
- *Survival*: a live cell, surrounded by two or three live cells, stays alive.

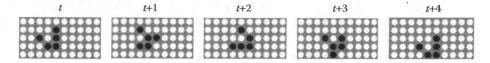

FIGURE 13.14 Evolution of a glider in the "Game of Life," reproduced by using "Mini-Life" model, Complexity Explorer project, http://complexityexplorer.org.

- *Loneliness*: a live cell, surrounded by less than two live neighbors, dies.
- *Overcrowding*: a live cell or a dead cell, surrounded by more than three live neighbors, dies or stays dead.

Conway formulated these "genetic laws" after a long period of careful experimentation to create interesting life-like behaviors. One interesting pattern is the "glider" represented in five consecutive time steps in Figure 13.14. It moves in a southeast direction indefinitely, as if it were a chemical wave.

Other intricate patterns have been discovered, such as the "glider gun," able to generate an infinite number of "gliders." Conway demonstrated that by assembling "glider guns," "gliders," and other structures, it is possible to carry out the fundamental logical operations: AND, OR, NOT. Therefore, the "Game of Life" can simulate a Turing machine that is a universal computer. It can run any program that can operate on a standard computer. However, it is very challenging to design initial configurations that can do non-trivial computations.

The most important message we receive from the study of cellular automata is that a simple system, ruled by simple laws, can give rise to complex phenomena. Therefore, cellular automata provide alternative models to interpret Complex Systems. Models based on cellular automata harness the parallelism of natural systems and are appropriate in highly nonlinear regimes where growth inhibition effects are relevant, and discrete threshold values emerge. The theory of cellular automata has been used in many fields. For example, it has been applied with success for the description of excitable media, such as the growth of dendritic crystals, snowflakes (Zhu and Hong 2001), the Belousov-Zhabotinsky reaction, and turbulent fluids (Markus and Hess 1990). Moreover, it has been used for the representation of many biological phenomena (Ermentrout and Edelstein-Keshet 1993), such as the formation of patterns, the pigmentation of many mollusk shells (Kusch and Markus 1996), and the uptake and loss of gases by plants through stomatal apertures (Peak et al. 2004).

Finally, cellular automata have been implemented in different ways; for example, by reaction-diffusion processes (Hiratsuka et al. 2001; Stone et al. 2008), nanometric quantum dots (Snider et al. 1999; Blair and Lent 2003) and memristors (Itoh and Chua 2009).

13.3.1.7 Artificial Intelligence, Fuzzy Logic, and Robots

Human intelligence may be conceived as the emergent property of the human nervous system (HNS). The nervous system is a complex network of billions of nerve cells organized in a structure where we can distinguish three main elements:[1] (1) the sensory system (SS), (2) the central nervous system (CNS), and (3) the effectors' system (ES) (Paxinos and Mai 2004).

1. The sensory system consists of many different specialized receptor cells, which either catch the stimuli coming from the external environment or monitor the inner state of the body. The receptor cells transduce the different kinds of stimuli in electrochemical signals that are sent to the CNS. The electrochemical signals contain the sensory information, and the sensory information may or may not lead to conscious awareness. If it does, it is called sensation.

---

[1] Traditionally, the human nervous system is presented as constituted by two parts: (a) the Central Nervous System (CNS) that is the brain and the spinal cord, and (b) the Peripheral Nervous System (PNS) made of the nerves carrying signals to and from the CNS.

2. The central nervous system consists of the brain and the spinal cord. The spinal cord is the highway of information. It lets the information coming from SS to reach the brain, and the information sent by the brain to reach the ES. The principal and fundamental roles of the brain are those of processing the information coming from the SS, memorizing it permanently or temporarily, and taking decisions. The brain extracts meaning from the world by mixing the different forms of sensory perception (Stein 2012).

3. Effectors' system consists of glands (both exocrine and endocrine) and muscles. Both of them receive instructions from the CNS.

It is interesting to compare the structure and working mechanisms of our nervous system with those of a von Neumann electronic computer. We may say that the brain plays the roles of both central processing unit and memory. The sensory system along with the effectors' cells play like the information exchanger of an electronic computer, and the spinal cord is the data and instructions bus. Apart from these formal similarities, there are sharp differences between the human nervous system and the von Neumann electronic computer. In fact, in any von Neumann computer the information is encoded by electrical signals, whereas in any brain through electrochemical signals. Moreover, in any von Neumann computer information is digital, whereas in HNS it is both digital and analog. In any von Neumann computer, we distinguish hardware and software. In brains, this distinction is not evident. The hardware of a computer is rigid, whereas any brain is flexible, and the synaptic connections among neurons are plastic. In a von Neumann computer, the operations are executed synchronously, whereas different regions of the brain work asynchronously. A brain is smaller and more efficient than a computer because it can achieve five or six orders of magnitude more computation for each joule of energy consumed (Ball 2012). The more significant *efficiency* of the brain is partly due to its 3D network architecture that is sharply different from the 2D grid-like structure of a von Neumann computer. A supercomputer is much faster in doing mathematical computations, and it has a much stronger memory for numerical data than any brain, even if we consider the brain of a mathematical genius. This evidence does not mean that we have reached the "Singularity," the point where machines surpass human intelligence, as envisioned by Kurzweil (2005). In fact, human intelligence does not reside solely on its computing rate and memory space, but on its ability to learn, understand scenarios, think creatively, react to unexpected events, recognize variable patterns, and compute with words. We, humans, can handle both accurate and vague information by computing with numbers and words. We can take decisions in complex situations when there are many intertwined variables, and accuracy and significance become two attributes of our statements, which are mutually exclusive (see Figure 13.15), in agreement with the Principle of Incompatibility (Zadeh 1973).

Therefore, it is evident that it is worthwhile trying to deeply understand the foundations and running mechanisms of the human intelligence and reproduce them artificially. The term "Artificial Intelligence" was coined by the American computer scientist John McCarthy. It was used in the title of a workshop that took place in the summer of 1956 at Dartmouth College in Hanover, New Hampshire (Moor 2006; Russell and Norvig 1995). The research on Artificial Intelligence has had cycles of success, misplaced optimism, and resulting cutbacks in enthusiasm and funding. Recently, it has received a renewed boost by the research initiatives called "The Decade of the Mind" started in 2007 (Albus et al. 2007), and "The Human Connectome Project" launched in 2009. A comprehensive mapping of the structural and functional connections, across all scales, from the micro-scale of individual synaptic connections between neurons to the macro-scale of brain regions and interregional pathways (Behrens and Sporn 2011) should facilitate the imitation of the human mind. The success in the imitation of the human mind will have a significant impact on science, medicine, economic growth, security and well-being.

For mimicking human intelligence, it is crucial to simulate the human power of computing with words and implement not only binary but also analog logic. These two tasks can be accomplished, at the same time, by turning to Fuzzy logic (Zadeh 2008). Fuzzy logic is based on the theory of Fuzzy sets. As we learned in Chapter 12, any Fuzzy set (Zadeh 1965) breaks the Law of Excluded-Middle. In fact, an item may belong to a Fuzzy set and its complement at the same time, with the

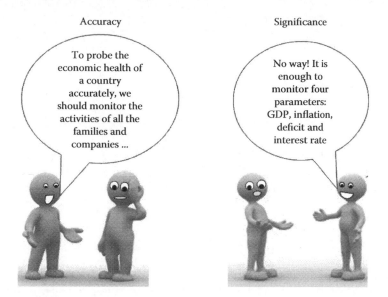

FIGURE 13.15 An illustration of the Incompatibility Principle. According to it, accuracy and significance are two mutually exclusive properties of the statements regarding Complex Systems.

same or different degrees of membership. The Law of Excluded-Middle is the foundation of the binary logic. In binary logic any variable is partitioned into two classical sets after fixing a threshold value: one set will include all the values below the threshold, whereas the other one will contain those above. In case of a positive logic convention, all the values of the first set become the binary 0, whereas those of the other set become the binary 1. The shape of a classical set is like that shown in Figure 13.16a. The degree of membership function, for such a set, changes discontinuously from 0

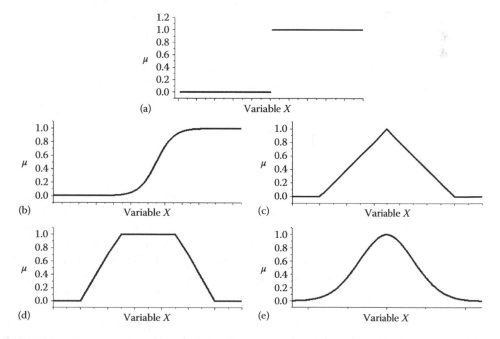

FIGURE 13.16 Shapes of the membership functions (μ) for a generic variable X. The case of a "classical" set: plot (a) Examples of Fuzzy sets are shown in (b) (sigmoidal shape), (c) (triangular shape), (d) (trapezoidal shape), and (e) (Gaussian shape) plots.

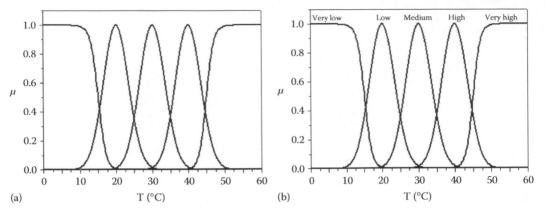

FIGURE 13.17 Graphical representation of the first two steps needed to build a FLS: granulation (a), and graduation (b) of the variable temperature (T) expressed in degree Celsius and taken as an example.

(below the threshold) to 1 (above the threshold). On the other hand, Fuzzy sets can have different shapes. They can be sigmoidal (b in Figure 13.16), triangular (c), trapezoidal (d), Gaussian (e), to cite a few. For a Fuzzy set, the degree of membership function changes smoothly from 0 to 1.

It derives that Fuzzy logic is an infinite-valued logic. Fuzzy logic is suitable to describe any complex non-linear cause and effect relation after the construction of a proper Fuzzy Logic System (FLS). The development of a FLS requires three fundamental steps. First, the granulation of every variable in Fuzzy sets (see Figure 13.17a). In other words, all the possible values of a variable are partitioned in a certain number of Fuzzy sets. The number, position, and shape of the Fuzzy sets are context-dependent. The second step is the graduation of all the variables: each Fuzzy set is labeled by a linguistic variable, often an adjective (see Figure 13.17b as an example for the variable temperature expressed in degree Celsius). These two steps imitate how our senses work (Gentili 2017). Physical and chemical sensations are epitomized by words in natural language. For instance, human perceptions of temperature are described by adjectives such as "freezing," "cold," "temperate," "warm," "hot," and so on. Visible radiations of different spectral compositions are described by words such as "green," "blue," "yellow," "red," and other words that we call colors. Savors are classified as salt, sweet, bitter, acid, umami. There are different words to describe the other sensations. These words of the natural language are cognitive granules originated by perceptions. Perceptions are constrained, noisy, and they generate clumps of Fuzzy-information (remember what we studied in Chapter 12).

The third final step required to build a FLS is the formulation of Fuzzy rules that describe the relations between input and output Fuzzy sets. Fuzzy rules are syllogistic statements of the type "IF…, THEN…" The "IF…" part is called the antecedent and involves the linguistic labels chosen for the input Fuzzy sets. The "THEN…" part is called the consequent and contains the linguistic labels chosen for the output Fuzzy sets. When we have multiple inputs, these are connected through the AND, OR, NOT operators (Mendel 1995). At the end of the three-step procedure, we have a FLS. In a FLS we distinguish three elements: A Fuzzifier, a Fuzzy Inference Engine, and a Defuzzifier (see graph a in Figure 13.18). A Fuzzifier is based on the partition of all the input variables in Fuzzy sets. A Fuzzifier transforms the crisp values of the input variables in degrees of membership to the input Fuzzy sets. The Fuzzy Inference Engine is based on the Fuzzy rules (see graph b in Figure 13.18). The IF-THEN rules are patches (Kosko 1994) covering the nonlinear function that is represented as the winding black trace of graph b in Figure 13.18. The higher the number of rules, the more accurate is the description. The Fuzzy rules are formulated by experts (according to the Mamdani's method), or they are determined by automatic learning mathematical techniques applied to a set of training experimental data (according to the Sugeno's method).[2] The Fuzzy Inference Engine turns on all the

---

[2] For more details about the methods of formulating Fuzzy rules, read the clear tutorial by Mendel (1995).

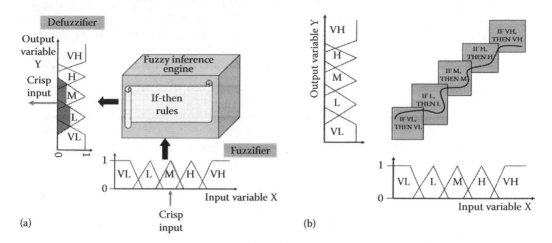

FIGURE 13.18 Graph a, on the left, shows the three elements of a FLS built according to the Mamdani's method (Mendel 1995). It consists of a Fuzzifier (based on Fuzzy sets of the input variable), a Fuzzy Inference Engine (based on the Fuzzy rules), and a Defuzzifier (based on the Fuzzy sets of the output variable). Graph b, on the right, shows the Fuzzy rules that, like patches, cover and describe the non-linear relation (represented by the winding black curve) between input and output variables.

rules that involve the Fuzzy sets activated by the crisp input values. Finally, the Defuzzifier is based on the Fuzzy sets of the output variables, and it transforms the collection of the output Fuzzy sets, activated by the rules, in crisp output values (see graph a of Figure 13.18). A complete FLS is an inferential tool to make predictions and take decisions about the nonlinear relation it refers to.

There are two main strategies to develop artificial intelligence: one is writing human-like intelligent programs running on computers or special-purpose hardware, and the other is neuromorphic engineering. For the first strategy, computer scientists are writing algorithms that can learn, analyze extensive data, and recognize patterns. At the same time, psychologists, biologists, and social scientists are giving information on human sensations, emotions, and intuitions. The merger of the two contributions provides algorithms that can easily communicate with us. Among the most promising algorithms, there are the artificial neural networks (remember Chapter 10, when we learned these algorithms for predicting chaotic time series) (Castelvecchi 2016). Recently, a program called AlphaGo Zero, based on an artificial neural network that learns through trial and error, has mastered the game of Go without any human data or guidance, and it has outperformed the skills of the best human players (Silver et al., 2017).

The second strategy for implementing Artificial Intelligence is neuromorphic engineering (the term was coined by California Institute of Technology electrical engineer Carver Mead in the 1980s). It implements surrogates of neurons through non-biological systems either for neuroprosthesis (Donoghue 2008) or to devise brain-like computing machines. Brain-like computing machines will exhibit the peculiar performances of human intelligence, such as learning, recognizing variable patterns, and computing with words as some programs have commenced doing. However, it is expected that brain-like computers will have the advantage of requiring much less power and occupying much less space than our best electronic supercomputers. Surrogates of neurons for brain-like computers can be implemented by using different strategies. A first strategy exploits conventional passive and active circuit elements, either analog (Mead 1989; Service 2014) or digital (Merolla et al. 2014). A second one grounds on circuits made of two-terminal devices with multiply-valued internal states that can be tuned in non-volatile or quasi-stable manners, keeping track of the devices' past dynamics (Ha and Ramanathan 2011; Di Ventra and Pershin 2013). A third approach uses phase-change devices wherein the actual membrane potential of the artificial neuron is stored in the form of the phase configuration of a chalcogenide-based material, which can undergo phase transition on a nanosecond timescale and at the nanometric level

(Tuma et al. 2016). Finally, a fourth methodology trusts in nonlinear chemical systems that can communicate through either continuous mass exchange (Marek and Stuchl 1975) or electro-chemical linkages (Hohmann et al. 1999; Zlotnik et al. 2016) or mechanical/light-induced pulsed release of chemicals (Taylor et al. 2015; Gentili et al. 2012; Horvath et al. 2012) or UV-visible radiation (Gentili et al. 2017a).

Artificial Intelligence is suitable to fulfill a long-cherished aspiration of humanity: that of designing machines that can help humans in both manual and mental activities, at the same time. Such devices, called robots, are programmable and potentially able to carry out many actions peculiar to humans (Russell and Norvig 1995). In fact, robots are usually able to perceive signals through sensors, plan activities through internal cognitive processes, and act through actuators, as the human nervous system does through its sensory cells, brain, and effector system. Traditionally, robots are made of electronic circuits, computer software, rigid mechanical parts and electric motors, and are designed primarily for accomplishing dangerous tasks, such as bombs disposal, deep ocean, and planetary exploration. Recently, the idea of developing robots grounded on wetware rather than on hardware is taking shape (Grančič and Štěpánek 2011; Hess and Ross 2017). A "Soft Robot," also called a "Chemical Robot," is thought as a molecular assembly that reacts autonomously to its environment through molecular sensors, makes decisions by its intrinsic chemical artificial neural networks, and performs actions upon its environment through molecular effectors (see Figure 13.19). Of course, it needs fuels to conduct its operations. Therefore, it is necessary to have a metabolic system inside of it. Chemical Robots could be easily miniaturized and implanted in living beings to interplay with cells or organelles. They are expected to play as auxiliary elements of the immune systems for biomedical applications (Hagiya et al. 2014).

13.3.1.8 Protein Computing

In Chapter 12, we have seen that the signaling network of a cell works as the "cellular brain." The neurons of the "cellular brain" are proteins. Many thousands of proteins, functionally connected to each other, carry information from the membrane to the genetic network or directly to the metabolic network (Bray 1995). Such system of interacting proteins acts as a neural network; the information of extracellular stimuli is collected by receptor proteins and transduced in the activities and concentrations of the inner cell proteins. The activity of any protein depends on its structure. Originally, it was taught that the interaction between proteins and substrates could be described only by the lock and key paradigm (Conrad 1992). Each type of enzyme works as a specific key, and its substrate is the lock. The fitting of the key to the lock is a sophisticated pattern recognition operation. The nonlinear process of recognition trusts in the three-dimensional structure of the protein. Nowadays, we are aware that many proteins or portions of proteins are structurally disordered in their native, functional states (Tompa 2010). This situation means that, quite often, a protein cannot be assumed to be just one rigid key. Instead, it might be described as a

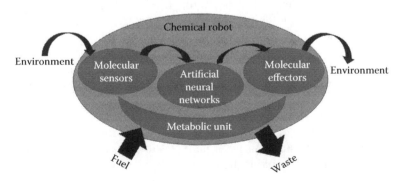

FIGURE 13.19 Essential modules for the design of a Chemical Robot: (I) Molecular sensors; (II) Chemical Artificial Neural Networks; (III) Molecular Effectors, and (IV) a Metabolic Unit.

pretty flexible key because the macromolecule is characterized by a dynamic disorder. Otherwise, it can be conceived as a bunch of keys because it exists as an ensemble of conformations. Static and dynamic structural disorder provides computational power to proteins because different conformations can work in parallel.

Some proteins are multifunctional. They exhibit phenomena of moonlighting. In analogy to moonlighting people who have multiple jobs, moonlighting proteins perform multiple autonomous, often unrelated, functions (Huberts and van der Klei 2010). An example of moonlighting protein is phosphoglucose isomerase (PGI), a protein that is both a cytoplasmic enzyme in glycolysis and an extracellular cytokine and growth factor. In other words, it can participate in both metabolic and signaling cellular events. In PGI, the enzyme active site is made of loops and helices from many parts of the protein, so that the active site domain is not separate from its receptor-binding domain (Jeffery 2014). At present, it is still difficult to assess how abundant moonlighting proteins are. However, moonlighting is a phenomenon that illustrates nature's ingenuity, and it can be exploited for computational reasons.

13.3.1.9 Amorphous Computing

In Chapter 9, we were stunned to investigate morphogenetic processes. For example, it is astonishing to observe how the cells in an embryo self-organize to form an organism. Under the point of view of Natural Computing, an embryo is a collection of cells that compute in parallel, based on a single program written in their DNA, and produces a globally coherent result.

The effort of mimicking natural morphogenetic processes for computational reasons has been named as Amorphous Computing by a team working at MIT in 1996 (Abelson et al. 2000). An amorphous computing system is an extensive collection of identically-programmed computational particles, sprinkled irregularly on a surface or mixed in a volume. They interact only locally because they communicate within a fixed distance, which is large compared to the dimension of the elements but small compared to the size of the entire system. The single elements compute asynchronously. The challenge of the amorphous computing is to write programs that when executed on each agent give rise to self-organization phenomena. The most promising way for implementing amorphous computing is by using cells. The goal is to organize cells into precise patterns that function as actuators or transform them in programmable delivery vehicles for pharmaceuticals or in chemical factories for the secretion of specific chemical components.

13.3.1.10 Building Models of Complex Systems: ODEs, Boolean Networks, and Fuzzy Cognitive Maps

Any Complex System involves many molecular components. Its behavior is intractable from the computational point of view. In fact, we cannot use the principles of quantum physics to predict the phenomenon accurately. Therefore, we need to develop a model. Such a model must have the structure of a network. The construction of a model is a multi-stage procedure as demonstrated by Le Novère (2015) for molecular and gene networks. The first stage requires the collection of data that are representative of the phenomenon we want to describe. The next step is to infer from the experimental data the biological entities that must be included in the model. The difficulty is to understand the components that are fundamental to the process we want to represent. The third stage consists in defining the interactions between the selected components, specifying the directionality and properties of the relationships. The model is complete by making a list of the constraints that represent the context of the analysis. Essential constraints are the initial conditions; they usually ground on conservation laws. After building the model, we need to test how it evolves, and if it reproduces the experimental evidence. For running the model, we can choose different methodologies. One methodology consists in writing the Ordinary Differential Equations (ODEs) for each component of the network and in solving the ODEs by numerical integration (read Appendices A and E for more details). Of course, we need to know the values of the parameters appearing in the equations (sometimes the values of the parameters are inferred by fitting the experimental time series).

Alternatively, we may use Boolean network models. In a Boolean network, a variable can take two states (ON or OFF, +1 or −1) and the edges are causal connections. An excitatory or positive (+1) edge from node A to node B means A causally increases B from −1 to +1. An inhibitory or negative (−1) edge from node A to B means A causally decreases B from +1 to −1.

TRY EXERCISES 13.2 AND 13.3

In certain situations, the use of discrete variables can be a crude approximation. In these cases, it is worth considering Fuzzy logic models. The electrical engineer Bart Kosko (1986) proposed Fuzzy Cognitive Maps (FCMs) for representing causal reasoning. The idea came after the political scientist Robert Axelrod (1976) introduced cognitive maps for representing social scientific knowledge. Cognitive maps are Boolean networks where nodes are variable concepts (like social stability), and edges are causal relationships among them. Since causality admits of degrees, and degrees are vague if expressed through words, Kosko had the brilliant idea of introducing continuous values of causality and variables. Since their invention, FCMs have been used in many fields for dealing with Complex Systems (Glykas 2010). In a FCM, each edge is associated with a number (i.e., weight), included in the [−1, +1] interval, which quantifies the degree of causal relationship. Values of −1 and +1 represent entirely negative and positive causal effects, respectively. A value of 0 denotes no causal effect. Traditionally, the values of the weights are fixed by experts, and they quantify the strength of the relationships, which often are expressed through natural language fuzzy terms, such as "strong," "medium," "weak." More recently, the weights are fixed through learning algorithms, such as artificial neural networks and genetic algorithms.

Any FCM is described by a square matrix, called "connection matrix," which stores all the weight values for edges between the variables defining rows and columns. A system with N nodes will be represented by $N \times N$ connection matrix C. A FCM is useful for predicting the dynamics of the Complex System it models. We start from a vector $N \times 1$ of initial values for all the variables: $V(t)$. Usually, they range in the normalized interval [0, 1]. Then, we select a function $f(x)$ that transforms the product of the connection matrix and the initial values vector, $C \times V(t)$, in a vector of values for the nodes at an instant later, $V(t+1) = f(C \times V(t))$. Three most commonly used transformation functions are (Stach et al. 2005):

1. The Heaviside function:

$$f(x) = \begin{cases} 0, & x \leq 0 \\ 1, & x > 0 \end{cases}$$ [13.4]

2. The trivalent function:

$$f(x) = \begin{cases} -1, & x \leq -0.5 \\ 0, & -0.5 < x < 0.5 \\ +1, & x \geq 0.5 \end{cases}$$ [13.5]

3. The logistic function:

$$f(x) = \frac{1}{1 + e^{-Dx}}$$ [13.6]

where D is a parameter that determines the shape of $f(x)$.

The possible dynamic scenarios that can be achieved by FCMs depend on the transformation function that is chosen. If we select a discrete transformation function, such as (1), the dynamics can evolve to a fixed point or a limit cycle. If, on the other hand, we select a continuous transformation function, such as (3), even chaotic attractors can be found.

13.3.1.11 Agent-Based Modeling

Some biological processes having living beings as protagonists require agent-based modeling to be understood and reproduced. An agent is an autonomous individual element that exhibits the capacity to achieve a goal or affect an outcome. Often, an agent operates within a population and interacts with other agents as well as a more passive environment. Many homogeneous agents, having limited individual capabilities, acting in parallel, without a centralized control, can exhibit a collective intelligent behavior called "Swarm Intelligence," which is better than the intelligence of the individuals. Examples of swarm intelligence are fish schooling and birds flocking, which we read in Chapter 12. Other examples are either ants that find the shortest path between their nest and a good source of food, bees that find the best source of nectar within the range of their hive, or termites that build their outstanding nests. Social insects, such as ants, termites, and bees communicate through stigmergy. The term stigmergy was introduced by the French zoologist Pierre-Paul Grassé (1959) from the Greek words $\sigma\tau\acute{\iota}\gamma\mu\alpha$ ("sign") and $\acute{\epsilon}\rho\gamma o\nu$ ("work," "action"). It refers to the notion that the action of a social insect leaves signs in the environment, signs that it and other insects sense, and that determine and incite their subsequent actions. For example, ants look for food sources by wandering randomly and laying pheromones in their trails. When an ant finds a suitable food source, it returns to its nest. Other ants setting out to seek food sense the pheromone laid down by their precursor, and this influences the path they take up. The collective behavior that emerges is autocatalytic (i.e., a behavior controlled by positive feedback) because the more the ants follow a trail, the more attractive that trail becomes for being followed. However, the evaporation of the pheromone assures that suboptimal paths discovered earlier do not impede to find better solutions, exploiting the power of each individual ant to make stochastic choices. The swarm intelligence of social insects has inspired two families of algorithms: (I) Ant Colony Optimization (ACO), and (II) Particle Swarm Optimization (PSO). ACO has been used to solve heuristically the Traveling Salesman Problem (Dorigo et al. 1996). An artificial ant releases pheromone on its path. The amount of pheromone laid down is inversely proportional to the overall length of the path. Longer routes will have much less pheromone than shorter routes. Longer routes become less attractive, and the longest routes can become wholly neglected after a while because pheromone evaporates over time. By repeating the travel many times, the artificial ants are able to find the shortest paths. ACO algorithm has also been proposed to solve other combinatorial optimization problems, such as the job-shop scheduling problem. A PSO algorithm starts with a swarm of particles, each representing a potential solution to a problem (Kennedy and Eberhart 1995). A particle has both a position and a velocity. Its position is the candidate solution it currently represents. Its velocity is a displacement vector in the search space, which will contribute toward a fruitful change in its location at the next iteration. Particles move through a multidimensional search space. The calculation of a new position for each particle is influenced by three contributions. First, the previous velocity. Second, the personal best: the particle remembers the best place it has encountered. Third, the global best: every particle in the swarm is aware of the best position that any particle has discovered. After a random initialization of positions and velocities, the evaluation of the fitness of the particles' current positions, and consequent initialization of the personal bests and global best, PSO proceeds in a remarkably straightforward manner. PSO algorithms have been used to solve various optimization problems. They have been applied to unsupervised learning, game learning, scheduling and planning applications, and as an alternative to backpropagation for training an artificial neural network.

TRY EXERCISE 13.4

What surprises biologists, economists, and social scientists is the observation that although living beings are selfish, there are notable examples of cooperation. Every living being is instinctively selfish because its primary purposes are those of fighting to survive and reproduce. Despite this fierce competition, there are many examples of the social organization driven by cooperation.

How can such cooperation arise among fundamentally selfish individuals? There are different mechanisms (Nowak 2006). Often, the donor and the recipient of an altruistic act are genetic relatives (according to the so-called "kin selection" mechanism). But, we observe cooperation also between unrelated individuals or even between members of different species. Cooperation may arise from two individuals, A and B, who repeatedly encounter, based on the *direct reciprocity* mechanism: at every meeting, every individual can choose to cooperate or defect. For example, if A cooperates now, B may cooperate next time. The idea might be to help someone who may later help you or metaphorically speaking "you scratch my back and I'll scratch yours." Hence, it might pay off to cooperate. What is a good strategy? Around 1950, during the Cold War, two mathematical game theorists, Merrill Flood and Melvin Drescher of the RAND Corporation,[3] invented the Prisoner's Dilemma as a model to investigate the cooperation dilemma. Two individuals, A and B, are arrested for committing a crime together and are questioned into separate rooms. Each prisoner is given the opportunity of either betraying the other by testifying against him or cooperating with the other by remaining silent. The offer is:

- If both A and B betray reciprocally, each of them serves three years in prison.
- If A betrays B, whereas B remains silent, A will be set free, whereas B will serve five years in prison. The opposite is also true.
- If both A and B remain silent, both of them will serve only one year in prison.

The options are synthetically presented in the payoff matrix of Table 13.1.

A and B must choose before knowing the other's decision. Both of them might reason as follows. "If my companion remains silent, it would be better for me to betray them, because I will be set free. On the other hand, if my companion betrays me, it is convenient that I betray him, too. Otherwise, he will be set free, and I will spend five years in prison." If the decision must be taken just once, there is no escape from this sad logic. Thus, they both betray, although they are aware that if both remained silent, they would serve just one year in prison. The pursuit of self-interest leads both of them to a poor outcome. The Prisoner's Dilemma can be easily transformed in a "donation game" in which cooperation corresponds to offering the other player a benefit b at a personal cost c with $b > c$. Defection means offering nothing. The payoff matrix is thus as Table 13.2.

In the 1980s, Robert Axelrod organized a series of Iterated Prisoner's Dilemma tournaments for computer programs to discover the best strategy. The winning strategy was the simplest of all, tit-for-tat, echoing the fundamental principle of *lex talionis*: take an eye for an eye (Axelrod 1984). This strategy always starts with cooperation; then it does whatever the other player has done in the

---

TABLE 13.1
Payoff Matrix for the Prisoner's Dilemma

A / B	A stays silent	A betrays
B stays silent	1 / 1	0 / 5
B betrays	5 / 0	3 / 3

---

[3] RAND Corporation is a nonprofit institution that helps improve policy and decision making on energy, education, health, justice, environment, and international and military affairs through research and analysis. The address of its website is https://www.rand.org/.

TABLE 13.2
Payoff Matrix for a Generalized Prisoner's Dilemma

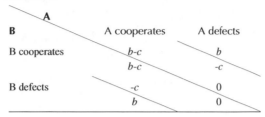

B	A cooperates	A defects
B cooperates	$b-c$ / $b-c$	b / $-c$
B defects	$-c$ / b	0 / 0

previous round: a cooperation for a cooperation, a defection for a defection. From the success of the tit-for-tat strategy, Axelrod extracted some lessons:

- *Be nice*: never be the first to defect.
- *Be retaliatory*: retaliate immediately when betrayed.
- *Be forgiving*: relent after a single cycle of punishment.
- *Be clear*: be transparent about your strategy, e.g., make it easy to infer.

Tit-for-tat is not a perfect strategy when there are accidental defections due to "trembling hands" or "blurred minds." In fact, it reacts to an accidental defection always with an immediate retaliation. An alternative strategy is generous-tit-for-tat that cooperates whenever the counterpart cooperates but sometimes cooperates although the other player has defected, admitting forgiveness (Nowak and Sigmund 1992). The tit-for-tat strategy is an efficient catalyst of cooperation in a society where nearly everybody is a defector, but once cooperation is established, the win-stay lose-shift strategy is better able to maintain it. Win-stay, lose-shift embodies the idea of repeating your previous move whenever you are doing well but changing otherwise (Nowak and Sigmund, 1993).

Sometimes humans cooperate even when they are not in direct contact. One person is in a position to help another, but there is no possibility for a direct reciprocation. For example, we help strangers who are in need by donating to charities that do not give us anything in return. What may fuel this kind of cooperation is reputation. Helping someone establishes a good reputation, which will be rewarded by others (Nowak and Sigmund 1998). People who are more helpful are more likely to receive help. This behavior is *indirect reciprocity*: "You scratch my back, and I'll scratch someone else's" or "I scratch your back, and someone else will scratch mine." Indirect reciprocity has cognitive demands. It is necessary that we remember our own interactions, and we must also monitor the ever-changing social network of the group. Language facilitates the acquisition of information and gossip. Indirect reciprocity is likely to be connected with the origin of moral and social norms (Nowak and Sigmund 2005).

A further improvement of the Prisoner's Dilemma model is the introduction of spatial structure. In the original model, there was no notion of space. In fact, it was equally likely for any player to encounter any other player. The mathematical biologist Martin Nowak envisioned that placing players on spatial structures or social networks imply that some individuals interact more often than others and have substantial effect on the evolution of cooperation. A cooperator pays a cost for each neighbor to receive a benefit. Defectors have no costs, and their neighbors receive no benefits. In this setting, cooperators form network clusters, where they help each other (Lieberman et al. 2005) and a group of cooperators is expected to be more successful than a group of defectors.

Computer simulations of idea models, such as the Prisoner's Dilemma, are sometimes the only available means of investigating Complex Systems. They are used to advance science and policy in many contexts. However, they describe a reality that is highly simplified. For example, the economy is a vast and complex set of arrangements and actions wherein economic agents

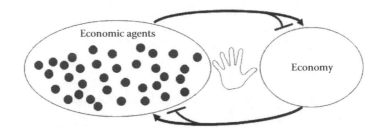

FIGURE 13.20 Scheme of the mutual relationships between the economic agents and the environment they originate. The bottom-up and top-down excitatory and inhibitory causations embody the Adam Smith's "Invisible Hand."

(producers, consumers, investors) constantly change their actions and strategies in response to the outcome they mutually create. The economy emerges from a continually developing set of technological innovations, institutions, and arrangements that induce further innovations, institutions, and arrangements (Arthur 2015). There are mutual "excitatory" and "inhibitory" effects between the economic agents and the economy they contribute to creating (see Figure 13.20). These bottom-up and top-down causation phenomena embody the Adam Smith's "Invisible Hand." The interacting agents self-organize, and the patterns they give rise, in turn, cause the interacting elements to change or adapt.

In complex situations, deductive rationality can break down. In fact, agents cannot rely upon the other agents to behave under perfect rationality, and so they are forced to guess their behavior. Agents are induced to formulate subjective beliefs about subjective beliefs. Objective, universal and well-defined assumptions then cease to apply, and rational deductive reasoning loses meaning. In the presence of complex, variable, and ill-defined situations humans look for patterns to construct internal models or hypotheses. We make localized deductions based on our current hypotheses. We test our temporary deductions by collecting feedback from the environment. We strengthen or weaken our trust in our hypothesis and models, discarding some of them when they fail to predict, and replacing them with new ones. Such mental behavior is inductive and can be modeled in a variety of ways. Usually, it leads to a rich psychological world in which agents' hypothesis or mental models compete for survival against other agents' hypothesis and mental models: a psychological world that is both complex and evolutionary (Arthur 2015).

TRY EXERCISE 13.5

13.3.2 Computing by Exploiting the Physicochemical Laws

All the phenomena that are investigated by science are causal. Any cause and effect relation can be described as a computation wherein the cause is the input, the effect is the output, and the law governing the transformation is the algorithm for the computation. The law can be either linear or non-linear. Therefore, different types of computation can be carried out depending on which physicochemical law we exploit (see Figure 13.21).

13.3.2.1 Thermodynamics

Any physicochemical system, pushed away from equilibrium by some temporary forces or maintained out-of-equilibrium constantly by some permanent forces, can be imagined being used as a computing machine. The laws that govern its response to the external forces rule its computational performances. The conditions required to process information are (I) the distinguishability of states within the system and (II) the capability for a controlled change of its states. If the system is merely a particle in two potential energy wells, like that shown in Figure 13.22, the location of the particle

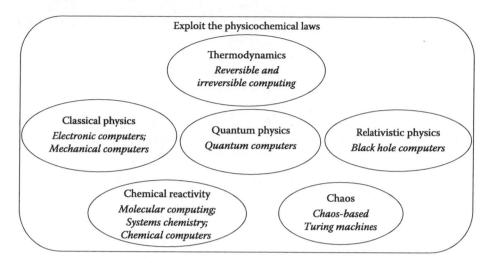

FIGURE 13.21   Types of computations that can be implemented by exploiting the natural physicochemical laws.

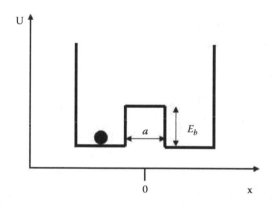

FIGURE 13.22   A particle (black circle) in two potential energy wells. The height of the energetic barrier is E_b whereas the barrier width is a.

in one of the two wells is said to be distinguishable if there is a very low probability of spontaneous transition from one well to the other. In other words, the probability of staying in the two wells must be different. If it is the same, i.e., 0.5, the two states are indistinguishable. The particle can spontaneously change well, and hence state, either going "classically" over the barrier or through quantum mechanical tunneling transitions.

The probability of a *classical* transition over barrier at temperature T is

$$P_{classical} = e^{-\left(\frac{E_b}{kT}\right)}$$

[13.7]

Imposing $P_{classical} = 0.5$, it derives that distinguishability is lost when the energetic barrier is $E_b \leq kTln2$. The barrier must have $E_b > kTln2$ to distinguish the two possible states.

From quantum mechanics, we know that when the barrier width a is small, the particle can tunnel through the energetic barrier E_b. A simple analytical form of tunneling probability for a particle of effective mass m through a rectangular barrier of width a is given by the Wentzel-Kramers-Brillouin approximation (Zhirnov et al. 2003):

$$P_{quantum} = e^{-\left(\frac{2\sqrt{2m}}{\hbar} a\sqrt{E_b}\right)}$$ [13.8]

Imposing $P_{quantum} = 0.5$, it derives that distinguishability is lost by quantum mechanical tunneling events when $a \leq (\hbar ln2)/(2\sqrt{2mE_b})$, i.e., when the barrier width is less than the limit fixed by the Heisenberg's uncertainty principle, which, for an electron, is:

$$\Delta x \geq \frac{\hbar}{\Delta p} = \frac{\hbar}{\sqrt{2mE_b}} = \frac{\hbar}{\sqrt{2mkTln2}} = 1.5 \text{ nm} \left(\text{at } 300 \text{ K}\right)$$ [13.9]

The total probability of transition by either "classical" or tunneling mechanism is:

$$P_{tot} = P_{classical} + P_{quantum} - \left(P_{classical}\right)\left(P_{quantum}\right)$$ [13.10]

The states in the two wells become indistinguishable when (Zhirnov et al. 2003)

$$E_b \approx kTln2 + \frac{\hbar^2(ln2)^2}{8ma^2}$$ [13.11]

The second term on the right part of the equation [13.11] becomes significant when $a < 5$ nm. If $a < 5$ nm, the barrier height must be $E_b > kTln2 + \left(\hbar^2(ln2)^2\right)/\left(8ma^2\right)$ to have two distinguishable states.

When we can distinguish states, we can make operations by switching from one state to the other in a controllable way. The logical operations of a computation can be reversible or irreversible. When we use elementary logic operations, such as AND, OR, etc., with two inputs and one output, or we erase a bit in the memory, we do irreversible logical operations. A logical operation is irreversible whenever the inputs cannot be reconstructed from the outputs.

When the opposite is true, we can infer the input from the output, and the information is preserved at every step, we are carrying out reversible logical operations. According to the Landauer's principle (Landauer 1961), any elementary irreversible logical operation determines a dissipation of energy equal to $kTln2$, and therefore an increase of entropy. Landauer's principle has been confirmed experimentally by measuring the amount of heat generated by erasure of a bit encoded by a single colloidal particle optically trapped in a modulated double-well potential (Bérut et al. 2012). However, recent theoretical studies (Maroney 2005; Gammaitoni et al. 2015), and experiments (López-Suárcz et al. 2016) have demonstrated that an irreversible logical operation can be performed in a thermo-dynamically reversible manner (i.e., very slowly) and generate an amount of heat smaller than $kTln2$. Logically irreversible operations do not always imply a dissipation of energy into the environment. Therefore, Landauer's principle holds strictly only for "information entropy," also called Shannon entropy or uncertainty H (remind what we studied in Chapter 2): any irreversible logical operation that determines the erasure of one bit, increases the "information entropy" of one bit.[4]

A computation, whether it is performed on an abacus or by an electronic computer or in a soup of chemicals, is a physicochemical process. How large is the computing machine? How much energy is required to perform an operation? How fast does the computer run? How frequently does it give the wrong result? The answers lie in the physical laws that govern the behavior of the computing machine.

13.3.2.2 Classical Physics

Initially, humans used fingers, small stones,[5] notched sticks, knotted strings, and other similar devices to make computations. People gradually realized that these methods of calculation could not go far

---

[4] Another example wherein "information entropy" pops up and that you might have encountered previously, is that associated with the process of mixing two ideal gases. The internal energy of the two gases does not change but the mixing determines a loss of information. *Try exercise 13.6.*

[5] The word calculation derives from the Latin "calculus" that means small stone.

enough to satisfy increasing computational needs. For counting with large numbers, they would have had to gather too many pebbles or other tools. Once the principle of the numerical base had been grasped, the usual pebbles were replaced with stones of various sizes to which different orders of units were assigned. For example, if a decimal system was used, the number 1 could be represented by a small stone, 10 by a larger one, 100 by a still larger one, and so on. It was a matter of time before someone developed first devices to facilitate calculation. Someone had the idea of placing pebbles or other objects in columns, assigning a distinct order of units to objects belonging to different columns and invented the counting board or abacus. Later, the arrangement of objects in columns was replaced with a more elegant one, wherein pebbles could slide along parallel rods. The abacus was really useful to make calculations with large numbers. A further significant contribution in this direction was contributed by the Scottish mathematician John Napier (1550–1617), who discovered that multiplication and division of numbers could be performed by addition and subtraction, respectively, of the logarithms of those numbers. Moreover, he designed his "Napier's bones," a manually-operated device that simplified calculations involving multiplication and division. In the 1620s, logarithmic scales were put to work by the English mathematician and Anglican minister William Oughtred (1574–1660) who developed the slide rule, where each number to be multiplied is represented by a length on a sliding ruler.

Although abacus-like devices and sliding rulers were significant computational aids, they still required direct physical and mental involvement of the operator with a consequent high risk of mistakes, especially after many hours of labor. An improvement was achieved by devising the first mechanical calculators. The history of mechanical computing machines is mostly the story of the numeral-toothed wheels and the devices that rotate them to register digital and tens-carry values (Chase 1980). The devices that rotate numeral wheels are grouped in (I) rotary or crank-type actuators (see Figure 13.23a) and (II) reciprocating or cam actuators (see Figure 13.23b).

In the mid-seventeenth century, Blaise Pascal (1623–1662) built a mechanical adding and subtracting machine called the Pascaline as an aid to his father Etienne, who was a high official in Basse-Normandie, and following a revolt over taxes, he reorganized the tax system of that area (Goldstine 1993). In the same century, Gottfried Wilhelm Leibniz (1646–1716) invented a device, now known as the Leibniz wheel, suitable for doing not only additions and subtractions but also multiplications and divisions. In the next two hundred years, mechanical calculating machines were mainly remarkable gadgets exhibiting human ingenuity, rather than concrete aids to calculation. Scientists, engineers, navigators, bankers, and actuaries who had to carry out complex computations, demanding high accuracy, could trust in the pen, paper, and numerical tables (such as trigonometric, multiplication, and logarithmic tables). Numerical tables were fruits of monotonous and fatiguing mental labor, and often they were full of errors. For this reason, in the first half of the nineteenth century, the English mathematician Charles Babbage (1791–1871) dreamt of designing automatic mechanical computers to compile more accurate numerical tables and

(a)

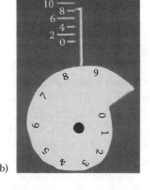

(b)

FIGURE 13.23 Two examples of digital value actuators: rotary or crank-type actuator (a), and reciprocating or cam actuator (b).

help people in routine calculations. He first designed a "Difference Machine" to compute polyno-mial functions. At the heart of his machine's operation was the use of gears to count. Babbage's Difference Machine was a single-purpose computer: it had only one program and one task. Later on, Babbage contrived a general-purpose computing machine by designing an "Analytical Engine." The Analytical Engine had the essential features and elements of a modern electronic computer but in a mechanical form. It was programmable using punched cards; it had a "Store," a memory where numbers and intermediate results could be held; and it had a separate "Mill," a mechanical processor where arithmetic logic unit and control flow were implemented in the form of condi-tional branching and loops. The "Mill" allowed for performing all the four fundamental arithmetic operations, plus comparison and optionally square roots. Finally, it had a variety of outputs includ-ing hardcopy printout, punched cards, and graph plotting. Unfortunately, Babbage never finished the construction of both his Difference Machine and his Analytical Engine. Probably, one reason was that the technology of those days was not reliable enough. In retrospect, we can notice some inherent problems with the use of mechanical devices for computing (Wolf 2017). One problem is that mechanical components are relatively heavy. Therefore, a substantial amount of energy is needed to make them work. Since the twentieth century, energy was generated by electric motors. Another problem is that the mechanical components are subject to friction and hence to damage in their connections.

The era of modern electronic computers began with a flurry of development before and during the Second World War. At first, mechanical computers were overtaken by electromechanical com-puters based on electric operated switches, called relays. These devices had a low operating speed and were eventually superseded by much faster all-electric computers, originally using vacuum tubes. At the same time, digital calculation replaced analog. The foundation of binary logic was laid by the English mathematician George Boole (1815–1864), who published *An Investigation of the Laws of Thought* in 1854. In his book, Boole investigated the fundamental laws of human reason-ing and gave expression to them in the symbolical language of a Calculus, now known as Boolean Algebra. In the 1930s, the American electrical engineer Claude Shannon (1916–2001) showed the correspondence between the concepts of Boolean logic and the electronic circuits implementing the logic gates. After the Second World War, the progress of computers continued, and the physicist John von Neumann, inspired among others by the ideas of Alan Turing (1912–1954), proposed his model that is still used in the current electronic computers that are based on transistors, as we learned in paragraph 13.2.

13.3.2.3 Computing with Subatomic Particles, Atoms, and Molecules

A strategy to make electronic computers increasingly faster and energetically efficient is to minia-turize the basic switching elements—the transistors. This strategy is called the top-down approach and the pace of its accomplishment is epitomized by Moore's law as we learned in paragraph 13.2. The American physicist Richard Feynman (1918–1988) paved the way for an alternative strategy, called the bottom-up approach in his seminal lecture titled "There is Plenty of Room at the Bottom," taken at the annual meeting of the American Physical Society at the end of 1959 (Feynman 1960). The bottom-up approach is based on the idea of manufacturing computers by assembling atoms and molecules. Matter at the atomic level does not behave classically but quantum-mechanically. Indeed, the subatomic particles, atoms, and molecules can be used to process a new kind of logic that is quantum logic. Quantum states of matter have different properties with respect to the classical states of matter. Therefore, quantum information is different from classical information. The ele-mentary unit of quantum information is the qubit or quantum bit (Schumacher 1995). A qubit is a quantum system that has two accessible states, labelled as $|0\rangle$ and $|1\rangle$ (note that the "bracket" notation $|\ \rangle$ means that whatever is contained in the bracket is a quantum-mechanical variable), and it can exist as a superposition of them. In other words, a qubit, $|\Psi\rangle$, is a linear combination of $|0\rangle$ and $|1\rangle$:

$$|\Psi\rangle = a|0\rangle + b|b\rangle \qquad [13.12]$$

wherein a and b are complex numbers, verifying the normalization condition

$$|a|^2 + |b|^2 = 1 \qquad [13.13]$$

The qubit can be described as a unit vector in a two-dimensional vector space, known as the Hilbert space. The states $|0\rangle$ and $|1\rangle$ are the computational basis states and form an orthonormal basis for the Hilbert space. The qubit can also be described by the following function (Chruściński 2006):

$$|\Psi\rangle = \cos\left(\frac{\theta}{2}\right)|0\rangle + e^{i\varphi}\sin\left(\frac{\theta}{2}\right)|1\rangle \qquad [13.14]$$

wherein θ and φ define a point on the unit three-dimensional sphere, called the Bloch sphere (see Figure 13.24). Represented on such a sphere, a classical bit could only lie on one of its poles, being the pure $|0\rangle$ and $|1\rangle$ states, respectively. A qubit can be implemented by any quantum two-level system. An example is the vertical and horizontal polarization of a single photon; other examples are the ground and excited states of an atom and the two spin states of a spin ½ particle within a magnetic field.

Since quantum states can be protected, manipulated, and transported, quantum information can be stored, processed, and conveyed. Therefore, it is possible to contrive quantum computers (Deutsch 1989; Feynman 1982). The execution of a program can be described by the Schrödinger equation. The latter applies only to isolated systems and is valid in the time interval separating a preparation (when the system is initialized) and a measurement (when the output is extracted from the system) (Wheeler and Zurek 1983). The Hamiltonian H generates a unitary evolution of $|\Psi\rangle$, which can be visualized as rotation of the qubit on the Bloch sphere. The qubit maintains its norm and the operation is reversible. When a calculation is ended, and a measurement is performed, the superposition $a|0\rangle + b|1\rangle$ behaves like $|0\rangle$ with probability $|a|^2$ and like $|1\rangle$ with probability $|b|^2$. The main difficulty in building a quantum computer comes from the fact that quantum states must continuously contend against insidious interactions with their environment (e.g., an atom colliding with another atom or a stray photon) triggering loss of coherence.

A quantum computer promises to be immensely powerful because it can be in multiple states at once. If it consists of N unmeasured qubits, it can be in an arbitrary superposition of up to 2^N different states simultaneously (the Hilbert space will be of 2^N dimensions). A classical computer could be

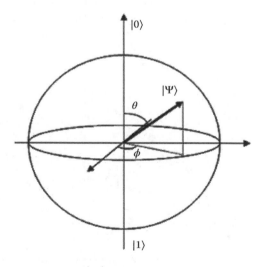

FIGURE 13.24 Representation of the qubit $|\Psi\rangle$ in the Bloch sphere.

only in one of the 2^N states at any one time (Bennet and DiVinceno 2000). For example, a quantum superposition of 300 particles would create as many configurations as there are atoms in the known Universe. The superposition can also involve the quantum states of physically separated particles if they are entangled (Plenio and Virmani 2007). Working with superimposed states can speed up not all, but some computations, such as the Shor's algorithm for factoring an n-digit number (Shor 1997). This feature makes possible some tasks, such as the absolute secrecy of communication between parties (i.e., quantum cryptography (Bennett et al. 1992)) that is impossible classically.

So far, different technologies have been proposed to implement quantum computations (DiVincenzo 2000; Brumfiel 2012). For instance, trapped ions in atomic physics, nuclear and electron magnetic resonance spectroscopies, quantum optics, loops of currents in superconducting circuits, and mesoscopic solid-state systems like arrays of quantum dots. In these technologies, the produced qubits have relaxation times sufficiently long to prevent quantum effects from decohering away before ending a computation.

Whenever too fast decoherent effects are unavoidable, the lure of quantum information vanishes. However, it is still possible to compute with molecules by processing "classical" logic. Since a qubit, $|\Psi\rangle$, can collapse into one of the two available states, $|0\rangle$ or $|1\rangle$, it seems evident that only crisp Boolean logic can be implemented at the atomic and molecular levels (see Figure 13.25). When the computation trusts in a chemical transformation, involving just two distinct parallel routes, every single molecule selects only one reaction path. Therefore, its behavior can be described in terms of Boolean algebra (de Silva 2013; Szaciłowski 2008). In the case of three-level, or k-level, quantum systems (with $k > 3$) that have three or k basis states, respectively, the decoherence gives the possibility of processing three-valued or multi-valued logics, also known as crisp logic wherein there are more than two truth values (Andréasson and Pischel 2015). The ability to make computation by molecules resides in their structures and reactivity (i.e., affinity). The order, the way the atoms of a molecule are linked, and their spatial distribution rule the intra- and inter-molecular interaction capabilities of the molecule itself, defining its potentiality of storing, processing and conveying information.

In practical terms, computation with single molecules is feasible by exploiting microscopic techniques that reach the atomic resolution, such as scanning tunneling microscopy (STM) or the atomic force field microscopy (AFM) (Joachim et al. 2000).

Alternatively, it is possible to process logic by using large assemblies of molecules (for instance, an Avogadro's number of computing elements). These extensive collections of molecules become macroscopic pieces of matter receiving inputs from the macroscopic world and giving back output signals to the surrounding macro-environment. Therefore, inputs and outputs become macroscopic variables whose values change in a continuous manner. They can be physical or chemical signals. Physical signals can be optical or electrical or magnetic or mechanical or thermal. Optical signals are particularly appealing because they can be easily focused and conveyed, and since they can be

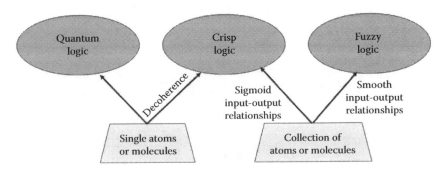

FIGURE 13.25 The kinds of logic that can be processed by using either single molecules or collection of atoms and molecules. (From Gentili, P.L., *ChemPhysChem.*, 12, 739–745, 2011a.)

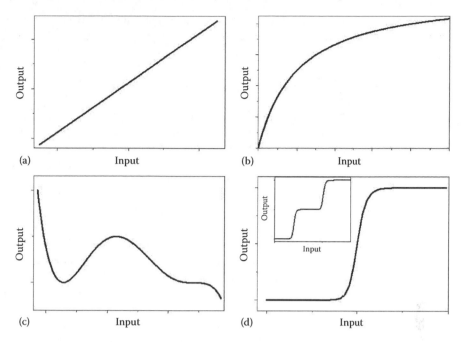

FIGURE 13.26 Examples of linear (a), smooth (b and c) and sigmoid (d) non-linear output-input functions. In the inset of graph D, a two-step sigmoid function is shown.

collected even by human eyes, they easily bridge the gap between the molecular and our macroscopic world. The relationships establishing between inputs and outputs have many shapes: they can be either linear or non-linear, and when they are non-linear, they can be either smooth or sigmoid (see Figure 13.26). Whenever the input-output relationship is sigmoid, having a steep slope around the inflection point, it is suitable to process crisp logic. In particular, if the input-output function is characterized by just one step, it is proper to implement binary logic. For this purpose, it is necessary to establish a threshold value and a logic convention for every input and output variable. The variables can assume merely high or low values that become digital 1 or 0, respectively, in the positive logic convention, whereas the negative logic convention reverses this relationship. On the other hand, if the input-output function is characterized by three or $k > 3$ steps, it is suitable to process three- or multi-valued logic, respectively. Whenever the input-output relationship is smooth, it is not adequate for crisp logic, but it is proper to implement infinite-valued logic, for instance, Fuzzy logic.[6] For this purpose, the entire domain of each variable referred to as the universe of discourse is divided into different Fuzzy sets whose shape and position define their membership functions (Gentili 2011b).

This method is not the only way for processing Fuzzy logic by molecules. Another approach considers the absorption bands of a compound as Fuzzy sets (Gentili et al. 2016), as we learned in Chapter 12 for the interpretation of the human color vision. When polychromatic radiation interacts with a compound, it belongs to more than one spectral Fuzzy set, with the same or different degrees of membership. The ensemble of degrees of membership of the radiation to the absorption bands of a compound is Fuzzy information encoded at the molecular level. The consequent photo-induced reactions process the collected Fuzzy information.

---

[6] This distinction in the use of sigmoid and smooth input-output functions holds in current electronic circuits, as well. In fact, when the electrical signals vary steeply, in sigmoid manner, they are used to process Boolean logic. On the other hand, the best strategy to implement Fuzzy logic is through analog electronic circuits that are based upon signals varying smoothly.

Even a third way of implementing Fuzzy logic by using molecules has been proposed, so far. It exploits compounds that exist as ensembles of many conformers (Gentili 2014b). The physical and chemical properties of such ensembles are context-dependent, like the properties of Fuzzy sets and the meaning of words in natural language. A set of conformers becomes a "Molecular Fuzzy set," and at the same time a "Word of the chemical language." A chemical reaction becomes an event of Fuzzy information processing, which is ruled by inter- and intra-molecular forces. The Fuzziness of a chemical compound becomes particularly important when we consider macromolecules, where we can have many conformational structures. For example, we already know that many proteins in their native, functional state, cannot be described adequately by a single conformation (Tompa and Fuxreiter 2007). Structural disorder becomes relevant in eukaryotic proteomes and correlates with important functions. Cellular processes become computational events of Fuzzy information. This feature should be considered when the method of logic modeling is used to describe molecular and gene networks in synthetic biology and the development of virtual organisms (Le Novère 2015; Macia and Sole 2014; Bavireddi et al. 2013).

A buzzing question in this research field is: how will the chemical computers look like? All the reasonable architectures thought so far, can be partitioned in two principal families: one family based on "interfacial hardware" and the other based on "wetware" (Gentili 2011c).

In the case of "interfacial hardware," the computations are carried out by molecules anchored to the surface of a solid phase. The computing molecules can be organic semiconductors, proteins or even DNA, to cite just a few examples. Organic semiconductors can behave like silicon-based semiconducting devices. Information is encoded through electric signals exchanged with the outside world through electrodes (Joachim et al. 2000). Proteins can compute through their ability to recognize a specific type of molecule and switching its state by making or breaking a precisely selected chemical bond. DNA derives its computing power from the hybridization reaction, i.e., the ability of nucleotides to bind together using Watson-Crick pairing. There is a code of "stick" or "don't stick," with DNA chains binding to form regions of double-stranded molecules or remaining free as regions of single-stranded DNA.

In the case of "wetware," soups of suitable chemicals process information through reactions, coupled or not with diffusion processes. These soups can work inside a test tube wherein computations are performed through perturbations coming from the outside world. Inside the test tube, we may think of putting a "Molecular Turing machine." The latter might be constituted by (I) a polymer that, as the tape of the Turing machine, has sites that can be modified chemically and reversibly, and (II) a polymer-manipulating catalyst, anchored to the macromolecular tape, which, as the head of the Turing machine, modifies the sites of the tape, controlled by external stimuli (such as light, electrons or acid/base chemicals) (Varghese et al. 2015). Alternatively, the chemical soups can operate in microfluidic systems structurally related to the pattern of the current electronic microchips. The microfluidic channels are the wires conveying chemical information, while logic operations are processed inside reaction chambers. In these systems, computation can also be pursued through chemical waves propagating across excitable chemical media. An alternative approach consists in devising open systems wherein properly combined chemical reactions mimic the signaling network of cells. Finally, it might be worthwhile trying to mimic a nervous system by implementing a network of distinct processing units, similar to neural networks. In a network of neurons, information is processed in two different kinds of spatial coordinates: a horizontal one and a vertical one. Along the vertical coordinate, information is processed hierarchically, from the molecular level (especially through conformations), to the mesoscopic one (through reaction-diffusion processes), up to the macroscopic one by the appearance of ordered structures playing the role of communication channels between the microscopic and the macroscopic worlds. Along the horizontal coordinate, information is processed and conveyed due to the chemical waves interconnecting spatially distant processing units. With a such complex computing system, it will be possible to devise a chemical computer more similar to the brain rather than to the current electronic computers.

13.3.2.4 The "Ultimate Laptop"

The physicist Seth Lloyd (2000) has calculated the physical limits of computation for an "ultimate laptop" operating at the limits of speed and memory space allowed by the laws of physics. According to the time-energy Heisenberg uncertainty principle extended by the Margolus and Levitin theorem, a quantum system in a state with average energy E takes time at least $\Delta t = \pi \hbar / 2E$ to evolve to an orthogonal state. If we assume that our "ultimate laptop" has a mass of 1 kg, it has an average energy of $E = mc^2 = 8.9874 \times 10^{16}$ J. Therefore, its maximum speed of computation is 5.4258×10^{50} operations per second. For the ultimate laptop, the rate grows by increasing its mass. Conventional laptops operate well below the ultimate laptop. In fact, in a conventional laptop, most of the energy is locked up in the mass of the materials, leaving only a tiny portion for performing calculations. Moreover, in a conventional laptop, many electrons are used to encode just one bit. From the physical perspective, such a computer operates in a highly redundant fashion. However, redundancy is required to reduce errors. A way of limiting redundancy is to compute with single subatomic particles.

The maximum amount of information, which the ultimate laptop can process and store, is determined by the total number of distinct physical states that are accessible. This number is, of course, related to the information entropy of the system. The calculation of the total entropy of 1 kg of matter contained in 1 L of volume would require complete knowledge of the dynamics of the elementary particles, quantum gravity, and so on. We cannot access all this information. However, its entropy can be estimated by assuming that the volume occupied by the laptop is a collection of modes of elementary particles with total energy E. It results (Lloyd 2000) that the amount of information that can be stored by the ultimate laptop is $\sim 10^{31}$ bits. This amount is much larger than the $\sim 10^{10}$ bits stored in a current laptop. This discrepancy is because conventional laptops use many degrees of freedom to store just one bit, assuring stability and controllability.

If the computation to be performed is highly parallelizable or requires many bits of memory, the volume of the computer should be large, and the energy available should be spread out evenly among the different parts of the computer. On the other hand, if the computation is highly serial and requires fewer bits of memory, the energy should be concentrated in a smaller volume. Ideally, the laptop can be compressed up to the black-hole limit (a black-hole of 1 kg has a "Schwarzschild radius" of 10^{27} m). Then, the computation becomes fully serial. A black-hole is suitable for computing because, according to the quantum mechanical picture, it is not entirely black. In fact, a black hole emits the so-called Hawking radiation that contains information about how it has been formed: what goes in does come out but in an altered form.

13.4 LAST CONCLUSIVE THOUGHTS AND PERSPECTIVES

We are at the end of this fascinating journey of discovering Complexity. We have learned that Complex Systems are (I) networks with many nodes (II) in out-of-equilibrium conditions, (III) which exhibit emergent properties. Complexity is "disorganized" (to cite a term used by Warren Weaver (1948) in his farsighted paper titled "Science and Complexity") when the networks are random. The probability theory and statistical mechanics are powerful tools for the description of the random networks. Complexity is "organized" when the networks are either regular, small-word, scale-free, modular or hierarchical. Their behavior can be investigated by analytical mathematical methods, such as that used by Marcus Covert and his colleagues at Stanford University for *Mycoplasma* cell's overall functions (Karr et al. 2012). However, if in a system of "Organized Complexity" there is an enormous number of factors that are all interrelated into an organic whole, an analytical approach becomes computationally intractable. In this latter case, we collect Big Data. But, then, the overflow of data must be transformed into information, knowledge, and finally wisdom (according to the DIKW pyramid). For this purpose, it is necessary to develop a model. Into the model, we are not expected to incorporate everything we know about a Complex System. In fact, we could originate a model that is too complicated to be understood. One can make a complicated model do anything one likes by fiddling with the parameters (suffice to think about the remarkable predictive power of Artificial Neural Networks). A model that predicts everything is not necessarily useful for understanding the phenomenon it refers to. A model should be an intentional symbolic simplification of a Complex System. The features of the model depend on which level we want to describe the behavior of a Complex System. As properly stated by the British statistician George Box (1919–2013), "All models are wrong, but some models are useful." The perfect model is not the model that best represents the world around us but, instead, is a model that in some ways exaggerates the aspects of the world we are most interested in and can help us win the challenges we are facing. For instance, can we predict when and where earthquakes occur? How much do human activities affect the climate? How can we save the biodiversity of our ecosystems? Which are the best strategies to guarantee a sustainable economic growth everywhere in the world? How can we guarantee social justice? Can we defeat cancer? Is it possible to slow down the aging of our bodies? And so on.

A useful methodology to build models for Complex Systems is represented in Figure 13.5. It assumes that any Complex System computes. Therefore, we must discern the inputs, outputs, and the computations that the system performs. Then, we need to develop algorithms that might carry out those computations. A suitable algorithm will become a predictive tool and a decision support system to try to win the Complexity Challenges related to that particular Complex System. Finally, we may contrive a mechanism for implementing the algorithm. If such a mechanism exists, it will contribute to the development of technology.

By constructing good models of Complex Systems, we should store the knowledge necessary to formulate a new theory and promote the expected third "gateway event" in the humankind journey to discovering the secrets of nature (remind Chapter 1). In fact, this new theory will probably apply equally well to the cell, the immune system, the brain, the ecosystem, human societies, the economy, and perhaps even the climate and the geology of our planet. Most likely, this new theory will have information as the pivotal variable. Information will be included not only in terms of quantity but also in terms of quality. The concepts of granulation and graduation of the variables, which are cornerstones of the theory of Fuzzy logic, could give clues on how to rationalize the quality of information.

To succeed in our effort of understanding Complex Systems and winning the Complexity Challenges, it is vital to support the formation of interdisciplinary research teams and promote the diffusion of interdisciplinary academic degrees focused on Complexity. In the interdisciplinary academic degrees in Complexity (see Figure 13.27), we should prepare the new generations of "Philo-physicists," i.e., scientists having the following attributes:

- Wonder for the beauty of nature.
- Curiosity for the unknown.
- Open-mindedness, multidisciplinary interests, and knowledge. It is fundamental to have polymath minds who approach problems by making analogies among ideas and concepts belonging to different disciplines.
- Dedication, patience, and perseverance in doing experiments and collecting data. In other words, patience in querying nature and catching its answers, by using the scientific method based on the rigorous laws of rational logic.
- High standards of personal honesty and love of truth when interpreting the answers of nature to our queries.
- Critical thinking, creativity, independence, and resourcefulness to think "out of the box" and formulate new ideas and theories.
- Ingenuity to contribute to the development of technology that exerts always a positive feedback action on the scientific enquiry.
- Resilience to recover from defeats and criticism.
- Awareness and excitement for having the possibility of contributing to the development of science and the improvement of human psycho-physical well-being.

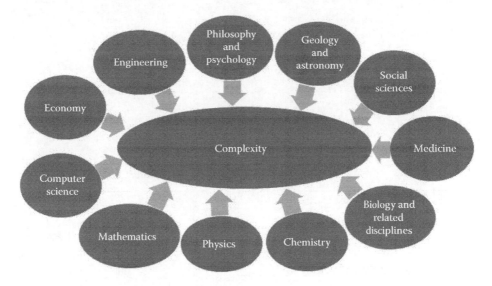

FIGURE 13.27 A multidisciplinary approach to face Complexity.

- Humbleness and wisdom. In fact, science alone cannot furnish a code of morals or a basis for aesthetics. Science cannot "furnish the yardstick for measuring, nor the motor for controlling, man's love of beauty and truth, his sense of value, or his convictions of faith" (Weaver, 1948).

The polymath figures formed at the interdisciplinary academic degrees in Complexity should work along with specialized scientists to build vibrant research teams. The lively multidisciplinary research teams will have a much higher potential for facing the Complexity Challenges than the potential of teams composed only of members specializing in one particular discipline. In fact, connections among different disciplines will blossom new great ideas and spark flashes of brilliance. However, a fundamental ingredient for success is the willingness of each member to sacrifice selfish short-term interests to bring about improvements for all.

13.5 LAST MOTIVATING SENTENCES PRONOUNCED BY "IMPORTANT PEOPLE"

I want to conclude this breathtaking journey exploring Complexity by citing some motivating sentences pronounced by eminent philosophers, scientists, and writers.

He who knows the happiness of understanding
has gained an infallible friend for life.
Thinking is to man
what flying is to birds.
Don't follow the example of a chicken
when you could be a lark.

Albert Einstein (1879–1955 AD)

Knowledge is like a sphere.
The greater it's volume
the larger it's contact with the unknown.

Blaise Pascal (1623–1662 AD)

The scientific man does not aim at an immediate result.
He does not expect that his advanced ideas will be readily taken up.
His work is like that of the planter—for the future.
His duty is to lay the foundation for those who are to come, and point the way.

Nikola Tesla (1856–1943 AD)

Learn from yesterday,
live for today,
hope for tomorrow.
The important thing is not to stop questioning.

Albert Einstein (1879–1955 AD)

One doesn't discover new lands
without consenting to lose sight of the shore.

André Gide (1869–1951 AD)

13.6   KEY QUESTIONS

- Which are the principal strategies to tackle the Complexity Challenges?
- Which are the fundamental elements of an electronic computer and how does it work?
- What is the future of the electronic computers?
- What is Natural Computing?
- Present the three-step methodology to study Complex Systems?
- Which are the essential attributes of a living being?
- Explain the basic idea of membrane computing.
- Explain the Adleman's idea and procedure of using DNA hybridization reaction to solve the Travelling Salesman Problem.
- Compare DNA with RNA computing.
- Which are the potentialities and limits of DNA as a memory?
- Which are the factors that rule biological evolution and are suitable to process information?
- Describe the essential features of the Human Immune System, which are used to make computations.
- Why are Cellular Automata good models of Complex Systems?
- Indicate what distinguishes a brain from a Von Neumann computer as far as their computational capabilities are concerned.
- How does Fuzzy logic mimic human ability to compute with words?
- Which are the routes for developing Artificial Intelligence?
- How do proteins compute?
- Describe some methodologies to model phenomena involving Complex Systems.
- Discuss the potentialities of those algorithms that are based on Swarm Intelligence.
- Present the Prisoner's dilemma and its usefulness.
- Which are the conditions required to compute with a particle in a force field that generates a potential energy profile with two wells?
- Is any irreversible computation an irreversible transformation?
- Describe the story of mechanical computing.
- What kind of logic can be processed by using subatomic particles, atoms and molecules?

13.7   KEY WORDS

Big Data; Transistor; Von Neumann architecture; Natural Computing; Artificial life; Systems biology; Systems chemistry; Synthetic biology; Artificial Intelligence; Agent; Prisoner's dilemma; Reversible and irreversible logical operations; Quantum logic; Fuzzy logic.

13.8 HINTS FOR FURTHER READING

- For those who are interested particularly in Systems Biology, Kitano (2002) suggests visiting two websites. First, the website of Systems Biology Markup Language (SBML) (http://sbml.org/Main_Page) that offers a free and open interchange format for computer models of biological processes. Second, the website of the CellML project (https://www.cellml.org/) whose purpose is to store and exchange computer-based mathematical models at a range of resolutions from the subcellular to organism level.
- For those interested in the theory of Complexity in Medicine, it is worthwhile reading the book edited by Sturmberg (2016).
- A survey on nucleic acid-based devices is the review by Krishnan and Simmel (2011). More about DNA origami and their use in chemical robots can be found in the review by Hong et al. (2017).
- More about the thermodynamics of computation in the review by Parrondo et al. (2015).
- To learn more about Information and have more cues to develop the new theory for understanding Complexity, it is really useful to read the book by von Baeyer (2004).
- For learning more about the origin of great new ideas, it is nice to read *Where Good Ideas Come From: The Natural History of Innovation* by Johnson (2010).
- For acquiring an interdisciplinary vision on Complex Systems, it is worthwhile reading *Humanity in a Creative Universe* by the Kauffman (2016). Kauffman sorts through the most significant questions and theories in biology, physics, and philosophy.

13.9 EXERCISES

13.1. According to the IBM website https://www.ibm.com/analytics/us/en/big-data/, Big Data will amount to 43 Trillion Gigabytes by 2020. If we imagine storing all these data in DNA, how much DNA do we need? The data density in bacterial DNA is 10^{19} bits/cm^3 (Extance 2016). The average density of DNA is 1.71 g/cm^3 (Panijpan 1977). Remember that 1 byte corresponds to 8 bits.

13.2. In Xenopus embryos, the cell cycle is driven by a protein circuit centered on the cyclin-dependent protein kinase CDK1 and the anaphase-promoting complex APC (Ferrell et al. 2011). The activation of CDK1 drives the cell into mitosis, whereas the activation of APC drives the cell back out of mitosis. A simple two-component model for the cell-cycle oscillator is shown in Figure 13.28.

Use a Boolean analysis, to predict the dynamics of the system.

13.3. A more complex model to describe the cell cycle of Xenopus embryo (see the previous exercise) consists of three components (Ferrell et al. 2011). Besides CDK1 and APC, there is a third protein that is Polo-like kinase 1 (Plk1), as shown in Figure 13.29.

Plk1 (P) is activated by CDK1 (C), and, in turn, it activates APC (A). APC inhibits CDK1. Use a Boolean analysis, to predict the dynamics of the system.

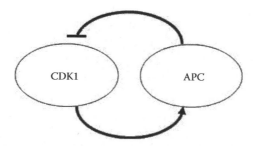

FIGURE 13.28 Mutual relationship between CDK1 and APC.

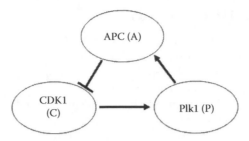

FIGURE 13.29   Model of the cell cycle for the Xenopus embryo.

13.4. Open NetLogo software (Wilensky 1999). Go to Models Library. Select Biology and open the "Ants" Model (Wilensky 1997). Fix "Populations" to 200 and "Diffusion rate" to 20.

   For each pheromone "evaporation-rate" value reported below, run the model five times. For each run, record the number of ticks it takes until all the food is eaten. Remember to click "setup" before each run. Then, average these five numbers. The result is the "average time taken" as a function of the evaporation rate value. What is the value of the evaporation-rates, reported below, which makes the ants quickest (on average) to eat all the food?

   A: "evaporation-rate" = 0
   B: "evaporation-rate" = 5
   C: "evaporation-rate" = 20

13.5. This exercise is an ideal model of inductive reasoning (Arthur 2015). One hundred people decide each week independently whether to go, or not to go, to a bar that offers entertainment on a certain night. Space is limited. The night is enjoyable if the bar is not too crowded, no more than sixty people. There is no sure way to tell the number of people coming in advance. Therefore, every agent goes if he expects fewer than sixty to show up or stays home if he expects more than sixty to go. There is no prior communication among the agents. The only information available is the numbers who went in the past M weeks. This model was inspired by the bar El Farol in Santa Fe, which offers Irish music on Thursday nights (Wilensky and Rand 2015). Let N be the number of strategies (or mental models) each agent has to make predictions. Let t be the current time (i.e., week). The previous weeks are thus $t-1, t-2$, and so on. Let $A(t-1)$ be the attendance at time $t-1, A(t-2)$ that at time $t-2$, and so on. Each strategy predicts the attendance at time t, $S(t)$, by the following equation:

$$S(t) = 100\left[w_1 A(t-1) + w_2 A(t-2) + \ldots + w_M A(t-M) + c \right]$$

where $w_i \in [-1, +1], \forall i = 1, \ldots, M$.

   The N strategies differ in the values of the coefficients w_i. The values of the coefficients w_i are randomly initialized. The attendance history (previous M time steps) is initialized at random, with values between 0 and 99. At each time step t, after making his decision and learning the current attendance $A(t)$, each agent evaluates his best current strategy S^* as the one that minimizes the error:

$$Error(S) = |S(t) - A(t)| + |S(t-1) - A(t-1)| + \ldots + |S(t-M) - A(t-M)|$$

S^* will be used by that agent to make his decision on the next round. If $S^*(t+1) > 60$ (i.e., the overcrowding threshold), the agent does not go; otherwise, he goes.

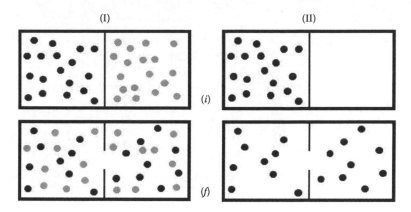

FIGURE 13.30 Mixing of two ideal gases (I) and free expansion of a perfect gas (II) where the state labeled as (i) and (f) represent the initial and final states, respectively.

Open NetLogo software (Wilensky 1999). Go to Models Library; select Social Science and open the "El Farol" Model (Rand and Wilensky 2007). Find a combination of memory-size and number of strategies that reduce the frequency and the extent of over-crowding. How does the attendance behave dynamically over time?

13.6. Determine the variation of the thermodynamic entropy and the information entropy for the following two processes: (I) mixing of two ideal gases and (II) free expansion of a perfect gas from volume V_i to $V_f = 2V_i$ (see Figure 13.30).

13.10 SOLUTIONS OF THE EXERCISES

13.1. The Big Data will amount to 4.3×10^{22} bytes $= 3.44 \times 10^{23}$ bits. If we imagine storing all these data in DNA, the volume needed would be

$$V(DNA) = \frac{3.44 \times 10^{23} \text{ bits}}{10^{19} \text{ bits/cm}^3} = 3.44 \times 10^4 \text{ cm}^3$$

The mass of DNA would be

$$m(DNA) = \frac{3.44 \times 10^4 \text{cm}^3}{1.71 \dfrac{\text{g}}{\text{cm}^3}} = 2 \times 10^4 \text{g} = 20 \text{ kg}$$

It is worthwhile noticing that the Big Data are constantly manipulated to extract useful information from them. Therefore, their storage in DNA is not the best choice because writing and reading information in DNA is not a fast process, yet. However, DNA requires much less volume than a hard disk and flash memory. In fact, the data density of a hard disk (based on magnetism) and flash memory (based on electric charges; the name "flash" refers to the fact that the data can be erased quickly) are one million and one thousand smaller than that of DNA, respectively (Extance 2016).

13.2. According to the Boolean analysis, the two variables, CDK1 and APC, are either entirely ON or entirely OFF. Then, the two-component system has four possible states, indicated in Figure 13.31.

Let us imagine that the system starts from a state wherein it prepares for the process of cell division: $CDK1_{OFF}$ and APC_{OFF} (state 1 in Figure). In the first discrete time step,

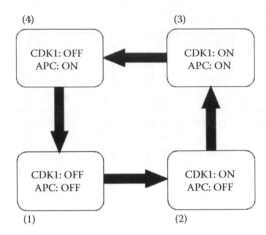

FIGURE 13.31   Possible binary states of CDK1 and APC.

$$
\begin{array}{cc}
 & \begin{array}{cc} \text{CDK1} & \quad \text{APC} \end{array} \\
\begin{array}{c} \text{CDK1} \\[3em] \\ \text{APC} \end{array} &
\left[\begin{array}{cc} 0 & -1 \\[4em] +1 & 0 \end{array}\right]
\end{array}
$$

FIGURE 13.32   Connection matrix for the CDK1-APC system.

CDK1 turns ON because APC is OFF: the system moves from the state 1 to the state 2. Then, the active CDK1 activates APC; thus, the system goes from the state 2 to the state 3. Now, the active APC inactivates CDK1, and the system goes from the state 3 to the state 4. Finally, in the absence of active CDK1, APC becomes inactive, and the system goes from the state 4 to the state 1, completing the cycle. The system will oscillate indefinitely. The transitions of the system can also be determined by using the "connection matrix" reported in Figure 13.32 and representing the state of the two-component system through 2×1 vectors. Here, -1 corresponds to OFF, whereas $+1$ to ON.

Ferrell et al. (2011) have demonstrated that if we use ODEs to describe the dynamics of this two-component system, we find that it gives damped oscillations to a stable steady state.

13.3. For the three-component system, there are 2^3 possible states (see Figure 13.33). If we start with the variables in their OFF state (i.e., the state 1), we get periodic cycling of six states (from 1 to 6). States 7 and 8 are not visited by the periodic cycle. However, they feed into the cycle (see grey arrows) according to the relations existing between the variables. Thus, no matter where the system starts, it will converge to the cycle traced by the black arrows. The cycle is stable.

The evolution of the system can also be calculated using the "connection matrix," shown in Figure 13.34, and 3×1 vectors to represent the states of the system.

The analysis of this three-component system by ODEs gives the same result: a stable limit cycle (Ferrell et al., 2011).

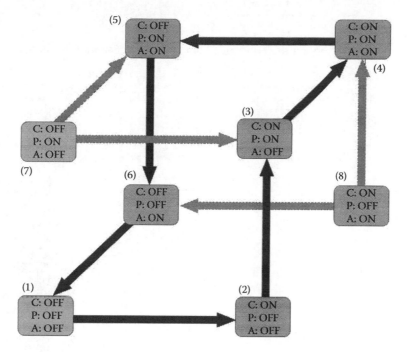

FIGURE 13.33 Possible states for the system consisting of CDK1 (C), Plk1 (P), and APC (A).

	CDK1	PlK1	APC
CDK1	0	0	−1
PlK1	+1	0	0
APC	0	+1	0

FIGURE 13.34 Connection matrix for the CDK1-P-APC system.

13.4. The results are listed in Table 13.3.

The best evaporation-rate is 5. When it is 0, ants have three broad paths to the three sources of food to follow. When the evaporation-rate is 20, ants can trust just in shreds of tracks towards the food, and most of them waste time by roving randomly throughout the entire space available.

13.5. By playing with the number of strategies and the memory size, different dynamics are observable. A list of results is reported in Table 13.4.

According to the results presented in Table 13.4, a good solution is when both memory size and number of strategies are fixed to ten. There are overcrowded outcomes, but the number of people does not exceed too much above sixty, and when there is no overcrowding, the number of people is close to sixty. The bar is never "desert," and this is good for the owner.

13.6. The first principle of thermodynamics allows us to write $dU = dq + dw$. Introducing the second law of thermodynamics, we have $dU = TdS - PdV$. Since the ideal gases are at constant temperature, their internal energies do not change. Therefore, $TdS = PdV$. Introducing

TABLE 13.3

Times Taken by the Ants to Eat the Food as a Function of the Evaporation Rate of Pheromone

Evaporation Rate	Time (ticks)					Average Time	Picture
0	600	790	729	624	775	703.6	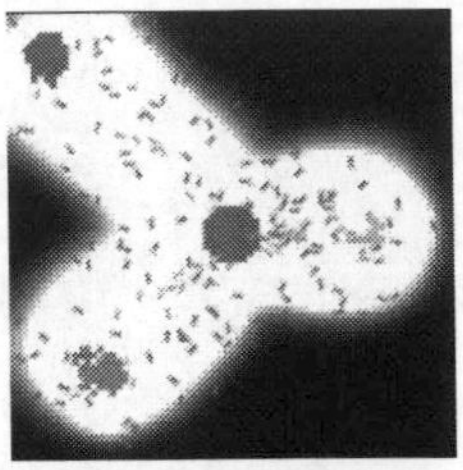
5	470	561	493	504	447	495	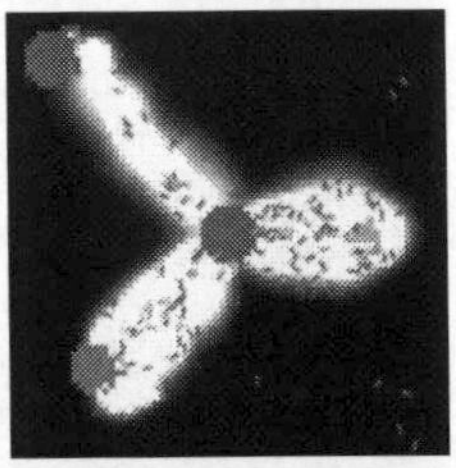
20	1031	1538	1625	1816	1598	1521.6	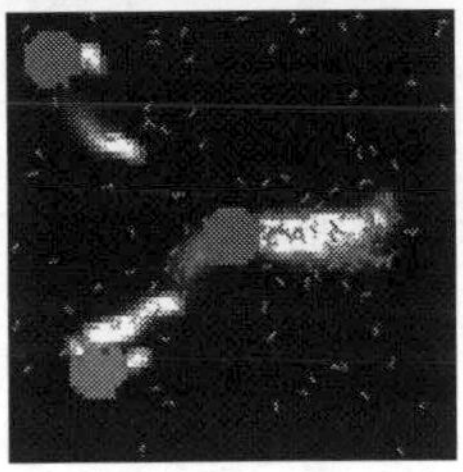

TABLE 13.4

Possible Dynamics of Attendance to the El Farol Bar as Functions of the Memory-size and the Number of Strategies

Memory-size	N° strategies	Dynamics	Attendance
1	1	Fixed point	>70
1	10	Oscillations	] ~85, ~25[
1	20	Oscillations	] ~90, ~10[
5	1	Fixed point or oscillations with small amplitudes	≥70
5	10	Chaotic	] ~70, ~45[
5	20	Chaotic	] ~90, ~20[
10	1	Chaotic (small amplitudes)	≥65
10	10	Chaotic	] ~67, ~50[
10	20	Chaotic	] ~70, ~40[

the equation of state for ideal gases, we obtain that $dS = (nR/V)dV$. For the mixing process (I), the total variation of the thermodynamic entropy is:

$$\Delta S = n_b R ln\left(\frac{V_{bf}}{V_{bi}}\right) + n_g R ln\left(\frac{V_{gf}}{V_{gi}}\right) = (n_b + n_g)Rln2$$

where the subscripts b and g refer to the black and gray gases, respectively.

If we consider the information entropy, the uncertainty (H) that we have about a gas contained in a volume V_i is $H_i = log_2 V_i$. Therefore, in the expansion from V_i to V_f, the increment of uncertainty and hence the variation of the information entropy is $\Delta H = log_2 \left(V_f / V_i \right)$. It is evident the strict relation between the two expressions:

$$(\Delta H)(n_b + n_r)Rln2 = \Delta S$$

Analogously, for the process of free expansion $(\Delta H)(n_b)Rln2 = \Delta S$.

Appendix A: Numerical Solutions of Differential Equations

No human investigation can become real science without going through mathematical people.

Leonardo da Vinci (1452–1519 AD)

In addressing the dynamics of a Complex System, the typical approach consists of the following steps. First, the key variables of the phenomenon are identified, together with the nature of their interactions. Second, differential equations describing the time evolution of the system are formulated. If the system is spatially homogeneous, the differential equations are ordinary (ODEs); otherwise they are partial (PDEs). Third, the differential equations are solved either analytically, graphically, or numerically (most non-linear differential equations cannot be solved analytically). Before the advent of the computer, numerical methods were impractical because they required an exhausting work of hand-calculation. With the implementation of always faster computers, all that has changed. Finally, after determining the fixed points, the stability properties of the steady state's solutions are probed.

Differential equations such as [A.1] are rules describing the time evolution of X as a function of X.

$$\frac{dX}{dt} = f(X) \qquad [A.1]$$

If only first order derivatives are involved, equation [A.1] is a first order differential equation. If the function $f(X)$ is continuous and smooth, i.e., differentiable in every point, then the solution to equation [A.1] exists and is unique.

In this Appendix A, a few methods of solving numerically differential equations, (such as [A.1]), are presented.

A.1 EULER'S METHOD

The most basic explicit method for the numerical integration of ordinary differential equations is Euler's Method. It is named after Leonhard Euler, a Swiss mathematician, and physicist of the eighteenth century. Euler's Method converts the rule [A.1] into values for X. Let us suppose that initially, i.e., for $t = t_0$, $X = X_0$. The derivative is equal to $\left(\frac{dX}{dt}\right)_{t_0} = f(X_0)$. Euler's method assumes that the derivative $\left(\frac{dX}{dt}\right)_{t_0}$ is constant over a certain time interval Δt. Therefore, the next value for X will be given by

$$X(t_0 + \Delta t = t_1) = X_1 = X_0 + f(X_0)\Delta t = X_0 + \left(\frac{dX}{dt}\right)_{t_0} \Delta t \qquad [A.2]$$

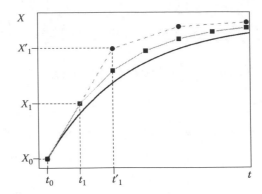

FIGURE A.1   Description of the function $X(t)$ by the Euler's method. The solutions shown as black circles are obtained by selecting a longer Δt than that chosen for the solutions represented as black squares.

Actually, the derivative $f(X)$ is constantly changing, but the assumption that it is constant over the time interval Δt becomes less and less wrong, as Δt becomes closer and closer to zero, i.e., infinitesimal. At time t_1, in X_1 location, the derivative is now equal to $f(X_1)$. We step forward to $X_2 = X(t_1 + \Delta t) = X_1 + f(X_1)\Delta t$, and then we iterate the procedure:

1. We calculate the derivative in X_n
2. We use the rate of change to determine the next value X_{n+1}:

$$X_{n+1} = X(t_n + \Delta t) = X_n + f(X_n)\Delta t \qquad \text{[A.3]}$$

until we have a satisfactory number of solutions.

A visualization of the possible solutions achievable by Euler's Method is shown in Figure A.1 where a non-linear function $X(t)$ is plotted as a continuous black curve. The squared and the circle dots represent the solutions of the differential equation [A.1] obtained by applying Euler's Method with two different time intervals. The black circles are obtained with a longer Δt. It is evident that the shorter the Δt, the more accurate the description of the rule.

A.2   ALTERNATIVE METHODS

An improvement on the Euler's Method was proposed by the German mathematicians Carl Runge and Martin Kutta at the beginning of the twentieth century. The Runge-Kutta Method samples the rate of change at several points along the interval Δt and averages them. In particular, the fourth-order Runge-Kutta Method, implemented in MATLAB through the command ode45, requires calculating the following four amounts

$$k_1 = f(X_n)\Delta t$$

$$k_2 = f\left(X_n + \frac{1}{2}k_1\right)\Delta t$$

$$k_3 = f\left(X_n + \frac{1}{2}k_2\right)\Delta t \qquad \text{[A.4]}$$

$$k_4 = f(X_n + k_3)\Delta t$$

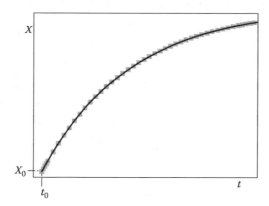

FIGURE A.2 The gray squares represent the numerical solution of the function depicted by a continuous black curve.[1] They describe the curve much better than the numerical results obtained by the Euler's Method (see Figure A.1).

to find X_{n+1} from X_n. In fact,

$$X_{n+1} = X_n + \frac{1}{6}\left(k_1 + 2k_2 + 2k_3 + k_4\right) \qquad [A.5]$$

This method gives much more accurate results than Euler's Method as it can be easily inferred by looking at Figure A.2.

Some differential equations may be more critical, and they may require an adaptive step size. In fact, there are methods where Δt is automatically adjusted on the fly: Δt is made small where the derivative changes rapidly and conversely, it is made large when the derivative changes slowly.

A.3 HINTS FOR FURTHER READINGS

More detailed information about differential equations and their numerical solutions can be found in many textbooks. See, for example, the book *Differential Equations* by Blanchard et al. (2006) and the book by Morton and Mayers (2005) for the numerical solution of partial differential equations.

---

[1] The syntax for the application of fourth-order Runge-Kutta method in MATLAB is the following one:

```
function dy = heating(t,y)
dy = zeros(1,1);
dy = 0.2*(20-y);
```

Then, you must paste in the Command Window the following string:

```
[t,y] = ode45('heating', [0 12], 5);
```

Appendix B: The Maximum Entropy Method

> The real voyage of discovery consists, not in seeking new landscapes, but in having new eyes.
>
> **Marcel Proust (1871–1922 AD)**

The Galilean scientific method is grounded on the collection of experimental data, their analysis, and interpretation. In the time-resolved spectroscopies, the time-evolving signal of a sample recorded upon pulse-excitation is a convolution of the pulse shape of the excitation (IRF) with the response of the molecular system ($N(t)$) (see Figure B.1).

After deconvolution, the experimental data should be expressed as a weighted infinite sum of exponentials:

$$N(t) = \int_0^\infty \alpha(\tau) e^{-t/\tau} d\tau \qquad [\text{B.1}]$$

Our computational task is to determine the spectrum of $\alpha(\tau)$ values having inevitably noisy data available. Mathematically, $\alpha(\tau)$ is the inverse Laplace transform of the measured signal deconvolved from the IRF. Although deconvolution is well-conditioned and relatively stable, inverting the Laplace transform is ill conditioned for a noisy signal. This statement means that small errors in the measurement of $R(t)$ can lead to huge errors in the reconstruction of $\alpha(\tau)$. In other words, the ill-conditioning leads to a multiplicity of allowable solutions. The ensemble of all possible shapes of $\alpha(\tau)$ is displayed as a rectangle labeled as set A in Figure B.2. Among all the elements of set A, only a subset of them agrees with the noisy data set shown on the left of Figure B.2. It is the set labeled as B, confined inside the dotted boundaries. Some of these spectra of set B are devoid of physical meaning (e.g., corresponding to negative concentrations). Therefore, set B is partitioned into two subsets: subset B" contains the unphysical spectra $\alpha(\tau)$ to be rejected, whereas subset B' contains the feasible spectra $\alpha(\tau)$. All the elements of subset B' fit the data and are physically allowable. Among them, we need to choose just one element. This operation is possible by maximizing some function of the spectra, $h[\alpha(\tau)]$. The function $h[\alpha(\tau)]$ should be chosen to introduce the fewest artifacts. For this purpose, we use the Maximum Entropy Method (MEM).

The roots of the Maximum Entropy Method reside in the probability theory. Statisticians are divided into two schools of thought regarding the interpretation of probability: the "classical," "frequentist," or "objective" school instituted by Fisher in the 1940s, and the "subjective" or "Bayesian" school, named after Thomas Bayes who proposed it around 1750 (Brochon 1994). The "objective" school of thought regards the probability of an event as an objective property of that event. It is possible to measure that probability by determining the frequency of positive events in an infinite number of trials. In calculating a probability distribution, the "objectivists" believe that they are making predictions which are in principle verifiable, just as they were dealing with deterministic events of classical mechanics. The question that an "objectivist" raises when he judges a probability distribution $p(x)$ is: "Does it correctly describe the observable fluctuations of x?"

The "subjective" school of thought regards the probability as personal degree of belief: the probability of an event is merely a formal expression of our expectation that the event will or did occur,

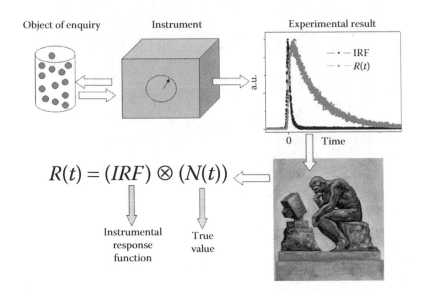

Object of enquiry   Instrument   Experimental result

$$R(t) = (IRF) \otimes (N(t))$$

Instrumental
response
function

True
value

FIGURE B.1 Schematic representation of the steps of a time-resolved measurement. The sample is the object of inquiry; the instrument allows to collect a time-resolved signal $R(t)$ that is a convolution of the Instrumental Response Function (IRF) and the molecular kinetics $N(t)$. The analysis of $R(t)$ is carried out by the operator with the help of some computational tools.

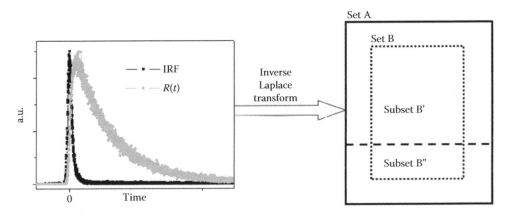

FIGURE B.2 Partition of all possible spectra $\alpha(\tau)$ contained in set A, into two ensembles: set B containing all the spectra agreeing with the experimental data and set not B ($\bar{B}$) containing the remaining elements. Set B is further partitioned into two subsets: B' containing the spectra with physical meaning and B" containing all the spectra devoid of physical significance.

based on whatever information is available. The test of a good subjective probability distribution $p(x)$ is: "Does it correctly represent our state of knowledge as to the value of x?" (Jaynes 1957). The Bayesian view of probability is the source of MEM.

The key aim of scientific data analysis is to determine the probability density of $h[\alpha(\tau)]$ given some data D, i.e., the conditional probability $\Pr(h|D)$. According to the Bayes' theorem, the probability of D and h, $\Pr(h,D)$ is given by

$$\Pr(h,D) = \Pr(h)Pr(D|h) = \Pr(D)\Pr(h|D)$$ [B.2]

The Bayes' theorem can be rearranged in the following form:

$$Pr(h|D) = \frac{Pr(h)\, Pr(D|h)}{Pr(D)}$$ [B.3]

The term $Pr(D|h)$ is the "likelihood:" it is the probability of obtaining some data D if the true spectrum h is known. In general, the likelihood expresses the instrumental response, its resolution and uncertainty. In the case of Gaussian noise statistics, the likelihood is

$$Pr(D|h) = \frac{e^{-\frac{1}{2}\chi^2}}{Z_l}$$ [B.4]

with

$$\frac{1}{2}\chi^2 = \frac{1}{2}(D - Rh)^T (\sigma^{-2})(D - Rh)$$ [B.5]

where:
    Rh is the calculated data from the spectrum h;
    σ^{-2} is the covariance matrix for the data;
    Z_l is the normalization factor.

The term $Pr(D)$ of equation [B.3] is the "plausibility" of the data based on our prior knowledge of the system. If the prior knowledge of the system maintains constant, this term is like a normalization factor (Z_D). Otherwise, it can be used to discriminate between models in which the data would have a different probability of occurring, for example, between a set of discrete or continuous distribution of exponential terms.

The term $Pr(h)$ is the "prior probability," and it codifies our knowledge about a possible spectrum before collecting experimental data. It is expressed as

$$Pr(h) \propto e^S$$ [B.6]

with S being the Information Entropy:

$$S(h) = -\int h(\alpha) \log\left[h(\alpha)\right] d\alpha$$ [B.7]

Finally, the term $Pr(h|D)$ is the "posterior probability." A single answer to the inverse Laplace transform problem can be obtained by maximizing $Pr(h|D)$. The "posterior probability" is

$$Pr(h|D) = \frac{e^{S - \frac{1}{2}\chi^2}}{Z_l Z_D} = \frac{e^Q}{Z_l Z_D}$$ [B.8]

Maximizing $Pr(h|D)$ means finding the maximum of the exponent $Q = S - (1/2)\chi^2$. Q is maximized when the Entropy S is maximized whereas χ^2 is minimized. MEM becomes a "tug of war" between maximizing the Entropy S and minimizing the χ^2 (see Figure B.3).

Mathematically, the Information Entropy has the important property that when maximizing it, no possibilities are excluded. In other words, maximizing S in the absence of the experimental data, we impose no correlations on the spectrum $\alpha(\tau)$; we choose the spectrum that is minimally structured or "maximally noncommittal." This concept can be illustrated by a funny numerical example: the so-called "Kangaroo problem" (see Figure B.4). The information available is that (1/3)

$$S(h) = \int [-h(\tau)\log h(\tau)]d\tau \qquad \chi^2 = \frac{1}{n^0_{data}} \sum_{i=1}^{n^0_{data}} \left(\frac{datum_i - fit_i}{\sigma_i}\right)^2$$

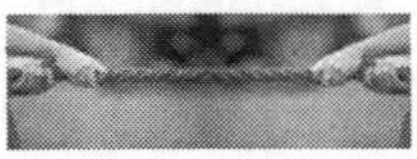

FIGURE B.3 The Maximum Entropy Method is a "tug of war" between the Information Entropy and the χ^2.

	Left-handed	
	True	False
Blue-eyed True	p_x	p_1
Blue-eyed False	p_2	p_3

	Left-handed	
	True	False
Blue-eyed True	x	$\frac{1}{3} - x$
Blue-eyed False	$\frac{1}{3} - x$	$\frac{1}{3} + x$

FIGURE B.4 The example of blue-eyed and left-handed kangaroos.

of kangaroos have blue eyes, and (1/3) are left-handed. The question is how many kangaroos are both left-handed and blue-eyed. Of course, the two properties are unconnected (as far as we know, unless one day we will discover that the genes responsible for blue-eyes are somehow correlated with the genes responsible for left-handedness). We have four possibilities shown in the left Table of Figure B.4. The normalization imposes that

$$p_x + p_1 + p_2 + p_3 = 1 \qquad [B.9]$$

We know that

$$p_x + p_1 = \left(\frac{1}{3}\right)$$
$$p_x + p_2 = \left(\frac{1}{3}\right) \qquad [B.10]$$

If we fix $p_x = x$, p_1, and p_2 will be $(1/3 - x)$, and p_3 will be $(1/3 + x)$, as shown in the table on the right of Figure B.4. According to the theory, the probability of two independent, uncorrelated events is given by the product of their single probabilities. In our case, it is expected that the probability to have left-handed and blue-eyed kangaroos is $(1/3)(1/3)$, i.e., $(1/9)$. Is there some function of the probabilities, p_i, which, when maximized, yields the maximally noncommittal result? The answer is yes, and it is just the Information Entropy function:

$$S = -\sum_i p_i \ln p_i = -\left[x \ln x + 2\left(\frac{1}{3} - x\right)\ln\left(\frac{1}{3} - x\right) + \left(\frac{1}{3} + x\right)\ln\left(\frac{1}{3} + x\right) \right] \qquad [B.11]$$

If we determine the maximum of S with respect to x, we achieve $x = 1/9$.

Try by yourself, just as an exercise.

Maximizing other functions of p_i imposes correlation between the eyes' color and handedness of kangaroos. For example, before the work by Shannon, other information measures had been proposed. One of them was $-\sum_i p_i^2$. Such function does not give the same result of the Shannon Entropy. In fact, maximizing $-\sum_i p_i^2$, we obtain $x = 1/12$, i.e., a negative correlation between the two uncorrelated properties: blue-eyes and left-handedness.

Before ending this Appendix, it is important to present another consideration about the "likelihood" function of equation [B.3]. When the experimental technique, chosen to collect the time-resolved signal, taints the data with "normal" noise, the Gaussian distribution is the right function, the nonlinear least-squares a good fitting procedure, and the minimization of χ^2 the right way to fit the data. However, there are techniques that are better described by the Poisson statistics. An example is the Time-Correlated Single Photon Counting technique. In such technique, the time scale is partitioned in n channels, and the luminescence intensity decay curve is built by collecting the number of independent photons emitted in each channel, $c_1, c_2, ..., c_n$. Such counts follow the Poisson statistics.

In statistics, there exist three probability distributions. The parent distribution is the Binomial one (see Figure B.5) representing the number of successes in a sequence of n-independent yes/no experiments, each of which yields the successful outcome with probability p, and the negative result with probability $q = 1 - p$. It is an asymmetric distribution unless $p = 1/2$. If we repeat the experiment n-times, the average value of the positive outcome is $\bar{v} = np$, and its standard deviation is $\sigma = \sqrt{npq}$. When the number of experiments is enormous, i.e., $n \to \infty$, the Binomial distribution becomes symmetric. It can be substituted by the Gaussian distribution (see Figure B.5) that is centered at $v = N$, which is also the mean value; σ is its standard deviation that defines the width of the distribution. When the number of experiments is huge, i.e., $n \to \infty$, and the probability of the positive result is tiny, i.e., $p \to 0$, the Binomial distribution can be replaced by the Poisson distribution (presented in Figure B.5), wherein μ is its average value of v, and $\sigma = \sqrt{\mu}$ is the standard deviation. The Poisson distribution describes appropriately all those experiments where random events occurring with small probabilities p are considered. Examples of such events are radioactive decays and the counting of emitted photons in a broad array of channels. Usually, the Poisson distribution is asymmetric, but when $\mu \to \infty$, it becomes symmetric, and it can be replaced by a Gaussian distribution having the same average and standard deviation.

In the case of the Single-Photon Counting Technique (Bajzer and Prendergast 1992), the probability of the observed counts c_i, $p\left(c_i; \alpha(\tau)\right)$, follows the Poisson distribution:

$$p\left(c_i; \alpha\right) = \frac{e^{-\langle c_i \rangle} \langle c_i \rangle^{c_i}}{c_i!} \qquad [B.12]$$

where $\langle c_i \rangle$ is the expected value of the number of counts in the i-th channel. Since the counts independent, the likelihood will be the joint probability

$$Pr(D|h) = \prod_i p(c_i; \alpha) \qquad [B.13]$$

The value $\langle c_i \rangle$ is modeled by the function Rh_i for $i = 1,...,n$. The best estimates of the Rh_i will be those values that maximize the likelihood and the entropy. Note that the log-likelihood function, $\ln\left[Pr(D|h)\right]$, attains its maximum for the same value of h, as the likelihood function. It is

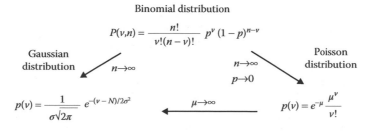

FIGURE B.5 Formula of the three distributions used in statistics to describe random events.

customary to determine the best h by minimizing $-\ln\lceil Pr(D|h)\rceil$, which is equivalent to minimizing the Poisson deviance

$$D(\alpha) = 2\sum_i \left\{ c_i ln\left[c_i / Rh_i \right] - \left[c_i - Rh_i \right] \right\}$$ [B.14]

The Poisson deviance is close to the Gaussian χ^2 when the counts are very high. This result is in agreement with the consideration that a Poisson distribution can be replaced by a Gaussian one when the average value in each channel goes to infinity.

B.1 HINTS FOR FURTHER READING

Further information regarding the Maximum Entropy Method applied to time-resolved spectroscopy, and image processing can be found on the website of Center for Molecular Modeling of the National Institutes of Health (https://cmm.cit.nih.gov/maxent/). It is possible to download the software MemExp by P. J. Steinbach, for the analysis of the kinetic data.

Appendix C: Fourier Transform of Waveforms

Computers are useless. They can only give you answers.

Pablo Picasso (1881–1973 AD)

Almost everything in the universe can be described as a waveform. Suffice to think that quantum particles are wave packets. A waveform is a curve showing the shape of a wave as a function of time or space or some other variables.

The Fourier Transform (Bracewell 1986) is a powerful mathematical tool that allows describing any waveform as a sum of sinusoids, i.e., as a sum of sine and cosine functions.

Let us remember the general expression and features of a sinusoid, like a sine function:

$$y = A\sin(2\pi\omega t + \varphi) \tag{C.1}$$

The function y has a frequency ω that is the inverse of the time it takes to complete one period of oscillation. The term A in equation [C.1] is the amplitude of y (see Figure C.1), and φ is its phase. In the bottom graph of Figure C.1, the phase difference ($\Delta\varphi$) between two sinusoids, having the same frequency and the same amplitude, is highlighted.

When we encounter a waveform like that depicted in Figure C.2, we ask ourselves: "If we describe the signal S as a sum of sinusoids having different frequencies, what frequencies are needed, and what is the distribution of the amplitudes and phases as the functions of the frequency?" This question has an answer because Fourier's Theorem states that any signal of zero mean value can be represented as a unique sum of sinusoids.

When we sample the signal S with a time resolution of Δt (see Figure C.2), and we collect N data, the total time window is $TW = N * \Delta t$. Inevitably, the smallest period we can infer from our set of data is $T_{MIN} = 2\Delta t$. In other words, the largest frequency we can measure in our series of sinusoids describing the data is $v_{MAX} = 1/2\Delta t$. If the original signal S contains periods shorter than $T_{MIN} = 2\Delta t$ or frequencies higher than v_{MAX}, our measurement is said to be aliased. If we want to evaluate shortest periods, we must reduce the time interval Δt. The longest period we can obtain unambiguously from a windowed trend is one that lasts the total sampling time, that is $T_{MAX} = N * \Delta t$; in other words, the smallest frequency we can infer is $v_{MIN} = 1/N\Delta t$. In addition, T_{MIN} defines our resolution in the determination of the period. Analogously, v_{MIN} defines the resolution in the frequency determination. It derives that the maximum number of the sinusoids in the Fourier Transform will be $N/2$. This statement makes sense because we have N data, and for the definition of each sinusoid we need to determine two parameters: its amplitude and its phase. For each frequency $v = n/N\Delta t$, with $n = 1, 2, ..., N/2$, its amplitude is

$$A_n = \sqrt{a_n^2 + b_n^2} \tag{C.2}$$

and its phase is

$$\varphi_n = -\arctan\left(\frac{b_n}{a_n}\right) \tag{C.3}$$

where the coefficients, a_n and b_n, are given in terms of the original signal $S(t)$ as

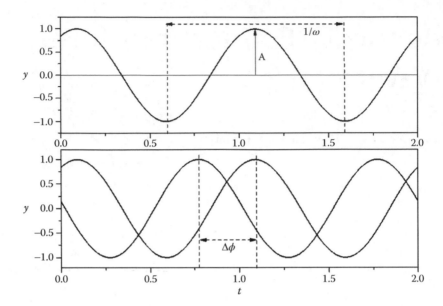

FIGURE C.1   Shape and features of the sinusoids.

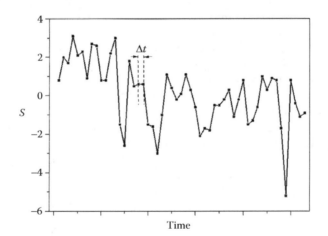

FIGURE C.2   An example of a waveform regarding the signal S as a function of time. Δt represents the sampling period.

$$a_n = \frac{2}{N}\sum_{i=1}^{N} S(i\Delta t)\cos\left(\frac{2\pi n i}{N}\right) \qquad b_n = \frac{2}{N}\sum_{i=1}^{N} S(i\Delta t)\sin\left(\frac{2\pi n i}{N}\right) \qquad [C.4]$$

Expressions [C.4], [C.2], and [C.3] are referred to as the Discrete Fourier Transform (DFT).

The original set of data $S(i\Delta t)$ is finally expressed in terms of the sum of the $N/2$ sinusoids:

$$S(i\Delta t) = \frac{A_0}{2} + \sum_{n=1}^{N/2} A_n \cos(2\pi v_n i\Delta t + \varphi_n) \qquad [C.5]$$

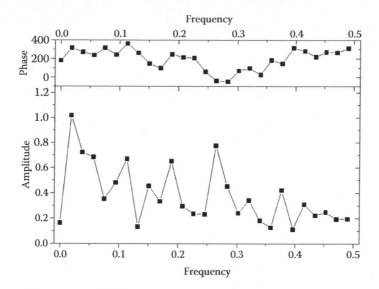

FIGURE C.3   Discrete Fourier Transform of the data plotted in Figure C.2.

Equation [C.5] is called the Inverse Discrete Fourier Transform. The term $A_0/2$ represents any mean value of the signal which is not accounted for in the other sinusoids. It comes from the definitions of a_n and b_n (cf. equation [C.4]) when $n = 0$, which corresponds to frequency 0.

When we decompose our signal S into its component sinusoids, the resulting graph reporting amplitude or phase as a function of frequency is referred to as its Fourier spectrum. In Figure C.3, we report the Fourier spectrum of the 53 data plotted in Figure C.2. It consists of 27 frequencies included between 0 and 0.5.

Usually, the DFT is computed by an algorithm known as the Fast Fourier Transform (FFT). The key advantage of the FFT over the DFT is that the operational complexity decreases from $O(N^2)$ for a DFT to $O(N\log_2(N))$ for the FFT.

Appendix D: Errors and Uncertainties in Laboratory Experiments

> Those sciences are vain and full of errors which are not born from experiment, the mother of certainty.

Leonardo da Vinci (1452–1519 AD)

Science is said to be exact not because it is based on infinitely exact assertions but because its rigorous methodology allows estimating the extent of the uncertainty associated with any determination.

D.1 ERRORS IN DIRECT MEASUREMENTS

Many scientific assertions are grounded in experiments. Leonardo da Vinci was used to say that scientific knowledge without experiments is vain and full of mistakes. However, we are aware that even when we perform experiments, we make errors.

Usually, an experiment involves four protagonists: the observer raising a question and looking for an answer; the object or the phenomenon that is inquired; an instrument, and the surrounding environment (see Figure D.1).

The Galilean experimental method requires that the object or the phenomenon under study is analyzed inside a laboratory where it is easier to control variables and parameters. When we face Complex Systems, this is not always feasible. Whether the environment is the laboratory or the real world, it often plays a relevant role. In fact, the environment can influence the phenomenon under study (especially if it obeys nonlinear laws), and it can also affect the performances of the observer, and the instrument. The use of an instrument is mandatory for the observer; the device allows making quantitative, objective, and reproducible determinations. Any measurement, however accurate and performed carefully, is tainted by errors, and the results have inevitable uncertainties. There are two principal sources of errors and uncertainties: one is the instrument that we use, and the other is the operator.[1] There are two types of sources of uncertainty: they are the systematic and the random errors.

D.2 SYSTEMATIC ERRORS

Any instrument determines the value of a variable with a definite degree of accuracy. When we choose an instrument to collect data, we must be aware of the error we make by using it. Almost all direct measurements require reading either a marked scale (e.g., the marked scale of a ruler, a chronometer, or a pipette) or a digital display (e.g., the digital display of a spectrophotometer).

When we read a marked scale, the main source of error is the interpolation between the scale markings. For instance, let us measure the length of the clip shown in Figure D.2 with a ruler graduated in millimeters. We might reasonably decide that the clip's length is undoubtedly closer to 3.2 cm than it is to 3.1 or 3.3 cm, but no more precise reading is possible. We conclude that our best estimate is 3.20 cm, and the uncertainty of our determination is in the range 3.15 and 3.25 cm.

---

[1] The contribution of the environment is included in those of the observer and instrument.

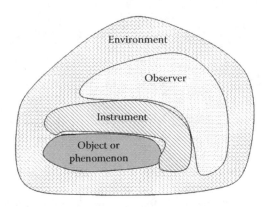

FIGURE D.1   Protagonists of a measurement.

FIGURE D.2   Determination of the length of a clip by a marked ruler.

The correct way to present the result of the measurement will be $l_{best} = (3.20 \pm 0.05)$ cm. More in general, when we measure a quantity x_{true}, we collect our best estimate x_{best} and we specify the range (Δx) within which we are confident x_{true} lies:

$$x_{best} \pm \Delta x \qquad\qquad\qquad\qquad [\text{D.1}]$$

The term Δx is the absolute error of our measurement, and it is usually stated with one significant figure. At most, with two significant figures, but no more. Moreover, the last significant figure in x_{best} must be of the same order of magnitude of the uncertainty, that is in the same decimal position. If we want to increase the accuracy of our determination and reduce Δx, we need a finer instrument. For instance, in the case of the length of the clip, we may use a Vernier caliper, like that shown in Figure D.3. First, we close the jaws lightly on the clip. Then, we measure its length. In the Vernier caliper, there are two marked scales: a fixed one and a sliding one. The fixed scale is identical to the scale of the ruler of Figure D.2. The separation between two consecutive marks is equal to 1 mm. The sliding scale has 20 marks. The left-most tick mark on the sliding scale will let us read from the fixed scale the number of whole millimeters that the jaws are opened. In Figure D.3, it is between 3.2 and 3.3 cm. For the estimation of the clip's length at the level of one-tenths of mm, we must determine the number of the first mark of the sliding scale that is aligned with a mark of the fixed scale. In Figure D.3, it is the seventh mark that corresponds to 0.30 mm. The uncertainty, as indicated onto the caliper, amounts to 0.05 mm. Therefore, the final best estimate of the clip's length is $l_{best} = (3.230 \pm 0.005)$ cm. Note that the use of the caliper has increased the accuracy of our determination of one order of magnitude.

The quality of a collected datum is expressed by the ratio of Δx over x_{best}, which is called the relative uncertainty (ε_{rel}):

$$\varepsilon_{rel} = \frac{\Delta x}{x_{best}} \qquad\qquad\qquad\qquad [\text{D.2}]$$

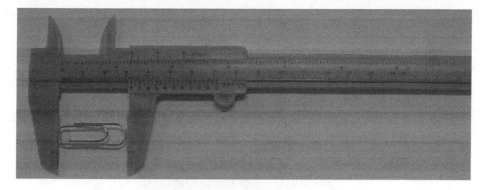

FIGURE D.3 Determination of the length of a clip by a Vernier caliper.

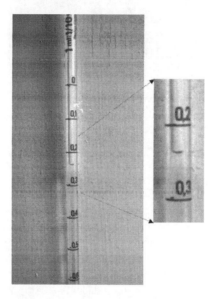

FIGURE D.4 Reading of a scale whose marks are rather far apart. The distance between two consecutive marks is (1.85 ± 0.05) cm.

If we multiply ε_{rel} by 100, we obtain the percentage relative error. The percentage relative error is $\varepsilon_{rel} = 1.56\%$ with the ruler, whereas it is $\varepsilon_{rel} = 0.155\%$ with the caliper. We confirm that the use of the caliper reduces the uncertainty of one order of magnitude.

Usually, the uncertainty in reading a scale when the marks are fairly close together is half the spacing. If the scale markings are rather far apart, like those of the pipette shown in Figure D.4, we might reasonably assume that we can read to one-fifth of a division. The volume of the liquid taken with the pipette of Figure D.4 is reasonably estimated to be (0.24 ± 0.02) mL.

The error we make when we read a value on a digital display is usually of one unit on the last significant figure, unless indicated differently by the specifications of the instrument we are using.

The errors described so far generated by the inherent inaccuracy of the instrument we use, are said systematic because they are always present, and their absolute extent (Δx) is constant and known, even before making the measurement.

An instrument may be a source of another type of systematic error that cannot be known unless we measure standard quantities, or we repeat our measurement by using different devices. Possible miscalibration of our equipment generates such a second type of systematic instrumental error.

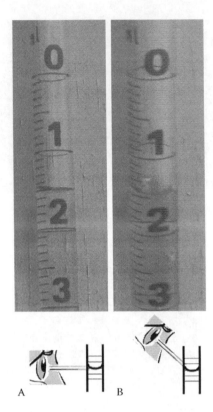

FIGURE D.5 Parallax error in reading a pipette's scale. Case A represents the correct viewing angle, whereas case B represents the incorrect one.

Moreover, the operator may introduce systematic errors in the measurements, as well. For instance, the right way of reading a volume on a scale of a pipette is to put our eyes at the height of the liquid's meniscus, like in situation A of Figure D.5. If we read the volume systematically by viewing the meniscus from above (see situation B of Figure D.5), we determine the volume constantly by defect. In situation B of Figure D.5, we will read (1.40 ± 0.05) mL rather than (1.50 ± 0.05) mL. The introduction of such uncertainty, called parallax error, might be even worse depending on the angle of viewing and the type of scale.

D.3   RANDOM UNCERTAINTIES

It may also occur that sometimes we read a scale assuming the right position, but in some others not. In this case, not all the measurements are affected by the same parallax error, but only some of them. The uncertainty introduced in such a way is not systematic but random.

The instrument itself may become a source of random errors. For instance, a chronometer may run faster or slower accidentally; a photomultiplier may change abruptly and temporarily its sensitivity to photons. Often, it is the environment that can contribute to the occurrence of random errors. In fact, fluctuations in the surrounding conditions may influence the performances of the instrument and the observer.

The only way for estimating the contribution of the random error is to repeat the same measurement many times. For example, imagine determining the lifetime of a long-lived phosphor by using a chronometer. A source of error is our reaction time in starting and stopping our clock: if we delay in starting or we anticipate in stopping, we underestimate the lifetime; if we delay in stopping or we anticipate in starting, we overestimate its lifetime. Since all these possibilities are equally likely, our measurements will spread statistically. We will have a set of results: $\tau_1, \tau_2, \ldots, \tau_n$ whose limiting distribution is Gaussian. The best estimate of τ_{true} is the center of the Gaussian distribution (read Box 1 of this Appendix), which is the average when $n \to \infty$:

$$\tau_{best} = \overline{\tau} = \frac{\sum_{i=1}^{n} \tau_i}{n} \qquad [D.3]$$

The uncertainty on the average[2] is the standard deviation of the mean:

$$\sigma_{\overline{\tau}} = \frac{\sqrt{\frac{1}{n-1} \sum_{i=1}^{n} (\tau_i - \overline{\tau})^2}}{\sqrt{n}} \qquad [D.4]$$

BOX 1   THE GAUSSIAN DISTRIBUTION

The form of the Gaussian function describing how the measurements of the τ_{true} value are distributed over the variable τ is $f(\tau) = \left(1/\sigma\sqrt{2\pi}\right)e^{-\frac{(\tau - \tau_{true})^2}{2\sigma^2}}$.

Why is the average the best estimate of τ_{true}?

If we collect n values $(\tau_1, \tau_2, \ldots, \tau_n)$, the probability of getting τ_1 is proportional to

$$f(\tau_1) \propto \frac{1}{\sigma\sqrt{2\pi}} e^{-\frac{(\tau_1 - \tau_{true})^2}{2\sigma^2}}$$

The probability of obtaining τ_2 is proportional to

$$f(\tau_2) \propto \frac{1}{\sigma\sqrt{2\pi}} e^{-\frac{(\tau_2 - \tau_{true})^2}{2\sigma^2}} \text{, and so on.}$$

The probability of getting τ_n is proportional to $f(\tau_n) \propto \left(1/\sigma\sqrt{2\pi}\right)e^{-\frac{(\tau_n - \tau_{true})^2}{2\sigma^2}}$.

The probability of observing the entire set of n values $(\tau_1, \tau_2, \ldots, \tau_n)$ is just the product of the probabilities of observing each of them when all the n measurements are independent events.

$$prob(\tau_1, \tau_2, \ldots, \tau_n) \propto f(\tau_1)f(\tau_2)\ldots f(\tau_n) = \left(\frac{1}{\sigma\sqrt{2\pi}}\right)^n e^{-\frac{\sum_{i=1}^{n}(\tau_i - \tau_{true})^2}{2\sigma^2}}$$

According to the principle of the Maximum Likelihood, the best estimate of τ_{true} is that value for which $prob(\tau_1, \tau_2, \ldots, \tau_n)$ is maximized. Obviously, $prob(\tau_1, \tau_2, \ldots, \tau_n)$ is maximum if the exponent $\sum_{i=1}^{n}(\tau_i - \tau_{true})^2/2\sigma^2$ is minimum. If we differentiate the exponent with respect to τ_{true} and we set the derivative equal to zero, we achieve that $\tau_{best} = \overline{\tau} = \sum_{i=1}^{n}\tau_i/n$.

(Continued)

---

[2] The uncertainty on the average value $\overline{\tau}$ is the standard deviation of the mean, whereas the uncertainty on the single determination τ_i is the standard deviation $\sigma = \sqrt{\frac{1}{n-1}\sum_{i=1}^{n}(\tau_i - \overline{\tau})^2}$.

BOX 1 (Continued)   THE GAUSSIAN DISTRIBUTION

By using the principle of the Maximum Likelihood, it is possible to also find the best estimate for σ, which is the width of the limiting distribution. If we differentiate $prob(\tau_1, \tau_2, ..., \tau_n)$ with respect to σ, and we set the derivative equal to zero, we confirm that σ_{best} is the standard deviation of the mean. In the figure within this box, there are two Gaussian distributions, both centered at the same τ_{true}, but having two different widths, i.e., two different σ values.

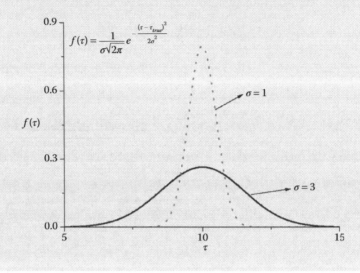

$$f(\tau) = \frac{1}{\sigma\sqrt{2\pi}} e^{-\frac{(\tau - \tau_{true})^2}{2\sigma^2}}$$

D.4   TOTAL UNCERTAINTY IN DIRECT MEASUREMENTS

In synthesis, all the direct measurements are subjected to both random and systematic errors (see Figure D.6). Some random errors are generated by the observer, whereas some others by the equipment. The unique strategy to cope with them is to repeat and repeat, as much as possible, the experimental determination. Also, the systematic errors may derive from both the observer and the instrument. The remedy consists in changing the operator or the experimental device. Whatever is the accuracy of a device, we cannot completely wipe out the systematic instrumental error due to the inherent uncertainty of any apparatus.

This means that our measurements of x_{true} can never be represented as dots along the x-axis (see Figure D.7) but as circles, whose area shrinks by upgrading our experimental apparatus. Figure D.7 shows four possible situations that may arise when we determine through direct measurements the value x_{true} of a variable. Case (1) occurs when the random error (ε_{random}) is small and when the only type of systematic error (ε_{sys}) is that generated by the inherent uncertainty of the instrument. In fact, the measurements are represented by small circles distributed symmetrically with respect to x_{true} and pretty close to it. Such determinations are said to be accurate and precise. Case (2) comes about when the random error is small, the inherent uncertainty is tiny, but our measurements are affected by miscalibration. In case (2) the determinations are still precise (ε_{random} is small like in case [1]) but they are inaccurate (the overall ε_{sys} is large). Case (3) refers to a situation where our measurements suffer from scarce precision, although they are accurate because the instrument has small inherent inaccuracy and there are no problems with miscalibrations. Case (4) represents the worse situation among all because the distribution and size of the circles along x-axis reveal that our determinations are imprecise (they are spread in a broad range of x values) and inaccurate (they are big circles, all located at values smaller than x_{true}).

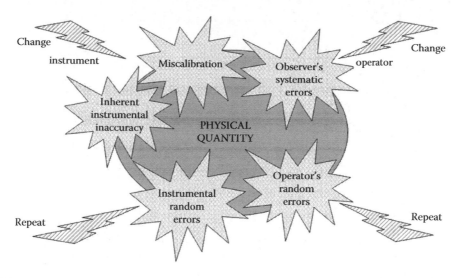

FIGURE D.6   Types of uncertainties in a direct measurement and their remedies.

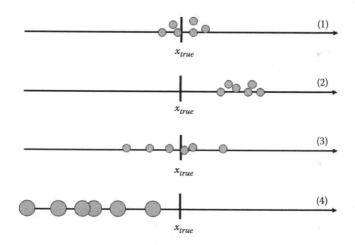

FIGURE D.7   Four possible situations we may encounter when we measure x_{true} more than once.

When we know both the random (ε_{random}) and the systematic uncertainty (ε_{sys}), and they are comparable, the reasonable estimation of the total uncertainty will be the sum of the two contributions:

$$x_{best} + \Delta x = \overline{x} + \varepsilon_{random} + \varepsilon_{sys} = \overline{x} + \sigma_{\overline{x}} + \varepsilon_{sys} \qquad [D.5]$$

From equations [D.5] and [D.4], it is evident that if we repeat many times our measurement, i.e., then $n \to \infty$, ε_{random} can become very tiny and negligible with respect to ε_{sys}. However, even if we repeat an infinite number of times our measurement, the total uncertainty in x_{best} cannot be smaller than ε_{sys} that is the inherent inaccuracy of our experimental device.

D.5   PROPAGATION OF THE UNCERTAINTIES

Many variables are measured indirectly. They need to be calculated from the values of variables that have been measured directly. For instance, the speed of a vehicle requires the direct determination of the traveled distance and the time spent; if we want to know the density of a material we need to measure its mass and its volume.

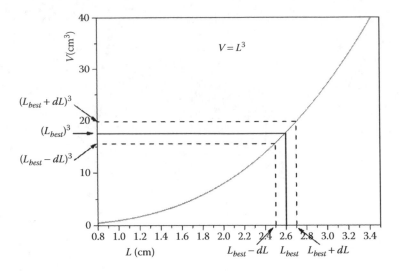

FIGURE D.8 The propagation of the L_{best} uncertainty in the determination of the volume.

How do the errors made in the direct measurements propagate in the estimation of the derived variables? Imagine having measured the length L of a cube's side directly and wanting to determine its volume. If $L_{best} = (2.6 \pm 0.1)$ cm, then the best estimate of the volume will be $V_{best} = (L_{best})^3 = 17.576$ cm^3. How much is the uncertainty in the volume? (Figure D.8)

If the uncertainty in L is small, the values of V for $L = L_{best} + dL$ and $L = L_{best} - dL$ are pretty close to V_{best}. The function $V(L)$, around V_{best}, can be approximated by a straight line. In other words, $V(L)$ can be described by a truncated Taylor's series expansion having V_{best} as its accumulation point:

$$V(L_{best} + dL) = V(L_{best}) + \left(\frac{dV}{dL}\right)_{L_{best}} dL = V_{best} + dV \qquad [D.6]$$

$$V(L_{best} - dL) = V(I_{best}) - \left(\frac{dV}{dL}\right)_{L_{best}} dL = V_{best} - dV \qquad [D.7]$$

It derives that the uncertainty in the volume is $dV = (dV/dL)_{L_{best}} dL = (3L_{best}^2)dL = 2.028$ cm^3. The final result is $V_{best} = (17.6 \pm 2.0)$ cm^3. This analysis of the error's propagation can be generalized. For any function z of the variable x that is measured directly with an uncertainty dx, the error will be:

$$dz = \left| \left(\frac{dz}{dx}\right)_{x_{best}} \right| dx \qquad [D.8]$$

Note that in equation [D.8], we have the absolute value of the first derivative because sometimes the first derivative may be negative, but the error never decreases when it propagates. If z is a function of two or more variables, $x_1, x_2, ..., x_n$, the uncertainty will be:

$$dz = \left| \left(\frac{dz}{dx_1}\right)_{x_{1,best}} \right| dx_1 + \left| \left(\frac{dz}{dx_2}\right)_{x_{2,best}} \right| dx_2 + ... + \left| \left(\frac{dz}{dx_n}\right)_{x_{n,best}} \right| dx_n \qquad [D.9]$$

The uncertainty in z determined through either the [D.8] or [D.9] formula is called the *maximum a priori absolute error*. In the absence of miscalibration errors, the z_{true} value will be surely within the range $\left[z_{best} + dz, z_{best} - dz \right]$. When the uncertainties in the variables x_1, x_2, ..., x_n that have been measured directly, are *independent* and *random*, we may use formula [D.10] for estimating the uncertainty in z.

$$dz = \sqrt{\left(\left| \left(\frac{dz}{dx_1} \right)_{x_1,best} \right| dx_1 \right)^2 + \left(\left| \left(\frac{dz}{dx_2} \right)_{x_2,best} \right| dx_2 \right)^2 + ... + \left(\left| \left(\frac{dz}{dx_n} \right)_{x_n,best} \right| dx_n \right)^2} \qquad [D.10]$$

Alternatively, we may calculate as many values of z as are those collected for the independent variables x_1, x_2, ..., x_n. Let us assume to have repeated m measurements for each variable. The best estimation of z and its uncertainty is the average $\left(\bar{z} = \left(\sum_{j=1}^{m} z_j \right) / m \right)$ and the standard deviation of the mean $\left(S_{\bar{z}} = \left(\sqrt{\frac{1}{m-1} \sum_{j=1}^{m} (z_j - \bar{z})^2} \right) / \sqrt{m} \right)$, respectively.

D.6   STUDENT'S *T*-DISTRIBUTION

We have already learned that when we repeat the measurement of a value of the variable x, n times, with $n \to \infty$, and the uncertainty is mainly random,[3] the data are well-described by a Gaussian function. The Gaussian function will be centered on the mean value $\bar{x}$, and its width will be the standard deviation of the mean, $\sigma_{\bar{x}}$. By exploiting the Gaussian distribution, we can calculate the probability that a further determination will be within the range $\bar{x} \pm g\sigma_{\bar{x}}$. Such probability is a function of the parameter g (see the values listed in Table D.1). For instance, the probability that a measurement is within the range $\left[\bar{x} + \sigma_{\bar{x}}, \bar{x} - \sigma_{\bar{x}} \right]$, with $g = 1$, amounts to 68.27%.

TABLE D.1

Percent Probability (Prob) That a Measurement is within the Range $\left[\bar{x} + g\sigma_{\bar{x}}, \bar{x} - g\sigma_{\bar{x}} \right]$ as a Function of g and Calculated under the Hypothesis That the Measurements are Described by a Gaussian distribution

g	Prob (%)
0.00	0.00
0.50	38.29
0.68	50.35
1.00	68.27
1.04	70.17
1.44	85.01
1.50	86.64
1.65	90.11
1.96	95.00
2.00	95.45
2.50	98.76
3.00	99.73

---

[3] In other words, $\varepsilon_{random} \gg \varepsilon_{sys}$.

Most often, we cannot repeat the same measurement an infinite number of times (i.e., many times), but just a few. In such cases, our data are not properly described by a Gaussian distribution but a Student's t-distribution. The Student's t-distribution is defined for the continuous random variable $t = (\bar{x} - \bar{x}_{Gauss}) / S_x$ where $\bar{x}_{Gauss}$ is the expectation value of the corresponding Gaussian distribution and $S_{\bar{x}}$ is the experimental standard deviation of the mean $\bar{x}$. The probability density function of the t variable is

$$f(t,v) = \frac{1}{\sqrt{\pi v}} \frac{\Gamma\left(\frac{v+1}{2}\right)}{\Gamma\left(\frac{v}{2}\right)} \left(1 + \frac{t^2}{v}\right)^{-\frac{v+1}{2}}$$ [D.11]

In equation [D.11], v is the number of the degrees of freedom that is equal to $m - 1$, if m is the finite number of times a measurement has been repeated; Γ is the gamma function. The function $f(t,v)$ is symmetric and bell-shaped (like the Gaussian distribution), but it has heavier tails than the Gaussian distribution; its expectation value is zero, and its variance is $v / (v - 2)$ for $v > 2$. If $v \to \infty$, the t-distribution approaches a Gaussian distribution centered at 0 and with a standard deviation $\sigma = 1$. The probability that the $(m + 1)$-th measurement is within the interval $\left[\bar{x} + tS_x, \bar{x} - tS_x\right]$ is a function of t and v; some values are reported in Table D.2. The last row of Table D.2 (with $v = \infty$) corresponds to the values defined for the Gaussian distribution.

We may use the Student's t-distribution for comparing the mean values of two independent sets of measurements, $\bar{x}_A$ and $\bar{x}_B$, and ask us if they belong to the same distribution. This test is suitable to ascertain if the experimental conditions have not changed in the two series of measurements or if the measurements of the same physical quantity carried out in different laboratories can be merged to give a single best estimate. First, we calculate the t-value through the following formula

$$t = \frac{\bar{x}_A - \bar{x}_B}{\sqrt{\dfrac{S_A^2}{n_A} + \dfrac{S_B^2}{n_B}}}$$ [D.12]

TABLE D.2

Values of the t Variable Expressed as a Function of the Percentage Probability (Prob) That a Further Measurement is within the Range $\left[\bar{x} + tS_x, \bar{x} - tS_x\right]$. The t-Value also Depends on the Degrees of Freedom $v = m-1$

Prob (%)

v	50	60	70	80	90	95	99
1	1.000	1.376	1.963	3.078	6.314	12.706	63.657
2	0.816	1.061	1.386	1.886	2.920	4.303	9.925
3	0.765	0.978	1.250	1.638	2.353	3.182	5.841
4	0.741	0.941	1.190	1.533	2.132	2.776	4.604
5	0.727	0.920	1.156	1.476	2.015	2.571	4.032
6	0.718	0.906	1.134	1.440	1.943	2.447	3.707
7	0.711	0.896	1.119	1.415	1.895	2.365	3.499
8	0.706	0.889	1.108	1.397	1.860	2.306	3.355
9	0.703	0.883	1.100	1.383	1.833	2.262	3.250
10	0.700	0.879	1.093	1.372	1.812	2.228	3.169
20	0.687	0.860	1.064	1.325	1.725	2.086	2.845
30	0.683	0.854	1.055	1.310	1.697	2.042	2.750
∞	0.674	0.842	1.036	1.282	1.645	1.960	2.576

In [D.12], S_A^2 and S_B^2 are the variances (i.e., the standard deviations S_A and S_B, both to the power of two) of the two series of measurements, A and B, whose numbers of elements are n_A and n_B, respectively. If the estimated t-value is larger than the expected t-value for a confidence level of the 95%, we may assert that the experimental conditions have been changed. Otherwise, the two series may be considered as belonging to the same distribution and may be used to determine a better estimate through the calculation of the weighted average:

$$x_{best} = \frac{\dfrac{\overline{x}_A}{S_A^2} + \dfrac{\overline{x}_B}{S_B^2}}{\dfrac{1}{S_A^2} + \dfrac{1}{S_B^2}}$$ [D.13]

The formula [D.13] can be generalized to more than two separate sets of measurements that are consistent among them. If the number of independent sets of measurements is k, $\overline{x}_1, \overline{x}_2, ..., \overline{x}_k$ are the corresponding mean values, and $S_1, S_2, ..., S_k$ are their standard deviations, the best estimate is

$$x_{best} = \frac{\sum_{i=1}^{k} w_i \overline{x}_i}{\sum_{i=1}^{k} w_i}$$ [D.14]

The terms $w_i = 1/S_i^2$ are called weights. The larger the standard deviation, the smaller the weight. The sets of measurements that are much less precise than the others contribute very much less to x_{best}. For example, if one set of measurements is three times less precise than the others, its weight is nine times smaller than the other weights. The uncertainty in the weighted average x_{best} is

$$S_x = \left(\sum_{i=1}^{k} w_i \right)^{-1/2}$$ [D.15]

D.7 LEAST-SQUARES FITTING

Sometimes the final goal of an experiment is the determination of the mathematical equation relating two variables. For instance, a chemist may be interested in determining the order of a chemical reaction with respect to a reagent X. In other words, the chemist may be interested in determining the exponent that appears in the equation $v = k[X]^{n_X}$. The chemist measures the reaction rates (v_1, v_2, ..., v_n) at different concentrations of X ($[X]_1$, $[X]_2$, ..., $[X]_n$) maintaining constant all the other experimental conditions. It is convenient to work with the logarithmic form of the kinetic equation ($v = k[X]^{n_X}$), i.e., $\log(v) = \log(k) + n_X \log[X]$. In fact, by plotting $\log(v_1)$, $\log(v_2)$, ..., $\log(v_n)$ vs. $\log[X]_1$, $\log[X]_2$, ..., $\log[X]_n$, the experimental data can be described by a straight line, $y = A + Bx$. The slope is $B = n_X$, and the intercept is $A = \log(k)$. The best estimates of the coefficients A and B are obtained by the linear regression, also called the least-squares method.

More in general, imagine having collected pairs of data (x_1, y_1), (x_2, y_2), ..., (x_n, y_n). The uncertainties on the x_i values are negligible with respect to those on the y_i values, and the uncertainties on the y_i values are of the random type. Each y_i value belongs to a Gaussian distribution centered on its true value that is $(A + Bx_i)$. Therefore, the probability of collecting y_i is:

$$P(y_i) \propto \frac{1}{\sigma_{y_i}} e^{-\frac{(y_i - A - Bx_i)^2}{\sigma_{y_i}^2}}$$ [D.16]

The probability of obtaining the complete set of independent measurements, y_1, y_2, ..., y_n, is the product of the probabilities of obtaining every single value:

$$P(y_1, y_2, \ldots, y_n) \propto \left(\prod_{i=1}^{n} \frac{1}{\sigma_{y_i}} \right) e^{-\sum_{i=1}^{n} \frac{(y_i - A - Bx_i)^2}{\sigma_{y_i}^2}} \qquad [\text{D.17}]$$

According to the principle of maximum likelihood, the best estimates of the A and B parameters are those that maximize the total probability, $P(y_1, y_2, \ldots, y_n)$, or minimize the exponent[4]

$$\chi^2 = \sum_{i=1}^{n} \frac{(y_i - A - Bx_i)^2}{\sigma_{y_i}^2} \qquad [\text{D.18}]$$

For the determination of these values, we differentiate χ^2 with respect to A and B and set the derivatives equal to zero:

$$\frac{\partial \chi^2}{\partial A} = -2 \sum_{i=1}^{n} w_i (y_i - A - Bx_i) = 0$$

$$\frac{\partial \chi^2}{\partial B} = -2 \sum_{i=1}^{n} w_i x_i (y_i - A - Bx_i) = 0 \qquad [\text{D.19}]$$

From the equations [D.19], wherein $w_i = 1/\sigma_{y_i}^2$, we achieve that the least-squares estimates of the coefficients A and B are:

$$A = \frac{\left(\sum_{i=1}^{n} w_i x_i^2 \right) \left(\sum_{i=1}^{n} w_i y_i \right) - \left(\sum_{i=1}^{n} w_i x_i \right) \left(\sum_{i=1}^{n} w_i x_i y_i \right)}{\left(\sum_{i=1}^{n} w_i \right) \left(\sum_{i=1}^{n} w_i x_i^2 \right) - \left(\sum_{i=1}^{n} w_i x_i \right)^2} \qquad [\text{D.20}]$$

$$B = \frac{\left(\sum_{i=1}^{n} w_i \right) \left(\sum_{i=1}^{n} w_i x_i y_i \right) - \left(\sum_{i=1}^{n} w_i x_i \right) \left(\sum_{i=1}^{n} w_i y_i \right)}{\left(\sum_{i=1}^{n} w_i \right) \left(\sum_{i=1}^{n} w_i x_i^2 \right) - \left(\sum_{i=1}^{n} w_i x_i \right)^2} \qquad [\text{D.21}]$$

When all the uncertainties in the y_i values are equal, the formula for the determination of A and B are like equations [D.20] and [D.21] but with all the weights $w_i = 1$. It may happen that we collect each pair of points (x_i, y_i) just once, and we estimate only their systematic uncertainty due to the equipment. In such cases, we still may use equations [D.20] and [D.21] to evaluate the A and B coefficients. Moreover, we may also determine a random-type uncertainty σ_y in the data $(y_1, y_2, \ldots, y_n)$ by the following formula:

$$\sigma_y = \sqrt{\frac{1}{n-2} \sum_{i=1}^{n} (y_i - A - Bx_i)^2} \qquad [\text{D.22}]$$

---

[4] The procedure of minimizing the sum of the squares appearing in the exponent χ^2 gives the name to the method: the least-squares fitting.

Equation [D.22] is based on the assumption that each y_i is normally distributed about its true value $A + Bx_i$ with width parameter σ_y.[5] After having found the uncertainty σ_y in the measured quantities $(y_1, y_2, ..., y_n)$, we can calculate the uncertainties in A and B by the error propagation formula. The result is:

$$\sigma_A = \sigma_y \sqrt{\frac{\sum_{i=1}^{n} x_i^2}{n\left(\sum_{i=1}^{n} x_i^2\right) - \left(\sum_{i=1}^{n} x_i\right)^2}} \qquad [D.23]$$

$$\sigma_B = \sigma_y \sqrt{\frac{n}{n\left(\sum_{i=1}^{n} x_i^2\right) - \left(\sum_{i=1}^{n} x_i\right)^2}} \qquad [D.24]$$

The extent to which a set of points (x_1, y_1), (x_2, y_2), ..., (x_n, y_n), supports a linear relation between the variables x and y is measured by the correlation coefficient r

$$r = \frac{\sum_{i=1}^{n} (x_i - \bar{x})(y_i - \bar{y})}{\left[\sum_{i=1}^{n} (x_i - \bar{x})^2 \sum_{i=1}^{n} (y_i - \bar{y})^2\right]^{1/2}} = \frac{\sigma_{xy}}{\sigma_x \sigma_y} \qquad [D.25]$$

The possible values of the correlation coefficients are in the range $-1 \leq r \leq +1$. If the points are correlated, $|\sigma_{xy}| \approx \sigma_x \sigma_y$ and $|r| \approx 1$. On the other hand, if the points are uncorrelated, then $\sigma_{xy} \approx 0$ and $r \approx 0$.

D.8 TEST χ^2

When we collect data about the value of a variable or a parameter, and we assume to describe them through a specific function or distribution, we might need a parameter that quantifies the agreement between the observed and the expected values. Such parameter exists, and it is the χ^2:

$$\chi^2 = \sum_{i=1}^{n} \left(\frac{\text{observed value}_i - \text{expected value}_i}{\text{standard deviation}}\right)^2 \qquad [D.26]$$

If the agreement is good, χ^2 will be of the order of the number of degrees of freedom (d), i.e., the number n of the collected data minus the number of the parameters calculated through the data. If the agreement is poor, χ^2 will be much larger than d. A slightly more convenient way to think about this test is to introduce the reduced chi-squared ($\tilde{\chi}^2$) defined as:

$$\tilde{\chi}^2 = \frac{\chi^2}{d} \qquad [D.27]$$

---

[5] The factor $(n - 2)$ appearing in equation [D.22] is analogous to the factor $(n - 1)$ of equation [D.4]. Both factors represent the degrees of freedom, which is the number of independent measurements minus the number of parameters calculated from those measurements. It is not difficult to remember the factors $(n - 1)$ and $(n - 2)$ because they are reasonable. In fact, if we collect just one datum regarding the value of a variable, we cannot calculate the standard deviation. Similarly, if we collect only two pairs of data, (x_1, y_1) and (x_2, y_2), we can always find a line that passes exactly through both points, and the least-squares method gives this line. With just two pairs of data, we cannot deduce anything about the reliability of the straight line.

Whatever is the number of the degrees of freedom, if we obtain a $\tilde{\chi}^2$ of the order of one or less, then we have no reason to doubt about our expected distribution or function. On the other hand, if we obtain a $\tilde{\chi}^2$ much larger than one, our expected distribution or function is unlikely to be correct.

D.9 CONCLUSION

Now it is clear why we say that science is exact. The rigorous treatment of the uncertainties in each measurement allows us to quantify how much we can trust in our scientific assertions. It is also evident that due to the random errors, many experiments are unique and unreproducible events. This statement is especially true when we deal with chaotic systems.

D.10 HINTS FOR FURTHER READING

More information about the error analysis in measurements can be found in the book titled *An Introduction to Error Analysis* by Taylor (1997), in *A Practical Guide to Data Analysis for Physical Science Students* by Lyons (1991), and in the *Statistical Treatment of Experimental Data* by Young (1962).

Appendix E: Errors in Numerical Computation

We are successful because we use the right level of abstraction.

Avi Wigderson (1956 AD–)

Calculations, just like experimental measurements, are subject to errors. In fact, any scientist, whatever is his/her expertise, introduces inevitable uncertainties not only when he/she performs an experiment but also when he/she computes using calculators, numerical methods, and algorithms. It is essential to have a notion of the errors that can be made in the computations.

E.1 ROUNDOFF ERRORS

Whatever is used as our computing machine, it can only represent numerical amounts using a limited number of significant figures. Thus, irrational numbers such as π, Napier's constant e, $\sqrt{2}$ cannot be represented exactly. The discrepancy between the real number and its representation in the computer is called *roundoff error*. Any real number N can be written with the floating-point method as

$$N = pb^q \qquad \text{[E.1]}$$

In [E.1], p is the mantissa, b is the base of the numeration system, and q is called the exponent or the characteristic.[1] In a calculator, the space available for the representation of any real number in binary digits is organized as shown in Figure E.1, wherein s represents the sign of the number.

In most of the modern binary calculators ($b = 2$), a number N is represented with either of two alternative different precisions, called single and double precision, respectively. In single precision, 32 bits are available to represent N: 23 bits are used to define the significant digits of p, 8 bits are used to store the exponent q, and one bit is used to store the sign s ("0" for positive numbers and "1" for negative numbers). In double precision, 64 bits are available with 52 digits used to represent the significant figures. For instance, the number π is represented as 3.141593 in the single precision, and as 3.141592653589793 in the double precision. Evidently, the use of the double precision reduces the effects of rounding error. However, it increases the run-time of the computation. There are two techniques of rounding. Let us assume that the maximum number of digits in the mantissa is n. The distance between two consecutive mantissas, $\bar{p}_1$ and $\bar{p}_2$, is equal to b^{-n}: $\bar{p}_2 = \bar{p}_1 + b^{-n}$. We may round a mantissa p, included between $\bar{p}_1$ and $\bar{p}_2$ (see the left part of Figure E.2) by chopping all the digits that are on the right of the n-th digit. Any mantissa p of Figure E.2 is substituted by $\bar{p}_1$. The roundoff error is

$$\left| p - \bar{p}_1 \right| < b^{-n} \qquad \text{[E.2]}$$

An alternative technique of rounding consists in adding $(1/2)b^{-n}$ to the mantissa p and then chop the result at the n-th digit. All the mantissas included in the range $(\bar{p}_1, \bar{p}_1 + 1/2b^{-n})$ are substituted by $\bar{p}_1$,

---

[1] The representation of the number $N = pb^q$ is defined normalized when $b^{-1} \le |p| < 1$. For instance, the representation of $a_1 = (12.73)$ is normalized when $a_{1n} = 0.1273 \times 10^2$, whereas that of $a_2 = (0.00098)$ is normalized when $a_{2n} = 0.98 \times 10^{-3}$.

FIGURE E.1 Representation of a real number in a calculator: s is the sign, q is the exponent, and p is the mantissa.

FIGURE E.2 On the left, a true mantissa p is included between the two consecutive mantissas, $\bar{p}_1$ and $\bar{p}_2$, rounded by the calculator. On the right, a true real number N is included between two consecutive numbers, $\bar{N}_1$ and $\bar{N}_2$, rounded by the calculator.

whereas all the mantissas included in the range $(\bar{p}_1 + 1/2b^{-n}, \bar{p}_2)$ are substituted by $\bar{p}_2$. Therefore, if we indicate with $\bar{p}$ the rounded mantissa, the roundoff error is

$$\left| p - \bar{p} \right| \le \frac{1}{2} b^{-n} \qquad [E.3]$$

If $N = pb^q$ is a real number whose rounded version is $\bar{N} = \bar{p}b^q$ (see the right part of Figure E.2), the absolute errors in $\bar{N}$ according to equations [E.2] and [E.3] are:

$$\left| N - \bar{N} \right| \le \begin{cases} b^{q-n} & [E.2] \\ \dfrac{1}{2} b^{q-n} & [E.3] \end{cases} \qquad [E.4]$$

whereas the relative errors are:

$$\frac{\left| N - \bar{N} \right|}{|N|} \le \begin{cases} b^{1-n} & [E.2] \\ \dfrac{1}{2} b^{1-n} & [E.3] \end{cases} \qquad [E.5]$$

We define the relative error [E.5] as the accuracy of the calculator (or machine accuracy, MA). It is equal to either $MA = b^{1-n}$ or $MA = (1/2)\, b^{1-n}$ depending on which rounding methodology is used in the calculator, if it is the first or the second one, respectively.

E.2 PROPAGATION OF THE ROUNDOFF ERROR

The arithmetic operations among numbers in the floating-point representation are not exact. For example, two floating-point numbers are added by first right-shifting the mantissa of the smaller one and simultaneously increasing its exponent until the two operands have the same exponent. Low-order bits belonging to the smaller operand are lost by this shifting. If the two operands differ too much in magnitude, then the smaller operand is effectively replaced by zero, since it is right-shifted to oblivion. Some of the well-known properties of the four arithmetic operations do not hold in a calculator. For example, whereas the commutative property for addition and multiplication continues to hold, the associative property does not.

Roundoff errors accumulate with increasing amounts of calculation. If we perform k arithmetic operations, and the roundoff errors come in randomly up and down, in the end, we have a total roundoff error of the order of $\sqrt{k}(MA)$, where the square root comes from a random-walk of the error. This statement is not true when the roundoff errors accumulate preferentially in one direction. In this case, the final error will be $k(MA)$, grown if it were a snowball.

The finite number of digits available to represent a number restricts the numerical ranges. Some operations can give rise to phenomena of overflow when the exponent of the output is larger than its maximum possible value or phenomena of underflow when the exponent of the output is smaller than its minimum possible value.

E.3 SOURCES OF ERRORS IN THE SOLUTION OF A NUMERICAL PROBLEM BY AN ALGORITHM IMPLEMENTED IN A COMPUTER

In general, a numerical problem has a certain number of input variables, x_1, x_2, ..., x_n, a certain number of output variables, y_1, y_2, ..., y_m, and some formulas that describe how the output variables depend on the input variables. For simplicity, we consider the case of one input and one output variable: $y = f(x)$.

If x is the true input value, when it is written as a machine number with the floating-point method, it can be tainted by a roundoff error. It becomes $\bar{x}$. The roundoff error is less or equal to the machine accuracy MA. The uncertainty in the input propagates into the output, according to the propagation formula we learned in Appendix D:

$$\bar{y} - y = \left| \frac{df(x)}{dx} \right| (\bar{x} - x) + \dots \qquad [\text{E.6}]$$

The relative error is:

$$\left| \frac{\bar{y} - y}{y} \right| \approx \left| \frac{df(x)}{dx} \right| \frac{|\bar{x} - x|}{|y|} = \frac{|x||f'(x)|}{|f(x)|} \frac{|\bar{x} - x|}{|x|} = \frac{|x||f'(x)|}{|f(x)|} (MA) = c(MA) \qquad [\text{E.7}]$$

The factor $c = (|x||f'(x)|)/|f(x)|$ is called the condition number of the function f at x. If c is not much larger than 1, the numerical problem is well-conditioned. On the other hand, if $c \gg 1$, the problem is ill-conditioned.[2] The output of the computation $\bar{y}$ is then approximated by the closest machine number $\bar{\bar{y}}$. Therefore, if we compare $\bar{\bar{y}}$ with the true value y, we obtain:

$$\left| \frac{\bar{\bar{y}} - y}{y} \right| \leq c(MA) + (MA) \qquad [\text{E.8}]$$

The ideal algorithm would give $\bar{y}$ as output. However, the perfect algorithm is usually impossible to implement. Many numerical algorithms compute "discrete" approximations to some desired "continuous" quantity. For example, an integral is evaluated numerically by computing a function at a discrete set of points, rather than at "every" point. The discrepancy between the true answer and the answer obtained in a practical calculation is called the *truncation error, T.* Truncation error would persist even on a hypothetical "perfect" computer that had an infinitely accurate representation and no roundoff error. The roundoff error is a characteristic of the computer hardware. The truncation

---

[2] For instance, if $f(x) = x^\alpha$, the condition number is $c = |x||f'(x)|/|f(x)| = |x||\alpha x^{\alpha-1}|/|x^\alpha| = |\alpha|$. The problem is ill-conditioned when $|\alpha| \gg 1$.

error, on the other hand, is entirely under the programmer's control. An algorithm is numerically stable if it yields in machine arithmetic a result $\tilde{y}$, such that

$$\left| \frac{\tilde{y} - y}{y} \right| \leq T + c(MA) + (MA) \qquad\qquad [E.9]$$

where T is of the order of MA.

E.4   HINTS FOR FURTHER READING

Who wants to learn more about errors in numerical computation can consult the books *Numerical Recipes* by Press et al. (2007), *Rounding Errors in Algebraic Processes* by Wilkinson (1994), and *Floating-Point Computation* by Sterbenz (1974).

References

Aaronson, S.; 2005, NP-complete problems and physical reality. *ACM SIGACT News* 36, 30–52.

Abarbanel, H. D. I.; 1996, *Analysis of Observed Chaotic Data*. Springer-Verlag, New York.

Abelson, H.; Allen, D.; Coore, D.; Hanson, C.; Homsy, G.; Knight, T. F.; Nagpal, R.; Rauch, E.; Sussman, G. J.; Weiss, R.; 2000, Amorphous computing. *Commun. ACM* 43, 74–82.

Adamatzky, A.; De Lacy Costello, B.; 2012, Reaction-diffusion computing. pp. 1897–1920 in *Handbook of Natural Computing*. Rozenberg, G.; Bäck, T.; Kok, J. N.; (Eds.), Springer-Verlag, Berlin, Germany.

Adleman, L. M.; 1994, Molecular computation of solutions to combinatorial problems. *Science* 266, 1021–1024.

Adleman, L. M.; 1998, Computing with DNA. *Sci. Am.* 279, 54–61.

Agladze, K. I.; Krinsky, V. I.; 1982, Multi-armed vortices in an active chemical medium. *Nature* 296, 424–426.

Aguilar, A.; 2009, *What Is Death? A Scientific, Philosophical and Theological Exploration of Life's End*. Libreria Editrice Vaticana, Vatican City.

Albert, R.; Barabási, A.-L.; 2002, Statistical mechanics of complex networks. *Rev. Mod. Phys.* 74, 47–97.

Albert, R.; Jeong, H.; Barabási, A.-L.; 2000, Error and attack tolerance of complex networks. *Nature* 406, 378–382.

Albus, J. S.; Bekey, G. A.; Holland, J. H.; Kanwisher, N. G.; Krichmar, J. L.; Mishkin, M.; Modha, D. S.; Raichle, M. E.; Shepherd G. M.; Tononi, G.; 2007, A proposal for a decade of the mind initiative. *Science* 317, 1321.

Alexander, S.; Orbach, R.; 1982, Density of states on fractals: <<fractons>>. *J. Phys. (Paris) Lett.* 43, L-625–L-631.

Alon, U.; 2007, *An Introduction to Systems Biology: Design Principles of Biological Circuits*. Chapman & Hall/CRC, Boca Raton, FL.

Amdahl, G. M.; 1967, Validity of the single processor approach to achieving large-scale computing capabilities. *AFIPS Conference Proc.* 30, 483–485.

Amir, Y.; Ben-Ishay, E.; Levner, D.; Ittah, S.; Abu-Horowitz, A.; Bachelet, I.; 2014, Universal computing by DNA origami robots in a living animal. *Nat. Nanotechnol.* 9, 353–357.

Amos, S. W.; James, M. R.; 2000, *Principles of Transistor Circuits*. 9th ed. Newnes Butterworth-Heinemann, Woburn, MA.

Anderson, J. D. Jr.; 2009, Governing equations of fluid dynamics. Chapter 2, pp. 15–51 in *Computational Fluid Dynamics*. Wendt, J. F.; (Ed.), Springer-Verlag, Berlin, Germany.

Andréasson, J.; Pischel, U.; 2015, Molecules with a sense of logic: A progress report. *Chem. Soc. Rev.* 44, 1053–1069.

Antal, T.; Droz, M.; Magnin, J.; Rácz, Z.; Zrinyi, M.; 1998, Derivation of the Matalon-Packter law for Liesegang patterns. *J. Chem. Phys.* 109, 9479–9486.

Arnold, R.; Showalter, K.; Tyson, J. J.; 1987, Propagation of chemical reactions in space. *J. Chem. Educ.* 64, 740–742.

Arthur, W. B.; 2015, *Complexity and the Economy*. Oxford University Press, New York.

Ashkenasy, G.; Hermans, T. M.; Otto, S.; Taylor, A. F.; 2017, Systems chemistry. *Chem. Soc. Rev.* 46, 2543–2554.

Ashkin, A.; 1997, Optical trapping and manipulation of neutral particles using lasers. *Proc. Natl. Acad. Sci. USA* 94, 4853–4860.

Atanasov, A. T.; 2014, Length of periods in the nasal cycle during 24-hours registration. *Open J. Biophys.* 4, 93–96.

Atkins, P.; de Paula, J.; 2010, *Atkins' Physical Chemistry*. 9th ed. Oxford University Press, Oxford.

Axelrod, R.; 1976, *Structure of Decision: The Cognitive Maps of Political Elites*. Princeton University Press, Princeton, NJ.

Axelrod, R.; 1984, *The Evolution of Cooperation*. Basic Books, New York.

Ayarpadikannan, S.; Lee, H. E.; Han, K.; Kim, H. S.; 2015, Transposable element-driven transcript diversification and its relevance to genetic disorders. *Gene* 558, 187–194.

Bajzer, Ž; Prendergast, F. G.; 1992, Maximum likelihood analysis of fluorescence data. *Methods Enzymol.* 210, 200–237.

Bak, P.; Paczuski, M.; 1995, Complexity, contingency, and criticality. *Proc. Natl. Acad. Sci. USA* 92, 6689–6696.

Bak, P.; Tang, C.; Wiesenfeld, K.; 1987, Self-organized criticality: An explanation of 1/f noise. *Phys. Rev. Let.* 59, 381–384.

Baker, R. E.; Schnell, S.; Maini, P. K.; 2006, A clock and wavefront mechanism for somite formation. *Dev. Biol.* 293, 116–126.

Balagaddé, F. K.; You, L.; Hansen, C. L.; Arnold, F. H.; Quake, S. R.; 2005, Long-term monitoring of bacteria undergoing programmed population control in a microchemostat. *Science* 309, 137–140.

Ball, P.; 2001, *The Self-Made Tapestry: Pattern Formation in Nature*. Oxford University Press, New York.

Ball, P.; 2009, *Nature's Patterns. A Tapestry in Three Parts. Branches.* Oxford University Press, New York.

Ball, P.; 2012, Computer engineering: Feeling the heat. *Nature* 492, 174–176.

Bancaud, A.; Huet, S.; Daigle, N.; Mozziconacci, J.; Beaudouin, J.; Ellenberg, J.; 2009, Molecular crowding affects diffusion and binding of nuclear proteins in heterochromatin and reveals the fractal organization of chromatin. *EMBO J.* 28, 3785–3798.

Bancaud, A.; Lavelle, C.; Huet, S.; Ellenberg, J.; 2012, A fractal model for nuclear organization: Current evidence and biological implications. *Nucleic Acids Res.* 40, 8783–8792.

Bandt, C.; Pompe, B.; 2002, Permutation entropy: A natural complexity measure for time series. *Phys. Rev. Lett.* 88, 174102.

Bánsági, T. Jr.; Vanag, V. K.; Epstein, I. R.; 2011, Tomography of reaction-diffusion microemulsions reveals three-dimensional Turing patterns. *Science* 331, 1309–1312.

Barabási, A.-L.; 2016, *Network Science.* Cambridge University Press, Cambridge, UK, freely available at http://barabasi.com/networksciencebook/.

Barabási, A.-L.; Albert, R.; 1999, Emergence of scaling in random networks. *Science* 286, 509–512.

Barabási, A.-L.; Oltvai, Z. N.; 2004, Network biology: Understanding the cell's functional organization. *Nat. Rev. Genet.* 5, 101–113.

Baranger, M.; 2000, *Chaos, Complexity, and Entropy.* New England Complex Systems Institute, http://www.necsi.edu/projects/baranger/ccc.html.

Barrio, R. A.; Varea, C.; Aragón, J. L.; Maini, P. K.; 1999, A two-dimensional numerical study of spatial pattern formation in interacting Turing systems. *Bull. Math. Biol.* 61, 483–505.

Bar-Yam, Y.; 2004, A mathematical theory of strong emergence using multiscale variety. *Complexity* 9, 15–24.

Bavireddi, H.; Bharate, P.; Kikkeri, R.; 2013, Use of Boolean and fuzzy logics in lactose glycocluster research. *Chem. Commun.* 49, 9185–9187.

Beasley, D.; 2000, Possible applications of evolutionary computation. pp. 4–19 in *Evolutionary Computation 1. Basic Algorithms and Operators.* Bäck, T.; Fogel, D. B.; Michalewicz, Z.; (Eds.), Institute of Physics Publishing, Bristol, UK.

Behrens, T. E. J.; Sporns, O.; 2011, Human connectomics. *Curr. Opin. Neurobiol.* 22, 1–10.

Beinhocker, E. D.; 2007, *The Origin of Wealth: Evolution, Complexity, and the Radical Remaking of Economics.* Harvard Business School Publishing, Boston, MA.

Belousov, B. P.; 1958, *A Periodic Reaction and Its Mechanism.* Sbornik Referatov po Radiatsionni Meditsine, Moscow: Medgiz, 145.

Ben-Jacob, E.; Garik, P.; 1990, The formation of patterns in non-equilibrium growth. *Nature* 343, 523–530.

Ben-Jacob, E.; Levine, H.; 2001, The artistry of nature. *Nature* 409, 985–986.

Benenson, Y.; Gil, B.; Ben-Dor, U.; Adar, R.; Shapiro, E.; 2004, An autonomous molecular computer for logical control of gene expression. *Nature* 429, 423–429.

Bennett, C. H.; 1987, Demons, engines and the second law. *Sci. Am.* 257, 108–116.

Bennett, C. H.; Bessette, F.; Brassard, G.; Salvail, L.; Smolin, J.; 1992, Experimental quantum cryptography. *J. Cryptol.* 5, 3–28.

Bennet, C. H.; DiVincenzo, D. P.; 2000, Quantum information and computation. *Nature* 404, 247–255.

Berberan-Santos, M. N.; Bodunov, E. N.; Valeur, B.; 2005, Mathematical functions for the analysis of luminescence decays with underlaying distributions 1. Kohlrausch decay function (stretched exponential). *Chem. Phys.* 315, 171–182.

Berger, B.; Leighton, T.; 1998, Protein folding in the hydrophobic-hydrophilic (HP) model is NP-complete. *J. Comput. Biol.* 5, 27–40.

Berridge, M. J.; Lipp, P.; Bootman, M. D.; 2000, The versatility and universality of calcium signalling. *Nat. Rev. Mol. Cell Biol.* 1, 11–21.

Bérut, A.; Arakelyan, A.; Petrosyan, A.; Ciliberto, S.; Dillenschneider, R.; Lutz, E.; 2012, Experimental verification of Landauer's principle linking information and thermodynamics. *Nature* 483, 187–189.

Bianconi, E.; Piovesan, A.; Facchin, F.; Beraudi, A.; Casadei, R.; Frabetti, F.; Vitale, L.; Pelleri, M. C.; Tassani, S.; Piva, F.; Perez-Amodio, S.; Strippoli, P.; Canaider, S.; 2013, An estimation of the number of cells in the human body. *Ann. Hum. Biol.* 40, 463–471.

Bishop, C. M.; 2006, *Pattern Recognition and Machine Learning*. Springer, Singapore.

Bissette, A. J.; Fletcher, S. P.; 2013, Mechanisms of autocatalysis. *Angew. Chem. Int. Ed.* 52, 2–30.

Björn, L. O.; (Ed.) 2015, *Photobiology. The Science of Light and Life*. 3rd ed. Springer, New York.

Blair, E. P.; Lent, C. S.; 2003, An architecture for molecular computing using quantum-dot cellular automata. *2003 Third IEEE Conference on Nanotechnology, 2003. IEEE-NANO 2003*, 2, 402–405.

Blanchard, P.; Devaney, R. L.; Hall, G. R.; 2006, *Differential Equations*. 3rd ed. Thomson Brooks/Cole, Belmont, CA.

Blankenship, R. E.; 2002, *Molecular Mechanisms of Photosynthesis*. Blackwell Science Publishers, Oxford, UK, Malden, MA.

Boccaletti, S.; Grebogi, C.; Lai, Y.-C.; Mancini, H.; Maza, D.; 2000, The control of chaos: Theory and applications. *Phys. Rep.* 329, 103–197.

Boccaletti, S.; Kurths, J.; Osipov, G.; Valladares, D. L.; Zhou, C. S.; 2002, The synchronization of chaotic systems. *Phys. Rep.* 366, 1–101.

Boissonade, J.; De Kepper, P.; 1980, Transitions from bistability to limit cycle oscillations. Theoretical analysis and experimental evidence in an open chemical system. *J. Phys. Chem.* 84, 501–506.

Bolotin, A.; 2014, Computational solution to quantum foundational problems. *Phys. Sci. Int. J.* 4 (8), 1145–1157.

Bonabeau, E.; Dorigo, M.; Theraulaz, G.; 1999, *Swarm Intelligence: From Natural to Artificial System*. Oxford University Press, New York.

Bonan, G. B.; Doney, S. C.; 2018, Climate, ecosystems, and planetary futures: The challenge to predict life in Earth system models. *Science* 359, eaam8328.

Boole, G.; 1854, *An Investigation of the Laws of Thought on Which Are Founded the Mathematical Theories of Logic and Probabilities*. Macmillan. Reprinted with corrections, Dover Publications, New York, 1958. (reissued by Cambridge University Press, 2009).

Bourzac, K.; 2012, Back to analogue. *Nature* 483, S34–S36.

Boyle, R. W.; 1987, *Gold. History and Genesis of Deposits*. Springer, Boston, MA.

Bracewell, R.; 1986, *The Fourier Transform and Its Applications*. McGraw-Hill, Boston, MA.

Bragard, J.; Velarde, M. G.; 1997, Bénard convection flows. *J. Non-Equilib. Thermodyn.* 22, 1–19.

Bray, D.; 1995, Protein molecules as computational elements in living beings. *Nature* 376, 307–312.

Bray, W. C.; 1921, A periodic reaction in homogeneous solution and its relation to catalysis. *J. Am. Chem. Soc.* 43, 1262–1267.

Bray, W. C.; Liebhafsky, H. A.; 1931, Reactions involving hydrogen peroxide, iodine and iodate ion. I. Introduction. *J. Am. Chem. Soc.* 53, 38–48.

Brent, R.; Bruck, J.; 2006, Can computers help to explain biology? *Nature* 440, 416–417.

Briggs, T. S.; Rauscher, W. C.; 1973, An oscillating iodine clock reaction. *J. Chem. Ed.* 50, 496.

Brillouin, L.; 1956, *Science and Information Theory*. Academic Press, New York.

Brochon, J.-C.; 1994, Maximum entropy method of data analysis in time-resolved spectroscopy. *Methods Enzymol.* 240, 262–311.

Brostow, W.; Hagg Lobland, H. E.; 2017, *Materials: Introduction and Applications*. John Wiley & Sons, Hoboken, NJ.

Brumfiel, G.; 2012, Quantum leaps. *Nature* 491, 322–324.

Budroni, M. A.; Lemaigre, L.; De Wit, A.; Rossi, F.; 2015, Cross-diffusion-induced convective patterns in microemulsion systems. *Phys. Chem. Chem. Phys.* 17, 1593–1600.

Burks, A. W.; Goldstine, H. H.; von Neumann, J.; 1963, Preliminary discussion of the logical design of an electronic computing instrument. Volume V, pp. 34–79 in *John von Neumann Collected works*. Taub, A. H.; (Ed.), The Macmillan, New York.

Bustamante, C.; Keller, D.; Oster, G.; 2001, The physics of molecular motors. *Acc. Chem. Res.* 34, 412–420.

Bustamante, C.; Liphardt, J.; Ritort, F.; 2005, The nonequilibrium thermodynamics of small systems. *Phys. Today* 58, 43–48.

Caldarelli, G.; Catanzaro, M.; 2012, *Networks. A Very Short Introduction*. Oxford University Press, Oxford.

Cañestro, C.; Yokoi, H.; Postlethwait, J. H.; 2007, Evolutionary developmental biology and genomics. *Nat. Rev. Genet.* 8, 932–942.

Cao, L.; 1997, Practical method for determining the minimum embedding dimension of a scalar time series. *Physica D* 110, 43–50.

Capra, F.; Luisi, P. L.; 2014, *The Systems View of Life. A Unifying Vision*. Cambridge University Press, New York.

Carroll, S. B.; 2005, *Endless Forms Most Beautiful: The New Science of Evo Devo*. Norton & Company, New York.

Cartwright, J. H. E.; Checa, A. G.; Escribano, B.; Sainz-Díaz, C. I.; 2012, Crystal growth as an excitable medium. *Phil. Trans. R. Soc. A.* 370, 2866–2876.

Castets, V. V.; Dulos, E.; Boissonade, J.; De Kepper, P.; 1990, Experimental evidence of a sustained standing Turing-type nonequilibrium chemical pattern. *Phys. Rev. Lett.* 64, 2953–2956.

Castelvecchi, D.; 2016, Can we open the black box of AI. *Nature* 538, 20–23.

Cavalli-Sforza, L. L.; 2001, *Genes, Peoples and Languages.* Penguin Books, London.

Cavin, R. K.; Lugli, P.; Zhirnov, V. V.; 2012, Science and engineering beyond Moore's law. *Proc. IEEE* 100, 1720–1749.

Carruthers, P.; Chamberlain, A.; 2000, *Evolution and the Human Mind: Modularity, Language, and Meta-Cognition.* Cambridge University Press, Cambridge, UK.

Cencini, M.; Cecconi, F.; Vulpiani, A.; 2009, *Chaos: From Simple Models to Complex Systems.* World Scientific Publishing, Singapore.

Chandrasekhar, S.; 1961, *Hydrodynamic and Hydromagnetic Stability.* Clarendon Press, Oxford.

Chaisson, E. J.; 2002, *Cosmic Evolution. The Rise of Complexity in Nature.* Harvard University Press, Cambridge, MA.

Chaitin, G.; 2012, *Proving Darwin. Making Biology Mathematical.* Pantheon Books, New York.

Chance, B.; Pye, E. K.; Ghosh, A. K.; Hess, B.; (Eds.) 1973, *Biological and Biochemical Oscillators.* Academic Press, New York.

Chase, G. C.; 1980, History of mechanical computing machinery. *Ann. Hist. Comput.* 2, 198–226.

Chock, P. B.; Stadtman, E. R.; 1977, Superiority of interconvertible enzyme cascades in metabolic regulation: Analysis of multicyclic systems. *Proc. Natl. Acad. Sci. USA* 74, 2766–2770.

Christensen, K.; Moloney, N. R.; 2005, *Complexity and Criticality.* Imperial College Press, London.

Chruściński, D.; 2006, Geometric aspect of quantum mechanics and quantum entanglement. *J. Phys. Conf. Series* 39, 9–16.

Chua, L. O.; 1971, Memristor: The missing circuit element. *IEEE Trans. Circuit Theory* 18, 5, 507–519.

Chua, L. O.; 1998, *CNN: A Paradigm for Complexity.* World Scientific, Singapore.

Church, G. M.; Gao, Y.; Kosuri, S.; 2012, Next-generation digital information storage in DNA. *Science* 337, 1628.

Cipra, B.; 2000, The Ising model is NP-complete. *SIAM News* 33, 6.

Citri, O.; Epstein, I. R.; 1986, Mechanism for the oscillatory Bromate-Iodide reaction. *J. Am. Chem. Soc.* 108, 357–363.

Clayton, P.; Davies, P.; (Eds.) 2006, *The Re-Emergence of Emergence: The Emergentist Hypothesis from Science to Religion.* Oxford University Press, New York.

Cline, D. B.; (Ed.) 1996, *The Physical Origin of Homochirality in Life.* American Institute of Physics, New York.

Coletti, G.; Scozzafava, R.; 2004, Conditional probability, fuzzy sets, and possibility: A unifying view. *Fuzzy Sets Syst.* 144, 1, 227–249.

Colussi, A. J.; Ghibaudi, E.; Yuan, Z.; Noyes, R. M.; 1990, Oscillatory oxidation of benzaldehyde by air. 1. Experimental observation. *J. Am. Chem. Soc.* 112, 8660–8670.

Coniglio, A.; De Arcangelis, L.; Herrmann, H. J.; 1989, Fractals and multifractals: applications in physics. *Physica A* 157, 21–30.

Conrad, M.; 1992, Molecular computing: The lock-key paradigm. *Computer* 25, 11–20.

Cronin, L.; Walker, S. I.; 2016, Beyond prebiotic chemistry. *Science* 352, 1174–1175.

Corning, P. A.; 2002, The Re-emergence of "emergence": A venerable concept in search of a theory. *Complexity* 7, 18–30.

Cross, M. C.; Hohenberg, P. C.; 1993, Pattern formation outside of equilibrium. *Rev. Mod. Phys.* 65, 851–1112.

Crutchfield, J. P.; 1994, The calculi of emergence: Computation, dynamics and induction. *Physica D* 75, 11–54.

Crutchfield, J. P.; Young, K.; 1989, Inferring statistical complexity. *Phys. Rev. Lett.* 63, 105–108.

Cubitt, T. S.; Eisert, J.; Wolf, M. M.; 2012, Extracting dynamical equations from experimental data is Np hard. *Phys. Rev. Lett.* 108, 120503.

Cui, Q.; Karplus, M.; 2008, Allostery and cooperativity revisited. *Protein Sci.* 17, 1295–1307.

Curie, P.; 1894, Sur la symétrie dans les phénomènes physiques, symétrie d'un champ électrique et d'un champ magnétique. *J. Phys. Theor. Appl.* 3, 393–415.

Cvitanovic, P.; 1989, *Universality in Chaos.* 2nd ed. Taylor and Francis Group, New York.

Dahlem, M. A.; Müller, S. C.; 2004, Reaction-diffusion waves in neuronal tissue and the window of cortical excitability. *Ann. Phys. (Leipzig),* 13, 442–449.

Darwin, C.; 1859, *On the Origin of Species by Means of Natural Selection, or the Preservation of Favoured Races in the Struggle for Life.* 1st ed. John Murray, London.

Das, J.; Busse, H.-G.; 1985, Long term oscillation in glycolysis. *J. Biochem.* 97, 719–727.

Dateo, C. E.; Orbán, M.; De Kepper, P.; Epstein, I. R.; 1982, Bistability and oscillations in the autocatalytic chlorite-iodide reaction in a stirred-flow reactor. *J. Am. Chem. Soc.* 104, 504–509.

Davies, M. L.; Halford-Maw, P. A.; Hill, J.; Tinsley, M. R.; Johnson, B. R.; Scott, S. K.; Kiss, I. Z.; Gáspár, V.; 2000, Control of chaos in combustion reaction. *J. Phys. Chem. A.* 104, 9944–9952.

Davies, P. C. W.; 2012, The epigenome and top-down causation. *Interface Focus* 2, 42–48.

de Castro, L. N.; 2007, Fundamentals of natural computing: An overview. *Phys. Life Rev.* 4, 1–36.

de Castro, L. N.; Timmis, J.; 2002, *Artificial Immune Systems: A New Computational Approach.* Springer-Verlag, London.

De Finetti, B.; Machi, A.; Smith, A.; 1992, *Theory of Probability: A Critical Introductory Treatment.* John Wiley & Sons, New York, NY.

De Kepper, P.; Epstein, I. R.; Kustin, K.; 1981a, A systematically designed homogeneous oscillating chemical reaction. The arsenite-iodate-chlorite system. *J. Am. Chem. Soc.* 103, 2133–2134.

De Kepper, P.; Epstein, I. R.; Kustin, K.; 1981b, Bistability in the oxidation of arsenite by iodate in a stirred flow reactor. *J. Am. Chem. Soc.* 103, 6121–6127.

Denbigh, K.; 1981, *The Principles of Chemical Equilibrium with Applications in Chemistry and Chemical Engineering.* Cambridge University Press, Cambridge, UK.

Deneke, V. E.; Di Talia, S.; 2018, Chemical waves in cell and developmental biology. *J. Cell Biol.* 217, 1193–1204.

Deneve, S.; Pouget, A.; 2004, Bayesian multisensory integration and cross-modal spatial links. *J. Physiol. Paris* 98, 249–258.

Dennard, R. H.; Gaensslen, F. H.; Yu, H.-N.; Rideout, L.; Bassous, E.; LeBlanc, A.; 1999, Design of ion-implanted MOSFET's with very small physical dimensions. *Proc. IEEE* 87, 4, 668–678.

de Silva, A. P.; 2013, *Molecular Logic-based Computation.* The Royal Society of Chemistry, Cambridge, UK.

Descartes, R.; 1664, Treatise of man. translated and commented by Steele Hall, T.; 2003, Prometheus Books, New York.

Deutsch, D.; 1989, Quantum computational networks. *Proc. R. Soc. London A* 425, 73–90.

Devlin, K.; 2002, *The Millennium Problems: The Seven Greatest Unsolved Mathematical Puzzles of Our Time.* Basic Books, New York, NY (USA).

Dewar, R. C.; 2009, *Maximum Entropy Production as an Inference Algorithm that Translates Physical Assumptions into Macroscopic Predictions: Don't Shoot the Messenger.* Entropy (Special Issue: What is Maximum Entropy Production and how should we apply it?) 11, 931–944. (doi:10.3390/e11040931)

Dewar, R. C.; Lineweaver, C. H.; Niven, R. K.; Regenauer-Lieb, K.; (Eds.) 2014, B*eyond the Second Law. Entropy Production and Non-Equilibrium Systems.* Springer-Verlag, Berlin, Germany.

Dewar, R. C.; Maritan, A.; 2014, A theoretical basis for maximum entropy production. Chapter 3, pp. 49–71 in *Beyond the Second Law.* Dewar, R. C.; Lineweaver, C. H.; Niven, R. K.; Regenauer-Lieb, K.; (Eds.), Springer-Verlag, Berlin, Germany.

Dillenschneider, R.; Lutz, E.; 2009, Memory erasure in small systems. *Phys. Rev. Lett.* 102, 210601.

Ditto, W. L.; Miliotis, A.; Murali, K.; Sinha, S.; Spano, M. L.; 2010, Chaogates: Morphing logic gates that exploit dynamical patterns. *Chaos* 20, 037107.

Ditto, W. L.; Rauseo, S. N.; Spano, M. L.; 1990, Experimental control of chaos. *Phys. Rev. Lett. A* 65, 3211–3214.

Di Ventra, M.; Pershin, Y. V.; 2013, The parallel approach. *Nat. Phys.* 9, 200–202.

DiVincenzo, D. P.; 2000, The physical implementation of quantum computation. *Fortschr. Phys.* 48, 771–783.

Donahue, B. S.; Abercrombie, R. F.; 1987, Free diffusion coefficient of ionic calcium in cytoplasm. *Cell Calcium* 8, 437–448.

Donoghue, J. P.; 2008, Bridging the brain to the world: A perspective on neural interface systems neuron. *Neuron* 60, 511–521.

Dorigo, M.; Maniezzo, V.; Colorni, A.; 1996, Ant system: Optimization by a colony of cooperating agents. *IEEE T. Syst. Man Cy. B* 26 (1), 29–41.

Dormann, D.; Vasiev, B.; Weijer, C. J.; 1998, Propagating waves control Dictyostelium discoideum morphogenesis. *Biophys. Chem.* 72, 21–35.

Dormann, D.; Weijer, C. J.; 2001, Propagating chemoattractant waves coordinate periodic cell movement in Dictyostelium slugs. *Development* 128, 4535–4543.

Driver, J.; Spense, C.; 2000, Multisensory perception: Beyond modularity and convergence. *Curr. Biol.* 10, R731–R735.

Duane, G. S.; Grabow, C.; Selten, F.; Ghill, M.; 2017, Introduction to focus issue: Synchronization in large networks and continuous media—data, models, and supermodels. *Chaos* 27, 126601.

Dufiet, V.; Boissonade, J.; 1992, Numerical studies of Turing patterns selection in a two-dimensional system. *Physica A* 188, 158–171.

Eckmann, J.-P.; Ruelle, D.; 1985, Ergodic theory of chaos and strange attractors. *Rev. Mod. Phys.* 57, 617–656.

Edblom, E. C.; Luo, Y.; Orbán, M.; Kustin, K.; Epstein, I. R.; 1989, Kinetics and Mechanism of the Oscillatory Bromate-Sulfite-Ferrocyanide Reaction. *J. Phys. Chem.* 93, 2722–2727.

Edblom, E. C.; Orbán, M.; Epstein, I. R.; 1986, A new iodate oscillator: The Landolt reaction with ferrocyanide in a CSTR. *J. Am. Chem. Soc.* 108, 2826–2830.

EIA (Energy Information Administration), 2016, Report Number: DOE/EIA-0484(2016).

Einstein, A.; 1916, translated in 1920, *Relativity: The Special and General Theory.* H. Holt and Company, New York.

Eisenstein, M.; 2013 Chronobiology: Stepping out of time. *Nature* 497, S10–S12.

Eiswirth, M.; Freund, A.; Ross, J.; 1991, Mechanistic classification of chemical oscillators and the role of species. *Adv. Chem. Phys.* 80, 127–199.

Elaydi, S. N.; 2005, *An introduction to Difference Equations.* 3rd ed. Springer, New York.

Eldredge, N.; Gould, S. J.; 1972, Punctuated equilibria: An alternative to phyletic gradualism. pp. 82–115 in *Models in paleobiology.* Schopf, T. J. M.; (Ed.), Freeman, Cooper, San Francisco, CA.

Elowitz, M. B.; Leibler, S.; 2000, A synthetic oscillatory network of transcriptional regulators. *Nature* 403, 335–338.

Elton, C.; Nicholson, M.; 1942, The ten-year cycle in numbers of the lynx in Canada. *J. Anim. Ecol.* 11 (2), 215–244.

Epstein, I. R.; Luo, Y.; 1991, Differential delay equations in chemical kinetics. Nonlinear models: The cross-shaped phase diagram and the Oregonator. *J. Chem. Phys.* 95, 244–254.

Epstein, I. R.; Pojman, J. A.; 1998, *An Introduction to Nonlinear Chemical Dynamics: Oscillations, waves, Patterns and Chaos.* Oxford University Press, New York.

Erdös, P.; Rényi, A.; 1960, On the evolution of random graphs. *Publ. Math. Inst. Hung. Acad. Sci.* 5, 17–61.

Erhart, J.; Sponar, S.; Sulyok, G.; Badurek, G.; Ozawa, M.; Hasegawa, Y.; 2012, Experimental demonstration of a universally valid error-disturbance uncertainty relation in spin measurements. *Nature* 8, 185–189.

Ermentrout, G. B.; Edelstein-Keshet, L.; 1993, Cellular automata approaches to biological modeling. *J. Theor. Biol.* 160, 97–133.

Ernst, M. O.; Banks, M. S.; 2002, Humans integrate visual and haptic information in a statistically optimal fashion. *Nature* 415, 429–433.

Evans, D. J.; Searles, D. J.; 2002, The Fluctuation theorem. *Adv. Phys.* 51, 7, 1529–1585.

Extance, A.; 2016, How DNA could store all the world's data. *Nature* 537, 22–24.

Falconer, K.; 1990, *Fractal Geometry: Mathematical Foundations and Applications.* John Wiley & Sons, Chichester, UK.

Farquhar, R.; Muhonen, D.; Davis, S. A.; 1985, Trajectories and orbital maneuvers for the ISEE-3/ICE comet mission. *J. Astronaut. Sci.* 33, 235–254.

Fausett, L. V.; 1994, *Fundamentals of Neural Networks: Architectures, Algorithms, and Applications.* Prentice Hall, Upper Saddle River, NJ.

Favaro, G.; di Nunzio, M. R.; Gentili, P. L.; Romani, A.; Becker, R. S.; 2007, Effects of the exciting wavelength and viscosity on the photobehavior of 9- and 9,10-bromoanthracenes. *J. Phys. Chem. A* 111, 5948–5953.

Fechner, A. T.; 1828, Über Umkehrungen der Polarität der einfachen Kette. *Schweigg. J.* 53, 61–76.

Feder, J.; 1988, *Fractals.* Plenum Press, New York.

Feigenbaum, M. J.; 1979, The universal metric properties of nonlinear transformations. *J. Stat. Phys.* 21, 669–706.

Feigenbaum, M. J.; 1980, Universal behaviour in nonlinear systems. *Los Alamos Sci.*, 1, 4–27.

Feldman, D. P.; 2012, *Chaos and Fractals: An Elementary Introduction.* Oxford University Press, Oxford.

Ferrell, J. E. Jr.; Ha, S. H.; 2014a, Ultrasensitivity Part I: Michaelian responses and zero-order ultrasensitivity. *Trends Biochem. Sci.* 39, 496–503.

Ferrell, J. E. Jr.; Ha, S. H.; 2014b, Ultrasensitivity Part II: Multisite phosphorylation, stoichiometric inhibitors, and positive feedback. *Trends Biochem. Sci.* 39, 556–569.

Ferrell, J. E. Jr.; Ha, S. H.; 2014c, Ultrasensitivity Part III: Cascades, bistable switches, and oscillators. *Trends Biochem. Sci.* 39, 612–618.

Ferrell, J. E. Jr.; Tsai, T. Y.-C.; Yang, Q.; 2011, Modelling the cell cycle: Why do certain circuits oscillate? *Cell* 144, 874–885.

Ferrell, J. E. Jr.; Xiong, W.; 2001, Bistability in cell signalling: How to make continuous processes discontinuous, and reversible processes irreversible. *Chaos* 11, 227–236.

Fewell, J. H.; 2003, Social insect networks. *Science* 301, 1867–1870.

Feynman, R. P.; 1960, There's plenty of room at the bottom. *Engineering and Science* 23, 22–36.

Feynman, R. P.; 1982, Simulating physics with computers. *Int. J. Theor. Phys.* 21, 467–488.

Feynman, R. P.; Leighton, R. B.; Sands, M.; 1963, *The Feynman Lectures on Physics*. Vol. 1, Chapter 46, Addison-Wesley, Reading, MA.

Field, R. J.; Körös, E.; Noyes, R. M.; 1972, Oscillations in chemical systems. II. Thorough analysis of temporal oscillation in the bromate-cerium-M4ic acid system. *J. Am. Chem. Soc.* 94, 8649–8664.

Field, R. J.; Noyes, R. M.; 1974a, Oscillations in chemical systems. IV. Limit cycle behavior in a model of a real chemical reaction. *J. Chem. Phys.* 60, 1877–1884.

Field, R. J.; Noyes, R. M.; 1974b, Oscillations in chemical systems. V. Quantitative explanation of band migration in the Belousov-Zhabotinskii reaction. *J. Am. Chem. Soc.* 96, 2001–2006.

Fife, P. C.; 1984, Propagator-controller systems and chemical patterns. pp. 76–88 in *Non-equilibrium Dynamics in Chemical Systems*. Vidal, C.; Pacault, A.; (Eds.), Springer-Verlag, Berlin, Germany.

Fisher, I.; 1933, The Debt-Deflation theory of great depressions. *Econometrica* 1, 337–357.

Fisher, R. A.; 1937, The wave of advance of advantageous gene. *Ann. Eugenics* 7, 355–369.

Foerster, P.; Müller, S. C.; Hess, B.; 1988, Curvature and propagation velocity of chemical waves. *Science* 241, 685–687.

Fortnow, L.; 2009, The status of the P versus NP problem. *Commun. ACM* 52, 78–86.

Foster, R. G.; Roenneberg, T.; 2008, Human responses to the geophysical daily, annual and lunar cycles. *Cur. Biol.* 18, R784–R794.

Franck, U. F.; 1985, Spontaneous temporal and spatial phenomena in physicochemical systems. pp. 2–12 in *Temporal Order*. Rensing, L.; Jaeger, N. I.; (Eds.), Springer, Berlin, Germany.

Fraser, A. M.; Swinney, H. L.; 1986, Independent coordinates for strange attractors from mutual information. *Phys. Rev. A* 33, 1134–1140.

Frerichs, G. A.; Thomson, R. C.; 1998, A pH-regulated chemical oscillator: The homogeneous system of hydrogen peroxide-sulfite-carbonate-sulfuric acid in a CSTR. *J. Phys. Chem. A* 102, 8142–8149.

Friedmann, H. S.; Zhou, H.; von der Heydt, R.; 2003, The coding of uniform color figures in monkey visual cortex. *J. Physiol. (Lond.)* 548, 593–613.

Frigg, R.; 2004, In what sense is the Kolmogorov-Sinai entropy a measure for chaotic behaviour?—Bridging the gap between dynamical systems theory and communication theory. *Brit. J. Phil. Sci.* 55, 411–434.

Fujii, T.; Rondelez, Y.; 2013, Predator-prey molecular ecosystems. *ACS Nano* 7, 27–34.

Furuta, A.; 2012, https://www.scientificamerican.com/article/heisenbergs-uncertainty-principle-is-not-dead/.

Gallistel, C. R.; King, A.; 2010, *Memory and the Computational Brain: Why Cognitive Science Will Transform Neuroscience*. Wiley- Blackwell, New York.

Gammaitoni, L.; Chiuchiú, D.; Madami, M.; Carlotti, G.; 2015, Towards zero-power ICT. *Nanotechnology* 26, 222001.

Gardner, M.; 1970, The fantastic combinations of John Conway's new solitaire game "life." *Sci. Am.* 223, 120–123.

Garey, M. R., Johnson, D. S.; 1979, *Computers and Intractability: A Guide to the Theory of NP-Completeness*. W. H. Freeman & Co., New York.

Garfinkel, A.; Spano, M. L.; Ditto, W. L.; Weiss, J. N.; 1992, Controlling cardiac chaos. *Science* 257, 1230–1235.

Gegenfurtner, K. R.; 2003, Cortical mechanisms of colour vision. *Nat. Rev. Neurosci.* 4, 563–572.

Gell-Mann, M.; 1994, *The Quark and the Jaguar*. Holt Paperbacks, Henry Holt and Company, LLC, New York.

Gentili, P. L.; 2011a, Molecular processors: From Qubits to fuzzy logic. *ChemPhysChem.* 12, 739–745.

Gentili, P. L.; 2011b, The fundamental Fuzzy logic operators and some complex Boolean logic circuits implemented by the chromogenism of a spirooxazine. *Phys. Chem. Chem. Phys.* 13, 20335–20344.

Gentili, P. L.; 2011c, Molecular fuzzy inference engines: Development of chemical systems to process fuzzy logic at the molecular level. *ICAART 2011: Proceedings of the 3rd International Conference on Agents and Artificial Intelligence* 1, 205–210.

Gentili, P. L.; 2013, Small steps towards the development of chemical artificial intelligent systems. *RSC Adv.* 3, 25523–25549.

Gentili, P. L.; 2014a, The human sensory system as a collection of specialized fuzzifiers: A conceptual framework to inspire new artificial intelligent systems computing with words. *J. Intel. & Fuzzy Sys.* 27, 2137–2151.

Gentili, P. L.; 2014b, The fuzziness of a chromogenic spirooxazine. *Dyes and Pigments* 110, 235–248.

Gentili, P. L.; 2017, A strategy to face complexity: The development of chemical artificial intelligence. pp. 151–160 in *Communications in Computer and Information Science, Vol. 708*. Rossi, F.; Piotto, S.; Concilio, S.; (Eds.), Springer International Publishing, Cham, Switzerland.

Gentili, P. L.; Dolnik, M.; Epstein, I. R.; 2014, "Photochemical oscillator": Colored hydrodynamic oscillations and waves in a photochromic system. *J. Phys. Chem. C* 118, 598–608.

Gentili, P. L.; Giubila, M. S.; Germani, R.; Romani, A.; Nicoziani, A.; Spalletti, A.; Heron, B. M.; 2017a, Optical communication among oscillatory reactions and photo-excitable systems: UV and visible radiation can synchronize artificial neuron models.*Angew. Chem. Int. Ed.* 56, 7535–7540.

Gentili, P. L.; Giubila, M. S.; Heron, B. M.; 2017b, Processing binary and fuzzy logic by chaotic time series generated by a hydrodynamic photochemical oscillator. *ChemPhysChem*. 18, 1831–1841.

Gentili, P. L.; Gotoda, H.; Dolnik, M.; Epstein, I. R.; 2015, Analysis and prediction of aperiodic hydrodynamic oscillatory time series by feed-forward neural networks, fuzzy logic, and a local nonlinear predictor. *Chaos* 25, 013104.

Gentili, P. L.; Horvath, V.; Vanag, V. K.; Epstein, I. R.; 2012, Belousov-Zhabotinsky "chemical neuron" as a binary and fuzzy logic processor. *Int. J. Unconv. Comput.* 8, 177–192.

Gentili, P. L.; Ortica, F.; Favaro, G.; 2008, Static and dynamic interaction of a naturally occurring photochromic molecule with bovine serum albumin studied by UV-visible absorption and fluorescence spectroscopy. *J. Phys. Chem. B* 112, 16793–16801.

Gentili, P. L.; Rightler, A. L.; Heron, B. M.; Gabbutt, C. D.; 2016, Extending human perception of electromagnetic radiation to the UV region through biologically inspired photochromic fuzzy logic (BIPFUL) systems. *Chem. Commun.* 52, 1474–1477.

Gershenfeld, N. A.; Weigend, A. S.; 1993, The future of time series: Learning and understanding. In *Time Series Prediction: Forecasting the Future and Understanding the Past.* Weigend, A. S.; Gershenfeld, N. A.; (Eds.), Addison-Wesley, Reading, MA.

Georgescu-Roegen, N.; 1971, *The Entropy Law and the Economic Process* Harvard University Press, Cambridge.

Georgescu-Roegen, N.; 1976, *Energy and Economic Myths* Pergamon Press, Oxford, United Kingdom.

GISTEMP Team, 2017: GISS Surface Temperature Analysis (GISTEMP). NASA Goddard Institute for Space Studies. Dataset accessed 2017-08-11 at https://data.giss.nasa.gov/gistemp/.

Glansdorff, P.; Prigogine, I.; 1954, Sur Les Propriétés Différentielles De La Production d'Entropie *Physica* 20, 773–780.

Glass, L.; 2001, Synchronization and rhythmic processes in physiology. *Nature* 410, 277–284.

Glass, L.; Mackey, M. C.; 1988, *From Clocks to Chaos: The Rhythms of Life.* Princeton University Press, Princeton, NJ.

Gleick, J.; 1987, *Chaos: Making a New Science.* Viking Penguin, New York.

Glykas, M.; (Ed.) 2010, *Fuzzy Cognitive Maps. Advances in Theory, Methodologies, Tools and Applications.* Springer-Verlag, Berlin, Germany.

Gierer, A.; Meinhardt, H.; 1972, A theory of biological formation. *Kybernetik* 12, 30–39.

Goehring, N. W.; Grill, S. W.; 2013, Cell polarity: Mechanochemical patterning. *Trends Cell Biol.* 23, 72–80.

Goldberger, A. L.; Amaral, L. A. N.; Glass, L.; Hausdorff, J. M.; Ivanov, P. Ch.; Mark, R. G.; Mietus, J. E.; Moody, G. B.; Peng, C.-K.; Stanley, H. E.; 2000, PhysioBank, PhysioToolkit, and PhysioNet: Components of a new research resource for complex physiologic signals. *Circulation* 101 (23), e215–e220 [Circulation Electronic Pages; http://circ.ahajournals.org/content/101/23/e215.full].

Goldbeter, A.; 1996, *Biochemical Oscillations and Cellular Rhythms.* Cambridge University Press, Cambridge, UK.

Goldbeter, A.; 2017, Dissipative structures and biological rhythms. *Chaos* 27, 104612.

Goldbeter, A.; Caplan, S. R.; 1976, Oscillatory enzymes. *Annu. Rev. Biophys. Bio.* 5, 449–476.

Goldbeter, A.; Pourquié, O.; 2008, Modeling the segmentation clock as a network of coupled oscillations in the Notch, Wnt and FGF signaling pathways. *J. Theor. Biol.* 252, 574–585.

Goldenfeld, N.; Woese, C.; 2011, Life is physics: Evolution as a collective phenomenon far from equilibrium. *Annu. Rev. Condens. Matter Phys.* 2, 375–399.

Goldreich, O.; 2008, *Computational Complexity. A Conceptual Perspective.* Cambridge University Press, New York.

Goldstein, H. H.; 1993, *The Computer from Pascal to von Neumann.* Princeton University Press, Princeton, NJ.

Goodey, N. M.; Benkovic, S. J.; 2008, Allosteric regulation and catalysis emerge via a common route. *Nat. Chem. Biol.* 4, 474–482.

Goodwin, R. M.; 1967, A growth cycle. pp. 54–58 in *Socialism, Capitalism and Economic Growth.* Feinstein, C. H.; (Ed.), Cambridge University Press, Cambridge.

Gorini, R.; 2003, Al-Haytham the man of experience. First steps in the science of vision. *JISHIM* 2, 53–55.

Gould, S. B.; 2012, Algae's complex origins. *Nature* 492, 46–48.

Grančič, P.; Štěpánek, F.; 2011, Chemical swarm robots. in *Handbook of Collective Robotics: Fundamentals and Challenges.* Kernbach, S.; (Ed.), Pan Stanford Publishing, Singapore.

Grassberger, P.; Procaccia, I.; 1983, Characterization of strange attractor. *Phys. Rev. Lett.* 50, 346–349.

Grassé, P. P.; 1959, La reconstruction du nid et les coordinations inter-individuelles chez Bellicositermes Natalensis et Cubitermes sp. La théorie de la stigmergie: essai d'interprétation du comportement des termites constructeurs. *Insectes Sociaux* 6, 41–84.

Gray, C. M.; 1994, Synchronous oscillations in neuronal systems: Mechanisms and functions. *J. Comput. Neurosci.* 1, 11–38.

Gregor, T.; Fujimoto, K.; Masaki, N.; Sawai, S.; 2010, The onset of collective behavior in social amoebae. *Science* 328, 1021–1025.

Grier, D. G.; 2003, A revolution in optical manipulation. *Nature* 424, 810–816.

de Groot, S. R.; Mazur, P.; 1984, *Nonequilibrium Thermodynamics* Dover Publications: New York.

Grzybowski, B. A.; 2009, *Chemistry in Motion: Reaction-Diffusion Systems for Micro- and Nanotechnology.* John Wiley & Sons, The Atrium, Southern Gate, Chichester, UK.

Grzybowski, B. A.; Wilmer, C. E.; Kim, J.; Browne, K. P.; Bishop, K. J. M.; 2009, Self-assembly: From crystals to cells *Soft Matter* 5, 1110–1128.

Gutenberg, B.; Richter, C. F.; 1956, Magnitude and energy of earthquakes. *Ann. Geophys.* 9, 1–15.

Gwinner, E.; 1986, *Circannual Rhythms. Endogenous Annual Clocks in the Organization of Seasonal Processes.* Springer-Verlag, Berlin, Germany.

Ha, S. D.; Ramanathan, S.; 2011, Adaptive oxide electronics: A review. *J. Appl. Phys.* 110, 071101.

Haeckel, E.; 1974, *Art Forms in Nature.* Dover Publications, New York.

Hagiya, M.; Konagaya, A.; Kobayashi, S.; Saito, H.; Murata, S.; 2014, Molecular robots with sensors and intelligence. *Acc. Chem. Res.* 47, 1681–1690.

Haken, H.; 1983, *Synergetics, an Introduction: Nonequilibrium Phase Transitions and Self-Organization in Physics, Chemistry and Biology.* 3rd edition. Springer-Verlag, New York.

Haken, H.; 2004, *Synergetics. Introduction and Advanced Topics.* Springer-Verlag, Berlin, Germany.

Haken, H.; 2006, *Information and Self-Organization: A Macroscopic Approach to Complex Systems.* Springer-Verlag, Berlin, Germany.

Halloy, J.; Lauzeral, J.; Goldbeter, A.; 1998, Modeling oscillations and waves of cAMP in Dictyostelium discoideum cells. *Biophys. Chem.* 72, 9–19.

Halsey, T. C., Jensen, M. H.; Kadanoff, L. P.; Procaccia, I.; Shraiman, B. I.; 1986, Fractal measures and their singularities: The characterization of strange sets. *Phys. Rev. A* 33, 1141–1151.

Hammes, G. G.; Wu, C. W.; 1974, Kinetics of allosteric enzymes. *Annu. Rev. Biophys. Bioeng.* 3, 1–33.

Hayashi, K.; Gotoda, H.; Gentili, P. L.; 2016, Probing and exploiting the chaotic dynamics of a hydrodynamic photochemical oscillator to implement all the basic binary logic functions. *Chaos* 26, 053102.

Hemphill, J.; Deiters, A.; 2013, DNA computation in mammalian cells: MicroRNA logic operations. *J. Am. Chem. Soc.* 135, 10512–10518.

Hess, H.; Ross, J. L.; 2017, Non-equilibrium assembly of microtubules: From molecules to autonomous chemical robots. *Chem. Soc. Rev.* 46, 5570–5587.

Hanna, A.; Saul, A.; Showalter, K.; 1982, Detailed studies of propagating fronts in the iodate oxidation of arsenous acid. *J. Am. Chem. Soc.* 104, 3838–3844.

Harris, M. P.; Williamson, S.; Fallon, J. F.; Meinhardt, H.; Prum, R. O.; 2005, Molecular evidence for an activator-inhibitor mechanism in development of embryonic feather branching. *Proc. Natl. Acad. Sci. USA* 102, 11734–11739.

Harrison, G. W.; 1979, Global stability of Predator-Prey interactions *J. Math. Biology* 8, 159–171.

Harrison, G. W.; 1995, Comparing Predator-Prey models to Luckinbill's experiment with didinium and paramecium *Ecology* 76, 357–374.

Harte, D.; 2001, *Multifractals. Theory and Applications.* Chapman & Hall/CRC, Boca Raton, FL.

Hartmann, E.; 1968, The 90-minute sleep-dream cycle. *Arch. Gen. Psychiatry* 18, 280–286.

Hartwell, L. H.; Hopfield, J. J.; Leibler, S.; Murray, A. W.; 1999, From molecular to modular cell biology. *Nature* 402, C47–C52.

Harvey, E. N.; 1957, *A History of Luminescence. From the Earliest Times until 1900.* American Philosophical Society, Philadelphia, PA.

Harvie, D.; 2000, Testing Goodwin: Growth cycles in ten OECD countries. *Camb. J. Econ.* 24 (3), 349–376.

Hassoun, M. H.; 1995, *Fundamentals of Artificial Neural Networks.* MIT Press, Cambridge, MA.

Hauser, M. J. B.; Strich, A.; Bakos, R.; Nagy-Ungvarai, Zs.; Müller, S. C.; 2002, pH oscillations in the hemin-hydrogen peroxide-sulfite reaction. *Faraday Discuss.* 120, 229–236.

Havlin, S.; Ben-Avraham, D.; 1987, Diffusion in disordered media. *Adv. Phys.* 36, 695–798.

Haynie, D. T.; 2001, *Biological Thermodynamics.* Cambridge University Press, New York.

Hegger, R.; Kantz, H.; Schreiber, T.; 1999, Practical implementation of nonlinear time series methods: The TISEAN package. *Chaos* 9, 413–435.

Heisenberg, W.; 1927, Über den anschaulichen Inhalt der quantentheoretischen Kinematik und Mechanik. *Zeitschrift für Physik* 43 (3–4), 172–198.

Helman, J. S.; Coniglio, A.; Tsallis, C.; 1984, Fractons and the fractal structure of proteins. *Phys. Rev. Lett.* 53, 1195–1197.

Henisch, H. K.; 1988, *Crystals in Gels and Liesegang Rings.* Cambridge University Press, Cambridge.

Hénon, M.; 1976, A two-dimensional mapping with a strange attractor. *Comm. Math. Phys.* 50, 69–77.

Hiratsuka, M.; Aoki, T.; Higuchi, T.; 2001, A model of reaction-diffusion cellular automata for massively parallel molecular computing. *Proceedings 31st IEEE International Symposium on Multiple-Valued Logic*, Warsaw, 2001, pp. 247–252.

Hodgkin, A. L.; Huxley, A. F.; 1939, Action potentials recorded from inside a nerve fibre. *Nature* 144, 710–711.

Hodgkin, A. L.; Huxley, A. F.; 1952, A quantitative description of membrane current and its application to conduction and excitation in nerve. *J. Physiol.* 117, 500–544.

Hoffmann, P. M.; 2012, *Life's Ratchet. How Molecular Machines Extract Order from Chaos.* Basic Books, New York.

Hohmann, W.; Kraus, M.; Schneider, F. W.; 1999, Pattern recognition by electrical coupling of eight chemical reactors. *J. Phys. Chem. A* 103, 7606–7611.

Holland, J. J.; 1975, *Adaptation in Natural and Artificial Systems.* MIT Press, Cambridge, MA.

Holling, C. S.; 1959, The components of predation as revealed by a study of small-mammal predation of the European pine sawfly. *Can. Entomol.* 91, 293–320.

Hong, F.; Zhang, F.; Liu, Y.; Yan, H.; 2017, DNA origami: Scaffolds for creating higher order structures. *Chem. Rev.* 117, 12584–12640.

Horvath, V.; Gentili, P. L.; Vanag, V. K.; Epstein, I. R.; 2012, Pulse-coupled chemical oscillators with time delay. *Angew. Chem. Int. Ed.* 51, 6878–6881.

Horváth, V.; Kurin-Csörgei, K.; Epstein, I. R.; Orbán, M.; 2010, Oscillatory concentration pulses of some divalent metal ions induced by a redox oscillator. *Phys. Chem. Chem. Phys.* 12, 1248–1252.

Horward, J.; Grill, S. W.; Bois, J. S.; 2011, Turing's next steps: The mechanochemical basis of morphogenesis. *Nat. Rev. Mol. Cell Biol.* 12, 392–398.

Hovers, E.; 2015, Tools go back in time. *Nature* 521, 294–295.

Hu, G.; Pojman, J. A.; Scott, S. K.; Wrobel, M. M.; Taylor, A. F.; 2010, Base-catalyzed feedback in the Urea-Urease reaction. *J. Phys. Chem. B* 114, 14059–14063.

Huberts, D. H. E. W.; van der Klei, I. J.; 2010, Moonlighting proteins: An intriguing mode of multitasking. *Biochim. Biophys. Acta* 1803, 520–525.

Hunt, E. R.; 1991, Stabilizing high-period orbits in a chaotic system: The diode resonator. *Phys. Rev. Lett.* 67, 1953–1955.

Huxley, J. S.; 1932, *Problems of Relative Growth.* Methuen & Co, London.

Ichimaru, Y.; Moody, G. B.; 1999, Development of the polysomnographic database on CD-ROM. *Psychiatry Clin. Neurosci.* 53, 175–177.

Idso, C. D.; Carter, R. M.; Singer, S. F.; 2013, *Climate Change Reconsidered II: Physical Science.* Published for the Nongovernmental International Panel on Climate Change (NIPCC). The Heartland Institute, Chicago, IL.

Idso, C. D.; Singer, S. F.; 2009, *Climate Change Reconsidered: 2009 Report of the Nongovernmental International Panel on Climate Change (NIPCC).* The Heartland Institute, Chicago, IL.

IPCC; 2013, Working group I Contribution to the Intergovernmental Panel on Climate Change Fifth Assessment Report: The Physical Science Basis. IPCC Fifth Assessment Report: Climate Change 2013. World Meteorological Organization, Geneva, Switzerland.

Ishida, K.; Matsumoto, S.; 1975, Non-equilibrium thermodynamics of temporally oscillating chemical reactions *J. Theor. Biol.* 52, 343–363.

Ishida, Y.; 2004, *Immunity-Based Systems. A Design Perspective.* Springer-Verlag, Berlin, Germany.

Itoh, M.; Chua, L. O.; 2009, Memristor cellular automata and memristor discrete-time cellular neural networks. *Int. J. Bif. Chaos* 19, 3605–3656.

Ivanov, P. Ch.; Nunes Amaral, L. A.; Goldberger, A. L.; Havlin, S.; Rosenblum, M. G.; Struzik, Z. R.; Stanley, H. E.; 1999, Multifractality in human heartbeat dynamics. *Nature* 399, 461–465.

Jaeger, W.; 1965, *Paideia: The Ideals of Greek Culture. I. Archaic Greece: The Mind of Athens.* Oxford University Press, New York.

Jain, A. K.; Duin, R. P. W.; Mao, J.; 2000, Statistical pattern recognition: A review. *IEEE T. Pattern Anal. Machine Intell.* 22, 4–37.

Jaynes, E. T.; 1957, Information theory and statistical mechanics. *Phys. Rev.* 106, 620–630.

Jeffery, C. J.; 2014, An introduction to protein moonlighting. *Biochem. Soc. Trans.* 42, 1679–1683.

Jennings, R. C.; Santabarbara, S.; Belgio, E.; Zucchelli, G.; 2014, The carnot efficiency and plant photosystems. *Biophysics* 59, 230–235.

Jerne, N. K.; 1974, Towards a network theory of the immune system. *Ann. Immunol. (Paris)* 125C, 373–389.

Jeschke, J. M.; Kopp, M.; Tollrian, R.; 2002, Predator functional responses: Discriminating between handling and digesting prey. *Ecol. Monogr.* 72, 95–112.

Ji, S.; 1999, The cell as the smallest DNA-based molecular computer. *BioSystems* 52, 123–133.

Joachim, C.; Gimzewski, J. K.; Aviram, A.; 2000, Electronics using hybrid-molecular and mono-molecular devices. *Nature* 408, 541–548.

Johnson, S.; 2010, *Where Good Ideas Come From: The Natural History of Innovation.* The Penguin Group, New York.

Johnson, E. N.; Hawken, M. J.; Shapley, R.; 2001, The spatial transformation of color in the primary visual cortex of the macaque monkey. *Nat. Neurosci.* 4, 409–416.

Jou, D.; Casas-Vázquez, J.; Lebon, G.; 2010, *Extended Irreversible Thermodynamics.* 4th ed. Springer, New York.

Juglar, C.; 1862, *Des Crises commerciales et leur retour periodique en France, en Angleterre, et aux Etats-Unis.* Guillaumin, Paris (France).

Jun, S.; Wright, A.; 2010, Entropy as the driver of Chromosome segregation. *Nat. Rev. Microbiol.* 8, 600–607.

Kádár, S.; Wang, J.; Showalter, K.; 1998, Noise-supported travelling waves in sub-excitable media. *Nature* 391, 770–772.

Kåhre, J.; 2002, *The Mathematical Theory of Information.* Kluwer Academic Publishers, Norwell, MA.

Kantz, H.; 1994, A robust method to estimate the maximal Lyapunov exponent of a time series. *Phys. Lett. A* 185, 77–87.

Kantz, H.; Kurths, J.; Mayer-Kress, G.; (Eds.) 1998, *Nonlinear Analysis of Physiological Data.* Springer-Verlag, Berlin, Germany.

Kantz, H.; Schreiber, T.; 2003, *Nonlinear Time Series Analysis.* 2nd ed. Cambridge University Press, Cambridge, UK.

Kapral, R.; Showalter, K.; (Eds.) 1995, *Chemical Waves and Patterns.* Springer-Science + Business Media, Dordrecht.

Karr, J. R.; Sanghvi, J. C.; Macklin, D. N.; Gutschow, M. V.; Jacobs, J. M.; Bolival, B. Jr.; Assad-Garcia, N.; Glass, J. I.; Covert, M. W.; 2012, A whole-cell computational model predicts phenotype from genotype. *Cell* 150, 389–401.

Kasdin, N. J.; 1995, Discrete simulation of colored noise and stochastic processes and $1/f^{\alpha}$ power law noise generation. *Proc. IEEE* 83, 802–827.

Kauffmann, S. A.; 1993, *The Origins of Order: Self-Organization and Selection in Evolution.* Oxford University Press, New York.

Kauffman, S. A.; 2016, *Humanity in a Creative Universe.* Oxford University Press, New York.

Kennedy, J.; Eberhart, R.; 1995, Particle swarm optimization. *Proceedings of IEEE International Joint Conference on Neural networks.* IEEE Press, Piscataway, 1942–1948.

Kennel, M. B.; Brown, R.; Abarbanel, H. D.; 1992, Determining embedding dimension for phase-space reconstruction using a geometrical construction. *Phys. Rev. A* 45, 3403–3411.

Kersten, D.; Mamassian, P.; Yuille, A.; 2004, Object perception as Bayesian inference. *Annu. Rev. Psychol.* 55, 271–304.

Kiprijanov, K. S.; 2016, Chaos and beauty in a beaker: The early history of the Belousov-Zhabotinsky reaction. *Ann. Phys. (Berlin)* 528 (3–4), 233–237.

Kirchhoff, G.; 1860, Ueber das Verhältniss zwischen dem Emissionsvermögen und dem Absorptionsvermögen der Körper für Wärme and Licht. *Annalen der Physik und Chemie* 109 (2), 275–301.

Kitano, H.; 2002, Computational systems biology. *Nature* 420, 206–210.

Kitchin, J.; 1923, Cycles and trends in economic factors *Rev. Econ. Stat.* 5 (1), 10–16.

Kleidon, A.; 2010, Life, hierarchy, and the thermodynamic machinery of planet Earth. *Phys. Life Rev.* 7, 424–460.

Klipp, E.; Liebermeister, W.; Wierling, C.; Kowald, A.; 2016, *Systems Biology: A Textbook.* Wiley-Blackwell, Weinheim (Germany).

Kolmogorov, A.; Petrovsky, I.; Piscounoff, N.; 1937, *Etude de l'équation de la diffusion avec croissance de la quantité de matière et son application à un problème biologique.* Moscow University, Bull. Math. 1, 1–25.

Kondepudi, D. K.; 2003, Theory of hierarchical chiral asymmetry. *Mendeleev Commun.* 13 (3), 128–129.

Kondepudi, D. K.; Asakura, K.; 2001, Chiral autocatalysis, spontaneous symmetry breaking, and stochastic behavior. *Acc. Chem. Res.* 34, 946–954.

Kondepudi, D. K.; Nelson, G. W.; 1985, Weak neutral currents and the origin of biomolecular chirality. *Nature* 314, 438–441.

Kondepudi, D. K.; Nelson, G. W.; 1984, Chiral-symmetry-breaking states and their sensitivity in nonequilibrium chimica systems. *Physica A* 125, 465–496.

Kondepudi, D.; Prigogine, I.; 1998, *Modern Thermodynamics. From Heat Engines to Dissipative Structures.* John Wiley & Sons, Chichester, UK.

Kondo, S.; Asai, R.; 1995, A reaction-diffusion wave on the skin of the marine angelfish Pomacanthus. *Nature* 376, 765–768.

Kondo, S.; Miura, T.; 2010, Reaction-diffusion model as a framework for understanding biological pattern formation. *Science* 329, 1616–1620.

Kondo, S.; Shirota, H.; 2009, Theoretical analysis of mechanisms that generate the pigmentation pattern of animals. *Semin. Cell Dev. Boil.* 20, 82–89.

Kondratiev, N. D.; 1935, The long waves in economic life. *Rev. Econ. Stat.* 17 (6), 105–115.

Kopell, N.; Howard, L. N.; 1973, Horizontal bands in the Belousov reaction. *Science* 180, 1171–1173.

Kopelman, R.; 1988, Fractal reaction kinetics. *Science* 241, 1620–1626.

Kopp, G.; Lean, J. L.; 2011, A new, lower value of total solar irradiance: Evidence and climate significance. *Geophys. Res. Lett.* 38, L01706.

Korotayev, A. V.; Tsirel, S. V.; 2010, A spectral analysis of world GDP dynamics: Kondratieff waves, kuznets swings, juglar and kitchin cycles in global economic development, and the 2008–2009 economic crisis. *Structure and Dynamics* 4 (1).

Korpimäki, E.; Krebs, C. J.; 1996, Predation and population cycles of small mammals *Bioscience* 46 (10), 754–764.

Koshland, D. E. Jr; 1980, Biochemistry of sensing and adaptation. *Trends Biochem. Sci.* 5, 297–302.

Koshland, D. E. Jr.; Goldbeter, A.; Stock, J. B.; 1982, Amplification and adaptation in regulatory and sensory systems. *Science* 217, 220–225.

Kosko, B.; 1986, Fuzzy cognitive maps. *Int. J. Man-Machine Studies* 24, 65–75.

Kosko, B.; 1994, Fuzzy systems as universal approximators. *IEEE T. Comput.* 43 (11), 1329–1333.

Kot, M.; 2001, *Elements of Mathematical Ecology.* Cambridge University Press, Cambridge, UK.

Kovács, K. M.; Rábai, Gy.; 2001, Large amplitude pH oscillation in the hydrogen peroxide-dithionite reaction in a flow reactor. *J. Phys. Chem. A* 105, 9183–9187.

Krishnan, Y.; Simmel, F. C.; 2011, Nucleic acid based molecular devices. *Angew. Chem. Int. Ed.*, 50, 3124–3156.

Kuhnert, L.; Agladze, K. I.; Krinsky, V. I.; 1989, Image processing using light-sensitive chemical waves. *Nature* 337, 244–247.

Kuramoto, Y.; 2003, *Chemical Oscillations, Waves, and Turbulence.* Dover Publications, Mineola, New York.

Kurin-Csörgei, K.; Epstein, I. R.; Orbán, M.; 2005, Systematic design of chemical oscillators using complexation and precipitation equilibria. *Nature* 433, 139–142.

Kurzweil, R.; 2005, The Singularity Is Near. Viking Penguin, London.

Kusch, I.; Markus, M.; 1996, Mollusc shell pigmentation: Cellular automaton simulations and evidence for undecidability. *J. Theor. Biol.* 178, 333–340.

Kuznets, S.; 1930, *Secular Movements in Production and Prices. Their Nature and Their Bearing upon Cyclical Fluctuations* Houghton Mifflin, Boston, MA.

Kwapień, J.; Drożdż, S.; 2012, Physical approach to complex systems. *Phys. Rep.* 515, 115–226.

Labrot, V.; De Kepper, P.; Boissonade, J.; Szalai, I.; Gauffre, F.; 2005, Wave patterns driven by chemomechanical instabilities in responsive gels. *J. Phys. Chem. B* 109, 21476–21480.

Laherrère, J.; Sornette, D.; 1998, Stretched exponential distributions in nature and economy: "fat tails" with characteristic scales. *Eur. Phys. J. B* 2, 525–539.

Lai, Y.-C.; Ye, N.; 2003, Recent developments in chaotic time series analysis. *Int. J. Bifurc. Chaos* 13, 1383–1422.

Landauer, R.; 1961, Irreversibility and heat generation in the computing process. *IBM J. Res. Develop.* 5, 183–191.

Landauer, R.; 1991, Information is physical. *Phys. Today* 44, 23–29.

Lander, A. D.; 2011, Pattern, growth, and control. *Cell* 144, 955–969.

Langer, J. S.; 1980, Instabilities and pattern formation in crystal growth. *Rev. Mod. Phys.* 52, 1–28.

Langton, C. G.; 1990, *Artificial Life.* Addison-Wesley Longman Publishing, Boston, MA.

Laplace, P.-S.; 1814, Essai Philosophique sur les Probabilités. Veuve Courcier, Paris. Translation: Laplace, P.-S.; 1902, *A Philosophical Essay on Probabilities*. John Wiley & Sons, New York; Chapman & Hall, London.

Laplante, J. P.; Micheau, J. C.; Gimenez, J.; 1984, Oscillatory fluorescence in irradiated solutions of 9,10-dimethylanthracene in chlorofom: A probe for the convective motion? *J. Phys. Chem.* 88, 4135–4137.

Laplante, J. P.; Pottier, R. H.; 1982, Study of the oscillatory behavior in irradiated 9,10-dimethylanthracene/chloroform solutions. *J. Phys. Chem.* 86, 4759–4766.

Lappa, M.; 2010, *Thermal Convection: Patterns, Evolution and Stability*. John Wiley & Sons Ltd, Chichester, UK.

Lathrop, R. H.; 1994, The protein threading problem with sequence amino acid interaction preferences is NP-complete. *Protein Eng.* 7, 1059–1068.

Lebedeva, M. I.; Vlachos, D. G.; Tsapatsis, M.; 2004, Bifurcation analysis of Liesegang Ring pattern formation. *Phys. Rev. Lett.* 92, 8, 088301.

Le Novère, N.; 2015, Quantitative and logic modelling of molecular and gene networks. *Nat. Rev. Genet.* 16, 146–158.

Leff, H. S.; Rex, A. F.; (Eds.) 1990, *Maxwell's Demon: Entropy, Information, Computing*. Princeton University Press, Princeton, NJ.

Lehn, J.-M.; 2013, *Perspectives in Chemistry-Steps towards Complex Matter*. Angew. Chem. Int. Ed. 52, 2836–2850.

Lehnertz, K.; Elger, C. E.; Arnhold, J.; Grassberger, P.; (Eds.) 2000, *Chaos in Brain?* World Scientific, Singapore.

Lekebusch, A.; Förster, A.; Schneider, F. W.; 1995, Chaos control in an enzymatic reaction. *J. Phys. Chem.* 99, 681–686.

Lenci, F.; 2008, *Basic Photomovement*. Photobiological Sciences Online (KC Smith, Ed.) American Society for Photobiology, http://www.photobiology.info/

Lengyel, I.; Epstein, I. R.; 1991, Modeling of Turing structures in the chlorite-iodide-malonic acid-starch reaction system. *Science* 251, 650–652.

Lengyel, I.; Epstein, I. R.; 1992, A chemical approach to designing Turing patterns in reaction-diffusion systems. *Proc. Natl. Acad. Sci. USA* 89, 3977–3979.

Lengyel, I.; Kádár, S.; Epstein, I. R.; 1993, Transient Turing structures in a gradient-free closed system. *Science* 259, 493–495.

Lengyel, I.; Rábai, G.; Epstein, I. R.; 1990, Experimental and modeling study of oscillations in the chlorine dioxide-iodine-malonic acid reaction. *J. Am. Chem. Soc.* 112, 9104–9110.

Lewis, S. L.; Maslin, M. A.; 2015, Defining the anthropocene *Nature* 519, 171–180.

Lewis, H. R.; Papadimitriou, C. H.; 1978, The efficiency of algorithms. *Sci. Am.* 238, 96–109.

Liaw, S. S.; Yang, C. C.; Liu, R. T.; Hong, J. T.; 2001, Turing model for the patterns of lady beetles. *Phys. Rev. E* 64, 041909.

Libchaber, A.; Laroche, C.; Fauve, S.; 1982, Period doubling cascade in mercury, a quantitative measurement. *J. Phys.-Paris*, 43, L-211–L-216.

Libchaber, A.; Maurer, J.; 1978, Local probe in a Rayleigh-Bénard experiment in liquid helium. *J. Phys.-Paris* 39, L-369–L-372.

Lide, D. R.; (Ed.) 2005, *CRC Handbook of Chemistry and Physics*. Internet Version 2005, <http://www.hbcpnetbase.com>, CRC Press, Boca Raton, FL.

Lieberman, E.; Hauert, C.; Nowak, M. A.; 2005, Evolutionary dynamics on graphs. *Nature* 433, 312–316.

Liesegang, R. E.; 1896, Über Einige Eigenschaften von Gallerten. *Naturwiss. Wochenschr.* 11, 353–362.

Linial, M.; Linial, N.; 1995, On the potential of molecular computing. *Science* 268, 481.

Lisman, J. E.; 1985, A mechanism for memory storage insensitive to molecular turnover: A bistable autophosphorylating kinase. *Proc. Natl. Acad. Sci. USA* 82, 3055–3057.

Liu, R. T.; Liaw, S. S.; Maini, P. K.; 2006, Two-stage Turing model for generating pigment patterns on the leopard and the jaguar. *Phys. Rev. E* 74, 011914-1-8.

Livio, M.; 2017, *Why? What Makes Us Curious*. Simon & Schuster, New York.

Lloyd, S.; 2000, Ultimate physical limits to computation. *Nature* 406, 1047–1054.

Lloyd, S.; 2006, *Programming the Universe. A Quantum Computer Scientist Takes on the Cosmos*. Vintage Books, London.

Lopes, R.; Betrouni, N.; 2009, Fractal and Multifractal analysis: A review. *Med. Image Anal.* 13, 634–649.

López-Suárez, M.; Neri, I.; Gammaitoni, L.; 2016, Sub-k_BT micro-electrochemical irreversible logic gate. *Nat. Comm.* 7, 12068.

Lorenz, E. N.; 1963, Deterministic nonperiodic flow. *J. Atmos. Sci.* 20, 130–141.

Luisi, P. L.; 2006, *The Emergence of Life: From Chemical Origins to Synthetic Biology.* Cambridge University Press, Cambridge, UK.

Luther, R.-L.; 1906, Räuemliche Fortpflanzung Chemischer Reaktionen. *Z. für Elektrochemie und angew. physikalische Chemie.* 12 (32), 506–600.

Lodish, H.; Berk, A.; Zipursky, S. L.; Matsudaira, P.; Baltimore, D.; Darnell, J.; 2000, *Molecular Cell Biology.* 4th ed. W. H. Freeman, New York.

Lotka, A. J.; 1910, Contribution to the theory of periodic reactions. *J. Phys. Chem.* 14, 271–274.

Lotka, A. J.; 1920a, Analytical note on certain rhythmic relations in organic systems. *Proc. Natl. Acad. Sci. USA* 6, 410–415.

Lotka, A. J.; 1920b, Undamped oscillations derived from the law of mass action. *J. Am. Chem. Soc.* 42, 1595–1599.

Lotka, A. J.; 1925, *Elements of Physical Biology* Williams and Wilkins Company, Baltimore, MD.

Luo, Y.; Epstein, I. R.; 1990, Feedback analysis of mechanisms for chemical oscillators. *Adv. Chem. Phys.* 79, 269–299.

Luo, Y.; Epstein, I. R.; 1991, A general model for pH oscillators. *J. Am. Chem. Soc.* 113, 1518–1522.

Luo, Y.; Orbán, M.; Kustin, K.; Epstein, I. R.; 1989, Mechanistic study of oscillations and bistability in the Cu(II)-catalyzed reaction between H_2O_2 and KSCN. *J. Am. Chem. Soc.* 111, 4541–4548.

Lyons, L.; 1991, *A Practical Guide to Data Analysis for Physical Science Students.* Cambridge University Press, Cambridge, UK.

Ma, W. J.; Beck, J. M.; Pouget, A.; 2008 Spiking networks for Bayesian inference and choice. *Curr. Opin. Neurobiol.* 18, 217–222.

MacDonald, M.; Jan, N.; 1986, Fractons and the fractal dimension of proteins. *Can. J. Phys.* 64, 1353–1355.

Mach, E.; 1984, *Contributions to the Analysis of the Sensations.* Open Court Publishing, Chicago, IL.

Macia, J.; Sole, R.; 2014, How to make a synthetic multicellular computer. *PLoS One* 9, e81248.

Maddox, J.; 2002, Maxwell's demon: Slamming the door. *Nature* 417, 903.

Maini, P. K.; Painter, K. J.; Nguyen Phong Chau, H.; 1997, Spatial pattern formation in chemical and biological system. *J. Chem. Soc. Faraday Trans.* 93, 3601–3610.

Mainzer, K.; 2007, *Thinking in Complexity.* 5th ed. Springer-Verlag, Berlin, Germany.

Mainzer, K.; Chua, L. O.; 2013, *Local Activity Principle.* Imperial College Press, London.

Malthus, T.; 1798, *An Essay on the Principle of Population* J. Johnson, London.

Mandelbrot, B. B.; 1967, How long is the coast of Britain? Statistical self-similarity and fractional dimension. *Science* 156, 636–638.

Mandelbrot, B. B.; 1982, *The Fractal Geometry of Nature.* Freeman and Company, New York.

Mandelbrot, B. B.; Hudson, R. L.; 2004, *The (Mis)Behaviour of Markets: A Fractal View of Risk, Ruin and Reward.* Profile Books, London.

Mandelbrot, B. B.; Passoja, D. E.; Paullay, A. J.; 1984, Fractal character of fracture surfaces of metals. *Nature* 308, 721–722.

Marek, M.; Stuchl, I.; 1975, Synchronization in two interacting oscillatory systems. *Biophys. Chem.* 3, 241–248.

Margolus, N.; Levitin, L. B.; 1998, The maximum speed of dynamical evolution. *Physica D* 120, 188–195.

Markov, I. L.; 2014, Limits on fundamental limits to computation. *Nature* 512, 147–154.

Markus, M.; Hess, B.; 1990, Isotropic cellular automaton for modelling excitable media. *Nature* 347, 56–58.

Maroney, O. J. E.; 2005, The (absence of a) relationship between thermodynamic and logical reversibility. *Studies Hist. Philos. Modern Phys.* 36, 355–374.

Marr, D.; 2010, *Vision. A Computational Investigation into the Human Representation and Processing of Visual Information.* The MIT Press, Cambridge, MA.

Marsden, J. E.; McCracken, M.; 1976, *The Hopf Bifurcation and Its Applications.* Springer-Verlag, New York.

Martyushev, L. M.; Seleznev, V. D.; 2006, Maximum entropy production principle in physics, chemistry and biology. *Phys. Rep.* 426, 1–45.

Maruyama, K.; Nori, F.; Vedral, V.; 2009, Colloquium: The physics of Maxwell's demon and information. *Rev. Mod. Phys.* 81, 1–23.

Marx, K.; 1867, *Das Capital.* Progress Publishers, Moscow, Russia.

Marx, V.; 2013, Biology: The big challenges of big data. *Nature* 498, 255–260.

Matsushita, M.; Ohgiwari, M.; Matsuyama, T.; 1993, Fractal growth and morphological change in bacterial colony formation. pp. 1–9 in *Growth Patterns in Physical Sciences and Biology.* Garcia-Ruiz, J. M.; Louis, E.; Meakin, P.; Sander, L. M.; (Eds.), Plenum Press, New York.

Maturana, H. R.; Varela, F. J.; 1980, Autopoiesis and cognition: The realization of the living. In *Boston Studies in the Philosophy of Science 42.* Cohen, R. S.; Wartofsky, M. V.; (Eds.), D. Reidel Publishing, Dordecht.

Maurer, J.; Libchaber, A.; 1979, Rayleigh-Bénard experiment in liquid helium; frequency locking and the onset of turbulence. *J. Phys.-Paris* 40, L-419–L-423.

Maxwell, J.; 1902, *Theory of Heat*. Longmans, Green, London.

May, Robert M.; 1998, The voles of Hokkaido. *Nature* 396, 409–410.

Mead, C.; 1989, *Analog VLSI and Neural Systems*. Addison-Wesley, Reading, MA.

Meinhardt, H.; 1982, *Models of Biological Pattern Formation*. Academic Press, London.

Meinhardt, H.; 2009, *The Algorithmic Beauty of Sea Shell*. 4th ed. Springer-Verlag, Heidelberg.

Meinhardt, H.; Gierer, A.; 2000, Pattern formation by local self-activation and lateral inhibition. *BioEssays* 22, 753–760.

Mendel, J. M.; 1995 Fuzzy logic systems for engineering: A tutorial. *Proc. IEEE* 83, 3, 345–377.

Merolla, P. A.; Arthur, J. V.; Alvarez-Icaza, R.; Cassidy, A. S.; Sawada, J.; Akopyan, F.; Jackson, B. L. et al.; 2014, Artificial brains: A million spiking-neuron integrated circuit with a scalable communication network and interface. *Science* 345, 668–673.

Merzbach, U. C.; Boyer, C. B.; 2011, *A History of Mathematics*. 3rd ed. John Wiley & Sons, Hoboken, NJ.

Miller, D. G.; 1960, Thermodynamics of irreversible processes: The experimental verification of the onsager reciprocal relations. *Chem. Rev.* 60, 15–37.

Miller, J. H.; Page, S. E.; 2007, *Complex Adaptive Systems. An Introduction to Computational Models of Social Life*. Princeton University Press, Princeton, NJ.

Milonni, P. W.; Eberly, J. H.; 2010, *Laser Physics*. John Wiley & Sons, Hoboken, NJ.

Mitchell, M.; 2009, *Complexity. A Guided Tour*. Oxford University Press, New York.

Miura, K.; Siegert, F.; 2000, Light affects cAMP signaling and cell movement activity in Dictyostelium discoideum. *Proc. Natl. Acad. Sci. USA* 97, 2111–2116.

Molina, M. G.; Medina, E. H.; 2010, Are there Goodwin employment-distribution cycles? International empirical evidence. *Cuadernos de Economía* 29 (53), 1–29.

Monasson, R.; Zecchina, R.; Kirkpatrick, S.; Selman, B.; Troyansky, L.; 1999, Determining computational complexity from characteristic 'phase transitions'. *Nature* 400, 133–137.

Montoya, J. M.; Pimm, S. L.; Solé, R. V.; 2006, Ecological networks and their fragility. *Nature* 442, 259–264.

Moor, J.; 2006, The Dartmouth college artificial intelligence conference: The next fifty years. *AI Magazine* 27 (4), 87–91.

Moore, G.; 1995, Lithography and the future of Moore's law. *Proc. SPIE* 2437, 1–8.

Morton, K. W.; Mayers, D. F.; 2005, *Numerical Solution of Partial Differential Equations*. 2nd ed. Cambridge University Press, New York.

Motlagh, H. N.; Wrabl, J. O.; Li, J.; Hilser, V. J.; 2014, The ensemble nature of allostery. *Nature* 508, 331–339.

Muller, L.; Piantoni, G.; Koller, D.; Cash, S. S.; Halgren, E.; Sejnowski, T. J.; 2016, Rotating waves during human sleep spindles organize global patterns of activity that repeat precisely through the night. *eLife* 5, e17267.

Murdoch, W. W.; 1977, Stabilizing effects of spatial heterogeneity in predator-prey systems *Theor. Popul. Biol.* 11, 252–273.

Murray, J. D.; 1988, How the leopard gets its spots. *Sci. Am.* 258, 80–87.

Murray, J. D.; 2002, *Mathematical Biology. I. An Introduction*. 3rd ed. Springer-Verlag, Berlin, Germany.

Murray, J. D.; 2003, *Mathematical Biology: II. Spatial Models and Biomedical Applications*. Springer-Verlag, Heidelberg (Germany).

Nakamura, T.; Mine, N.; Nakaguchi, E.; Mochizuki, A.; Yamamoto, M.; Yashiro, K.; Meno C.; Hamada, H.; 2006, Generation of robust left-right asymmetry in the mouse embryo requires a self-enhancement and lateral-inhibition system. *Dev. Cell* 11, 495–504.

NASA website at https://solarsystem.nasa.gov/planets/sun/facts, visited the 8th of August 2017.

Newman, M. E. J.; 2010, *Networks: An Introduction*. Oxford University Press, New York.

Nicolis, G.; Portnow, J.; 1973, Chemical oscillations. *Chem. Rev.* 73, 365–384.

Nicolis, G.; Prigogine, I.; 1977, *Self-Organization in Nonequilibrium Systems: From Dissipative Structures to Order through Fluctuations*. John Wiley & Sons, New York.

Nisbet, E. G.; Sleep, N. H.; 2001, The habitat and nature of early life. *Nature* 409, 1083–1091.

Nishida, T. Y.; 2004, An application of P systems: A new algorithm for NP-complete optimization problems. vol. V, pp. 109–112 in *Proceedings of the 8th World Multi-Conference on Systemics, Cybernetics and Informatics*. Callaos N. et al.; (Eds.). (SCI 2002 ISAS 2002, Ext. Vol. XX, Orlando, FL.

Nitsan, I.; Drori, S.; Lewis, Y. E.; Cohen, S.; Tzlil, S.; 2016, Mechanical communication in cardiac cell synchronized beating. *Nat. Phys.* 12, 472–477.

Noble, D.; 2006, *The Music of Life: Biology Beyond the Genome*. Oxford University Press, New York.

Nordhaus, W. D.; 1975, The political business cycles *Rev. Econ. Stud.* 42, 169–190.

Novák, B.; Tyson, J. J.; 2008, Design principles of biochemical oscillators. *Nat. Rev. Mol. Cell Bio.* 9, 981–991.

Nowak, M. A.; 2006, Five rules for the evolution of cooperation. *Science* 314, 1560–1563.

Nowak, M. A.; Sigmund, K.; 1992, Tit for tat in heterogeneous population. *Nature* 355, 250–253.

Nowak, M. A.; Sigmund, K.; 1993, A strategy of win-stay, lose-shift that outperforms tit-for-tat in the Prisoner's Dilemma game. *Nature* 364, 56–58.

Nowak, M. A.; Sigmund, K.; 1998, Evolution of indirect reciprocity by image scoring. *Nature* 393, 573–577.

Nowak, M. A.; Sigmund, K.; 2005, Evolution of indirect reciprocity. *Nature* 437, 1291–1298.

Noyes, R. M.; Field, R. J.; Körös, E. J.; 1972, Oscillations in chemical systems. I. Detailed mechanism in a system showing temporal oscillations. *J. Am. Chem. Soc.* 94, 1394–1395.

O'Brien, E. L.; Itallie, E. V.; Bennett, M. R.; 2012, Modelling synthetic gene oscillators. *Math. Biosci.* 236, 1–15.

Okazaki, N.; Rábai, Gy.; Hanazaki, I.; 1999, Discovery of novel bromate-sulfite pH oscillator with Mn^{2+} or MnO_4^- as a negative feedback species. *J. Phys. Chem. A* 103, 10915–10920.

Olson, J. M.; Rosenberger, F.; 1979, Convective instabilities in a closed vertical cylinder heated from below. Part 1. Monocomponent gases. *J. Fluid Mech.* 92, 609–629.

Onsager, L.; 1931a, Reciprocal relations in irreversible processes. I. *Phys Rev.* 37 (4), 405–426.

Onsager, L.; 1931b, Reciprocal relations in irreversible processes. II. *Phys Rev.* 38 (12), 2265–2279.

Orbán, M.; 1986, Oscillations and bistability in the Cu(II)-catalyzed reaction between H_2O_2 and KSCN. *J. Am. Chem. Soc.* 108, 6893–6898.

Orbán, M.; Epstein, I. R.; 1985, A new halogen-free chemical oscillator: The reaction between sulfide ion and hydrogen peroxide in a CSTR. *J. Am. Chem. Soc.* 107, 2302–2305.

Orbán, M.; Epstein, I. R.; 1987, Chemical oscillator in group VIA: The Cu(II)-catalyzed reaction between hydrogen peroxide and thiosulfate ion. *J. Am. Chem. Soc.* 109, 101–106.

Orbán, M.; Epstein, I. R.; 1992, A new type of oxyhalogen oscillator: The bromite-iodide reaction in a CSTR. *J. Am. Chem. Soc.* 114, 1252–1256.

Orbán, M.; Epstein, I. R.; 1994, Simple and complex pH oscillations and bistability in the phenol-perturbed bromite-hydroxylamine reaction. *J. Phys. Chem.* 98, 2930–2935.

Orbán, M.; Epstein, I. R.; 1995, A new bromite oscillator: Large amplitude pH in the bromite-thiosulfate-phenol flow system. *J. Phys. Chem.* 99, 2358–2362.

Orbán, M.; Körös, E.; Noyes, R. M.; 1979, Chemical oscillations during the uncatalyzed reaction of aromatic compounds with bromate. 2. A plausible skeleton mechanism. *J. Phys. Chem.* 83, 3056–3057.

Orbán, M.; Kurin-Csörgei, K.; Epstein, I. R.; 2015, pH-regulated chemical oscillators. *Acc. Chem. Res.* 48, 593–601.

Orbán, M.; Kurin-Csörgei, K.; Rábai, G.; Epstein, I. R.; 2000, Mechanistic studies of oscillatory copper(II) catalyzed oxidation reactions of sulfur compounds. *Chem. Eng. Sci.* 55, 267–273.

Osborne, A. R.; Provenzale, A.; 1989, Finite correlation dimension for stochastic systems with power-law spectra. *Physica D* 35, 357–381.

Ostwald, W.; 1899, Periodisch veraenderliche Reaktionsgeschwindigkeiten. *Phys. Zeitschr.* 8, 87–88.

Otto, S. P.; Day, T.; 2007, *A Biologist's Guide to Mathematical Modeling in Ecology and Evolution.* Princeton University Press, Princeton, NJ.

Oyster, C. W.; 1999, *The Human Eye, Structure and Function.* Sinauer Associates, Sunderland, MA.

Ozawa, H.; Shimokawa, S.; 2014, The time evolution of entropy production in nonlinear dynamic systems. Chapter 6, pp. 113–128 in *Beyond the Second Law.* Dewar, R. C.; Lineweaver, C. H.; Niven, R. K.; Regenauer-Lieb, K.; (Eds.), Springer-Verlag, Berlin, Germany.

Pääbo, S.; Poinar, H.; Serre, D.; Jaenicke-Desprès, V.; Hebler, J.; Rohland, N.; Kuch, M.; Krause, J.; Vigilant, L.; Hofreiter, M.; 2004, Genetic analyses from ancient DNA. *Annu. Rev. Genet.* 38, 645–679.

Padirac, A.; Fujii, T.; Estévez-Torres, A.; Rondelez, Y.; 2013, Spatial waves in synthetic biochemical networks. *J. Am. Chem. Soc.* 135, 14586–14592.

Paladin, G.; Vulpiani, A.; 1987, Anomalous scaling laws in multifractal objects. *Phys. Rep.* 156, 147–225.

Palva, S.; Palva, J. M.; 2007, New vistas for α-frequency band oscillations. *Trends Neurosci.* 30, 150–158.

Panijpan, B.; 1977, The buoyant density of DNA and the G + C content. *J. Chem. Edu.* 54, 172–173.

Papadimitriou, C. H.; 1994, *Computational Complexity.* Addison-Wesley, Reading, MA.

Parrondo, J. M.; Horowitz, J. M.; Sagawa, T.; 2015, Thermodynamics of information. *Nat. Phys.* 11, 131–139.

Paun, G.; 2002, *Membrane Computing: An Introduction.* Spinger-Verlag, Berlin, Germany.

Paxinos, G.; Mai, J. K.; 2004, *The Human Nervous System.* 2nd ed. Academic Press/Elsevier, San Diego, CA.

Peak, D.; West, J. D.; Messinger, S. M.; Mott, K. A.; 2004, Evidence for complex, collective dynamics and emergent, distributed computation in plants. *Proc. Natl. Acad. Sci. USA* 101, 918–922.

Peitgen, H. O.; Jürgens, H.; Saupe, D.; 1992a, *Fractals for the Classroom. Part One. Introduction to Fractals and Chaos*. Springer-Verlag, New York.

Peitgen, H. O.; Jürgens, H.; Saupe, D.; 1992b, *Fractals for the Classroom. Part Two. Complex Systems and Mandelbrot Set*. Springer-Verlag, New York.

Pegas, K. L.; Edelweiss, M. I.; Cambruzzi, E.; Zettler, C. G.; 2010, Liesegang rings in xanthogranulomatous pyelonephritis: A case report. *Pathol. Res. Int.* 602523, 1–3.

Pellitero, M. A.; Lamsfus, C. Á.; Borge, J.; 2013, The Belousov-Zhabotinskii reaction: Improving the oregonator model with the Arrhenius equation. *J. Chem. Edu.* 90, 82–89.

Perelson, A. S.; Oster, G. F.; 1979, Theoretical studies of clonal selection: Minimal antibody repertoire size and reliability of self-nonself discrimination. *J. Theor. Biol.* 81, 645–670.

Petrov, V.; Gáspár, V.; Masere, J.; Showalter, K.; 1993, Controlling chaos in the Belousov-Zhabotinsky reaction. *Nature* 361, 240–243.

Petrovskii, S. V.; Li, B.-L.; 2006, *Exactly Solvable Models of Biological Invasion*. Chapman and Hall/CRC, Boca Raton, FL.

Planck, M.; 1914, *The Theory of Heat Radiation*. Masius, M. (transl.), 2nd ed, P. Blakiston's Son & Co., Philadelphia, PA.

Plenio, M. B.; Virmani, S.; 2007, An introduction to entanglement measures. *Quant. Inf. Comp.* 7, 1–51.

Plosser, C. I.; 1989, Understanding real business cycles. *J. Econ. Perspect.* 3, 51–77.

Poincaré, H.; 1908, *Science and Method*. Thomas Nelson and Sons, London, Edinburgh, Dublin and New York.

Pojman, J. A.; Craven, R.; Leard, D. C.; 1994, Chemical oscillations and waves in the physical chemistry lab. *J. Chem. Edu.* 71, 84–90.

Popper, K. R.; 1972, Of clouds and clocks. pp. 206–255 in *Objective Knowledge. An Evolutionary Approach*. revised edition, Oxford University Press, Oxford.

Poros, E.; Horváth, V.; Kurin-Csörgei, K.; Epstein, I. R.; Orbán, M.; 2011, Generation of pH-oscillations in closed chemical systems: Method and applications. *J. Am. Chem. Soc.* 133, 7174–7179.

Pouget, A.; Deneve, S.; Duhamel, J.-R.; 2002, A computational perspective on the neural basis of multisensory spatial representations. *Nat. Rev. Neurosci.* 3, 741–747.

Press, W. H.; Teukolsky, S. A.; Vetterling, W. T.; Flannery, B. P.; 2007, *Numerical Recipes*. 3rd ed. Cambridge University Press, New York.

Prigogine, I.; 1968, *Introduction to Thermodynamics of Irreversible Processes* John Wiley & Sons, New York.

Prigogine, I.; 1978, Time, structure and fluctuations *Science* 201, 777–785.

Prigogine, I.; Lefever, R.; 1968, Symmetry breaking instabilities in dissipative systems. II. *J. Chem. Phys.* 48, 1695–1700.

Prigogine, I.; Stengers, I.; 1984, *Order Out of Chaos. Man's New Dialogue with Nature*. Bantam Books, Inc. New York.

Prypsztejn, H. E.; Mulford, D. R.; Stratton, D.; 2005, Chemiluminescent oscillating demonstrations: The chemical buoy, the lighting wave, and the ghostly cylinder. *J. Chem. Edu.* 82, 53–54.

Puhl, A.; Nicolis, G.; 1987, Normal form analysis of multiple bifurcations in incompletely mixed chemical reactors. *J. Chem. Phys.* 87, 1070–1078.

Purcell, O.; Savery, N. J.; Grierson, C. S.; di Bernardo, M.; 2010, A comparative analysis of synthetic genetic oscillators. *J. R. Soc. Interface* 7, 1503–1524.

Qiu, M.; Khisamutdinov, E.; Zhao, Z.; Pan, C.; Choi, J.-W.; Leontis, N. B.; Guo, P.; 2013, RNA nanotechnology for computer design and in vivo computation. *Phil. Trans. R. Soc. A* 371, 20120310.

Quinlan, M. E.; 2016, Cytoplasmic streaming in the Drosophila oocyte. *Annu. Rev. Cell Dev. Biol.* 32, 173–195.

Rábai, Gy.; 1997, Period-doubling routing to chaos in the hydrogen peroxide-sulfur(IV)-hydrogen carbonate flow system. *J. Phys. Chem. A* 101, 7085–7089.

Rábai, Gy.; 2011, pH-oscillations in a closed chemical system of $CaSO_3-H_2O_2-HCO_3^-$. *Phys. Chem. Chem. Phys.* 13, 13604–13606.

Rábai, Gy.; Beck, M. T.; 1988, Exotic kinetic phenomena and their chemical explanation in the iodate-sulfite-thiosulfate system. *J. Phys. Chem.* 92, 2804–2807.

Rábai, Gy.; Beck, M. T.; Kustin, K.; Epstein, I. R.; 1989a, Sustained and damped pH oscillation in the periodate-thiosulfate reaction in a continuous-flow stirred tank reactor. *J. Phys. Chem.* 93, 2853–2858.

Rábai, Gy.; Epstein, I. R.; 1989, Oxidation of hydroxylamine by periodate in a continuous-flow stirred tank reactor: A new pH oscillator. *J. Phys. Chem.* 93, 7556–7559.

Rábai, Gy.; Epstein, I. R.; 1990, Large amplitude pH oscillation in the oxidation of hydroxylamine by iodate in a continuous-flow stirred tank reactor. *J. Phys. Chem.* 94, 6361–6365.

Rábai, Gy.; Hanazaki, I.; 1999, Chaotic pH oscillations in the hydrogen peroxide-thiosulfate-sulfite flow system. *J. Phys. Chem. A* 103, 7268–7273.

Rábai, Gy.; Kustin, K.; Epstein, I. R.; 1989b, A systematically designed pH oscillator: The hydrogen peroxide-sulfite-ferrocyanide reaction in a continuous flow stirred tank reactor. *J. Am. Chem. Soc.* 111, 3870–3874.

Rábai, Gy.; Nagy, Z. V.; Beck, M. T.; 1987, Quantitative description of the oscillatory behavior of the iodate-sulfite-thiourea system in CSTR. *React. Kinet. Catal. Lett.* 33, 23–29.

Rábai, Gy.; Orbán, M.; Epstein, I. R.; 1990, Design of pH-regulated oscillators. *Acc. Chem. Res.* 23, 258–263.

Rácz, Z.; 1999, Formation of Liesegang patterns. *Physica A* 274, 50–59.

Rand, W.; Wilensky, U.; 2007, *NetLogo El Farol model*.http://ccl.northwestern.edu/netlogo/models/ElFarol. Center for Connected Learning and Computer-Based Modeling, Northwestern Institute on Complex Systems, Northwestern University, Evanston, IL.

Raspopovic, J.; Marcon, L.; Russo, L.; Sharpe, J.; 2014, Digit patterning is controlled by a Bmp-Sox9-Wnt Turing network modulated by morphogen gradients. *Science* 345, 566–570.

Ravasz, E.; Barabási, A.-L.; 2003, Hierarchical organization in complex networks. *Phys. Rev. E* 67, 026112.

Ravasz, E.; Somera, A. L.; Mongru, D. A.; Oltvai, Z. N.; Barabási, A.-L.; 2002, Hierarchical organization of modularity in metabolic networks. *Science* 297, 1551–1555.

Reactome project: http://www.reactome.org/, visited the 14th of August 2017.

Reale, G.; 1987, *A History of Ancient Philosophy from the Origins to Socrates*. State University of New York Press, Albany, NY.

Reinhardt, D.; Pesce, E.-R.; Stieger, P.; Mandel, T.; Baltensperger, K.; Bennett, M.; Traas, J.; Friml, J.; Kuhlemeier, C.; 2003, Regulation of phyllotaxis by polar auxin transport. *Nature* 426, 255–260.

Reusser, E. J.; Field, R. J.; 1979, The transition from phase waves to trigger waves in a model of the Zhabotinskii reaction. *J. Am. Chem. Soc.* 101, 1063–1071.

Reynolds, C. W.; 1987, Flocks, herds, and schools: A distributed behavioral model. *Comp. Graph.* 21 (4), 25–34.

Rigney, D. R.; Goldberger, A. L.; Ocasio, W. C.; Ichimaru, Y.; Moody, G. B.; Mark, R. G.; 1993, Multi-channel physiological data: Description and analysis. pp. 105–129 in *Time Series Prediction: Forecasting the Future and Understanding the Past*. Weigend, A. S.; Gershenfeld, N. A.; (Eds.), Addison-Wesley, Reading, MA.

Roederer, J. G.; 2003, On the concept of information and its role in nature. *Entropy* 5, 3–33.

Romero, E.; Novoderezhkin, V. I.; van Grondelle, R.; 2017, Quantum design of photosynthesis for bio-inspired solar-energy conversion. *Nature* 543, 355–365.

Rosenstein, M. T.; Collins, J. J.; De Luca, C. J.; 1993, A practical method for calculating largest Lyapunov exponents from small data sets. *Physica D* 65, 117–134.

Ross, J.; Müller, S. C.; Vidal, C.; 1988, Chemical waves. *Nature* 240, 460–465.

Ross, J.; Vlad, M. O.; 1999, Nonlinear kinetics and new approaches to complex reaction mechanisms. *Ann. Rev. Phys. Chem.* 50, 51–78.

Rossi, F.; Vanag, V. K.; Epstein, I. R.; 2011, Pentanary cross-diffusion in water-in-oil microemulsions loaded with two components of the Belousov-Zhabotinsky reaction. *Chem. Eur. J.* 17, 2138–2145.

Rothemund, P. W. K.; 2006, Folding DNA to create nanoscale shapes and patterns. *Nature* 440, 297–302.

Rothschild, W. G.; 1998, *Fractals in Chemistry*. John Wiley & Sons, New York.

Roy, R.; Murphy, T. W. Jr.; Maier, T. D.; Gills, Z.; Hunt, E. R.; 1992, Dynamical control of a chaotic laser: Experimental stabilization of a globally coupled system. *Phys. Rev. Lett.* 68, 1259–1262.

Rozenberg, G.; Bäck, T.; Kok, J. N.; 2012, *Handbook of Natural Computing*. Springer-Verlag, Berlin, Germany.

Rumelhart, D. E.; Durbin, R.; Golden, R.; Chauvin Y.; 1993, Backpropagation: The basic theory. Chapter 1, pp. 1–24 in *Backpropagation: Theory, Architectures and Applications*. Chauvin, Y.; Rumelhart, D. E.; (Eds.), Lawrence Erlbaum, Hillsdale, NJ.

Runge, F. F.; 1855, *Der Bildungstrieb der Stoffe veranschaulicht in selbständig gewachsenen Bildern*. Selbstverlag, Oranienburg (Germany).

Ruoff, P.; Varga, M.; Körös, E.; 1988, How bromate oscillators are controlled. *Acc. Chem. Res.* 21, 326–332.

Russell, S. J.; Norvig, P.; 1995, *Artificial Intelligence: A Modern Approach*. 3rd ed. Prentice-Hall, Englewood Cliffs, NJ.

Sabry Aly, M. M.; Gao, M.; Hills, G.; Lee, C.-S.; Pitner, G.; Shulaker, M. M.; Wu, T. F. et al.; 2015, Energy-efficient abundant-data computing: The N3XT 1,000x. *Computer* 48, 12, 24–33.

Salawitch, R. J.; Canty, T. P.; Hope, A. P.; Tribett, W. R.; Bennett, B. F.; 2017, *Paris Climate Agreement: Beacon of Hope*. Springer Open. Cham, Switzerland.

Samuelson, P. A.; 1939, Interactions between the multiplier analysis and the principle of acceleration. *Rev. Econ. Stat.* 21 (2), 75–78.

Samuelson, P. A.; Nordhaus, W. D.; 2004, *Economics* 19th edition, McGraw-Hill, New York.

Sander, L. M.; 1986, Fractal growth processes. *Nature* 322, 789–793.

Sardet, C.; Roegiers, F.; Dumollard, R.; Rouviere, C.; McDougall, A.; 1998, Calcium waves and oscillations in eggs. *Biophys. Chem.* 72, 131–140.

Sattar, S.; Epstein, I. R.; 1990, Interaction of luminol with the oscillating system H_2O_2-KSCN-$CuSO_4$-NaOH. *J. Phys. Chem.* 94, 275–277.

Savageau, M. A.; 1979, Allometric morphogenesis of complex systems: Derivation of the basic equations from first principles. *Proc. Natl. Acad. Sci. USA* 76, 6023–6025.

Savageau, M. A.; 1995, Michaelis-Menten mechanism reconsidered: Implications of fractal kinetics. *J. Theor. Biol.* 176, 115–124.

Scheres, B.; van der Putten, W. H.; 2017, The plant perceptron connects environment to development. *Nature* 543, 337–345.

Schiff, S. J.; Jerger, K.; Duong, D. H.; Chang, T.; Spano, M. L.; Ditto, W. L.; 1994, Controlling chaos in the brain. *Nature* 370, 615–620.

Schilpp, P. A.; 1979, *Albert Einstein: Autobiographical Notes.* A Centennial Edition. Open Court Publishing Company, La Salle, IL.

Schirmer, T.; Evans, P. R.; 1990, Structural basis of the allosteric behaviour of phosphofructokinase. *Nature* 343, 140–145.

Schmidt-Nielsen, K.; 1984, *Scaling: Why Is Animal Size So Important?* Cambridge University Press, New York.

Schnakenberg, J.; 1979, Simple chemical reaction systems with limit cycle behaviour. *J. Theor. Biol.* 81, 389–400.

Schneider, J. T.; 2012, Perfect stripes from a general Turing model in different geometries. Thesis for the degree in Master of Science in Mathematics, Boise State University, Boise, ID.

Schöll, E.; Schuster, H. G.; (Eds.) 2008, *Handbook of Chaos Control.* 2nd ed. John Wiley & Sons, Weinheim, Germany.

Schreiber, T.; Schmitz, A.; 1996, Improved surrogate data for nonlinearity tests. *Phys. Rev. Lett.* 77, 635–638.

Schrödinger, E.; 1944, *What Is life?* Cambridge University Press, Cambridge, UK.

Schumacher, B., 1995, Quantum coding. *Phys. Rev. A* 51, 2738–2747.

Schweitzer, F.; Fagiolo, G.; Sornette, D.; Vega-Redondo, F.; Vespignani, A.; White, D. R.; 2009, Economic networks: The new challenges. *Science* 325, 422–425.

Schwierz, F.; 2010, Graphene transistors. *Nat. Nanotechnol.* 5, 487–496.

Schwille, P.; 2017, How simple could life be? *Angew. Chem. Int. Ed.* 56, 10998–11002.

Scott, S. K.; Showalter, K.; 1992, Simple and complex propagating reaction-diffusion fronts. *J. Phys. Chem.* 96, 8702–8711.

Semenov, S. N.; Kraft, L. J.; Ainla, A.; Zhao, M.; Baghbanzadeh, M.; Campbell, V. E.; Kang, K.; Fox, J. M.; Whitesides, G. M.; 2016, Autocatalytic, bistable, oscillatory networks of biologically relevant organic reactions. *Nature* 537, 656–660.

Semenov, S. N.; Wong, A. S. Y.; van der Made, R. M.; Postma, S. G. J.; Groen, J.; van Roekel, H. W. H.; de Greef, T. F. A.; Huck, W. T. S.; 2015, Rational design of functional and tunable oscillating enzymatic networks. *Nat. Chem.* 7, 160–165.

Service, R. F.; 2014, The brain chip. *Science* 345, 614–616.

Shannon, C.; 1948, A mathematical theory of communication. *Bell Syst. Tech. J.* 27 (3), 379–423.

Shea, J. J.; 2011, Refuting a myth about human origins. *Am. Sci.* 99, 128–135.

Shenker, O. R.; 1994, Fractal geometry is not the geometry of nature. *Stud. Hist. Phil. Sci. A* 25, 967–981.

Shigesada, N.; Kawasaki, K.; 1997, *Biological Invasions: Theory and Practice.* Oxford University Press, Oxford.

Shinbrot, T.; Grebogi, C.; Ott, E.; Yorke, J. A.; 1993, Using small perturbations to control chaos. *Nature* 363, 411–417.

Shinbrot, T.; Grebogi, C.; Wisdom, J.; Yorke, J. A.; 1992, Chaos in a double pendulum. *Am. J. Phys.* 60 (6), 491–499.

Shiva, S. G.; 2008, *Computer Organization, Design, and Architecture.* 4th ed. CRC Press, Boca Raton, FL.

Shor, P. W.; 1997, Polynomial time algorithms for prime factorization and discrete logarithms on a quantum computer. *SIAM J. Comput.* 26, 1484–1509.

Short, M. B.; Brantingham, P. J.; Bertozzi, A. L.; Tita, G. E.; 2010, Dissipation and displacement of hotspots in reaction-diffusion models of crime. *Proc. Natl. Acad. Sci. USA* 107, 3961–3965.

Showalter, K.; Tyson, J. J.; 1987, Luther's 1906 discovery and analysis of chemical waves. *J. Chem. Edu.* 64, 742–744.

Silver, D.; Schrittwieser, J.; Simonyan, K.; Antonoglou, I.; Huang, A.; Guez, A.; Hubert, T. et al.; 2017, Mastering the game of Go without human knowledge. *Nature* 550, 354–359.

Singer, J.; Wang, Y.-Z.; Bau, H. H.; 1991, Controlling a chaotic system. *Phys. Rev. Lett.* 66, 1123–1125.

Sipser, M.; 1992, The history and status of the P versus NP question. *Proc. ACM STOC* 603–618.

Skellam, J. G.; 1951, Random dispersal in theoretical populations. *Biometrika* 38, 196–218.

Small, M.; Judd, K.; 1998, Detecting nonlinearity in experimental data. *Int. J. Bifurcat. Chaos* 8, 1231–1244.

Smith, A.; 1776, *The Wealth of Nations* Reprint Edition, 1994. Random House, New York.

Snell, B.; 1982, *The Discovery of the Mind in Greek Philosophy and Literature.* Dover Publications, New York.

Snider, G. L.; Orlov, A. O.; Amlani, I.; Zuo, X.; Bernstein, G. H.; Lent, C. S.; Merz, J. L.; Porod, W.; 1999, Quantum-dot cellular automata: Review and recent experiments. *J. Appl. Phys.* 85, 4283–4285.

Sparrow, C.; 1982, The Lorenz equations: Bifurcations, chaos, and strange attractors. *Appl. Math. Sci.* Vol. 41, Springer-Verlag, New York.

Stach, W.; Kurgan, L.; Pedrycz, W.; Reformat, M.; 2005, Genetic learning of fuzzy cognitive maps. *Fuzzy Set Syst.* 153, 371–401.

Stadtman, E. R.; Chock, P. B.; 1977, Superiority of interconvertible enzyme cascades in metabolic regulation: Analysis of multicyclic systems. *Proc. Natl. Acad. Sci. USA* 74, 2766–2770.

Stahel, W. R.; 2016, Circular economy. *Nature* 531, 435–438.

Stanley, H. E.; Meakin, P.; 1988, Multifractal phenomena in physics and chemistry. *Nature* 335, 405–409.

Stein, B. E.; (Ed.) 2012, *The New Handbook of Multisensory Processing.* MIT Press, Cambridge, MA.

Stein, B. E.; Meredith, M. A.; 1993, *The Merging of the Senses.* MIT Press, Cambridge, MA.

Stephanopoulos, G.; 2012, Synthetic biology and metabolic engineering. *ACS Synth. Biol.* 1, 514–525.

Sterbenz, P. H.; 1974, *Floating-Point Computation.* Prentice-Hall, Englewood Cliffs, NJ.

Stone, C.; Toth, R.; de Lacy Costello, B.; Bull, L.; Adamatzky, A.; 2008, Coevolving cellular automata with memory for chemical computing: Boolean logic gates in the BZ Reaction. pp. 579–588 in *Parallel Problem Solving from Nature: PPSN X. PPSN 2008.* Rudolph, G.; Jansen, T.; Beume, N.; Lucas, S.; Poloni, C.; (Eds.), Lecture Notes in Computer Science, vol. 5199, Springer-Verlag, Berlin, Germany.

Stong, C. L.; 1962, Growing crystals in silica gel mimics natural mineralization. *Sci. Am.* 206, 3, 155–164.

Strassmann, J. E.; Queller, D. C.; 2011, Evolution of cooperation and control of cheating in a social microbe. *Proc. Natl. Acad. Sci. USA* 108, 10855–10862.

Stricker, J.; Cookson, S.; Bennett, M. R.; Mather, W. H.; Tsimring, L. S.; Hasty, J.; 2008, A fast, robust and tunable synthetic gene oscillator. *Nature* 456, 516–519.

Strogatz, S. H.; 1994, *Nonlinear Dynamics and Chaos.* 1st ed. Westview Press, Perseus Books Publishing, Boulder, CO.

Strogatz, S. H.; 2001, Exploring complex networks. *Nature* 410, 268–276.

Strogatz, S.; 2004, *Sync* Penguin Books, London.

Sturmberg, J. P.; 2016, *The Value of Systems and Complexity Sciences for Healthcare.* Springer-Verlag, Berlin, Germany.

Subramanian, N.; Torabi-Parizi, P.; Gottschalk, R. A.; Germain, R. N.; Dutta, B.; 2015, Network representations of immune system complexity. *WIREs Syst. Biol. Med.* 7, 13–38.

Sugihara, G.; May, R. M.; 1990, Nonlinear forecasting as a way of distinguishing chaos from measurement error in time series. *Nature* 344, 734–741.

Sugihara, G.; May, R.; Ye, H.; Hsieh, C.-H.; Deyle, E.; Fogarty, M.; Munch, S.; 2012, Detecting causality in complex ecosystems. *Science* 338, 496–500.

Suykens, J. A. K.; Vandewalle, J.; (Eds.) 1998, *Nonlinear Modeling: Advanced Black-Box Techniques.* Kluwer Academic Publishers, Dordrecht, the Netherlands.

Swaab, D. F.; Van Someren, E. J.; Zhou, J. N.; Hofman, M. A.; 1996, Biological rhythms in the human life cycle and their relationship to functional changes in the suprachiasmatic nucleus. *Prog. Brain Res.* 111, 349–368.

Szaciłowski, K.; 2008, Digital information processing in molecular systems. *Chem. Rev.* 108, 3481–3548.

Szathmáry, E.; Számadó, S.; 2008, Being human: Language—A social history of words. *Nature* 456, 40–41.

Szilard, L.; 1964, On the decrease of entropy in a thermodynamic system by the intervention of intelligent beings. *Behav. Sci.* 9 (4), 301–310.

Szent-Györgyi, A.; 1972, Dionysians and apollonians. *Science* 176, 966.

Takens, F.; 1981, Detecting strange attractors in turbulence. pp. 366–381 in *Dynamical Systems and Turbulence.* Rand, D. A.; Young, L.-S.; (Eds.), Springer-Verlag, Berlin, Germany.

Tang, L.; Lv, H.; Yang, F.; Yu, L.; 2015, Complexity testing techniques for time series data: A comprehensive literature review. *Chaos Soliton. Fract.* 81, 117–135.

Taylor, J. R.; 1997, *An Introduction to Error Analysis. The Study of Uncertainties in Physical Measurements.* University Science Books, Sausalito, CA.

Taylor, A. F.; Tinsley, M. R.; Showalter, K.; 2015, Insights into collective cell behaviour from populations of coupled chemical oscillators. *Phys. Chem. Chem. Phys.* 17, 20047–20055.

Theiler, J.; Eubank, S.; Longtin, A.; Galdrikian, B.; Farmer, J. D.; 1992, Testing for nonlinearity in time series: The method of surrogate data. *Physica D* 58, 77–94.

Theraulaz, G.; Bonabeau, E.; Nicolis, S. C.; Sole, R. V.; Fourcassie, V.; Blanco, S.; Fournier, R. et al.; 2002, Spatial patterns in ant colonies. *Proc. Natl. Acad. Sci. USA* 99, 9645–9649.

Thomas, D.; 1975, Artificial enzyme membranes, transport, memory, and oscillatory phenomena. pp. 115–150 in *Analysis and Control of Immobilized Enzyme Systems.* Thomas, D.; Kernevez, J.-P.; (Eds.), Springer-Verlag, Berlin, Germany-New York.

Thompson, D. W.; 1917, *On Growth and Form* Cambridge University Press, Cambridge, UK.

Thomson, W., 1852, On a universal tendency in nature to the dissipation of mechanical energy. *Proceedings of the Royal Society of Edinburgh for April 19, 1852,* also Philosophical Magazine, Oct. 1852. [This version from Mathematical and Physical Papers, vol. i, art. 59, pp. 511–514]

Timmis, J.; Andrews, P.; Owens, N.; Clark, E.; 2008, An interdisciplinary perspective on artificial immune systems. *Evol. Intell.* 1, 5–26.

Tinsley, M. R.; Collison, D.; Showalter, K.; 2013, Propagating precipitation waves: Experiments and modelling. *J. Phys. Chem. A* 117, 12719–12725.

Tompa, P.; 2002, Intrinsically unstructured proteins. *Trends Biochem. Sci.* 27, 527–533.

Tompa, P.; 2010, *Structure and Function of Intrinsically Disordered Proteins.* Chapman & Hall/CRC Taylor & Francis Group, Boca Raton, FL.

Tompa, P.; 2012, On the supertertiary structure of proteins. *Nat. Chem. Biol.* 8, 597–600.

Tompa, P.; Fuxreiter, M.; 2007, Fuzzy complexes: Polymorphism and structural disorder in protein-protein interactions. *Trends Biochem. Sci.* 33, 2–8.

Tóth, R.; Walliser, R. M.; Lagzi, I.; Boudoire, F.; Düggelin, M.; Braun, A.; Housecroft, C. E.; Constable, E. C.; 2016, Probing the mystery of Liesegang band formation: Revealing the origin of self-organized dual-frequency micro and nanoparticle arrays. *Soft Matter.* 12, 8367–8374.

Toyabe, S.; Sagawa, T.; Ueda, M.; Muneyuki, E.; Sano, M.; 2010, Experimental demonstration of information-to-energy conversion and validation of the generalized Jarzynski equality. *Nat. Phys.* 6, 988–992.

Tsiairis, C. D.; Aulehla, A.; 2016, Self-organization of embryonic genetic oscillators into spatiotemporal wave patterns. *Cell* 164, 656–667.

Tuma, T.; Pantazi, A.; Le Gallo, M.; Sebastian, A.; Eleftheriou, E.; 2016, Stochastic phase-change neurons. *Nat. Nanotechnol.* 11, 693–699.

Turing, A. M.; 1936, On computable numbers, with an application to the Entscheidungsproblem. *Proc. London Math. Soc.* 42, 230–265.

Turing, A. M.; 1952, The chemical basis of morphogenesis. *Philos. Trans. Roy. Soc. London Ser. B* 237, 37–72.

Tuur, S. M.; Nelson, A. M.; Gibson, D. W.; Neafie, R. C.; Johnson, F. B.; Mostofi, F. K.; Connor, D. H.; 1987, Liesegang rings in tissue: How to distinguish Liesegang rings from the giant kidney worm, Dioctophyma renale. *Am. J. Surg. Pathol.* 11, 598–605.

Tyson, J. J.; Keener, J. P.; 1988, Singular Perturbation theory of traveling waves in excitable media (a review). *Physica D* 32, 327–361.

Utsu, T.; Ogata, Y.; Matsu'ura, R. S.; 1995, The centenary of the Omori formula for a Decay law of aftershock activity. *J. Phys. Earth* 43, 1–33.

Vanag, V. K.; Epstein, I. R.; 2001a, Pattern formation in a tunable medium: The Belousov-Zhabotinsky reaction in an Aerosol OT microemulsion. *Phys. Rev. Lett.* 87, 228301-1/4.

Vanag, V. K.; Epstein, I. R.; 2001b, Inwardly rotating spiral waves in a reaction-diffusion system. *Science* 294, 835–837.

Vanag, V. K.; Epstein, I. R.; 2002, Packet waves in a reaction-diffusion system. *Phys. Rev. Lett.* 88, 088303-1/4.

Vanag, V. K.; Epstein, I. R.; 2003, Dash waves in a reaction-diffusion system. *Phys. Rev. Lett.* 90, 098301-1/4.

Vanag, V. K.; Epstein, I. R.; 2009, Cross-diffusion and pattern formation in reaction-diffusion systems. *Phys. Chem. Chem. Phys.* 11, 897–912.

Van Hook, S. J.; Schatz, M. F.; 1997, Simple demonstrations of pattern formation. *Phys. Teach.* 35, 391–395.

Van Horn, K. S.; 2003, Constructing a logic of plausible inference: A guide to Cox's theorem. *Int. J. Approx. Reason.* 34, 3–24.

Varga, M.; Györgyi, L.; Körös, E.; 1985, Bromate oscillators: Elucidation of the source of bromide ion and modification of the chemical mechanism. *J. Am. Chem. Soc.* 107, 4780–4781.

Varghese, S.; Elemans, J. A. A. W.; Rowan, A. E.; Nolte, R. J. M.; 2015, Molecular computing: Paths to chemical Turing machines. *Chem. Sci.* 6, 6050–6058.

Velarde, M. G.; Normand, C.; 1980, Convection. *Sci. Am.* 243, 92–108.

Vickery, H. B.; 1950, The origin of the word protein. *Yale J. Biol. Med.* 22 (5), 387–393.

Vicsek, T.; 1988, Construction of a radial Hele-Shaw cell. pp. 82 in *Random Fluctuations and Pattern Growth: Experiments and Models.* Stanley H. E.; Ostrowsky, N.; (Eds.), Kluwer Academic, Dordrecht, the Netherlands.

Viswanath, D.; 2004, The fractal property of the Lorenz attractor. *Physica D* 190, 115–128.

Volpert, V.; Petrovskii, S.; 2009, Reaction-diffusion waves in biology. *Phys. Life Rev.* 6, 267–310.

Volterra, V.; 1926, Fluctuations in the abundance of a species considered mathematically *Nature* 118, 558–560.

von Baeyer, H. C.; 2004, *Information. The New Language of Science.* Harvard University Press, Cambridge, MA.

von Bertalanffy, L.; 1969, *General System Theory: Foundations, Development, Applications.* G. Braziller, New York.

von Neumann, J.; 1966, *Theory of Self Reproducing Automata.* Burks, A.; (Ed.), University of Illinois Press, Urbana and London.

Waldrop, M. M.; 2016, More than Moore. *Nature* 530, 144–147.

Walker, S. I.; Davies, P. C. W.; Ellis, G. F. R.; 2017, *From Matter to Life: Information and Causality.* Cambridge University Press, Cambridge, UK.

Walsh, C. T.; Tu, B. P.; Tang, Y.; 2018, Eight kinetically stable but thermodynamically activated molecules that power cell metabolism. *Chem. Rev.* 118, 1460–1494.

Wang, G. M.; Sevick, E. M.; Mittag, E.; Searles, D. J.; Evans, D. J.; 2002, Experimental demonstration of violations of the second law of thermodynamics for small systems and short time scales. *Phys. Rev. Lett.* 89, 050601.

Wang, Q.; Schoenlein, R. W.; Peteanu, L. A.; Mathies, R. A.; Shank, C. V.; 1994, Vibrationally coherent photochemistry in the femtosecond primary event of vision. *Science* 266, 422–424.

Watanabe, M.; Kondo, S.; 2012, Changing clothes easily: Connexin4108 Regulates Skin Pattern Variation. *Pigment Cell Melanoma Res.* 25, 326–330.

Watson, J. D.; Crick, F. H. C.; 1953, Molecular structure of nucleic acids. *Nature* 171, 737–738.

Watts, D. J.; Strogatz, S. H.; 1998, Collective dynamics of 'small-world' networks. *Nature* 393, 440–442.

Watzl, M.; Münster, A. F.; 1995, Turing-like spatial patterns in a polyacrylamide-methylene blue-sulfide-oxygen system. *Chem. Phys. Lett.* 242, 273–278.

Weaver, W.; 1948, Science and complexity. *Am. Sci.* 36, 536–544.

Weiss, M. C.; Sousa, F. L.; Mrnjavac, N.; Neukirchen, S.; Roettger, M.; Nelson-Sathi, S.; Martin, W. F.; 2016, The physiology and habitat of the last universal common ancestor. *Nat. Microbiol.* 1, 16116.

Wessel, A.; 2009, What is epigenesis? Or Gene's place in development. *Hum. Ontogenet.* 3, 35–37.

West, G.; 2017, *Scale: The Universal Laws of Growth, Innovation, Sustainability, and the Pace of Life in Organisms, Cities, Economies, and Companies.* Penguin Press, New York.

Wheeler, J. A.; Zurek, W. H.; (Eds.) 1983, *Quantum Theory and Measurement.* Princeton University, Princeton, NJ.

Whewell, W.; 1834, On the connexion of the physical sciences. *Quarterly Review* LI, 54–68.

Whitaker, M.; 2006, Calcium at fertilization and in early development. *Physiol. Rev.* 86, 25–88.

Whitesides, G. M.; Grzybowski, B.; 2002, Self-assembly at all scales. *Science* 295, 2418–2421.

Whitfield, J. D.; Love, P. J.; Aspuru-Guzik, A.; 2013, Computational complexity in electronic structure. *Phys. Chem. Chem. Phys.* 15, 397–411.

Wiener, N.; 1948, *Cybernetics: Or Control and Communication in the Animal and the Machine.* MIT Press, Cambridge, MA.

Wilensky, U.; 1997, *NetLogo Ants model.* http://ccl.northwestern.edu/netlogo/models/Ants. Center for Connected Learning and Computer-Based Modeling, Northwestern University, Evanston, IL.

Wilensky, U.; 1998, *NetLogo Flocking model.* http://ccl.northwestern.edu/netlogo/models/Flocking. Center for Connected Learning and Computer-Based Modeling, Northwestern University, Evanston, IL.

Wilensky, U.; 1999, *NetLogo.* http://ccl.northwestern.edu/netlogo/. Center for Connected Learning and Computer-Based Modeling, Northwestern University, Evanston, IL.

Wilensky, U.; 2005, *NetLogo Small Worlds model.* http://ccl.northwestern.edu/netlogo/models/SmallWorlds. Center for Connected Learning and Computer-Based Modeling, Northwestern University, Evanston, IL.

Wilensky, U.; Rand, W.; 2015, *Introduction to Agent-Based Modeling: Modeling Natural, Social and Engineered Complex Systems with NetLogo.* MIT Press, Cambridge, MA.

Wilkinson, J. H.; 1994, *Rounding Errors in Algebraic Processes.* Dover Publications, New York.

Winfree, A. T.; Strogatz, S. H.; 1984, Organizing centres for three-dimensional chemical waves. *Nature* 311, 611–615.

Witten, T. A. Jr.; Sander, L. M.; 1981, Diffusion-limited aggregation, a kinetic critical phenomenon. *Phys. Rev. Lett.* 47, 1400–1403.

Witten, T.; Sander, L.; 1983, Diffusion-limited aggregation. *Phys. Rev. B* 27, 5686–5697.

Wolf, M.; 2017, *The Physics of Computing*. Elsevier, Cambridge, MA.

Wolfram, S.; 2002, *A New Kind of Science*. Wolfram Media, Champaign, IL.

Wolpert, L.; 2008, *The Triumph of the Embryo*. Dover Publications, Mineola, New York.

Wolpert, L.; 2011, Positional information and patterning revisited. *J. Theor. Biol.* 269, 359–365.

Würger, A.; 2014, Do thermal diffusion and Dufour coefficients satisfy Onsager's reciprocity relation? *Eur. Phys. J. E* 37, 96 1–11.

Xepapadeas, A.; 2010, The spatial dimension in environmental and resource economics. *Environ. Dev. Econ.* 15, 747–758.

Xiong, W.; Ferrell, J. E. Jr.; 2003, A positive-feedback-based bistable "memory module" that governs a cell fate decision. *Nature* 426, 460–465.

Yamaguchi, M.; Yoshimoto, E.; Kondo, S.; 2007, Pattern regulation in the stripe of zebrafish suggests an underlying dynamic and autonomous mechanism. *Proc. Natl. Acad. Sci. USA* 104 (12), 4790–4793.

Yang, J. J.; Strukov, D. B.; Stewart, D. R.; 2013, Memristive devices for computing. *Nat. Nanotechnol.* 8, 13–24.

Young, H. D.; 1962, *Statistical Treatment of Experimental Data*. McGraw-Hill Book Company, New York.

Young, J. K.; 2012, *Hunger, Thirst, Sex, and Sleep. How the Brain Controls Our Passions*. Rowan & Littlefield Publishers, Lanham, MD.

Yule, G. U.; 1927, On a method of investigating periodicities in disturbed series, with special reference to Wolfer's sunspot numbers. *Philos. Trans. R. Soc. London, Ser. A* 226, 267–298.

Zadeh, L. A.; 1965, Fuzzy sets. *Inform. Control* 8, 338–353.

Zadeh, L. A.; 1973, Outline of a new approach to the analysis of complex systems and decision processes. *IEEE T. Syst. Man Cyb.* 3, 28–44.

Zadeh, L. A.; 2008, Toward human level machine intelligence-Is it achievable? The need for a paradigm shift. *IEEE Comput. Intell. M.* 3, 11–??

Zhirnov, V. V.; Cavin, R. K.; Hutchby, J. A.; Bourianoff, G. I.; 2003, Limits to binary logic switch scaling – A Gedanken model. *Proc. IEEE* 91, 1934–1939.

Zhu, M. F.; Hong, C. P.; 2001, A modified cellular automaton model for the simulation of dendritic growth in solidification alloy. *ISIJ Int.* 41, 436–445.

Zlotnik, A.; Nagao, R.; Kiss, I. Z.; Li, Jr-S.; 2016, Phase-selective entrainment of nonlinear oscillator ensembles. *Nat. Commun.* 7, 10788.

Zuo, R.; Wang, J.; 2016, Fractal/Multifractal modeling of geochemical data: A review. *J. Geochem. Explor.* 164, 33–41.

Index

Note: Page numbers in italic and bold refer to figures and tables respectively.